CLYMER®

YAMAHA

OUTBOARD SHOP MANUAL
2-225 HP 2-STROKE • 1984-1989
(includes Jet Drives)

The World's Finest Publisher of Mechanical How-To Manuals

INTERTEC PUBLISHING

P.O. Box 12901, Overland Park, Kansas 66282-2901

Copyright ©1990 Intertec Publishing

FIRST EDITION
First Printing July, 1986

SECOND EDITION
Updated to include 1987 models
First Printing December, 1987
Second Printing January, 1989

THIRD EDITION
Updated to include 1988-1989 models
First Printing July, 1990
Second Printing November, 1991
Third Printing February, 1993
Fourth Printing May, 1994
Fifth Printing September, 1995
Sixth Printing March, 1997
Seventh Printing June, 1999

Printed in U.S.A.

CLYMER and colophon are registered trademarks of Intertec Publishing.

ISBN: 0-89287-498-8

Library of Congress: 89-46359

Tools shown in Chapter Two courtesy of Thorsen Tool, Dallas, Texas. Test equipment shown in Chapter Two courtesy of Dixson, Inc., Grand Junction, Colorado.

Technical illustrations courtesy of Yamaha Motor Corporation, U.S.A., Cypress, California. Additional technical illustrations by Mitzi McCarthy and Diana Kirkland. With thanks to Camborac Inc., Fullerton, California.

COVER: Photographed by Michael Brown Photographic Productions, Los Angeles, California.

Intertec Book Division

President Raymond E. Maloney
Vice President, Book Division Ted Marcus

The following books and guides are published by Intertec Publishing.

CLYMER SHOP MANUALS
Boat Motors and Drives
Motorcycles and ATVs
Snowmobiles
Personal Watercraft

ABOS/INTERTEC/CLYMER BLUE BOOKS AND TRADE-IN GUIDES
Recreational Vehicles
Outdoor Power Equipment
Agricultural Tractors
Lawn and Garden Tractors
Motorcycles and ATVs
Snowmobiles and Personal Watercraft
Boats and Motors

AIRCRAFT BLUEBOOK-PRICE DIGEST
Airplanes
Helicopters

AC-U-KWIK DIRECTORIES
The Corporate Pilot's Airport/FBO Directory
International Manager's Edition
Jet Book

I&T SHOP SERVICE MANUALS
Tractors

INTERTEC SERVICE MANUALS
Snowmobiles
Outdoor Power Equipment
Personal Watercraft
Gasoline and Diesel Engines
Recreational Vehicles
Boat Motors and Drives
Motorcycles
Lawn and Garden Tractors

Contents

CHAPTER FIVE

TIMING, SYNCHRONIZING AND ADJUSTING 109

CHAPTER SIX

FUEL SYSTEM 159

CHAPTER SEVEN

IGNITION AND ELECTRICAL SYSTEMS. 203

CHAPTER EIGHT

POWER HEAD. 228

CHAPTER NINE

GEARCASE AND DRIVE SHAFT HOUSING 308

Quick Reference Data

TUNE-UP SPECIFICATIONS (YAMAHA 2)

Firing order	Single cylinder
Spark plug	
Type	NGK B5HS
Gap	0.020-0.024 in.
Torque	14 ft.-lb. (20 N•m)
Static WOT (wide open throttle)	0.044 ±0.0047 in. BTDC
Breaker point gap	0.014 in.
Full throttle rpm	4,000-5,000 rpm
Idle rpm	1,150-1,250 rpm

TUNE-UP SPECIFICATIONS (YAMAHA 3)

Firing order	Single cylinder
Spark plug	
Type	NGK B6HS-10
Gap	0.035-0.039 in.
Torque	14 ft.-lb. (20 N•m)
Maximum timing	18° ±1° BTDC
Idle timing	18° ±1° BTDC
Full throttle rpm	4,500-5,500 rpm
Idle rpm	1,500-1,250 rpm

TUNE-UP SPECIFICATIONS (YAMAHA 4 AND 5)

Firing order	Single cylinder
Spark plug	
Type	NGK B7HS
Gap	0.020-0.024 in.
Torque	18 ft.-lb. (25 N•m)
Maximum timing	28° ±3° BTDC
Idle timing	5° ±2° BTDC
Full throttle rpm	4,500-5,500 rpm
Idle rpm	1,100-1,200 rpm
Trolling rpm	950-1,050 rpm

TUNE-UP SPECIFICATIONS (YAMAHA 6 AND 8)

Firing order	Alternate
Spark plug	
Type	NGK B7HS-10
Gap	0.035-0.039 in.
Torque	14 ft.-lb. (20 N•m)
Maximum timing	35° ±1 BTDC
Idle timing	Automatically set when maximum timing is correct
Full throttle rpm	4,500-5,500 rpm
Idle rpm	850-950 rpm
Trolling rpm	750-850 rpm

TUNE-UP SPECIFICATIONS (YAMAHA 9.9 AND 15)

Firing order	Alternate
Spark plug	
Type	NGK B7HS-10
Gap	0.035-0.039 in.
Torque	14 ft.-lb. (20 N•m)
Static WOT (wide open throttle)	0.166 ±0.011 in.
Dynamic WOT (wide open throttle)	30° ±1 BTDC
Static idle timing	0.005 ±0.002 in.
Dynamic idle timing	5° ±1° BTDC
Full throttle rpm	4,500-5.500 rpm
Idle rpm	850-950 rpm
Trolling rpm	670-770 rpm

TUNE-UP SPECIFICATIONS (YAMAHA 25)

Firing order	Alternate
Spark plug	
Type	
1984	NGK B7HS
1985-on	NGK B7HS-10
Gap	
1984	0.020-0.024 in.
1985-on	0.035-0.039 in.
Torque	
1984-1987	14 ft.-lb. (20 N•m)
1988-1989	18 ft.-lb. (25 N•m)
Dynamic WOT (wide open throttle)	25° BTDC
Full throttle rpm	
1984-1987	4,500-5,500 rpm
1988-1989	5,000-6,000 rpm
Idle rpm	
1984-1987	850-950 rpm
1988-1989	700-800 rpm
Trolling rpm	
1984-1987	750-850 rpm
1988-1989	600-700 rpm

TUNE-UP SPECIFICATIONS (YAMAHA 40 AND 50 [includes Pro 50])

Firing order	1-2-3
Spark plug	
Type	
40 hp	NGK B7HS-10
50 hp	NGK B8HS-10
Gap	0.035-0.039 in.
Torque	14 ft.-lb. (20 N•m)
Dynamic WOT (wide open throttle)	25° ±1° BTDC
Dynamic idle timing	5° ±1° ATDC
Pickup timing	1° ±1° ATDC
Full throttle rpm	4,500-5,500 rpm
Idle rpm	750-850 rpm
Trolling rpm	550-650 rpm

TUNE-UP SPECIFICATIONS (YAMAHA 30)

Firing order	
1984-1986	Alternate
1987	1-2-3
Spark plug	
Type	NGK B7HS-10
Gap	0.035-0.039 in.
Torque	14 ft.-lb. (20 N•m)
Dynamic WOT (wide open throttle)	25° BTDC
Full throttle rpm	4,500-5,500 rpm
Idle rpm	
1984-1986	850-950 rpm
1987	700-800 rpm
Trolling rpm	
1984-1986	750-850 rpm
1987	600-700 rpm

TUNE-UP SPECIFICATIONS (YAMAHA 70)

Firing order	1-2-3
Spark plug	
Type	NGK B8HS-10
Gap	0.035-0.039 in.
Torque	18 ft.-lb. (25 N•m)
Dynamic WOT (wide open throttle)	20° ±1° BTDC
Dynamic idle timing	
1984	4° ±1° ATDC
1985-on	7° ±1° ATDC
Pickup timing	
1984	NA
1985-on	7° ATDC
Full throttle rpm	4,500-5,500 rpm
Idle rpm	750-850 rpm
Trolling rpm	550-650 rpm

TUNE-UP SPECIFICATIONS (YAMAHA 90)

Firing order	1-2-3
Spark plug	
Type	NGK B8HS-10
Gap	0.035-0.039 in.
Torque	18 ft.-lb. (25 N•m)
Dynamic WOT (wide open throttle)	22° ±1° BTDC
Dynamic idle timing	
1984	5° ±1° ATDC
1985-on	10° ±1° ATDC
Pickup timing	
1984	5° ATDC
1985-on	10° ±1° ATDC
Full throttle rpm	4,500-5,500 rpm
Idle rpm	750-850 rpm
Trolling rpm	550-650 rpm

TUNE-UP SPECIFICATIONS (YAMAHA 115 AND 130)

Firing order	1-2-3-4
Spark plug	
Type	
115 hp	NGK B8HS-10
130 hp	NGK B9HS-10
Gap	0.035-0.039 in.
Torque	14 ft.-lb. (20 N•m)
Dynamic WOT (wide open throttle)	
115 hp	25° ±1° BTDC
130 hp	23° ±1° BTDC
Dynamic idle timing	5° ±1° ATDC
Pickup timing	4° ±1° ATDC
Full throttle rpm	4,500-5,500 rpm
Idle rpm	700-800 rpm
Trolling rpm	600-700 rpm

TUNE-UP SPECIFICATIONS (YAMAHA V6)

Firing order	1-2-3-4-5-6
Spark plug	
Type	
150 hp	NGK B8HS-10
175-200 hp	NGK B8HS-10
220 hp	NGK BR8HS-10
225 Excel	NGK BR9HS-10
Gap	0.035-0.039 in.
Torque	14 ft.-lb. (20 N•m)
Dynamic WOT (wide open throttle)	
150 hp	24° ±1° BTDC
150 hp Pro V	28° ±1° BTDC
175-200 hp	22° ±1° BTDC
220 hp	26° ±2° BTDC
225 hp Excel	22° ±1° BTDC
Dynamic idle timing	
1984 150-200 hp	5° ±1° ATDC
1985-on 150-200 hp	7° ±1° ATDC
220 hp	5° ±2° ATDC
225 hp Excel	6° ±1° ATDC
Pickup timing	
1984 150-200 hp	4° ±1° ATDC
1985-on 150 and 200 hp	6° ±1° ATDC
175 hp	7° ±1° ATDC
220 hp	5° ±2° ATDC
225 hp Excel	6° ±1° ATDC
Full throttle rpm	
150-200 hp	4,500-5,500 rpm
220 hp and 225 hp Excel	4,800-5,800 rpm
Idle rpm	
1984 150 hp	795-845 rpm
1984 175-200 hp	775-825 rpm
1984 220 hp	755-805 rpm
1985-on 150-175 hp	775-825 rpm
1985-on 200 hp	675-725 rpm
1985-on 220 hp and 225 hp Excel	725-775 rpm

(continued)

Trolling rpm	
1984 (all)	600-650 rpm
1985-on 150-200 hp	550-600 rpm)
1985-on 220 hp and 225 hp Excel	575-625 rpm
Initial idle screw	
adjustment[1]	
150 hp w/carburetor	
#6G400	2 1/8-2 5/8 turns
#6G401	1-1 1/2 turns
#6G402	7/8-1 3/8 turns
#6G403	1-1 1/2 turns
150 hp Pro V w/carburetor	
#6J900	
Starboard	3/4-1 1/4 turns
Port	1 1/4-1 3/4 turns
175 hp w/carburetor	
#6G500	1 7/8-2 3/8 turns
#6G501	1-1 1/2 turns
#6G502	
Starboard	7/8-1 3/8 turns
Port	1 1/2-2 turns
#6G503	1 1/8-1 5/8 turns
200 hp w/carburetor	
#6G600	1 7/8-2 3/8 turns
#6G601	1 1/8-1 5/8 turns
#6G602	
Starboard	7/8-1 3/8 turns
Port	1 1/4-1 3/4 turns
#6G603	
Starboard	5/8-1 1/8 turns
Port	1 3/8-1 7/8 turns
220 hp w/carburetor	
#6G700	1-1 1/2 turns
#6G701	1 1/4-1 3/4 turns
#6G702	
Starboard	1/2-1 turn
Port	1 3/8-1 7/8 turns
225 hp Excel w/carburetor	
#6K700	3/4-1 1/4 turns

1. Lightly seat the idle mixture screw, then back it out the specified number of turns as an initial adjustment point.

Introduction

This Clymer shop manual covers Yamaha 2 to 225 hp 2-stroke outboard engines from 1984-1989—including jet drives. It does not cover high-performance, counter-rotating, commercial or 4-stroke models. Step-by-step instructions guide you through jobs ranging from simple maintenance to complete overhaul.

This manual can be used by anyone from a first time do-it-yourselfer to a professional mechanic. Easy to read type, detailed drawings and clear photographs give you all the information you need to do the work right.

Having a well-maintained engine will increase your enjoyment of your boat as well as assure your safety when offshore. Keep this shop manual handy and use it often. It can save you hundreds of dollars in maintenance and repair bills and make yours a reliable top-performing boat.

Chapter One

General Information

This detailed, comprehensive manual contains complete information on maintenance, tune-up, repair and overhaul. Hundreds of photos and drawings guide you through every step-by-step procedure.

Troubleshooting, tune-up, maintenance and repair are not difficult if you know what tools and equipment to use and what to do. Anyone not afraid to get their hands dirty, of average intelligence and with some mechanical ability, can perform most of the procedures in this book. See Chapter Two for more information on tools and techniques.

A shop manual is a reference. You want to be able to find information fast. Clymer books are designed with you in mind. All chapters are thumb tabbed and important items are indexed at the end of the book. All procedures, tables, photos, etc., in this manual assume that the reader may be working on the machine or using this manual for the first time.

Keep this book handy in your tool box. It will help you to better understand how your machine runs, lower repair and maintenance costs and generally increase your enjoyment of your marine equipment.

MANUAL ORGANIZATION

This chapter provides general information useful to marine owners and mechanics.

Chapter Two discusses the tools and techniques for preventive maintenance, troubleshooting and repair.

Chapter Three describes typical equipment problems and provides logical troubleshooting procedures.

Following chapters describe specific systems, providing disassembly, repair, assembly and adjustment procedures in simple step-by-step form. Specifications concerning a specific system are included at the end of the appropriate chapter.

NOTES, CAUTIONS AND WARNINGS

The terms NOTE, CAUTION and WARNING have specific meanings in this manual. A NOTE provides additional information to make a step or procedure easier or clearer. Disregarding a NOTE could cause inconvenience, but would not cause damage or personal injury.

A CAUTION emphasizes areas where equipment damage could result. Disregarding a CAUTION could cause permanent mechanical damage; however, personal injury is unlikely.

A WARNING emphasizes areas where personal injury or even death could result from negligence. Mechanical damage may also occur. WARNINGS *are to be taken seriously*. In some cases, serious injury or death has resulted from disregarding similar warnings.

TORQUE SPECIFICATIONS

Torque specifications throughout this manual are given in foot-pounds (ft.-lb.) and either Newton meters (N·m) or meter-kilograms (mkg). Newton meters are being adopted in place of meter-kilograms in accordance with the International Modernized Metric System. Existing torque wrenches calibrated in meter-kilograms can be used by performing a simple conversion: move the decimal point one place to the right. For example, 4.7 mkg = 47 N·m. This conversion is accurate enough for mechanics' use even though the exact mathematical conversion is 3.5 mkg = 34.3 N·m.

ENGINE OPERATION

All marine engines, whether 2- or 4-stroke, gasoline or diesel, operate on the Otto cycle of intake, compression, power and exhaust phases.

4-stroke Cycle

A 4-stroke engine requires two crankshaft revolutions (4 strokes of the piston) to complete the Otto cycle. **Figure 1** shows gasoline 4-stroke engine operation. **Figure 2** shows diesel 4-stroke engine operation.

2-stroke Cycle

A 2-stroke engine requires only 1 crankshaft revolution (2 strokes of the piston) to complete the Otto cycle. **Figure 3** shows gasoline 2-stroke engine operation. Although diesel 2-strokes exist, they are not commonly used in light marine applications.

FASTENERS

The material and design of the various fasteners used on marine equipment are not arrived at by chance or accident. Fastener design determines the type of tool required to work with the fastener. Fastener material is carefully selected to decrease the possibility of physical failure or corrosion. See *Galvanic Corrosion* in this chapter for more information on marine materials.

Threads

Nuts, bolts and screws are manufactured in a wide range of thread patterns. To join a nut and bolt, the diameter of the bolt and the diameter of the hole in the nut must be the same. It is just as important that the threads on both be properly matched.

The best way to determine if the threads on two fasteners are matched is to turn the nut on the bolt (or the bolt into the threaded hole in a piece of equipment) with fingers only. Be sure both pieces are clean. If much force is required, check the thread condition on each fastener. If the thread condition is good but the fasteners jam, the threads are not compatible.

Four important specifications describe every thread:

  a. Diameter.
  b. Threads per inch.
  c. Thread pattern.
  d. Thread direction.

Figure 4 shows the first two specifications. Thread pattern is more subtle. Italian and British

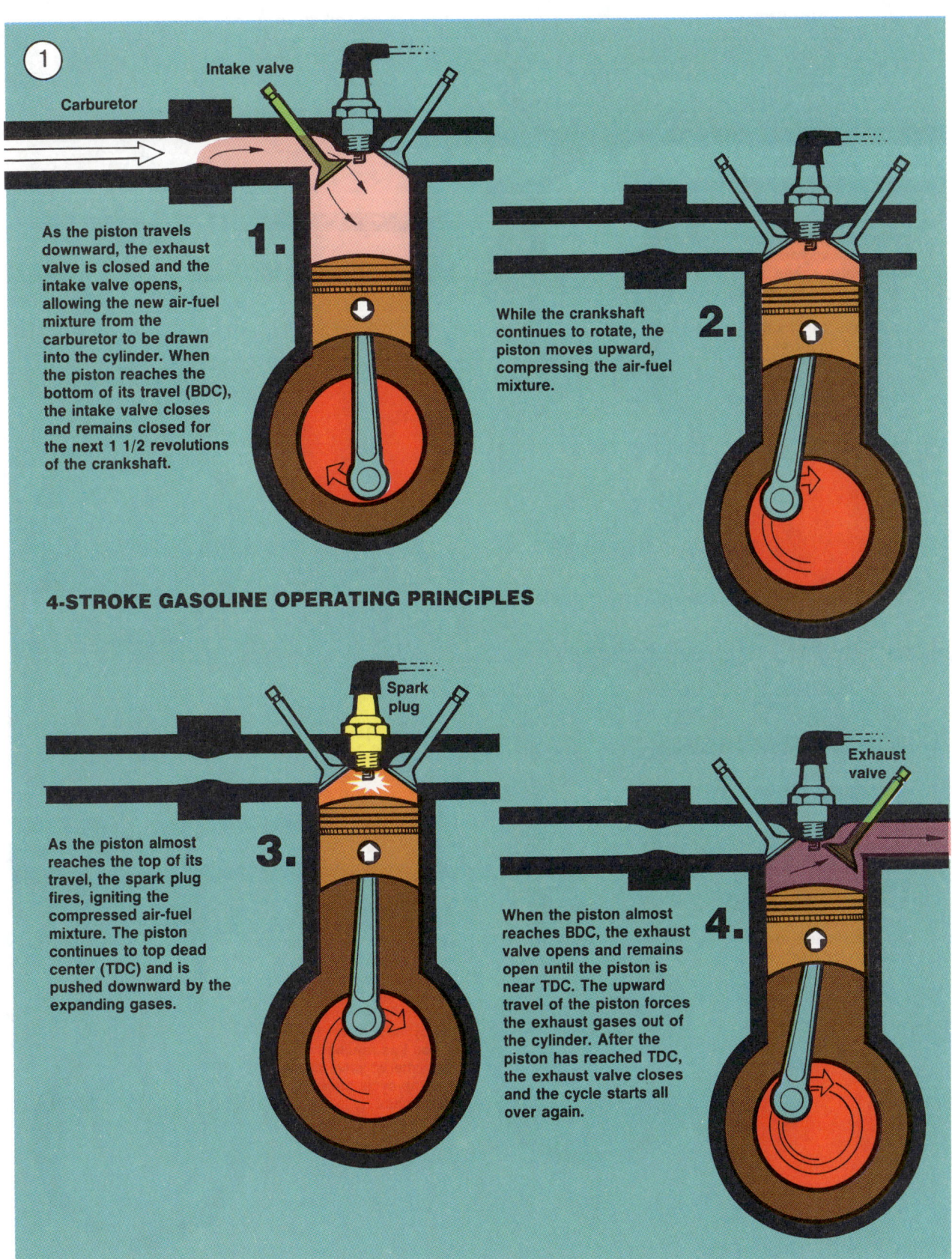

①

Intake valve

Carburetor

1. As the piston travels downward, the exhaust valve is closed and the intake valve opens, allowing the new air-fuel mixture from the carburetor to be drawn into the cylinder. When the piston reaches the bottom of its travel (BDC), the intake valve closes and remains closed for the next 1 1/2 revolutions of the crankshaft.

2. While the crankshaft continues to rotate, the piston moves upward, compressing the air-fuel mixture.

4-STROKE GASOLINE OPERATING PRINCIPLES

Spark plug

3. As the piston almost reaches the top of its travel, the spark plug fires, igniting the compressed air-fuel mixture. The piston continues to top dead center (TDC) and is pushed downward by the expanding gases.

Exhaust valve

4. When the piston almost reaches BDC, the exhaust valve opens and remains open until the piston is near TDC. The upward travel of the piston forces the exhaust gases out of the cylinder. After the piston has reached TDC, the exhaust valve closes and the cycle starts all over again.

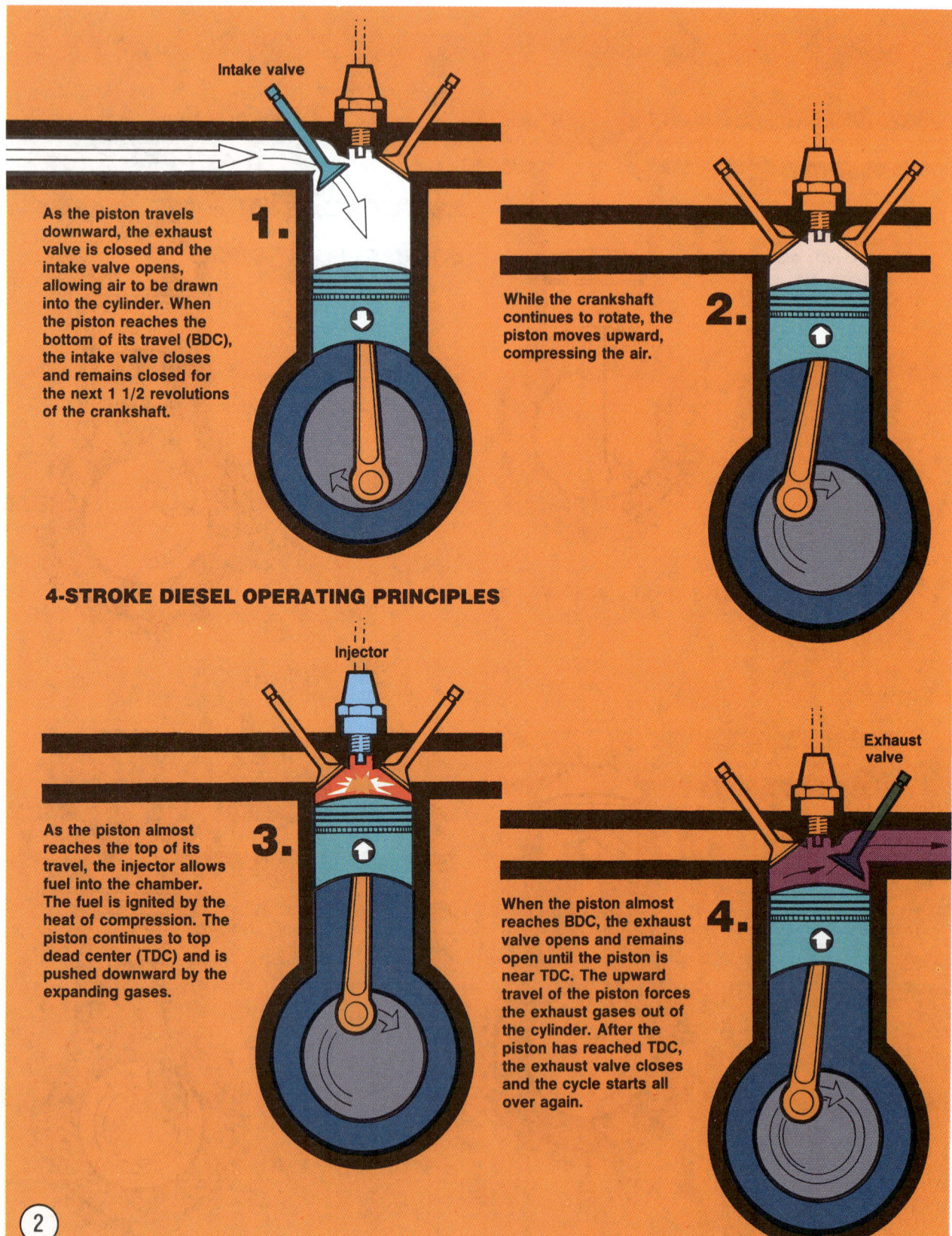

Intake valve

1. As the piston travels downward, the exhaust valve is closed and the intake valve opens, allowing air to be drawn into the cylinder. When the piston reaches the bottom of its travel (BDC), the intake valve closes and remains closed for the next 1 1/2 revolutions of the crankshaft.

2. While the crankshaft continues to rotate, the piston moves upward, compressing the air.

4-STROKE DIESEL OPERATING PRINCIPLES

Injector

3. As the piston almost reaches the top of its travel, the injector allows fuel into the chamber. The fuel is ignited by the heat of compression. The piston continues to top dead center (TDC) and is pushed downward by the expanding gases.

Exhaust valve

4. When the piston almost reaches BDC, the exhaust valve opens and remains open until the piston is near TDC. The upward travel of the piston forces the exhaust gases out of the cylinder. After the piston has reached TDC, the exhaust valve closes and the cycle starts all over again.

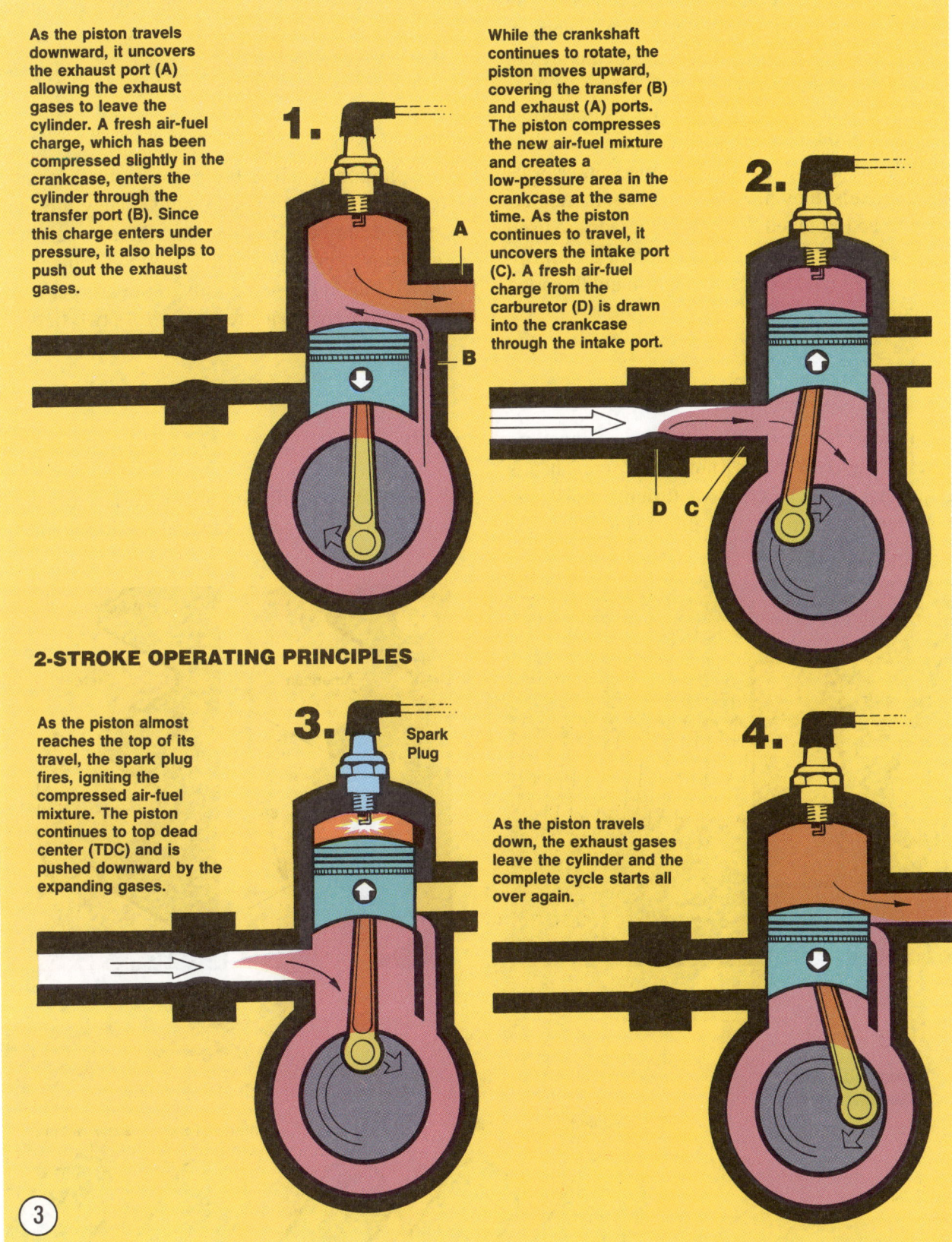

As the piston travels downward, it uncovers the exhaust port (A) allowing the exhaust gases to leave the cylinder. A fresh air-fuel charge, which has been compressed slightly in the crankcase, enters the cylinder through the transfer port (B). Since this charge enters under pressure, it also helps to push out the exhaust gases.

1.

A

B

While the crankshaft continues to rotate, the piston moves upward, covering the transfer (B) and exhaust (A) ports. The piston compresses the new air-fuel mixture and creates a low-pressure area in the crankcase at the same time. As the piston continues to travel, it uncovers the intake port (C). A fresh air-fuel charge from the carburetor (D) is drawn into the crankcase through the intake port.

2.

D C

2-STROKE OPERATING PRINCIPLES

As the piston almost reaches the top of its travel, the spark plug fires, igniting the compressed air-fuel mixture. The piston continues to top dead center (TDC) and is pushed downward by the expanding gases.

3. Spark Plug

As the piston travels down, the exhaust gases leave the cylinder and the complete cycle starts all over again.

4.

standards exist, but the most commonly used by marine equipment manufacturers are American standard and metric standard. The threads are cut differently as shown in **Figure 5**.

Most threads are cut so that the fastener must be turned clockwise to tighten it. These are called right-hand threads. Some fasteners have left-hand threads; they must be turned counterclockwise to be tightened. Left-hand threads are used in locations where normal rotation of the equipment would tend to loosen a right-hand threaded fastener.

Machine Screws

There are many different types of machine screws. **Figure 6** shows a number of screw heads requiring different types of turning tools (see Chapter Two for detailed information). Heads

are also designed to protrude above the metal (round) or to be slightly recessed in the metal (flat) (**Figure 7**).

Bolts

Commonly called bolts, the technical name for these fasteners is cap screw. They are normally described by diameter, threads per inch and length. For example, 1/4-20 × 1 indicates a bolt 1/4 in. in diameter with 20 threads per inch, 1 in. long. The measurement across two flats on the head of the bolt indicates the proper wrench size to be used.

Nuts

Nuts are manufactured in a variety of types and sizes. Most are hexagonal (6-sided) and fit

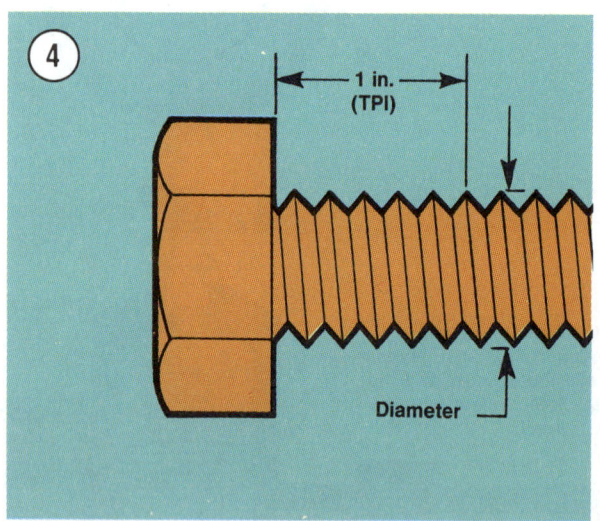

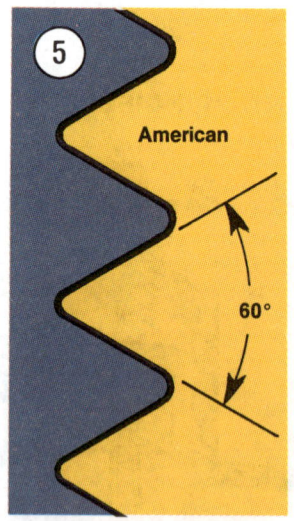

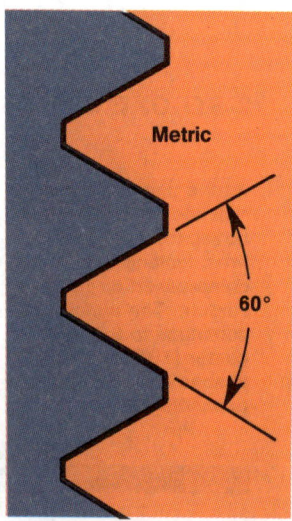

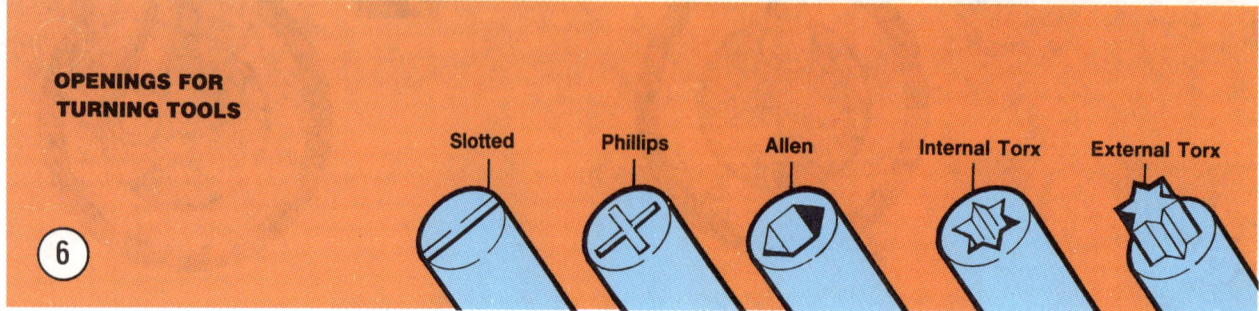

on bolts, screws and studs with the same diameter and threads per inch.

Figure 8 shows several types of nuts. The common nut is usually used with a lockwasher. Self-locking nuts have a nylon insert that prevents the nut from loosening; no lockwasher is required. Wing nuts are designed for fast removal by hand. Wing nuts are used for convenience in non-critical locations.

To indicate the size of a nut, manufacturers specify the diameter of the opening and the threads per inch. This is similar to bolt specification, but without the length dimension. The measurement across two flats on the nut indicates the proper wrench size to be used.

Washers

There are two basic types of washers: flat washers and lockwashers. Flat washers are simple discs with a hole to fit a screw or bolt. Lockwashers are designed to prevent a fastener from working loose due to vibration, expansion and contraction. **Figure 9** shows several types of lockwashers. Note that flat washers are often used between a lockwasher and a fastener to provide a smooth bearing surface. This allows the fastener to be turned easily with a tool.

Cotter Pins

Cotter pins (**Figure 10**) are used to secure special kinds of fasteners. The threaded stud

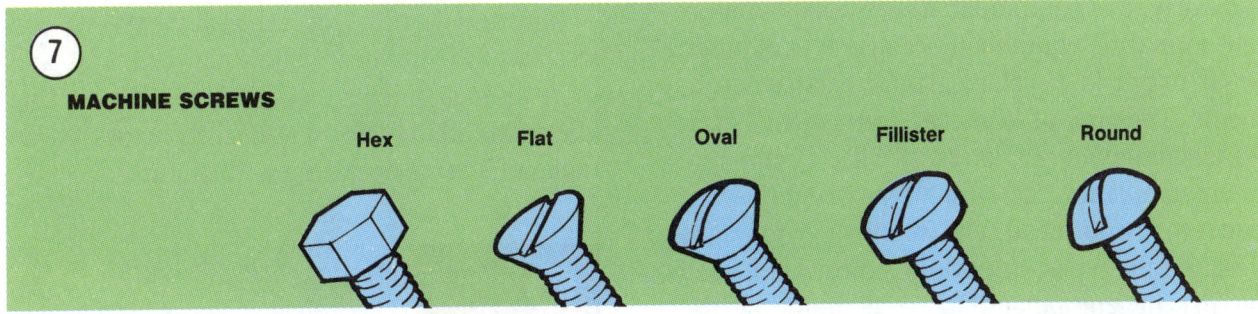

7 MACHINE SCREWS

Hex   Flat   Oval   Fillister   Round

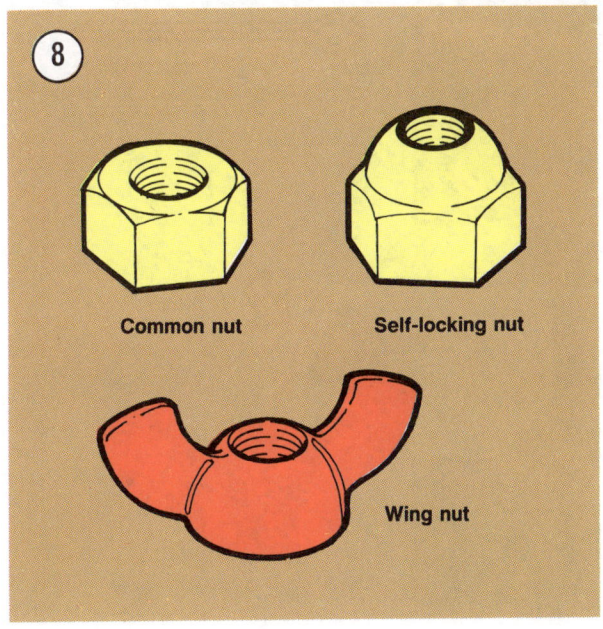

8

Common nut   Self-locking nut

Wing nut

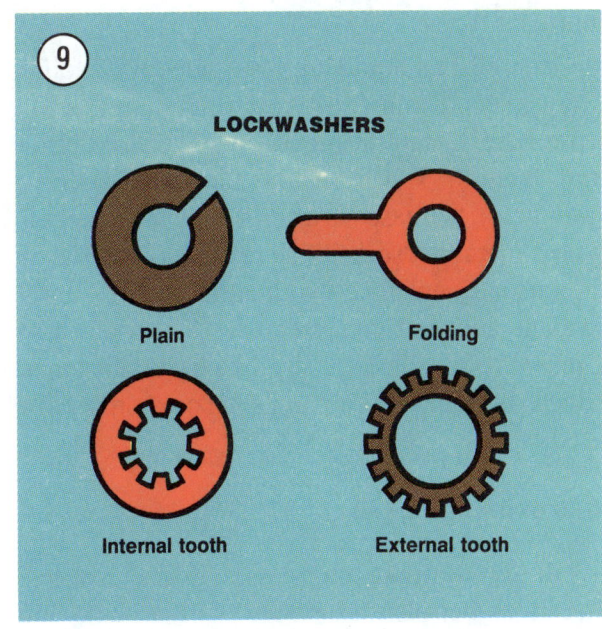

9 LOCKWASHERS

Plain   Folding

Internal tooth   External tooth

must have a hole in it; the nut or nut lock piece has projections that the cotter pin fits between. This type of nut is called a "Castellated nut." Cotter pins should not be reused after removal.

Snap Rings

Snap rings can be of an internal or external design. They are used to retain items on shafts (external type) or within tubes (internal type). Snap rings can be reused if they are not distorted during removal. In some applications, snap rings of varying thickness can be selected to control the end play of parts assemblies.

LUBRICANTS

Periodic lubrication ensures long service life for any type of equipment. It is especially important to marine equipment because it is exposed to salt or brackish water and other harsh environments. The *type* of lubricant used is just as important as the lubrication service itself; although, in an emergency, the wrong type of lubricant is better than none at all. The following paragraphs describe the types of lubricants most often used on marine equipment. Be sure to follow the equipment manufacturer's recommendations for lubricant types.

Generally, all liquid lubricants are called "oil." They may be mineral-based (including petroleum bases), natural-based (vegetable and animal bases), synthetic-based or emulsions (mixtures). "Grease" is an oil which is thickened with a metallic "soap." The resulting material is then usually enhanced with anticorrosion, antioxidant and extreme pressure (EP) additives. Grease is often classified by the type of thickener added; lithium and calcium soap are commonly used.

4-stroke Engine Oil

Oil for 4-stroke engines is graded by the American Petroleum Institute (API) and the So-

ciety of Automotive Engineers (SAE) in several categories. Oil containers display these ratings on the top or label (**Figure 11**).

API oil grade is indicated by letters, oils for gasoline engines are identified by an "S" and oils for diesel engines are identified by a "C." Most modern gasoline engines require SF or SG graded oil. Automotive and marine diesel engines use CC or CD graded oil.

Viscosity is an indication of the oil's thickness, or resistance to flow. The SAE uses numbers to indicate viscosity; thin oils have low numbers and thick oils have high numbers. A "W" after the number indicates that the viscosity testing was done at low temperature to simulate cold weather operation. Engine oils fall into the 5W-20W and 20-50 range.

Multi-grade oils (for example, 10W-40) are less viscous (thinner) at low temperatures and more viscous (thicker) at high temperatures. This allows the oil to perform efficiently across a wide range of engine operating temperatures.

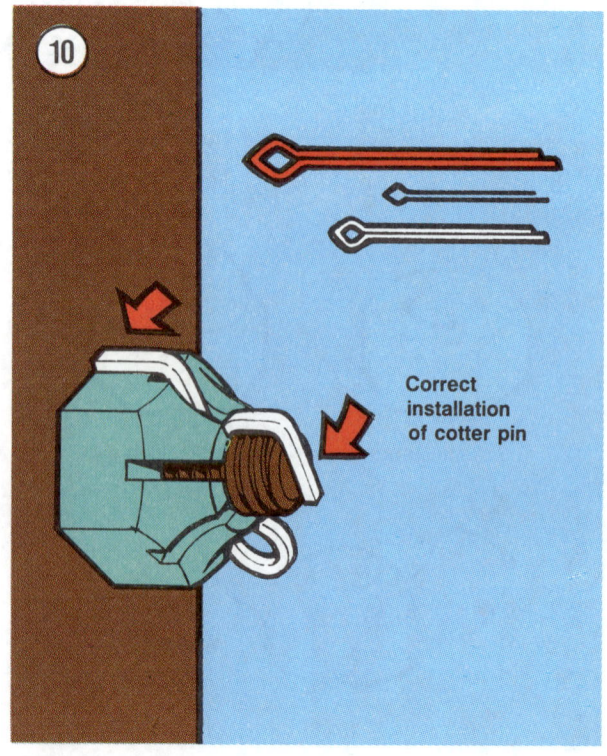

Correct installation of cotter pin

2-stroke Engine Oil

Lubrication for a 2-stroke engine is provided by oil mixed with the incoming fuel-air mixture. Some of the oil mist settles out in the crankcase, lubricating the crankshaft and lower end of the connecting rods. The rest of the oil enters the combustion chamber to lubricate the piston, rings and cylinder wall. This oil is then burned along with the fuel-air mixture during the combustion process.

Engine oil must have several special qualities to work well in a 2-stroke engine. It must mix easily and stay in suspension in gasoline. When burned, it can't leave behind excessive deposits. It must also be able to withstand the high temperatures associated with 2-stroke engines.

The National Marine Manufacturer's Association (NMMA) has set standards for oil used in 2-stroke, water-cooled engines. This is the NMMA TC-W (two-cycle, water-cooled) grade (**Figure 12**). The oil's performance in the following areas is evaluated:

   a. Lubrication (prevention of wear and scuffing).
   b. Spark plug fouling.
   c. Preignition.
   d. Piston ring sticking.
   e. Piston varnish.
   f. General engine condition (including deposits).
   g. Exhaust port blockage.
   h. Rust prevention.
   i. Mixing ability with gasoline.

In addition to oil grade, manufacturers specify the ratio of gasoline to oil required during break-in and normal engine operation.

Gear Oil

Gear lubricants are assigned SAE viscosity numbers under the same system as 4-stroke engine oil. Gear lubricant falls into the SAE 72-250 range (**Figure 13**). Some gear lubricants are multi-grade; for example, SAE 85W-90.

Three types of marine gear lubricant are generally available: SAE 90 hypoid gear lubricant is designed for older manual-shift units; Type C gear lubricant contains additives designed for electric shift mechanisms; High viscosity gear lubricant is a heavier oil designed to withstand the shock loading of high-performance engines or units subjected to severe duty use. Always use a gear lubricant of the type specified by the unit's manufacturer.

Grease

Greases are graded by the National Lubricating Grease Institute (NLGI). Greases are graded by number according to the consistency of the grease; these ratings range from No. 000 to No. 6, with No. 6 being the most solid. A typical multipurpose grease is NLGI No. 2 (**Figure 14**). For specific applications, equipment manufacturers may require grease with an additive such as molybdenum disulfide (MOS_2).

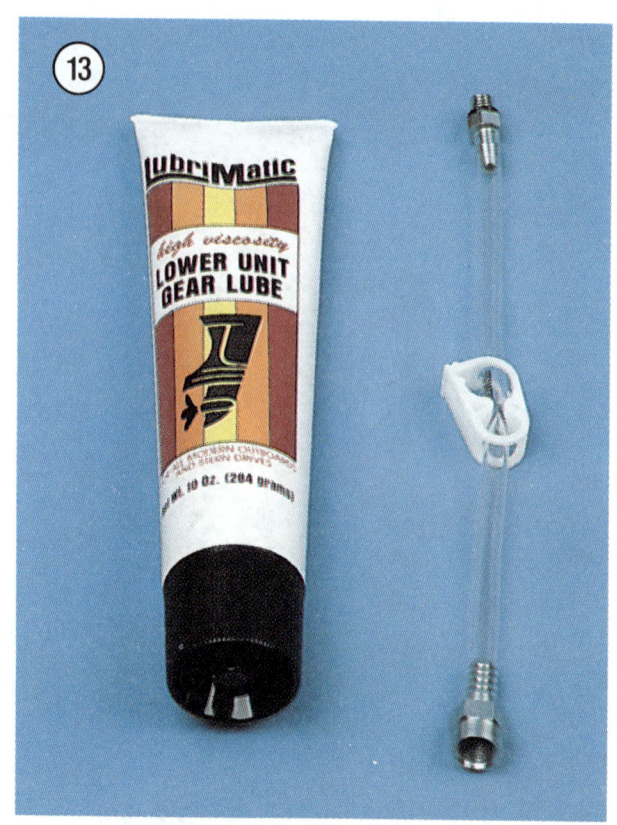

GASKET SEALANT

Gasket sealant is used instead of pre-formed gaskets on some applications, or as a gasket dressing on others. Two types of gasket sealant are commonly used: room temperature vulcanizing (RTV) and anaerobic. Because these two materials have different sealing properties, they cannot be used interchangeably.

RTV Sealant

This is a silicone gel supplied in tubes (**Figure 15**). Moisture in the air causes RTV to cure. Always place the cap on the tube as soon as possible when using RTV. RTV has a shelf life of one year and will not cure properly when the shelf life has expired. Check the expiration date

on RTV tubes before using and keep partially used tubes tightly sealed. RTV sealant can generally fill gaps up to 1/4 in. (6.3 mm) and works well on slightly flexible surfaces.

Applying RTV Sealant

Clean all gasket residue from mating surfaces. Surfaces should be clean and free of oil and dirt. Remove all RTV gasket material from blind attaching holes because it can create a "hydraulic" effect and affect bolt torque.

Apply RTV sealant in a continuous bead 2-3 mm (0.08-0.12 in.) thick. Circle all mounting holes unless otherwise specified. Torque mating parts within 10 minutes after application.

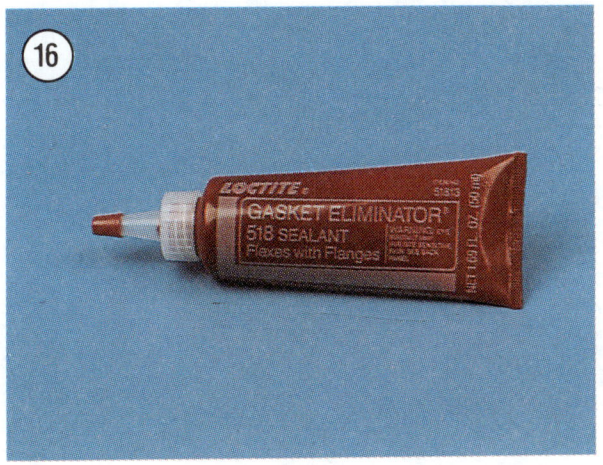

Anaerobic Sealant

This is a gel supplied in tubes (**Figure 16**). It cures only in the absence of air, as when squeezed tightly between two machined mating surfaces. For this reason, it will not spoil if the cap is left off the tube. It should not be used if one mating surface is flexible. Anaerobic sealant is able to fill gaps up to 0.030 in. (0.8 mm) and generally works best on rigid, machined flanges or surfaces.

Applying Anaerobic Sealant

Clean all gasket residue from mating surfaces. Surfaces must be clean and free of oil and dirt. Remove all gasket material from blind attaching holes, as it can cause a "hydraulic" effect and affect bolt torque.

Apply anaerobic sealant in a 1 mm or less (0.04 in.) bead to one sealing surface. Circle all mounting holes. Torque mating parts within 15 minutes after application.

GALVANIC CORROSION

A chemical reaction occurs whenever two different types of metal are joined by an electrical conductor and immersed in an electrolyte. Electrons transfer from one metal to the other through the electrolyte and return through the conductor.

The hardware on a boat is made of many different types of metal. The boat hull acts as a conductor between the metals. Even if the hull is wooden or fiberglass, the slightest film of water (electrolyte) within the hull provides conductivity. This combination creates a good environment for electron flow (**Figure 17**). Unfortunately, this electron flow results in galvanic corrosion of the metal involved, causing one of the metals to be corroded or eaten away by the process. The amount of electron flow (and, therefore, the amount of corrosion) depends on several factors:

  a. The types of metal involved.

  b. The efficiency of the conductor.

  c. The strength of the electrolyte.

Metals

The chemical composition of the metals used in marine equipment has a significant effect on the amount and speed of galvanic corrosion. Certain metals are more resistant to corrosion than others. These electrically negative metals are commonly called "noble;" they act as the cathode in any reaction. Metals that are more subject to corrosion are electrically positive; they act as the anode in a reaction. The more noble metals include titanium, 18-8 stainless steel and nickel. Less noble metals include zinc, aluminum and magnesium. Galvanic corrosion becomes more severe as the difference in electrical potential between the two metals increases.

In some cases, galvanic corrosion can occur within a single piece of metal. Common brass is a mixture of zinc and copper, and, when immersed in an electrolyte, the zinc portion of the mixture will corrode away as reaction occurs between the zinc and the copper particles.

Conductors

The hull of the boat often acts as the conductor between different types of metal. Marine equipment, such as an outboard motor or stern drive unit, can also act as the conductor. Large masses of metal, firmly connected together, are more efficient conductors than water. Rubber mountings and vinyl-based paint can act as insulators between pieces of metal.

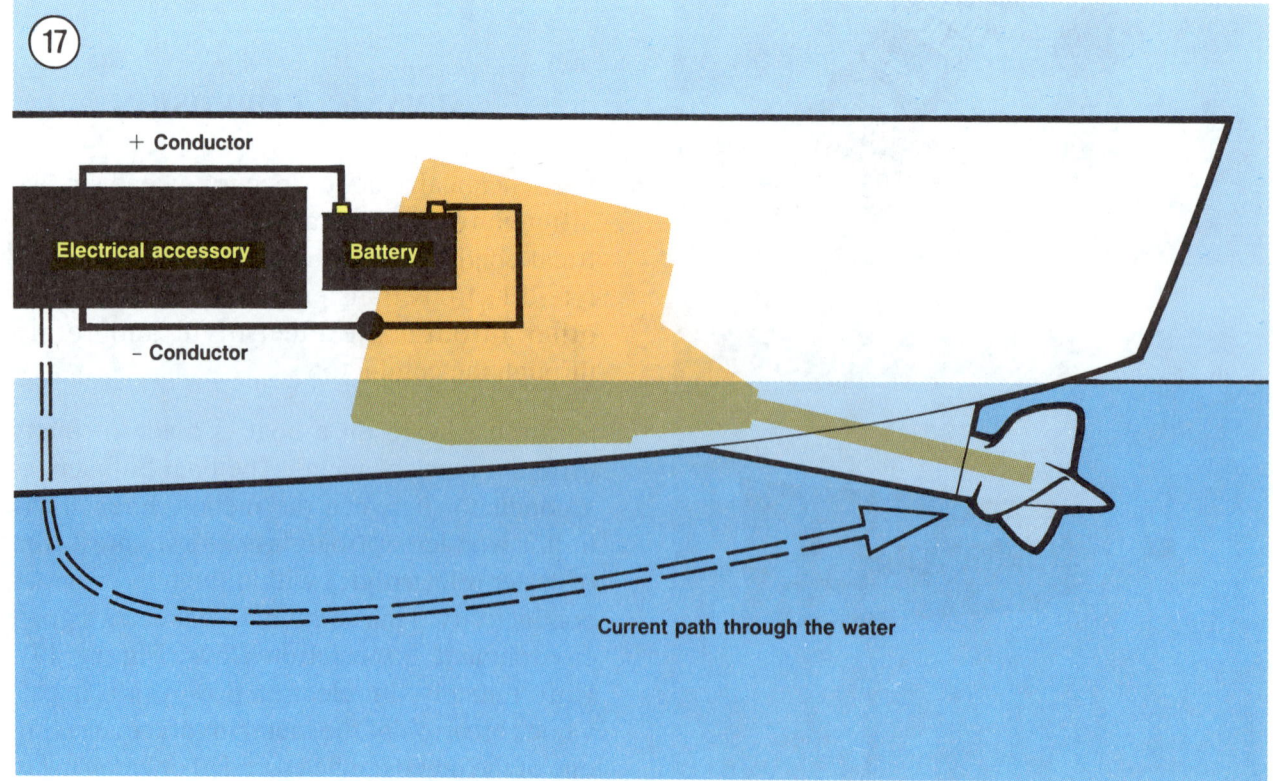

(17)

+ Conductor

Electrical accessory   Battery

– Conductor

Current path through the water

Electrolyte

The water in which a boat operates acts as the electrolyte for the galvanic corrosion process. The better a conductor the electrolyte is, the more severe and rapid the corrosion.

Cold, clean freshwater is the poorest electrolyte. As water temperature increases, its conductivity increases. Pollutants will increase conductivity; brackish or saltwater is also an efficient electrolyte. This is one of the reasons that most manufacturers recommend a freshwater flush for marine equipment after operation in saltwater, polluted or brackish water.

PROTECTION FROM GALVANIC CORROSION

Because of the environment in which marine equipment must operate, it is practically impossible to totally prevent galvanic corrosion. There are several ways by which the process can be slowed. After taking these precautions, the next step is to "fool" the process into occurring only where *you* want it to occur. This is the role of sacrificial anodes and impressed current systems.

Slowing Corrosion

Some simple precautions can help reduce the amount of corrosion taking place outside the hull. These are *not* a substitute for the corrosion protection methods discussed under *Sacrificial Anodes* and *Impressed Current Systems* in this chapter, but they can help these protection methods do their job.

Use fasteners of a metal more noble than the part they are fastening. If corrosion occurs, the larger equipment will suffer but the fastener will be protected. Because fasteners are usually very small in comparison to the equipment being fastened, the equipment can survive the loss of material. If the fastener were to corrode instead of the equipment, major problems could arise.

Keep all painted surfaces in good condition. If paint is scraped off and bare metal exposed, corrosion will rapidly increase. Use a vinyl- or plastic-based paint, which acts as an electrical insulator.

Be careful when using metal-based antifouling paints. These should not be applied to metal parts of the boat, outboard motor or stern drive unit or they will actually react with the equipment, causing corrosion between the equipment and the layer of paint. Organic-based paints are available for use on metal surfaces.

Where a corrosion protection device is used, remember that it must be immersed in the electrolyte along with the rest of the boat to have any effect. If you raise the power unit out of the water when the boat is docked, any anodes on the power unit will be removed from the corrosion cycle and will not protect the rest of the equipment that is still immersed. Also, such corrosion protection devices must not be painted because this would insulate them from the corrosion process.

Any change in the boat's equipment, such as the installation of a new stainless steel propeller, will change the electrical potential and could cause increased corrosion. Keep in mind that when you add new equipment or change materials, you should review your corrosion protection system to be sure it is up to the job.

Sacrificial Anodes

Anodes are usually made of zinc, a far from noble metal. Sacrificial anodes are specially designed to do nothing but corrode. Properly fastening such pieces to the boat will cause them to act as the anode in *any* galvanic reaction that occurs; any other metal present will act as the cathode and will not be damaged.

Anodes must be used properly to be effective. Simply fastening pieces of zinc to your boat in random locations won't do the job.

You must determine how much anode surface area is required to adequately protect the equipment's surface area. A good starting point is provided by Military Specification MIL-A-818001, which states that one square inch of new anode will protect either:

a. 800 square inches of freshly painted steel.
b. 250 square inches of bare steel or bare aluminum alloy.
c. 100 square inches of copper or copper alloy.

This rule is for a boat at rest. When underway, more anode area is required to protect the same equipment surface area.

The anode must be fastened so that it has good electrical contact with the metal to be protected. If possible, the anode can be attached directly to the other metal. If that is not possible, the entire network of metal parts in the boat should be electrically bonded together so that all pieces are protected.

Good quality anodes have inserts of some other metal around the fastener holes. Otherwise, the anode could erode away around the fastener. The anode can then become loose or even fall off, removing all protection.

Another Military Specification (MIL-A-18001) defines the type of alloy preferred that will corrode at a uniform rate without forming a crust that could reduce its efficiency after a time.

Impressed Current Systems

An impressed current system can be installed on any boat that has a battery. The system consists of an anode, a control box and a sensor. The anode in this system is coated with a very noble metal, such as platinum, so that it is almost corrosion-free and will last indefinitely. The sensor, under the boat's waterline, monitors the potential for corrosion. When it senses that

corrosion could be occurring, it transmits this information to the control box.

The control box connects the boat's battery to the anode. When the sensor signals the need, the control box applies positive battery voltage to the anode. Current from the battery flows from the anode to all other metal parts of the boat, no matter how noble or non-noble these parts may be. This battery current takes the place of any galvanic current flow.

Only a very small amount of battery current is needed to counteract galvanic corrosion. Manufacturers estimate that it would take two or three months of constant use to drain a typical marine battery, assuming the battery is never recharged.

An impressed current system is more expensive to install than simple anodes but, considering its low maintenance requirements and the excellent protection it provides, the long-term cost may actually be lower.

PROPELLERS

The propeller is the final link between the boat's drive system and the water. A perfectly

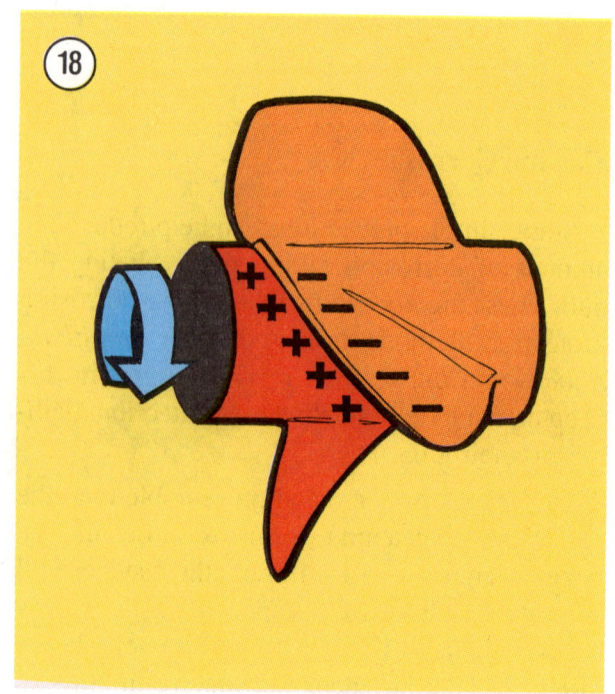

maintained engine and hull are useless if the propeller is the wrong type or has been allowed to deteriorate. Although propeller selection for a specific situation is beyond the scope of this book, the following information on propeller construction and design will allow you to discuss the subject intelligently with your marine dealer.

How a Propeller Works

As the curved blades of a propeller rotate through the water, a high-pressure area is created on one side of the blade and a low-pressure area exists on the other side of the blade (**Figure 18**). The propeller moves toward the low-pressure area, carrying the boat with it.

Propeller Parts

Although a propeller may be a one-piece unit, it is made up of several different parts (**Figure 19**). Variations in the design of these parts make different propellers suitable for different jobs.

The blade tip is the point on the blade farthest from the center of the propeller hub. The blade

tip separates the leading edge from the trailing edge.

The leading edge is the edge of the blade nearest to the boat. During normal rotation, this is the area of the blade that first cuts through the water.

The trailing edge is the edge of the blade farthest from the boat.

The blade face is the surface of the blade that faces away from the boat. During normal rotation, high pressure exists on this side of the blade.

The blade back is the surface of the blade that faces toward the boat. During normal rotation, low pressure exists on this side of the blade.

The cup is a small curve or lip on the trailing edge of the blade.

The hub is the central portion of the propeller. It connects the blades to the propeller shaft (part of the boat's drive system). On some drive systems, engine exhaust is routed through the hub; in this case, the hub is made up of an outer and an inner portion, connected by ribs.

The diffuser ring is used on through-hub exhaust models to prevent exhaust gases from entering the blade area.

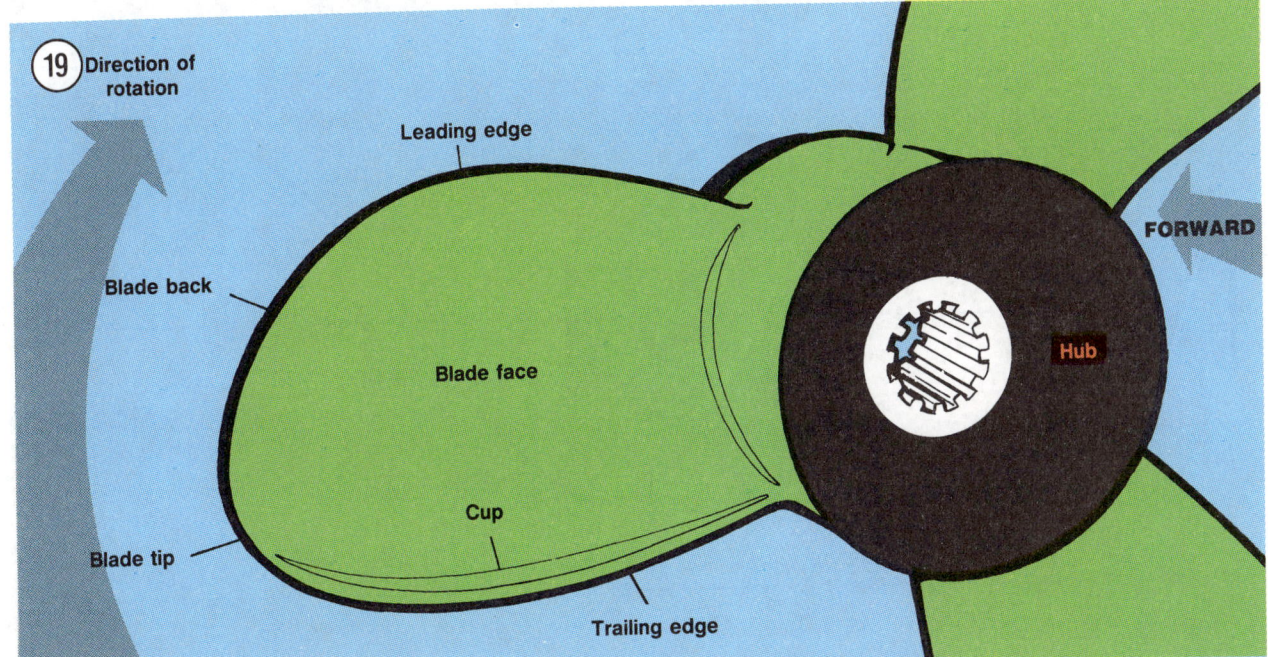

(19) Direction of rotation

Leading edge

FORWARD

Blade back

Hub

Blade face

Cup

Blade tip

Trailing edge

Propeller Design

Changes in length, angle, thickness and material of propeller parts make different propellers suitable for different situations.

Diameter

Propeller diameter is the distance from the center of the hub to the blade tip, multiplied by

2. That is, it is the diameter of the circle formed by the blade tips during propeller rotation (**Figure 20**).

Pitch and rake

Propeller pitch and rake describe the placement of the blade in relation to the hub (**Figure 21**).

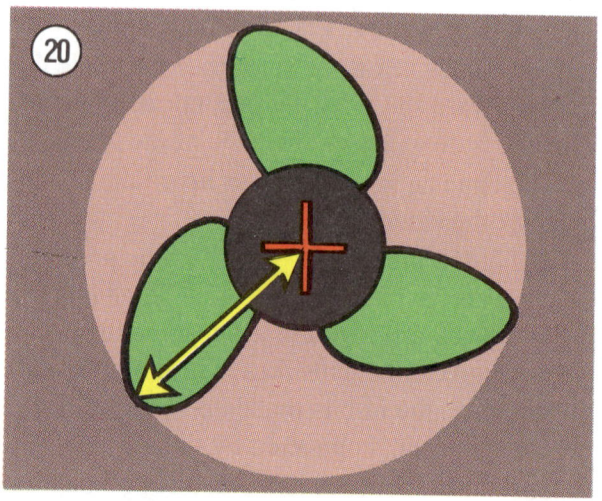

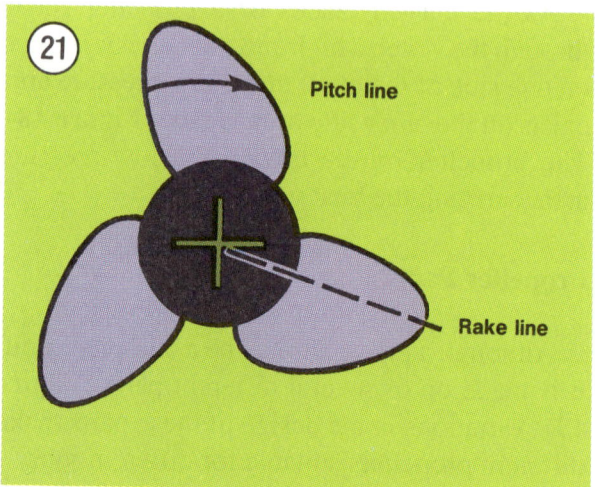

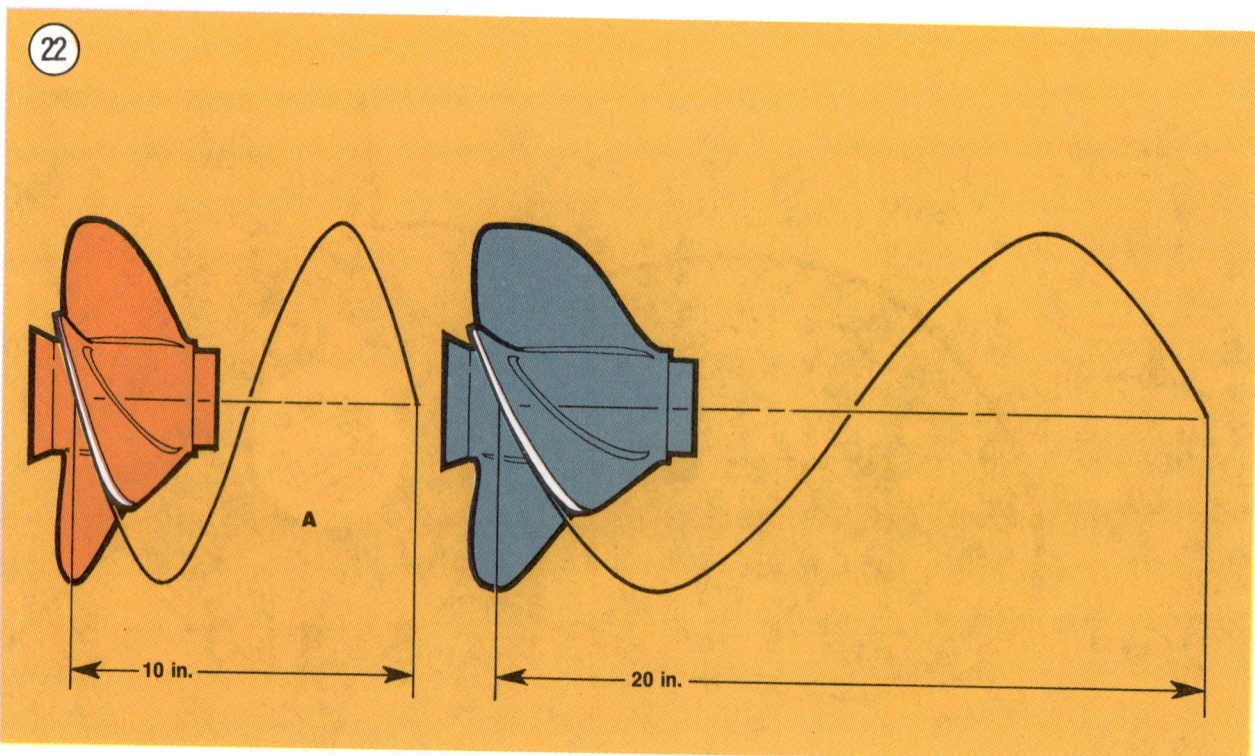

Pitch is expressed by the theoretical distance that the propeller would travel in one revolution. In A, **Figure 22**, the propeller would travel 10 inches in one revolution. In B, **Figure 22**, the propeller would travel 20 inches in one revolution. This distance is only theoretical; during actual operation, the propeller achieves about 80% of its rated travel.

Propeller blades can be constructed with constant pitch (**Figure 23**) or progressive pitch (**Figure 24**). Progressive pitch starts low at the leading edge and increases toward to trailing edge. The propeller pitch specification is the average of the pitch across the entire blade.

Blade rake is specified in degrees and is measured along a line from the center of the hub to the blade tip. A blade that is perpendicular to the hub (A, **Figure 25**) has 0° of rake. A blade that is angled from perpendicular (B, **Figure 25**) has a rake expressed by its difference from perpen-

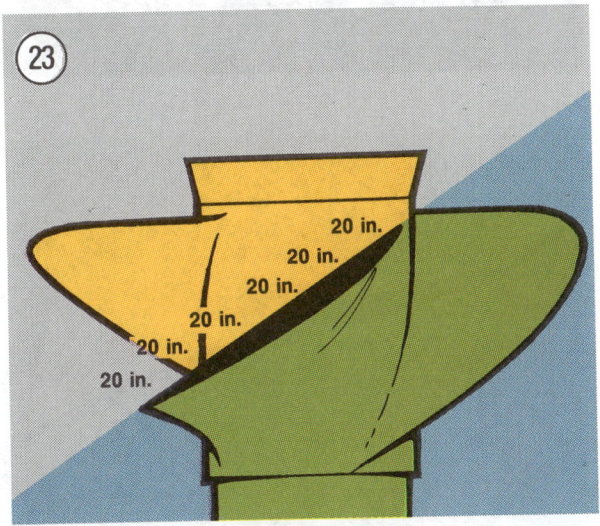

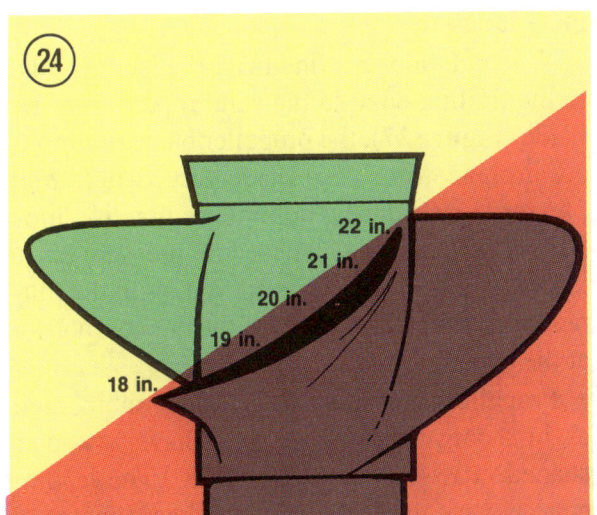

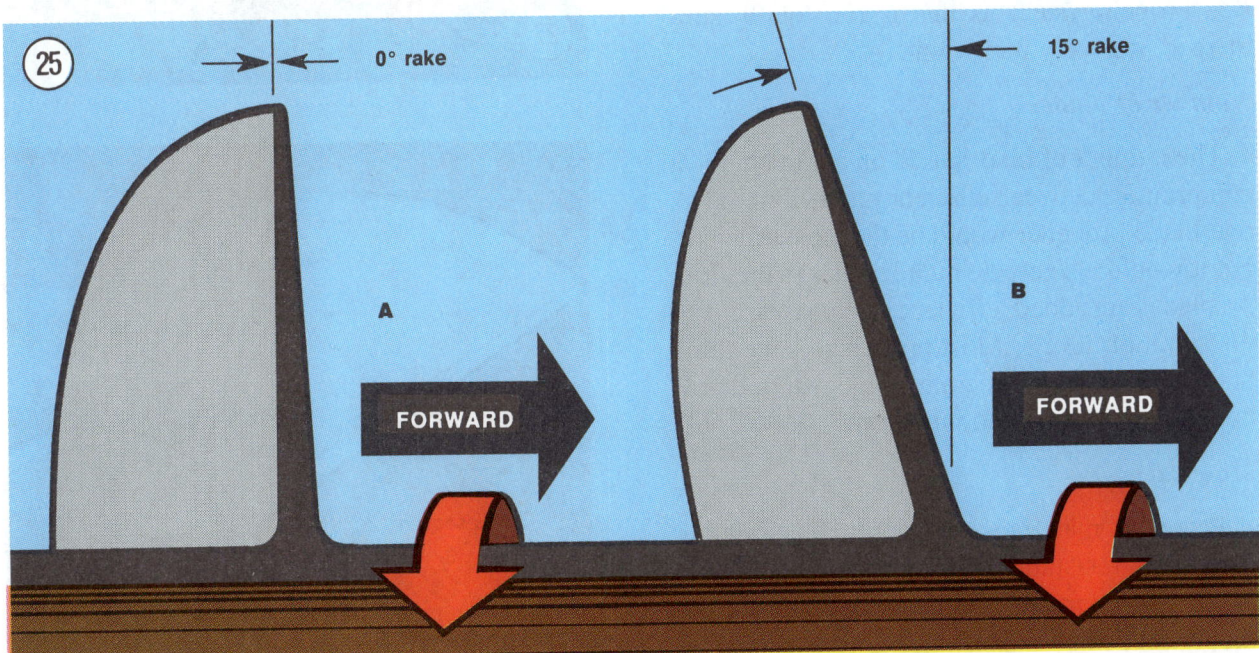

dicular. Most propellers have rakes ranging from 0-20°.

Blade thickness

Blade thickness is not uniform at all points along the blade. For efficiency, blades should be as thin as possible at all points while retaining enough strength to move the boat. Blades tend to be thicker where they meet the hub and thinner at the blade tip (**Figure 26**). This is to support the heavier loads at the hub section of the blade. This thickness is dependent on the strength of the material used.

When cut along a line from the leading edge to the trailing edge in the central portion of the blade (**Figure 27**), the propeller blade resembles an airplane wing. The blade face, where high pressure exists during normal rotation, is almost flat. The blade back, where low pressure exists during normal rotation, is curved, with the thinnest portions at the edges and the thickest portion at the center.

Propellers that run only partially submerged, as in racing applications, may have a wedge-shaped cross-section (**Figure 28**). The leading edge is very thin; the blade thickness increases toward the trailing edge, where it is the thickest. If a propeller such as this is run totally submerged, it is very inefficient.

Number of blades

The number of blades used on a propeller is a compromise between efficiency and vibration. A one-blade propeller would be the most efficient, but it would also create high levels of vibration. As blades are added, efficiency decreases, but so do vibration levels. Most propellers have three blades, representing the most practical trade-off between efficiency and vibration.

Material

Propeller materials are chosen for strength, corrosion resistance and economy. Stainless steel, aluminum and bronze are the most commonly used materials. Bronze is quite strong but

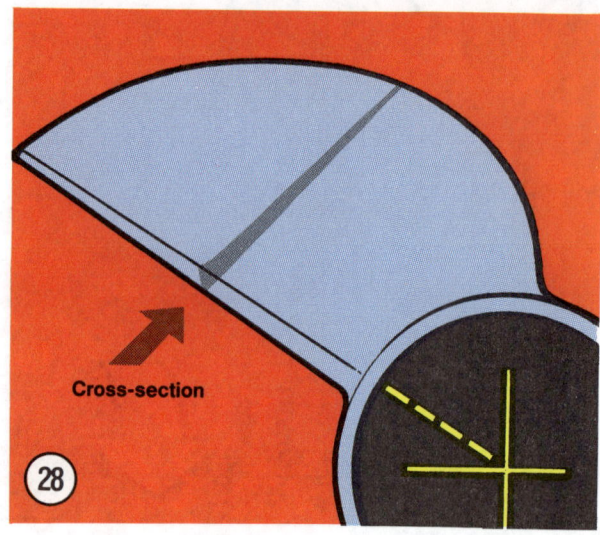

Cross-section

rather expensive. Stainless steel is more common than bronze because of its combination of strength and lower cost. Aluminum alloys are the least expensive but usually lack the strength of steel. Plastic propellers may be used in some low horsepower applications.

Direction of rotation

Propellers are made for both right-hand and left-hand rotation although right-hand is the most commonly used. When seen from behind the boat in forward motion, a right-hand propeller turns clockwise and a left-hand propeller turns counterclockwise. Off the boat, you can tell the difference by observing the angle of the blades (**Figure 29**). A right-hand propeller's blades slant from the upper left to the lower right; a left-hand propeller's blades are the opposite.

Cavitation and Ventilation

Cavitation and ventilation are *not* interchangeable terms; they refer to two distinct problems encountered during propeller operation.

To understand cavitation, you must first understand the relationship between pressure and the boiling point of water. At sea level, water will boil at 212° F. As pressure increases, such as within an engine's closed cooling system, the boiling point of water increases—it will boil at some temperature higher than 212° F. The opposite is also true. As pressure decreases, water will boil at a temperature lower than 212° F. If pressure drops low enough, water will boil at typical ambient temperatures of 50-60° F.

We have said that, during normal propeller operation, low-pressure exists on the blade back. Normally, the pressure does not drop low enough for boiling to occur. However, poor blade design

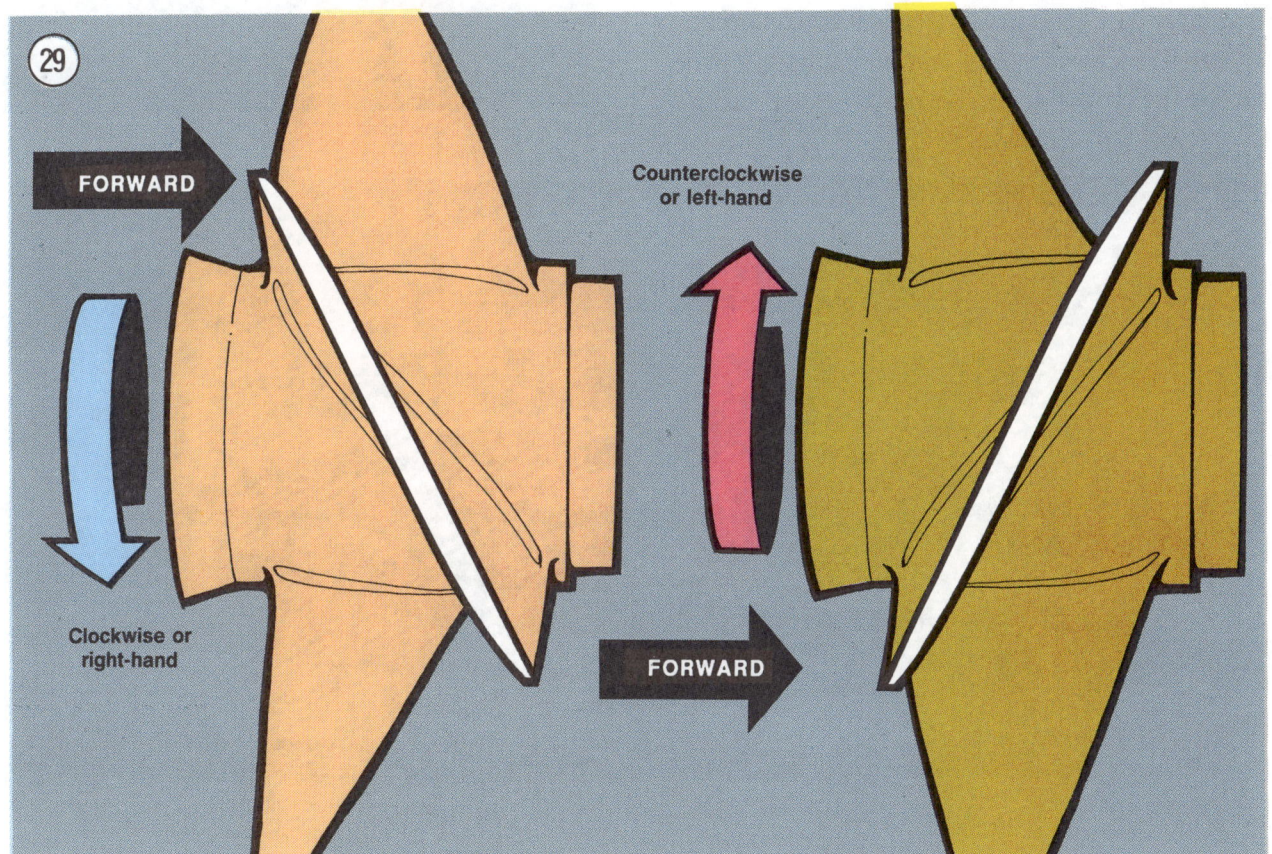

or selection, or blade damage can cause an unusual pressure drop on a small area of the blade (**Figure 30**). Boiling can occur in this small area. As the water boils, air bubbles form. As the boiling water passes to a higher pressure area of the blade, the boiling stops and the bubbles collapse. The collapsing bubbles release enough energy to erode the surface of the blade.

This entire process of pressure drop, boiling and bubble collapse is called "cavitation." The damage caused by the collapsing bubbles is called a "cavitation burn." It is important to remember that cavitation is caused by a decrease in pressure, *not* an increase in temperature.

Ventilation is not as complex a process as cavitation. Ventilation refers to air entering the blade area, either from above the surface of the water or from a through-hub exhaust system. As the blades meet the air, the propeller momentarily over-revs, losing most of its thrust. An added complication is that as the propeller over-revs, pressure on the blade back decreases and massive cavitation can occur.

Most pieces of marine equipment have a plate above the propeller area designed to keep surface air from entering the blade area (**Figure 31**). This plate is correctly called an "antiventilation plate," although you will often *see* it called an "anticavitation plate." Through hub exhaust systems also have specially designed hubs to keep exhaust gases from entering the blade area.

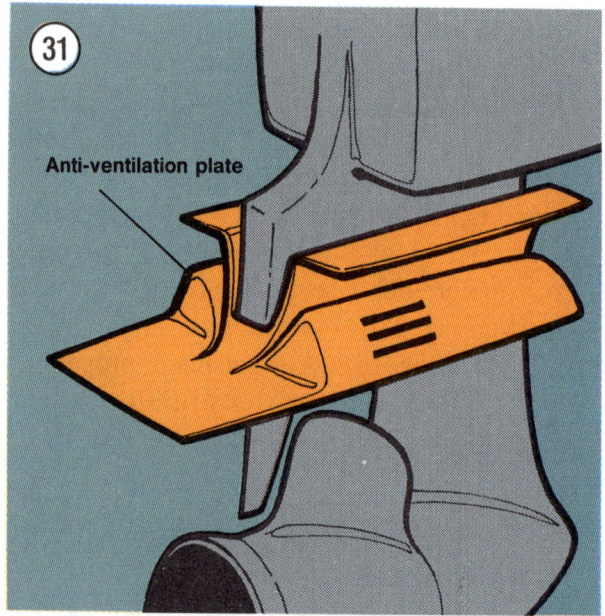

Anti-ventilation plate

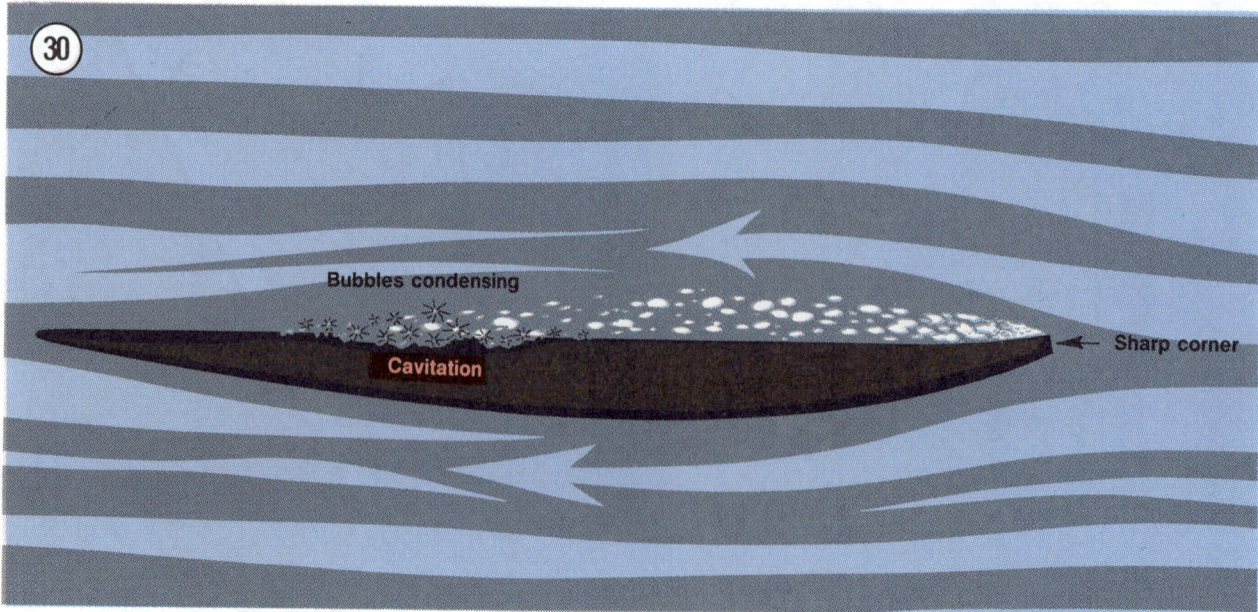

Bubbles condensing

Cavitation

Sharp corner

Chapter Two

Tools and Techniques

This chapter describes the common tools required for marine equipment repairs and troubleshooting. Techniques that will make your work easier and more effective are also described. Some of the procedures in this book require special skills or expertise; in some cases, you are better off entrusting the job to a dealer or qualified specialist.

SAFETY FIRST

Professional mechanics can work for years and never suffer a serious injury. If you follow a few rules of common sense and safety, you too can enjoy many safe hours servicing your marine equipment. If you ignore these rules, you can hurt yourself or damage the equipment.

1. Never use gasoline as a cleaning solvent.
2. Never smoke or use a torch near flammable liquids, such as cleaning solvent. If you are working in your home garage, remember that your home gas appliances have pilot lights.
3. Never smoke or use a torch in an area where batteries are being charged. Highly explosive hydrogen gas is formed during the charging process.

4. Use the proper size wrenches to avoid damage to fasteners and injury to yourself.
5. When loosening a tight or stuck fastener, think of what would happen if the wrench should slip. Protect yourself accordingly.
6. Keep your work area clean, uncluttered and well lighted.
7. Wear safety goggles during all operations involving drilling, grinding or the use of a cold chisel.
8. Never use worn tools.
9. Keep a Coast Guard approved fire extinguisher handy. Be sure it is rated for gasoline (Class B) and electrical (Class C) fires.

BASIC HAND TOOLS

A number of tools are required to maintain marine equipment. You may already have some of these tools for home or car repairs. There are also tools made especially for marine equipment repairs; these you will have to purchase. In any case, a wide variety of quality tools will make repairs easier and more effective.

Keep your tools clean and in a tool box. Keep them organized with the sockets and related drives together, the open end and box wrenches together, etc. After using a tool, wipe off dirt and grease with a clean cloth and place the tool in its correct place.

The following tools are required to perform virtually any repair job. Each tool is described and the recommended size given for starting a tool collection. Additional tools and some duplications may be added as you become more familiar with the equipment. You may need all standard U.S. size tools, all metric size tools or a mixture of both.

Screwdrivers

The screwdriver is a very basic tool, but if used improperly, it will do more damage than good. The slot on a screw has a definite dimension and shape. A screwdriver must be selected to conform with that shape. Use a small screwdriver for small screws and a large one for large screws or the screw head will be damaged.

Two types of screwdriver are commonly required: a common (flat-blade) screwdriver (**Figure 1**) and Phillips screwdrivers (**Figure 2**).

Screwdrivers are available in sets, which often include an assortment of common and Phillips blades. If you buy them individually, buy at least the following:

    a. Common screwdriver—5/16 × 6 in. blade.
    b. Common screwdriver—3/8 × 12 in. blade
    c. Phillips screwdriver—size 2 tip, 6 in. blade.

Use screwdrivers only for driving screws. Never use a screwdriver for prying or chiseling. Do not try to remove a Phillips or Allen head screw with a common screwdriver; you can damage the head so that the proper tool will be unable to remove it.

Keep screwdrivers in the proper condition and they will last longer and perform better. Always keep the tip of a common screwdriver in good condition. **Figure 3** shows how to grind the tip to the proper shape if it becomes damaged. Note the parallel sides of the tip.

Pliers

Pliers come in a wide range of types and sizes. Pliers are useful for cutting, bending and crimping. They should never be used to cut hardened objects or to turn bolts or nuts. **Figure 4** shows several types of pliers.

Each type of pliers has a specialized function. General purpose pliers are used mainly for holding things and for bending. Locking pliers are used as pliers or to hold objects very tightly, like a vise. Needlenose pliers are used to hold or bend small objects. Adjustable or slip-joint pliers can

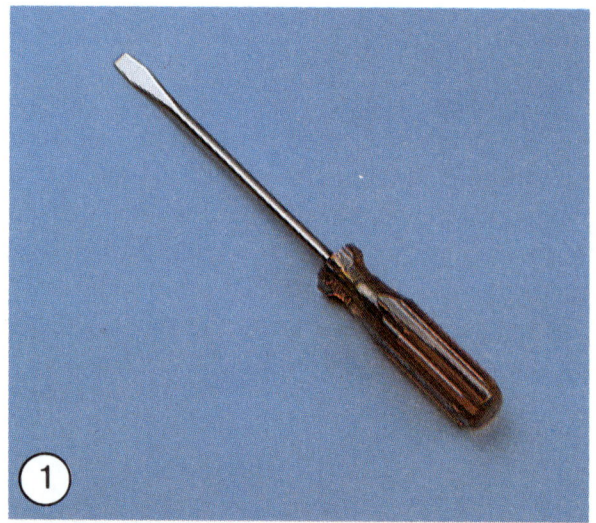

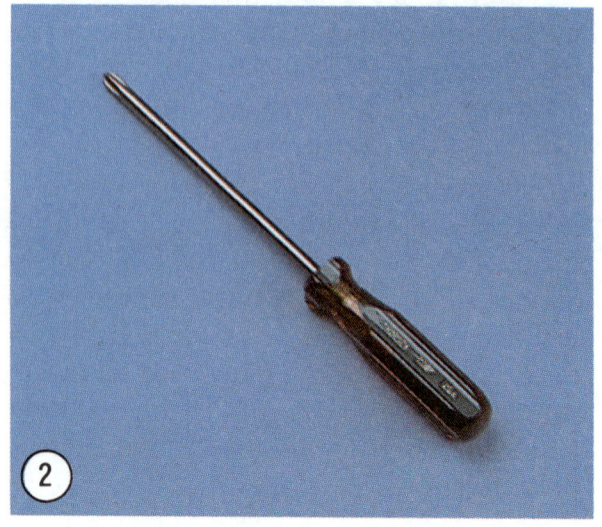

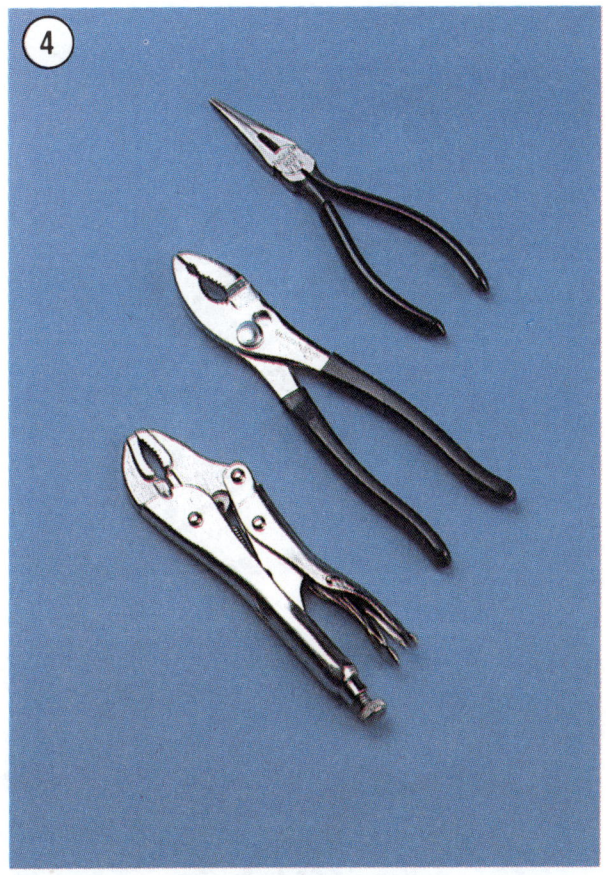

be adjusted to hold various sizes of objects; the jaws remain parallel to grip around objects such as pipe or tubing. There are many more types of pliers. The ones described here are the most commonly used.

Box and Open-end Wrenches

Box and open-end wrenches are available in sets or separately in a variety of sizes. See **Figure 5** and **Figure 6**. The number stamped near the end refers to the distance between two parallel flats on the hex head bolt or nut.

Box wrenches are usually superior to open-end wrenches. An open-end wrench grips the nut on only two flats. Unless it fits well, it may slip and round off the points on the nut. The box wrench grips all 6 flats. Both 6-point and 12-point openings on box wrenches are available. The 6-point gives superior holding power; the 12-point allows a shorter swing.

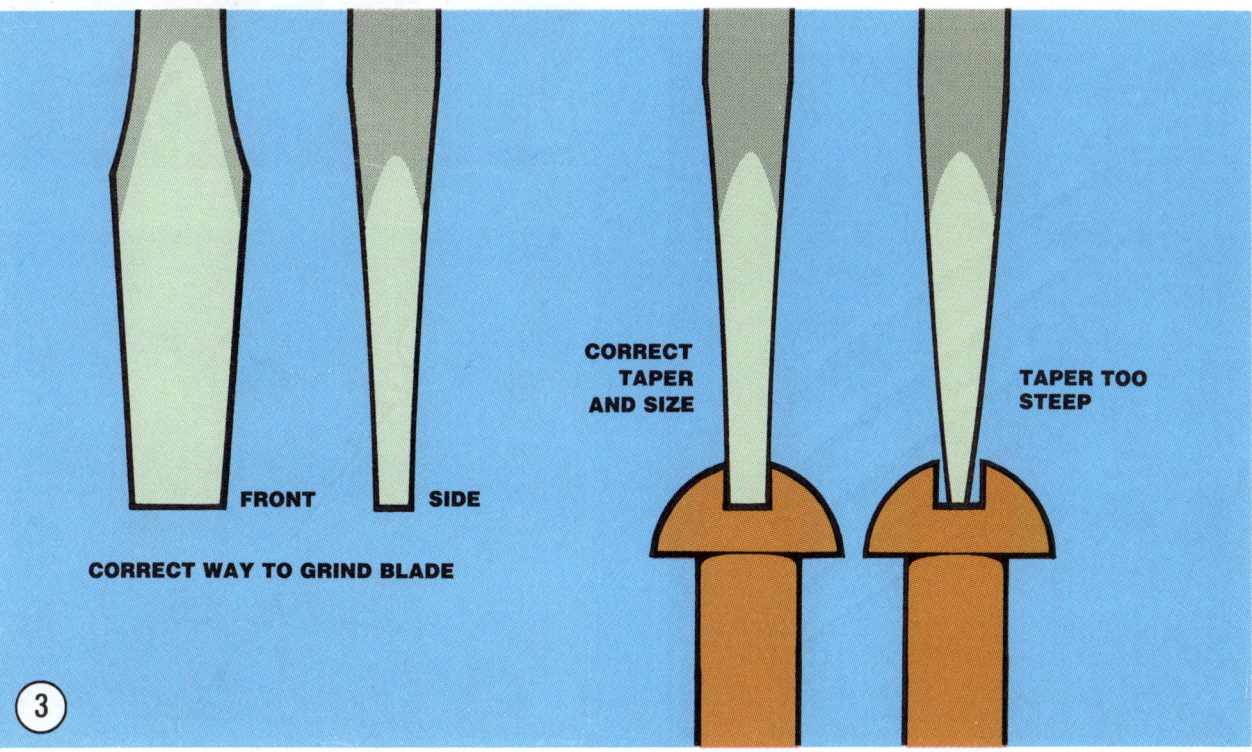

FRONT   SIDE

CORRECT WAY TO GRIND BLADE

CORRECT TAPER AND SIZE

TAPER TOO STEEP

Combination wrenches, which are open on one side and boxed on the other, are also available. Both ends are the same size.

Adjustable Wrenches

An adjustable wrench can be adjusted to fit nearly any nut or bolt head. See **Figure 7**. However, it can loosen and slip, causing damage to the nut and maybe to your knuckles. Use an adjustable wrench only when other wrenches are not available.

Adjustable wrenches come in sizes ranging from 4-18 in. overall. A 6 or 8 in. wrench is recommended as an all-purpose wrench.

Socket Wrenches

This type is undoubtedly the fastest, safest and most convenient to use. See **Figure 8**. Sockets, which attach to a suitable handle, are available with 6-point or 12-point openings and use 1/4, 3/8 and 3/4 inch drives. The drive size indicates

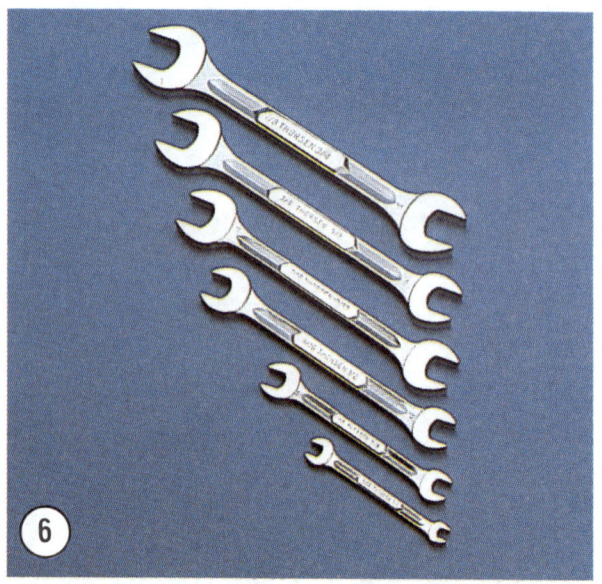

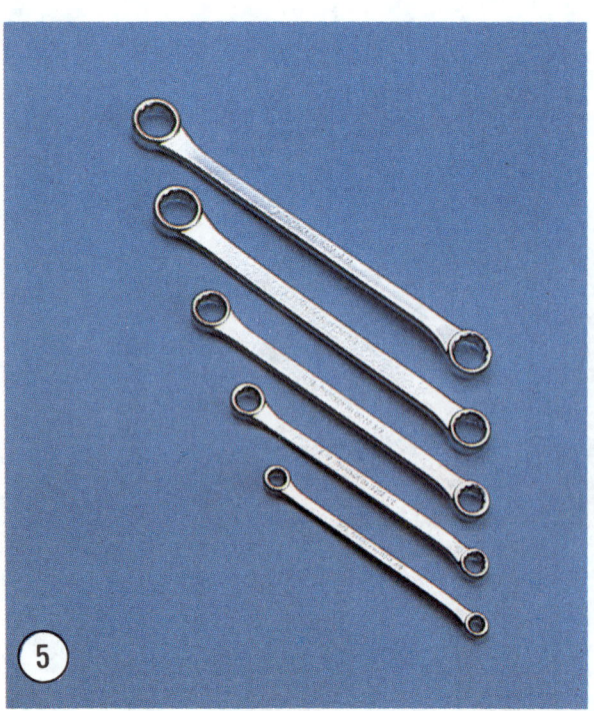

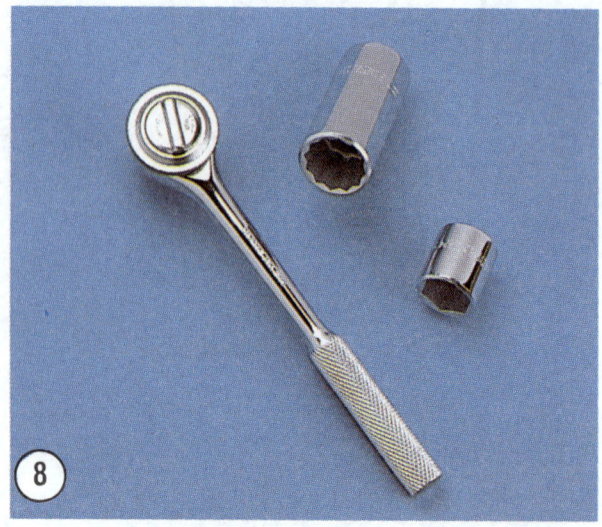

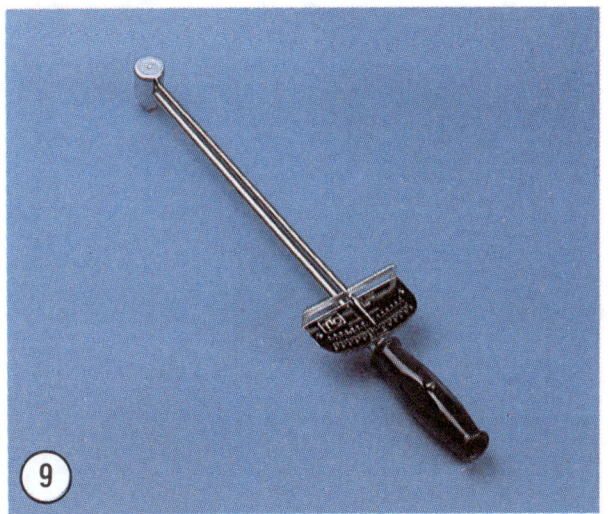

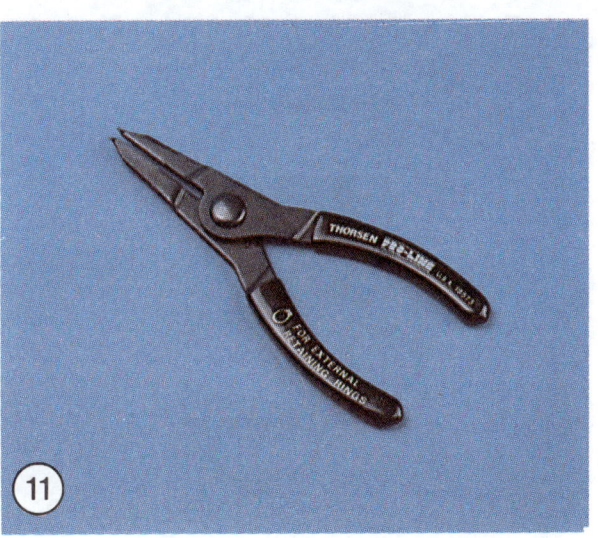

the size of the square hole that mates with the ratchet or flex handle.

Torque Wrench

A torque wrench (**Figure 9**) is used with a socket to measure how tight a nut or bolt is installed. They come in a wide price range and with either 3/8 or 1/2 in. square drive. The drive size indicates the size of the square drive that mates with the socket. Purchase one that measures up to 150 ft.-lb. (203 N•m).

Impact Driver

This tool (**Figure 10**) makes removal of tight fasteners easy and eliminates damage to bolts and screw slots. Impact drivers and interchangeable bits are available at most large hardware and auto parts stores.

Circlip Pliers

Circlip pliers (sometimes referred to as snap-ring pliers) are necessary to remove circlips. See **Figure 11**. Circlip pliers usually come with several different size tips; many designs can be switched from internal type to external type.

Hammers

The correct hammer is necessary for repairs. Use only a hammer with a face (or head) of rubber or plastic or the soft-faced type that is filled with buckshot (**Figure 12**). These are sometimes necessary in engine tear-downs. *Never* use a metal-faced hammer as severe damage will result in most cases. You can always produce the same amount of force with a soft-faced hammer.

Feeler Gauge

This tool has either flat or wire measuring gauges (**Figure 13**). Wire gauges are used to measure spark plug gap; flat gauges are used for all other measurements. A non-magnetic (brass) gauge may be specified when working around magnetized parts.

Other Special Tools

Some procedures require special tools; these are identified in the appropriate chapter. Unless otherwise specified, the part number used in this book to identify a special tool is the marine equipment manufacturer's part number.

Special tools can usually be purchased through your marine equipment dealer. Some can be made locally by a machinist, often at a much lower price. You may find certain special tools at tool rental dealers. Don't use makeshift tools if you can't locate the correct special tool; you will probably cause more damage than good.

TEST EQUIPMENT

Multimeter

This instrument (**Figure 14**) is invaluable for electrical system troubleshooting and service. It combines a voltmeter, an ohmmeter and an ammeter into one unit, so it is often called a VOM.

Two types of multimeter are available, analog and digital. Analog meters have a moving needle with marked bands indicating the volt, ohm and amperage scales. The digital meter (DVOM) is ideally suited for troubleshooting because it is easy to read, more accurate than analog, contains internal overload protection, is auto-ranging (analog meters must be recalibrated each time the scale is changed) and has automatic polarity compensation.

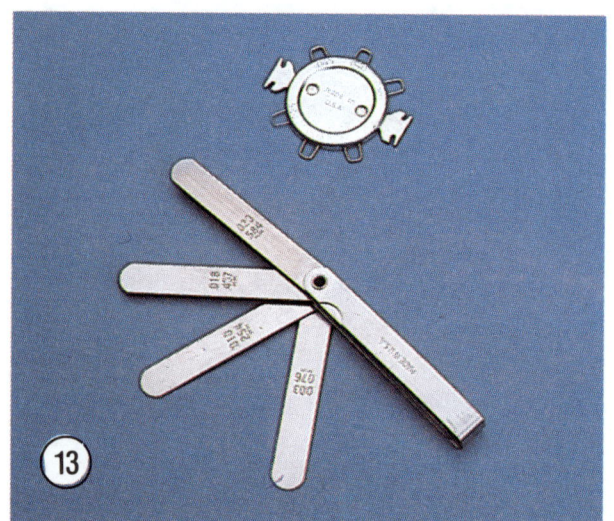

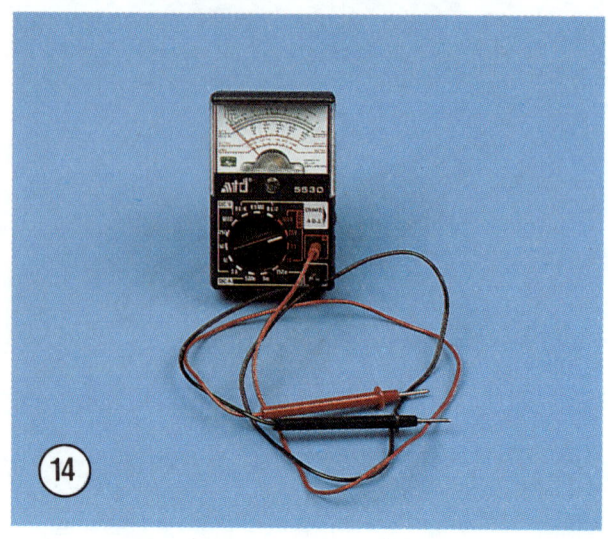

2

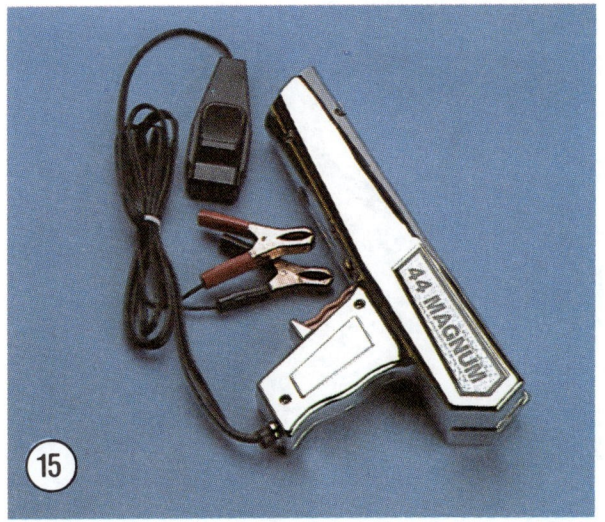

15

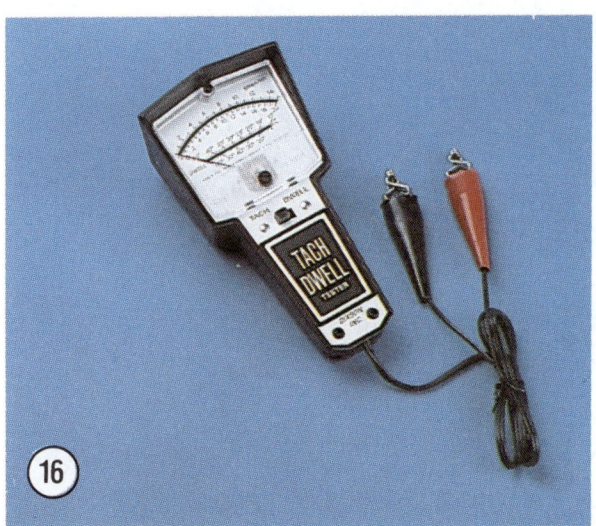

16

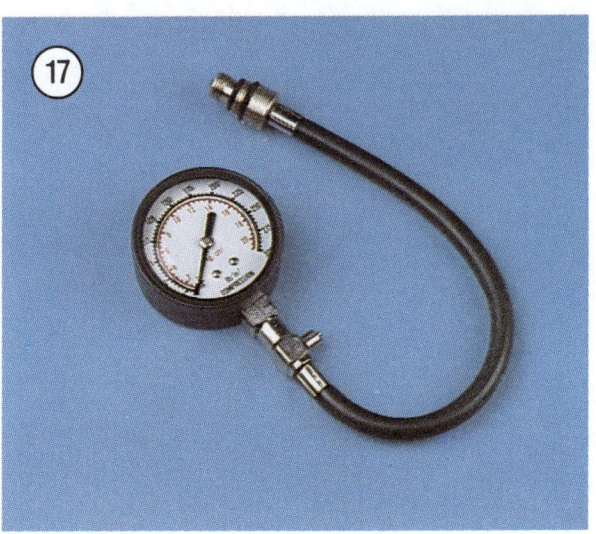

17

Strobe Timing Light

This instrument is necessary for dynamic tuning (setting ignition timing while the engine is running). By flashing a light at the precise instant the spark plug fires, the position of the timing mark can be seen. The flashing light makes a moving mark appear to stand still opposite a stationary mark.

Suitable lights range from inexpensive neon bulb types to powerful xenon strobe lights. See **Figure 15**. A light with an inductive pickup is best because it eliminates any possible damage to ignition wiring.

Tachometer/Dwell Meter

A portable tachometer is necessary for tuning. See **Figure 16**. Ignition timing and carburetor adjustments must be performed at the specified idle speed. The best instrument for this purpose is one with a low range of 0-1000 or 0-2000 rpm and a high range of 0-6000 rpm. Extended range (0-6000 or 0-8000 rpm) instruments lack accuracy at lower speeds. The instrument should be capable of detecting changes of 25 rpm on the low range.

A dwell meter is often combined with a tachometer. Dwell meters are used with breaker point ignition systems to measure the amount of time the points remain closed during engine operation.

Compression Gauge

This tool (**Figure 17**) measures the amount of pressure present in the engine's combustion chamber during the compression stroke. This indicates general engine condition. Compression readings can be interpreted along with vacuum gauge readings to pinpoint specific engine mechanical problems.

The easiest type to use has screw-in adapters that fit into the spark plug holes. Press-in rubber-tipped types are also available.

Vacuum Gauge

The vacuum gauge (**Figure 18**) measures the intake manifold vacuum created by the engine's intake stroke. Manifold and valve problems (on 4-stroke engines) can be identified by interpreting the readings. When combined with compression gauge readings, other engine problems can be diagnosed.

Some vacuum gauges can also be used as fuel pressure gauges to trace fuel system problems.

Hydrometer

Battery electrolyte specific gravity is measured with a hydrometer (**Figure 19**). The specific gravity of the electrolyte indicates the battery's state of charge. The best type has automatic temperature compensation; otherwise, you must calculate the compensation yourself.

Precision Measuring Tools

Various tools are needed to make precision measurements. A dial indicator (**Figure 20**), for example, is used to determine run-out of rotating parts and end play of parts assemblies. A dial indicator can also be used to precisely measure piston position in relation to top dead center; some engines require this measurement for ignition timing adjustment.

Vernier calipers (**Figure 21**) and micrometers (**Figure 22**) are other precision measuring tools used to determine the size of parts (such as piston diameter).

Precision measuring equipment must be stored, handled and used carefully or it will not remain accurate.

SERVICE HINTS

Most of the service procedures covered in this manual are straightforward and can be performed by anyone reasonably handy with tools.

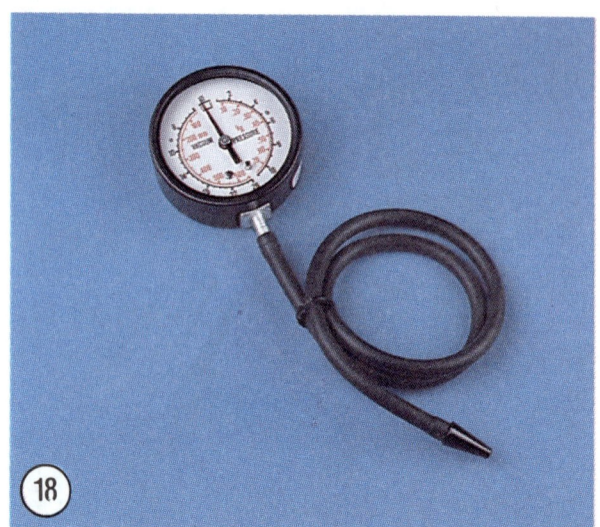

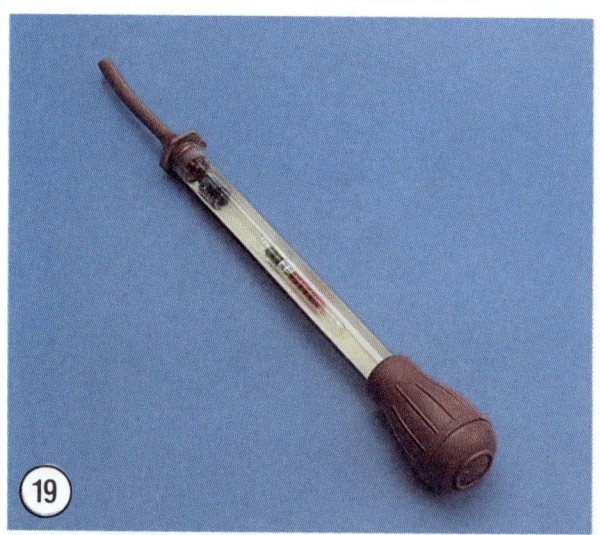

It is suggested, however, that you consider your own skills and toolbox carefully before attempting any operation involving major disassembly of the engine or gearcase.

Some operations, for example, require the use of a press. It would be wiser to have these performed by a shop equipped for such work, rather than trying to do the job yourself with makeshift equipment. Other procedures require precise measurements. Unless you have the skills and

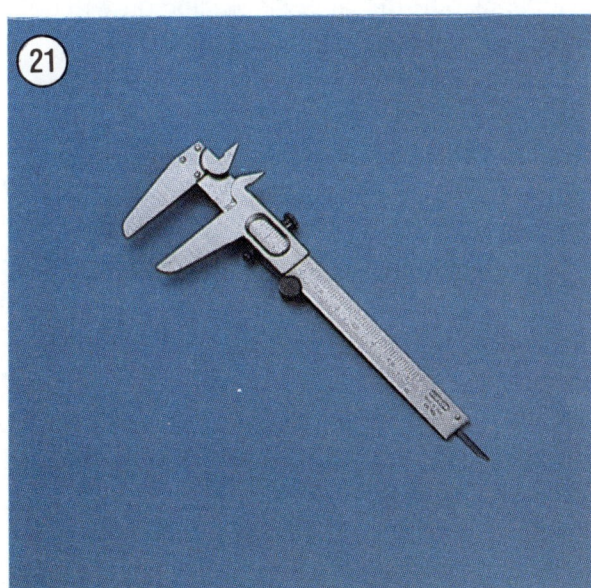

equipment required, it would be better to have a qualified repair shop make the measurements for you.

Preparation for Disassembly

Repairs go much faster and easier if the equipment is clean before you begin work. There are special cleaners, such as Gunk or Bel-Ray Degreaser, for washing the engine and related parts. Just spray or brush on the cleaning solution, let it stand, then rinse away with a garden hose. Clean all oily or greasy parts with cleaning solvent as you remove them.

> *WARNING*
> *Never use gasoline as a cleaning agent. It presents an extreme fire hazard. Be sure to work in a well-ventilated area when using cleaning solvent. Keep a Coast Guard approved fire extinguisher, rated for gasoline fires, handy in any case.*

Much of the labor charged for repairs made by dealers is for the removal and disassembly of other parts to reach the defective unit. It is frequently possible to perform the preliminary operations yourself and then take the defective unit in to the dealer for repair.

If you decide to tackle the job yourself, read the entire section in this manual that pertains to it, making sure you have identified the proper one. Study the illustrations and text until you have a good idea of what is involved in completing the job satisfactorily. If special tools or replacement parts are required, make arrangements to get them before you start. It is frustrating and time-consuming to get partly into a job and then be unable to complete it.

Disassembly Precautions

During disassembly of parts, keep a few general precautions in mind. Force is rarely needed to get things apart. If parts are a tight fit, such as a bearing in a case, there is usually a tool designed to separate them. Never use a screwdriver to pry apart parts with machined surfaces (such as cylinder heads and crankcases). You will mar the surfaces and end up with leaks.

Make diagrams (or take an instant picture) wherever similar-appearing parts are found. For example, head and crankcase bolts are often not the same length. You may think you can remember where everything came from, but mistakes are costly. There is also the possibility you may be sidetracked and not return to work for days or even weeks. In the interval, carefully laid out parts may have been disturbed.

Cover all openings after removing parts to keep small parts, dirt or other contamination from entering.

Tag all similar internal parts for location and direction. All internal components should be reinstalled in the same location and direction from which removed. Record the number and thickness of any shims as they are removed. Small parts, such as bolts, can be identified by placing them in plastic sandwich bags. Seal and label them with masking tape.

Wiring should be tagged with masking tape and marked as each wire is removed. Again, do not rely on memory alone.

Protect finished surfaces from physical damage or corrosion. Keep gasoline off painted surfaces.

Assembly Precautions

No parts, except those assembled with a press fit, require unusual force during assembly. If a part is hard to remove or install, find out why before proceeding.

When assembling two parts, start all fasteners, then tighten evenly in an alternating or crossing pattern if no specific tightening sequence is given.

When assembling parts, be sure all shims and washers are installed exactly as they came out.

Whenever a rotating part butts against a stationary part, look for a shim or washer. Use new gaskets if there is any doubt about the condition of the old ones. Unless otherwise specified, a thin coat of oil on gaskets may help them seal effectively.

Heavy grease can be used to hold small parts in place if they tend to fall out during assembly. However, keep grease and oil away from electrical components.

High spots may be sanded off a piston with sandpaper, but fine emery cloth and oil will do a much more professional job.

Carbon can be removed from the cylinder head, the piston crown and the exhaust port with a dull screwdriver. *Do not* scratch either surface. Wipe off the surface with a clean cloth when finished.

The carburetor is best cleaned by disassembling it and soaking the parts in a commercial carburetor cleaner. Never soak gaskets and rubber parts in these cleaners. Never use wire to clean out jets and air passages; they are easily damaged. Use compressed air to blow out the carburetor *after* the float has been removed.

Take your time and do the job right. Do not forget that the break-in procedure on a newly rebuilt engine is the same as that of a new one. Use the break-in oil recommendations and follow other instructions given in your owner's manual.

SPECIAL TIPS

Because of the extreme demands placed on marine equipment, several points should be kept in mind when performing service and repair. The following items are general suggestions that may improve the overall life of the machine and help avoid costly failures.

1. Unless otherwise specified, use a locking compound, such as Loctite Threadlocker, on all bolts and nuts, even if they are secured with lockwashers. Be sure to use the specified grade

of thread locking compound. A screw or bolt lost from an engine cover or bearing retainer could easily cause serious and expensive damage before its loss is noticed.

When applying thread locking compound, use a small amount. If too much is used, it can work its way down the threads and stick parts together that were not meant to be stuck together.

Keep a tube of thread locking compound in your tool box; when used properly, it is cheap insurance.

2. Use a hammer-driven impact tool to remove and install screws and bolts. These tools help prevent the rounding off of bolt heads and screw slots and ensure a tight installation.

3. When straightening the fold-over type lockwasher, use a wide-blade chisel, such as an old and dull wood chisel. Such a tool provides a better purchase on the folded tab, making straightening easier.

4. When installing the fold-over type lockwasher, always use a new washer if possible. If a new washer is not available, always fold over a part of the washer that has not been previously folded. Reusing the same fold may cause the washer to break, resulting in the loss of its locking ability and a loose piece of metal adrift in the engine.

When folding the washer, start the fold with a screwdriver and finish it with a pair of pliers. If a punch is used to make the fold, the fold may be too sharp, thereby increasing the chances of the washer breaking under stress.

These washers are relatively inexpensive and it is suggested that you keep several of each size in your tool box for repairs.

5. When replacing missing or broken fasteners (bolts, nuts and screws), always use authorized replacement parts. They are specially hardened for each application. The wrong 50-cent bolt could easily cause serious and expensive damage.

6. When installing gaskets, always use authorized replacement gaskets *without* sealer, unless designated. Many gaskets are designed to swell when they come in contact with oil. Gasket sealer will prevent the gaskets from swelling as intended and can result in oil leaks. Authorized replacement gaskets are cut from material of the precise thickness needed. Installation of a too thick or too thin gasket in a critical area could cause equipment damage.

MECHANIC'S TECHNIQUES

Removing Frozen Fasteners

When a fastener rusts and cannot be removed, several methods may be used to loosen it. First, apply penetrating oil, such as Liquid Wrench or WD-40 (available at any hardware or auto supply store). Apply it liberally and allow it penetrate for 10-15 minutes. Tap the fastener several times with a small hammer; do not hit it hard enough to cause damage. Reapply the penetrating oil if necessary.

For frozen screws, apply penetrating oil as described, then insert a screwdriver in the slot and tap the top of the screwdriver with a hammer. This loosens the rust so the screw can be removed in the normal way. If the screw head is too chewed up to use a screwdriver, grip the head with locking pliers and twist the screw out.

Avoid applying heat unless specifically instructed because it may melt, warp or remove the temper from parts.

Remedying Stripped Threads

Occasionally, threads are stripped through carelessness or impact damage. Often the threads can be cleaned up by running a tap (for internal threads on nuts) or die (for external threads on bolts) through threads. See **Figure 23**.

Removing Broken Screws or Bolts

When the head breaks off a screw or bolt, several methods are available for removing the remaining portion.

If a large portion of the remainder projects out, try gripping it with vise-grip pliers. If the projecting portion is too small, file it to fit a wrench or cut a slot in it to fit a screwdriver. See **Figure 24**.

If the head breaks off flush, use a screw extractor. To do this, centerpunch the remaining portion of the screw or bolt. Drill a small hole in the screw and tap the extractor into the hole. Back the screw out with a wrench on the extractor. See **Figure 25**.

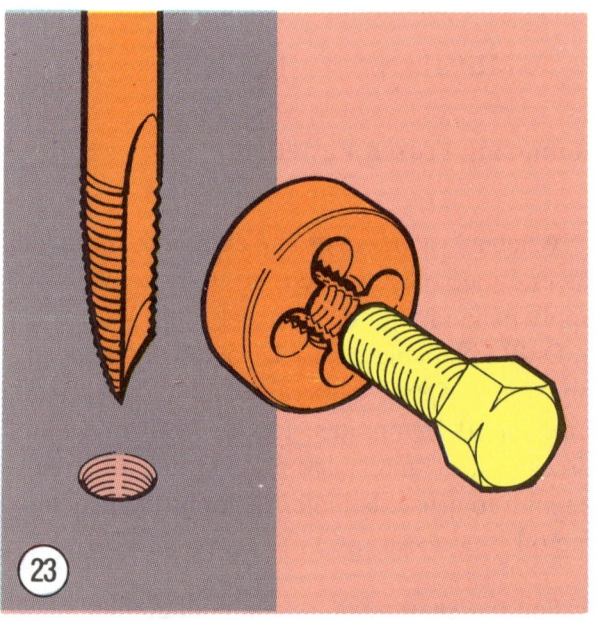

(23)

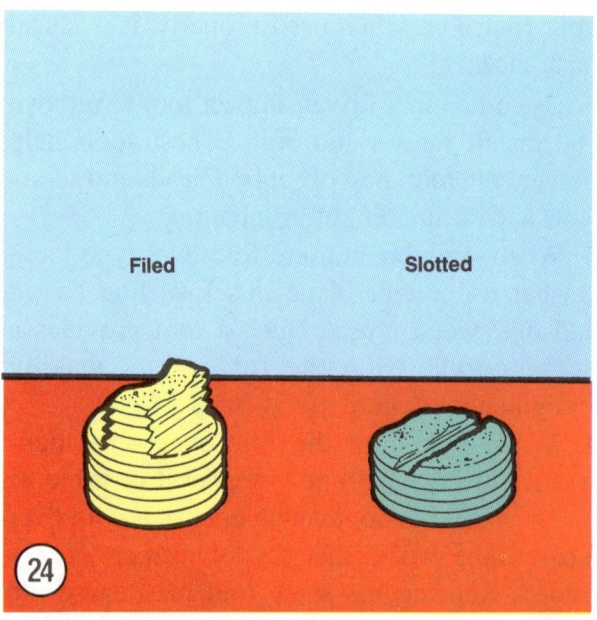

Filed   Slotted

(24)

(25)

Center punch   Drill hole   Tap extractor into hole   Remove screw

Chapter Three

Troubleshooting

Troubleshooting is a relatively simple matter when it is done logically. The first step in any troubleshooting procedure is to define the symptoms as closely as possible and then localize the problem. Subsequent steps involve testing and analyzing those areas which could cause the symptoms. A haphazard approach may eventually solve the problem, but it can be very costly in terms of wasted time and unnecessary parts replacement.

Proper lubrication, maintenance and periodic tune-ups as described in Chapter Four will reduce the necessity for troubleshooting. Even with the best of care, however, an outboard motor is prone to problems which will require troubleshooting.

This chapter contains brief descriptions of each operating system and troubleshooting procedures to be used. **Tables 1-3** present typical starting, ignition and fuel system problems with their probable causes and solutions. **Tables 1-7** are at the end of the chapter.

OPERATING REQUIREMENTS

Every outboard motor requires 3 basic things to run properly: an uninterrupted supply of fuel and air in the correct proportions, proper ignition at the right time and adequate compression. If any of these are lacking, the motor will not run.

The electrical system is the weakest link in the chain. More problems result from electrical malfunctions than from any other source. Keep this in mind before you blame the fuel system and start making unnecessary carburetor adjustments.

If a motor has been sitting for any length of time and refuses to start, check the condition of the battery first to make sure it has an adequate charge, then look to the fuel delivery system. This includes the gas tank, fuel pump, fuel lines and carburetor(s). Rust may have formed in the tank, obstructing fuel flow. Gasoline deposits may have gummed up carburetor jets and air passages. Gasoline tends to lose its potency after standing for long periods. Condensation may contaminate it with water. Drain the old gas and try starting with a fresh tankful.

STARTING SYSTEM

Description

Yamaha 9.9 hp and larger outboard motors whose model designation contains the letter "E" following the model number (Yamaha 40EN) use an electric starter motor (**Figure 1**). An electric start option is available for 6 and 8 hp engines.

The starter motor is mounted vertically on the engine. When battery current is supplied to the starter motor, its pinion gear is thrust upward to engage the teeth on the engine flywheel. Once the engine starts, the pinion gear disengages from the flywheel. This is similar to the method used in cranking an automotive engine.

The electric starting system requires a fully charged battery to provide the large amount of current required to operate the starter motor. The battery may be charged externally or by a charging coil on the magneto base (6-70 hp) or alternator stator (90-225 hp) which will keep the battery charged while the engine is running.

The starting circuit on all outboards covered in this manual equipped with an electric starting system consists of the battery, a key ignition or push-button starter switch, stop switch, an interlock or neutral start switch, starter motor, starter relay (**Figure 2**) to carry the heavy electrical current to the motor, choke solenoid, a fuse (on some models) and connecting wiring.

Depressing the starter button or turning the key ignition switch to the START position allows current to flow from the battery

through the relay coil. The relay contacts close, allowing current to flow from the battery to the starter motor.

An interlock or neutral safety switch in the remote control box prevents current flow to the starter motor if the shift mechanism is not in NEUTRAL. Models without a remote control box have a mechanical interlock in the rewind starter. This device is connected to the interlock switch by a cable.

The stop switch shorts out the magneto base or stator charge coil(s). The choke solenoid electrically moves the choke valve linkage to open the choke for starting. **Figure 3** is a schematic of a typical Yamaha electrical system showing the starting and stop circuits.

CAUTION
Do not operate an electric starter motor continuously for more than 10 seconds. Allow the motor to cool for at least 2 minutes between attempts to start the engine.

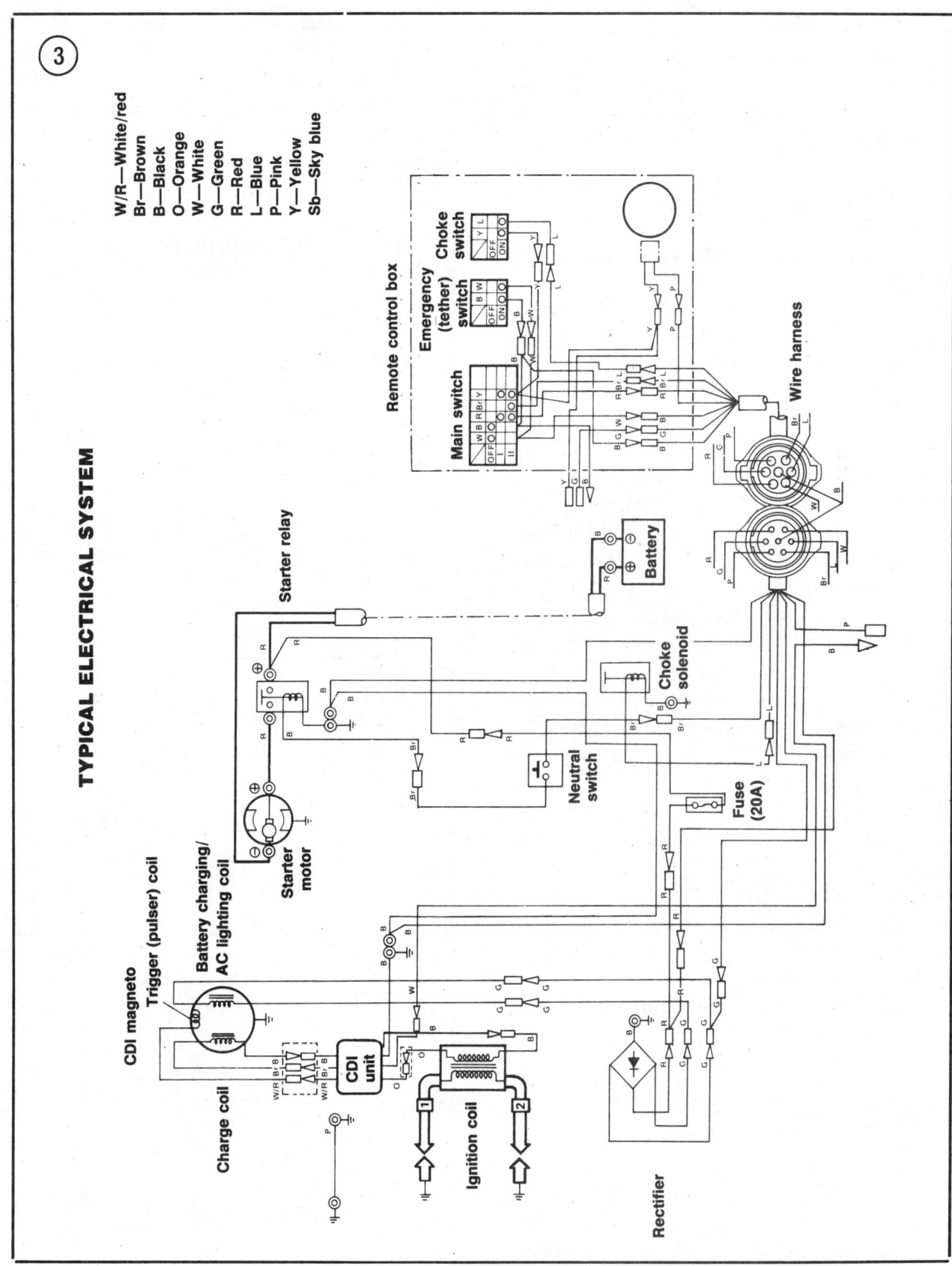

Starting Difficulties

Two-stroke engines, especially older or well-used models, are often plagued by hard starting and generally poor running for which there seems to be no good cause. Carburetion and ignition are satisfactory and a compression test shows all is well in the engine's upper end.

What a compression test does not show is a lack of primary compression. The crankcase in a 2-stroke engine must be alternately under pressure and vacuum. After the piston closes the intake port, further downward movement of the piston causes the trapped mixture to be pressurized so it can rush quickly into the cylinder when the scavenging ports are opened. Upward piston movement creates a vacuum in the crankcase, enabling air-fuel mixture to be drawn in from the carburetor.

If the crankshaft seals or case gaskets leak, the crankcase cannot hold pressure or vacuum and proper engine operation becomes impossible. Any other source of leakage, such as defective cylinder base gaskets or porous or cracked crankcase castings, will result in the same conditions.

Such engines suffering from hard starting should be checked for pressure leaks with a small brush and a solution of soap suds. The following is a list of possible leakage points in the engine:

    a. Crankshaft seals.
    b. Spark plug threads.
    c. Cylinder head joint.
    d. Cylinder base joint.
    e. Carburetor mounting flange(s).
    f. Crankcase joint.

Troubleshooting Preparation

If the following procedures do not locate the cause of the problem, refer to **Table 1** for more extensive testing. Before troubleshooting the starting circuit, make sure that:

    a. The battery is fully charged.
    b. Battery cables are the proper size and length. Replace cables that are undersize or relocate the battery to shorten the distance between battery and starter relay.
    c. The shift mechanism is in NEUTRAL position and the emergency switch lock plate or lanyard is properly installed on remote control models.
    d. All electrical connections are clean and tight.
    e. The wiring harness is in good condition, with no worn or frayed insulation or loose harness sockets.
    f. The electrical circuit fuse (if so equipped) is in good condition.
    g. The fuel system is filled with an adequate supply of fresh gasoline that has been properly mixed with Yamalube Two-cycle oil (for outboards). See Chapter Four.

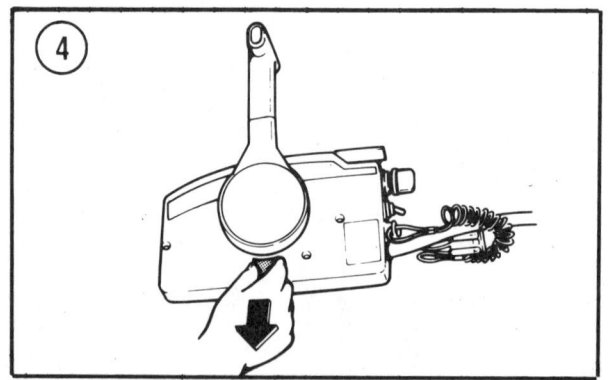

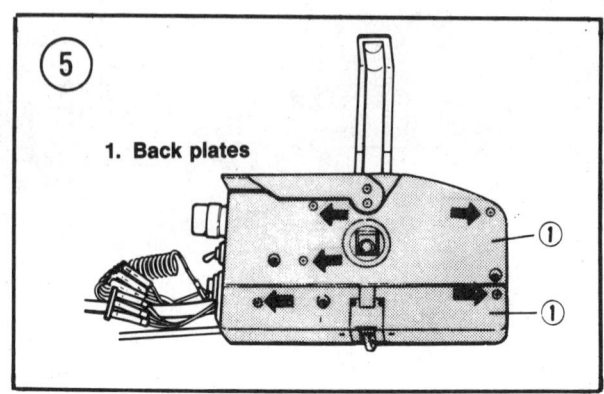

1. Back plates

h. The spark plugs are in good condition and properly gapped.

i. The ignition system is correctly timed, synchronized and adjusted. See Chapter Five.

Troubleshooting is intended only to isolate a malfunction to a certain component. If further bench testing is deemed necessary, remove the suspected component and have it tested by an authorized service center. Refer to Chapter Seven for electrical component removal and installation procedures.

Starter Relay Resistance Test

See Chapter Four.

Push-Button Starter Switch
Continuity Test

1. Disconnect the red and black leads at the starter push-button.
2. Connect an ohmmeter between the disconnected push-button leads.
3. Depress the push-button while watching the meter needle. If the needle does not deflect (indicating continuity) when the button is depressed and return to its original position when released, replace the starter push-button.

Engine Stop Switch
Continuity Test

1. Disconnect the white and black leads at the engine stop switch.

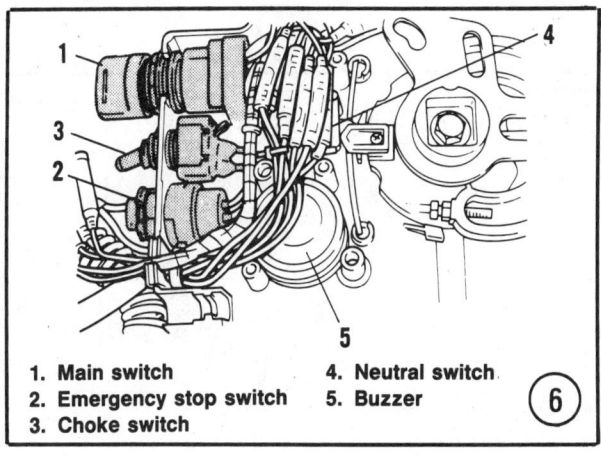

1. **Main switch**
2. **Emergency stop switch**
3. **Choke switch**
4. **Neutral switch**
5. **Buzzer**

6

2. Connect an ohmmeter between the disconnected stop switch leads.
3. Depress the stop switch while watching the meter needle. If the needle does not deflect (indicating continuity) when the switch is depressed and return to its original position when released, replace the stop switch.

Neutral Start Switch
Continuity Test
(25-30 hp)

1. Make sure the shift lever is in NEUTRAL.
2. Disconnect the neutral start switch leads at the starter relay and wiring harness.
3. Connect an ohmmeter between the disconnected start switch leads. The meter should indicate continuity.
4. Move the shift lever from NEUTRAL to FORWARD, then back to NEUTRAL and into REVERSE while watching the meter needle. The meter should show continuity only in NEUTRAL. If continuity is shown in FORWARD or REVERSE, or no continuity is shown in NEUTRAL, replace the start switch.

Choke Solenoid
Resistance Test

See Chapter Four.

Remote Control Box Switches

The warning buzzer, ignition, choke, emergency stop, neutral start and power trim/tilt switches are located in the remote control box on models so equipped.

1. Remove the remote control box from its mounting bracket.
2. Remove the cover from the lower side of the box (**Figure 4**).
3. Loosen the control box back plate screws (arrows, **Figure 5**). Remove the back plates.
4. Loosen the fasteners holding the switch(es) to be tested. Remove the switch(es) or warning buzzer. See **Figure 6**.

3

5. To test the ignition switch:
  a. Connect an ohmmeter between the white and black switch leads (**Figure 7**). There should be continuity when the switch is OFF.
  b. Connect the ohmmeter between the red and yellow switch leads (**Figure 8**). Turn the ignition switch first to the ON and then to the START position. There should be continuity at both positions.
  c. Connect the ohmmeter between the red and brown switch leads (**Figure 9**). There should be continuity with the switch in the START position.

6. To test the emergency stop switch—Connect an ohmmeter between the stop switch leads. The meter should show continuity with the lock plate removed and no continuity when the lock plate is properly installed. See **Figure 10**.

7. To test the choke switch—Connect an ohmmeter between the choke switch leads. The meter should show continuity with the switch ON and no continuity with the switch OFF. See **Figure 11**.

8. To test the neutral start switch—Connect an ohmmeter between the neutral switch leads. The meter should show continuity with the switch plunger depressed and no continuity when the plunger is released. See **Figure 12**.

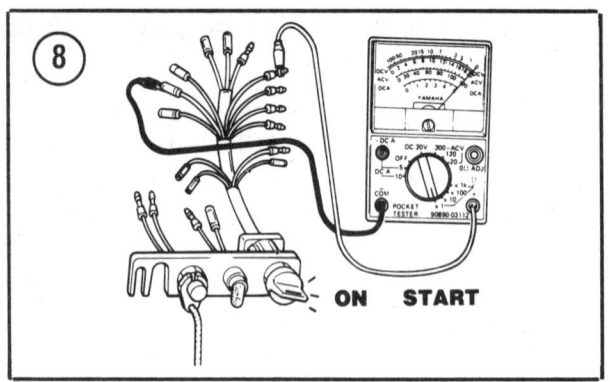

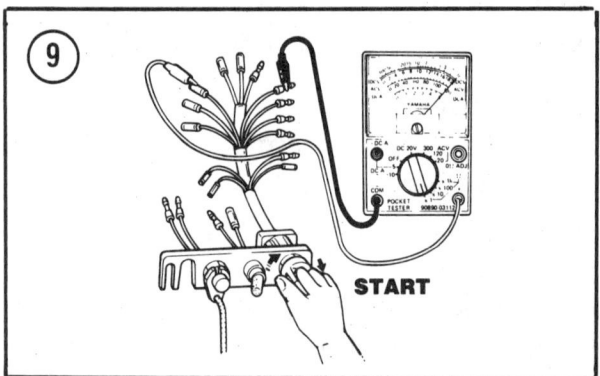

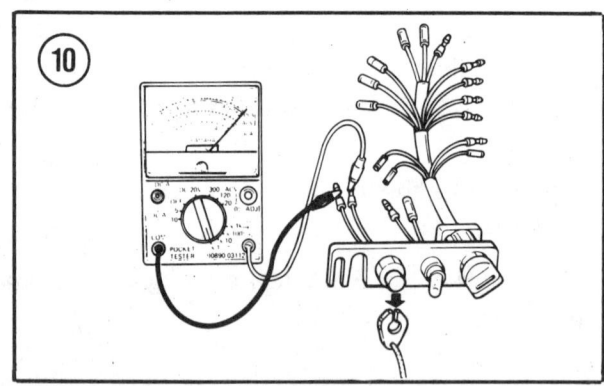

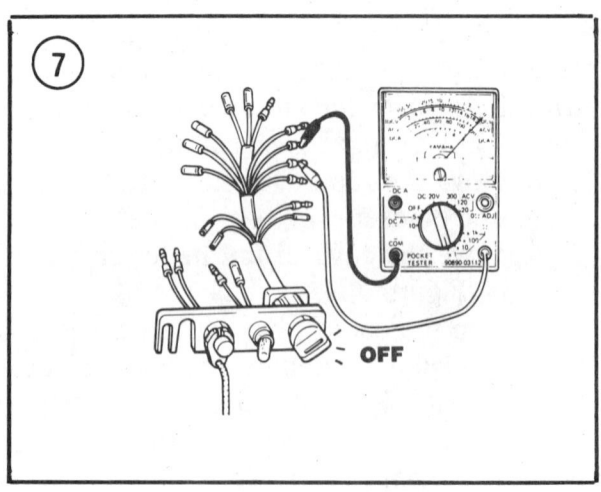

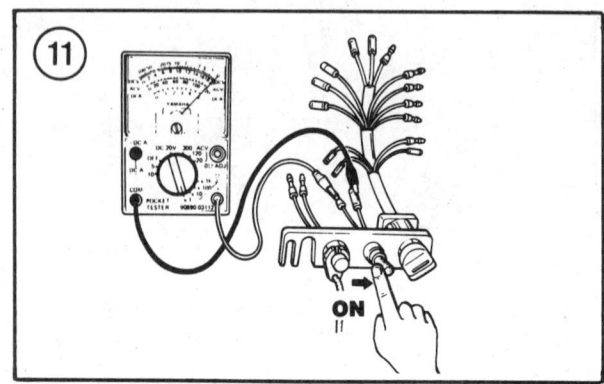

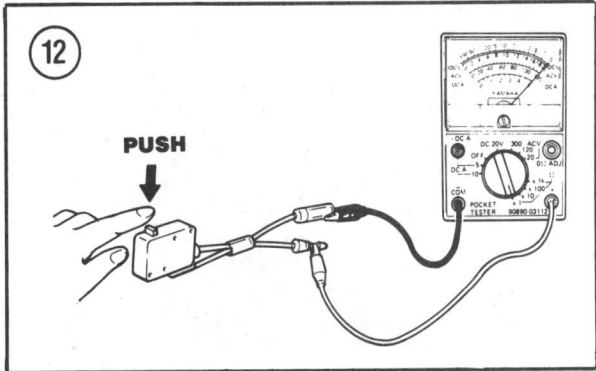

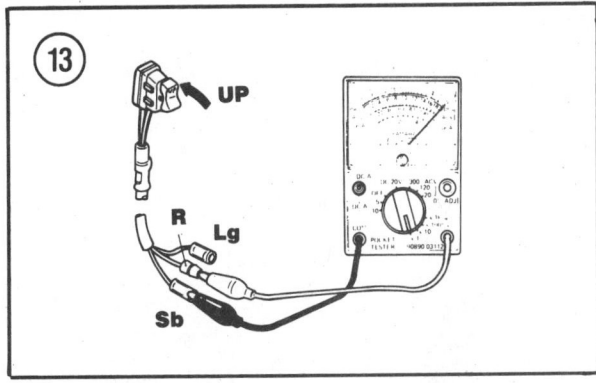

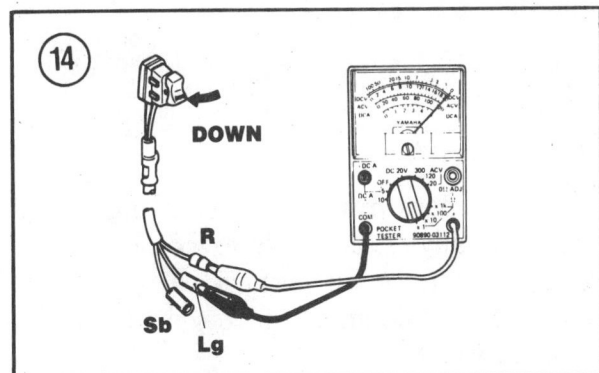

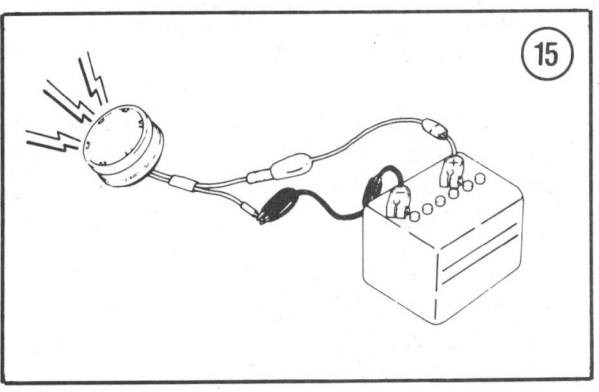

9. To test the power trim/tilt switch:
   a. Connect an ohmmeter between the red and sky blue switch leads (**Figure 13**). There should be continuity when the switch is pushed upward and no continuity when it is released.
   b. Connect the ohmmeter between the red and light green switch leads (**Figure 14**). There should be continuity when the switch is pushed downward and no continuity when it is released.

10. To test the warning buzzer—Connect the buzzer leads to a 12-volt battery with jumper leads (**Figure 15**). Replace the buzzer if it does not emit a steady sound.

11. Replace any switch that does not perform as specified in this procedure.

12. Reverse Steps 1-4.

CHARGING SYSTEM

Description

The charging system consists of the magneto base on 9.9-70 hp (**Figure 16**) or an alternator stator on 90-225 hp (**Figure 17**) containing one or more coils wound on a laminated core, a series of permanent magnets located within the flywheel rim

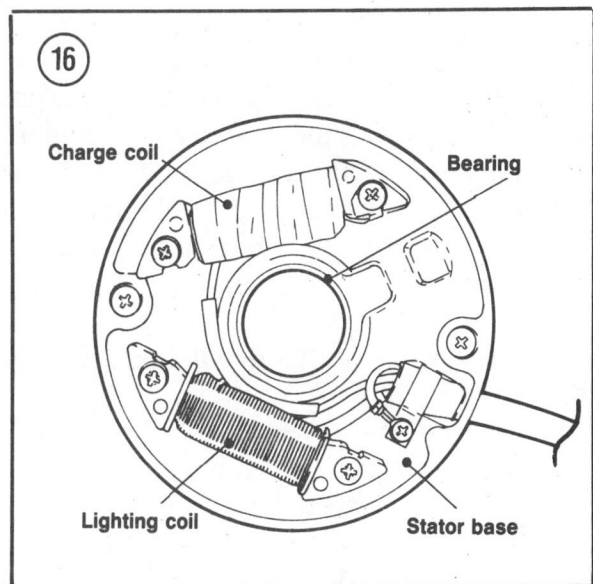

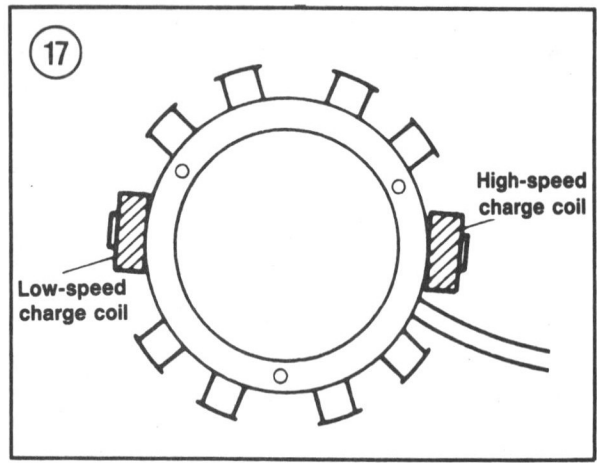

Low-speed
charge coil

High-speed
charge coil

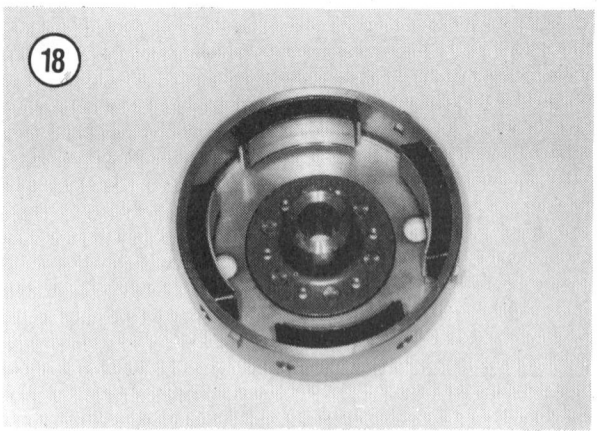

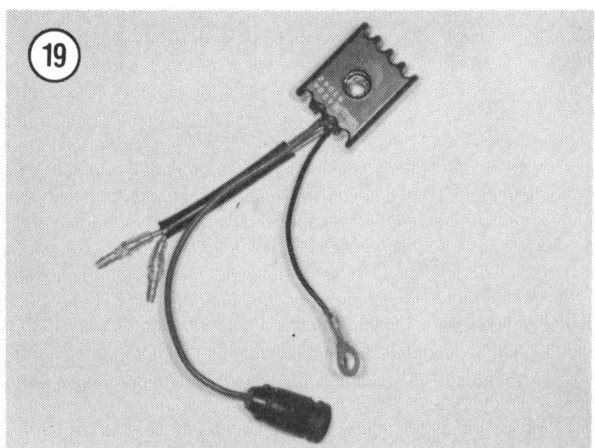

(**Figure 18**), a rectifier (**Figure 19**) or rectifier/regulator to change alternating current (AC) to direct current (DC), the starter relay, battery and connecting wiring. **Figure 3** is a schematic of a typical Yamaha electrical system showing the charging circuit.

> *NOTE*
> *The AC lighting coil system used on 6-25 hp manual start models consists of a coil mounted on the magneto base and a flywheel with permanent magnets in the rim. The AC current produced is sent directly to AC accessories. An AC lighting coil can be tested with the same procedures specified for a battery charging coil.*

A malfunction in the charging system generally causes the battery to remain undercharged. Since the magneto base or stator is located underneath the flywheel and is thus protected, it is more likely that the battery, rectifier, starter relay or connecting wiring will cause problems.

The following conditions will cause rectifier damage:

    a. Battery leads reversed.

    b. Running the engine with the battery leads disconnected.

    c. A broken wire or loose connection resulting in an open circuit.

Troubleshooting

Before performing any charging circuit tests, visually check the following.

1. Make sure the battery cables are properly connected. If polarity is reversed, check for a damaged rectifier.

> *NOTE*
> *A damaged rectifier will generally have a discolored or a burned appearance.*

2. Carefully inspect all wiring between the magneto base or stator and battery for damaged or deteriorated insulation and corroded, loose or disconnected connections. Replace wiring or clean and tighten connections as required.

3. Check battery condition. Clean and recharge as required.

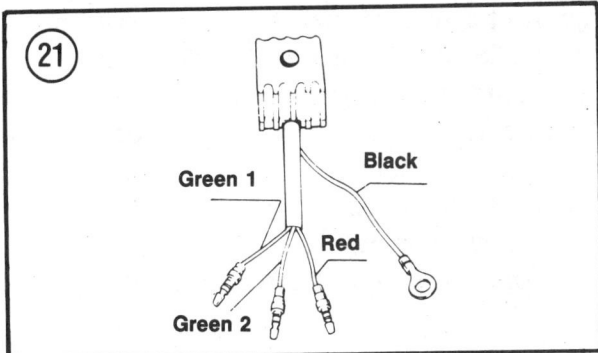

Battery Charging Coil/ Lighting Coil Output Test

CAUTION
*The engine must be provided with an adequate supply of water while performing this procedure. The use of a flushing device is **not** recommended. Place the engine in a test tank or perform the test with the boat in the water.*

1. Install the engine in a test tank and connect a remote fuel tank to the fuel inlet.
2. Make sure the battery is fully charged.
3. Remove the engine cover.
4. Connect a tachometer according to manufacturer's instructions.
5. Start the engine and warm to normal operating temperature.
6. Install an ampere gauge (**Figure 20**) over the green charging coil leads.
7. Watch the ampere gauge and gradually increase engine speed to approximately 5,000 rpm and note the gauge reading. If it is not 6

amps (6-70 hp), 10 amps (1984-1985 90-220 hp and 1986-on 90-130 hp) or 15 amps (1986-on 150-225 hp), replace the battery charging coil.

Battery Charging Coil/ Lighting Coil Resistance Test

1. Disconnect the negative battery cable, if so equipped.
2. Remove the engine cover.

NOTE
On some 1986 and later models, the 2 green lighting coil leads to the rectifier/regulator terminals have been changed to 1 green lead with a black terminal eye and 1 green/white lead with a white terminal eye. Note their positioning before disconnecting and reconnect to the proper terminals.

3. Disconnect the 2 green battery charging coil leads at their bullet connectors or at the rectifier or rectifier/regulator terminals.
4. Connect an ohmmeter between the 2 disconnected leads. With the ohmmeter set on the R×1 scale, note the reading and compare to **Table 4**. If not within specifications, replace the battery charging coil. See Chapter Seven.

Rectifier or Rectifier/ Regulator Test

A rectifier or rectifier/regulator can be tested without removal from the engine. Disconnect the battery leads at the battery before performing this procedure.
1. Remove the engine cover.

NOTE
If the rectifier or rectifier/regulator is installed on the CDI unit bracket, the CDI unit cover must be removed to provide access for testing.

2A. Rectifier—Disconnect the red and 2 green leads at their bullet connectors.

3

Disconnect the black ground lead. See **Figure 21**.

2B. Rectifier/regulator—Disconnect the red, black and 2 green leads at the rectifier/regulator terminals. See **Figure 22**.

3. Set the ohmmeter on the R×1 scale.

4. Connect the red ohmmeter test lead to the black rectifier lead or terminal. Connect the black ohmmeter test lead alternately to the red and 2 green leads or terminals. The ohmmeter should show continuity.

5. Connect the black ohmmeter test lead to the black rectifier lead or terminal. Connect the red ohmmeter test lead alternately to the red and 2 green leads or terminals. The ohmmeter should not show continuity.

6. Connect the red ohmmeter test lead to the red rectifier lead or terminal. Connect the black ohmmeter test lead alternately to the black and 2 green leads or terminals. The ohmmeter should show continuity.

7. Connect the black ohmmeter test lead to the red rectifier lead or terminal. Connect the black ohmmeter test lead alternately to the red and 2 green leads or terminals. The ohmmeter should not show continuity.

NOTE
The ohmmeter's internal battery polarity may cause the test results to turn out exactly the reverse of those specified. If this occurs, the rectifier or rectifier/regulator is good and need not be replaced.

8. Replace the rectifier or rectifier/regulator if the ohmmeter readings are not as specified in Steps 4-7.

Voltage Regulator Test

CAUTION
*The engine must be provided with an adequate supply of water while performing this procedure. The use of a flushing device is **not** recommended. Place the engine in a test tank or perform the test with the boat in the water.*

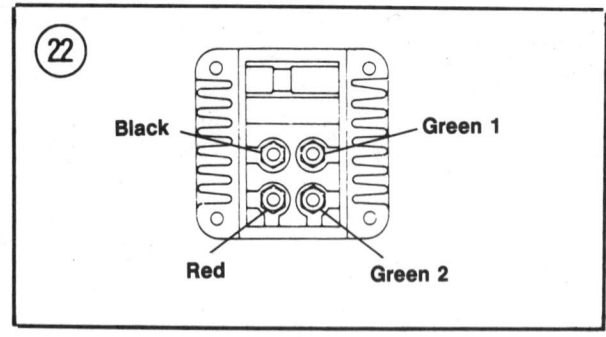

1. Install the engine in a test tank and connect a remote fuel tank to the fuel inlet.

2. Make sure the battery is fully charged.

3. Remove the engine cover.

4. Connect a tachometer according to manufacturer's instructions.

5. Start the engine and warm to normal operating temperature.

6. Connect a voltmeter across the battery terminals.

7. Gradually increase engine speed to approximately 4,200-4,500 rpm and note the voltmeter reading. If it is not 14-15 volts, replace the voltage regulator.

IGNITION SYSTEM

The wiring harness used between the ignition switch and engine is adequate to handle the electrical needs of the outboard. It will *not* handle the electrical needs of accessories. Whenever an accessory is added, run new wiring between the battery and accessory, installing a separate fuse block on the instrument panel.

If the ignition switch requires replacement, *never* install an automotive-type switch. A marine-type switch must always be used.

Description

Variations of two different ignition systems are used on Yamaha outboards. See Chapter Seven for a full description. For the purposes of troubleshooting, the ignition systems can be divided into 2 basic types:

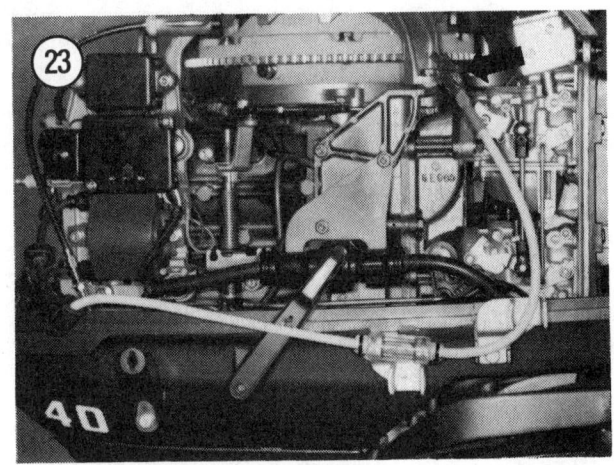

1. A flywheel magneto breaker-point ignition (2 hp only).
2. A flywheel magneto or alternator stator capacitor discharge (breakerless) ignition (CDI).

General troubleshooting procedures are provided in **Table 2**.

Precautions

Several precautions should be strictly observed to avoid damage to the ignition system.

1. Do not reverse the battery connections. This reverses polarity and can damage the rectifier, rectifier/regulator or CDI unit on CDI ignitions.
2. Do not "spark" the battery terminals with the battery cable connections to check polarity.
3. Do not disconnect the battery cables with the engine running.
4. Do not crank the engine if the CDI unit is not grounded to the engine.
5. Do not touch or disconnect any ignition components when the engine is running, while the ignition switch is ON or while the battery cables are connected.
6. If you must run an engine that has a CDI ignition without the battery connected to the harness, disconnect the magneto base or stator rectifier leads and tape them separately.

Troubleshooting Preparation (All Ignition Systems)

> *NOTE*
> *To test the wiring harness for poor connections in Step 1, bend the molded rubber connector while checking each wire for resistance.*

1. Check the wiring harness and all plug-in connections to make sure that all terminals are free of corrosion, all connectors are tight and the wiring insulation is in good condition.
2. Check all electrical components that are grounded to the engine for a good ground.
3. Make sure that all ground wires are properly connected and that the connections are clean and tight.
4. Check remainder of the wiring for disconnected wires and short or open circuits.
5. Check the fuse (if so equipped) to make sure it is not defective.
6. Make sure there is an adequate supply of fresh and properly mixed fuel available to the engine.
7. Check the battery condition (if so equipped). Clean terminals and recharge battery, if necessary.
8. Check spark plug cable routing. Make sure the cables are properly connected to their respective spark plugs.
9. Remove all spark plugs, keeping them in order. Check the condition of each plug. See Chapter Four.
10. Install a spark tester between the plug wire and a good ground to check for spark at each cylinder. See **Figure 23**. If a spark tester is not available, reconnect the proper plug cable to one plug. Lay the plug against the cylinder head so its base makes a good connection and crank the engine. If there is no spark or only a weak one, check for loose connections at the coil and battery. If all external wiring connections are good, the problem is most likely in the ignition system.

3

COMPONENT TESTING

An ignition analyzer should be used for accurate testing of the breaker points, condenser, charge coil and ignition coil. A Merc-O-Tronic ignition analyzer can be used but Yamaha recommends the use of an Electro-tester and its Pocket Tester (part No. YU-3112). For ignition coil tests, Yamaha recommends the Electro-tester for electric start models and its ignition coil tester (part No. YU-33261) for manual start models, although a standard spark tester can be used. All test equipment can be purchased through your Yamaha dealer. Each analyzer includes detailed instructions for component testing as well as complete component specifications according to engine model and year of manufacture. The procedures given here are general in nature to acquaint you with component testing. Refer to the instructions provided with the particular test equipment to be used for the exact procedure.

Ohmmeter readings should be made when the engine is cold. Readings taken on a hot engine will show increased resistance caused by engine heat and result in unnecessary parts replacement without solving the basic problem. When switching between ohmmeter scales, always cross the test leads and zero the needle to assure a correct reading.

BREAKER POINT IGNITION TESTING

Breaker Point Test

1. Remove the engine cowling.
2. Remove the flywheel. See Chapter Eight.
3. Disconnect the breaker point leads from the magneto base.
4. Connect one analyzer test lead to the breaker arm. Connect the other test lead to the breaker point screw terminal.
5. Set the analyzer controls according to manufacturer's instructions.
6. If the breaker points are good, the analyzer needle will rest in the OK segment.

7. If the analyzer needle does not fall within the specified segment on the scale, clean the points with alcohol and recheck the analyzer leads to make sure that the connections are tight before discarding the points. The low current present in this test makes clean points and proper connections very important.

Condenser Tests

1. Remove the engine cowling.
2. Remove the flywheel. See Chapter Eight.
3. Disconnect the condenser lead from the breaker point set.
4. Connect one analyzer test lead to the condenser lead. Connect the other test lead to the magneto base.

> *WARNING*
> *High voltage is involved in a condenser leakage test. Handle the analyzer leads carefully and turn the analyzer switch to DISCHARGE before disconnecting the leads from the condenser.*

5. Set the analyzer controls according to manufacturer's instructions and check the condenser for leakage, resistance and capacity.
6. Compare the results in Step 5 with the specifications provided by the analyzer manufacturer (Yamaha specifies capacity at 0.25 microfarads ± 10 percent and resistance at 5K ohms). Replace the condenser if it fails any of the 3 tests.

Ignition Coil Test

This test checks the primary and secondary windings of the coil for circuit continuity.

1. Remove the engine cowling.
2. Disconnect the white primary coil lead. See arrow, **Figure 24**.
3. Disconnect the secondary coil lead at the spark plug.
4. Remove the flywheel. See Chapter Eight.
5. Connect an ohmmeter between the primary coil lead and the ignition coil body. Set the ohmmeter on the R×1 scale. The meter should show a resistance of 1.06 ohms ±10 percent at 68° F (20° C).
6. Connect the ohmmeter between the secondary coil lead and the ignition coil body. Set the ohmmeter on the R×1,000 scale. It should show a resistance of 6050 ohms ±10 percent at 68° F (20° C).
7. If the resistance values are not as specified in Step 5 and Step 6, replace the ignition coil. See Chapter Seven.

CDI IGNITION TESTING

Charge Coil Resistance Test

1. Disconnect the negative battery cable, if so equipped.
2. Remove the engine cover.
3. Remove the CDI unit cover as required and disconnect the charge coil leads at the CDI unit or connector as follows:
    a. 3-30 and 70 hp—Brown and black leads.
    b. 40 and 50 hp—Brown and blue leads.
    c. 90 hp—Brown, blue, red/blue leads.
    d. 115-225 hp—Brown, red, black/red and blue leads.

4A. 3-70 hp—Connect an ohmmeter between the disconnected leads. Replace the charge coil if the reading is not within specifications (**Table 5**). See Chapter Seven.

4B. 90-225 hp—Connect an ohmmeter between one set of the disconnected leads.

Take the reading and repeat this step to test the other set of leads. Replace the alternator stator if either reading is not within specifications (**Table 5**). See Chapter Seven.

Trigger Coil Resistance Test

NOTE
The trigger coil may also be referred to as a pulser coil.

1. Disconnect the negative battery cable, if so equipped.
2. Remove the engine cover.
3. Remove the CDI unit cover as required and disconnect the trigger coil leads at the CDI unit or connector as follows:
    a. 3-5 hp—White/green and black (low speed) and white/red and black (high speed).
    b. 6-30 (2-cyl.) hp—White/red and black leads.
    c. 30 (3 cyl.)-70 hp—White/red, white/black, white/green and black leads.
    d. 40-130 hp—White/red, white/black, white/green, white/yellow, white/blue and white/brown leads.
    e. 150-225 hp—White/red, white/black, white/green, white/yellow, white/blue and white/brown leads.

4A. 3-5 hp—Connect an ohmmeter between each set of disconnected leads. Replace the appropriate trigger coil if the reading is not within specifications (**Table 6**). See Chapter Seven.

4B. 6-30 (2-cyl.) hp—Connect an ohmmeter between the disconnected leads. Replace the trigger coil if the reading is not within specifications (**Table 6**). See Chapter Seven.

4C. 30 (3 cyl.)-70 hp—Connect the black ohmmeter test lead to the black connector terminal. Alternately probe the other 3 terminals of the connector with the red test lead and note each reading. Replace any trigger coil that does not fall within specifications (**Table 6**). See Chapter Seven.

4D. 90-130 hp—Connect an ohmmeter between the white/black and white/green leads. Take the reading and repeat this step to test the other set of leads. Replace the alternator stator if either reading is not within specifications (**Table 6**). See Chapter Seven.

4E. 150-225 hp—Connect an ohmmeter between the white/red and white/green leads and take the reading. Repeat this step, connecting the ohmmeter between the white/black and white/blue leads, then between the white/yellow and white/brown leads. Replace the alternator stator if any reading is not within specifications (**Table 6**). See Chapter Seven under "Charging Coil Replacement, V4 & V6".

Ignition Coil Resistance Test

1. Disconnect the negative battery cable, if so equipped.

2. Remove the engine cover.

3. Remove the coil(s) to be tested from the engine. See Chapter Seven.

4. Remove the plug cap from each spark plug lead.

5. Set the ohmmeter on the R×1 scale and connect it between the primary and ground leads. Note the reading.

6. Set the ohmmeter on the R×1,000 scale.

7A. Single spark plug lead—Connect the ohmmeter between the ground lead and spark plug lead. Note the reading.

7B. Dual spark plug leads—Connect the ohmmeter between the spark plug leads and note the reading.

8. Compare the readings obtained in Step 5 and Step 7 with specifications (**Table 7**). Replace the coil if not within specifications.

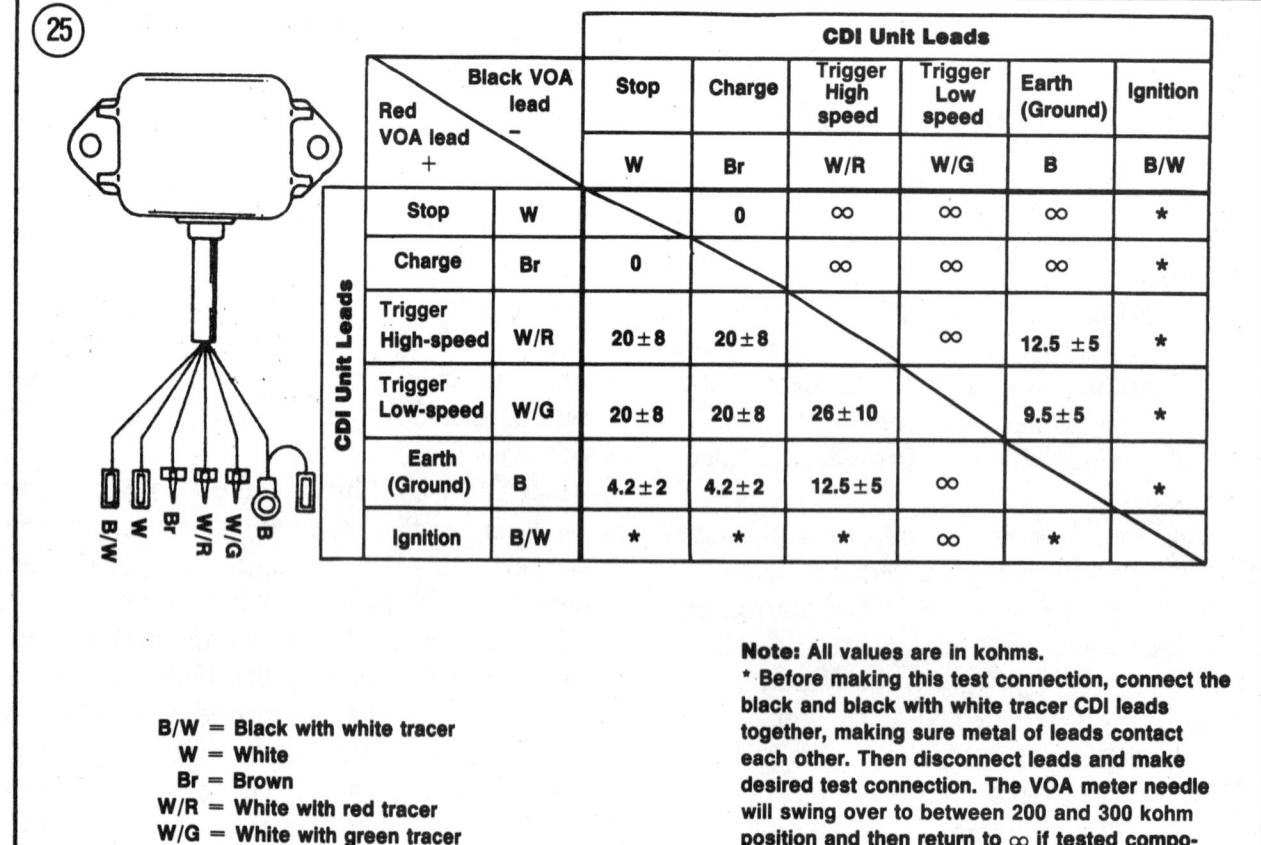

			CDI Unit Leads					
Red VOA lead + \ Black VOA lead −			Stop	Charge	Trigger High speed	Trigger Low speed	Earth (Ground)	Ignition
			W	Br	W/R	W/G	B	B/W
Stop	W			0	∞	∞	∞	*
Charge	Br		0		∞	∞	∞	*
Trigger High-speed	W/R		20±8	20±8		∞	12.5 ±5	*
Trigger Low-speed	W/G		20±8	20±8	26±10		9.5±5	*
Earth (Ground)	B		4.2±2	4.2±2	12.5±5	∞		*
Ignition	B/W		*	*	*	∞	*	

Note: All values are in kohms.
* Before making this test connection, connect the black and black with white tracer CDI leads together, making sure metal of leads contact each other. Then disconnect leads and make desired test connection. The VOA meter needle will swing over to between 200 and 300 kohm position and then return to ∞ if tested component is functioning properly.

B/W = Black with white tracer
W = White
Br = Brown
W/R = White with red tracer
W/G = White with green tracer
B = Black

CDI Unit Resistance Test

The resistance values provided by Yamaha are based on the use of its Pocket Tester (part No. YU-3112). If another ohmmeter is used, the readings obtained may not agree with those specified due to internal resistance of the ohmmeter. When switching between ohmmeter scales, always cross the test leads and zero the needle to assure a correct reading.

3-5 hp

1. Remove the CDI unit. See Chapter Seven.
2. Connect the black and black/white CDI leads together for 10-20 seconds to discharge the condenser.
3. Set the Pocket Tester on the R×1,000 scale.
4. Refer to **Figure 25** for test connections and values. Make each connection (repeat Step 2 after each connection before making the next one) and compare the meter reading to the stated value. If any of the meter readings differ from the stated values, replace the CDI unit.

6-15 hp

1. Remove the CDI unit. See Chapter Seven.
2. Connect the black and orange CDI leads together for 10-20 seconds to discharge the condenser.
3. Set the Pocket Tester on the R×1,000 scale.
4. Refer to **Figure 26** for test connections and values. Make each connection (repeat Step 2 after each connection before making the next one) and compare the meter reading to the stated value. If any of the meter readings differ from the stated values, replace the CDI unit.

25 and 30 hp (2-cylinder)

1. Remove the CDI unit. See Chapter Seven.
2. Connect the black and orange CDI leads together for 10-20 seconds to discharge the condenser.

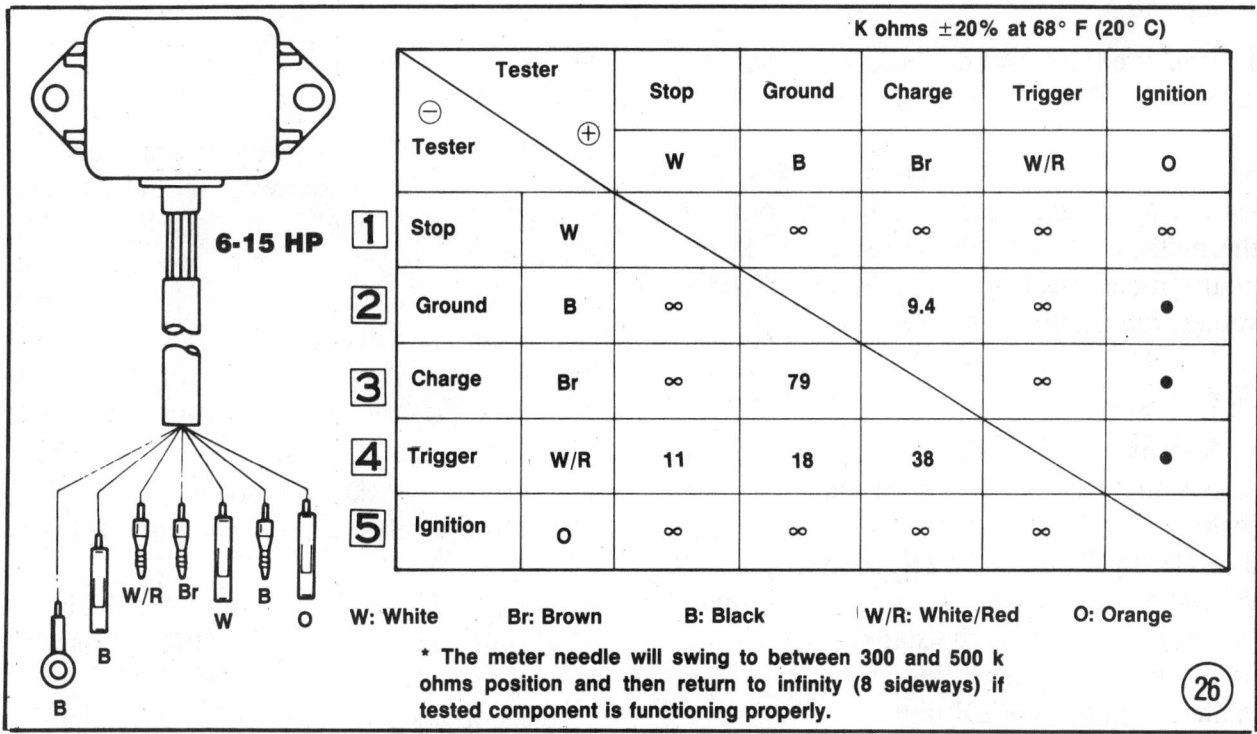

		Tester (−)	Stop	Ground	Charge	Trigger	Ignition
	Tester (+)		W	B	Br	W/R	O
1	Stop	W		∞	∞	∞	∞
2	Ground	B	∞		9.4	∞	●
3	Charge	Br	∞	79		∞	●
4	Trigger	W/R	11	18	38		●
5	Ignition	O	∞	∞	∞	∞	

K ohms ±20% at 68° F (20° C)

W: White   Br: Brown   B: Black   W/R: White/Red   O: Orange

* The meter needle will swing to between 300 and 500 k ohms position and then return to infinity (8 sideways) if tested component is functioning properly.

26

3. Set the Pocket Tester on the R×1,000 scale.

4. Refer to **Figure 27** for test connections and values. Make each connection (repeat Step 2 after each connection before making the next one) and compare the meter reading to the stated value. If any of the meter readings differ from the stated values, replace the CDI unit.

30 hp (3-cylinder)

1. Remove the CDI unit. See Chapter Seven.
2. Connect the black and orange CDI leads together for 10-20 seconds to discharge the condenser.
3. Set the Pocket Tester on the R×1,000 scale.
4. Refer to **Figure 28** for test connections and values. Make each connection (repeat Step 2 after each connection before making the next one) and compare the meter reading to the stated value. If any of the meter readings differ from the stated values, replace the CDI unit.

40 and 50 hp

1. Remove the CDI unit. See Chapter Seven.
2. Set the Pocket Tester on the R×1,000 scale.
3. Refer to **Figure 29** for test connections and values. Make each connection and compare the meter reading to the stated value. If any of the meter readings differ from the stated values, replace the CDI unit.

70 hp

1. Remove the CDI unit. See Chapter Seven.
2. Set the Pocket Tester on the R×1,000 scale.
3. Refer to **Figure 30** for test connections and values. Make each connection and compare the meter reading to the stated value. If any of the meter readings differ from the stated values, replace the CDI unit.

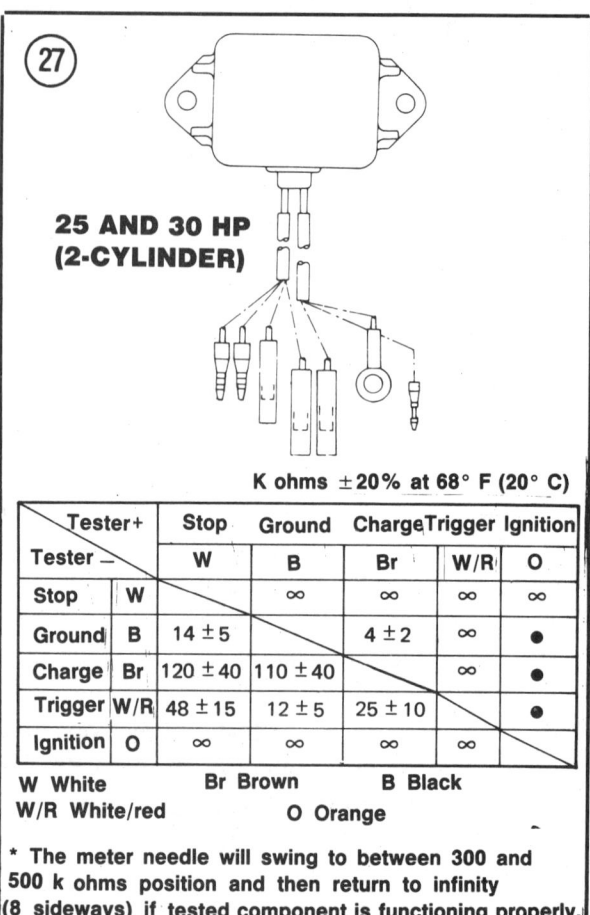

25 AND 30 HP (2-CYLINDER)

K ohms ±20% at 68° F (20° C)

Tester+		Stop	Ground	Charge	Trigger	Ignition
Tester −		W	B	Br	W/R	O
Stop	W		∞	∞	∞	∞
Ground	B	14 ±5		4 ±2	∞	●
Charge	Br	120 ±40	110 ±40		∞	●
Trigger	W/R	48 ±15	12 ±5	25 ±10		●
Ignition	O	∞	∞	∞	∞	

W White   Br Brown   B Black
W/R White/red   O Orange

* The meter needle will swing to between 300 and 500 k ohms position and then return to infinity (8 sideways) if tested component is functioning properly.

90 hp

Two different electrical circuits are used on 1984 models. Type A circuits are found on outboards with the following serial numbers: L450101 to L450400 and UL850101 to UL850150. Type B circuits are found on outboards with serial numbers L450401 and above or UL850151 and above. Type B circuits use a diode installed in the wiring between the oil level sensor and the thermoswitch.

1. Remove the CDI unit cover.
2. Set the Pocket Tester on the R×1,000 scale.
3. Refer to **Figure 31** (1984 Type A), **Figure 32** (1984 Type B) or **Figure 33** (1985-on) for test connections and values. Make each connection and compare the meter reading to

30 HP (3-CYLINDER)

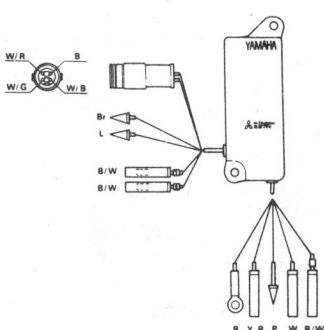

(+) / (−)	Pulser			Charge		Grond	Ignition			Stop	Thermo switch	Overheat
	W/R	W/B	W/G	Br	L	B	B/W 1	B/W 2	B/W 3	W	P	Y/R
W/R		∞	∞	∞	∞	∞	34^{+30}_{-15}	∞	∞	∞	∞	∞
W/B	∞		∞	∞	∞	∞	∞	34^{+30}_{-15}	∞	∞	∞	∞
W/G	∞	∞		∞	∞	∞	∞	∞	34^{+30}_{-15}	∞	∞	∞
Br	$*1000^{+\infty}_{-500}$	$*1000^{+\infty}_{-500}$	$*1000^{+\infty}_{-500}$		$*400^{+400}_{-200}$	$*400^{+400}_{-200}$	$*1000^{+\infty}_{-500}$	$*1000^{+\infty}_{-500}$	$*1000^{+\infty}_{-500}$	$*140^{+100}_{-40}$	$*1000^{+\infty}_{-500}$	$*1000^{+\infty}_{-500}$
L	16±6	16±6	16±6	4±2		0	4±2	4±2	4±2	62^{+60}_{-30}	13±6	4.3±2
B	16±6	16±6	16±6	4±2	0		4±2	4±2	4±2	62^{+60}_{-30}	13±6	4.3±2
B/W 1	∞	∞	∞	∞	∞	∞		∞	∞	∞	∞	∞
B/W 2	∞	∞	∞	∞	∞	∞	∞		∞	∞	∞	∞
B/W 3	∞	∞	∞	∞	∞	∞	∞	∞		∞	∞	∞
W	∞	∞	∞	∞	∞	∞	∞	∞	∞		∞	∞
P	∞	∞	∞	∞	∞	∞	∞	∞	∞	∞		∞
Y/R	∞	∞	∞	∞	∞	∞	∞	∞	∞	∞	∞	

∞··· **No continuity**

The asterisk (*) indicated that the tester needle should swing toward "O" and slowly swing back to stay at the specified value.

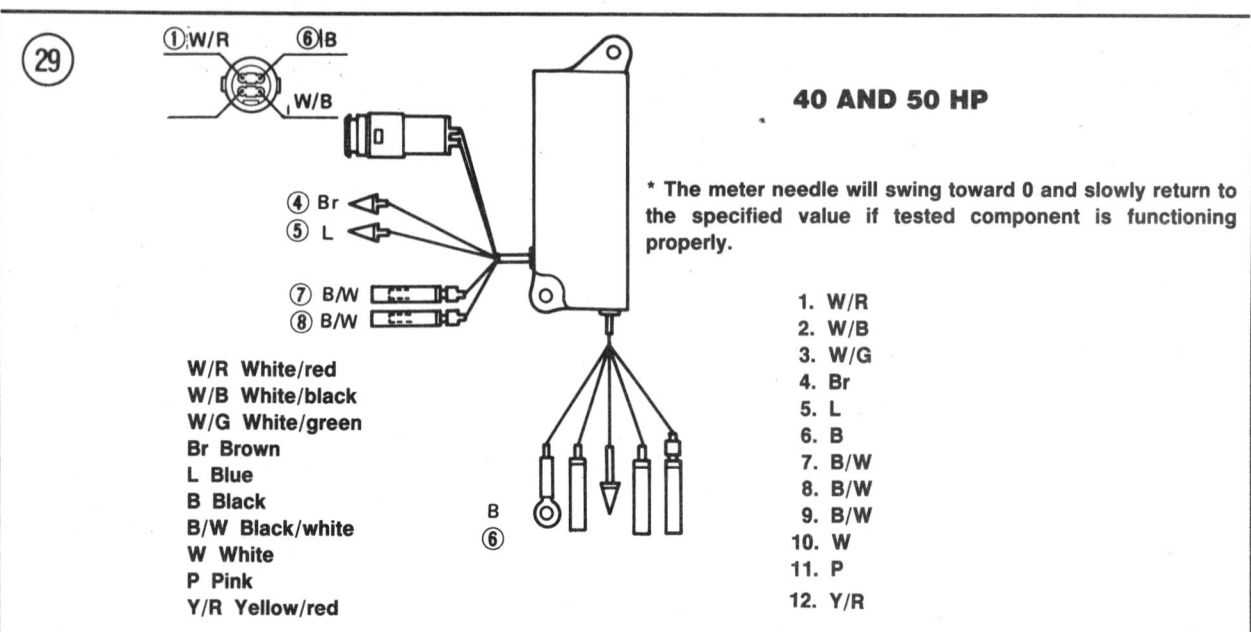

40 AND 50 HP

* The meter needle will swing toward 0 and slowly return to the specified value if tested component is functioning properly.

W/R White/red
W/B White/black
W/G White/green
Br   Brown
L   Blue
B   Black
B/W Black/white
W   White
P   Pink
Y/R Yellow/red

1. W/R
2. W/B
3. W/G
4. Br
5. L
6. B
7. B/W
8. B/W
9. B/W
10. W
11. P
12. Y/R

(−) \ (+)	Trigger ① W/R	Trigger ② W/B	Trigger ③ W/G	Charge ④ Br	Charge ⑤ L	Ground ⑥ B	Ignition ⑦ B/W	Ignition ⑧ B/W	Ignition ⑨ B/W	Stop ⑩ W	Thermo switch ⑪ P	Over-heat ⑫ Y/R
① W/R		∞	∞	∞	∞	∞	34^{+30}_{-15}	∞	∞	∞	∞	∞
② W/B	∞		∞	∞	∞	∞	∞	34^{+30}_{-15}	∞	∞	∞	∞
③ W/G	∞	∞		∞	∞	∞	∞	∞	34^{+30}_{-15}	∞	∞	∞
④ Br	$*1000^{+000}_{-500}$	$*1000^{+000}_{-500}$	$*1000^{+000}_{-500}$		$*400^{+400}_{-200}$	$*400^{+400}_{-200}$	$*1000^{+000}_{-500}$	$*1000^{+000}_{-500}$	$*1000^{+000}_{-500}$	$*140^{+100}_{-40}$	$*1000^{+000}_{-500}$	$*1000^{+000}_{-500}$
⑤ L	16±6	16±6	16±6	4±2		0	4±2	4±2	4±2	62^{+60}_{-30}	13±6	4.3±2
⑥ B	16±6	16±6	16±6	4±2	0		4±2	4±2	4±2	62^{+60}_{-30}	13±6	4.3±2
⑦ B/W	∞	∞	∞	∞	∞	∞		∞	∞	∞	∞	∞
⑧ B/W	∞	∞	∞	∞	∞	∞	∞		∞	∞	∞	∞
⑨ B/W	∞	∞	∞	∞	∞	∞	∞	∞		∞	∞	∞
⑩ W	∞	∞	∞	∞	∞	∞	∞	∞	∞		∞	∞
⑪ P	∞	∞	∞	∞	∞	∞	∞	∞	∞	∞		∞
⑫ Y/R	∞	∞	∞	∞	∞	∞	∞	∞	∞	∞	∞	

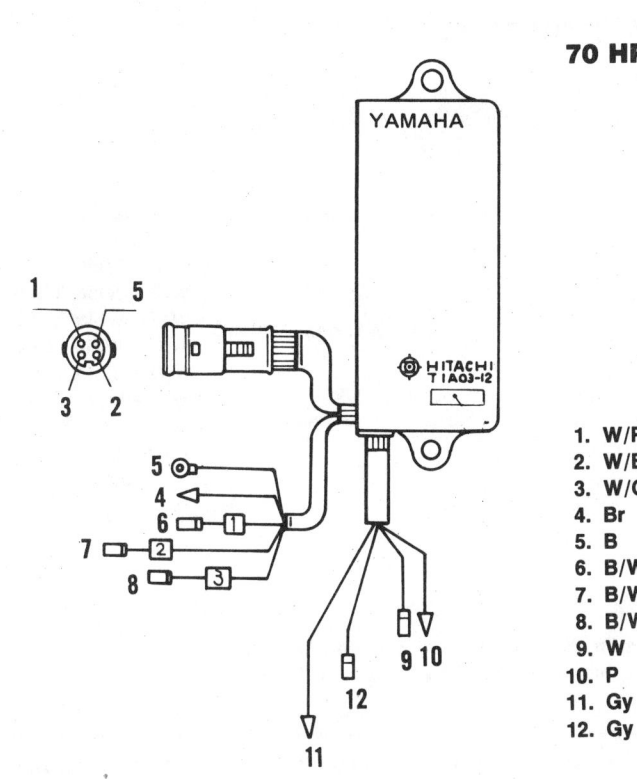

70 HP

W/R—White/red
W/B—White/black
W/G—White/green
Br—Brown
B—Black
B/W—Black/white
W—White
P—Pink
Gy—Gray

3

1. W/R
2. W/B
3. W/G
4. Br
5. B
6. B/W
7. B/W
8. B/W
9. W
10. P
11. Gy
12. Gy

⊕ ⊖		Trigger			Charge	Ground	Ignition			Stop	Thermo switch	Over-rev.	
		1 W/R	2 W/B	3 W/G	4 Br	5 B	6 B/W	7 B/W	8 B/W	9 W	10 P	11 Gy	12 Gy
1	W/R		75	80	49	28	28	55	55	4.5	40	32	∞
2	W/B	79		80	49	28	55	28	55	4.5	40	32	∞
3	W/G	80	80		49	28	55	55	28	4.5	40	32	∞
4	Br	46	46	46		*11	19	19	18	8.2	26	30	∞
5	B	29	29	29	4.1		3.9	3.9	3.9	7.5	29	19	∞
6	B/W	∞	∞	∞	∞	∞		∞	∞	∞	∞	∞	∞
7	B/W	∞	∞	∞	∞	∞	∞		∞	∞	∞	∞	∞
8	B/W	∞	∞	∞	∞	∞	∞	∞		∞	∞	∞	∞
9	W	47	46	46	26	17	26	26	26		20	16	∞
10	P	∞	∞	∞	∞	∞	∞	∞	∞	∞		∞	∞
11	Gy	∞	∞	∞	∞	∞	∞	∞	∞	∞	∞		∞
12	Gy	∞	∞	∞	∞	∞	∞	∞	∞	∞	∞	∞	

* The meter needle will swing toward 0 and slowly return to the specified value if tested component is functioning properly.

(31)

90 HP (1984 TYPE A)

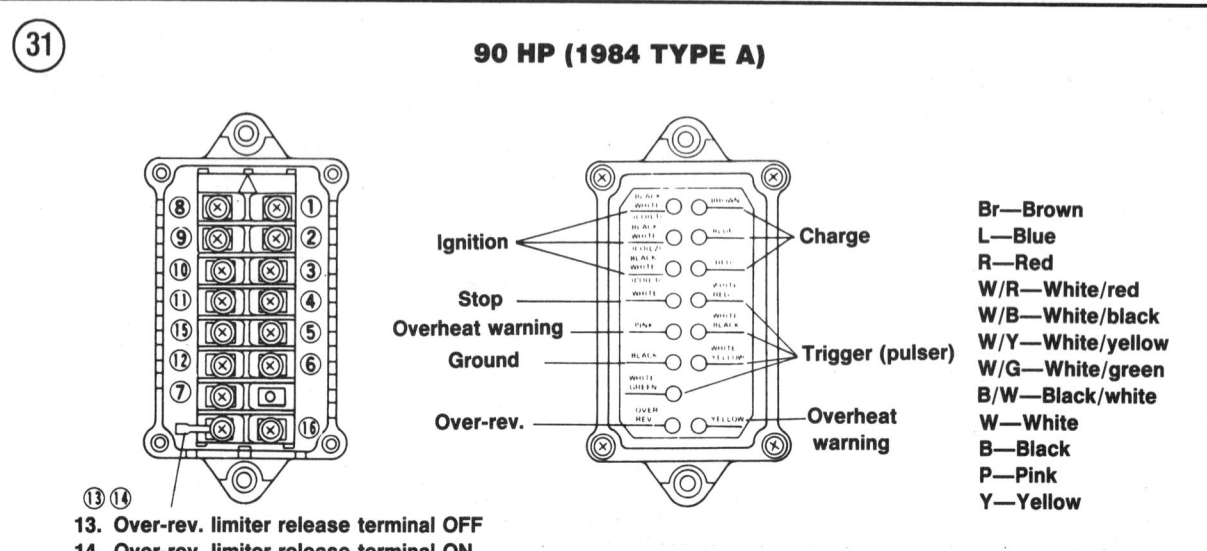

Ignition
Charge
Stop
Overheat warning
Ground
Trigger (pulser)
Over-rev.
Overheat warning

Br—Brown
L—Blue
R—Red
W/R—White/red
W/B—White/black
W/Y—White/yellow
W/G—White/green
B/W—Black/white
W—White
B—Black
P—Pink
Y—Yellow

⑬ ⑭
13. Over-rev. limiter release terminal OFF
14. Over-rev. limiter release terminal ON

* The meter needle will swing toward 0 and slowly return to the specified value if tested component is functioning properly.

⊕ / ⊖	Charge			Trigger				Ignition			Stop	Ground	Over rev.		Overheat	
	① Br	② L	③ R	④ W/R	⑤ W/B	⑥ W/Y	⑦ W/G	⑧ COIL1 B/W	⑨ COIL2 B/W	⑩ COIL3 B/W	⑪ W	⑫ B	⑬ OFF	⑭ ON	⑮ P	⑯ Y
① Br		∞	400~600	400~600	∞	400~600	320~480	∞	∞	∞	400~600	320~480	900~1100	400~600	400~600	12~19
② L	3~6		900~1100	900~1100	∞	900~1100	400~600	∞	∞	∞	900~1100	400~600	900~1100	900~1100	900~1100	24~38
③ R	112~168	∞		*112~168	∞	*112~168	*56~84	∞	∞	∞	*144~216	*56~84	200~300	46~70	72~108	320~480
④ W/R	36~54	∞	72~108		∞	36~54	20~32	∞	∞	∞	3~6	20~32	53~81	104~156	144~216	104~156
⑤ W/B	36~54	∞	72~108	36~54		36~54	20~32	∞	∞	∞	3~6	20~32	53~81	104~156	144~216	104~156
⑥ W/Y	36~54	∞	72~108	36~54	∞		20~32	∞	∞	∞	3~6	20~32	53~81	104~156	144~216	104~156
⑦ W/G	3~5	∞	9~15	3~5	∞	3~5		∞	∞	∞	8~14	0	6~10	15~23	53~81	24~36
⑧ COIL1 B/W	9~15	∞	27~41	9~15	∞	9~15	3~5		∞	∞	24~36	3~5	17~27	44~66	112~168	54~82
⑨ COIL2 B/W	9~15	∞	27~41	9~15	∞	9~15	3~5	∞		∞	24~36	3~5	17~27	44~66	112~168	54~82
⑩ COIL3 B/W	9~15	∞	27~41	9~15	∞	9~15	3~5	∞	∞		24~36	3~5	17~27	44~66	112~168	54~82
⑪ W	68~102	∞	104~156	65~99	∞	65~99	46~70	∞	∞	∞		46~70	88~132	120~180	160~240	120~180
⑫ B	3~5	∞	9~15	3~5	∞	3~5	0	∞	∞	∞	8~14		6~10	15~23	53~81	24~36
⑬ Over.Rev (OFF)	∞	∞	∞	∞	∞	∞	∞	∞	∞	∞	∞	∞		∞	∞	∞
⑭ Over Rev (ON)	∞	∞	∞	∞	∞	∞	∞	∞	∞	∞	∞	∞	∞		∞	∞
⑮ P	∞	∞	∞	∞	∞	∞	∞	∞	∞	∞	∞	∞	∞	∞		∞
⑯ Y	∞	∞	∞	∞	∞	∞	∞	∞	∞	∞	∞	∞	∞	∞	∞	

3

90 HP (1984 TYPE B)

(−) \ (+)	Charge ① Br	② L	③ R	Trigger ④ W/R	⑤ W/B	⑥ W/Y	⑦ W/G	Ignition ⑧ COIL1 B/W	⑨ COIL2 B/W	⑩ COIL3 B/W	Stop ⑪ W	Ground ⑫ B	Over rev. ⑬ OFF	⑭ ON	Overheat ⑮ P	⑯ Y
① Br		∞	400~600	600~900	∞	600~900	400~600	∞	∞	∞	600~900	320~480	900~1100	400~600	400~600	12~19
② L	3~6		900~1100	900~1100	∞	900~1100	900~1100	∞	∞	∞	900~1100	400~600	900~1100	900~1100	900~1100	24~38
③ R	*200~300	∞		*200~300	∞	*200~300	*80~120	∞	∞	∞	*200~300	*400~600	200~300	46~70	72~108	600~900
④ W/R	36~54	∞	72~108		∞	36~54	20~32	∞	∞	∞	3~6	20~32	53~81	*96~144	*112~168	104~156
⑤ W/B	36~54	∞	72~108	36~54		36~54	20~32	∞	∞	∞	3~6	20~32	53~81	104~156	144~216	104~156
⑥ W/Y	36~54	∞	72~108	36~54	∞		20~32	∞	∞	∞	3~6	20~32	53~81	104~156	144~216	104~156
⑦ W/G	3~5	∞	9~15	3~5	∞	3~5		∞	∞	∞	8~14	0	6~10	15~23	53~81	24~36
⑧ COIL1 B/W	9~15	∞	27~41	9~15	∞	9~15	3~5		∞	∞	24~36	3~5	17~27	44~66	112~168	54~82
⑨ COIL2 B/W	9~15	∞	27~41	9~15	∞	9~15	3~5	∞		∞	24~36	3~5	17~27	44~66	76~114	54~82
⑩ COIL3 B/W	9~15	∞	27~41	9~15	∞	9~15	3~5	∞	∞		24~36	3~5	17~27	44~66	76~114	54~82
⑪ W	68~102	∞	104~156	65~99	∞	65~99	46~70	∞	∞	∞		46~70	88~132	120~180	76~114	120~180
⑫ B	3~5	∞	9~15	3~5	∞	3~5	0	∞	∞	∞	8~14		6~10	15~23	53~81	24~36
⑬ Over.Rev. (OFF)	∞	∞	∞	∞	∞	∞	∞	∞	∞	∞	∞	∞		∞	∞	∞
⑭ Over.Rev. (ON)	∞	∞	∞	∞	∞	∞	∞	∞	∞	∞	∞	∞	∞		∞	∞
⑮ P	∞	∞	∞	∞	∞	∞	∞	∞	∞	∞	∞	∞	∞	∞		∞
⑯ Y	∞	∞	∞	∞	∞	∞	∞	∞	∞	∞	∞	∞	∞	∞	∞	

* The meter needle will swing toward 0 and slowly return to the specified value if tested component is functioning properly.

33

90 HP (1985-ON)

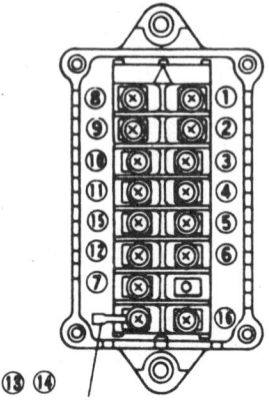

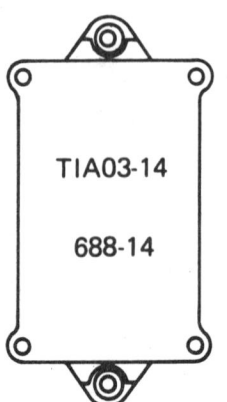

TIA03-14

688-14

1. Brown
2. Blue
3. Red
4. White/red
5. White/black
6. White/yellow
7. White/green
8. Black/white
9. Black/white
10. Black/white
11. White
12. Black
13. Over-rev. ON
14. Over-rev. OFF
15. Pink
16. Yellow

13 & 14. Over-rev. limiter release terminal

		Charge			Pulser				Ignition			Stop	Ground	Over rev.		Overheat	
		(1) Br	(2) L	(3) R	(4) W/R	(5) W/B	(6) W/Y	(7) W/G	(8) Coil 1 B/W	(9) Coil 2 B/W	(10) Coil 3 B/W	(11) W	(12) B	(13) ON	(14) OFF	(15) P	(16) Y
(1)	Br		∞	700~1300	525~975	∞	525~975	350~650	∞	∞	∞	525~975	350~650	700~1300	350~650	350~650	12~19
(2)	L	3.5~5.3		700~1300	700~1300	∞	700~1300	700~1300	∞	∞	∞	700~1300	700~1300	700~1300	700~1300	700~1300	24~37
(3)	R	*175~325	∞		*175~325	∞	*175~325	*70~130	∞	∞	∞	*175~325	*70~650	*350~650	48~72	72~108	525~975
(4)	W/R	38~58	∞	77~143		∞	38~58	21~31	∞	∞	∞	3.4~5.2	21~31	56~84	*84~156	*98~182	98~182
(5)	W/B	38~58	∞	77~143	38~58		38~58	21~31	∞	∞	∞	3.4~5.2	21~31	56~84	98~182	112~208	98~182
(6)	W/Y	39~59	∞	77~143	38~58	∞		21~31	∞	∞	∞	3.4~5.2	21~31	56~84	98~182	112~208	98~182
(7)	W/G	3.5~5.3	∞	10~16	3.4~5.2	∞	3.5~5.3		∞	∞	∞	9~14	0	6~10	16~24	53~81	25~38
(8)	Coil 1 B/W	10~16	∞	32~48	10~16	∞	10~16	3.2~4.8		∞	∞	25~38	3.2~4.8	18~27	48~72	76~114	60~98
(9)	Coil 2 B/W	10~16	∞	32~48	10~16	∞	10~16	3.2~4.8	∞		∞	25~38	3.2~4.8	18~27	48~72	76~114	54~98
(10)	Coil 3 B/W	10~16	∞	32~48	10~16	∞	10~16	3.2~4.8	∞	∞		25~38	3.2~4.8	18~27	48~72	76~114	60~98
(11)	W	68~102	∞	98~182	68~102	∞	68~102	48~72	∞	∞	∞		48~72	77~143	112~208	140~260	105~195
(12)	B	3.5~5.3	∞	10~16	3.4~5.2	∞	3.5~5.3	0	∞	∞	∞	9~14		6~10	16~24	53~81	25~38
(13)	Over rev. (ON)	∞	∞	∞	∞	∞	∞	∞	∞	∞	∞	∞	∞		∞	∞	∞
(14)	Over rev. (OFF)	∞	∞	∞	∞	∞	∞	∞	∞	∞	∞	∞	∞	∞		∞	∞
(15)	P	∞	∞	∞	∞	∞	∞	∞	∞	∞	∞	∞	∞	∞	∞		∞
(16)	Y	∞	∞	∞	∞	∞	∞	∞	∞	∞	∞	∞	∞	∞	∞	∞	

* The meter needle will swing toward 0 and slowly return to the specified value if tested component is functioning properly.

the stated value. If any of the meter readings differ from the stated values, replace the CDI unit.

115-130 hp

1. Remove the CDI unit cover.
2. Set the Pocket Tester on the R×1,000 scale.

> *NOTE*
> *Disconnect the lead wire from the CDI over-rev terminal and measure resistance to find the OFF value, then reconnect the wire and measure resistance again to find the ON value.*

3. Refer to **Figure 34** or **Figure 35** for 1984 models and **Figure 36** (1985-on) for test connections and values. Make each connection and compare the meter reading to the stated value. If any of the meter readings differ from the stated values, replace the CDI unit.

150-200 hp

1. Remove the CDI unit cover.
2. Set the Pocket Tester on the R×1,000 scale.
3. Refer to **Figure 37** (1984), **Figure 38** (1985) or **Figure 39** (1986-on) for test connections and values. Make each connection and compare the meter reading to the stated value. If any of the meter readings differ from the stated value, replace the CDI unit.

CDI Control Unit Resistance Test (150-200 hp Only)

This unit contains the circuitry for the over-rev and overheat warning system. The 40-130 hp models incorporate the circuitry in the CDI unit; the 220 hp and 225 hp systems are controlled by the YMIS computer.

The resistance values provided by Yamaha are based on the use of its Pocket Tester (part No. YU-3112). If another ohmmeter is used, the readings obtained may not agree with those specified due to internal resistance of the ohmmeter.

1. Remove the CDI unit cover.
2. Disconnect the control unit leads at the CDI unit.
3. Set the Pocket Tester on the R×100 scale.
4. Refer to **Figure 40** for connections and test values. Make each connection and compare the meter reading to the stated value. If any of the meter readings differ from the stated values, replace the control unit.

Oil Pump Control Unit Resistance Test

The resistance values provided by Yamaha are based on the use of its Pocket Tester (part No. YU-3112). If another ohmmeter is used, the readings obtained may not agree with those specified due to internal resistance of the ohmmeter.

40-50 hp

1. Disconnect the oil pump control unit leads.
2. Set the Pocket Tester on the R×1,000 scale.
3. Refer to **Figure 41** for connections and test values. Make each connection and compare the meter reading to the stated value. If any of the meter readings differ from the stated values, replace the oil pump control unit.

1984-1985 70 and 90 hp

1. Disconnect the oil pump control unit leads.
2. Set the Pocket Tester on the R×1,000 scale.
3. Refer to **Figure 42** for connections and test values. Make each connection and compare

3

**TYPE A AND B
OIL LEVEL
WARNING SYSTEM
(1984 115 HP)**

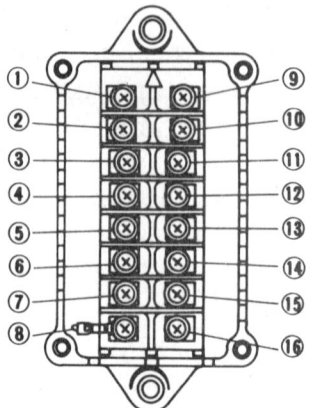

1. Brown
2. Blue
3. Red
4. Black/red
5. White/red
6. White/black
7. White/yellow
8. White/green
9. Black/white
10. Black/white
11. Black/white
12. Black/white
13. White
14. Pink
15. Black
16. Over-rev.

* The meter needle will swing toward 0 and slowly return to the specified value if tested component is functioning properly.

Tester (+) / Tester (−)	Charge				Trigger				Ignition				Stop	Ground		Over-rev.	
	Br	L	R	B/R	W/R	W/B	W/Y	W/G	Coil 1 (B/W)	Coil 2 (B/W)	Coil 3 (B/W)	Coil 4 (B/W)	W	P	B	OR	OR LE104
① Br		120~180	160~240	120~180	120~180	120~180	120~180	120~180	120~180	120~180	120~180	120~180	160~240	200~300	80~120	160~240	∞
② L	3.6~5.4		*120~180	*72~108	*72~108	*72~108	*72~108	*72~108	*72~108	*72~108	*72~108	*72~108	120~180	72~108	*40~60	48~72	∞
③ R	160~240	120~180		120~180	120~180	120~180	120~180	120~180	120~180	120~180	120~180	120~180	160~240	160~240	80~120	160~240	∞
④ B/R	*120~180	*72~108	3.6~5.4		*72~108	*72~108	*72~108	*72~108	*72~108	*72~108	*72~108	*72~108	120~180	72~108	40~60	48~72	∞
⑤ W/R	96~134	44~66	96~134	44~66		44~66	44~66	44~66	28~42	44~66	44~66	44~66	12~18	*160~240	24~36	*129~180	∞
⑥ W/B	120~180	72~108	120~180	72~108	72~108		72~108	72~108	72~108	32~48	72~108	72~108	12~18	*160~240	40~60	*160~240	∞
⑦ W/Y	130~180	72~108	120~180	72~108	72~108	72~108		72~108	72~108	72~108	32~48	72~108	12~18	*160~240	40~60	*160~240	∞
⑧ W/G	120~180	72~108	120~180	72~108	72~108	72~108	72~108		72~108	72~108	72~108	32~48	12~18	*160~240	40~60	*160~240	∞
⑨ Coil 1 (B/W)	∞	∞	∞	∞	∞	∞	∞	∞		∞	∞	∞	∞	∞	∞	∞	∞
⑩ Coil 2 (B/W)	∞	∞	∞	∞	∞	∞	∞	∞	∞		∞	∞	∞	∞	∞	∞	∞
⑪ Coil 3 (B/W)	∞	∞	∞	∞	∞	∞	∞	∞	∞	∞		∞	∞	∞	∞	∞	∞
⑫ Coil 4 (B/W)	∞	∞	∞	∞	∞	∞	∞	∞	∞	∞	∞		∞	∞	∞	∞	∞
⑬ W	∞	∞	∞	∞	∞	∞	∞	∞	∞	∞	∞	∞		∞	∞	∞	∞
⑭ P	∞	∞	∞	∞	∞	∞	∞	∞	∞	∞	∞	∞	∞		∞	∞	∞
⑮ B	12~18	3.6~5.4	12~18	3.6~5.4	3.6~5.4	3.6~5.4	3.6~5.4	3.6~5.4	3.6~5.4	3.6~5.4	3.6~5.4	3.6~5.4	24~36	*56~84		*20~30	∞
⑯ OR	∞	∞	∞	∞	∞	∞	∞	∞	∞	∞	∞	∞	∞	∞	∞		∞
⑯ OR LE104	∞	∞	∞	∞	∞	∞	∞	∞	∞	∞	∞	∞	∞	∞	∞	∞	

3

TYPE C OIL LEVEL WARNING SYSTEM (1984 115 HP)

Tester (−) \ Tester (+)	Charge				Trigger				Ignition				Stop	Ground		Over-rev.	
	Br	L	R	B/R	W/R	W/B	W/Y	W/G	Coil 1 (B/W)	Coil 2 (B/W)	Coil 3 (B/W)	Coil 4 (B/W)	W	P	B	OR	OR LE104
① Br		120~180	160~240	120~180	120~180	120~180	120~180	120~180	120~180	120~180	120~180	120~180	160~240	160~240	80~120	160~240	∞
② L	3.6~5.4		*120~180	*72~108	*72~108	*72~108	*72~108	*72~108	*72~108	*72~108	*72~108	*72~108	120~180	72~108	*40~60	48~72	∞
③ R	160~240	120~180		120~180	120~180	120~180	120~180	120~180	120~180	120~180	120~180	120~180	160~240	160~240	80~120	160~240	∞
④ B/R	*120~180	*72~108	3.6~5.4		*72~108	*72~108	*72~108	*72~108	*72~108	*72~108	*72~108	*72~108	120~180	72~108	*40~60	48~72	∞
⑤ W/R	96~134	44~66	96~134	64~96		62~94	64~96	64~96	36~48	60~90	60~90	60~90	12~18	176~264	36~54	128~192	∞
⑥ W/B	120~180	72~108	120~180	72~108	72~108		72~108	72~108	72~108	32~48	72~108	72~108	12~18	200~300	40~60	*160~240	∞
⑦ W/Y	130~180	72~108	120~180	72~108	72~108	72~108		72~108	72~108	72~108	32~48	72~108	12~18	200~300	40~60	*160~240	∞
⑧ W/G	120~180	72~108	120~180	72~108	72~108	72~108	72~108		72~108	72~108	72~108	32~48	12~18	200~300	40~60	*160~240	∞
⑨ Coil 1 (B/W)	∞	∞	∞	∞	∞	∞	∞	∞		∞	∞	∞	∞	∞	∞	∞	∞
⑩ Coil 2 (B/W)	∞	∞	∞	∞	∞	∞	∞	∞	∞		∞	∞	∞	∞	∞	∞	∞
⑪ Coil 3 (B/W)	∞	∞	∞	∞	∞	∞	∞	∞	∞	∞		∞	∞	∞	∞	∞	∞
⑫ Coil 4 (B/W)	∞	∞	∞	∞	∞	∞	∞	∞	∞	∞	∞		∞	∞	∞	∞	∞
⑬ W	∞	∞	∞	∞	∞	∞	∞	∞	∞	∞	∞	∞		∞	∞	∞	∞
⑭ P	∞	∞	∞	∞	∞	∞	∞	∞	∞	∞	∞	∞	∞			∞	∞
⑮ B	12~18	3.6~5.4	12~18	3.6~5.4	3.6~5.4	3.6~5.4	3.6~5.4	3.6~5.4	3.6~5.4	3.6~5.4	3.6~5.4	3.6~5.4	24~36	30~46		19~29	∞
⑯ OR	∞	∞	∞	∞	∞	∞	∞	∞	∞	∞	∞	∞	∞	∞	∞		∞
⑯ OR LE104	∞	∞	∞	∞	∞	∞	∞	∞	∞	∞	∞	∞	∞	∞	∞	∞	

* The meter needle will swing toward 0 and slowly return to the specified value if tested component is functioning properly.

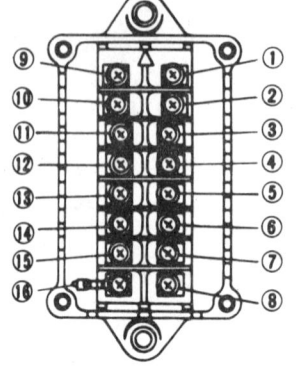

36

115 HP (1985-ON) AND 130 HP

1. Brown
2. Blue
3. Red
4. Black/red
5. White/red
6. White/black
7. White/yellow
8. White/green
9. Black/white
10. Black/white
11. Black/white
12. Black/white
13. White
14. Pink
15. Black
16. Over-rev.

(−) → (+) ↓		Charge				Trigger				Ignition				Stop	Ground		Over-rev	
		Br	L	R	B/R	W/R	W/B	W/Y	W/G	Coil 1 (B/W)	Coil 2 (B/W)	Coil 3 (B/W)	Coil 4 (B/W)	W	P	B	OR	OR LE104
①	Br		77~143	105~195	77~143	77~143	77~143	77~143	84~156	77~143	77~143	77~143	77~143	112~208	140~260	68~102	105~195	∞
②	L	3.6~5.4		*98~182	*64~96	*64~96	*64~96	*60~90	*64~96	*64~96	*64~96	*64~96	*64~96	*105~195	60~90	*37~57	48~72	∞
③	R	105~195	77~143		77~143	77~143	77~143	77~143	84~156	77~143	77~143	77~143	77~143	112~208	140~260	68~102	140~260	∞
④	B/R	*91~169	*64~96	3.6~5.4		*64~96	*64~96	*64~96	*64~96	*64~96	*64~96	*64~96	*64~96	98~182	60~90	*37~57	48~72	∞
⑤	W/R	91~169	60~90	91~169	60~90		60~90	56~84	60~90	29~43	56~84	56~84	56~84	12~18	122~227	38~54	112~208	∞
⑥	W/B	98~182	64~96	98~182	64~96	64~96		60~90	64~96	60~90	29~44	60~90	60~90	12~18	140~260	37~57	119~221	∞
⑦	W/Y	98~182	64~96	98~182	64~96	64~96	64~96		64~96	60~90	60~90	29~44	56~84	12~18	140~260	37~57	119~221	∞
⑧	W/G	98~182	64~96	98~182	64~96	64~96	64~96	60~90		60~90	60~90	60~90	29~44	12~18	140~260	37~57	119~221	∞
⑨	Coil 1 (B/W)	∞	∞	∞	∞	∞	∞	∞	∞		∞	∞	∞	∞	∞	∞	∞	∞
⑩	Coil 2 (B/W)	∞	∞	∞	∞	∞	∞	∞	∞	∞		∞	∞	∞	∞	∞	∞	∞
⑪	Coil 3 (B/W)	∞	∞	∞	∞	∞	∞	∞	∞	∞	∞		∞	∞	∞	∞	∞	∞
⑫	Coil 4 (B/W)	∞	∞	∞	∞	∞	∞	∞	∞	∞	∞	∞		∞	∞	∞	∞	∞
⑬	W	∞	∞	∞	∞	∞	∞	∞	∞	∞	∞	∞	∞		∞	∞	∞	∞
⑭	P	∞	∞	∞	∞	∞	∞	∞	∞	∞	∞	∞	∞	∞		∞	∞	∞
⑮	B	11.8~17.8	3.7~5.5	11.8~17.8	3.6~5.5	3.7~5.5	3.7~5.5	3.2~4.9	3.7~5.5	3.0~4.6	3.2~4.8	3.0~4.6	3.1~4.7	24~36	28~42		17~26	∞
⑯	OR	∞	∞	∞	∞	∞	∞	∞	∞	∞	∞	∞	∞	∞	∞	∞		∞
⑯	OR LE104	∞	∞	∞	∞	∞	∞	∞	∞	∞	∞	∞	∞	∞	∞	∞	∞	

* The meter needle will swing toward 0 and slowly return to the specified value if tested component is functioning properly.

150-200 HP (1984)

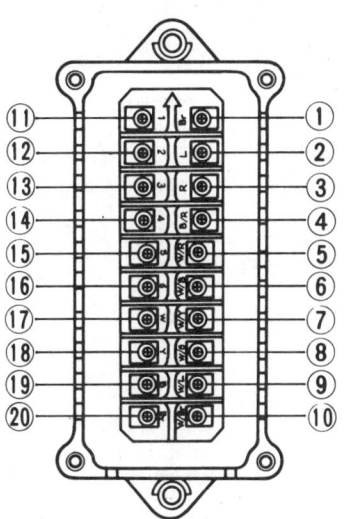

1. Brown
2. Blue
3. Red
4. Black/red
5. White/red
6. White/black
7. White/yellow
8. White/green
9. Black/blue
10. Black/brown
11. Ignition 1
12. Ignition 2
13. Ignition 3
14. Ignition 4
15. Ignition 5
16. Ignition 6
17. White
18. Yellow
19. Black
20. Brown

3

− +		Charge				Trigger						Ignition						Stop	Ground & control		
		Br	L	R	B/R	W/R	W/B	W/Y	W/G	W/L	W/Br	1	2	3	4	5	6	W	Y	B	Br
1	Br		0.6	100~∞	100~∞	100~∞	100~∞	100~∞	100~∞	100~∞	100~∞	∞	∞	∞	∞	∞	∞	100~∞	26	2.5	∞
2	L	100~∞		100~∞	100~∞	100~∞	100~∞	100~∞	100~∞	100~∞	100~∞	∞	∞	∞	∞	∞	∞	100~∞	6.8	0.6	∞
3	R	100~∞	100~∞		0.6	100~∞	100~∞	100~∞	100~∞	100~∞	100~∞	∞	∞	∞	∞	∞	∞	100~∞	26	2.5	∞
4	B/R	100~∞	100~∞	100~∞		100~∞	100~∞	100~∞	100~∞	100~∞	100~∞	∞	∞	∞	∞	∞	∞	100~∞	6.8	0.6	∞
5	W/R	100~∞	100~∞	100~∞	100~∞		100~∞	100~∞	100~∞	100~∞	100~∞	∞	∞	∞	∞	∞	∞	100~∞	6.8	0.6	∞
6	W/B	100~∞	100~∞	100~∞	100~∞	100		100~∞	100~∞	100~∞	100~∞	∞	∞	∞	∞	∞	∞	100~∞	6.8	0.6	∞
7	W/Y	100~∞	100~∞	100~∞	100~∞	100~∞	100~∞		100~∞	100~∞	100~∞	∞	∞	∞	∞	∞	∞	100~∞	6.8	0.6	∞
8	W/G	100~∞	100~∞	100~∞	100~∞	100~∞	100~∞	100~∞		100~∞	100~∞	∞	∞	∞	∞	∞	∞	100~∞	6.8	0.6	∞
9	W/L	100~∞	100~∞	100~∞	100~∞	100~∞	100~∞	100~∞	100~∞		100~∞	∞	∞	∞	∞	∞	∞	100~∞	6.8	0.6	∞
10	W/Br	100~∞	100~∞	100~∞	100~∞	100~∞	100~∞	100~∞	100~∞	100~∞		∞	∞	∞	∞	∞	∞	100~∞	6.8	0.6	∞
11	1	100~∞	100~∞	100~∞	100~∞	45	100~∞	100~∞	100~∞	100~∞	100~∞		∞	∞	∞	∞	∞	100~∞	6	0.5	∞
12	2	100~∞	100~∞	100~∞	100~∞	100~∞	45	100~∞	100~∞	100~∞	100~∞	∞		∞	∞	∞	∞	100~∞	6	0.5	∞
13	3	100~∞	100~∞	100~∞	100~∞	100~∞	100~∞	45	100~∞	100~∞	100~∞	∞	∞		∞	∞	∞	100~∞	6	0.5	∞
14	4	100~∞	100~∞	100~∞	100~∞	100~∞	100~∞	100~∞	45	100~∞	100~∞	∞	∞	∞		∞	∞	100~∞	6	0.5	∞
15	5	100~∞	100~∞	100~∞	100~∞	100~∞	100~∞	100~∞	100~∞	45	100~∞	∞	∞	∞	∞		∞	100~∞	6	0.5	∞
16	6	100~∞	100~∞	100~∞	100~∞	100~∞	100~∞	100~∞	100~∞	100~∞	45	∞	∞	∞	∞	∞		100~∞	6	0.5	∞
17	W	100~∞	100~∞	100~∞	100~∞	0.6	0.6	0.6	0.6	0.6	0.6	∞	∞	∞	∞	∞	∞		17	1.6	∞
18	Y	100~∞	100~∞	100~∞	100~∞	100~∞	100~∞	100~∞	100~∞	100~∞	100~∞	∞	∞	∞	∞	∞	∞	100~∞		10	∞
19	B	100~∞	100~∞	100~∞	100~∞	100~∞	100~∞	100~∞	100~∞	100~∞	100~∞	∞	∞	∞	∞	∞	∞	100~∞	2.5		∞
20	Br	0.6	2.5	100~∞	100~∞	100~∞	100~∞	100~∞	100~∞	100~∞	100~∞	∞	∞	∞	∞	∞	∞	100~∞	75	14	

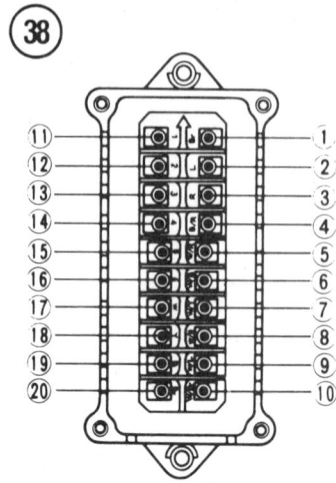

38

150-200 HP (1985)

1. Brown	11. Ignition 1
2. Blue	12. Ignition 2
3. Red	13. Ignition 3
4. Black/red	14. Ignition 4
5. White/red	15. Ignition 5
6. White/black	16. Ignition 6
7. White/yellow	17. White
8. White/green	18. Yellow
9. Black/blue	19. Black
10. Black/brown	20. Brown

Tester (−) → Tester (+) ↓		Charge				Pulser						Ignition						Stop	Ground & control		
		Br	L	R	B/R	W/R	W/B	W/Y	W/G	W/L	W/Br	1	2	3	4	5	6	W	Y	B	Br
1	Br		0.6	100~∞	100~∞	100~∞	100~∞	100~∞	100~∞	100~∞	100~∞	∞	∞	∞	∞	∞	∞	100~∞	26	2.5	∞
2	L	100~∞		100~∞	100~∞	100~∞	100~∞	100~∞	100~∞	100~∞	100~∞	∞	∞	∞	∞	∞	∞	100~∞	6.8	0.6	∞
3	R	100~∞	100~∞		0.6	100~∞	100~∞	100~∞	100~∞	100~∞	100~∞	∞	∞	∞	∞	∞	∞	100~∞	26	2.5	∞
4	B/R	100~∞	100~∞	100~∞		100~∞	100~∞	100~∞	100~∞	100~∞	100~∞	∞	∞	∞	∞	∞	∞	100~∞	6.8	0.6	∞
5	W/R	100~∞	100~∞	100~∞	100~∞		100~∞	100~∞	100~∞	100~∞	100~∞	∞	∞	∞	∞	∞	∞	100~∞	6.8	0.6	∞
6	W/B	100~∞	100~∞	100~∞	100~∞	100~∞		100~∞	100~∞	100~∞	100~∞	∞	∞	∞	∞	∞	∞	100~∞	6.8	0.6	∞
7	W/Y	100~∞	100~∞	100~∞	100~∞	100~∞	100~∞		100~∞	100~∞	100~∞	∞	∞	∞	∞	∞	∞	100~∞	6.8	0.6	∞
8	W/G	100~∞	100~∞	100~∞	100~∞	100~∞	100~∞	100~∞		100~∞	100~∞	∞	∞	∞	∞	∞	∞	100~∞	6.8	0.6	∞
9	W/L	100~∞	100~∞	100~∞	100~∞	100~∞	100~∞	100~∞	100~∞		100~∞	∞	∞	∞	∞	∞	∞	100~∞	6.8	0.6	∞
10	W/Br	100~∞	100~∞	100~∞	100~∞	100~∞	100~∞	100~∞	100~∞	100~∞		∞	∞	∞	∞	∞	∞	100~∞	6.8	0.6	∞
11	1	100~∞	100~∞	100~∞	100~∞	45	100~∞	100~∞	100~∞	100~∞	100~∞		∞	∞	∞	∞	∞	100~∞	6	0.5	∞
12	2	100~∞	100~∞	100~∞	100~∞	100~∞	45	100~∞	100~∞	100~∞	100~∞	∞		∞	∞	∞	∞	100~∞	6	0.5	∞
13	3	100~∞	100~∞	100~∞	100~∞	100~∞	100~∞	45	100~∞	100~∞	100~∞	∞	∞		∞	∞	∞	100~∞	6	0.5	∞
14	4	100~∞	100~∞	100~∞	100~∞	100~∞	100~∞	100~∞	45	100~∞	100~∞	∞	∞	∞		∞	∞	100~∞	6	0.5	∞
15	5	100~∞	100~∞	100~∞	100~∞	100~∞	100~∞	100~∞	100~∞	45	100~∞	∞	∞	∞	∞		∞	100~∞	6	0.5	∞
16	6	100~∞	100~∞	100~∞	100~∞	100~∞	100~∞	100~∞	100~∞	100~∞	45	∞	∞	∞	∞	∞		100~∞	6	0.5	∞
17	W	100~∞	100~∞	100~∞	100~∞	0.6	0.6	0.6	0.6	0.6	0.6	∞	∞	∞	∞	∞	∞		17	1.6	∞
18	Y	100~∞	100~∞	100~∞	100~∞	100~∞	100~∞	100~∞	100~∞	100~∞	100~∞	∞	∞	∞	∞	∞	∞	100~∞		10	∞
19	B	100~∞	100~∞	100~∞	100~∞	100~∞	100~∞	100~∞	100~∞	100~∞	100~∞	∞	∞	∞	∞	∞	∞	100~∞	2.5		∞
20	Br	100~∞	100~∞	0.6	2.5	100~∞	100~∞	100~∞	100~∞	100~∞	100~∞	∞	∞	∞	∞	∞	∞	100~∞	75	14	

☐ *Revised

150-200 HP (1986-ON)

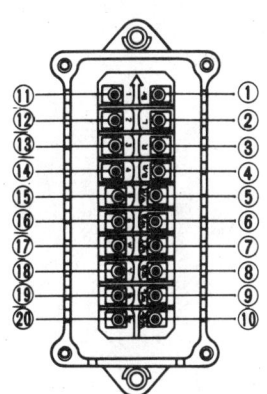

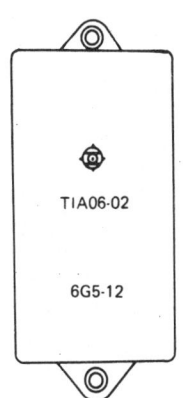

TIA06-02

6G5-12

CDI unit test by Yamaha pocket tester

Unit: kΩ ± 20%

(+) \ (−)	Charge ① Br	② L	③ R	④ B/R	Pulser ⑤ W/R	⑥ W/B	⑦ W/Y	⑧ W/G	⑨ W/L	⑩ W/Br	Ignition ⑪ 1	⑫ 2	⑬ 3	⑭ 4	⑮ 5	⑯ 6	Stop ⑰ W	Ground & control ⑱ Y	⑲ B	⑳ Br
① Br		0.56	100~∞	100~∞	100~∞	100~∞	100~∞	100~∞	100~∞	100~∞	∞	∞	∞	∞	∞	∞	100~∞	27	2.1	∞
② L	100~∞		100~∞	100~∞	100~∞	100~∞	100~∞	100~∞	100~∞	100~∞	∞	∞	∞	∞	∞	∞	100~∞	7	0.56	∞
③ R	100~∞	100~∞		0.56	100~∞	100~∞	100~∞	100~∞	100~∞	100~∞	∞	∞	∞	∞	∞	∞	100~∞	27	2.1	∞
④ B/R	100~∞	100~∞	100~∞		100~∞	100~∞	100~∞	100~∞	100~∞	100~∞	∞	∞	∞	∞	∞	∞	100~∞	7	0.56	∞
⑤ W/R	100~∞	100~∞	100~∞	100~∞		100~∞	100~∞	100~∞	100~∞	100~∞	∞	∞	∞	∞	∞	∞	100~∞	7	0.56	∞
⑥ W/B	100~∞	100~∞	100~∞	100~∞	100~∞		100~∞	100~∞	100~∞	100~∞	∞	∞	∞	∞	∞	∞	100~∞	7	0.56	∞
⑦ W/Y	100~∞	100~∞	100~∞	100~∞	100~∞	100~∞		100~∞	100~∞	100~∞	∞	∞	∞	∞	∞	∞	100~∞	7	0.56	∞
⑧ W/G	100~∞	100~∞	100~∞	100~∞	100~∞	100~∞	100~∞		100~∞	100~∞	∞	∞	∞	∞	∞	∞	100~∞	7	0.56	∞
⑨ W/L	100~∞	100~∞	100~∞	100~∞	100~∞	100~∞	100~∞	100~∞		100~∞	∞	∞	∞	∞	∞	∞	100~∞	7	0.56	∞
⑩ W/Br	100~∞	100~∞	100~∞	100~∞	100~∞	100~∞	100~∞	100~∞	100~∞		∞	∞	∞	∞	∞	∞	100~∞	7	0.56	∞
⑪ 1	100~∞	100~∞	100~∞	100~∞	40	100~∞	100~∞	100~∞	100~∞	100~∞		∞	∞	∞	∞	∞	100~∞	7	0.56	∞
⑫ 2	100~∞	100~∞	100~∞	100~∞	100~∞	40	100~∞	100~∞	100~∞	100~∞	∞		∞	∞	∞	∞	100~∞	7	0.56	∞
⑬ 3	100~∞	100~∞	100~∞	100~∞	100~∞	100~∞	40	100~∞	100~∞	100~∞	∞	∞		∞	∞	∞	100~∞	7	0.56	∞
⑭ 4	100~∞	100~∞	100~∞	100~∞	100~∞	100~∞	100~∞	40	100~∞	100~∞	∞	∞	∞		∞	∞	100~∞	7	0.56	∞
⑮ 5	100~∞	100~∞	100~∞	100~∞	100~∞	100~∞	100~∞	100~∞	40	100~∞	∞	∞	∞	∞		∞	100~∞	7	0.56	∞
⑯ 6	100~∞	100~∞	100~∞	100~∞	100~∞	100~∞	100~∞	100~∞	100~∞	40	∞	∞	∞	∞	∞		100~∞	7	0.56	∞
⑰ W	100~∞	100~∞	100~∞	100~∞	0.56	0.56	0.56	0.56	0.56	0.56	∞	∞	∞	∞	∞	∞		20	1.6	∞
⑱ Y	100~∞	100~∞	100~∞	100~∞	100~∞	100~∞	100~∞	100~∞	100~∞	100~∞	∞	∞	∞	∞	∞	∞	100~∞		12	∞
⑲ B	100~∞	100~∞	100~∞	100~∞	100~∞	100~∞	100~∞	100~∞	100~∞	100~∞	∞	∞	∞	∞	∞	∞	100~∞	2.8		∞
⑳ Br	100~∞	100~∞	0.56	2.2	100~∞	100~∞	100~∞	100~∞	100~∞	100~∞	∞	∞	∞	∞	∞	∞	100~∞	75	13	

∞ : No continuity

 Revised

3

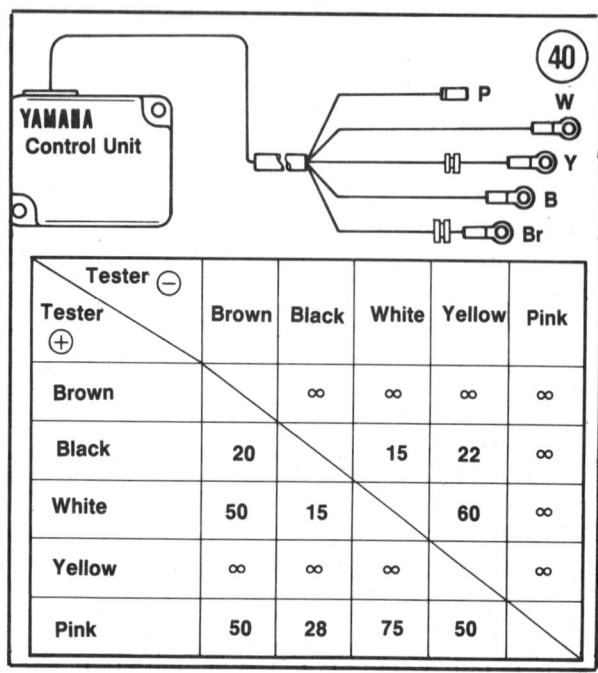

Tester (−) \ Tester (+)	Brown	Black	White	Yellow	Pink
Brown		∞	∞	∞	∞
Black	20		15	22	∞
White	50	15		60	∞
Yellow	∞	∞	∞		∞
Pink	50	28	75	50	

the meter reading to the stated value. If any of the meter readings differ from the stated values, replace the oil pump control unit.

1986-on 70 and 90 hp

1. Disconnect the oil pump control unit leads.
2. Set the Pocket Tester on the R×1,000 scale.
3. Refer to **Figure 43** for connections and test values. Make each connection and compare the meter reading to the stated value. If any of the meter readings differ from the stated value, replace the oil pump control unit.

115-130 hp

1. Disconnect the oil pump control unit leads.
2. Set the Pocket Tester on the R×1,000 scale.
3. Refer to **Figure 44** for connections and test values. Make each connection and compare the meter reading to the stated value. If any of the meter readings differ from the stated values, replace the oil pump control unit.

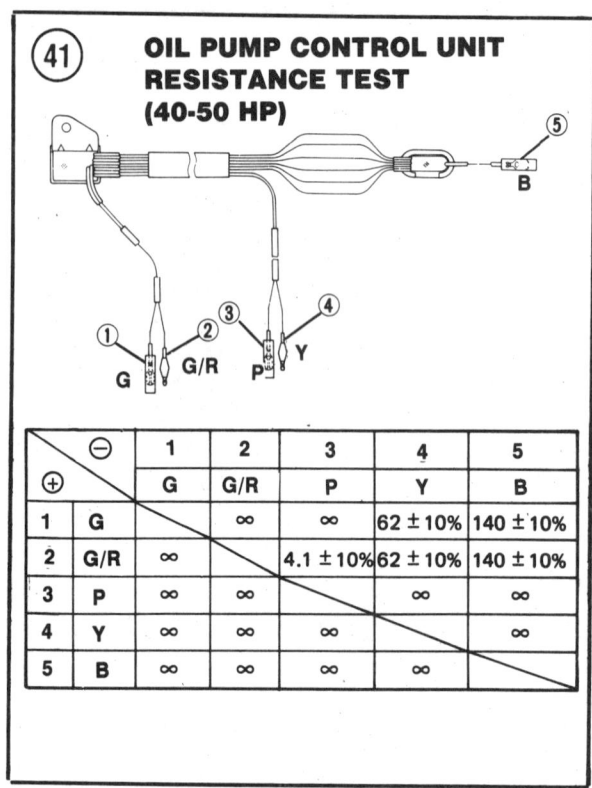

OIL PUMP CONTROL UNIT RESISTANCE TEST (40-50 HP)

(−) / (+)		1	2	3	4	5
		G	G/R	P	Y	B
1	G		∞	∞	62 ± 10%	140 ± 10%
2	G/R	∞		4.1 ± 10%	62 ± 10%	140 ± 10%
3	P	∞	∞		∞	∞
4	Y	∞	∞	∞		∞
5	B	∞	∞	∞	∞	

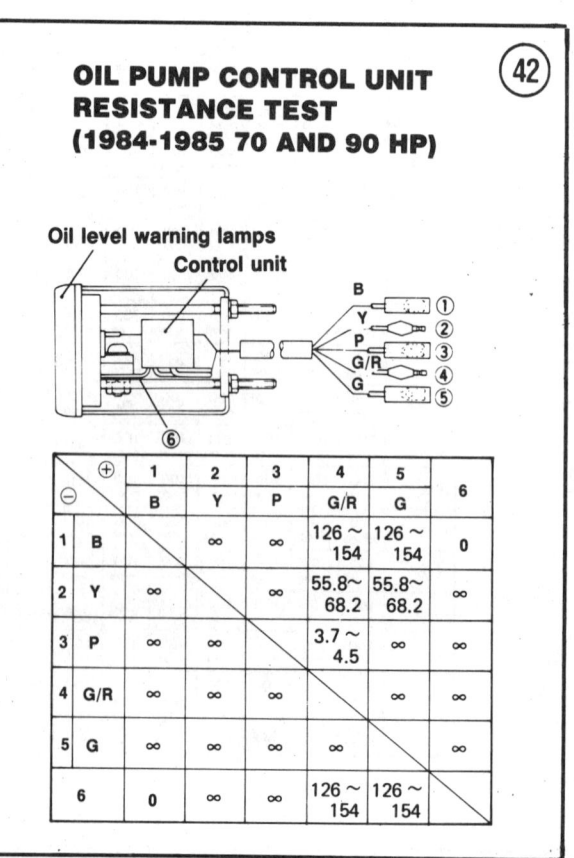

OIL PUMP CONTROL UNIT RESISTANCE TEST (1984-1985 70 AND 90 HP)

(−) / (+)		1	2	3	4	5	6
		B	Y	P	G/R	G	
1	B		∞	∞	126 ~ 154	126 ~ 154	0
2	Y	∞		∞	55.8~ 68.2	55.8~ 68.2	∞
3	P	∞	∞		3.7 ~ 4.5	∞	∞
4	G/R	∞	∞	∞		∞	∞
5	G	∞	∞	∞	∞		∞
6		0	∞	∞	126 ~ 154	126 ~ 154	

150-225 hp

1. Disconnect the oil pump control unit leads.

2. Set the Pocket Tester on the R×1,000 scale.

3. Refer to **Figure 45** for connections and test values. Make each connection and compare the meter reading to the stated value. If any of the meter readings differ from the stated values, replace the oil pump control unit.

Thermoswitch Continuity Test

The resistance values provided by Yamaha are based on the use of its Pocket Tester (part No. YU-3112). If another ohmmeter is used, the readings obtained may not agree with those specified due to internal resistance of the ohmmeter.

1. Remove the engine cover.

2. Disconnect and remove the thermoswitch from the power head. See **Figure 46** (typical).

3. Pour some water in a container that can be heated. Suspend a thermometer in the container.

4. Connect the Pocket Tester to the thermoswitch leads and suspend the tip of the

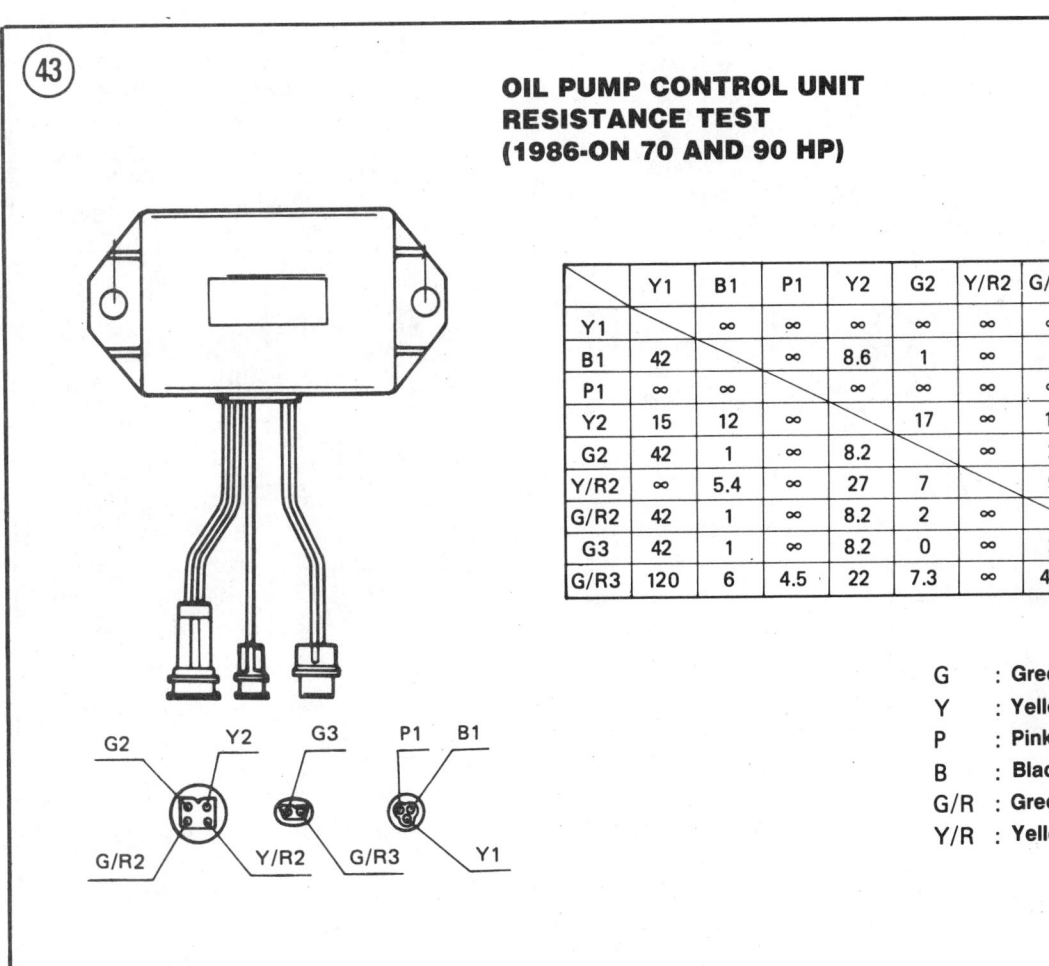

(43)

**OIL PUMP CONTROL UNIT
RESISTANCE TEST
(1986-ON 70 AND 90 HP)**

Unit: kΩ ± 20%

	Y1	B1	P1	Y2	G2	Y/R2	G/R2	G3	G/R3
Y1		∞	∞	∞	∞	∞	∞	∞	∞
B1	42		∞	8.6	1	∞	1	1	∞
P1	∞	∞		∞	∞	∞	∞	∞	∞
Y2	15	12	∞		17	∞	17	17	∞
G2	42	1	∞	8.2		∞	2	0	∞
Y/R2	∞	5.4	∞	27	7		9	7	∞
G/R2	42	1	∞	8.2	2	∞		2	∞
G3	42	1	∞	8.2	0	∞	2		∞
G/R3	120	6	4.5	22	7.3	∞	4.5	7.3	

G   : Green
Y   : Yellow
P   : Pink
B   : Black
G/R : Green/Red
Y/R : Yellow/Red

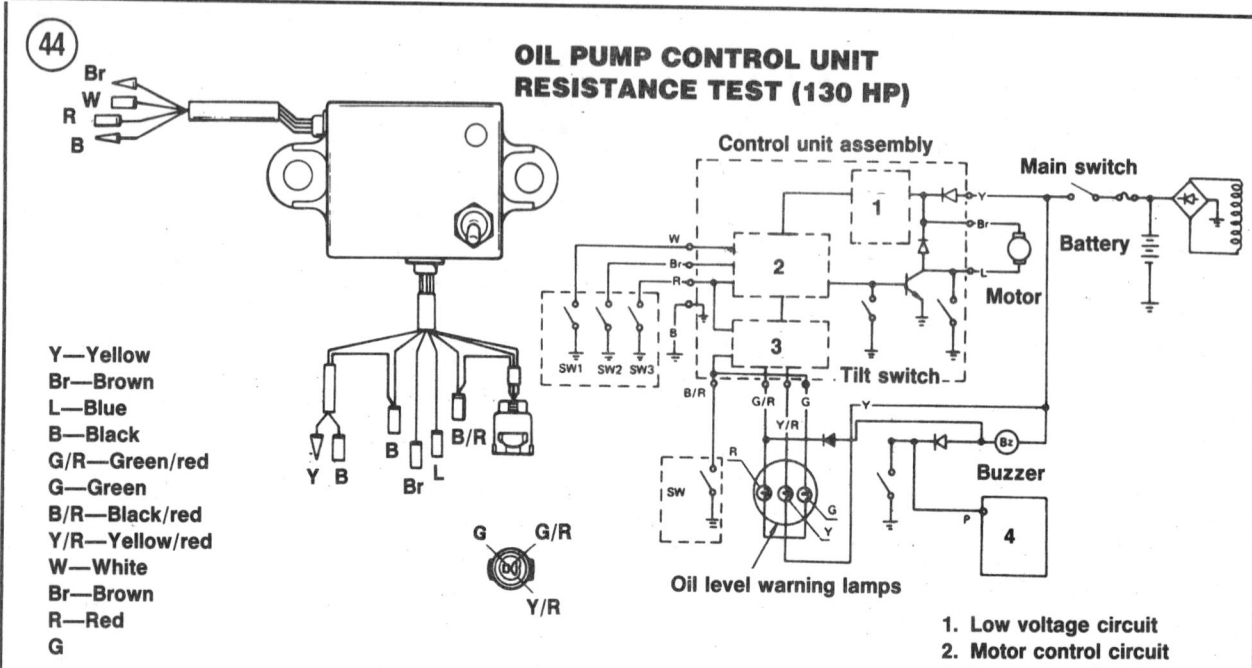

OIL PUMP CONTROL UNIT RESISTANCE TEST (130 HP)

Y—Yellow
Br—Brown
L—Blue
B—Black
G/R—Green/red
G—Green
B/R—Black/red
Y/R—Yellow/red
W—White
Br—Brown
R—Red
G

1. Low voltage circuit
2. Motor control circuit
3. Lamp control circuit
4. CDI unit

		To main switch		To feed pump motor			Grnd	To lamps			To oil level switch		
								Red	Green	Yellow	SW₁	SW₂	SW₃
⊕ Tester	(−)	① Y	①*1 Y	② Br	③ L	③*2 L	④ B	⑤ G/R	⑥ B/R	⑦ Y/R	⑧ W	⑨ Br	⑩ R
To main switch	① Y			3.2~4.8	12~18	4.8~7.2	4.8~7.2	16~24	16~24	16~24	16~24	16~24	16~24
	①*1 Y			3.2~4.8	11.2~16.8	4~6	4~6	16~24	16~24	12~18	16~24	16~24	16~24
To feed pump motor	② Br	∞	∞		4.8~7.2	1.6~2.4	1.6~2.4	8~12	6.4~9.6	4.8~7.2	8~12	8~12	8~12
	③ L	∞	∞	3.2~4.8			4.8~7.2	16~24	16~24	16~24	16~24	16~24	16~24
	③*2 L	∞	∞	1.6~2.4			0	8~12	8~12	3.2~4.8	8~12	8~12	8~12
Ground	④ B	∞	∞	1.6~2.4	3.2~4.8	0		8~12	8~12	3.2~4.8	8~12	8~12	8~12
To Lamps Red	⑤ G/R	∞	∞	∞	∞	∞	∞		∞	∞	∞	∞	0
To Lamps Green	⑥ G B/R	∞	∞	∞	∞	∞	∞	∞		∞	∞	∞	∞
To Lamps Yellow	⑦ Y/R	∞	∞	∞	∞	∞	∞	∞	∞		∞	∞	∞
To oil level switch SW₁	⑧ W	∞	∞	8~12	16~24	8~12	8~12	16~24	16~24	16~24		16~24	16~24
To oil level switch SW₂	⑨ Br	∞	∞	8~12	16~24	8~12	8~12	16~24	16~24	16~24	16~24		16~24
To oil level switch SW₃	⑩ R	∞	∞	∞	∞	∞	∞	0	∞	∞	∞	∞	

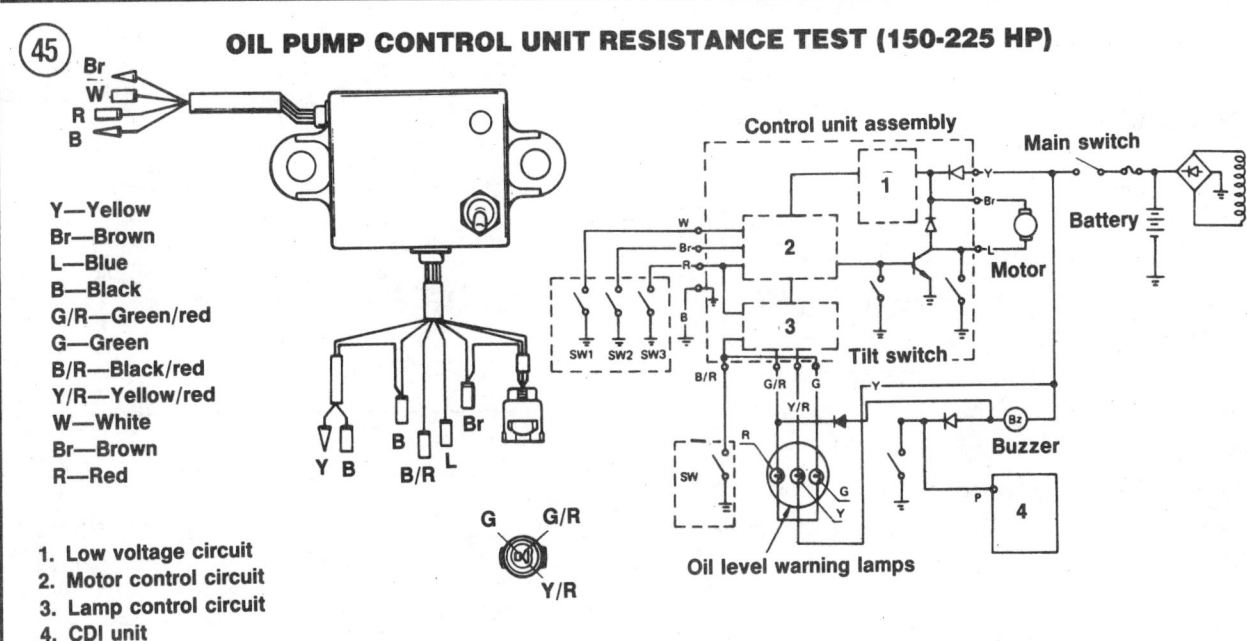

OIL PUMP CONTROL UNIT RESISTANCE TEST (150-225 HP)

Y—Yellow
Br—Brown
L—Blue
B—Black
G/R—Green/red
G—Green
B/R—Black/red
Y/R—Yellow/red
W—White
Br—Brown
R—Red

1. Low voltage circuit
2. Motor control circuit
3. Lamp control circuit
4. CDI unit

	Tester ⊖	To main switch		To feed pump motor			Grnd	To lamps			To oil level switch		
								Red	Green	Yellow	SW₁	SW₂	SW₃
⊕ Tester		① Y	①*1 Y	② Br	③ L	③*2 L	④ B	⑤ G/R	⑥ B/R	⑦ Y/R	⑧ W	⑨ Br	⑩ R
To main switch	① Y			3.2~4.8	12~18	4.8~7.2	4.8~7.2	16~24	16~24	16~24	16~24	16~24	16~24
	①*1 Y			3.2~4.8	11.2~16.8	4~6	4~6	16~24	16~24	12~18	16~24	16~24	16~24
To feed pump motor	② Br	∞	∞		4.8~7.2	1.6~2.4	1.6~2.4	8~12	6.4~9.6	4.8~7.2	8~12	8~12	8~12
	③ L	∞	∞	3.2~4.8			4.8~7.2	16~24	16~24	16~24	16~24	16~24	16~24
	③*2 L	∞	∞	1.6~2.4			0	8~12	8~12	3.2~4.8	8~12	8~12	8~12
Ground	④ B	∞	∞	1.6~2.4	3.2~4.8	0		8~12	8~12	3.2~4.8	8~12	8~12	8~12
To Lamps	Red ⑤ G/R	∞	∞	∞	∞	∞	∞		∞	∞	∞	∞	0
	Green ⑥ G B/R	∞	∞	∞	∞	∞	∞	∞		∞	∞	∞	∞
	Yellow ⑦ Y/R	∞	∞	∞	∞	∞	∞	∞	∞		∞	∞	∞
To oil level switch	SW₁ ⑧ W	∞	∞	8~12	16~24	8~12	8~12	16~24	16~24	16~24		16~24	16~24
	SW₂ ⑨ Br	∞	∞	8~12	16~24	8~12	8~12	16~24	16~24	16~24	16~24		16~24
	SW₃ ⑩ R	∞	∞	∞	∞	∞	∞	0	∞	∞	∞	∞	

3

thermoswitch in the water as it is being heated. See **Figure 47**. Do not submerge the thermoswitch in the water as the readings will be incorrect.

5. No continuity should be shown until the water temperature reaches:

    a. 40-50 hp—199°F (93°C).

    b. 70-90 hp—189°±37°F (87°±3°C).

    c. 115-200 hp—190°±37°F (88°±3°C)

6. When the water reaches the temperature specified in Step 5, the meter should show continuity. Allow the water to reach the boiling point, then remove the heat and allow it to cool down. The meter should continue to show continuity until the water temperature cools to:

    a. 40-50 hp—181°F (83° C).

    b. 70-200 hp—153°±45°F (67°±7° C).

7. When the water cools to the temperature specified in Step 6, the meter should show no continuity.

8. If the thermoswitch does not provide the readings as specified, replace it.

9. Reinstall the thermoswitch in the power head and reconnect its leads. Install the engine cover.

YMIS Component Testing

V6 Special and Excel engines equipped with the YMIS ignition use a thermo sensor, crank position sensor, throttle position sensor and knock sensor not found on other models. The following procedures will determine whether or not the sensors are functioning properly.

Resistance values are based on the use of its Pocket Tester (part No. YU-3112). If another ohmmeter is used, the readings obtained may not agree with those specified due to internal resistance of the ohmmeter.

Thermo sensor

1. Remove the engine cover.

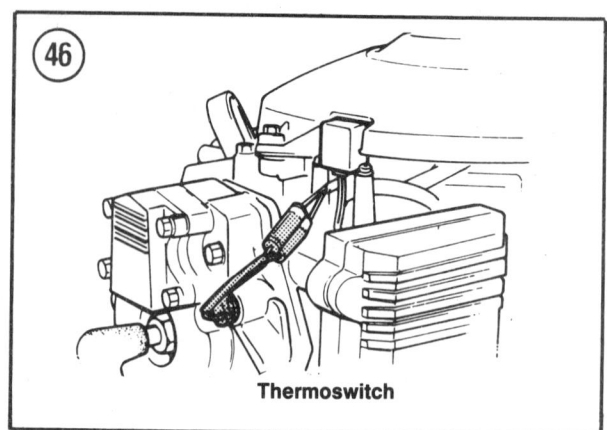

Thermoswitch

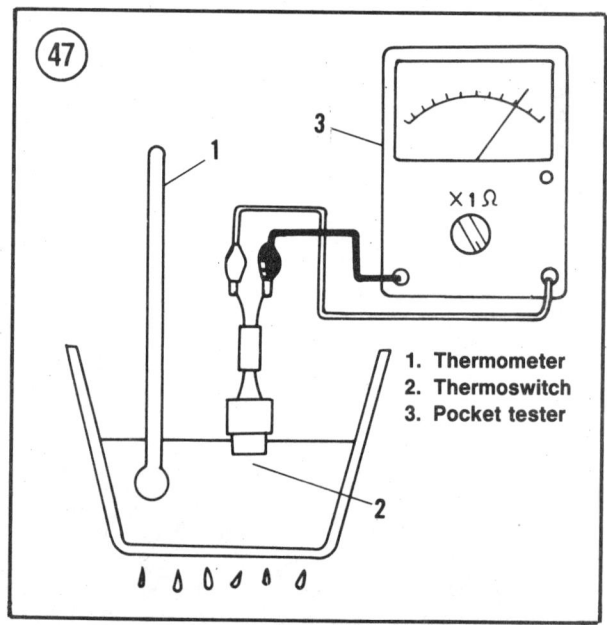

1. Thermometer
2. Thermoswitch
3. Pocket tester

NOTE
Do not confuse the thermo sensor with the thermo switch installed in each cylinder head. The thermo sensor connects to the YMIS box; the thermo switches connect to the CDI unit.

2. Disconnect the thermo sensor leads from the YMIS box. See **Figure 48** for sensor location.

3. Connect the Pocket Tester to the thermo sensor leads and test the sensor resistance and amperage values. If either value is not within specifications, replace the thermo sensor.

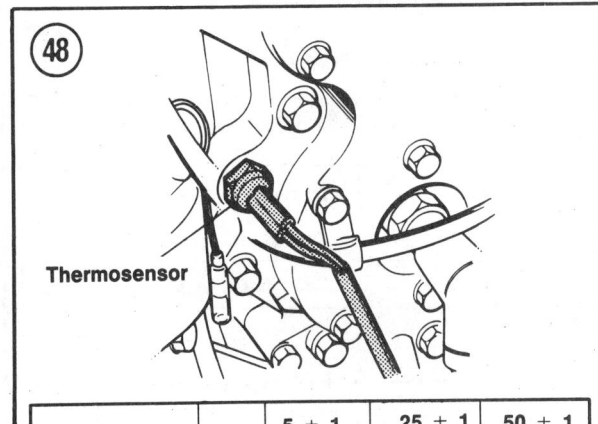

Thermosensor

Atmosphere temp.	°C (°F)	5 ± 1 (41 ± 33.8)	25 ± 1 (77 ± 33.8)	50 ± 1 (122 ± 33.8)
Measurement amp.	mA	Below 1	Below 1	Below 1
Terminal resistance (Black-black)	kΩ	20.6-26.4	76.5-93.7	25.2-30.8

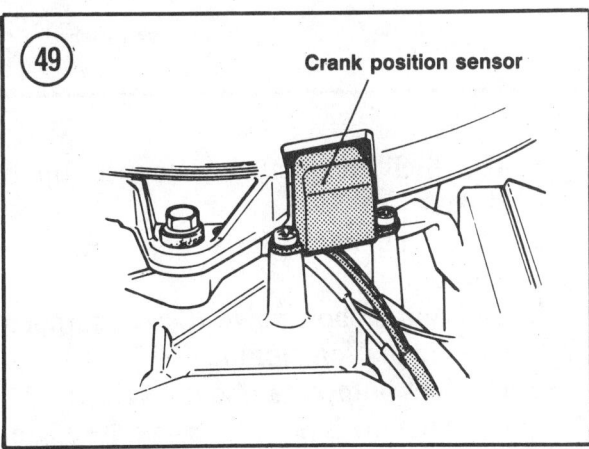

Crank position sensor

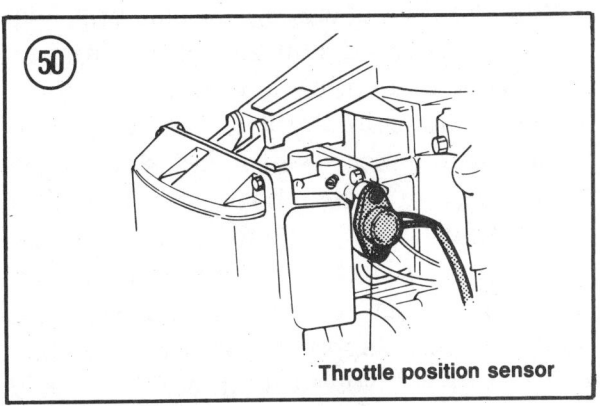

Throttle position sensor

3

Crank position sensor

1. Remove the engine cover.
2. Disconnect the crank position sensor connector from the YMIS box. See **Figure 49** for sensor location.
3. Connect the Pocket Tester to the crank position sensor connector terminals and test the sensor resistance value. If meter does not read 710 ohms ± 10 percent at 68° F (20° C), replace the crank position sensor. See *Disassembly, V4 and V6*, Chapter Eight.

Throttle position sensor

1. Remove the engine cover.
2. Disconnect the throttle position sensor connector from the YMIS box. See **Figure 50** for sensor location.
3. Set the Pocket Tester on the R×1 ohm position.
4. Connect the Pocket Tester to the black and white wire terminals in the crank position sensor connector and note the reading. Refer to **Figure 51** for terminal location and resistance values.
5. Set the ohmmeter on the R×1,000 ohm position and repeat Step 4 to test first the red and white, then the red and black terminals. Refer to **Figure 51** for terminal location and resistance values.
6. If a correct reading is not obtained at each of the 3 terminal connections, replace the throttle position sensor. See Chapter Five.

Knock sensor

1. Remove the engine cover.
2. Disconnect the knock sensor lead at the sensor. See **Figure 52** for sensor location in the cylinder head.
3. Connect the Pocket Tester between the sensor lead terminal and a good engine ground. Replace the sensor if continuity is shown.

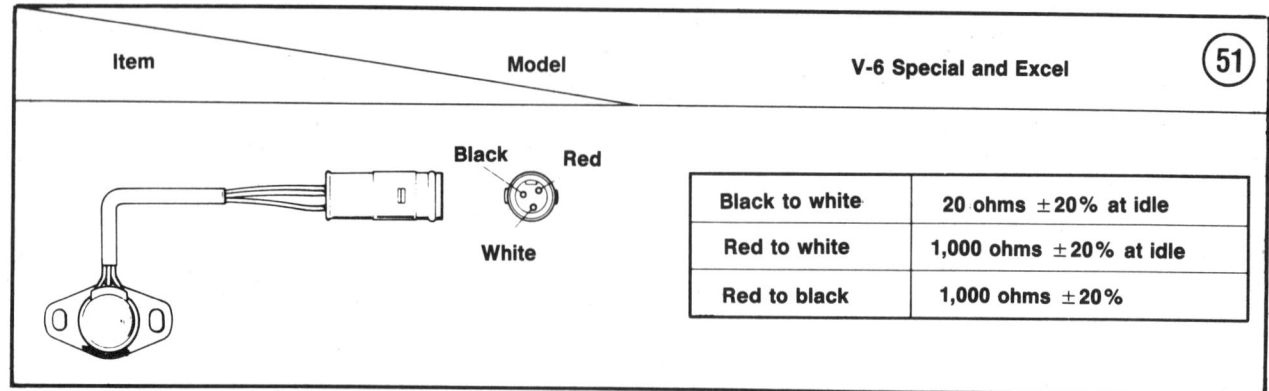

Item	Model	V-6 Special and Excel	(51)

Black to white	20 ohms ±20% at idle
Red to white	1,000 ohms ±20% at idle
Red to black	1,000 ohms ±20%

FUEL SYSTEM

Many outboard owners automatically assume that the carburetor is at fault when the engine does not run properly. While fuel system problems are not uncommon, carburetor adjustment is seldom the answer. In many cases, adjusting the carburetor only compounds the problem by making the engine run worse.

Fuel system troubleshooting should start at the gas tank and work through the system, reserving the carburetor(s) as the final point. Most fuel system problems result from an empty fuel tank, sour fuel, a plugged fuel filter or a malfunctioning fuel pump or anti-siphon valve. **Table 3** provides a series of symptoms and causes that can be useful in localizing fuel system problems.

Troubleshooting

As a first step, check the fuel flow. Remove the fuel tank cap and look into the tank. If there is fuel present, disconnect and ground the spark plug lead(s) as a safety precaution. Disconnect the fuel line at the carburetor and place it in a suitable container to catch any discharged fuel. See if gas flows freely from the line when the primer bulb is squeezed.

If there is no fuel flow from the line:
a. The fuel petcock may be shut off or blocked by rust or foreign matter.

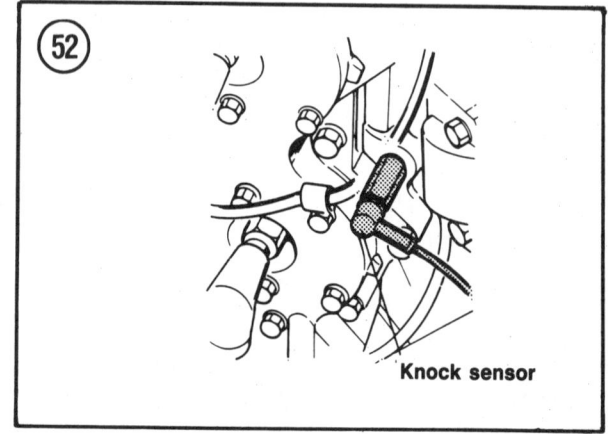

Knock sensor

b. The fuel line may be stopped up or kinked.
c. A primer bulb check valve may be defective.
d. The anti-siphon valve (if so equipped) may be malfunctioning.
e. The fuel pump may be defective.

If a good fuel flow is present, crank the engine 10-12 times to check fuel pump operation. A pump that is operating satisfactorily will deliver a good, constant flow of fuel from the line. If the amount of flow varies from pulse to pulse, the fuel pump is probably failing.

In accordance with industry safety standards, late-model boats with a built-in fuel tank will have some form of anti-siphon device installed between the tank outlet and engine fuel inlet. This device is designed to shut the fuel supply off in case the boat capsizes or is involved in an accident. Quite often, the malfunction of such devices leads the owner to replace a fuel pump in the belief that it is defective.

Anti-siphon devices can malfunction in one of the following ways:

a. Anti-siphon valve: orifice in valve is too small or clogs easily; valve sticks in closed or partially closed position; valve fluctuates between open and closed position; thread sealer, metal filing or other debris clogs orifice or lodges in the relief spring.

b. Solenoid-operated fuel shut-off valve: solenoid fails with valve in closed position; solenoid malfunctions, leaving valve in partially closed position.

c. Manually-operated fuel shut-off valve: valve is left in completely closed position; valve is not fully opened.

The easiest way to determine if the anti-siphon valve is defective is to bypass it by operating the engine with a remote fuel supply such as an outboard fuel tank.

Carburetor chokes can also present problems. A choke that sticks open will show up as a hard starting problem; one that sticks closed will result in a flooding condition.

During a hot engine shut-down, the fuel bowl temperature can rise above 200° F, causing the fuel inside to boil. While marine carburetors are vented to atmosphere to prevent this problem, there is a possibility that

some fuel will percolate over the high-speed nozzle.

A leaking inlet needle and seat or a defective float will allow an excessive amount of fuel into the intake manifold. Pressure in the fuel line after the engine is shut down forces fuel past the leaking needle and seat. This raises the fuel bowl level, allowing fuel to overflow into the manifold.

A defective bleed line or bleed line check valve may cause fuel starvation in one or more cylinders.

Excessive fuel consumption may not necessarily mean an engine or fuel system problem. Marine growth on the boat's hull, a bent or otherwise damaged propeller or a fuel line leak can cause an increase in fuel consumption. These areas should all be checked *before* blaming the carburetor.

ENGINE TEMPERATURE AND OVERHEATING

Proper engine temperature is critical to good engine operation. An engine that runs too hot will be damaged internally. One that operates too cool will not run smoothly or efficiently.

A variety of problems can cause engine overheating. Some of the most commonly encountered are a defective thermostat, a low output or defective water pump, damaged or mispositioned water passage restrictors or even engine flashing (material left over from the manufacturing process) in the cylinder head casting water discharge passage that was not removed during manufacture.

Troubleshooting

Engine temperature can be checked with the use of Markal thermomelt sticks available at your local marine dealer. This heat-sensitive stick looks like a large crayon (**Figure 53**) and will melt on contact with a metal surface at a specific temperature.

Two thermomelt sticks are required to properly check a Yamaha outboard: a 125° F (52° C) stick and a 163° F (73° C) stick. The stick should not be applied to the center of the cylinder head, as this area may normally run hotter than 163° F.

The test is most efficient when carried out on a motor operating on a boat in the water. If necessary to perform the test using a test tank, run the engine at 3,000 rpm for a minimum of 5 minutes to assure that it is at operating temperature. Make sure inlet water temperature is below 80°degrees F (26° C) and perform the test as follows:

1. Mark the cylinder water jacket with each stick. The mark will appear similar to a chalk mark. Make sure sufficient material is applied to the metal surface.

2. With the engine at operating temperature and running at idle in FORWARD gear, the 125° F stick mark should melt. If it does not melt on thermostat-equipped models, the thermostat is stuck open and the engine is running cold.

3. With the engine at operating temperature and running at full throttle in FORWARD gear, the 163° F stick mark should not melt. If it does, the power head is overheating. Look for a defective water pump, clogged or leaking cooling system. On thermostat-equipped models, the thermostat may be stuck closed.

ENGINE

Engine problems are generally symptoms of something wrong in another system, such as ignition, fuel or starting. If properly maintained and serviced, the engine should experience no problems other than those caused by age and wear.

Overheating and Lack of Lubrication

Overheating and lack of lubrication cause the majority of engine mechanical problems. Outboard motors create a great deal of heat and are not designed to operate at a standstill for any length of time. Using a spark plug of the wrong heat range can burn a piston. Incorrect ignition timing, a defective water pump or thermostat, a propeller that is too large (over-propping) or an excessively lean fuel mixture can also cause the engine to overheat.

Preignition

Preignition is the premature burning of fuel and is caused by hot spots in the combustion chamber (**Figure 54**). The fuel actually ignites before it is supposed to. Glowing deposits in the combustion chamber, inadequate cooling or overheated spark plugs can all cause preignition. This is first noticed in the form of

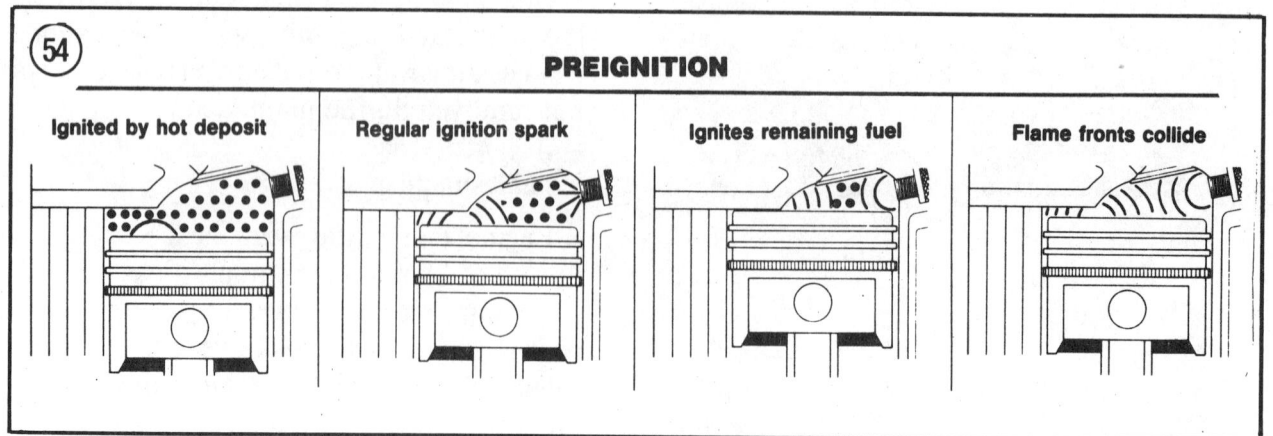

PREIGNITION

| Ignited by hot deposit | Regular ignition spark | Ignites remaining fuel | Flame fronts collide |

a power loss but will eventually result in extensive damage to the internal parts of the engine because of higher combustion chamber temperatures.

Detonation

Commonly called "spark knock" or "fuel knock," detonation is the violent explosion of fuel in the combustion chamber prior to the proper time of combustion (**Figure 55**). Severe damage can result. Use of low octane gasoline is a common cause of detonation.

Even when high octane gasoline is used, detonation can still occur if the engine is improperly timed. Other causes are over-advanced ignition timing, lean fuel mixture at or near full throttle, inadequate engine cooling, cross-firing of spark plugs, excessive accumulation of deposits on piston and combustion chamber or the use of a prop that is too large (over-propping).

Since outboard motors are noisy, engine knock or detonation is likely to go unnoticed by owners, especially at high engine rpm when wind noise is also present. Such inaudible detonation, as it is called, is usually the cause when engine damage occurs for no apparent reason.

Poor Idling

A poor idle can be caused by improper carburetor adjustment, incorrect timing or ignition system malfunctions. Check the gas cap vent for an obstruction.

Misfiring

Misfiring can result from a weak spark or a dirty spark plug. Check for fuel contamination. If misfiring occurs only under heavy load, as when accelerating, it is usually caused by a defective spark plug. Run the motor at night to check for spark leaks along the plug wire and under spark plug cap or use a spark leak tester.

> *WARNING*
> *Do not run engine in a dark garage to check for spark leak. There is considerable danger of carbon monoxide poisoning.*

Water Leakage in Cylinder

The fastest and easiest way to check for water leakage in a cylinder is to check the spark plugs. Water will clean a spark plug. If one of the plugs on a multi-cylinder engine is clean and the others are dirty, there is most likely a water leak in the cylinder with the clean plug.

3

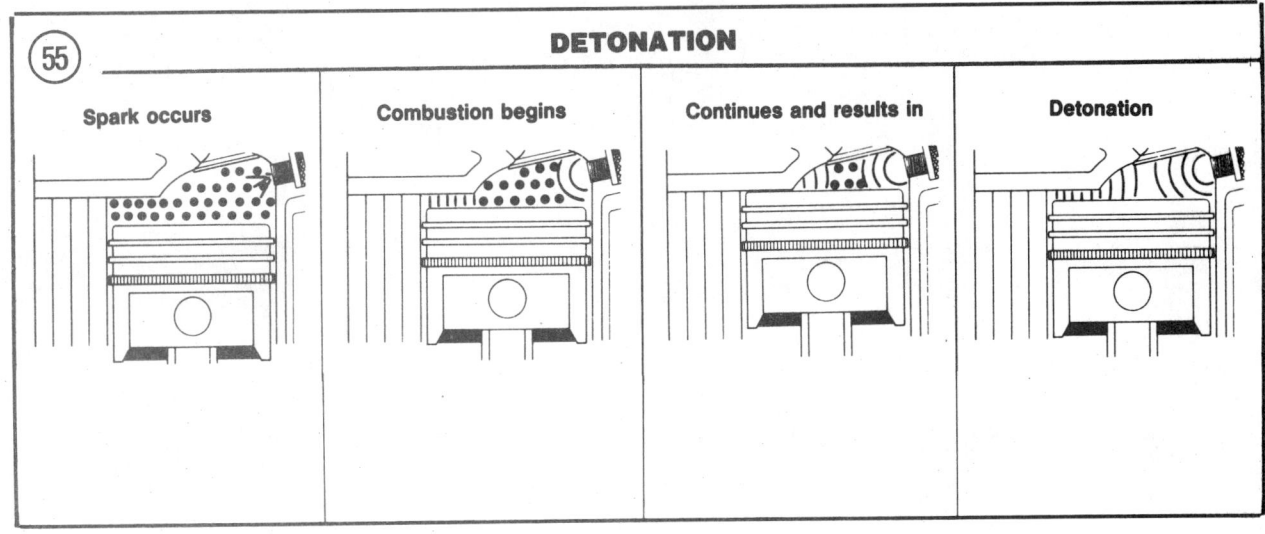

55 — **DETONATION**

| Spark occurs | Combustion begins | Continues and results in | Detonation |

To remove all doubt, install a dirty plug in each cylinder. Run the engine in a test tank or on the boat in water for 5-10 minutes. Shut the engine off and remove the plugs. If one plug is clean and the others are dirty (or if all plugs are clean), a water leak in the cylinder(s) is the problem.

Flat Spots

If the engine seems to die momentarily when the throttle is opened and then recovers, check for a dirty main jet in the carburetor, water in the fuel or an excessively lean mixture.

Power Loss

Several factors can cause a lack of power and speed. Look for air leaks in the fuel line or fuel pump, a clogged fuel filter or a choke/throttle valve that does not operate properly. Check ignition timing.

A piston or cylinder that is galling, incorrect piston clearance or a worn/sticky piston ring may be responsible. Look for loose bolts, defective gaskets or leaking machined mating surfaces on the cylinder head, cylinder or crankcase. Also check the crankcase oil seal; if worn, it can allow gas to leak between cylinders.

Piston Seizure

This is caused by one or more pistons with incorrect bore clearances, piston rings with an improper end gap, the use of an oil/fuel mixture containing less than 1 part oil to 50 parts of gasoline or an oil of poor quality, a spark plug of the wrong heat range or incorrect ignition timing. Overheating from any cause may result in piston seizure.

Excessive Vibrations

Excessive vibrations may be caused by loose motor mounts, worn bearings or a generally poor running motor.

Engine Noises

Experience is needed to diagnose accurately in this area. Noises are difficult to differentiate and even harder to describe. Deep knocking noises usually mean main bearing failure. A slapping noise generally comes from a loose piston. A light knocking noise during acceleration may be a bad connecting rod bearing. Pinging should be corrected immediately or damage to the piston will result. A compression leak at the head-to-cylinder joint will sound like a rapid on-off squeal.

Table 1 STARTER TROUBLESHOOTING

Trouble	Cause	Remedy
Pinion does not move when starter is turned on	Blown fuse	Replace fuse.
	Pinion rusted to armature shaft	Remove, clean or replace as required.
	Series coil or shunt broken or shorted	Replace coil or shunt.
	Loose switch connections	Tighten connections.
	Rusted or dirty plunger	Clean plunger.
Pinion meshes with ring gear but starter does not run	Worn brushes or brush springs touching armature	Replace brushes or brush springs.
	Dirty or burned commutator	Clean or replace as required.
	Defective armature field coil	Replace armature.
	Worn or rusted armature shaft bearing	Replace bearing.
Starter motor runs at full speed before pinion meshes with ring gear	Worn pinion sleeve	Replace sleeve.
	Pinion does not stop in correct position	Replace pinion.
Pinion meshes with gear and motor starts but engine does not crank	Defective overrunning clutch	Replace overrunning clutch
Starter motor does not stop when turned off after engine has started	Rusted or dirty plunger	Clean or replace plunger
Starter motor has low no-load speed and high current draw	Armature may be dragging on pole shoes from bent shaft, worn bearings or loose pole shoes	Replace shaft or bearings and/or tighten pole shoes.
	Tight or dirty bearings	Loosen or clean bearings.
High current draw with no armature rotation	A direct ground switch, @ terminal or @ brushes or field connections	Replace defective parts.
	Frozen shaft bearings which prevent armature from rotating	Loosen, clean or replace bearings.

(continued)

3

Table 1 STARTER TROUBLESHOOTING (continued)

Trouble	Cause	Remedy
Starter motor has grounded armature or field winding	Current passes through armature first, then to ground field windings	Disconnect grounded leads, then locate any abnormal grounds in starter motor.
Starter motor fails to operate and draws no current and/or high resistance	Open circuit in fields or armature, @ connections or brushes or between brushes and commutator	Repair or adjust broken or weak brush springs, worn brushes, high insulation between commutator bars or a dirty, gummy or oily commutator.
Low no-load and a low current draw and low developed torque	High resistance in starter motor	Close "open" field winding on unit which has 2 or 3 circuits in starter motor (unit in which current divides as it enters, taking 2 or 3 parallel paths).
High free speed and high current draw	Shorted fields in starter motor	Install new fields and check for improved performance (fields normally have very low resistance, thus it is difficult to detect shorted fields, since difference in current draw between normal starter motor field windings would not be very great).
Excessive voltage drop	Cables too small	Install larger cables to accomodate high current draw.
High circuit resistance	Dirty connections	Clean connections.
Field and/or armature is burned or lead is thrown out of commutator due to excess leakage	Starter motor has grounded armature or field winding	Raise grounded brushes from commutator and insulate them with cardboard. Use Magneto Analyzer (part No. C-91-25213) (Selector No. 3) and test points to check between insulated terminal or starter motor and starter motor frame (remove ground connection of shunt coils on motors with this feature). If analyzer shows resistance (meter needle moves to right), there is a ground. Raise other brushes from armature and check armature and fields separately to locate ground.

(continued)

Table 1 STARTER TROUBLESHOOTING (continued)

Trouble	Cause	Remedy
Starter does not operate	Run-down battery	Check battery with hydrometer. If reading is below 1.230, recharge or replace battery.
	Poor contact @ terminals	Remove terminal clamps. Scrape terminals and clamps clean and tighten bolts securely.
	Wiring or key switch	Coat with sealer to protect against further corrosion.
	Starter solenoid	Check for resistance between: (a) positive (+) terminal of battery and large input terminal of starter solenoid, (b) large wire @ top of starter motor and negative (-) terminal of battery and (c) small terminal of starter solenoid and positive battery terminal. Key switch must be in START position. Repair all defective parts.
	Starter motor	With a fully charged battery, connect a negative (-) jumper wire to upper terminal on side of starter motor and a positive jumper to large lower terminal of starter motor. If motor still does not operate, remove for overhaul or replacement.
Starter turns over too slowly	Low battery or poor contact @ battery terminal	See "Starter does not operate."
	Poor contact @ starter solenoid or starter motor	Check all terminals for looseness and tighten all nuts securely.
	Starter mechanism	Disconnect positive (+) battery terminal. Rotate pinion gear in disengaged position. Pinion gear and motor should run freely by hand. If motor does not turn over easily, clean starter and replace all defective parts.
	Starter motor	See "Starter does not operate."
Starter spins freely but does not engage engine	Low battery or poor contact @ battery terminal	See "Starter does not operate."
	Poor contact @ starter solenoid or starter motor	See "Starter does not operate."
	Dirty or corroded pinion drive	Clean thoroughly and lubricate the spline underneath the pinion with water-resistant grease

(continued)

3

Table 1 STARTER TROUBLESHOOTING (continued)

Trouble	Cause	Remedy
Starter does not engage freely	Pinion or flywheel gear	Inspect mating gears for excessive wear. Replace all defective parts.
	Small anti-drift spring	If drive pinion interferes with flywheel gear after engine has started, inspect anti-drift spring located under pinion gear. Replace all defective parts. NOTE: If drive pinion tends to stay engaged in flywheel gear when starter motor is in idle position, start motor @ 1/4 throttle to allow starter pinion gear to release flywheel ring gear instantly.
Starter keeps on spinning after key is turned ON	Key not fully returned	Check that key has returned to normal ON position from START position. Replace switch if key constantly stays in START position.
	Starter solenoid	Inspect starter solenoid to see if contacts have become stuck in closed position. If starter does not stop running with small yellow lead disconnected from starter solenoid, replace starter solenoid.
	Wiring or key switch	Inspect all wires for defects. Open remote control box and inspect wiring @ switches. Repair or replace all defective parts.
Wires overheat	Battery terminals improperly connected	Check that negative marking on harness matches that of battery. If battery is connected improperly, red wire to rectifier will overheat.
	Short circuit in wiring system	Inspect all connections and wires for looseness or defects. Open remote control box and inspect wiring @ switches.
	Short circuit in choke solenoid	Repair or replace all defective parts. Check for high resistance. If blue choke wire heats rapidly when choke is used, choke solenoid may have internal short. Replace if defective.
	Short circuit in starter relay	If starter relay lead overheats, there may be internal short (resistance) in starter relay. Replace if defective.
	Low battery voltage	Battery voltage is checked with an ampere-volt tester when battery is under a starting load. Battery must be recharged if it registers under 9.5 volts. If battery is below specified hydrometer reading of 1.230, it will not turn engine fast enough to start it.

Table 2 IGNITION TROUBLESHOOTING

Symptom	Probable cause
Engine won't start, but fuel and spark are good	Defective or dirty spark plugs Spark plug gap set too wide Improper spark timing Shorted "kill" or stop button Air leaks into fuel pump Broken piston ring(s) Cylinder head, crankcase or cylinder sealing faulty Worn crankcase oil seal
Engine misfires @ idle	Incorrect spark plug gap Defective, dirty or loose spark plugs Spark plugs of incorrect heat range Cracked distributor cap Leaking or broken high tension wires Weak armature magnets Defective coil or condenser Defective ignition switch Spark timing out of adjustment
Engine misfires at high speed	See "Engine misfires @ idle." Coil breaks down Coil shorts through insulation Spark plug gap too wide Wrong type spark plugs Too much spark advance
Engine backfires through exhaust	Cracked spark plug insulator Carbon path in distributor cap Improper timing Crossed spark plug wires
Engine backfires through carburetor	Improper ignition timing
Engine preignition	Spark advanced too far Incorrect type spark plug Burned spark plug electrodes
Engine noises (knocking at power head)	Spark advanced too far
Ignition coil fails	Extremely high voltage Moisture formation Excessive heat from engine
Spark plugs burn and foul	Incorrect type plug Fuel mixture too rich Inferior grade of gasoline Overheated engine Excessive carbon in combustion chambers
Ignition causing high fuel consumption	Incorrect spark timing Leaking high tension wires Incorrect spark plug gap Fouled spark plugs Incorrect spark advance Weak ignition coil Preignition

Table 3 FUEL SYSTEM TROUBLESHOOTING

Symptom	Probable cause
No fuel @ carburetor	No gas in tank Air vent in gas cap not open Air vent in gas cap clogged Fuel tank sitting on fuel line Fuel line fittings not properly connected to engine or fuel tank Air leak @ fuel connection Fuel pickup clogged Defective fuel pump
Flooding @ carburetor	Choke out of adjustment High float level Float stuck Excessive fuel pump pressure Float saturated beyond buoyancy
Rough operation	Dirt or water in fuel Reed valve open or broken Incorrect fuel level in carburetor bowl Carburetor loose @ mounting flange Throttle shutter not closing completely Throttle shutter valve installed incorrectly Carburetor backdraft jets plugged (if so equipped)
Carburetor spit-back @ idle	Chipped or broken reed valve(s)
Engine misfires @ high speed	Dirty carburetor Lean carburetor adjustment Restriction in fuel system Low fuel pump pressure
Engine backfires	Poor quality fuel Air-fuel mixture too rich or too lean Improperly adjusted carburetor
Engine preignition	Excessive oil in fuel Inferior grade of gasoline Lean carburetor mixture
Spark plugs burn and foul	Fuel mixture too rich Inferior grade of gasoline
High gas consumption: **Flooding or leaking**	Cracked carburetor casting Leaks @ line connections Defective carburetor bowl gasket High float level Plugged vent hole in cover Loose needle and seat Defective needle valve seat gasket Worn needle valve and seat Foreign matter clogging needle valve Worn float pin or bracket Float binding in bowl High fuel pump pressure

(continued)

Table 3 FUEL SYSTEM TROUBLESHOOTING (continued)

Symptom	Probable cause
Overrich mixture	Choke lever stuck
	High float level
	High fuel pump pressure
Abnormal speeds	Carburetor out of adjustment
	Too much oil in fuel

Table 4 BATTERY CHARGING (LIGHTING) COIL RESISTANCE SPECIFICATIONS

Model	ohms
6-15 hp	0.4 ±10 percent @ 68° F (20° C)
25-30 hp	0.287 ±20 percent @ 68° F (20° C)
40-50 hp	
Standard	0.287 ±20 percent @ 68° F (20° C)
Optional	0.270 ±20 percent @ 68° F (20° C)
70 hp	0.45 ±20 percent @ 68° F (20° C)
90-130 hp	0.6 ±10 percent @ 68° F (20° C)
150-225 hp	
1984-on	0.6 ±10 percent @ 68° F (20° C)

Table 5 CHARGE COIL RESISTANCE SPECIFICATIONS

Model	ohms
3 hp	275 ±10 percent @ 68° F (20° C)
4 and 5 hp	275 ±20 percent @ 68° F (20° C)
6-15 hp	90 ±10 percent @ 68° F (20° C)
25 hp	134 ±10 percent @ 68° F (20° C)
30 hp	
2-cylinder	134 ±10 percent @ 68° F (20° C)
3-cylinder	205 ±10 percent @ 68° F (20° C)
40 and 50 hp	205 ±10 percent @ 68° F (20° C)
70 hp	165 ±10 percent @ 68° F (20° C)
90 hp	
Brown-blue leads	850 ±10 percent @ 68° F (20° C)
Blue-red leads	120 ±10 percent @ 68° F (20° C)
115-130 hp	
Brown-red leads	1,050 ±20 percent @ 68° F (20° C)
Black/red and blue leads	127 ±20 percent @ 68° F (20° C)
150-225 hp	
1984-1985	
Brown-red leads	1,050 ±20 percent @ 68° F (20° C)
Black/red and blue leads	127 ±20 percent @ 68° F (20° C)
1986-on	
Brown-red leads	1,050 ±20 percent @ 68° F (20° C)
Black/red and blue leads	24 ±20 percent @ 68° F (20° C)

Table 6 TRIGGER COIL RESISTANCE SPECIFICATIONS

Model	ohms
3-5 hp	
Low-speed	210 ±20 percent @ 68° F (20° C)
High-speed	33 ±10 percent @ 68° F (20° C)
6-15 hp	102 ±10 percent @ 68° F (20° C)
25 hp	14 ±10 percent @ 68° F (20° C)
30 hp	
2-cylinder	14 ±10 percent @ 68° F (20° C)
3-cylinder	346 ±10 percent @ 68° F (20° C)
40 and 50 hp	346 ±10 percent @ 68° F (20° C)
70 hp	130 ±10 percent @ 68° F (20° C)
90 hp	380 ±10 percent @ 68° F (20° C)
115-225 hp	360 ±20 percent @ 68° F (20° C)

Table 7 IGNITION COIL RESISTANCE SPECIFICATIONS

Model	Primary (ohms)	Secondary (ohms)
2 hp	1.06[1]	6,050[1]
3 hp	0.10[2]	2,700[2]
4 and 5 hp	0.25[2]	2,500[2]
6 and 8 hp	0.40[1]	3,500[1]
9.9 and 15 hp	1.50[2]	3,500[2]
25 hp	0.09[3]	3,500[3]
30 hp		
2-cylinder	0.09[3]	3,500[3]
3-cylinder	0.54[3]	6,300[3]
40 and 50 hp	0.54[3]	6,300[3]
70 hp	0.22[1]	4,800[1]
90 hp	0.25[1]	2,500[1]
115-225 hp	0.25[2]	2,500[2]

1. ±10% @ 68° F (20° C).
2. ±20% @ 68° F (20° C).
3. ±15% @ 68° F (20° C).

Chapter Four

Lubrication, Maintenance and Tune-up

4

The modern 2-stroke outboard motor delivers more power and performance than ever before, with higher compression ratios, new and improved electrical systems and other design advances. Proper lubrication, maintenance and tune-ups have thus become increasingly important as ways in which you can maintain a high level of performance, extend engine life and extract the maximum economy of operation.

You can do your own lubrication, maintenance and tune-ups if you follow the correct procedures and use common sense. You should read and follow the information provided by Yamaha in the Owner's Manual accompanying your outboard. This booklet is a good source of operating and maintenance information pertaining to your particular model. If you have misplaced or lost your Owner's Manual, or if you purchased the outboard used, obtain a replacement manual from your Yamaha dealer.

The following information is based on recommendations from Yamaha that will help you keep your 2-stroke outboard motor operating at its peak performance level.

Tables 1-4 are at the end of the chapter.

LUBRICATION

Proper Fuel Selection

Two-stroke engines are lubricated by mixing oil with the fuel. The various components of the engine are thus lubricated as the fuel-oil mixture passes through the crankcase and cylinders. Since outboard fuel serves the dual function of producing ignition and distributing the lubrication, the use of low octane marine white gasolines should be avoided. Such gasolines also have a tendency to cause ring sticking and port plugging.

Yamaha recommends the use of any gasoline with a minimum posted pump octane rating of 84 with its 2-50 hp models. A minimum posted pump octane rating of 86 is recommended for its 70-200 hp models. The 220 hp V6 Special and 225 hp V6 Excel models, however, requires an octane rating of 89 to prevent engine knock and assure proper operation.

Sour Fuel

Fuel should not be stored for more than 60 days (under ideal conditions). Gasoline forms gum and varnish deposits as it ages. Such fuel will cause starting problems. Yamalube Fuel Conditioner or another good grade of gasoline stabilizer and conditioner additive may be used to prevent gum and varnish formation during storage or prolonged periods of non-use but it is always better to drain the tank in such cases. Always use fresh gasoline when mixing fuel for your outboard.

Gasohol

Gasoline blended with alcohol and sold for marine use is widely available, although it is not legally required to be labeled as such in many states. A mixture of 10 percent ethyl alcohol and 90 percent unleaded gasoline is called gasohol. Yamaha does *not* recommend gasohol for use in its 2-stroke outboards. Testing to date has found that its use can cause a major deterioration of the fuel system and possible engine damage under certain operating conditions.

Fuels with an alcohol content tend to absorb moisture from the air. When the moisture content of the fuel reaches approximately one percent, it combines with the alcohol and separates from the fuel. This separation does not normally occur when gasohol is used in an automobile, as the tank is generally emptied within a few days after filling it.

The problem does occur in marine use, however, because boats often remain idle between start-ups for days or even weeks. This length of time permits separation to take place. The water-alcohol mixture settles at the bottom of the fuel tank where the fuel pickup carries it into the fuel line to the carburetor(s). Since outboard motors will not run on this mixture, it is necessary to drain the fuel tank, flush out the fuel system with clean gasoline and then remove, clean and reinstall the spark plugs before the engine can be started.

Some methods of blending alcohol with gasoline now make use of "cosolvents" as a suspension agent to prevent the water-alcohol from separating from the gasoline. Regardless of the method used, however, alcohol mixed with gasoline in any manner can cause numerous and serious problems with an outboard motor and fuel system.

Continued use of fuels containing alcohol can "melt" the fuel level indicator lens in portable fuel tanks. Many late-model replacement tanks now contain an alcohol-resistant lens.

Other than premature and costly failure of fuel system components, the major danger of using gasoline blended with alcohol in a 2-stroke outboard motor is that a shot of the water-alcohol mix may be picked up and sent to one of the carburetors of a multicylinder engine. Since this mixture contains no oil, it will wash oil off the bore of any cylinder it enters. The other carburetor(s) receiving good fuel-oil mixture will keep the engine running while the cylinder(s) receiving the water-alcohol mixture can suffer internal damage.

The problem of gasoline blended with alcohol has become so prevalent around the United States that Miller Tools (32615 Park Lane, Garden City, MI 48135) now offers an Alcohol Detection Kit (part No. C-4846) so that owners and mechanics can determine the quality of fuel being used.

The detection procedure is performed with water as a reacting agent. However, if cosolvents have been used as a suspension agent in alcohol blending, the test will not show the presence of alcohol unless ethylene glycol (automotive antifreeze) is used instead of water as a reacting agent. It is suggested that a gasoline sample be tested twice using the detection kit: first with water and then with ethylene glycol (automotive antifreeze).

The procedure cannot differentiate between types of alcohol (ethanol, methanol, etc.) nor is it considered to be absolutely accurate from a scientific standpoint, but it is accurate enough to determine whether or not there is sufficient alcohol in the fuel to cause the user to take precautions. Maintaining a close watch on the quality of fuel used can save hundreds of dollars in marine engine and fuel system repairs.

Recommended Fuel Mixture

The Yamaha Precision Blend oil injection system (**Figure 1**) is used with the 1988-on 25 hp, 1987-on 30 hp and all 40 hp electric start and larger engines. A mechanical pump

driven by the crankshaft automatically injects oil into the intake manifold between the reed valves and carburetor at a variable ratio from approximately 200:1 at idle to 100:1 on 25-50 hp models at full throttle and 50:1 on 70-225 hp models at full throttle. **Figure 2** shows the typical engine-mounted components of this system.

A sensor monitoring the injection system sounds a buzzer and illuminates a gauge warning lamp when the oil tank requires replenishment. Engine rpm is also reduced at this time on some models. These engines should be run with a 50:1 pre-mix during the break-in period or after a long period of storage as specified in your owner's manual.

To replenish the system, remove the engine cover. Remove the oil reservoir cap and pour in the required amount of Yamalube Two-cycle Lubricant (for outboards). See Chapter Twelve for system operation and service procedures.

With all other Yamaha 2-stroke engines covered in this manual, use the specified gasoline and mix with Yamalube Two-cycle Lubricant (for outboards) in the following ratios:

① OIL INJECTION SYSTEM (PART ONE)

Warning lamp
Top cowling
Oil tank cap
Oil tank
Oil tank sensor
Oil injection pump
Remote control box
Battery
Gasoline tank
Buzzer (EH model)

CAUTION
Do not, under any circumstances, use multigrade or other high detergent automotive oils, or oils containing metallic additives. Such oils are harmful to 2-stroke engines. Since they do not mix properly with gasoline, do not burn as 2-stroke oils do and leave an ash residue, their use may result in piston scoring, bearing failure or other engine damage.

Thoroughly mix one 8-ounce can of Yamalube Two-cycle Lubricant (for outboards) with each 6.3 gallons of gasoline in your fuel tank. This provides the recommended 100:1 mixture.

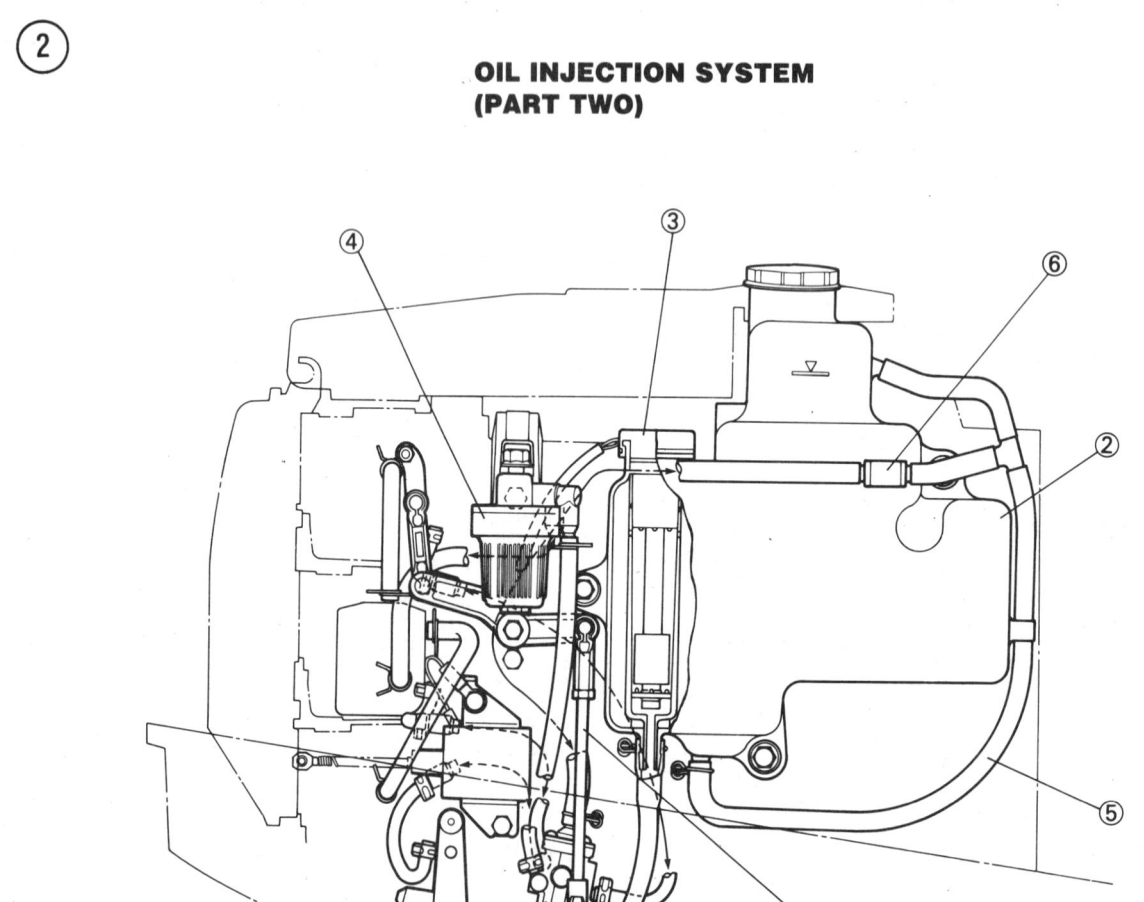

**OIL INJECTION SYSTEM
(PART TWO)**

1. Oil injection pump
2. Oil tank
3. Oil level gauge
4. Fuel filter
5. Drain hose
6. Check valve
7. Oil pump control link

30H	30EH
Manual start	Electric start
Manual control	Remote control
Precision blend system	Precision blend system

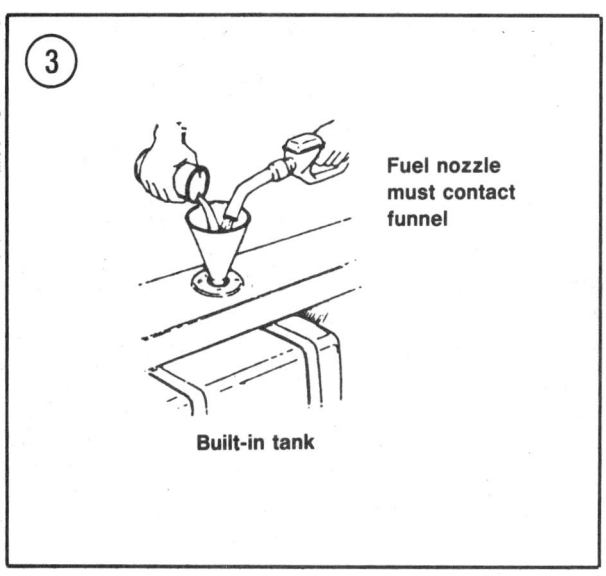

Fuel nozzle must contact funnel

Built-in tank

CAUTION
*There are a number of oil products on the market which specify use at 100:1. They are **not** BIA-TC-W approved and should **not** be used.*

If Yamalube Two-cycle Lubricant (for outboards) is not available, any high-quality 2-stroke oil intended for outboard use may be substituted provided the oil meets BIA rating TC-W and specifies so on the container. Follow the manufacturer's mixing instructions on the container but do not exceed a 100:1 ratio (after break-in).

Correct Fuel Mixing

Mix the fuel and oil outdoors or in a well-ventilated indoor location. Mix the fuel directly in the portable tank.

WARNING
Gasoline is an extreme fire hazard. Never use gasoline near heat, sparks or flame. Do not smoke while mixing fuel.

Using less than the specified amount of oil can result in insufficient lubrication and serious engine damage. Using more oil than specified causes spark plug fouling, erratic carburetion, excessive smoking and rapid

carbon accumulation which can cause preignition.

Cleanliness is of prime importance. Even a very small particle of dirt can cause carburetion problems. Always use fresh gasoline. Gum and varnish deposits tend to form in gasoline stored in a tank for any length of time. Use of sour fuel can result in carburetor problems and spark plug fouling.

Above 32° F (10° C)

Measure the required amounts of gasoline and Yamalube Two-cycle Lubricant (for outboards) accurately. Pour the specified amount of oil into the portable tank and add one-half of the gasoline to be mixed. Replace the tank filler cap and mix the fuel by tipping the tank on its side and back to an upright position several times. Remove the tank cap and add the balance of the gasoline, then mix again.

If a built-in tank is used, insert a large metal filter funnel in the tank filler neck. Slowly pour the Yamalube Two-cycle Lubricant (for outboards) into the funnel at the same time the tank is being filled with gasoline. See **Figure 3**.

Below 32° F (0° C)

Measure the required amounts of gasoline and Yamalube Two-cycle Lubricant (for outboards) accurately. Pour about one gallon of gasoline in the tank and then add the required amount of oil. Replace the tank filler cap and shake the tank to thoroughly mix the fuel and oil. Remove the cap and add the balance of the gasoline.

If a built-in tank is used, insert a large metal filter funnel in the tank filler neck. Mix the required amount of Yamalube Two-cycle Lubricant (for outboards) with one gallon of gasoline in a separate container. Slowly pour the mixture into the funnel at the same time the tank is being filled with gasoline.

4

Consistent Fuel Mixtures

The carburetor idle adjustment is sensitive to fuel mixture variations which result from the use of different oils and gasolines or from inaccurate measuring and mixing. This may require readjustment of the idle needle. To prevent the necessity for constant readjustment of the carburetor from one batch of fuel to the next, always be consistent. Prepare each batch of fuel exactly the same as previous ones.

Pre-mixed fuels sold at some marinas are not recommended for use in Yamaha 2-stroke outboards, since the quality and consistency of pre-mixed fuels can vary greatly. The possibility of engine damage resulting from use of an incorrect fuel mixture outweighs the convenience offered by pre-mixed fuel.

Gearcase Lubrication

Change the gearcase lubricant after the first 10 hours of operation. Check the lubricant level after the first 50 hours. Replace the lubricant at 100 hour intervals or at least once per season. Use Yamalube Gearcase Lube.

CAUTION
Do not use regular automotive grease in the gearcase. Its expansion and foam characteristics are not suitable for marine use.

Gearcase Lubricant Check

To assure a correct level check, the engine must be in the upright position and not run for at least 2 hours before performing this procedure. Refer to **Figure 4A** for this procedure.
1. Remove the engine cover and disconnect the spark plug lead(s) as a safety precaution to prevent any accidental starting of the engine.
2. Locate and loosen the lubricant drain plug on the right side of the gear housing (just

above the skeg). Allow a small amount of lubricant to drain. If there is water in the gearcase, it will drain before the lubricant. Retighten the plug securely.
3. If water is noted in Step 2, retighten the plug securely and pressure test the gearcase to determine if a seal has failed or if the water is simply condensation in the gearcase. See Chapter Nine.
4. Remove the oil level plug above the anti-cavitation plate. Do not lose the accompanying washer. The lubricant should be level with the bottom rim of the plug hole.

CAUTION
Never lubricate the gearcase without first removing the oil level plug, as the injected lubricant displaces air which must be allowed to escape. The gearcase cannot be completely filled otherwise.

5. If the lubricant level is low, place a suitable container under the gearcase and remove the drain plug.
6. Inject lubricant into the drain plug hole until excess fluid flows out the oil level plug hole.

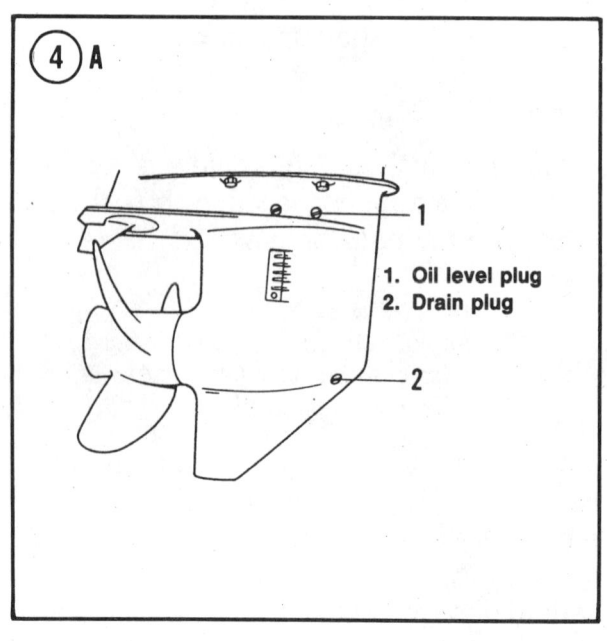

1. Oil level plug
2. Drain plug

7. Drain about one fluid ounce of lubricant to allow for expansion.

8. Install the oil level screw, then the drain plug. Be sure the washers are in place under the head of each plug, so that water will not leak past the threads into the housing.

9. Wipe any excess lubricant off the gearcase exterior. Let gearcase stand upright for a minimum of 1/2 hour, then remove the oil level plug and check the lubricant level. Top up if necessary, then reinstall oil level plug and washer.

Gearcase Lubriant Change

Refer to **Figure 4A** for this procedure.

1. Remove the engine cover and disconnect the spark plug lead(s) as a safety precaution to prevent any accidental starting of the engine.

2. Place a container under the drain plug and remove it. Remove the oil level plug. Drain the lubricant from the gearcase.

NOTE
If the lubricant is creamy in color or metallic particles are found in Step 3, remove and disassemble the gearcase to determine and correct the cause of the problem.

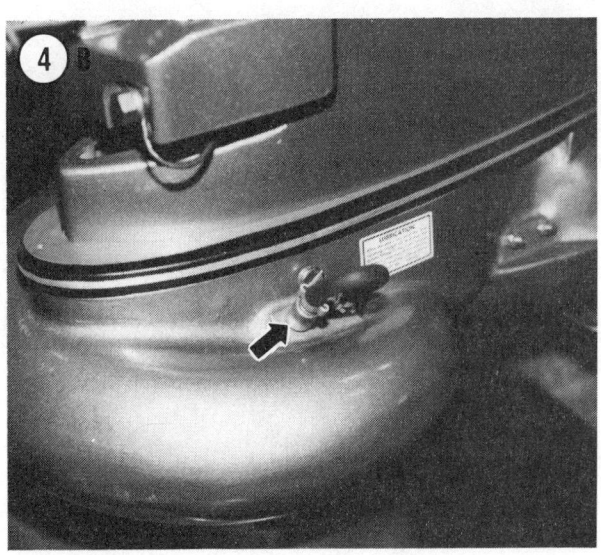

3. Wipe a small amount of lubricant on a finger and rub the finger and thumb together. Check for the presence of metallic particles in the lubricant. Note the color of the lubricant. A white or creamy color indicates water in the lubricant. Check the drain container and magnetic drain plug (used on some models) for metallic particles or signs of water separation from the lubricant.

4. Perform Steps 6-9 of *Gearcase Lubricant Check* in this chapter.

Jet Drive Bearing Lubrication

The jet drive bearing(s) should be lubricated after *each* operating period, after every 10 hours of operation and prior to storage. In addition, after every 50 hours of operation additional grease should be pumped into the bearing(s) to purge any moisture. The bearing(s) is/are lubricated by first removing the cap on the end of the excess grease hose from the grease fitting on the side of the jet drive. See **Figure 4B**. Use a grease gun and inject grease into the fitting until grease exits from the capped end of the excess grease hose. Use only Yamalube All-purpose Marine grease or a NLGI No.1 rated grease to lubricate bearing(s).

Note the color of the grease being expelled from the excess grease hose. During the break-in period some discoloration of the grease is normal. After the break-in period, if the grease starts to turn a dark or dirty grey, then the jet drive assembly should be disassembled as outlined under *JET DRIVE* in *Chapter Ten* and the seals and bearing inspected and replaced as needed. If moisture is noted being expelled from excess grease hose, then the jet drive should be disassembled and the seals replaced and the bearing(s) inspected and replaced if needed.

Other Lubrication Points

Refer to **Figures 5-11** (typical) and **Table 1** for other lubrication points, frequency of lubrication and lubricant to be used.

In addition to these lubrication points, some motors may also have grease fittings provided at critical points where bearing surfaces are not externally exposed. See **Figure 12** (typical). Remove the protective rubber cap from such fittings (A, **Figure 12**) and lubricate at least once each season with an automotive type grease gun and Yamalube All-purpose Marine Grease until grease can be seen at an external point (B, **Figure 12**).

> *CAUTION*
> *When lubricating the steering cable on models so equipped, make sure its core is fully retracted into the cable housing. Lubricating the cable while extended can cause a hydraulic lock to occur.*

Salt Water Corrosion of Gearcase Bearing Housing or Cap

Salt water corrosion that is allowed to build up unchecked can eventually split the gearcase and destroy the lower unit. If the motor is used in salt water, remove the propeller, cover nuts and bearing housing or cap at least once a year after the initial

10-hour inspection. Clean all corrosive deposits and dried-up lubricant from each end of the housing or cap.

Install new O-rings on the bearing housing or cap and wipe the outer diameter of the housing at A and B, **Figure 13** with

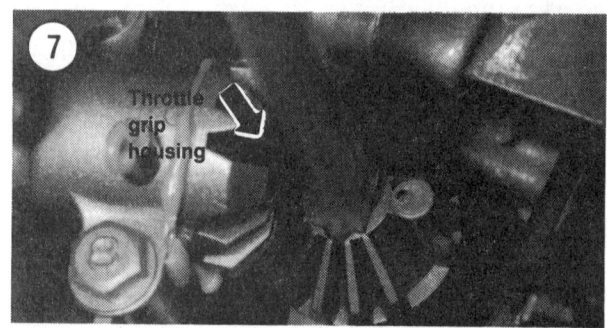

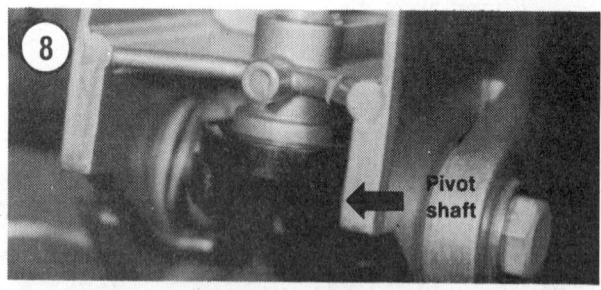

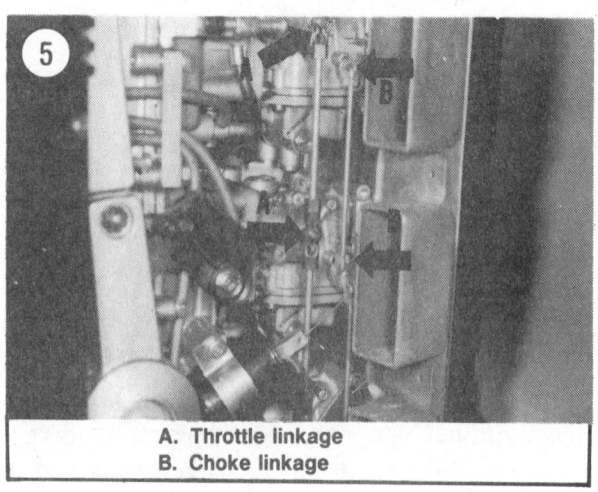

A. Throttle linkage
B. Choke linkage

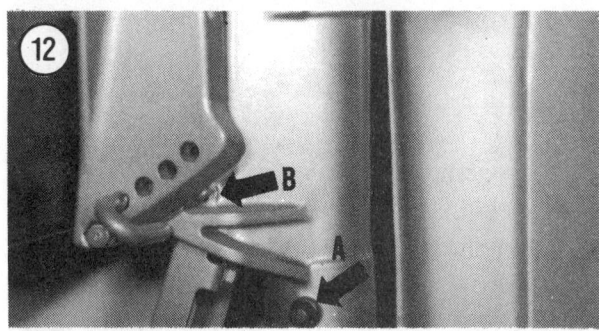

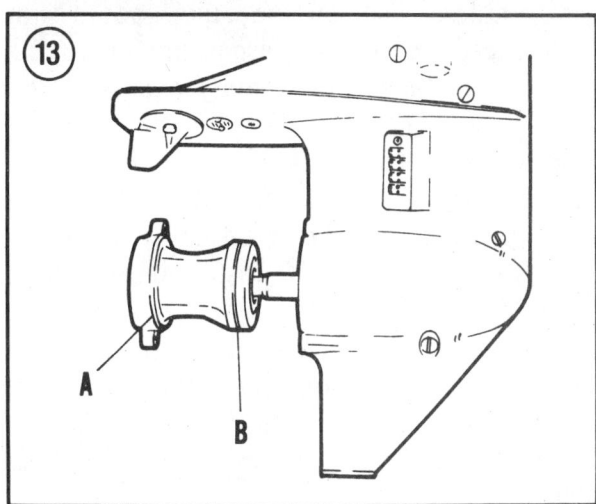

Yamalube All-purpose Marine Grease. Install the housing or cap and tighten bolts to specifications (Chapter Nine). Lubricate the propeller shaft splines with the same grease and reinstall the propeller.

STORAGE

The major consideration in preparing an outboard motor for storage is to protect it from rust, corrosion and dirt. Use the following procedure with your Yamaha outboard.

1. Remove the engine cover.

2A. Gearcase models—Operate the motor in a test tank or attach a flush device. See **Figure 14** (typical). Start the engine and run at fast idle until warmed up.

2B. Jet drive models—Operate the motor with a cooling system flush device attached as outlined in Chapter 10.

3. Gearcase models—Shift the outboard motor into NEUTRAL.

4. Disconnect the fuel line and run the engine at fast idle while pouring about 2 ounces of Yamalube Two-cycle Lubricant (for outboards) into each carburetor throat until the engine stalls out.

5. Remove spark plug(s) as described in this chapter. Pour about one ounce of Yamalube Two-cycle Lubricant (for outboards) into each spark plug hole. Slowly rotate flywheel by hand several times to distribute the oil throughout the cylinder(s). Reinstall spark plugs.

6. Service the portable fuel tank and filter (strainer) as follows:

   a. Disconnect the fuel line at the fuel meter housing on the tank. Remove the attaching screws and pull the housing assembly from the tank.

   b. Clean the fine wire mesh filter (strainer) at the bottom of the housing assembly suction pipe by rinsing it in clean benzine.

   c. Pour approximately one quart of clean benzine in the tank and slosh it about to clean the interior. Pour out the benzine and discard safely.

   d. Reinstall fuel meter housing to fuel tank and connect the fuel line.

7. Service the engine fuel filter as described in this chapter.

8A. Gearcase models—Drain and refill gearcase as described in this chapter. Check condition of oil level and drain plug gaskets. Replace as required.

8B. Jet drive models—Lubricate jet drive bearing(s) as described in this chapter.

9. Refer to **Figures 5-12** and **Table 1** as appropriate and lubricate motor at all specified points.

10. Clean the motor, including all accessible power head parts. Remove all dirt, grease and scum with a good quality marine cleaner. Coat with a good marine-type wax. Install the engine cover and wipe a thin film of clean engine oil on all painted surfaces.

11. Gearcase models—Remove the propeller and lubricate propeller shaft splines with Yamalube All-purpose Marine Lubricant. Reinstall the propeller.

CAUTION
Make certain that all water drain holes in the gearcase are free and open to allow water to drain out. Water expands as it freezes and can crack the gearcase or water pump.

12. Drain the system completely to prevent damage from freezing.

13. Store the motor upright in a dry and well-ventilated area.

14. Service the battery (if so equipped) as follows:

   a. Disconnect the negative battery cable, then the positive battery cable.

   b. Remove all grease, corrosion and dirt from the battery surface.

   c. Check the electrolyte level in each battery cell and top up with distilled water, if necessary. Fluid level in each cell should not be higher than 3/16 in. above the perforated baffles.

   d. Lubricate the terminal bolts with grease or petroleum jelly.

CAUTION
A discharged battery can be damaged by freezing.

   e. With the battery in a fully-charged condition (specific gravity 1.260-1.275), store in a dry place where the temperature will not drop below freezing.

   f. Recharge the battery every 45 days or whenever the specific gravity drops below 1.230. Before charging, cover the plates with distilled water, but not more than 3/16 in. above the perforated baffles. The charge rate should not exceed 6 amps. Discontinue charging when the specific gravity reaches 1.260 at 80° F (27° C).

   g. Before placing the battery back into service after storage, remove the excess grease from the terminals, leaving a small amount on. Install battery in a fully-charged state.

15. If an outboard equipped with oil injection is stored for more than one season, use a 50:1 pre-mix to start the engine and follow the *Break-in Procedure* provided in Chapter Thirteen.

ANTI-CORROSION MAINTENANCE

1. Flush the cooling system with fresh water as described in this chapter after each time the motor is used in salt water. Wash exterior with fresh water.

2. Dry exterior of motor and apply Yamaha zinc primer (part No. LUB-84PNT-PR-MR) over any paint nicks and scratches. Let primer flash off, then apply Yamaha Silver Touch-up Paint (part No. LUB-84PNT-SL-VR). If Yamaha paints are not available, use a suitable tin anti-fouling paint; do not use paints containing mercury or copper. Do not paint sacrificial anodes or trim tab.

3. Spray powerhead and all electrical connections with a good quality corrosion and rust preventative.

4. Check condition of sacrificial anodes and trim tab. Replace any anodes that are less than half their original size. Replace trim tab, if damaged.

5. Lubricate more frequently than specified in **Table 1**. If used consistently in salt water, reduce lubrication intervals by one-half.

COMPLETE SUBMERSION

An outboard motor which has been lost overboard should be recovered as quickly as possible. If the motor was running when submerged, disassemble and clean it immediately—any delay will result in rust and corrosion of internal components once it has been removed from the water. If the motor was not running and appears to be undamaged mechanically with no abrasive dirt or silt inside, take the following emergency steps immediately.

1. Wash the outside of the motor with clean water to remove weeds, mud and other debris.

2. Remove the engine cover.

3. If recovered from salt water, flush motor completely with fresh water.

4. Remove the spark plug(s) as described in this chapter.

CAUTION
Do not force the motor if it does not turn over freely by hand in Step 5. This may be an indication of internal damage such as a bent connecting rod or broken piston.

5. Drain as much water as possible from the power head by placing the motor in a horizontal position. Manually rotate the flywheel with the spark plug hole(s) facing downward.

6. Pour alcohol into the carburetor throat(s) to displace water. Manually rotate the flywheel at least a full turn, then position the motor so you can pour alcohol into the spark plug hole(s). Manually rotate the flywheel another full turn.

7. Dry and reinstall the spark plug(s).

8. Dry all ignition components.

9. Drain the fuel lines and carburetor(s).

10. On models with an integral fuel tank, drain the tank and flush with fresh gasoline until all water has been removed.

CAUTION
If there is a possibility that sand may have entered the power head, do not try to start the motor or severe internal damage may occur.

11. Try starting the motor with a fresh fuel source. If motor will start, let it run at least one hour to eliminate any water remaining inside.

CAUTION
If it is not possible to disassemble and clean the motor immediately in Step 12, resubmerge the power head in water to prevent rust and corrosion formation until such time as it can be properly serviced.

12. If the motor will not start in Step 11, try to diagnose the cause as fuel, electrical or mechanical and correct the problem. If the engine cannot be started within 2 hours, disassemble, clean and oil all parts thoroughly as soon as possible.

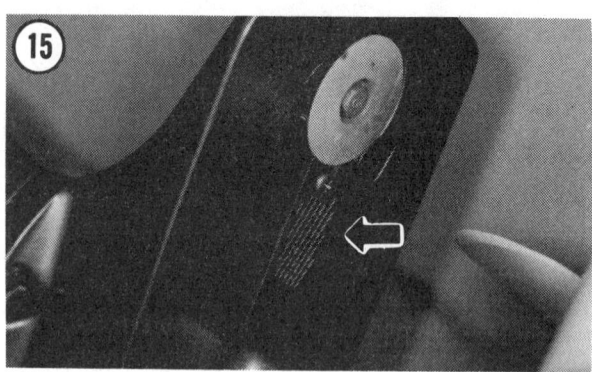

ENGINE FLUSHING

Gearcase Models

Periodic engine flushing will prevent salt or silt deposits from accumulating in the water passageways. This procedure should also be performed whenever an outboard motor is operated in salt water or polluted water.

Keep the engine in an upright position during and after flushing. This prevents water from passing into the power head through the drive shaft housing and exhaust ports during the flushing procedure. It also eliminates the possibility of residual water being trapped in the drive shaft housing or other passageways.

Some Yamaha outboards have the water intake located on the underside of the anti-ventilation plate (**Figure 15**). These models require the use of flushing devices other than a flush-test unit. See your Yamaha dealer for the proper flushing device.

1. Attach a flushing device to the front of the gearcase according to manufacturer's instructions. See **Figure 14** (typical).
2. Connect a garden hose between a water tap and the flushing device.
3. Open the water tap partially—do not use full pressure.
4. Shift into NEUTRAL, then start engine. Do not run engine above idle speed.
5. Adjust water flow so that there is a slight loss of water around the rubber cups of the flushing device.
6. Check the engine to make sure that water is being discharged from the pilot or

"tell-tale" nozzle. If it is not, stop the engine immediately and determine the cause of the problem.

> *CAUTION*
> *Flush the engine for at least 5 minutes if used in salt water.*

7. Flush engine until discharged water is clear. Stop engine.
8. Close water tap and remove flushing device from gearcase.

Jet Drive Models

The cooling system can become blocked by sand and salt deposits if it is not flushed occasionally. Clean the cooling system after each use in salt water.

> *NOTE*
> *If the outboard motor was produced prior to January 1987, refer to **Cooling System Cleaning** under **Chapter Ten: Jet Drives** for procedures to install a cooling system flush port.*

1. Remove plug and gasket from port side of jet drive housing to gain access to flush passage.
2. Install adapter (part No. 6EO-28193-00-94) into flush passage.

3. Connect a suitable fresh water supply to adapter and turn on to full pressure (maximum output).

CAUTION
When the outboard motor is running, make sure a stream of water is noted being discharged from the motor's tell-tale outlet. If not, stop the motor immediately and diagnose the problem.

CAUTION
Do not operate the outboard motor, when connected to the flush adapter, at a high rpm.

4. Start the motor and allow the fresh water to circulate for approximately 15 minutes.
5. Stop the motor, turn off the auxillary water supply and disconnect the water supply from the adapter.
6. Remove the adapter.
7. Replace gasket if damaged is noted, then install on plug.
8. Install plug and gasket into jet drive housing flush passage and securely tighten.

NOTE
To flush jet drive impeller and intake housing, direct a fresh water supply into the intake housing area.

TUNE-UP

A tune-up consists of a series of inspections, adjustments and parts replacements to compensate for normal wear and deterioration of outboard engine components. Regular tune-ups are important for power, performance and economy. Yamaha recommends that its outboards be serviced every 6 months or 100 hours of operation (whichever comes first) after the first 50 hours of operation. See **Table 1**. If subjected to limited use, the engine should be tuned at least once a year.

Since proper outboard engine operation depends upon a number of interrelated system functions, a tune-up consisting of only one or two corrections will seldom give lasting results. For best results, a thorough and systematic procedure of analysis and correction is necessary.

Prior to performing a tune-up, it is a good idea to flush the engine as described in this chapter and check for satisfactory water pump operation.

The tune-up sequence recommended by Yamaha includes the following:
   a. Compression check.
   b. Spark plug service.
   c. Gearcase and water pump check.
   d. Fuel system service.
   e. Ignition system service.
   f. Battery, starter motor and solenoid check (if so equipped).
   g. Wiring harness check.
   h. Timing, synchronization and adjustment.
   i. Performance test (on boat).

Any time the fuel or ignition systems are adjusted or defective parts replaced, the engine timing, synchronization and adjustment *must* be checked. These procedures are described in Chapter Five. Perform the timing, synchronization and adjustment procedure for your engine *before* running the performance test.

Compression Check

An accurate cylinder compression check gives a good idea of the condition of the basic working parts of the engine. It is also an important first step in any tune-up, as an engine with low or unequal compression between cylinders *cannot* be satisfactorily tuned. Any compression problem discovered during this check must be corrected before continuing with the tune-up procedure.

1. With the engine warm, disconnect the spark plug lead(s) and remove the plug(s) as described in this chapter.

2. Ground the spark plug lead(s) to the engine to disable the ignition system.

NOTE
The top cylinder is the No. 1 cylinder on inline engines. On V4 or V6 engines, the top cylinder on the starboard bank is the No. 1 cylinder.

3. Connect the compression tester to the top spark plug hole according to manufacturer's instructions (**Figure 16**).

4. Make sure the throttle is set to the wide open position, then crank the engine through at least 4 compression strokes. Record the gauge reading.

5. Repeat Step 3 and Step 4 on each remaining cylinder of multicylinder engines.

NOTE
In order to maintain proper cylinder temperature on 3-cylinder and V6 engines, Yamaha varies the cylinder head volume on some models. A V6 will normally show the highest compression on cylinders No. 1 and No. 2, with cylinders No. 5 and No. 6 giving the lowest reading. Three-cylinder engines follow the same progression, with the highest reading in No. 1 and the lowest in No. 3. J and later-model 175-200 hp engines, however, will produce the highest readings in cylinders No. 3 and No. 4 due to an increase in the volume of cylinders No. 1 and No. 2 over that of the previous N- and K-models.

The actual readings are not as important as the differences in readings when interpreting the results. A variation of more than 15 psi between 2 cylinders indicates a problem with the lower reading cylinder, such as worn or sticking piston rings and/or scored pistons or cylinders. In such cases, pour a tablespoon of engine oil into the suspect cylinder and repeat

Step 3 and Step 4. If the compression is raised significantly (by 10 psi in an older engine) the rings are worn and should be replaced.

Many outboard owners are plagued by hard starting and generally poor running for which there seems to be no good cause. Carburetion and ignition check out satisfactorily and a compression test may show that everything is well in the engine's upper end. With everything apparently pointing to a sound engine, they focus on the propeller as the cause of the problem and start swapping props, often with disastrous results.

What a compression test does *not* show is lack of primary compression. In a 2-stroke engine, the crankcase must be alternately under high pressure and low pressure. After the piston closes the intake port, further downward movement of the piston causes the entrapped mixture to be pressurized so that it can rush quickly into the cylinder when the scavenging ports are opened. Upward piston movement creates a lower pressure in the crankcase, enabling fuel-air mixture to pass in from the carburetor.

When the crankshaft seals or case gaskets leak, the crankcase cannot hold pressure and proper engine operation becomes impossible. Any other source of leakage, such as defective cylinder base gaskets or a porous or cracked crankcase casting will result in the same conditions.

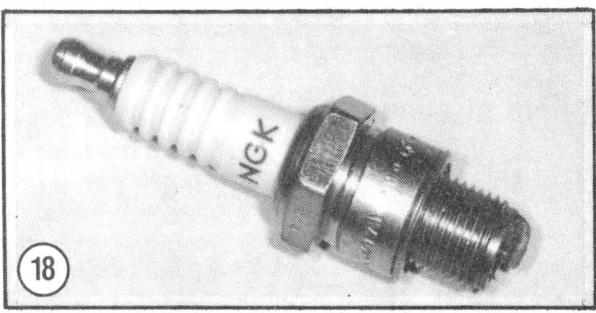

If the power head shows signs of overheating (discolored or scorched paint) but the compression test turns up nothing abnormal, check the cylinder(s) visually through the transfer ports for possible scoring. A cylinder can be slightly scored and still deliver a relatively good compression reading. In such a case, it is also a good idea to double-check the water pump operation as a possible cause for overheating.

Spark Plugs

Yamaha outboards are equipped with NGK spark plugs selected for average use conditions. Under adverse use conditions, the recommended spark plug may foul or overheat. In such cases, check the ignition and carburetion systems to make sure they are operating correctly. If no defect is found, replace the spark plug with one of a hotter or colder heat range as required.

NOTE
*Some NGK spark plug gap recommendations differ from those provided by Yamaha. **Table 2** contains both sets of gap recommendations.*

Table 2 contains the recommended spark plugs for all models covered in this book. **Table 3** provides a cross-reference for use when NGK spark plugs are not available.

Spark Plug Removal

CAUTION
Whenever the spark plugs are removed, dirt around them can fall into the plug holes. This can cause engine damage that is expensive to repair.

1. Blow out any foreign matter from around the spark plugs with compressed air. Use a compressor if you have one. If you do not, use a can of compressed inert gas, available from photo stores.
2. Disconnect the spark plug leads by twisting the lead cap back and forth on the plug insulator while pulling outward. See **Figure 17**. Pulling on the lead instead of the cap may cause internal damage to the lead.
3. Remove the plugs with an appropriate size spark plug socket or box end wrench. Keep the plugs in order so you know which cylinder they came from.
4. Examine each spark plug. See **Figure 18** (typical). Compare its condition with **Figure 19**. Spark plug condition indicates engine condition and can warn of developing trouble.
5. Check each plug for make and heat range. All should be of the same make and number or heat range.
6. Discard the plugs. Although they could be cleaned and reused if in good condition, they seldom last very long. New plugs are inexpensive and far more reliable.

Gapping Plugs

New plugs should be carefully gapped to ensure a reliable, consistent spark. Use a

4

SPARK PLUG ANALYSIS

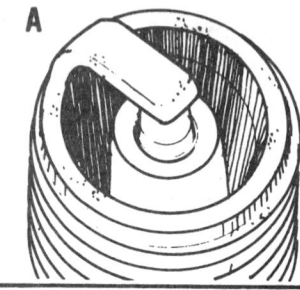

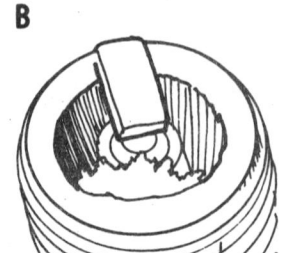

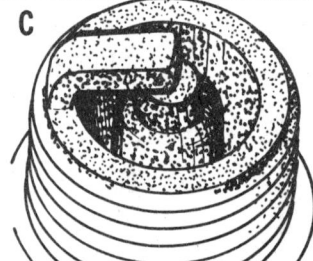

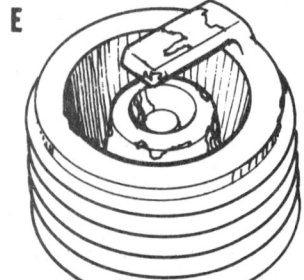

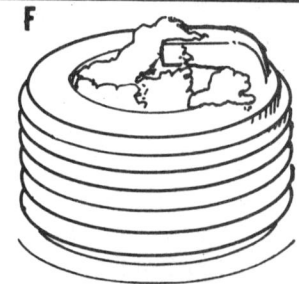

A. **Normal**—Light tan to gray color of insulator indicates correct heat range. Few deposits are present and the electrodes are not burned.

B. **Core bridging**—These defects are caused by excessive combustion chamber deposits striking and adhering to the firing end of the plug. In this case, they wedge or fuse between the electrode and core nose. They originate from the piston and cylinder head surfaces. Deposits are formed by one or more of the following:

a. Excessive carbon in cylinder.

b. Use of non-recommended oils.

c. Immediate high-speed operation after prolonged trolling.

d. Improper fuel-oil ratio.

C. **Wet fouling**—Damp or wet, black carbon coating over entire firing end of plug. Forms sludge in some engines. Caused by one or more of the following:

a. Spark plug heat range too cold.

b. Prolonged trolling.

c. Low-speed carburetor adjustment too rich.

d. Improper fuel-oil ratio.

e. Induction manifold bleed-off passage obstructed.

f. Worn or defective breaker points.

D. **Gap bridging**—Similar to core bridging, except the combustion particles are wedged or fused between the electrodes. Causes are the same.

E. **Overheating**—Badly worn electrodes and premature gap wear are indicative of this problem, along with a gray or white "blistered" appearance on the insulator. Caused by one or more of the following:

a. Spark plug heat range too hot.

b. Incorrect propeller usage, causing engine to lug.

c. Worn or defective water pump.

d. Restricted water intake or restriction somewhere in the cooling system.

F. **Ash deposits or lead fouling**—Ash deposits are light brown to white in color and result from use of fuel or oil additives. Lead fouling produces a yellowish brown discoloration and can be avoided by using unleaded fuels.

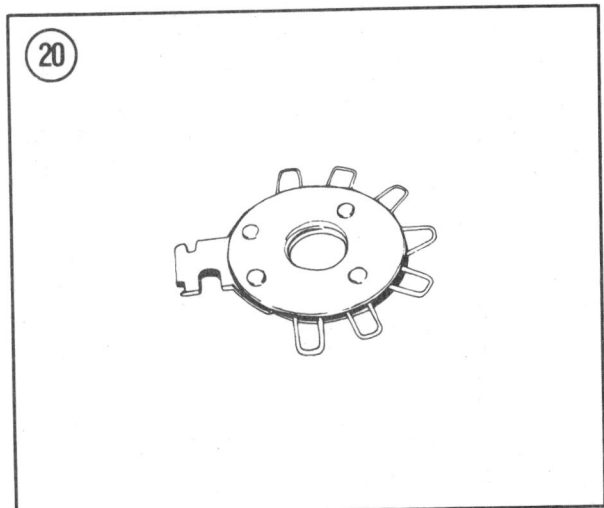

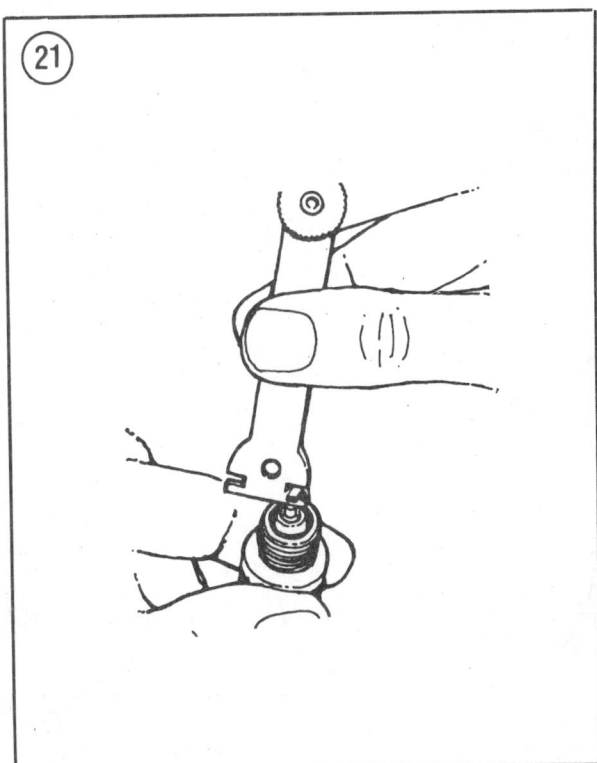

4

special spark plug tool with a round gauge. See **Figure 20** for one common type.

1. Remove the plugs and gaskets from the boxes. Install the gaskets.

NOTE
Some plug brands may have small end pieces that must be screwed on before the plugs can be used.

2. Insert the appropriate size feeler gauge (**Table 2**) between the electrodes. If the gap is correct, there will be a slight drag as the wire is pulled through. If there is no drag, or if the wire will not pull through, bend the side electrode with the gapping tool (**Figure 21**) to change the gap. Remeasure with the wire gauge.

NOTE
Never try to close the electrode gap by tapping the spark plug on a solid surface. This can damage the plug internally. Always use the gapping and adjusting tool to open or close the gap.

Spark Plug Installation

Improper installation of spark plugs is one of the most common causes of poor spark plug performance in outboard engines. The gasket on the plug must be fully compressed against a clean plug seat in order for heat transfer to take place effectively. This requires close attention to proper tightening during installation.

1. Inspect the spark plug hole threads and clean them with a thread chaser (**Figure 22**). Wipe cylinder head seats clean before installing the new plugs.

2. Screw each plug in by hand until it seats. Very little effort is required. If force is necessary, the plug is cross-threaded. Unscrew it and try again.

3. Tighten the spark plugs. If you have a torque wrench, tighten to 10-15 ft.-lb. If not, seat the plug finger-tight on the gasket, then tighten an additional 1/4-1/2 turn with a wrench.

NOTE
If damage to the spark plug lead insulation noted in Step 4 is close to the spark plug cap, it may be possible to remove the cap, cut the damaged end off and reinstall the spring on the lead to accept the cap. This will avoid the necessity of replacing an otherwise good coil.

4. Inspect each spark plug lead and cap before reconnecting it to its cylinder (**Figure 23**). If lead insulation is damaged or deteriorated, install a new coil and plug lead assembly.
5. If the cap is cracked or deteriorated, remove it from the coil lead (**Figure 24**). Wipe the spring on the end of the coil lead with a light coat of Yamalube All-purpose Marine Grease and install a new cap.
6. Push wire cap onto plug terminal and make sure it seats fully.

Water Pump Check

A faulty water pump or one that performs below specifications can result in extensive engine damage. Thus, it is a good idea to replace the water pump impeller, seals and gaskets once a year or whenever the gearcase (see Chapter Nine) or jet drive (see Chapter Ten) is removed for service.

While many outboard owners depend upon the visual sign provided by the operation of the pilot or "tell-tale" hole, installation of a water pressure meter (available from your Yamaha dealer for 40 hp and larger engines) is a far more effective check of water pump operation. A sending unit (part No. 688-83667-00-00) is permanently installed in the cylinder head and connects to a gauge (part No. 688-83661-40-00) for use in the boat.

After installing a pressure meter, note the gauge reading the first time the engine is run. If the pressure reading falls off during subsequent engine use, service the water pump as soon as possible. See Chapter Nine or Chapter Ten.

Fuel System Service

The clearance between the carburetor and choke shutter should not be greater than 0.015 in. when the choke is closed or a hard starting condition will result. When changing from one brand of gasoline to another, it may

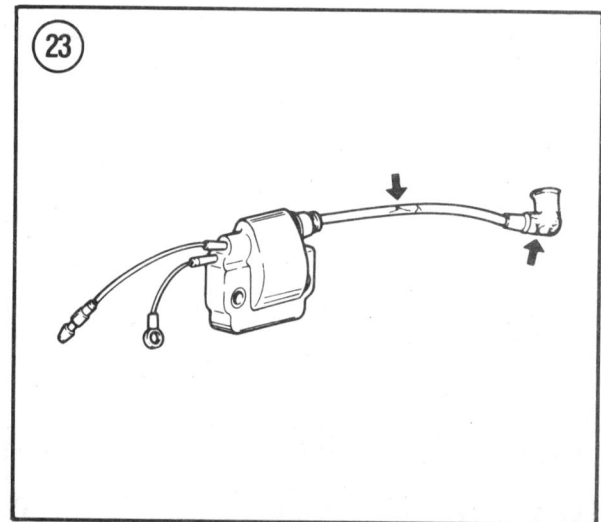

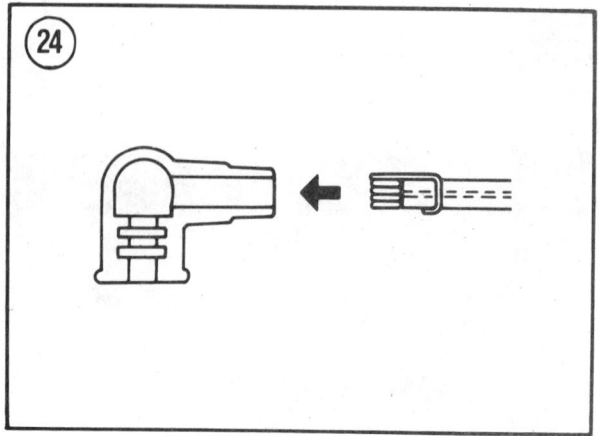

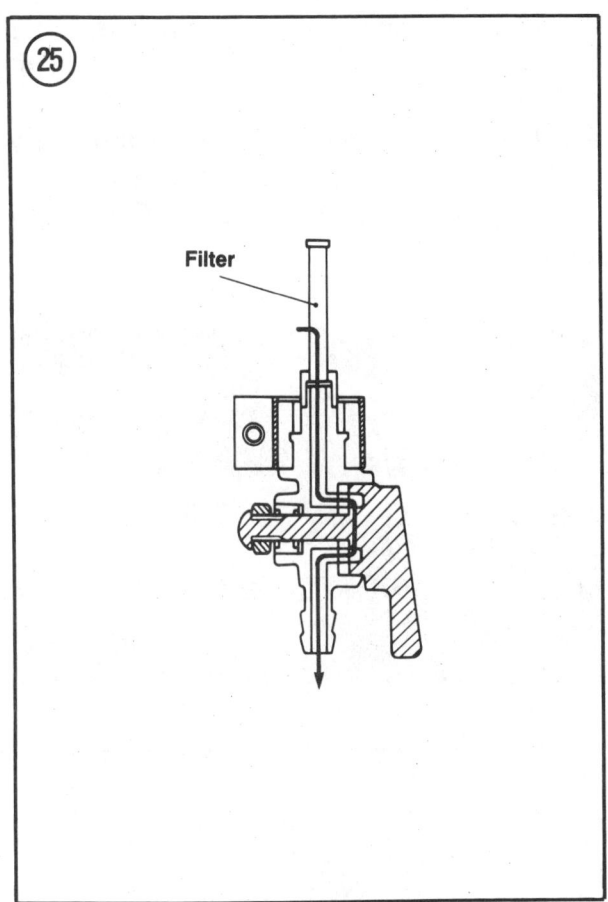

Filter

be necessary to readjust the carburetor idle mixture needle slightly (1/4 turn) to accomodate the variations in volatility.

Fuel Lines

1. Visually check all fuel lines for kinks, leaks, deterioration or other damage.
2. Disconnect fuel lines and blow out with compressed air to dislodge any contamination or foreign material.
3. Coat fuel line fittings sparingly with Type 2 Permatex and reconnect the lines.

Engine Fuel Filter Service

The 2 hp, 3 hp and 4 hp models have an integral fuel tank and utilize gravity fuel feed instead of a fuel pump. See Chapter Six. A filter screen (**Figure 25**) is installed in the fuel tank petcock (**Figure 26** shows the 4 hp model; the 2 hp and 3 hp models are similar).

Other models may use either an inline filter (**Figure 27**) or a canister filter (**Figure 28**).

> *WARNING*
> *Gasoline is recommended as a cleaning solvent in the following procedures. Work in a well-ventilated area away from any source of ignition. Keep a fire extinguisher handy.*

Integral tank filter

Refer to **Figure 29** (typical) for this procedure.
1. Drain and remove the fuel tank to provide working room. See Chapter Six.
2. Loosen the fuel petcock clamp screw or nut. Remove the petcock from the clamp and tank, then disconnect the hose.
3. Clean the petcock and filter assembly in gasoline and blow dry with compressed air. If excessively dirty or contaminated with water, discard and install a new filter.
4. Installation is the reverse of removal.

4

Inline filter

1. Place a clean cloth under the inline filter to absorb fuel in the line when the filter is removed.

2. Slide each hose retaining clip off the filter nipple with a pair of pliers and disconnect the hoses from the filter. See **Figure 30**.

3. Clean the filter assembly in gasoline. If excessively dirty or contaminated with water, discard and install a new filter.

4. Reinstall the hoses on the filter nipples. Make sure embossed arrow on filter points in the direction of fuel flow (**Figure 31**).

5. Slide the retaining clips on each hose over the nipple to assure a leak-free connection.

6. Check fuel filter installation for leakage by priming fuel system with fuel line primer bulb.

7. Remove and safely discard cloth used to catch fuel.

Canister filter

> *NOTE*
> *It may be possible to service the filter without removing the entire canister assembly from the engine on some installations. If so, perform Steps 3-7 and Step 11.*

1A. Metal clips—Slide each hose retaining clip off the filter assembly cover nipples with a pair of pliers. Disconnect the hoses from the cover and plug to prevent leakage.

1B. Plastic clips—Unsnap the plastic clips holding the inlet and outlet hoses to the filter assembly cover. Disconnect and plug the hoses to prevent leakage.

2. Remove the nut holding the filter cover to its mounting bracket (**Figure 32**). Remove the cover and canister assembly.

3. Unscrew the canister from the filter assembly cover. Remove the filter element from the canister.

4. Drain the canister and wipe the inside dry with a clean lint-free cloth or paper towel.

5. Remove and discard the cover O-ring and gasket, if used.

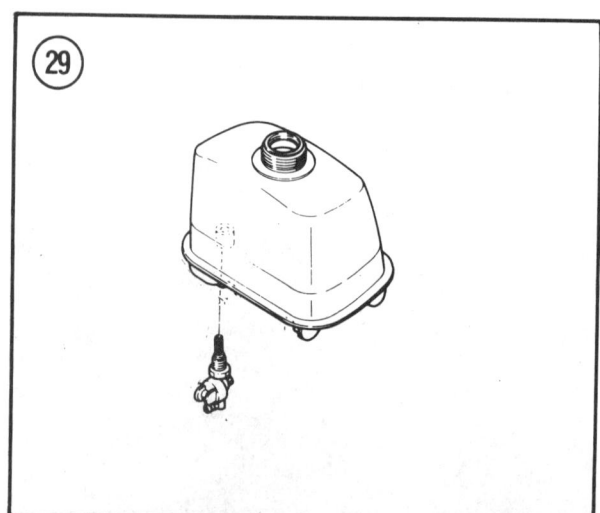

6. Clean the filter element in gasoline. If excessively dirty or contaminated with water, discard and install a new element.

7. Reinstall the element in the canister. Fit a new O-ring and gasket (if used) in position, then thread the canister on the filter assembly cover and tighten securely.

8. Reinstall the cover and canister assembly to its mounting bracket. Position cover so its nipples align properly with the inlet and outlet hoses, then tighten the nut.

9. Connect the inlet and outlet hoses to the canister cover.

10. Position the plastic clips on the hoses and snap into place. See Chapter Six.

11. Check fuel filter installation for leakage by priming fuel system with fuel line primer bulb.

Fuel Pump

The fuel pump does not generally require service during a tune-up. However, if the engine has more than 100 hours on it since the fuel pump was last serviced, it is a good idea to remove and disassemble the pump, inspect each part carefully for wear or damage and reassemble it with a new diaphragm. See Chapter Six.

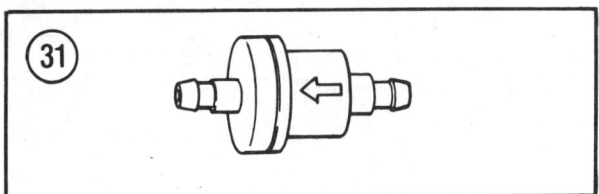

Fuel pump diaphragms are fragile and one that is defective often produces symptoms that are diagnosed as an ignition system problem. A common malfunction results from a tiny pinhole or crack in the diaphragm caused by an engine backfire. This defect allows gasoline to enter the crankcase and wet-foul the spark plug at idle speed, causing hard starting and engine stall at low rpm. The problem disappears at higher speeds, as fuel quantity is limited. Since the plug is not fouled by excess fuel at higher speeds, it fires normally.

Pressure test

NOTE
Fuel pump pressure cannot be tested on integral fuel pump carburetors.

1A. On single fuel pump installations, use a tee-fitting to connect a fuel pressure gauge into the line between the carburetor(s) and fuel pump.

1B. If equipped with twin fuel pumps, connect the pressure gauge into the line between the 2 pumps at the lower pump to test the lower pump. To test the upper pump, disconnect the fuel inlet line at the lower pump and the line between the 2 pumps at the upper pump. Move the disconnected fuel inlet line from the lower pump to the upper pump and connect it to the pump with the tee and pressure gauge in place of the line between the 2 pumps.

2. Before starting the engine, loosen the fuel tank vent cap to relieve any pressure in the system. With the engine in a test tank or on the boat in the water, the fuel pump pressure should be at least:

   a. 1 psi at 600 rpm.
   b. 1.5 psi at 1,500-3,000 rpm.
   c. 2.5 psi at 4,500 rpm.

If not, rebuild the fuel pump with a new diaphragm and gaskets. See Chapter Six.

Breaker Point Ignition
System Service

The Yamaha 2 hp model uses a breaker point ignition; all other models use a CDI ignition.

The condition and gap of the breaker points will greatly affect engine operation. Burned or badly oxidized points will allow little or no current to pass. A gap that is too narrow will not allow the coil to build up sufficient voltage and will result in a weak spark. An excessive point gap will allow the points to open before the primary current reaches its maximum.

While slightly pitted points can be dressed with a file, this should be done only as a temporary measure, as the points may arc after filing. Oxidized, dirty or oily points can be cleaned with alcohol but new points are inexpensive and always preferable for efficient engine operation.

The condenser absorbs the surge of high voltage from the coil and prevents current from arcing across the points when they open. Condensers can be tested with a condenser tester but are also inexpensive and should be replaced as a matter of course whenever new breaker points are installed.

NOTE
Breaker points must be adjusted correctly. An error in gap of 0.0015 in. will change engine timing by as much as one degree.

Breaker point sets are installed on the stator base under the flywheel and are set to 0.014 ± 0.002 in. After establishing the breaker point gap, ignition timing should be checked. See Chapter Five.

CAUTION
Always rotate the crankshaft in a clockwise direction in the following procedures. If rotated more than 180° in a counterclockwise direction, the water pump impeller may be damaged.

Breaker Point Replacement

Refer to **Figure 33** for this procedure.
1. Disconnect the negative battery cable, if so equipped.
2. Remove the 10 screws holding the engine cowling. Remove the cowling.
3. Remove the flywheel. See Chapter Eight.
4. Remove the screw holding the breaker point set to the stator base (A, **Figure 33**).
5. Disconnect the coil and condenser leads (B, **Figure 33**) at the breaker point set. Remove the breaker point set.
6. Remove the screw holding the condenser to the stator base (C, **Figure 33**). Remove the condenser.
7. Install a new breaker point set on the stator base. Make sure the pivot point on the bottom of the point set engages the hole in the stator base. Install but do not tighten the hold-down screw.
8. Install a new condenser on the stator base and tighten the attaching screw securely, then

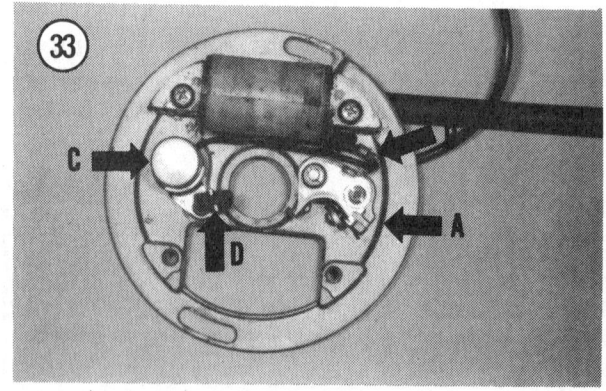

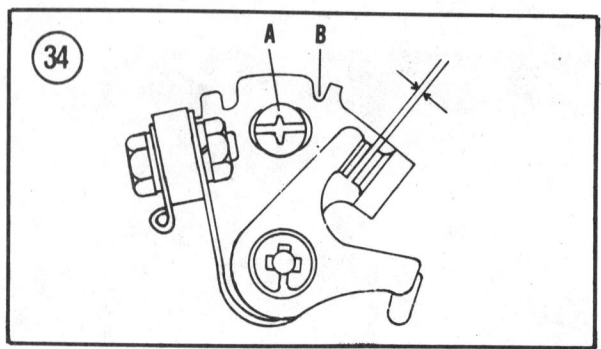

connect the coil and condenser leads to the breaker point set.

CAUTION
Do not over-lubricate the breaker cam in Step 9. Excessive lubrication will cause premature point set failure.

9A. If equipped with a felt lubrication wick (D, **Figure 33**), gently squeeze the wick to see if it is dry. If dry, lubricate with 1-2 drops of 30W engine oil.

9B. If no felt wick is used, lightly lubricate the breaker cam with cam grease.

10. Adjust the breaker point gap as described in this chapter.

11. Reverse Steps 1-3 to complete installation.

Breaker Point Adjustment

The following procedure is used when new breaker points have been installed. Slots are provided in the flywheel rotor for checking and adjusting the point gap without flywheel removal. When checking or adjusting old breaker points, remove the engine cowling and proceed to Step 3.

1. Install the flywheel nut on the crankshaft and rotate the stator base to the wide-open throttle position.

2. Place a wrench on the flywheel nut and rotate the crankshaft clockwise until the breaker point rubbing block rests on a high point on the cam (the points will be wide open).

3. Loosen the point set hold-down screw (A, **Figure 34**). Insert a screwdriver in the adjusting notch and move the point set base to obtain a gap of 0.014 in. when measured with a flat feeler gauge. The gap is correct when the feeler gauge offers a slight drag as it is slipped between the points. When the gap is correct, tighten the hold-down screw securely and recheck the point gap.

Battery and Starter Motor Check (Electric Start Models Only)

1. Check the battery's state of charge. See Chapter Seven.

2. Connect a voltmeter between the starter motor positive terminal (**Figure 35**) and ground.

3. Turn ignition switch to START and check voltmeter scale.

4A. If voltage exceeds 9.5 volts and the starter motor does not operate, replace the motor.

4B. If voltage is less than 9.5 volts, recheck battery and connections. Charge battery, if necessary, and repeat procedure.

Starter Relay Test (Electric Start Models Only)

NOTE
*The size and shape of the starter relay is reduced on 1985 and later 9.9-30 hp models. See **Figure 36**.*

1. Disconnect the negative battery lead, then the positive battery lead.

2. Disconnect the brown and black starter relay leads.

3. Connect an ohmmeter between the disconnected relay leads (**Figure 37**). If the meter does not read 3 ±0.3 ohms (9.9-50 hp) or 3 ±0.6 ohms (70-225 hp) at 68° F (20° C), replace the starter relay.

4. Connect a voltmeter or a test lamp between the starter relay terminals. Use

jumper leads to connect the brown relay lead to the positive battery terminal and the black relay lead to the negative battery terminal. See **Figure 38**.

5. The relay should click and the meter needle deflect or the test lamp light when the relay leads are connected to the battery in Step 4.

   a. If the relay clicks but the needle does not deflect (test lamp does not light) in

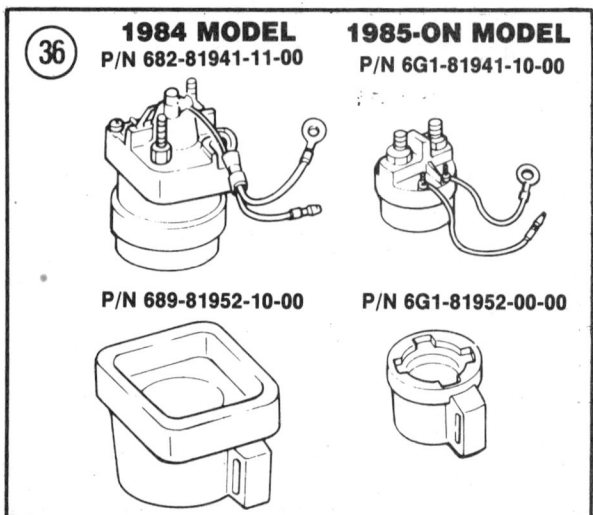

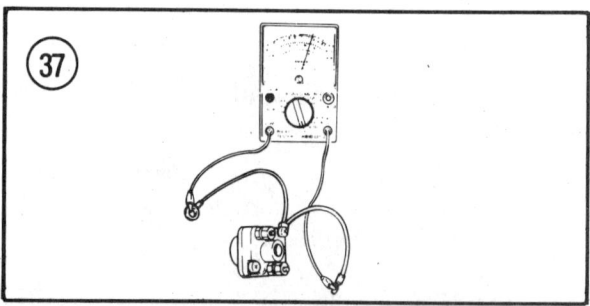

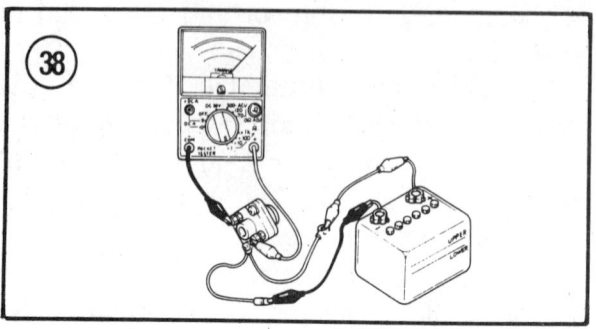

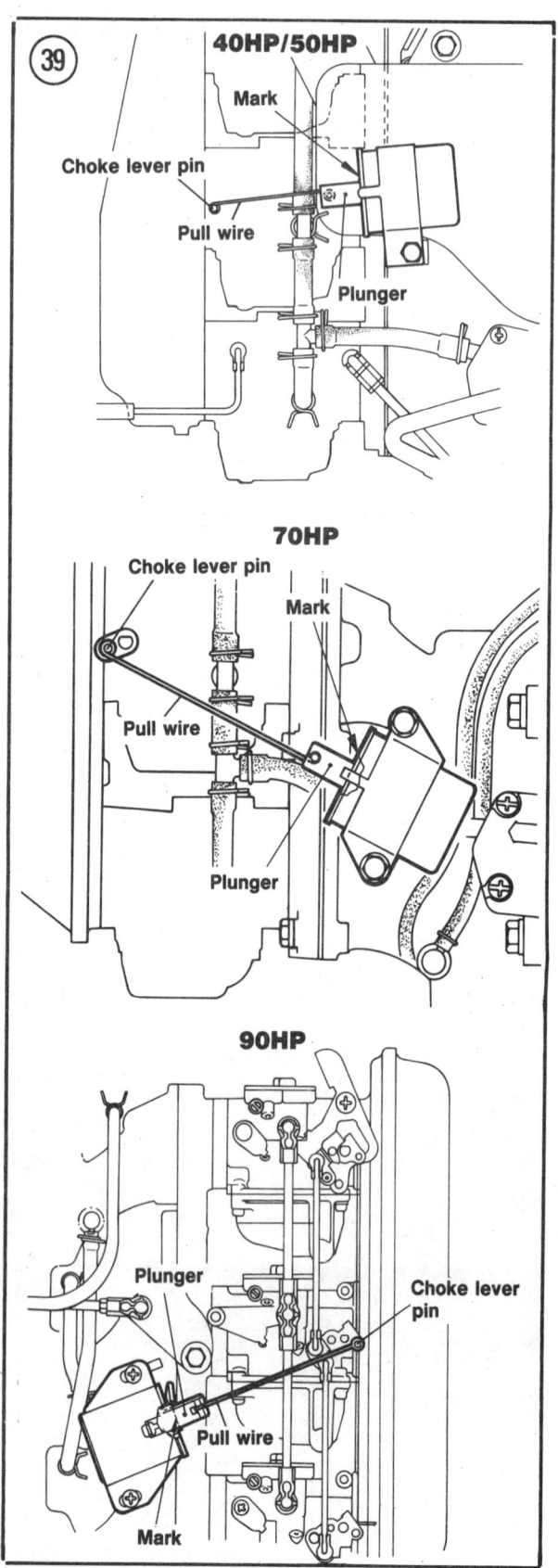

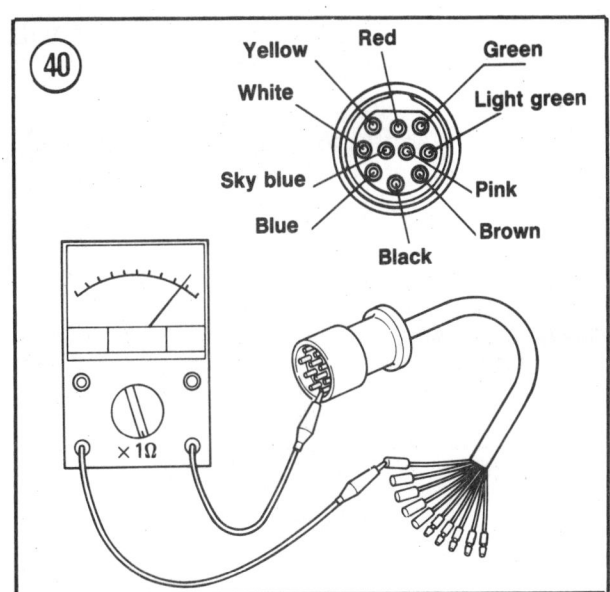

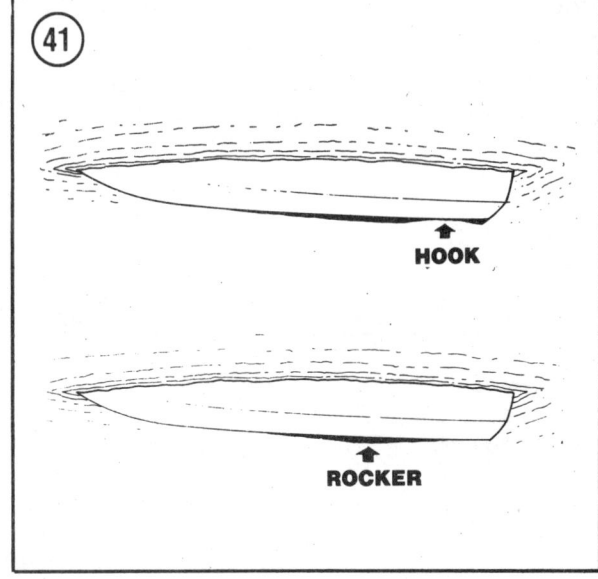

Step 4, the relay contacts are defective. Replace the starter relay.

b. If there is no relay click or needle deflection (test lamp does not light) in Step 4, the relay coil is defective. Replace the starter relay.

Choke Solenoid Resistance Check (Electric Start Models Only)

The choke solenoid may be located on the port or starboard side of the engine. See **Figure 39** for typical installations.

1. Disconnect the choke solenoid lead at the bullet connector.

2. Connect an ohmmeter between the disconnected lead and a good engine ground.

3. If the reading obtained in Step 2 is not within specifications (**Table 4**), replace the choke solenoid.

Wiring Harness Check

1. Check the wiring harness for signs of frayed or chafed insulation.

2. Check for loose connections between the wires and terminal ends.

3. Check harness connector for bent electrical prongs.

4. Check harness connector and prongs for signs of corrosion. Clean as required.

5. If harness is suspected of contributing to electrical malfunction, disconnect the harness at the engine and switch panel. Check the continuity of all wires with an ohmmeter or the Yahama pocket tester. See **Figure 40** (typical). Replace the harness if any wire shows no continuity.

Engine Synchronization and Adjustment

See Chapter Five.

Performance Test (On Boat)

Before performance testing the engine, make sure that the boat bottom is cleaned of all marine growth and that there is no evidence of a "hook" or "rocker" (**Figure 41**) on the bottom. Any of these conditions will reduce performance considerably.

The boat should be performance tested with an average load and with the motor tilted at an angle that will allow the boat to ride on an even keel. If equipped with an adjustable trim tab, it should be properly adjusted to allow the boat to steer in either direction with equal ease.

Check engine rpm at full throttle. If not within the maximum rpm range for the motor as specified in Chapter Five, check the

propeller pitch. A high pitch propeller will reduce rpm while a lower pitch prop will increase it.

Readjust the idle mixture and speed under actual operating conditions as required to obtain the best low-speed engine performance.

Table 1 LUBRICATION & MAINTENANCE [1]

Frequency	Item	Lubricant type [2]
After each use	Check for loose nuts, bolts and spark plugs	—
	Check propeller, shear pin and cotter pin condition	—
	Make sure cooling water runs out of exhaust ports while cruising	—
	Grease jet drive bearing(s)	A
After first 10 hours or 1 month	Check throttle link operation	—
	Check shift mechanism operation	—
	Check throttle grip/housing	—
	Check choke lever	—
	Check swivel bracket	—
After first 10 and 50 hour intervals, (1 and 3 months), then every 100 hours (6 months) or once per season	Clean and flush fuel tank	—
	Service fuel filter	—
	Check head bolt torque	—
	Check water pump operation	—
	Check sacrifical anode condition; replace as required	—
	Lubricate steering pivot shaft	A
After first 10 hours or 1 month, then every 100 hours (6 months) or once per season	Replace spark plugs	—
	Check all condition of all fuel line hoses	—
	Check and adjust carburetor	—
	Check and adjust ignition timing	—
	Check and adjust ignition and throttle synchronization	—
	Check and adjust throttle	—
	Lubricate shift lever housing	A
	Lubricate throttle gear housing	
	Lubricate clamp bolts or screws	A
	Lubricate tilt mechanism	A
	Change gearcase lubricant	B
	Lubricate propeller shaft [3]	A
After first 50 hours (3 months), then every 100 hours (6 months) or once per season	Lubricate throttle link	A
	Lubricate shift mechanism	A
	Lubricate throttle grip/housing	A
	Lubricate choke lever	A
	Lubricate swivel bracket	A
	(continued)	

Table 1 LUBRICATION & MAINTENANCE[1] (continued)

Frequency	Item	Lubricant type [2]
Every 100 hours (6 months) or once per season	Lubricate oil injection pump return spring	C
	Replace water pump impeller	
	Lubricate recoil starter [4]	A
	Check power trim/tilt fluid level	D
	Check battery condition	
Every 200 hours or once per season	Check neutral start system operation	
	Check neutral opening stop adjustment	
	Check and adjust choke solenoid	
	Check oil injection pump link adjustment	
	Check oil injection system operation	
Once per season	Clean/paint exterior	E

1. Complete list may not apply to all models. Perform only those items which apply to your model. Reduce service interval by one-half if used in salt water.
2. Lubricant legend:
  A. Yamalube All-purpose marine grease.
  B. Yamalube Gearcase lubricant.
  C. Yamalube Two-cycle outboard lubricant.
  D. Yamalube Power Trim/tilt fluid.
  E. Yamaha spray paint.
3. And whenever propeller is removed or changed.
4. 2 hp only.

Table 2 RECOMMENDED SPARK PLUGS

Engine	NGK plug type	Gap (in.)[1]
2 hp	B5HS	0.020-0.024 in.
3 hp	B6HS-10	0.035-0.039 in.
4 hp	B7HS	0.020-0.024 in.
5 hp	B7HS	0.020-0.024 in.
6 hp	B7HS-10	0.035-0.039 in.
8 hp	B7HS-10	0.035-0.039 in.
9.9 hp	B7HS-10	0.035-0.039 in.
15 hp	B7HS-10	0.035-0.039 in.
25 hp		
1984	B7HS	0.020-0.024 in.
1985-on	B7HS-10	0.035-0.039 in.
30 hp	B7HS-10	0.035-0.039 in.
40 hp	B7HS-10	0.035-0.039 in.
50 hp	B8HS-10	0.035-0.039 in.
70 hp	B8HS-10	0.035-0.039 in.
90 hp	B8HS-10	0.035-0.039 in.
115 hp	B8HS-10	0.035-0.039 in.
130 hp	B9HS-10	0.035-0.039 in.
150 hp	B8HS-10	0.035-0.039 in.
175 hp	B8HS-10	0.035-0.039 in. (continued)

4

Table 2 RECOMMENDED SPARK PLUGS (continued)

Engine	NGK plug type	Gap (in.)[1]
200 hp	B8HS-10	0.035-0.039 in.
220 hp	BR8HS-10	0.035-0.039 in.
225 hp	BR9HS-10	0.035-0.039 in.

1. NGK recommends a gap of 0.028 in. for B5HS and B7HS plugs and a gap of 0.040 in. for all plugs designated −10 when used in Yamaha outboards.

Table 3 SPARK PLUG CROSS-REFERENCE CHART *

NGK	Champion	AC
B5HS	L9J, L10	45F, M45FF, 46F
B7HS	L4J, L82, L82C	42F, S42F, M42FF
B6HS-10	—	—
B7HS-10	—	—
B8HS-10	—	—
B9HS-10	L77JC4, L77J4	M40FFX
BR8HS-10	—	—
BR9HS-10	QL77J4	—

* The cross-referenced spark plugs are not exact replacements for original plug heat range and should be used only as a temporary replacement.

Table 4 CHOKE SOLENOID RESISTANCE SPECIFICATIONS

Model	ohms
25-30 (Prior to 1988) hp	7.2-8.8
25-30 (After 1987) hp	3.6-4.4
40-90 hp	3.4-4.0
115-220 hp	2.9-4.4
225 hp	3.4-4.0

Chapter Five

Timing, Synchronizing and Adjusting

If an engine is to deliver its maximum efficiency and peak performance, the ignition system must be timed and carburetor operation synchronized with the ignition. The engine must be timed and synchronized as the final step of a tune-up or whenever the fuel or ignition systems are serviced or adjusted.

Procedures for timing, synchronizing and adjustment on Yamaha outboards differ according to model and ignition system. This chapter is divided into self-contained sections dealing with particular models/ignition systems for fast and easy reference. Each section specifies the appropriate procedure and sequence to be followed and provides the necessary tune-up data. Read the general information at the beginning of the chapter and then select the section pertaining to your outboard.

Tables 1-13 are at the end of the chapter.

ENGINE TIMING AND SYNCHRONIZATION

As engine rpm increases, the ignition system must fire the spark plug(s) more rapidly. Proper ignition timing synchronizes spark plug firing with engine speed.

As engine speed increases, the carburetor must provide an increased amount of fuel for combustion. Synchronizing is the process of timing the carburetor operation to the ignition (and thereby the engine speed).

The Yamaha 2 hp is equipped with a breaker point ignition and is static-timed by measuring the amount of piston travel relative to breaker point opening and closing. Correct timing thus depends upon a correct breaker point gap setting.

Required Equipment

Static timing of Yamaha engines requires the use of an accurate dial indicator to determine top dead center (TDC) of the piston before making any timing adjustment. TDC is determined by removing the spark plug and installing the dial indicator in the plug opening. See **Figure 1**.

> *NOTE*
> *Yamaha cylinder heads use offset spark plug holes, that is, the spark plug hole is **not** perpendicular to the piston. For a true reading, the dial indicator must be installed so that the indicator plunger **is** perpendicular to the piston. When properly installed, the indicator plunger will be off-center in the spark plug hole.*

Dynamic engine timing uses a timing light connected to the No. 1 spark plug lead. See **Figure 2**. As the engine is cranked or operated, the light flashes each time the spark plug fires. When the light is pointed at the moving flywheel, the mark on the flywheel appears to stand still. The flywheel mark should align with the stationary timing pointer or decal on the engine.

A tachometer connected to the engine is used to determine engine speed during idle and high-speed adjustments.

> *CAUTION*
> *Never operate the engine without water circulating through the gearcase to the engine. This will damage the water pump and the gearcase and can cause engine damage.*

Some form of water supply is required whenever the engine must be operated during the procedure. The use of a test tank and test wheel is the most convenient method. While the procedures may be carried out with the boat in the water, checking engine timing while speeding across open water is neither easy nor safe. The use of a flushing device is *not* recommended if the engine must be run above idle speed.

Yamaha recommends that a test wheel (**Figure 3**) be substituted for the propeller

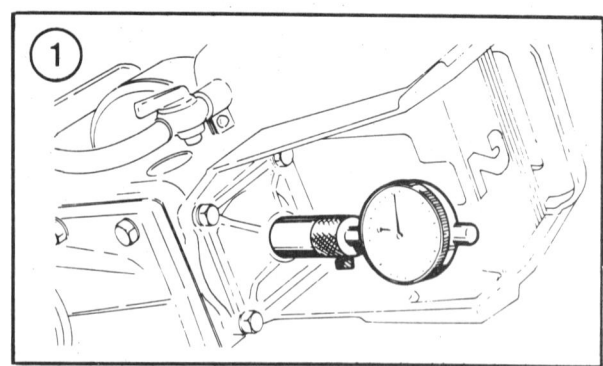

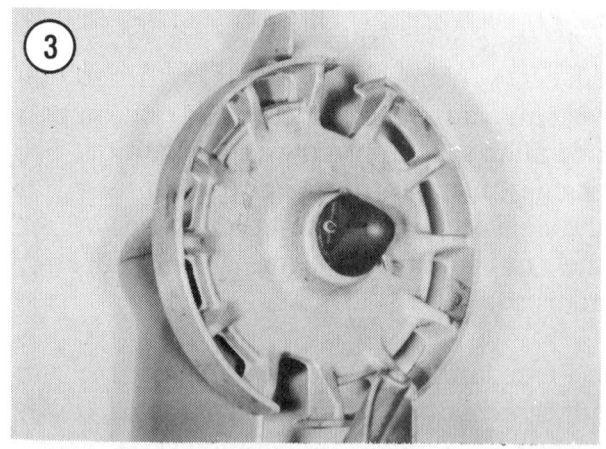

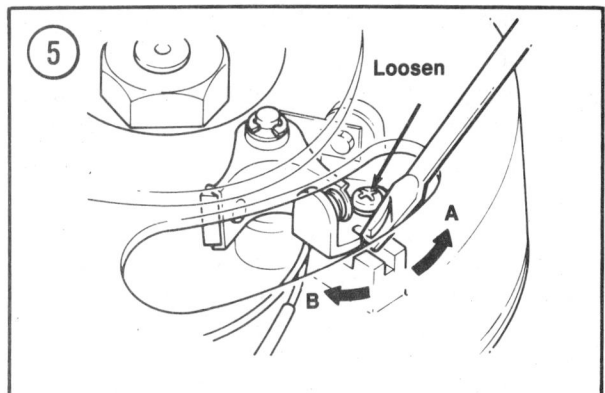

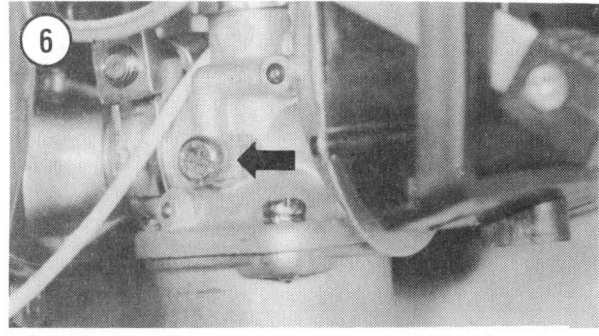

with some models to put a load on the propeller shaft and prevent engine damage from excessive rpm. Test wheel recommendations provided by Yamaha are given in **Table 1**.

YAMAHA 2

The flywheel magneto on this model is designed to provide automatic spark advance. Timing adjustments are made by setting the breaker point gap to specifications.

1. Remove the 10 screws holding the engine cowling. Remove the cowling.
2. Remove the flywheel. See Chapter Eight.
3. Disconnect the spark plug lead and remove the spark plug. See Chapter Four.
4. Disconnect the white stop switch lead at the connector (**Figure 4**).
5. Make sure the breaker points are set to specifications (**Table 2**). See Chapter Four.
6. Install a dial indicator in the spark plug hole (**Figure 1**).

7. Rotate the flywheel clockwise until the dial indicator shows that the piston has reached top dead center (TDC). This is the point at which the indicator needle reverses its direction of movement as the flywheel is rotated. Set the indicator gauge to zero.
8. Connect the red test lead of a low-reading ohmmeter to the white magneto assembly lead disconnected in Step 4. Connect the black test lead to a good engine ground.
9. Slowly rotate the magneto base counterclockwise until the ohmmeter needle moves, indicating that the breaker points have closed. The dial indicator should show movement at a point 0.044 ± 0.0047 in. BTDC.

   a. If the meter needle moves before passing 0.039 in. BTDC, loosen the breaker point hold-down screw and adjust the point gap to 0.014-0.016 in. See **Figure 5**.

   b. If the meter needle moves after passing 0.048 in. BTDC, loosen the breaker point hold-down screw and adjust the point gap to 0.012-0.014 in. See **Figure 5**.

   c. If correct ignition timing cannot be obtained by adjusting the point gap, replace the breaker points. See Chapter Four.

10. With the engine in a test tank or on the boat in the water, turn the idle speed screw (**Figure 6**) clockwise until it lightly seats, then back it out 3 full turns.
11. Connect a tachometer to the engine according to manufacturer's instructions.
12. Start the engine, place in FORWARD gear and warm to normal operating temperature.
13. Adjust the idle speed screw as required to bring the idle speed within specifications (**Table 2**).
14. Shut the engine off, remove the test equipment and install the engine cowling.

YAMAHA 3, 4 AND 5

Timing Inspection

The ignition system used on these models is designed to provide automatic spark advance. Ignition timing is non-adjustable. The following dynamic timing procedure will determine if timing is incorrect.

1. Install the engine in a test tank.
2. Remove the engine cover.
3. Connect a tachometer and timing light to the engine according to manufacturer's instructions. See **Figure 7**.
4. Start the engine and adjust the throttle to run at a speed under 1,700 rpm.
5. Aim the timing light at the magneto base window nearest the front of the motor. If low-speed timing is correct, the flywheel timing mark will be seen in the window indicating the specified timing (**Table 3**, Yamaha 3 and **Table 4**, Yamaha 4 and 5). See **Figure 8**.
6. Increase engine speed to 4,500-5,000 rpm.
7. Aim the timing light at the magneto base window nearest the rear of the motor.
   a. If high-speed timing is correct, the flywheel timing mark will be seen in the window indicating the specified timing (**Table 3**). See **Figure 9**.
   b. If no timing mark is seen in the window, either the high-speed pulser coil or CDI unit is defective. Refer to Chapter Three for test procedures and Chapter Seven for replacement procedures.
8. Bring the throttle back to idle and depress the stop switch.
9. Remove the test equipment.
10. Install the engine cover.

Carburetor Adjustment

Refer to **Figure 10A** for 3 hp models and **Figure 10B** for 4 and 5 hp models.
1. Install the engine in a test tank.
2. Remove the engine cover.

3. Connect a tachometer to the engine according to manufacturer's instructions.
4. Turn the idle mixture screw (A) clockwise until it lightly seats, then back it out 1 1/4 turns (3 hp), 1 7/8 turns (4 hp) or 1 5/8 turns (5 hp).

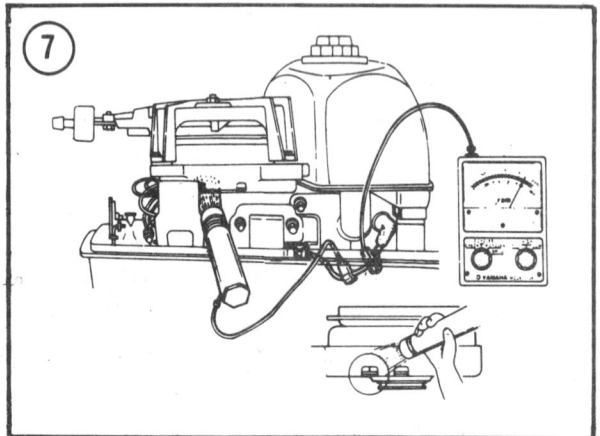

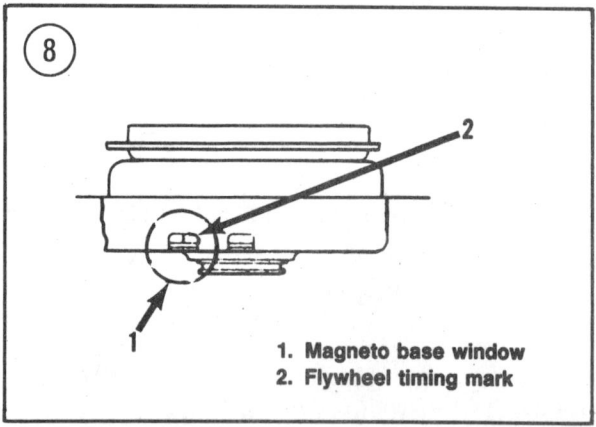

1. Magneto base window
2. Flywheel timing mark

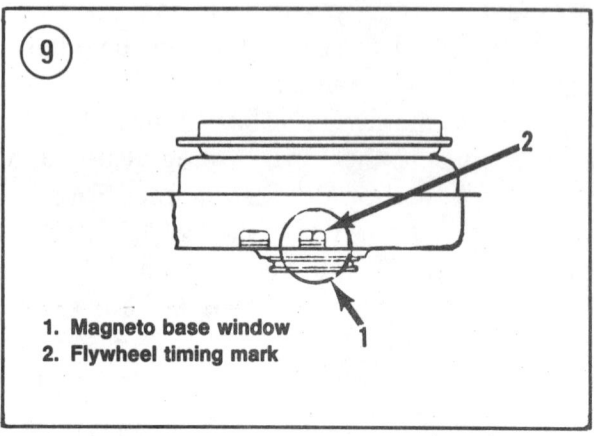

1. Magneto base window
2. Flywheel timing mark

5. Start the engine, place in FORWARD gear and warm to normal operating temperature.

6. Adjust the idle speed screw (B) as required until the lowest possible rpm is obtained.

7. Slowly adjust the idle mixture screw counterclockwise until the engine starts to load up, then turn the screw clockwise until the engine fires evenly and rpm increases.

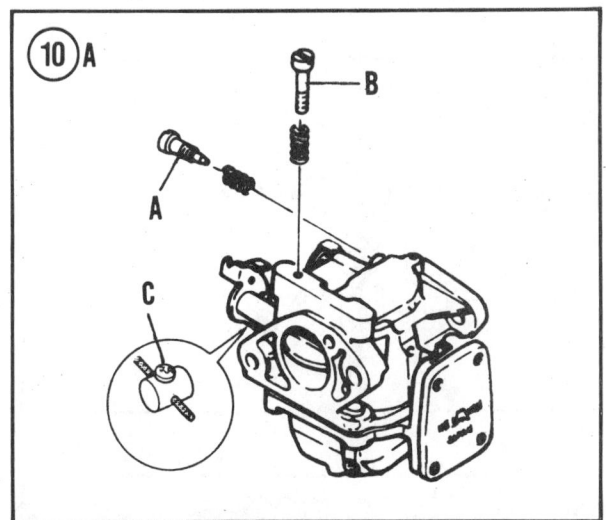

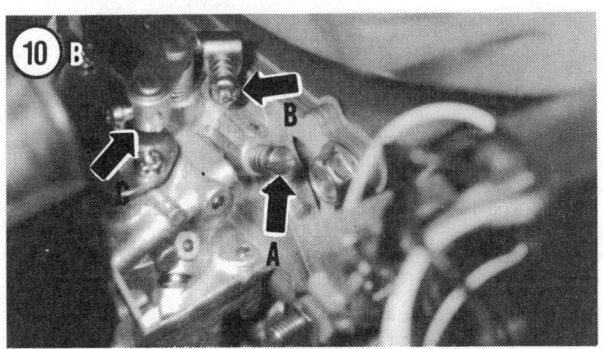

8. Repeat Step 6 and Step 7 until the smoothest idle is obtained at the lowest possible rpm.

9. Rapidly open and close the throttle. If the engine hesitates during acceleration, turn the idle mixture screw clockwise slightly until acceleration is smooth.

10. Return engine to idle rpm and recheck idle speed. Readjust idle speed screw if necessary to bring idle speed within specifications (**Table 3** or **Table 4**).

11. Shut the engine off. Remove the tachometer.

12. Rotate the throttle grip to open and close the throttle several times. With the throttle grip set at the FAST or wide-open position, the throttle valve wide-open stop should touch the carburetor stop. If it does not, loosen the cable screw (C). Adjust the cable length so that it projects 0.12-0.16 in. with the throttle valve wide open.

YAMAHA 6 AND 8

Timing Adjustment

Correctly adjusting the full-advance timing on this model automatically adjusts the full retard timing.

1. Install the engine in a test tank.

2. Remove the engine cover.

3. Connect a tachometer and timing light to the engine according to manufacturer's instructions.

4. Disconnect the magneto base link rod from the control lever ball stud. See **Figure 11**.

> *WARNING*
> *The remaining adjustment steps require movement of the magneto base by hand while the engine is running. Work slowly and carefully to avoid contact with the moving flywheel or serious personal injury may result.*

5. Start the engine and manually move the magneto base to the full-advance position.

6. Aim the timing light at the magneto timing pointer (**Figure 12**). If the pointer does not align with the specified mark (**Table 5**) on the flywheel timing indicator (**Figure 13**), loosen the bolt holding the magneto base stop (**Figure 14**). Slowly move the magneto base until the indicator mark and timing plate align (**Figure 13**) under the timing light.

7. Working carefully to avoid moving the magneto base, move the magneto base stop until it contacts the full-advance side stop on the magneto base, then tighten the stop bolt. See **Figure 15**.

8. Recheck timing with the timing light. If it is still incorrect, repeat Step 5 and Step 6 as required. When timing is correct, shut the engine off.

9. Reverse Steps 1-4 to complete timing adjustment procedure.

Throttle Linkage Adjustment

1. Remove the engine cover.

2. Disconnect the magneto base link rod from the control lever ball stud. See **Figure 11**.

3. Rotate the throttle grip to the wide-open (FAST) position (**Figure 16**).

4. Loosen the locknut on the tight-side (outer) throttle wire. Turn the adjusting nut until the carburetor throttle lever contacts its stop, then tighten the locknut. See **Figure 17**.

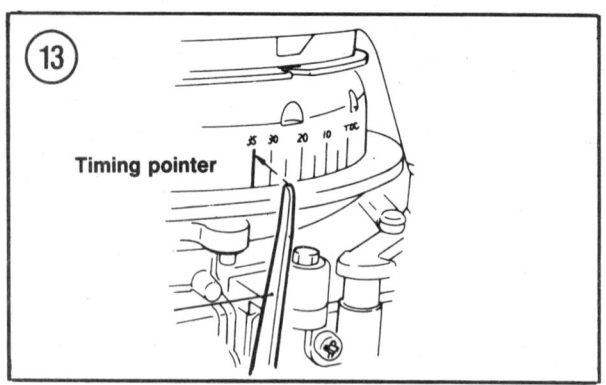

Timing pointer

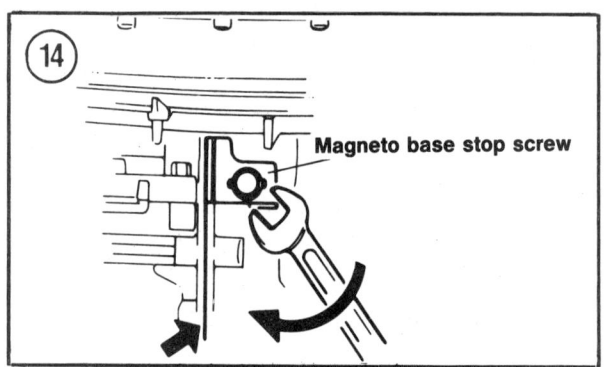

Magneto base stop screw

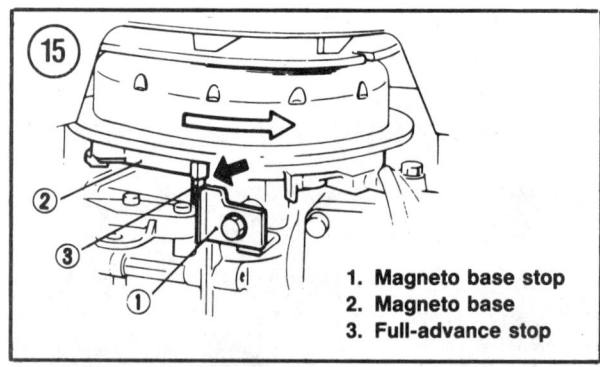

1. Magneto base stop
2. Magneto base
3. Full-advance stop

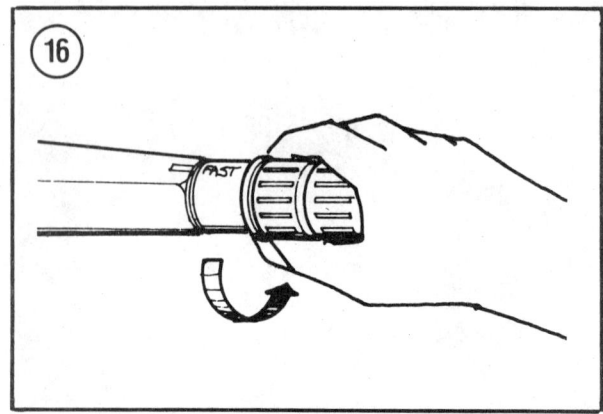

5. Rotate the magneto base until the full-advance side stop contacts the magneto base stop (**Figure 15**).

6. Adjust the plastic snap-on connector on the end of the link rod until it can be reconnected to the control lever ball stud

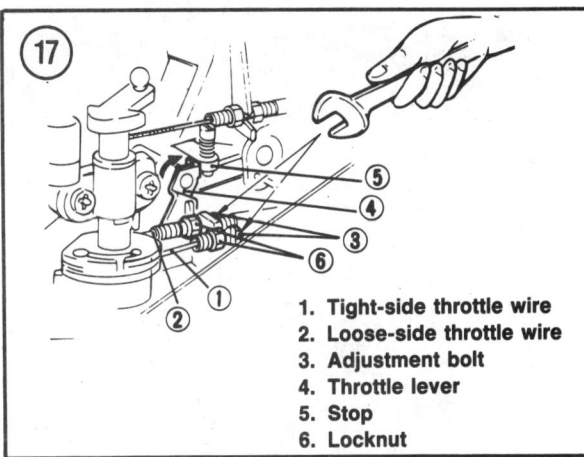

1. Tight-side throttle wire
2. Loose-side throttle wire
3. Adjustment bolt
4. Throttle lever
5. Stop
6. Locknut

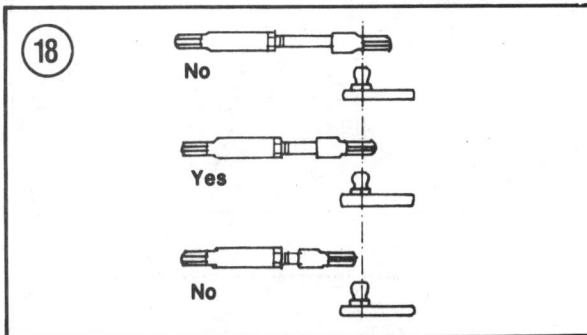

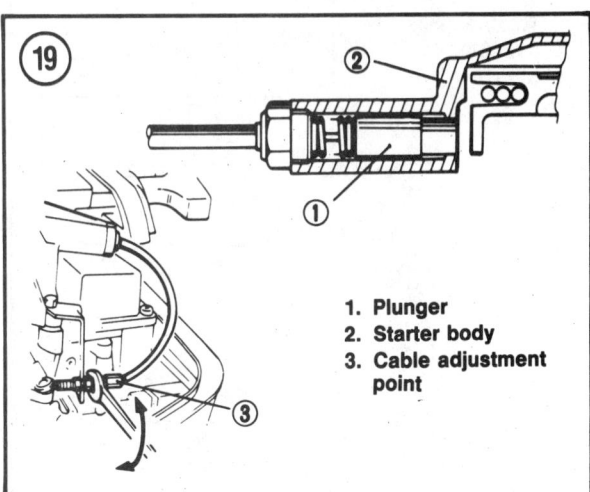

1. Plunger
2. Starter body
3. Cable adjustment point

without changing the position of the linkage or magneto base. See **Figure 18**.

7. Operate throttle twist grip. The throttle lever should contact its stop (**Figure 17**) and the magneto base full-advance side stop should contact the magneto base stop (**Figure 15**) when the twist grip is set to the FAST position (**Figure 16**). If either condition is not met, readjust as required.

8. Now rotate the twist grip to its SLOW position. If the throttle opening limiter cable plunger is not flush with the surface of the starter housing, turn the adjustment nut as required. See **Figure 19**.

Carburetor Adjustment

1. Install the engine in a test tank with the engine cover off.

2. Connect a tachometer according to manufacturer's instructions.

3. Turn the idle mixture screw (A, **Figure 20**) clockwise until it lightly seats, then back it out 1-1 1/2 turns.

4. Start the engine, place in FORWARD gear and warm to normal operating temperature.

5. Adjust the throttle stop screw (B, **Figure 20**) as required until the lowest possible rpm is obtained.

6. Slowly adjust the idle mixture screw counterclockwise until the engine starts to load up, then turn the screw clockwise until the engine fires evenly and rpm increases.

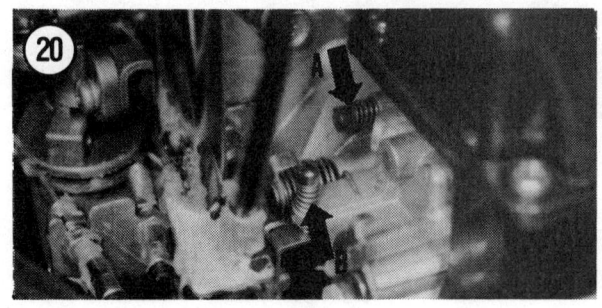

5

7. Repeat Step 5 and Step 6 until the smoothest idle is obtained at the lowest possible rpm.

8. Rapidly open and close the throttle. If the engine hesitates during acceleration, turn the idle mixture screw clockwise slightly until acceleration is smooth.

9. Return engine to idle rpm and recheck idle speed. Readjust throttle stop screw if necessary to bring idle speed within specifications (**Table 5**).

10. Shut the engine off. Remove the tachometer and install the engine cover.

YAMAHA 9.9 AND 15

Timing Adjustment

1. Remove the engine cover.

2. Move the shift lever to the FORWARD position.

3. Disconnect the magneto base link rod from the control lever ball stud. See **Figure 21**.

4. Disconnect the spark plug leads and remove the spark plugs. See Chapter Four.

5. Install a dial indicator in the No. 1 (top) spark plug hole. See **Figure 22**.

6. Rotate the flywheel clockwise until the dial indicator shows that the piston has reached top dead center (TDC). This is the point at which the indicator needle reverses its direction as the flywheel is rotated.

1. Magneto base
2. Timing mark

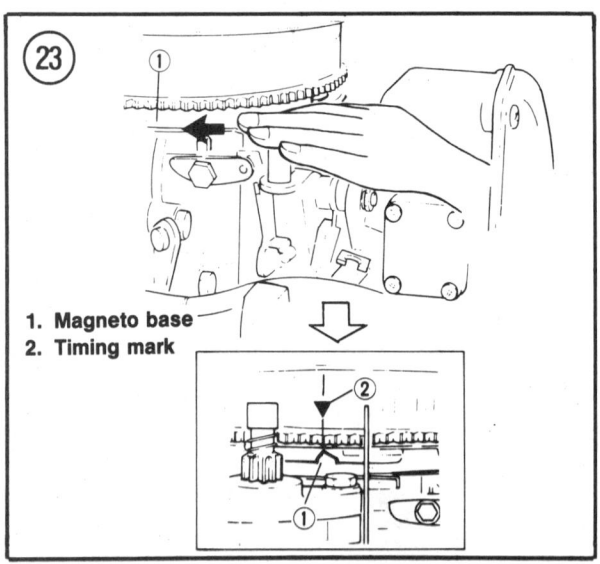

1. Throttle cam
2. Cam bolt
3. Stop

7. Rotate the flywheel counterclockwise until the dial indicator reads 0.166 ±0.011 in. (equivalent to 30° ±1° BTDC).

8. Slowly move the magneto base until its timing mark aligns with the flywheel timing mark. See **Figure 23**.

9. Loosen the bolts holding the throttle cam/magneto stop bracket to the magneto

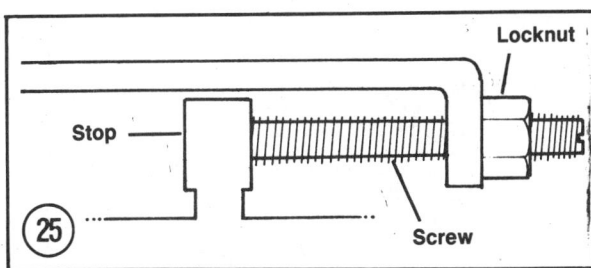

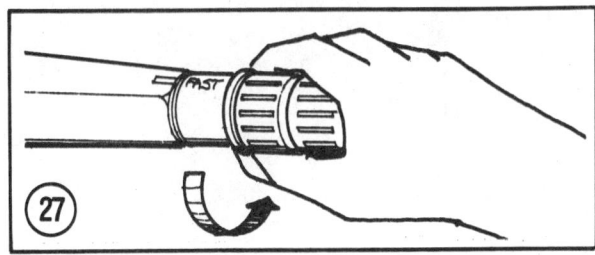

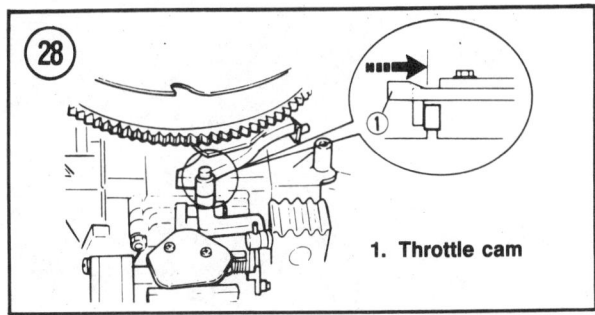

1. Throttle cam

base. With magneto base and flywheel aligned as described in Step 8, move the throttle cam/magneto stop bracket until it just touches the engine block stop and tighten the bracket bolts snugly. See **Figure 24**.

10. Once the full-advance timing is set, repeat Step 6.

11. Rotate the flywheel counterclockwise until the dial indicator reads 0.005 ±0.002 in. (equivalent to 5° ±1° BTDC).

12. Rotate the magneto base clockwise until the idle timing adjustment screw on the bracket stop just touches the engine block stop (**Figure 25**).

13. Loosen the adjustment screw locknut. Hold adjustment screw against engine block stop and adjust screw (**Figure 26**) until magneto base timing mark aligns with 5° BTDC mark on flywheel. Tighten the locknut.

14. Rotate the throttle grip to the FAST position (**Figure 27**).

15. Move the throttle cam to the full advance position (**Figure 28**).

16. Loosen the locknut and adjust the plastic snap-on connector on the end of the link rod so that it can be reconnected to the control lever ball stud without changing the position of the linkage or magneto base. See **Figure 29**.

17. Install the link rod on the ball stud and tighten the locknut. See **Figure 21**.

18. Operate the throttle twist grip. The throttle cam/magneto stop should touch the

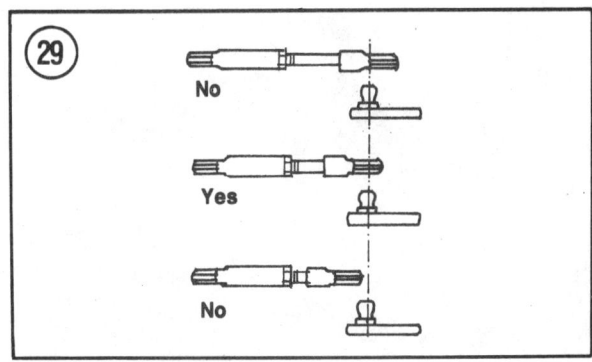

5

engine stop with the grip set to FAST and the idle timing adjustment screw should touch the engine stop with the grip set to SLOW. If either one or both conditions are not met, readjust as required.

19. Remove the dial indicator. Install the spark plug and connect the spark plug lead. See Chapter Four.

Wide-open Throttle Linkage Adjustment

1. Remove the engine cover.
2. Move the shift lever to the FORWARD position.
3. Rotate the throttle grip to the FAST position (**Figure 27**).
4. Loosen the throttle link setscrew (**Figure 30**).
5. Depress the throttle link until the throttle lever touches the stop with the throttle cam roller touching the throttle cam, then tighten the setscrew. See **Figure 31**.
6. Return the shift lever to the NEUTRAL position and install the engine cover.

Throttle Linkage Neutral Adjustment

1. Place the engine in a test tank.
2. Remove the engine cover.
3. Move the shift lever to the FORWARD position.
4. Depress the magneto control lever and hold it against its stop (**Figure 32**).

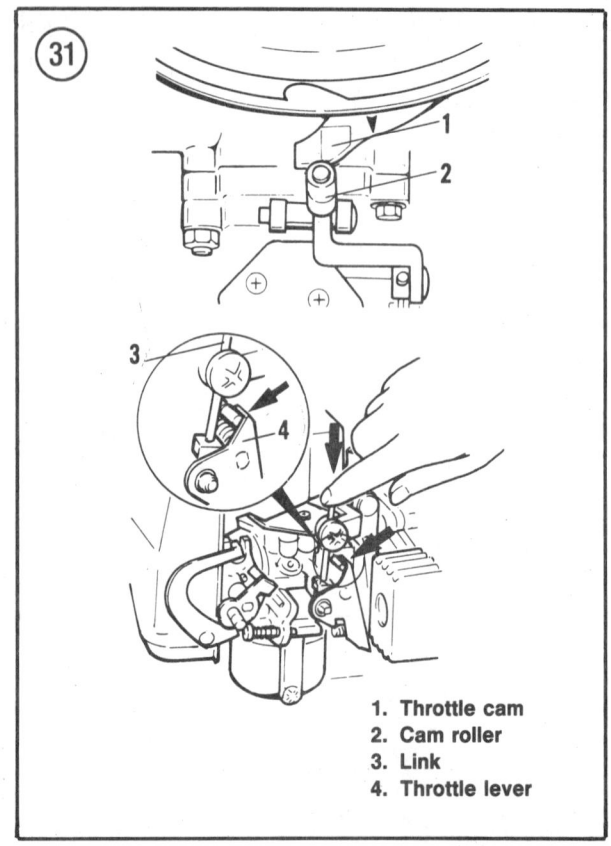

1. Throttle cam
2. Cam roller
3. Link
4. Throttle lever

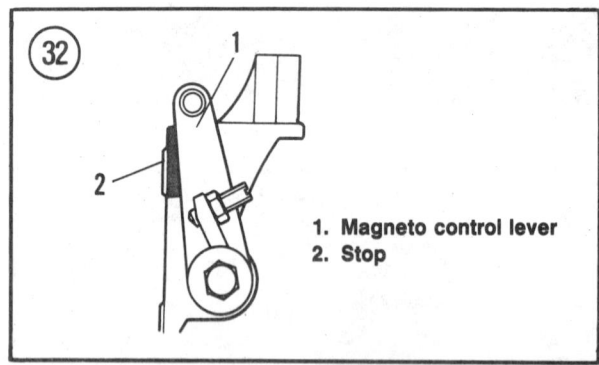

1. Magneto control lever
2. Stop

5. Disconnect the magneto base link rod from the control lever ball stud. See **Figure 33**.

6. Loosen the locknut. Adjust the plastic snap-on connector on the end of the link rod so that it can be reconnected to the control

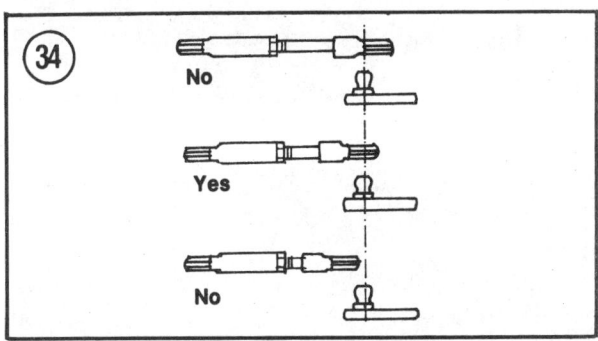

lever ball stud without changing the position of the linkage or magneto base. See **Figure 34**.

7. Install the link rod on the ball stud and tighten the locknut.

8. Connect a tachometer according to manufacturer's instructions.

9. Loosen the locknut on the magneto control lever adjustment screw (**Figure 35**).

10. Start the engine and adjust the throttle grip to obtain an engine speed of 1,300 ± 300 rpm. Turn the adjustment screw (**Figure 35**) until it touches the stop, then tighten the locknut.

11. Reverse Steps 1-3 to complete procedure.

Carburetor Adjustment

1. Install the engine in a test tank and remove the engine cover.

2. Connect a tachometer according to manufacturer's instructions.

3. Turn the idle mixture screw (A, **Figure 36**) clockwise until it lightly seats, then back it out 1 1/4-1 3/4 turns (9.9) or 1 1/8-1 5/8 turns (15).

4. Start the engine, place in FORWARD gear and warm to normal operating temperature.

5. Adjust the throttle stop screw (B, **Figure 36**) as required until the lowest possible rpm is obtained.

6. Slowly adjust the idle mixture screw counterclockwise until the engine starts to load up, then turn the screw clockwise until the engine fires evenly and rpm increases.

7. Repeat Step 5 and Step 6 until the smoothest idle is obtained at the lowest possible rpm.

8. Rapidly open and close the throttle. If the engine hesitates during acceleration, turn the

idle mixture screw clockwise slightly until acceleration is smooth.

9. Return engine to idle rpm and recheck idle speed. Readjust throttle stop screw if necessary to bring idle speed within specifications (**Table 6**).

10. Shut the engine off. Remove the tachometer and install the engine cover.

YAMAHA 25 (1984-1987)

Timing Adjustment

Correctly adjusting the full-advance timing on this model automatically adjusts the full retard timing.

1. Remove the engine cover.

2. Disconnect the magneto base link rod from the control lever ball stud. See A, **Figure 37**.

3. Disconnect the spark plug leads and remove the spark plugs. See Chapter Four.

4. Install a dial indicator in the No. 1 (top) spark plug hole. See **Figure 38** (typical).

5. Rotate the flywheel clockwise until the dial indicator shows that the piston has reached top dead center (TDC). This is the point at which the indicator needle reverses its direction as the flywheel is rotated.

6. Check timing pointer alignment with flywheel timing scale (**Figure 39**). If pointer is not properly aligned, recheck Step 5 to make sure piston is at TDC. If this is not the problem, loosen the timing pointer bolt and move the pointer as required to align with the flywheel TDC mark, then tighten the pointer bolt.

7. Remove the dial indicator.

8. Rotate the flywheel clockwise to align the timing pointer with the 25° BTDC mark on the flywheel timing scale. See **Figure 40**. Hold flywheel in this position.

9. Slowly move the magneto base counterclockwise to align the stamped timing mark on the base with the TDC mark on the flywheel timing scale (**Figure 41**).

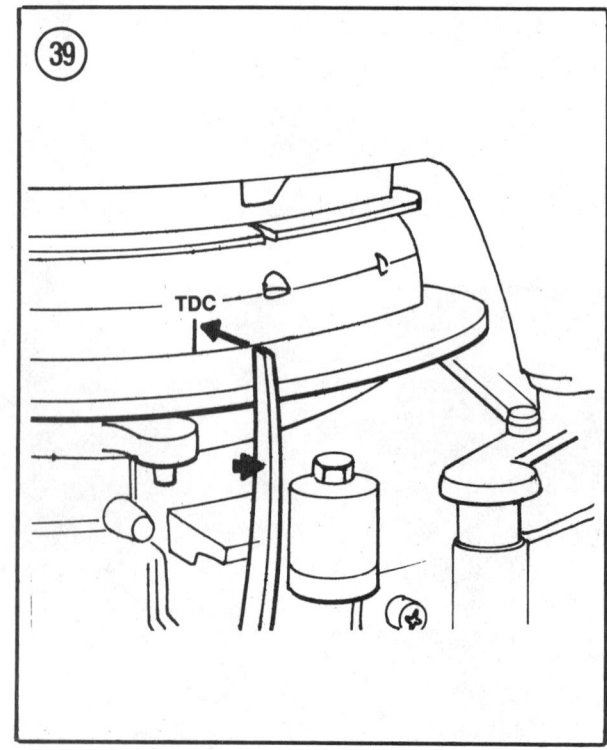

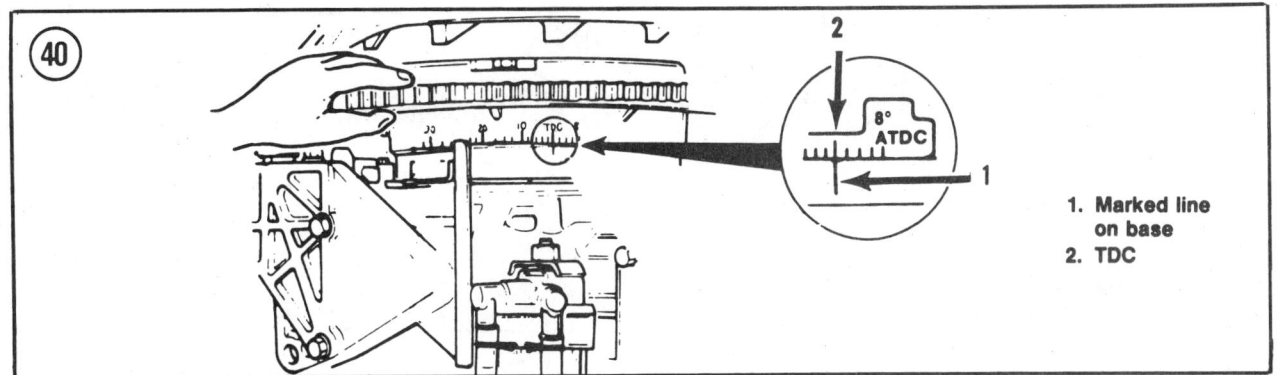

1. Marked line on base
2. TDC

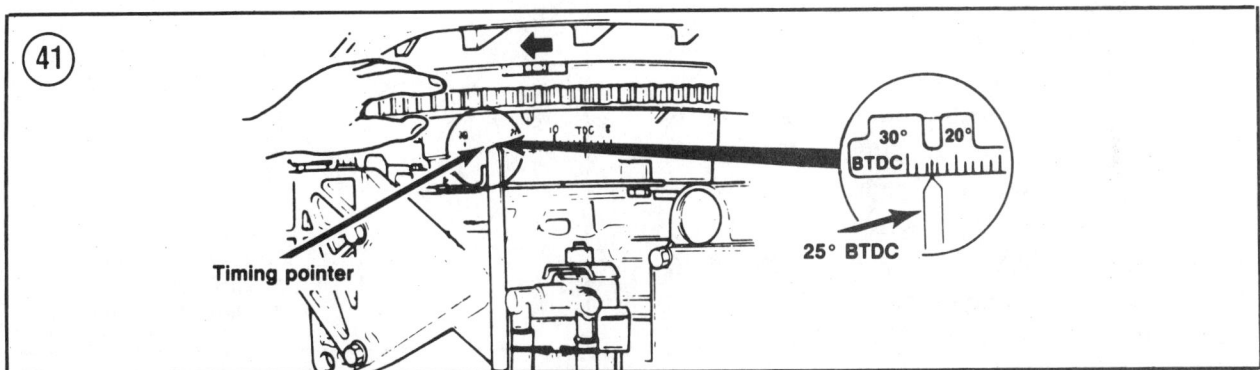

Timing pointer

25° BTDC

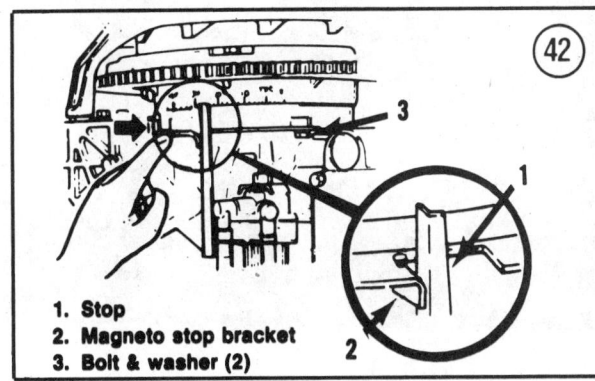

1. Stop
2. Magneto stop bracket
3. Bolt & washer (2)

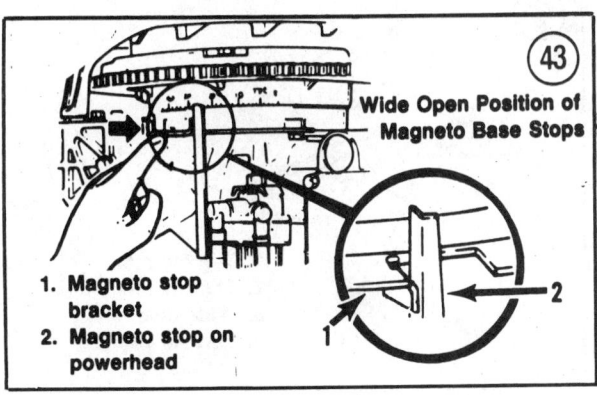

Wide Open Position of Magneto Base Stops

1. Magneto stop bracket
2. Magneto stop on powerhead

10. The magneto stop bracket cap should touch the engine stop. If it does not, loosen the bolt holding the magneto stop bracket. With the timing marks aligned as in Step 9, move the bracket until it contacts the engine stop, then tighten the bolt. See **Figure 42**.

Wide-open Throttle Linkage Adjustment

1. Disconnect the magneto base (A), throttle cam lever (B) and throttle control lever (C) linkage at the points shown in **Figure 37**. Loosen the locknut on each link rod.

2. Slowly rotate the magneto base by hand to its wide open position as shown in **Figure 43**.

3. Shift engine into FORWARD and rotate twist grip to FAST position (**Figure 44**).

4. Adjust the plastic snap-on connector on the end of the magneto base link rod until link rod length is 3.11 ±0.02 in. Reconnect link to control lever ball stud and tighten the locknut.

5. Depress the carburetor throttle arm until its stop touches the wide open throttle stop. Hold throttle arm in this position and and rotate throttle cam clockwise until it touches the throttle arm. See **Figure 45**.

6. Adjust the plastic snap-on connector on the end of the throttle cam lever link rod until link rod length is 6.71 in. Reconnect link to control lever ball stud and tighten the locknut.

7. Loosen the upper throttle cable locknuts. Turn adjusting nut until all slack is removed from the upper cable, then tighten the locknuts. Loosen the lower throttle cable locknuts. Turn adjusting nut until there is 0.04-0.08 in. slack in the lower cable, then tighten the locknuts. See **Figure 46**.

8. Open and close the throttle several times with the twist grip to make sure the magneto base stop touches the crankcase at the wide open and closed positions. If not, repeat Steps 4-7 as required.

9. Open and close the throttle several times with the twist grip to make sure the carburetor throttle arm touches the wide open throttle stop and throttle adjustment screw at the wide open and closed positions. If not, repeat Steps 5-7 as required.

10. Pull the throttle control lever link rod upward until the lever touches the stop in the bottom of the cowling. See arrow, **Figure 47**. Adjust the plastic snap-on connector on the end of the throttle control lever link rod until link rod length is 3.7 ±0.2 in. Reconnect link to control lever ball stud and tighten the locknut.

Carburetor Adjustment

1. Install the engine in a test tank with the engine cover off.

2. Connect a tachometer according to manufacturer's instructions.

3. Turn the idle mixture screw (A, **Figure 48**) clockwise until it lightly seats, then back it out 1 1/4-1 3/4 turns.

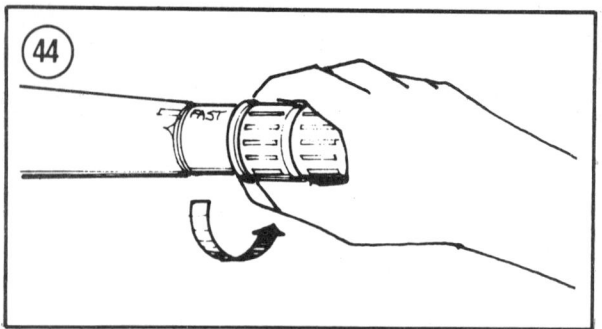

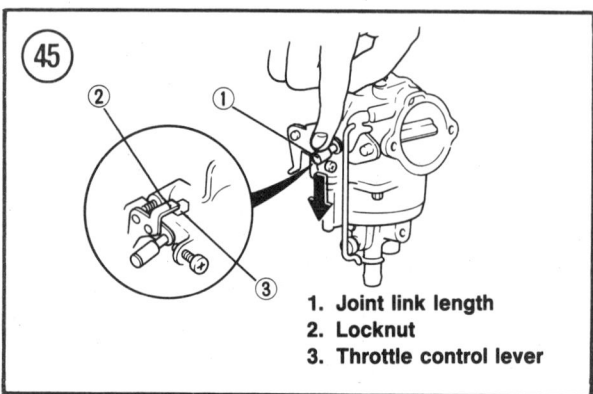

1. Joint link length
2. Locknut
3. Throttle control lever

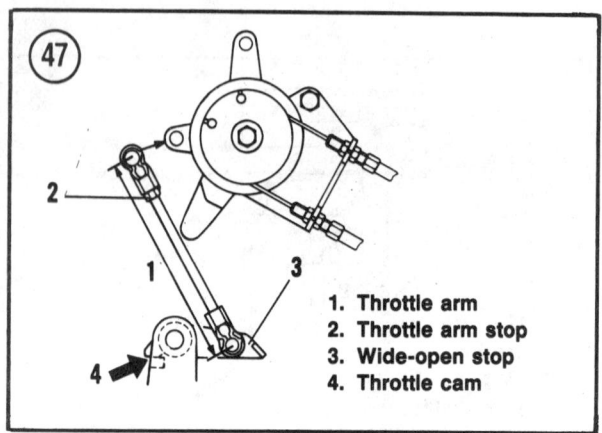

1. Throttle arm
2. Throttle arm stop
3. Wide-open stop
4. Throttle cam

4. Start the engine, place in FORWARD gear and warm to normal operating temperature.

5. Adjust the throttle stop screw (B, **Figure 48**) as required until the lowest possible rpm is obtained.

6. Slowly adjust the idle mixture screw counterclockwise until the engine starts to

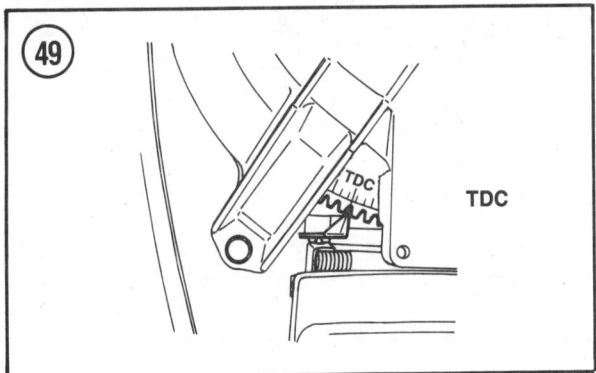

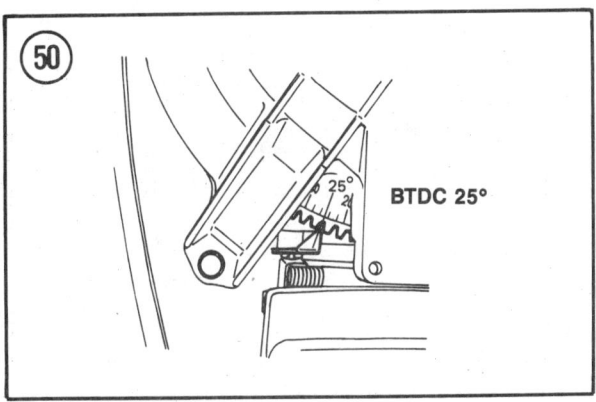

load up, then turn the screw clockwise until the engine fires evenly and rpm increases.

7. Repeat Step 5 and Step 6 until the smoothest idle is obtained at the lowest possible rpm.

8. Rapidly open and close the throttle. If the engine hesitates during acceleration, turn the idle mixture screw clockwise slightly until acceleration is smooth.

9. Return engine to idle rpm and recheck idle speed. Readjust throttle stop screw if necessary to bring idle speed within specifications (**Table 6**).

10. Shut the engine off. Remove the tachometer and install the engine cover.

YAMAHA 25 (1988-ON)

Timing Adjustment

1. Remove the engine cover.

2. Disconnect the spark plug leads and remove the spark plugs. See Chapter Four.

3. Install a dial indicator in the No. 1 (top) spark plug hole. See **Figure 38** (typical).

4. Rotate the flywheel clockwise until the dial indicator shows that the piston has reached top dead center (TDC). This is the point at which the indicator needle reverses its direction as the flywheel is rotated.

5. Check timing pointer alignment with the flywheel timing scale (**Figure 49**). If pointer is not properly aligned, recheck Step 4 to make sure piston is at TDC. If this not the problem, loosen the timing plate setscrew and move the plate as required to align with the flywheel TDC mark, then tighten the setscrew securely.

6. Remove the dial indicator.

7. Rotate the flywheel clockwise to align the timing plate with the 25° BTDC mark on the flywheel timing scale. See **Figure 50**.

8. Move the magneto control lever to the right (full advance) until it touches the stopper.

9. If the timing marks on the magneto base plate and flywheel do not align as shown in **Figure 51**, disconnect the link rod from the control lever ball stud. Loosen the locknut and adjust the plastic snap-on connector on the end of the link rod until it can be reconnected to the ball stud without changing the position of the linkage or timing marks. Reconnect the link rod to the ball stud and tighten the locknut.

10. Rotate the flywheel clockwise to align the timing plate with the 7° ATDC mark on the flywheel timing scale. See **Figure 52**.

11. Move the magneto control lever to the left (full retard) until it touches the stopper bolt (**Figure 53**).

12. If the timing marks do not align as shown in **Figure 53**, loosen the stopper bolt locknut and adjust the bolt length as required to align the marks, then tighten the locknut.

Carburetor Linkage Adjustment

1. Loosen throttle stop screw (**Figure 54**) until clearance is noted between end of screw and throttle valve stop.

2. Close the throttle valves.

> *NOTE*
> *The throttle lever screws have a left-hand thread. Turn the screws clockwise to loosen and counterclockwise to tighten.*

3. If throttle valves do not completely close, loosen No. 1 cylinder (top) carburetor throttle lever screw (A, **Figure 55**).

4. Completely loosen adjustment screw (B, **Figure 55**) on accelerator lever rod.

5. Make sure No. 1 cylinder (top) carburetor throttle valve is closed, then tighten throttle lever screw (A, **Figure 55**).

6. Make sure No. 1 cylinder (top) and No. 2 cylinder (bottom) carburetor throttle valves are closed, then tighten adjustment screw (B, **Figure 55**).

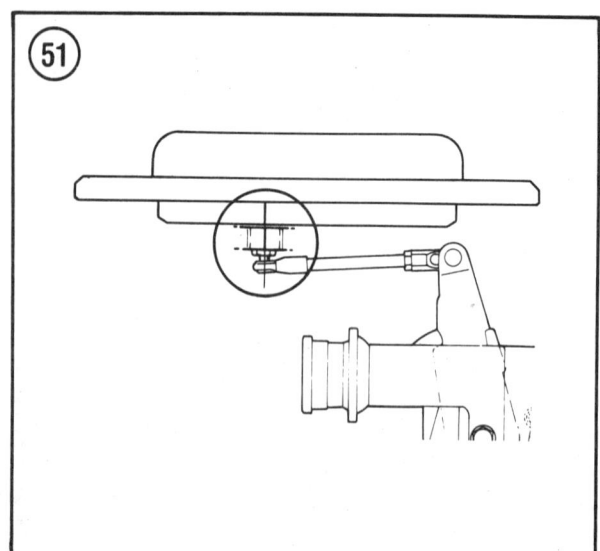

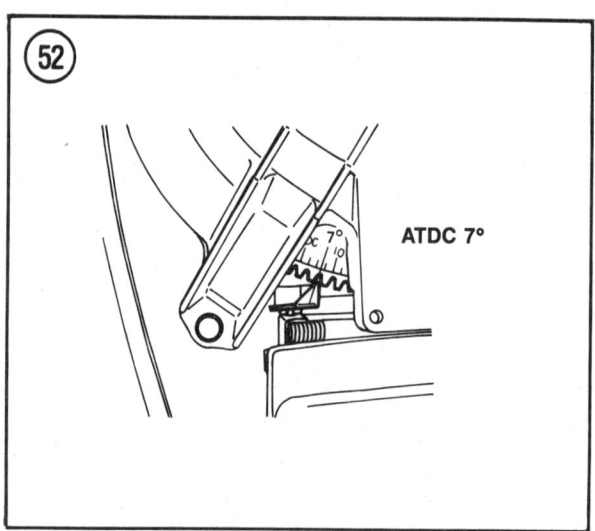

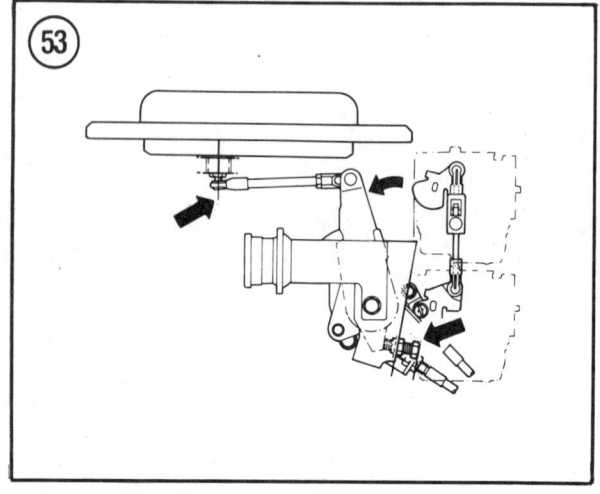

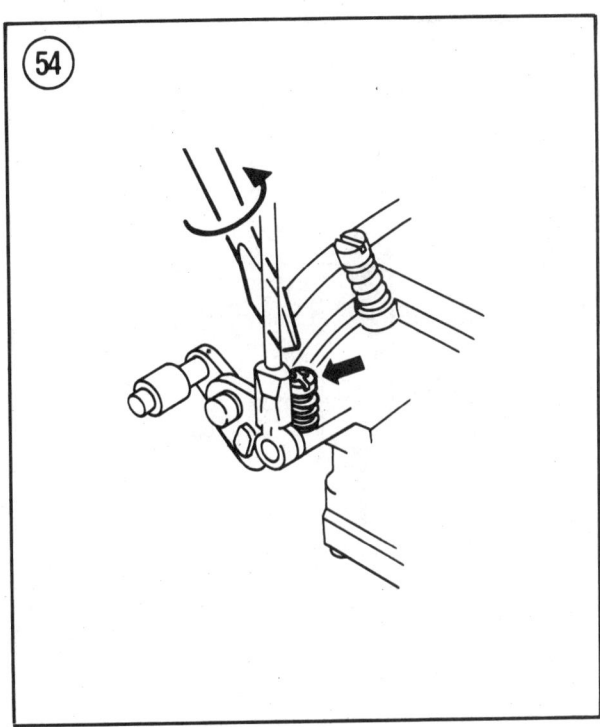

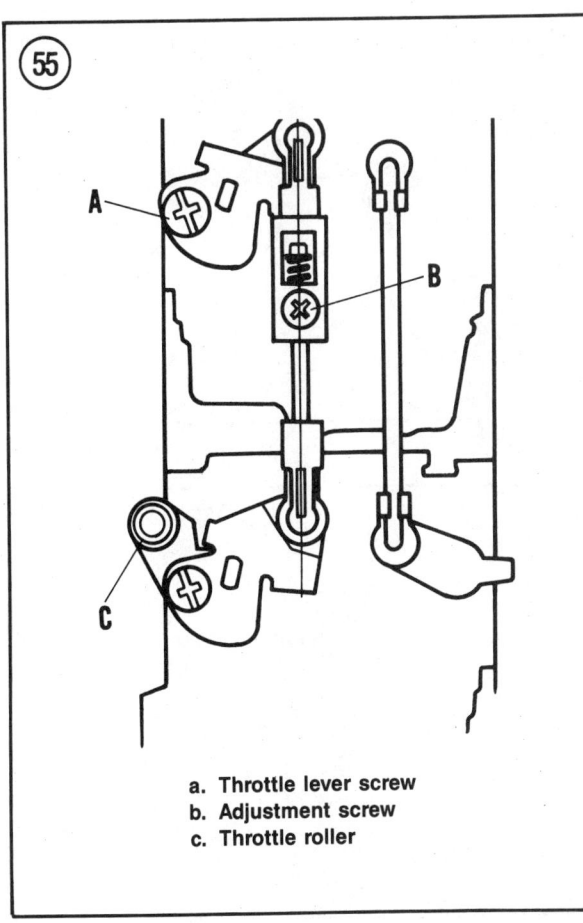

a. Throttle lever screw
b. Adjustment screw
c. Throttle roller

Pickup Timing Adjustment

1. Install the engine in a test tank.
2. Remove the engine cover.
3. Connect a tachometer and timing light to the engine according to manufacturer's instructions.
4. Start the engine. Check and adjust engine idle speed as outlined in following *Carburetor Adjustment*.
5. Aim the timing light at the timing plate and manually move the magneto control lever until 1° ATDC mark on the flywheel is aligned with the timing plate.
6. At this point, throttle roller (C, **Figure 55**) should just lightly contact accelerator cam. No throttle valve movement should be noted.

NOTE
The throttle lever screws have a left-hand thread. Turn the screws clockwise to loosen and counterclockwise to tighten.

7. To adjust, ensure correct ignition timing. Loosen No. 2 cylinder (bottom) carburetor throttle lever screw, then depress throttle roller (C, **Figure 55**) with light pressure until roller just contacts accelerator cam and no throttle valve movement is noted.
8. Tighten No. 2 cylinder (bottom) carburetor throttle lever screw.

Carburetor Adjustment

1. Install the engine in a test tank.
2. Remove the engine cover.
3. Connect a tachometer to the engine according to manufacturer's instructions.
4. Turn the idle mixture screw (**Figure 56**) on each carburetor clockwise until it lightly seats, then back it out 1 1/4-2 3/4 turns.
5. Start the engine, place in FORWARD gear and warm to normal operating temperature.

5

6. Adjust the throttle stop screw (**Figure 54**) as required until engine trolling rpm is between 600-700.

7. Slowly adjust each idle mixture screw (**Figure 56**) counterclockwise 1/8 turn at a time until the engine starts to load up, then turn the screws clockwise 1/16 turn at a time until the engine fires evenly and rpm increases.

8. Repeat Step 6 and Step 7 until the smoothest idle is obtained at the lowest possible rpm.

9. Rapidly open and close the throttle. If the engine hesitates during acceleration, turn each idle mixture screw clockwise slightly until acceleration is smooth.

10. Return the engine to trolling speed, then shift into NEUTRAL and check idle speed. Readjust the throttle stop screw (**Figure 54**) if necessary to bring the idle speed within specifications (**Table 7**).

11. Shut the engine off. Remove the tachometer. Install the engine cover.

Throttle Cable Adjustment (Manual Model)

1. Shift into FORWARD gear.

2. Rotate the throttle grip toward the wide-open throttle position as far as possible. At this position, the center of the throttle roller should align with the wide-open throttle mark on the throttle cam. See **Figure 57**.

3. If the marks do not align in Step 2, loosen the adjusting bolt locknut on the pull side of the throttle wire. Turn the adjusting bolt until all slack in the wire is removed, then tighten the locknut.

4. Loosen the adjusting bolt locknut on the opposite side of the throttle wire. Turn the adjusting bolt until the wire has approximately 0.12 in. (3 mm) slack, then tighten the locknut. See **Figure 58**.

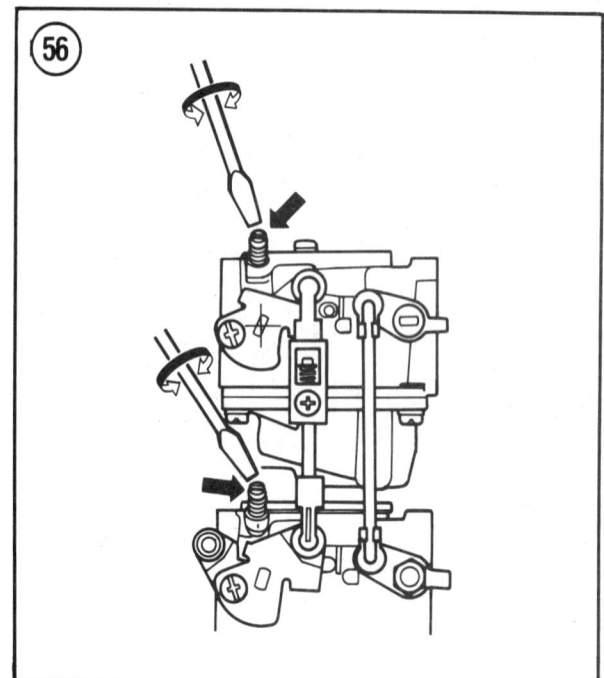

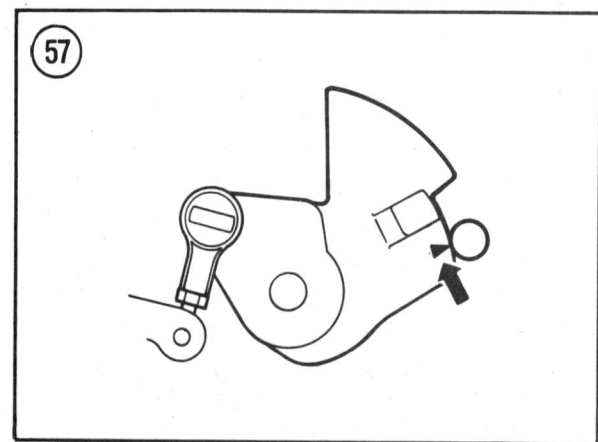

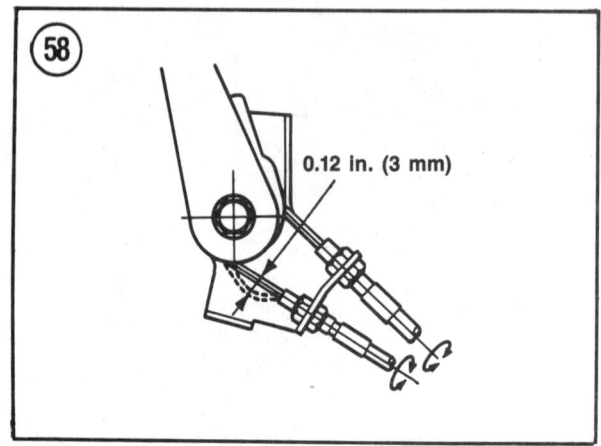

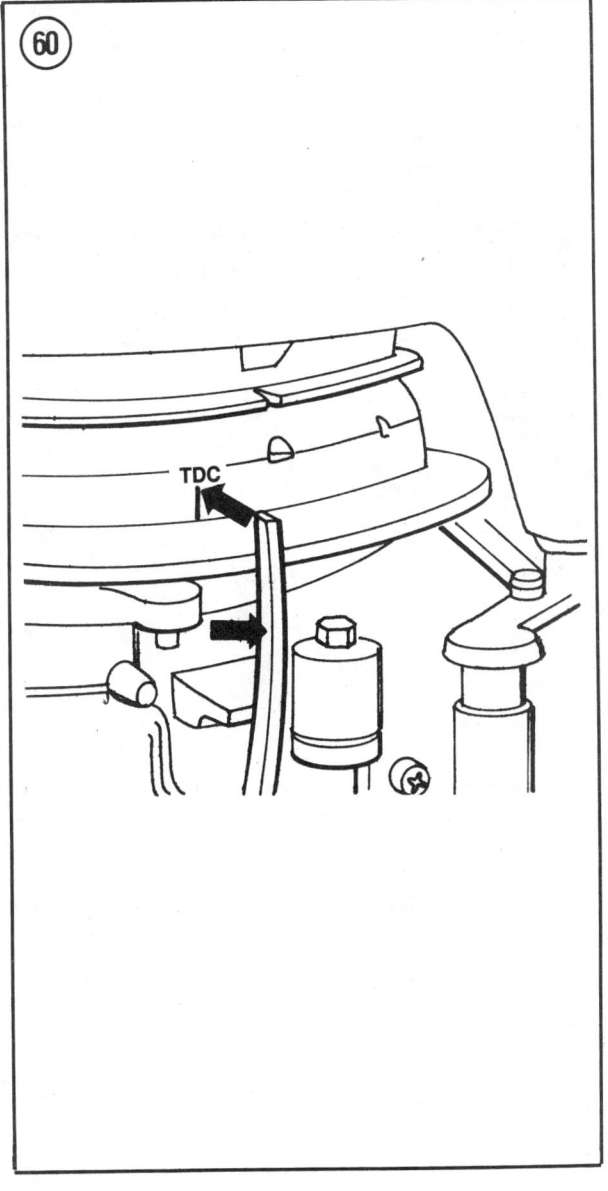

Oil Pump Link Adjustment

1. Rotate the throttle toward the wide-open throttle position as far as possible. At this position, the oil pump lever should contact the wide-open stop.

2. If not, disconnect the link rod from the oil pump lever ball stud. Rotate oil pump lever against wide-open stop. Loosen the locknut and adjust the plastic snap-on connector on the end of the link rod until it can be reconnected to the ball stud without changing the throttle or oil pump lever position. Reconnect the link rod to the ball stud and tighten the locknut.

YAMAHA 30 (1984-1986)

Timing Adjustment

Correctly adjusting the full-advance timing on this model automatically adjusts the full retard timing.

1. Remove the engine cover.

2. Disconnect the magneto base link rod from the control lever ball stud. See **Figure 59**.

3. Disconnect the spark plug leads and remove the spark plugs. See Chapter Four.

4. Install a dial indicator in the No. 1 (top) spark plug hole. See **Figure 38** (typical).

5. Rotate the flywheel clockwise until the dial indicator shows that the piston has reached top dead center (TDC). This is the point at which the indicator needle reverses its direction as the flywheel is rotated.

6. Check timing pointer alignment with flywheel timing scale (**Figure 60**). If pointer is not properly aligned, recheck Step 5 to make sure piston is at TDC. If this is not the problem, loosen the timing pointer bolt and move the pointer as required to align with the flywheel TDC mark, then tighten the pointer bolt.

5

7. Remove the dial indicator.

8. Rotate the flywheel clockwise to align the timing pointer with the 25° BTDC mark on the flywheel timing scale. See **Figure 61**. Hold flywheel in this position.

9. Slowly move the magneto base counterclockwise to align the stamped timing mark on the base with the TDC mark on the flywheel timing scale (**Figure 62**).

10. The magneto stop bracket cap should touch the engine stop. If it does not, loosen the bolt holding the magneto stop bracket. With the timing marks aligned as in Step 9, move the bracket until it contacts the engine stop, then tighten the bolt. See **Figure 63**.

Wide-open Throttle
Linkage Adjustment

1. With the timing marks aligned and the magneto stop bracket touching the engine

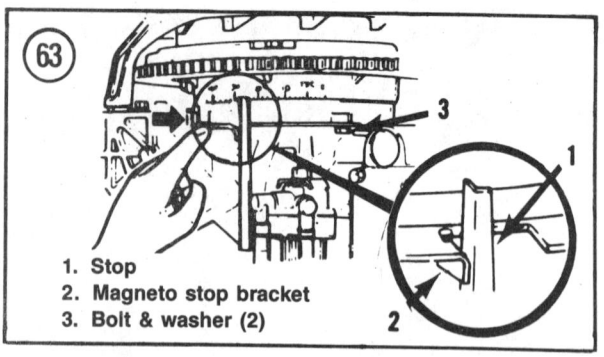

1. Stop
2. Magneto stop bracket
3. Bolt & washer (2)

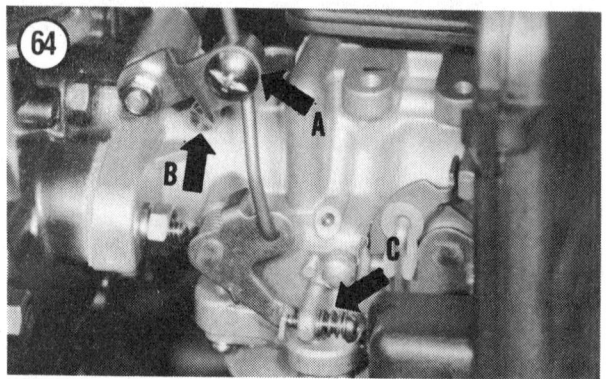

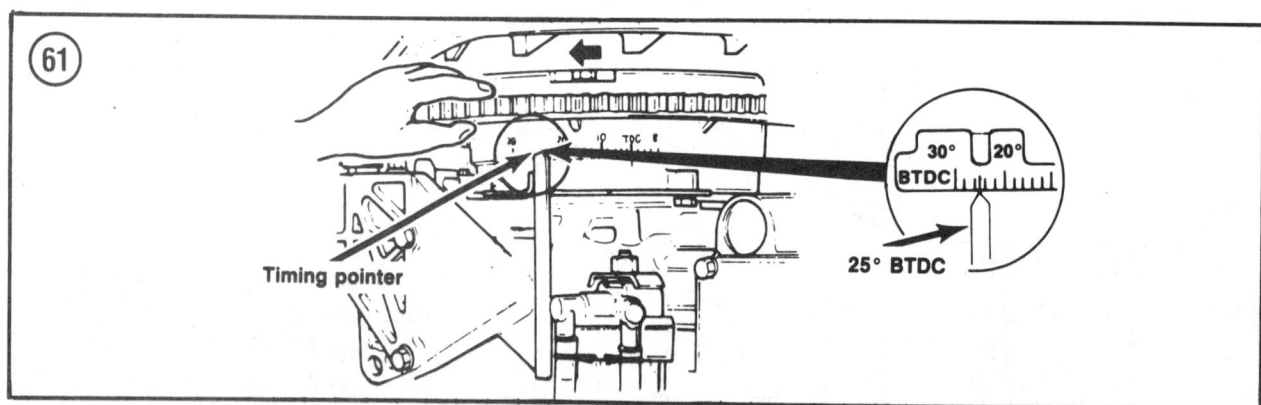

Timing pointer

30° 20°
BTDC

25° BTDC

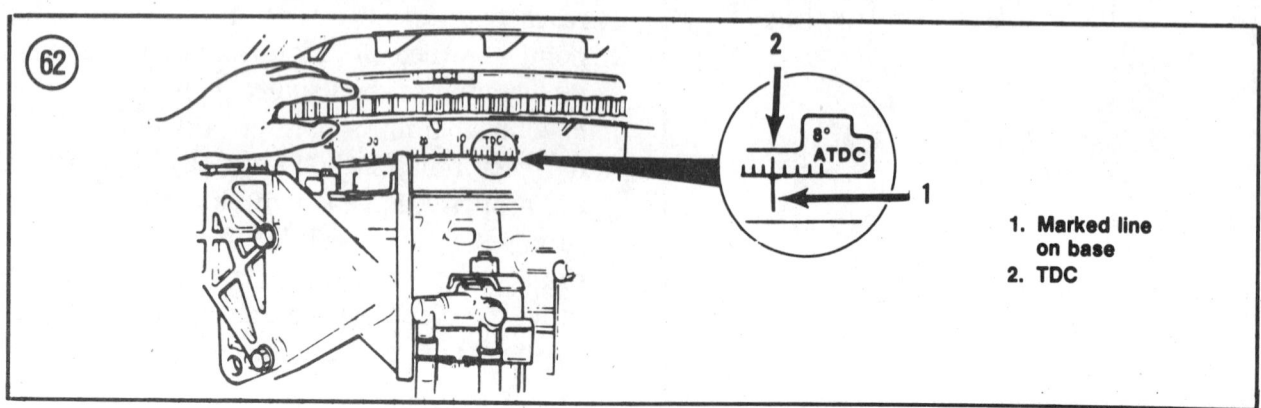

2
8°
ATDC
1

1. Marked line
   on base
2. TDC

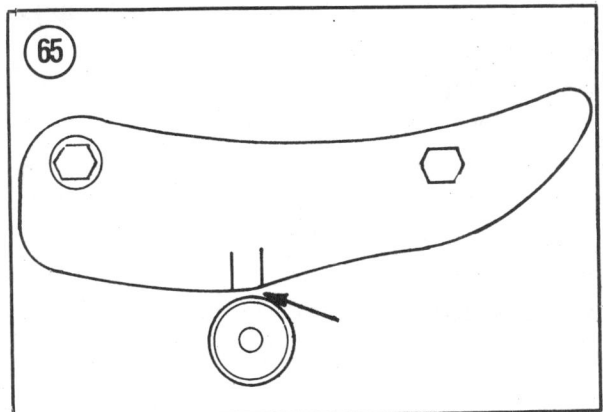

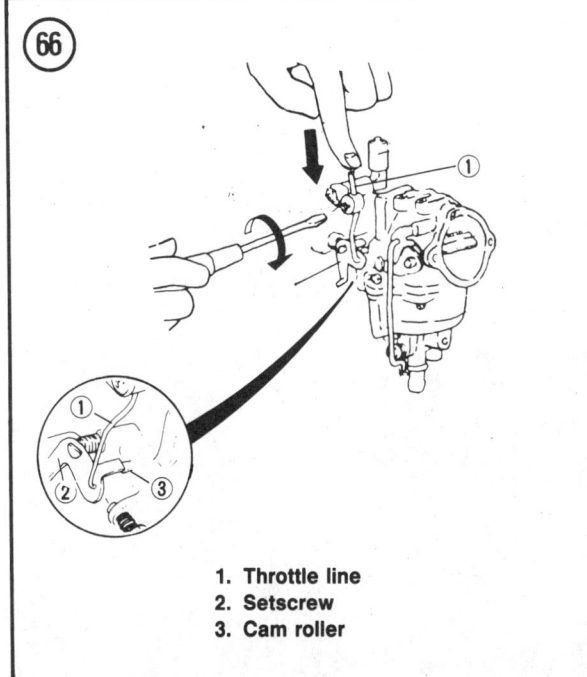

1. Throttle line
2. Setscrew
3. Cam roller

stop (see Step 9 and Step 10 of *Timing Adjustment* in this section), loosen the setscrew holding the throttle link (A, **Figure 64**).

2. Loosen the throttle cam bolts and move the cam as required to center its alignment mark with the throttle cam roller (**Figure 65**). Tighten the throttle cam bolts snugly.

3. With the throttle cam roller and cam mark aligned as described in Step 2, push the throttle link until the throttle arm stop touches the wide open throttle stop then tighten the setscrew. See **Figure 66**.

4. Move the shift lever to the FORWARD position.

5. Disconnect the magneto base link rod (A, **Figure 67**) and throttle lever link rod (B, **Figure 67**) from the throttle control lever ball studs. Loosen the locknut on each link rod.

6. Slowly rotate the magneto base by hand to its wide open position as shown in **Figure 68**.

7. Adjust the plastic snap-on connector on the end of the magneto base link rod until it can be reconnected to the control lever ball stud without changing the position of the linkage or magneto base. Install on ball stud and tighten the locknut.

8. Push the throttle lever link rod downward to position the lower control lever against its

5

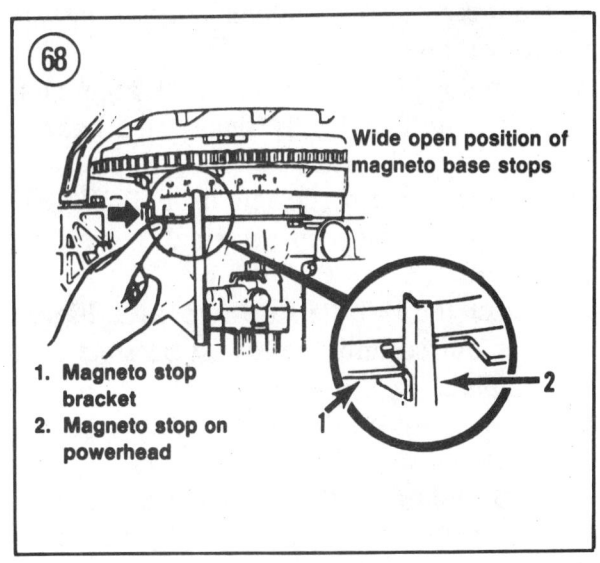

Wide open position of magneto base stops

1. Magneto stop bracket
2. Magneto stop on powerhead

stop on the bottom of the cowling. Make sure the magneto base is at its full advance position. Adjust the plastic snap-on connector on the end of the throttle lever link rod until it can be reconnected to the upper control lever ball stud without changing the position of the linkage or magneto base. Install on ball stud and tighten the locknut. See **Figure 69**.

Throttle Linkage
Neutral Adjustment

1. Place the engine in a test tank.
2. Remove the engine cover.
3. Connect a tachometer according to manufacturer's instructions.
4. Loosen the locknut on the magneto control lever adjustment screw (**Figure 70**).
5. Start the engine and run in NEUTRAL. Adjust the throttle grip to obtain an engine speed of 3,200 ± 300 rpm. Turn the adjustment screw until it touches the throttle control lever under the magneto control lever, then tighten the locknut.
6. Reverse Steps 1-3 to complete procedure.

Carburetor Adjustment

1. Install the engine in a test tank with the engine cover off.
2. Connect a tachometer according to manufacturer's instructions.
3. Turn the idle mixture screw (B, **Figure 64**) clockwise until it lightly seats, then back it out 1 1/4-1 3/4 turns.
4. Start the engine, place in FORWARD gear and warm to normal operating temperature.
5. Adjust the throttle stop screw (C, **Figure 64**) as required until the lowest possible rpm is obtained.
6. Slowly adjust the idle mixture screw counterclockwise until the engine starts to load up, then turn the screw clockwise until the engine fires evenly and rpm increases.

7. Repeat Step 5 and Step 6 until the smoothest idle is obtained at the lowest possible rpm.
8. Rapidly open and close the throttle. If the engine hesitates during acceleration, turn the idle mixture screw clockwise slightly until acceleration is smooth.
9. Return engine to idle rpm and recheck idle speed. If it is not within specifications (**Table 8**) readjust throttle stop screw.

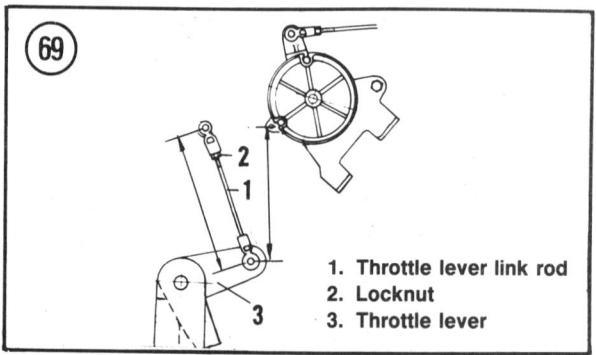

1. Throttle lever link rod
2. Locknut
3. Throttle lever

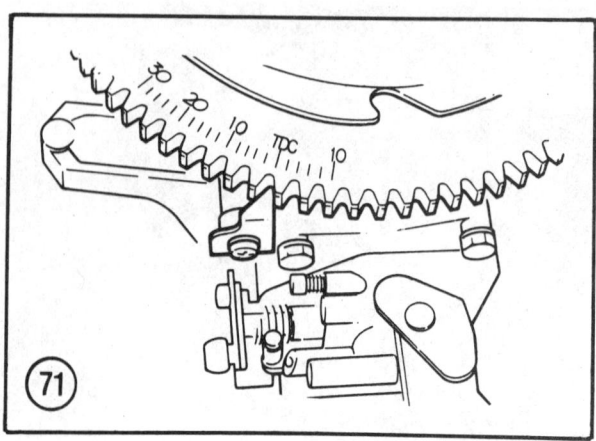

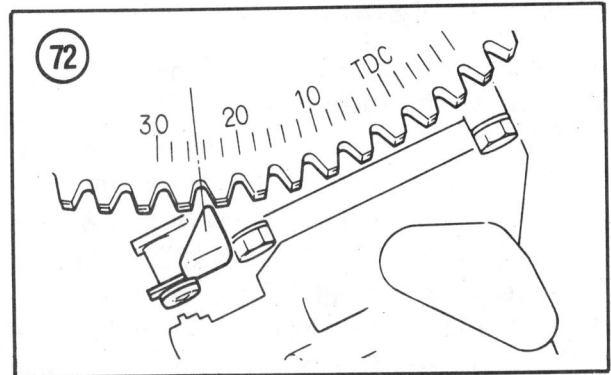

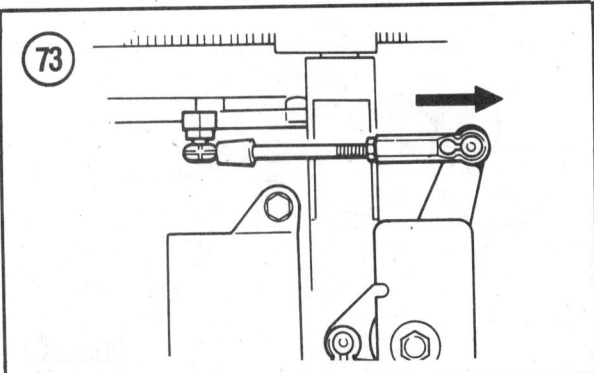

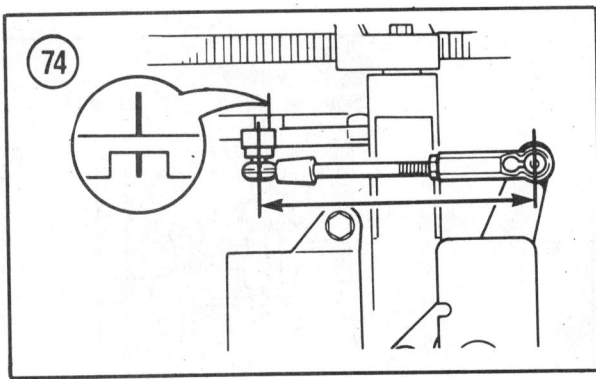

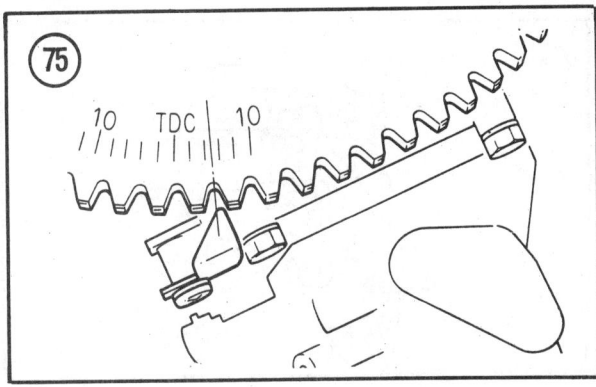

10. Shut the engine off. Remove the tachometer and install the engine cover.

YAMAHA 30 (1987-ON)

Timing Adjustment

1. Remove the engine cover.

2. Disconnect the spark plug leads and remove the spark plugs. See Chapter Four.

3. Install a dial indicator in the No. 1 (top) spark plug hole. See **Figure 38** (typical).

4. Rotate the flywheel clockwise until the dial indicator shows that the piston has reached top dead center (TDC). This is the point at which the indicator needle reverses its direction as the flywheel is rotated.

5. Check timing pointer alignment with the flywheel timing scale (**Figure 71**). If pointer is not properly aligned, recheck Step 4 to make sure piston is at TDC. If this is not the problem, loosen the timing plate setscrew and move the plate as required to align with the flywheel TDC mark, then tighten the setscrew securely.

6. Remove the dial indicator.

7. Rotate the flywheel clockwise to align the timing plate with the 25° BTDC mark on the flywheel timing scale. See **Figure 72**.

8. Move the magneto control lever to the right (full advance) until it touches the stopper (**Figure 73**).

9. If the timing marks do not align as shown in **Figure 74**, disconnect the link rod from the control lever ball stud. Loosen the locknut and adjust the plastic snap-on connector on the end of the link rod until it can be reconnected to the ball stud without changing the position of the linkage or timing marks. Reconnect the link rod to the ball stud and tighten the locknut.

10. Rotate the flywheel clockwise to align the timing plate with the 5° ATDC mark on the flywheel timing scale. See **Figure 75**.

5

11. Move the magneto control lever to the left (full retard) until it touches the stopper bolt (**Figure 76**).

12. If the timing marks do not align as shown in **Figure 76**, loosen the stopper bolt locknut and adjust the bolt length as required to align the marks, then tighten the locknut.

Carburetor Linkage Adjustment

1. Loosen the idle adjustment screw on the upper carburetor (**Figure 77**) and close the throttle valve.

> *NOTE*
> *The throttle lever screws have a left-hand thread. Turn the screws clockwise to loosen and counterclockwise to tighten.*

2. Loosen the throttle lever screws for the No. 1 and No. 3 carburetors. See **Figure 78**.

3. Make sure all carburetor throttle valves are completely closed. Lightly depress the No. 2 carburetor throttle roller (arrow, **Figure 79**) and tighten the No. 1 and No. 3 throttle lever screws.

Pickup Timing Adjustment

1. Install the engine in a test tank.
2. Remove the engine cover.
3. Connect a tachometer and timing light to the engine according to manufacturer's instructions.
4. Start the engine. Aim the timing light at the timing plate and manually advance the magneto control lever (arrow, **Figure 80**) until the throttle cam just contacts the throttle cam roller. Ignition timing should be $2° \pm 1°$ ATDC. If not, stop the engine and repeat Steps 7-9 under *Timing Adjustment* in this chapter.
5. Loosen the No. 2 carburetor throttle lever screw. Lightly depress the No. 2 carburetor throttle roller (arrow, **Figure 81**) and tighten the throttle lever screw.

Carburetor Adjustment

1. Install the engine in a test tank.
2. Remove the engine cover.
3. Connect a tachometer to the engine according to manufacturer's instructions.
4. Turn the idle mixture screw (**Figure 82**) on each carburetor clockwise until it lightly seats, then back it out 1-1 1/2 turns.
5. Start the engine, place in FORWARD gear and warm to normal operating temperature.

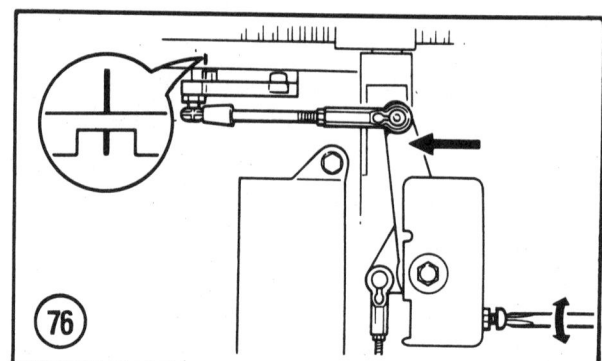

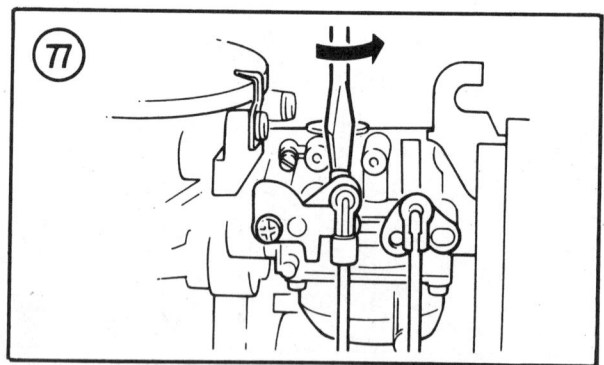

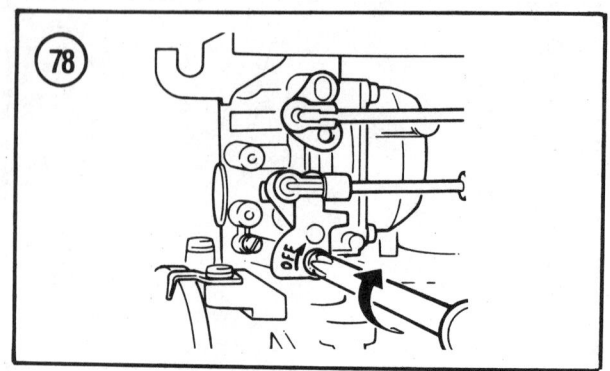

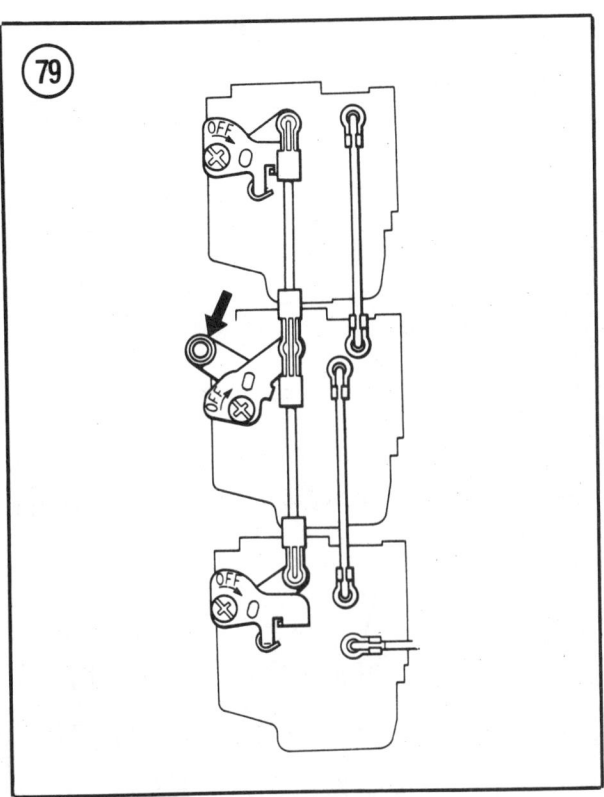

6. Adjust the throttle stop screw (**Figure 83**) as required until the lowest possible rpm is obtained.

7. Slowly adjust each idle mixture screw counterclockwise 1/8 turn at a time until the engine starts to load up, then turn the screws clockwise 1/16 turn at a time until the engine fires evenly and rpm increases.

8. Repeat Step 6 and Step 7 until the smoothest idle is obtained at the lowest possible rpm.

9. Rapidly open and close the throttle. If the engine hesitates during acceleration, turn each idle mixture screw clockwise slightly until acceleration is smooth.

10. Return engine to idle rpm and recheck idle speed. Readjust the throttle stop screw if necessary to bring the idle speed within specifications (**Table 8**).

11. Shut the engine off. Remove the tachometer. Install the engine cover.

5

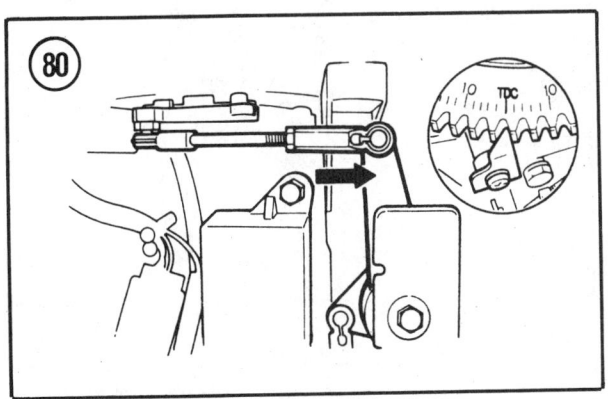

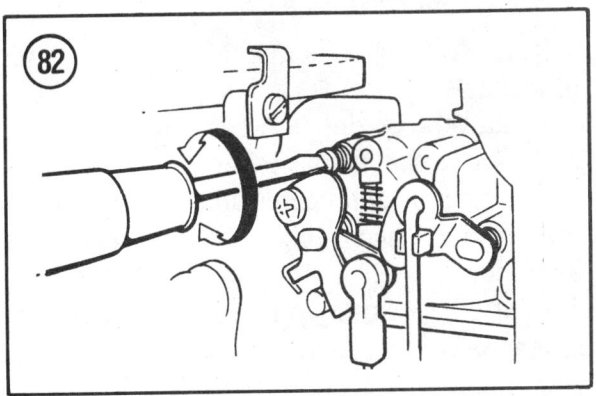

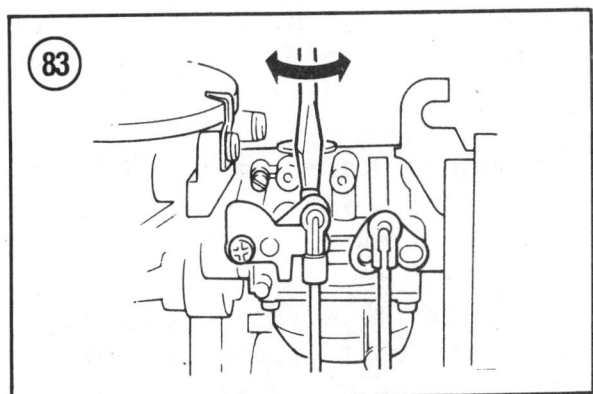

Throttle Cable Adjustment (Manual Model)

1. Shift into FORWARD gear.

2. Rotate the throttle grip toward the wide-open throttle position as far as possible. At this position, the center of the throttle roller should align with the wide-open throttle mark on the throttle cam. See **Figure 84**.

3. If the marks do not align in Step 2, loosen the adjusting bolt locknut on the pull side of the throttle wire. Turn the adjusting bolt until all slack in the wire is removed, then tighten the locknut.

4. Loosen the adjusting bolt locknut on the opposite side of the throttle wire. Turn the adjusting bolt until the wire has approximately 0.12 in. (3 mm) slack, then tighten the locknut. See **Figure 85**.

YAMAHA 40 AND 50 (1984-1988)

Timing Adjustment

Refer to **Figure 86** and **Figure 87** for this procedure.

1. Install the engine in a test tank.

2. Remove the engine cover.

3. Connect a tachometer and timing light to the engine according to manufacturer's instructions.

4. Start the engine and manually move the magneto control lever to the closed throttle position.

5. Aim the timing light at the magneto pointer and turn the full retard screw (A, **Figure 86**) until the timing indicator aligns with the specified timing mark on the flywheel (**Table 9**).

6. Loosen the locknut on the full advance screw (A, **Figure 87**) and manually move the magneto control lever to the wide open throttle position.

7. Aim the timing light at the magneto pointer and turn the full advance screw (A, **Figure 87**) until the timing indicator aligns

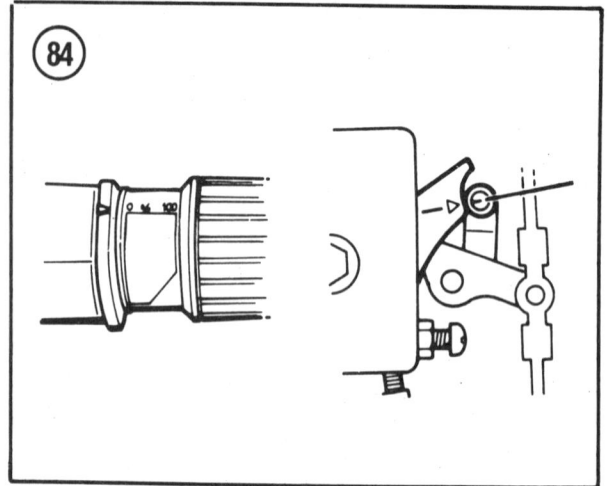

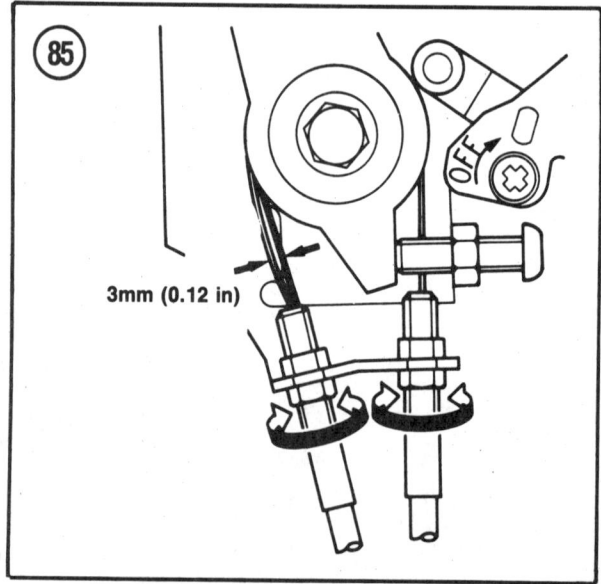

3mm (0.12 in)

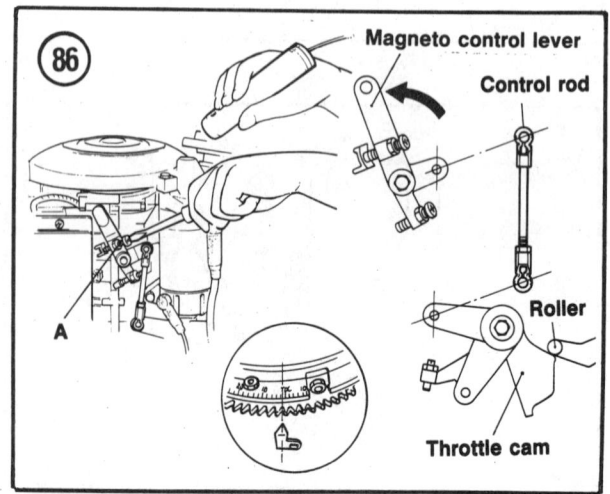

Magneto control lever

Control rod

Roller

Throttle cam

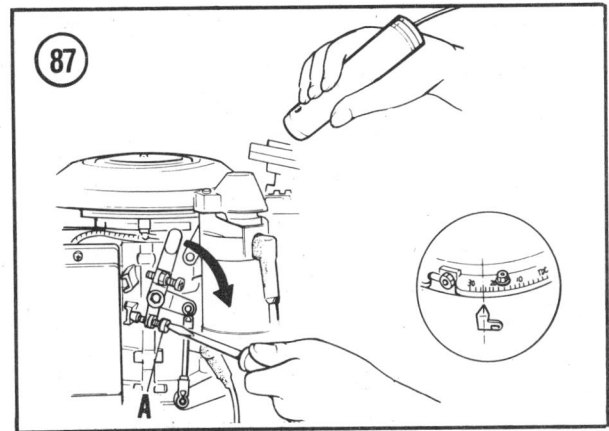

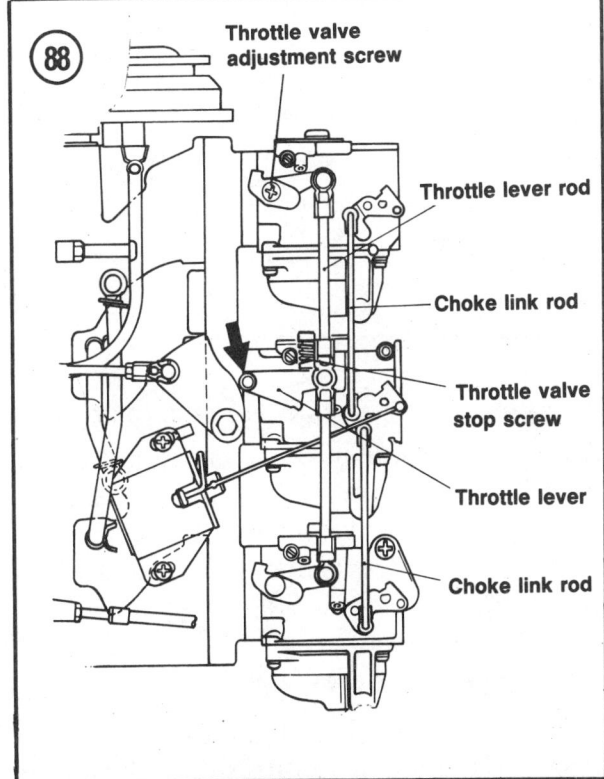

Throttle valve
adjustment screw

Throttle lever rod

Choke link rod

Throttle valve
stop screw

Throttle lever

Choke link rod

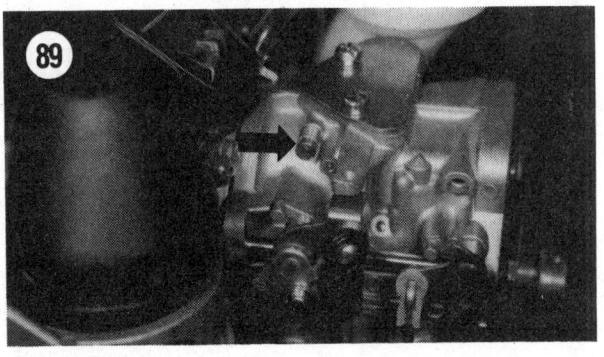

with the specified timing mark on the flywheel (**Table 9**).

8. Once the timing is correct, check the engine speed on the tachometer. It should be between 750-850 rpm. If not, adjust the carburetor throttle stop screw to bring engine speed into specifications. See *Carburetor Adjustment* in this section.

9. Shut the engine off. With the magneto control lever in the closed throttle position, the throttle valve roller should contact the accelerator cam. If it does not, disconnect the link rod at the control lever. Adjust the plastic snap-on connector on the end of the link rod until it can be reconnected to the control lever ball stud with the cam and roller in contact.

10. Remove the timing light and tachometer. Install the engine cover.

Carburetor Linkage Adjustment

Refer to **Figure 88** for this procedure.
1. Loosen the throttle valve stop screw.

NOTE
The throttle valve adjustment screws have a left-hand thread. Turn the screw clockwise to loosen and counterclockwise to tighten.

2. Loosen the throttle valve adjustment screws for the No. 1 and No. 2 carburetors.
3. Make sure all carburetor throttle valves are completely closed. Depress the throttle lever (arrow, **Figure 88**) and tighten the adjustment screws.

Carburetor Adjustment

1. Install the engine in a test tank.
2. Remove the engine cover.
3. Connect a tachometer to the engine according to manufacturer's instructions.
4. Turn the idle mixture screw (arrow, **Figure 89**) on each carburetor clockwise until it

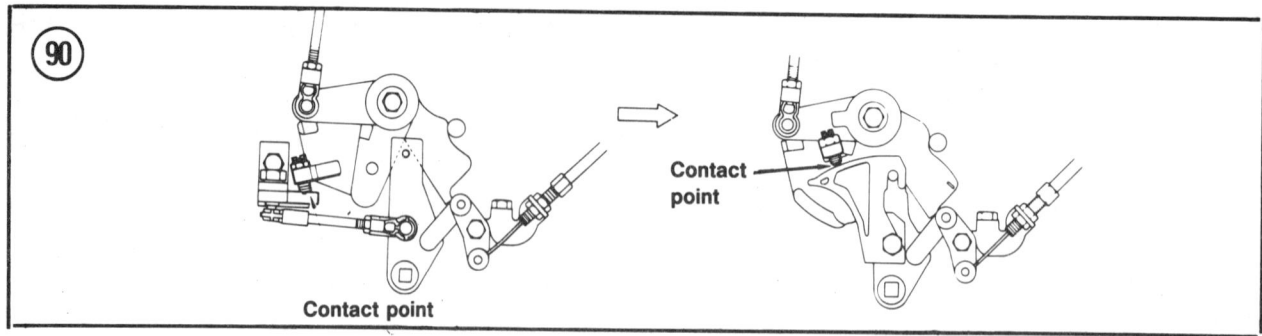

Contact point

Contact point

lightly seats, then back it out 1 3/8-1 7/8 turns.

5. Start the engine, place in FORWARD gear and warm to normal operating temperature.

6. Adjust the throttle stop screw (**Figure 88**) as required until the lowest possible rpm is obtained.

7. Slowly adjust each idle mixture screw counterclockwise 1/8 turn at a time until the engine starts to load up, then turn the screws clockwise 1/16 turn at a time until the engine fires evenly and rpm increases.

8. Repeat Step 6 and Step 7 until the smoothest idle is obtained at the lowest possible rpm.

9. Rapidly open and close the throttle. If the engine hesitates during acceleration, turn each idle mixture screw clockwise slightly until acceleration is smooth.

10. Return engine to idle rpm and recheck idle speed. Readjust throttle stop screw if necessary to bring idle speed within specifications (**Table 9**).

11. Shut the engine off. Remove the tachometer.

Throttle Linkage
Neutral Adjustment

1. Place the engine in a test tank.
2. Remove the engine cover.
3. Connect a tachometer according to manufacturer's instructions.
4. Loosen the locknut on the neutral opening adjustment screw, if so equipped.

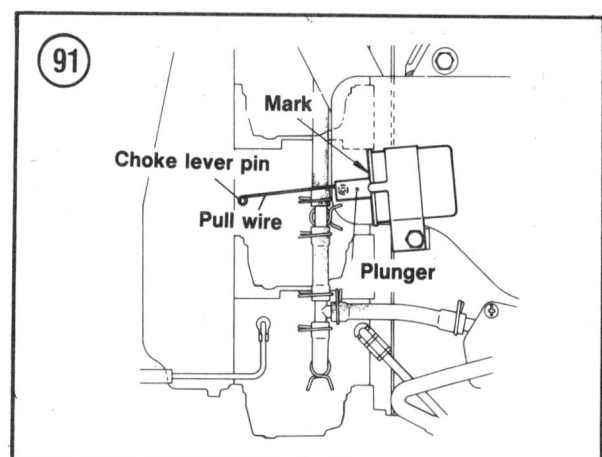

Mark

Choke lever pin

Pull wire

Plunger

5. Start the engine and run in NEUTRAL. Adjust the throttle to obtain an engine speed of 3,200 ±300 rpm. Turn the adjustment screw (**Figure 90**) until it touches the neutral opening stop, then tighten the locknut (if so equipped).

Choke Solenoid Adjustment

With the choke valves fully open, the groove on the end of the solenoid plunger should align with the end of the solenoid (**Figure 91**). If it does not, loosen the solenoid attaching bolt and move the solenoid in the proper direction to align the groove, then tighten the attaching bolt.

Oil Pump Link Adjustment

Loosen the throttle valve stop screw (**Figure 88**) and close the throttle valve completely. The punch mark on the pump

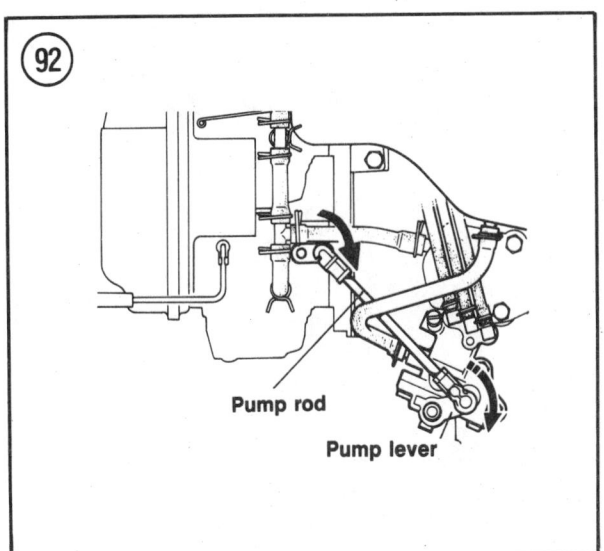

Pump rod

Pump lever

lever should align with the mark on the pump body. If it does not, disconnect the pump link and adjust the connector as required. See **Figure 92**. Reinstall the link and readjust the throttle valve stop screw.

YAMAHA 40 AND 50 (1989)

The 1989 models are equipped with an electronic ignition advance system in place of the earlier mechanical ignition advance system used on 1984-1988 models. See **Figure 93**. Adjust length of rod (1) to alter idle timing. Screw (2) adjust wide-open throttle (WOT) stop.

5

Charge coil

Pulser coil

Lighting coil

To rectifier

Spark plug cap

Ignition coil

Stay

CDI unit

1

2

Control lever

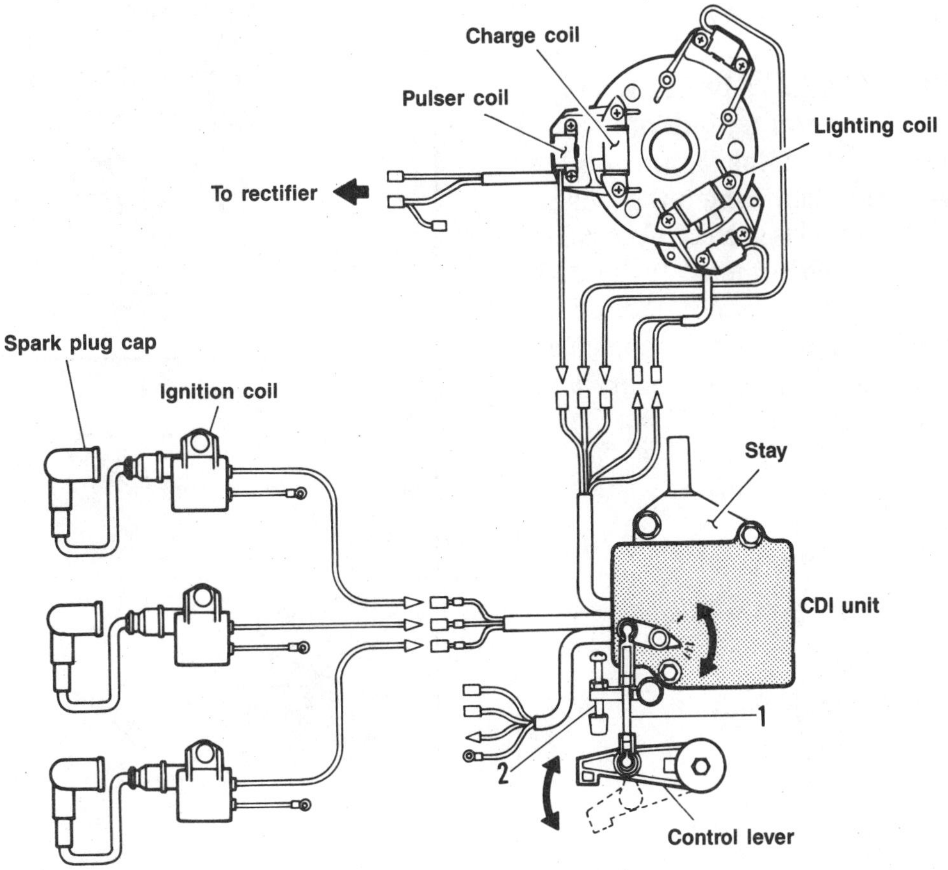

YAMAHA 70

Timing Adjustment

1. Install the engine in a test tank.
2. Remove the engine cover.
3. Connect a tachometer and timing light to the engine according to manufacturer's instructions.
4. Start the engine and manually move the magneto control lever to the closed throttle position (**Figure 94**).
5. Aim the timing light at the magneto pointer and turn the full retard screw (A, **Figure 95**) until the timing indicator aligns with the specified timing mark on the flywheel (**Table 10**).
6. Loosen the locknut on the full advance screw and manually move the magneto control lever to the wide open throttle position (**Figure 96**).
7. Aim the timing light at the magneto pointer and turn the full advance screw (A, **Figure 97**) until the timing indicator aligns with the specified timing mark on the flywheel (**Table 10**), then tighten the locknut.
8. Once the timing is correct, check the engine speed on the tachometer. It should be between 750-850 rpm. If not, adjust the carburetor throttle stop screw to bring engine speed into specifications. See *Carburetor Adjustment* in this section.

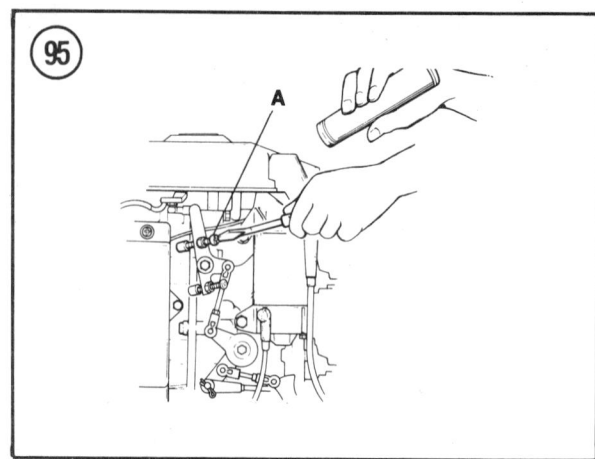

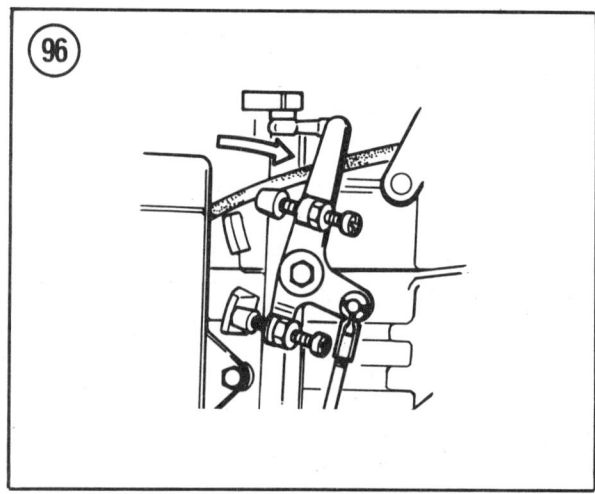

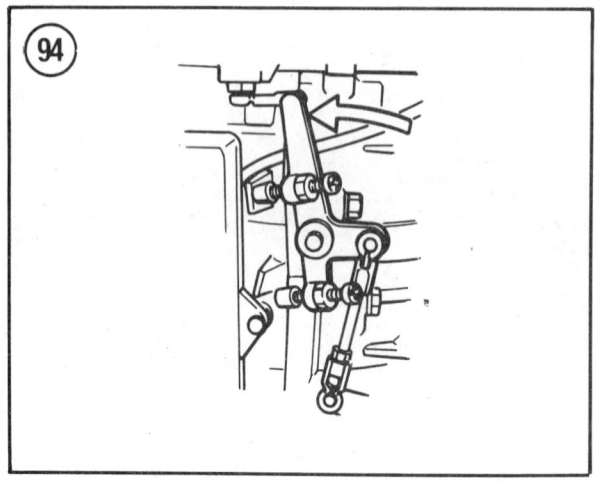

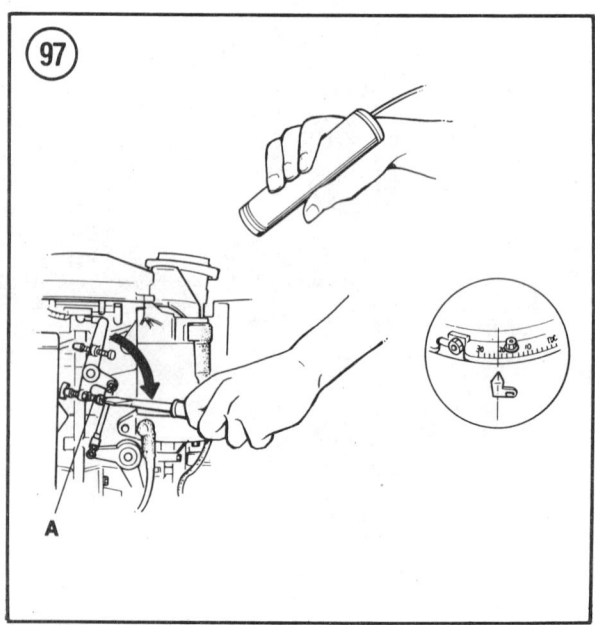

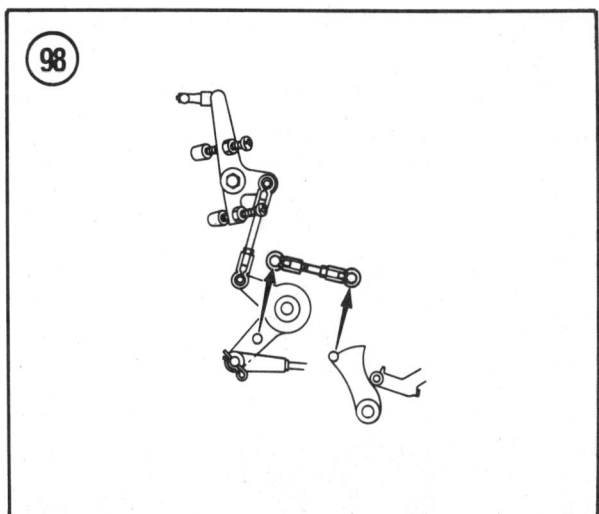

9. Shut the engine off. With the magneto control lever in the closed throttle position, the throttle valve roller should contact the accelerator cam. If it does not, disconnect the link rod at the control lever. See **Figure 98**. Adjust the plastic snap-on connector on the end of the link rod until it can be reconnected to the control lever ball stud with the cam and roller in contact.

10. Remove the timing light and tachometer. Install the engine cover.

Carburetor Linkage Adjustment

1. Loosen the throttle valve stop screw (middle carburetor).

NOTE
The throttle valve adjustment screws have a left-hand thread. Turn the screw clockwise to loosen and counter-clockwise to tighten.

2. Loosen the throttle valve adjustment screws for the No. 1 and No. 2 carburetors. See A, **Figure 99**.

3. Make sure all carburetor throttle valves are completely closed. Depress the throttle lever (B, **Figure 99**) and tighten the adjustment screws.

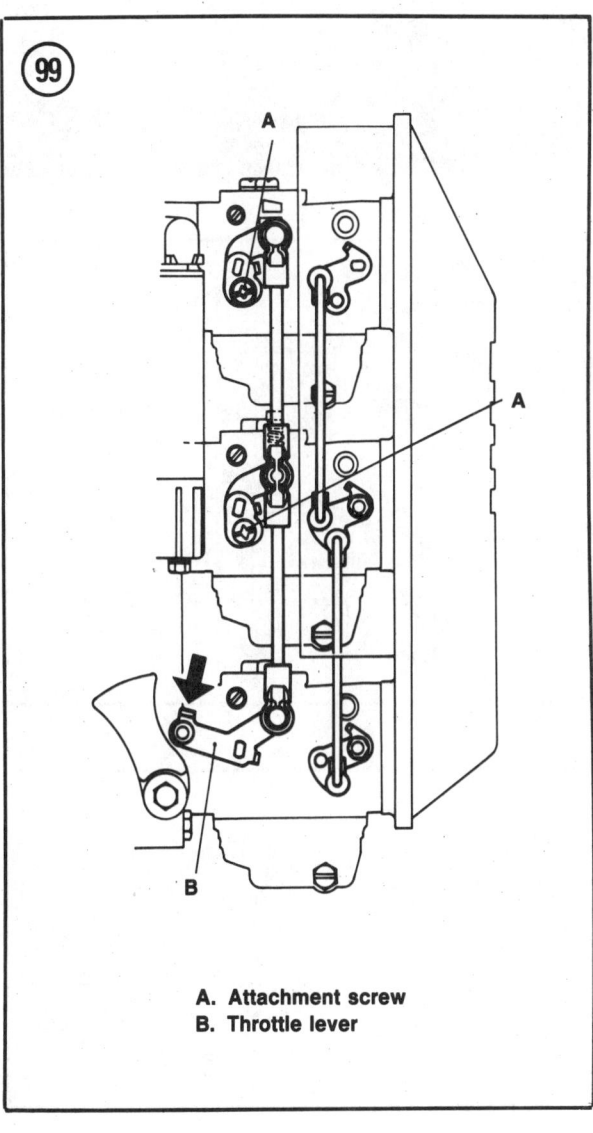

A. Attachment screw
B. Throttle lever

Carburetor Adjustment

1. Install the engine in a test tank.
2. Remove the engine cover.
3. Connect a tachometer to the engine according to manufacturer's instructions.

4. Turn the idle mixture screw (arrow, **Figure 100**) on each carburetor clockwise until it lightly seats, then back it out 1 5/8-2 1/8 turns.

5. Start the engine, place in FORWARD gear and warm to normal operating temperature.

6. Adjust the throttle stop screw (**Figure 88**) as required until the lowest possible rpm is obtained.

7. Slowly adjust each idle mixture screw counterclockwise 1/8 turn at a time until the engine starts to load up, then turn the screws clockwise 1/16 turn at a time until the engine fires evenly and rpm increases.

8. Repeat Step 6 and Step 7 until the smoothest idle is obtained at the lowest possible rpm.

9. Rapidly open and close the throttle. If the engine hesitates during acceleration, turn each idle mixture screw clockwise slightly until acceleration is smooth.

10. Return engine to idle rpm and recheck idle speed. Readjust throttle stop screw if necessary to bring idle speed within specifications (**Table 10**).

11. Shut the engine off. Remove the tachometer.

Choke Solenoid Adjustment

With the choke valves fully open, the groove on the end of the solenoid plunger should align with the end of the solenoid (**Figure 101**). If it does not, loosen the solenoid attaching bolts and move the solenoid in the proper direction to align the groove, then tighten the attaching bolts.

Oil Pump Link Adjustment

Loosen the throttle valve stop screw (**Figure 88**) and close the throttle valve completely. The punch mark on the pump lever should align with the mark on the pump

body (**Figure 102**). If it does not, disconnect the pump link and adjust the connector as required. Reinstall the link and readjust the throttle valve stop screw.

YAMAHA 90

Timing Adjustment

1. Install the engine in a test tank.

2. Remove the engine cover.

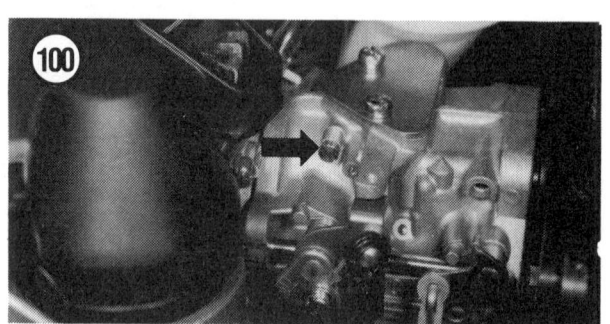

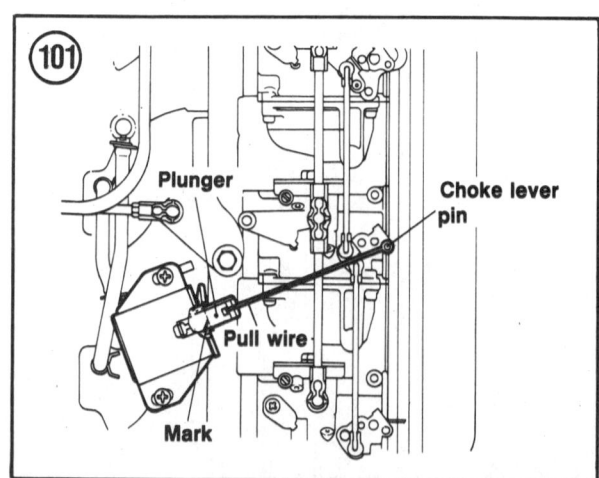

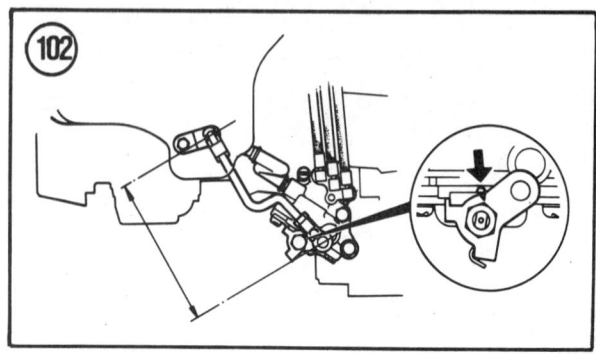

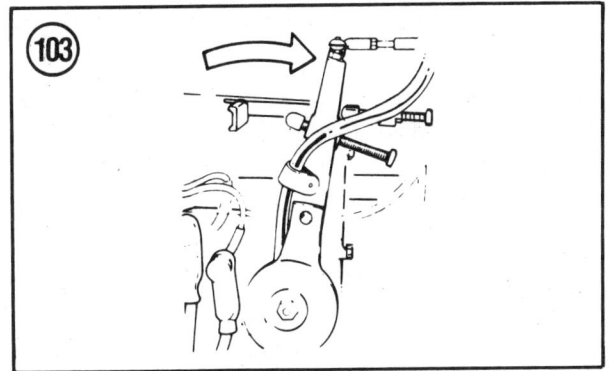

3. Connect a tachometer and timing light to the engine according to manufacturer's instructions.

4. Start the engine and manually move the magneto control lever to the closed throttle position (**Figure 103**).

5. Aim the timing light at the magneto pointer and turn the full retard screw (A, **Figure 104**) until the timing indicator aligns with the specified timing mark on the flywheel (**Table 11**).

6. Move the magneto control lever to the wide open throttle position (**Figure 105**).

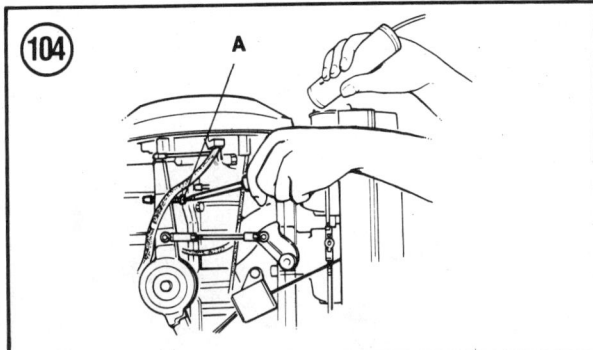

7. Loosen the locknut on the full advance screw (A, **Figure 106**). Aim the timing light at the magneto pointer and turn the full advance screw until the timing indicator aligns with the specified timing mark on the flywheel (**Table 11**), then tighten the locknut.

8. Once the timing is correct, check the engine speed on the tachometer. It should be between 750-850 rpm. If not, adjust the carburetor throttle stop screw to bring engine speed into specifications. See *Carburetor Adjustment* in this section.

9. Move the throttle cam until it barely touches the throttle valve roller. The throttle valve should not open. If it does, disconnect the link rod at the control lever. Adjust the plastic snap-on connector on the end of the link rod until it can be reconnected to the control lever ball stud with the cam and roller in contact. See **Figure 107**.

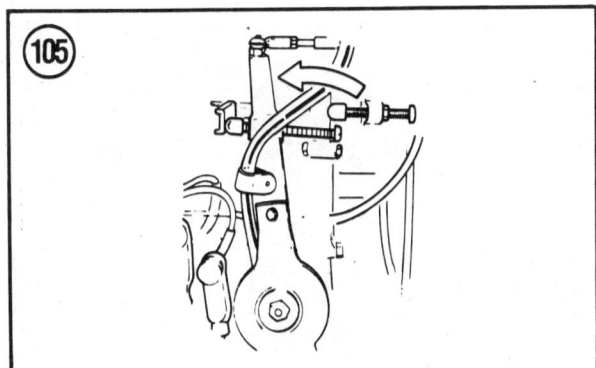

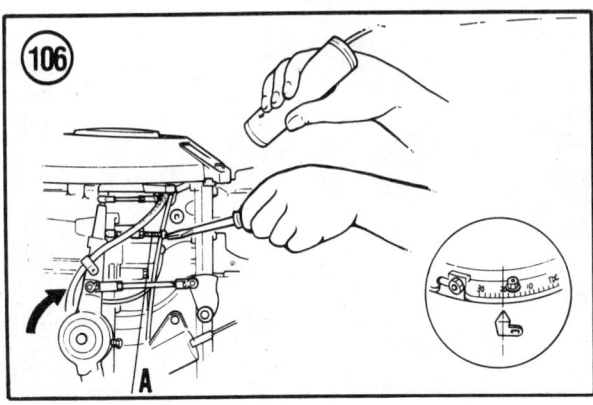

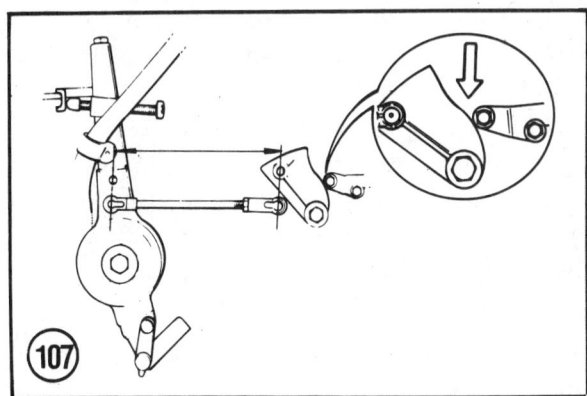

10. Shut the engine off. Disconnect the timing light and tachometer. Install the engine cover.

Carburetor Linkage Adjustment

1. Loosen the throttle valve stop screw (A, **Figure 108**) and close the throttle valve.

> *NOTE*
> *The throttle valve adjustment screws have a left-hand thread. Turn the screw clockwise to loosen and counterclockwise to tighten.*

2. Loosen the throttle valve adjustment screws for the No. 1 and No. 3 carburetors. See B, **Figure 108**.

3. Make sure all carburetor throttle valves are completely closed. Depress the throttle lever (arrow, **Figure 108**) and tighten the adjustment screws.

Carburetor Adjustment

1. Install the engine in a test tank.
2. Remove the engine cover.
3. Connect a tachometer to the engine according to manufacturer's instructions.
4. Turn the idle mixture screw (arrow, **Figure 109**) on each carburetor clockwise until it lightly seats, then back it out 1 5/8-2 1/8 turns.
5. Start the engine, place in FORWARD gear and warm to normal operating temperature.
6. Adjust the throttle stop screw (C, **Figure 109**) as required until the lowest possible rpm is obtained.
7. Slowly adjust each idle mixture screw counterclockwise 1/8 turn at a time until the engine starts to load up, then turn the screws clockwise 1/16 turn at a time until the engine fires evenly and rpm increases.
8. Repeat Step 6 and Step 7 until the smoothest idle is obtained at the lowest possible rpm.

9. Rapidly open and close the throttle. If the engine hesitates during acceleration, turn each idle mixture screw clockwise slightly until acceleration is smooth.
10. Return engine to idle rpm and recheck idle speed. Readjust throttle stop screw if necessary to bring idle speed within specifications (**Table 11**).
11. Shut the engine off. Remove the tachometer. Install the engine cover.

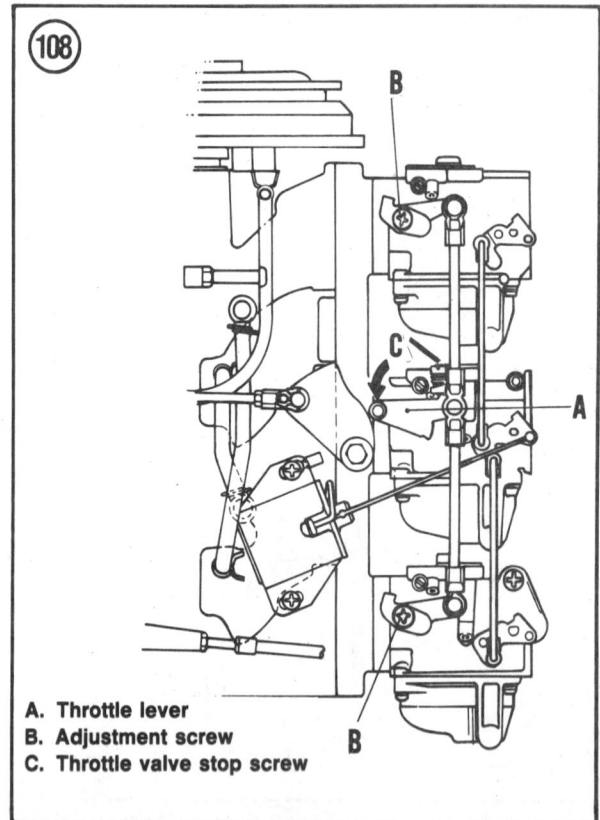

A. Throttle lever
B. Adjustment screw
C. Throttle valve stop screw

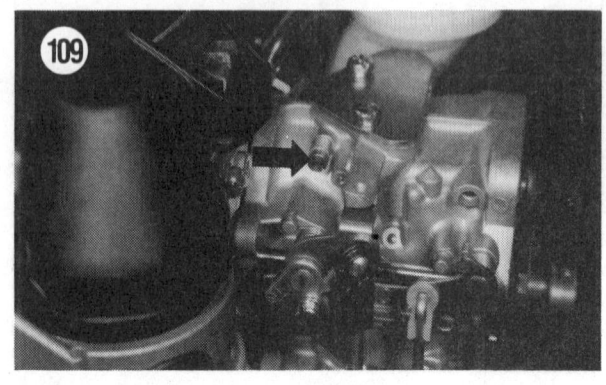

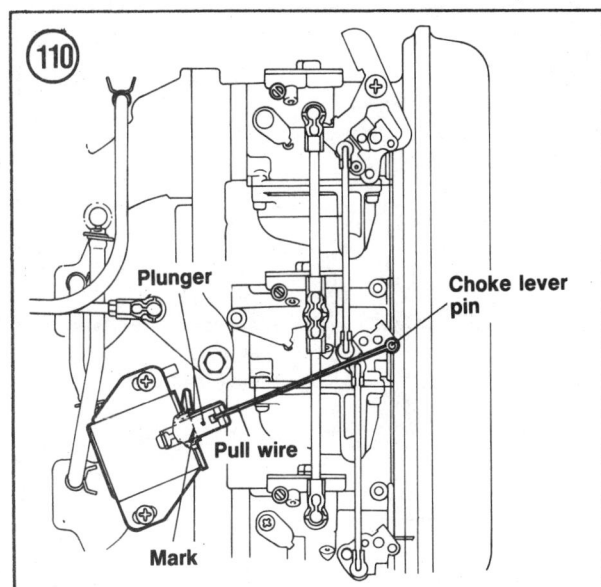

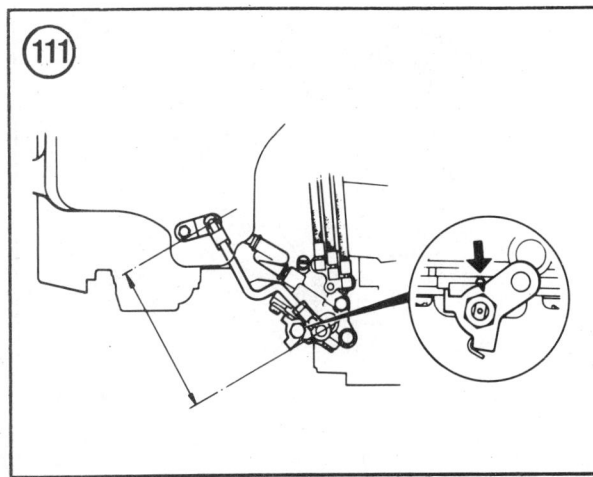

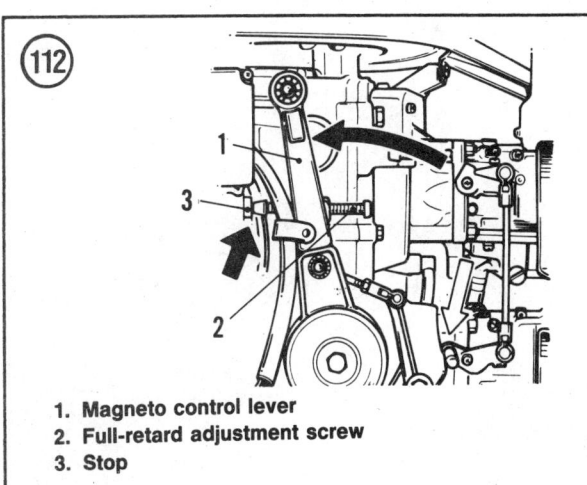

1. **Magneto control lever**
2. **Full-retard adjustment screw**
3. **Stop**

Choke Solenoid Adjustment

With the choke valves fully open, the groove on the end of the solenoid plunger should align with the end of the solenoid (**Figure 110**). If it does not, loosen the solenoid attaching bolts and move the solenoid in the proper direction to align the groove, then tighten the attaching bolts.

Oil Pump Link Adjustment

Loosen the throttle valve stop screw (C, **Figure 108**) and close the throttle valve completely. The punch mark on the pump lever should align with the mark on the pump body (**Figure 111**). If it does not, disconnect the pump link and adjust the connector as required. Reinstall the link and readjust the throttle valve stop screw.

YAMAHA 115 AND 130

Timing Adjustment

1. Install the engine in a test tank.
2. Remove the engine cover.
3. Connect a tachometer and timing light to the engine according to manufacturer's instructions.
4. Start the engine and manually move the magneto control lever to the closed throttle position until the full retard screw touches its stop (**Figure 112**).
5. Aim the timing light at the magneto pointer and turn the full retard screw until the timing indicator aligns with the specified timing mark on the flywheel (**Table 12**). See **Figure 113**.
6. Move the magneto control lever to the wide open throttle position until the full advance screw touches its stop (**Figure 113**).
7. Loosen the locknut on the full advance screw. Aim the timing light at the magneto pointer and turn the full advance screw until the timing indicator aligns with the specified

5

timing mark on the flywheel (**Table 12**), then tighten the locknut. See **Figure 115**.

8. Once the timing is correct, check the engine speed on the tachometer. It should be between 750-850 rpm. If not, adjust the carburetor throttle stop screw to bring engine speed into specifications. See *Carburetor Adjustment* in this section.

9. Aim the timing light at the timing plate and move the magneto control lever until the specified pickup timing mark (**Table 12**) aligns with the timing plate.

10. Move the throttle cam until it barely touches the throttle valve roller without opening the throttle valve. The center of the throttle valve roller should intersect the stamped mark on the throttle cam (**Figure 116**). If it does not, perform Step 11 and Step 12. If it does, omit Step 11 and Step 12.

11. Disconnect the throttle link between the control lever and throttle cam (**Figure 117**).

12. Loosen the throttle lever adjustment screw enough to align the roller and cam, then tighten the screw.

13. Adjust the plastic snap-on connector on the end of the link rod until it can be reconnected to the control lever ball stud with the cam and roller in contact.

14. Shut the engine off. Disconnect the timing light and tachometer. Install the engine cover.

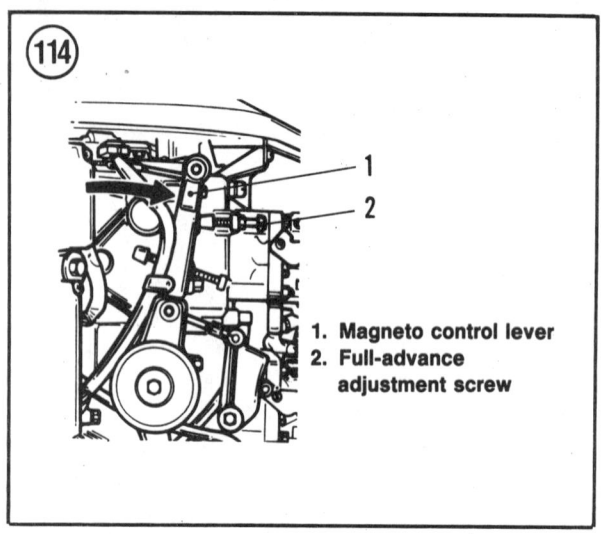

1. Magneto control lever
2. Full-advance adjustment screw

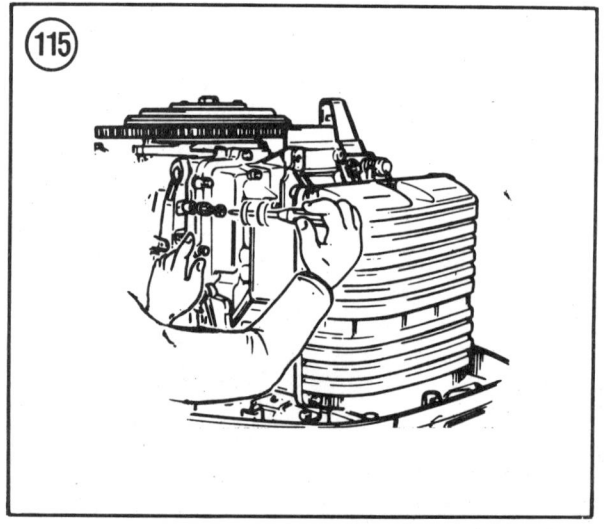

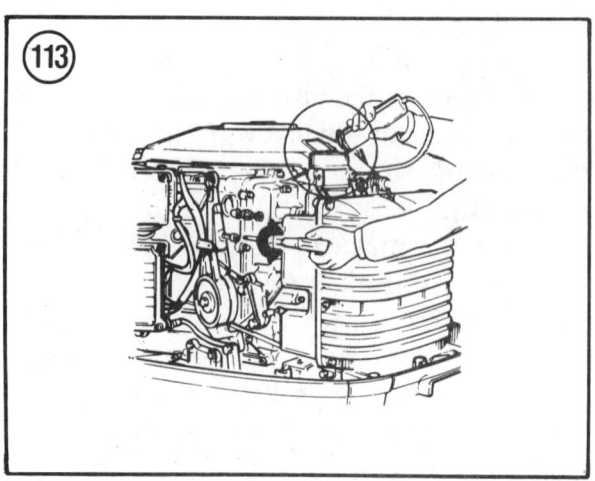

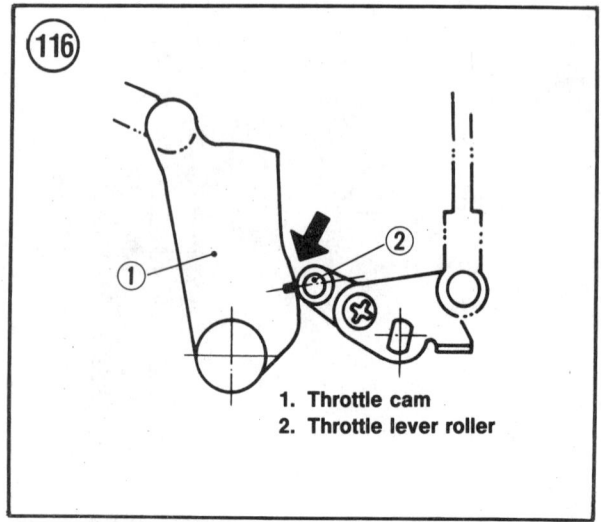

1. Throttle cam
2. Throttle lever roller

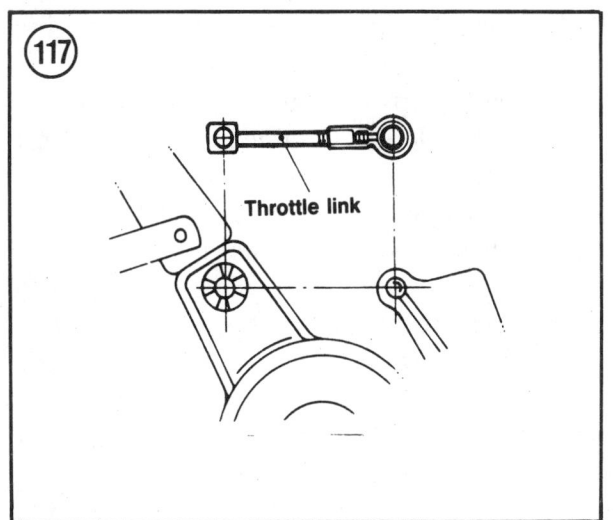

Throttle link

Carburetor Linkage Adjustment

1984

1. Loosen the upper carburetor throttle valve stop screw (**Figure 118**) until it does not touch the stop.

> *NOTE*
> *The throttle valve adjustment screws have a left-hand thread. Turn the screws clockwise to loosen and counterclockwise to tighten.*

2. Loosen the throttle valve adjustment screw on each carburetor. See **Figure 119**.

3. There should be no play between the throttle link and throttle arm ball stud, allowing the upper and lower throttle valves to operate together. If there is, remove and adjust the throttle link as required and reinstall the link. See **Figure 120**.

4. Tighten the throttle valve adjustment screws.

5. Tighten the upper carburetor throttle valve stop screw until it touches the stop, then tighten an additional 1 1/2 turns. See **Figure 121**.

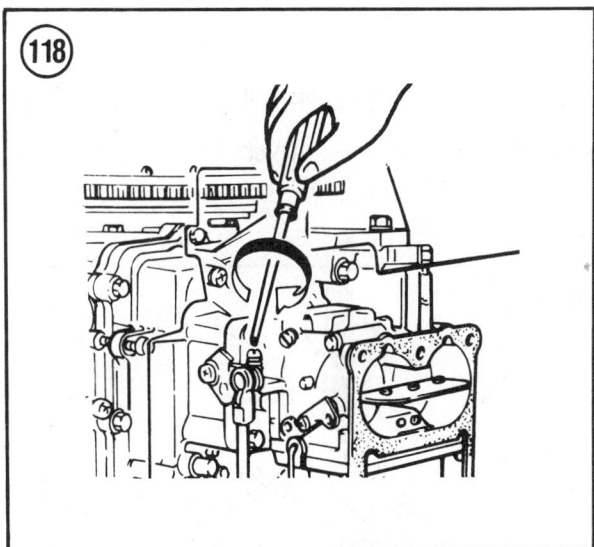

1985-on

Spring-loaded levers are used to automatically synchronize the throttle valves when the adjustment screw is loosened. In addition, the throttle valve stop screw has been relocated from the upper to the lower carburetor. See **Figure 122**.

1. Loosen the throttle valve stop screw completely. See **Figure 122** or A, **Figure 123**. Make sure the throttle valve is fully closed.

NOTE
The throttle valve adjustment screw has a left-hand thread. Turn the screw clockwise to loosen and counter-clockwise to tighten.

2. Loosen the throttle valve adjustment screw on the upper carburetor (B, **Figure 123**) until levers 1 and 2 are free.
3. Depress lever 1 (arrow, **Figure 123**) and tighten the throttle valve adjustment screw (B, **Figure 123**).
4. Adjust the throttle valve stop screw (A, **Figure 123**) to set idle speed to specifications.

Carburetor Adjustment

1. Install the engine in a test tank.
2. Remove the engine cover.
3. Connect a tachometer to the engine according to manufacturer's instructions.
4. Turn the idle mixture screw (A, **Figure 124**) on each carburetor clockwise until it lightly seats, then back it out 1 7/8-2 3/8 turns.
5. Start the engine, place in FORWARD gear and warm to normal operating temperature.
6. Adjust the throttle stop screw as required until the lowest possible rpm is obtained. See B, **Figure 124** (1984) or A, **Figure 123** (1985-on).
7. Slowly adjust each idle mixture screw counterclockwise 1/8 turn at a time until the engine starts to load up, then turn the screws

clockwise 1/16 turn at a time until the engine fires evenly and rpm increases.
8. Repeat Step 6 and Step 7 until the smoothest idle is obtained at the lowest possible rpm.
9. Rapidly open and close the throttle. If the engine hesitates during acceleration, turn each idle mixture screw clockwise slightly until acceleration is smooth.
10. Return engine to idle rpm and recheck idle speed. Readjust throttle stop screw if necessary to bring idle speed within specifications (**Table 12**).
11. Shut the engine off. Remove the tachometer. Install the engine cover.

Choke Solenoid Adjustment

With the choke valves fully open, the groove on the end of the solenoid plunger

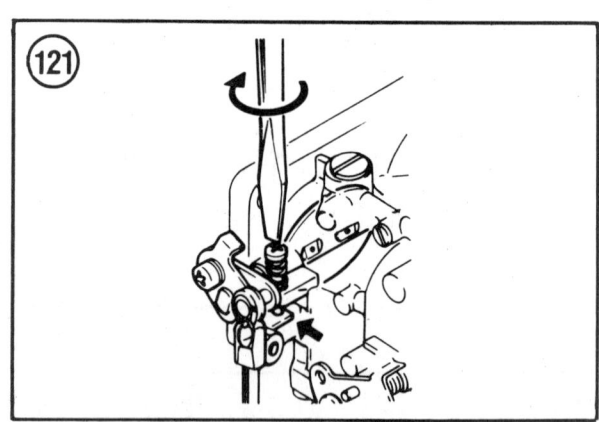

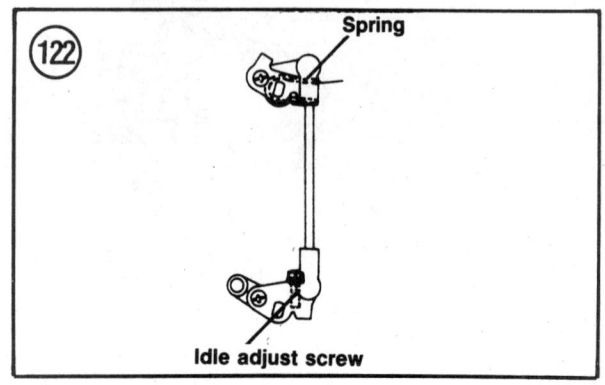

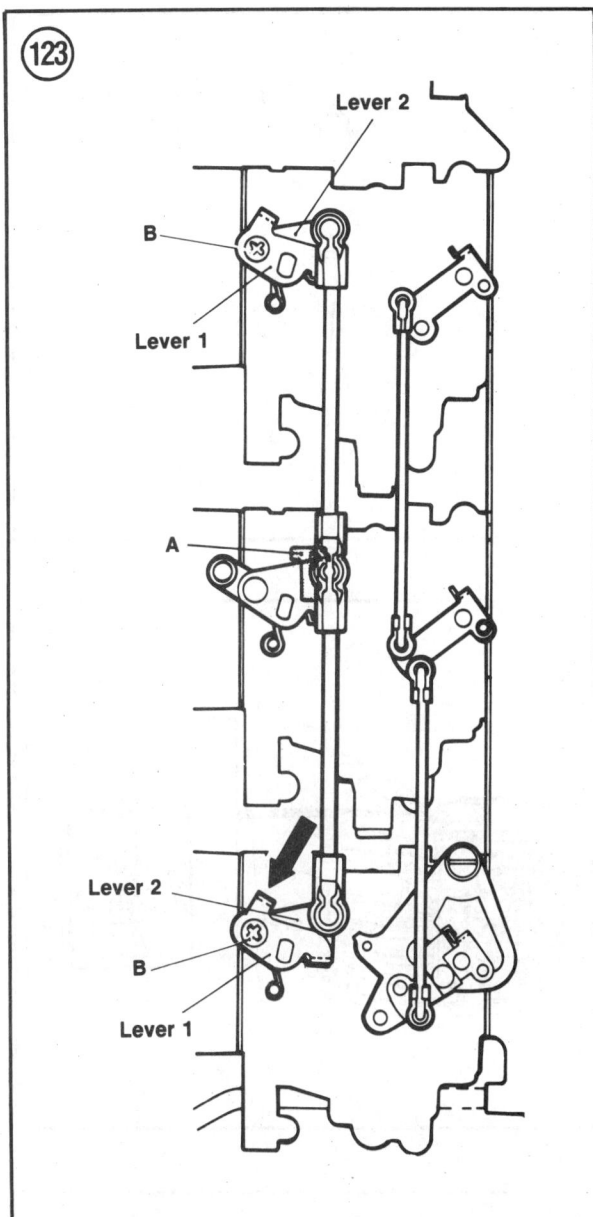

should align with the end of the solenoid. If it does not, loosen the solenoid attaching bolts and move the solenoid in the proper direction to align the groove, then tighten the attaching bolts.

Oil Pump Link Adjustment

Loosen the throttle valve stop screw. See B, **Figure 124** (1984) or A, **Figure 123** (1985-on). Close the throttle valve completely. The punch mark on the pump lever should align with the mark on the pump body (**Figure 125**). If it does not, disconnect the pump link and adjust the connector as required. Reinstall the pump link and readjust the throttle valve stop screw.

YAMAHA 150, 175, 200, 220 AND 225

Timing Adjustment

> *NOTE*
> *The 220 hp and 225 hp models are equipped with the Yamaha Micro-computer Ignition System (YMIS) and does not require spark advance adjustment. The ignition timing mark is used to confirm automatic advance by the YMIS module.*

1. Install the engine in a test tank.
2. Remove the engine cover.

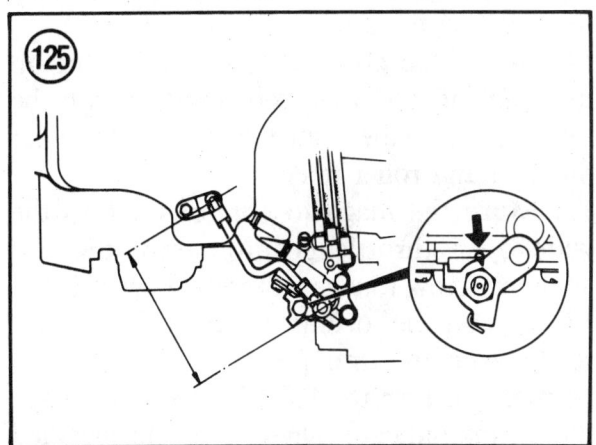

3. Connect a tachometer and timing light to the engine according to manufacturer's instructions.

4. Start the engine and manually move the magneto control lever to the closed throttle position until the full retard screw touches its stop (**Figure 126**).

5. Aim the timing light at the magneto pointer and turn the full retard screw until the timing indicator aligns with the specified timing mark on the flywheel (**Table 13**). See **Figure 127** (typical).

6. Once the timing is correct, check the engine speed on the tachometer. If not within specifications (**Table 13**), adjust the carburetor throttle stop screw to bring engine speed into specifications. See *Carburetor Adjustment* in this section.

7. Aim the timing light at the timing plate and move the magneto control lever until the specified pickup timing mark (**Table 13**) aligns with the timing plate.

8. Move the throttle cam until it barely touches the throttle valve roller without opening the throttle valve. The center of the throttle valve roller should intersect the stamped mark on the throttle cam (**Figure 128**).

9. If it does not, disconnect the throttle link between the control lever and throttle cam (**Figure 129**).

10. Loosen the throttle lever adjustment screw enough to align the roller and cam, then tighten the screw.

11. Adjust the plastic snap-on connector on the end of the link rod until it can be reconnected to the control lever ball stud with the cam and roller in contact.

12. Move the magneto control lever to the wide open throttle position until the full advance screw touches its stop (**Figure 130**).

13. Loosen the locknut on the full advance screw. Aim the timing light at the magneto pointer and turn the full advance screw until the timing indicator aligns with the specified

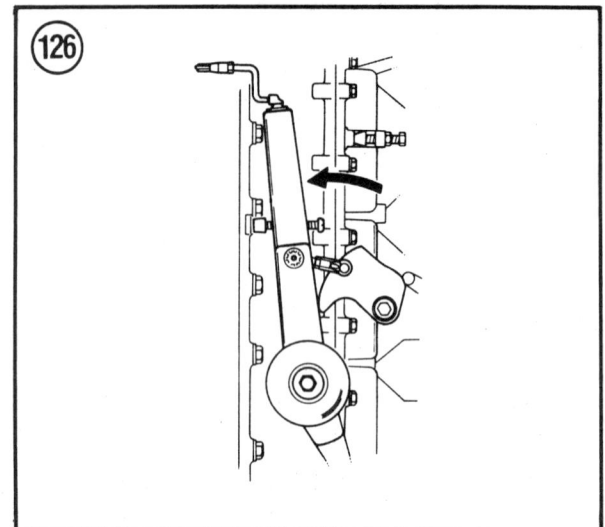

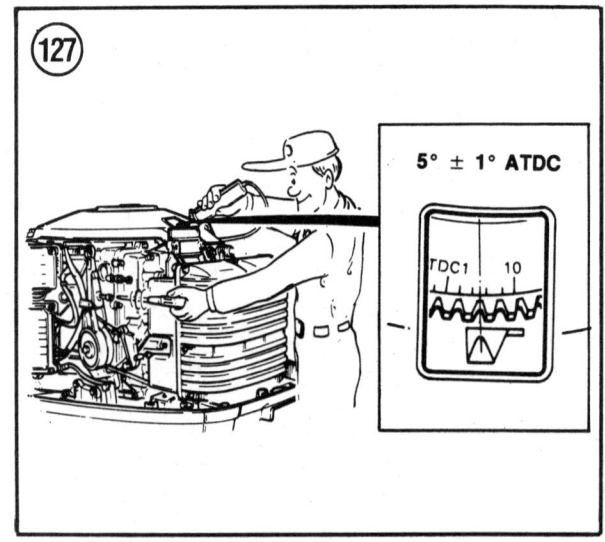

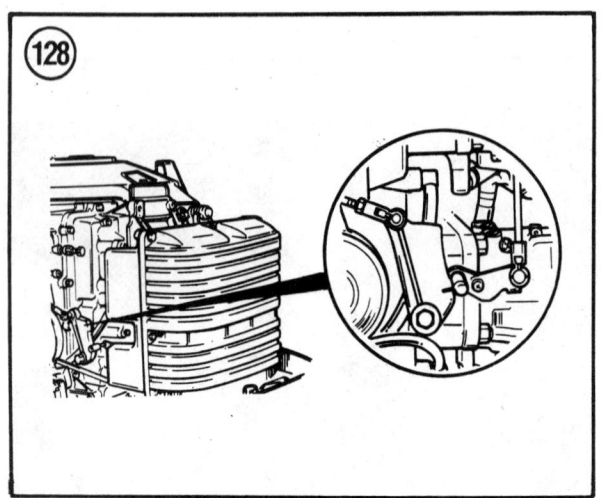

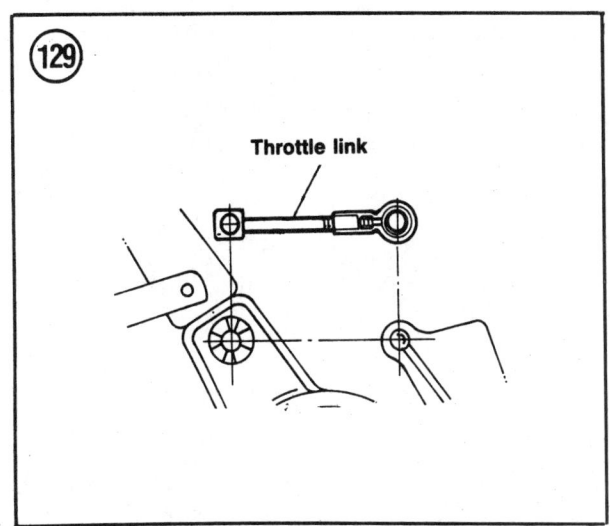

Throttle link

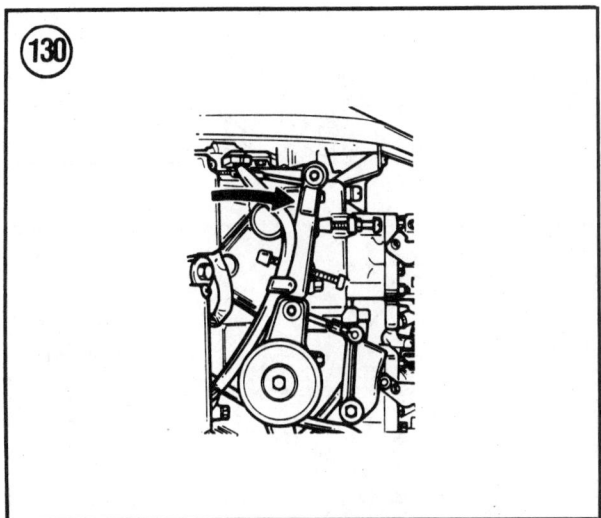

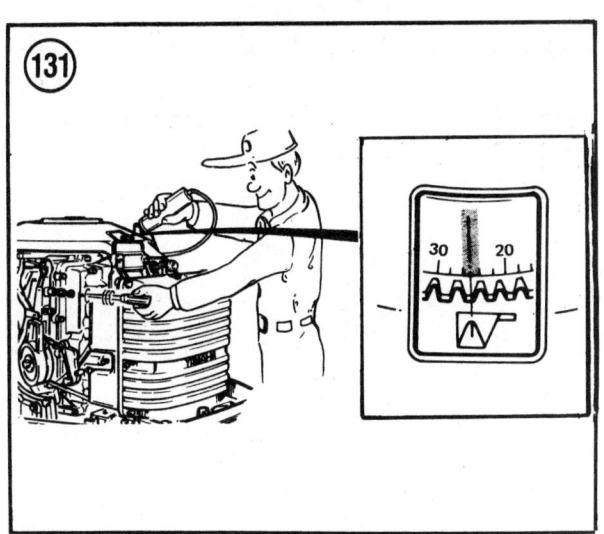

timing mark on the flywheel (**Table 13**), then tighten the locknut. See **Figure 131**.

14. Shut the engine off. Disconnect the timing light and tachometer. Install the engine cover.

Carburetor Linkage Adjustment

1984

1. Loosen the upper carburetor throttle valve stop screw (**Figure 132**) until it does not touch the stop.

> NOTE
> *The throttle valve adjustment screw has a left-hand thread. Turn the screw clockwise to loosen and counter-clockwise to tighten.*

2. Loosen the throttle valve adjustment screw on each carburetor. See **Figure 133**.

3. There should be no play between the throttle link and throttle arm ball stud, allowing the upper and lower throttle valves to operate together. If there is, remove and adjust the throttle link as required and reinstall the link. See **Figure 134**.

4. Tighten the throttle valve adjustment screws.

5. Tighten the upper carburetor throttle valve stop screw until it touches the stop, then tighten an additional 1 1/2 turns. See **Figure 135**.

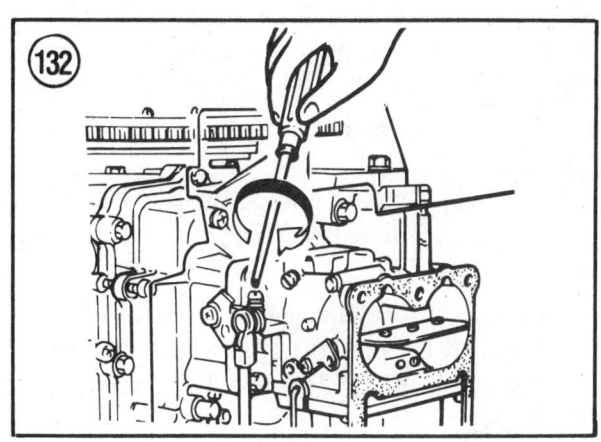

1985-on

Spring-loaded levers are used to automatically synchronize the throttle valves when the adjustment screw is loosened. In addition, the throttle valve stop screw has been relocated from the upper to the middle carburetor. See **Figure 136**.

1. Loosen the throttle valve stop screw completely. See **Figure 136** or A, **Figure 137**. Make sure the throttle valve is fully closed.

> *NOTE*
> *The throttle valve adjustment screws have a left-hand thread. Turn the screw clockwise to loosen and counter-clockwise to tighten.*

2. Loosen the throttle valve adjustment screws on the upper and lower carburetors (B, **Figure 137**) until levers 1 and 2 are free at each carburetor.
3. Depress lever 1 at each carburetor and tighten the throttle valve adjustment screws (B, **Figure 137**).
4. Adjust the throttle valve stop screw (A, **Figure 137**) to set idle speed to specifications.

Carburetor Adjustment

1. Install the engine in a test tank.
2. Remove the engine cover.
3. Connect a tachometer to the engine according to manufacturer's instructions.
4. Turn the idle mixture screw (A, **Figure 138**) on each carburetor clockwise until it lightly seats, then back it out as specified in **Table 13**.
5. Start the engine, place in FORWARD gear and warm to normal operating temperature.
6. Adjust the throttle stop screw as required until the lowest possible rpm is obtained. See B, **Figure 138** (1984) or A, **Figure 137** (1985-on).

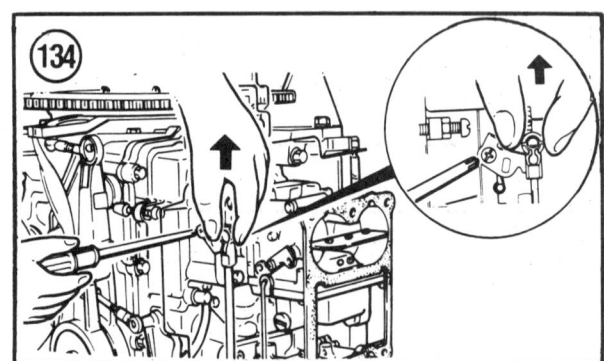

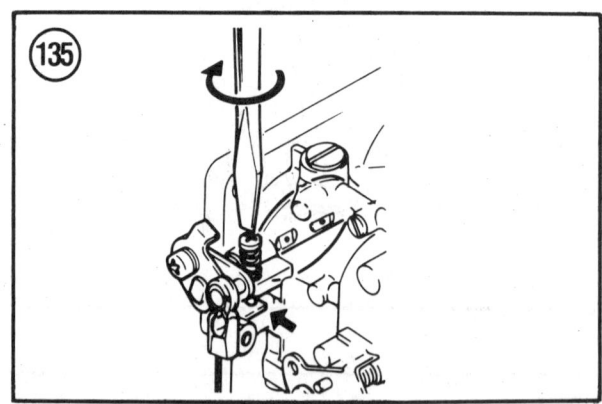

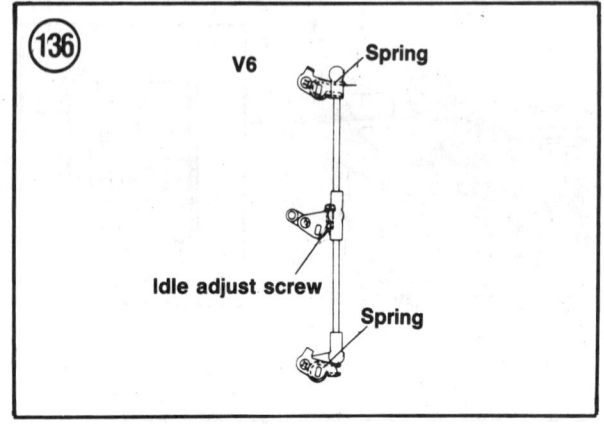

7. Slowly adjust each idle mixture screw counterclockwise 1/8 turn at a time until the engine starts to load up, then turn the screws clockwise 1/16 turn at a time until the engine fires evenly and rpm increases.

8. Repeat Step 6 and Step 7 until the smoothest idle is obtained at the lowest possible rpm.

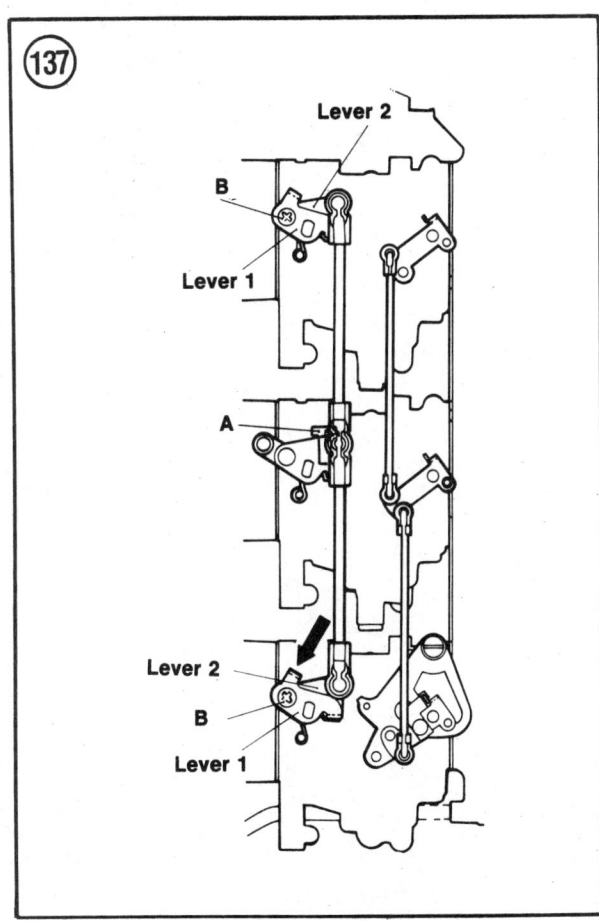

9. Rapidly open and close the throttle. If the engine hesitates during acceleration, turn each idle mixture screw clockwise slightly until acceleration is smooth.

10. Return engine to idle rpm and recheck idle speed. Readjust throttle stop screw if necessary to bring idle speed within specifications (**Table 13**).

11. Shut the engine off. Remove the tachometer. Install the engine cover.

5

Choke Solenoid Adjustment

With the choke valves fully open, the groove on the end of the solenoid plunger should align with the end of the solenoid (**Figure 139**). If it does not, loosen the solenoid attaching bolts and move the solenoid in the proper direction to align the groove, then tighten the attaching bolts.

Oil Pump Link Adjustment

Loosen the throttle valve stop screw. See B, **Figure 138** (1984) or A, **Figure 137** (1985-on). Close the throttle valve completely. The punch mark on the pump lever should align with the mark on the pump body (**Figure 140**). If it does not, disconnect the pump link and adjust the connector as required. Reinstall the pump link and readjust the throttle valve stop screw.

If the oil pump does not have a punch mark, disconnect the link between the pump and carburetor at the pump. Loosen the throttle stop screw. See B, **Figure 138** (1984) or A, **Figure 137** (1985-on). Move the oil pump stroke adjustment arm to its fully-closed position. Adjust the length of the link as required. Reconnect the link and adjust the throttle linkage as described in this chapter.

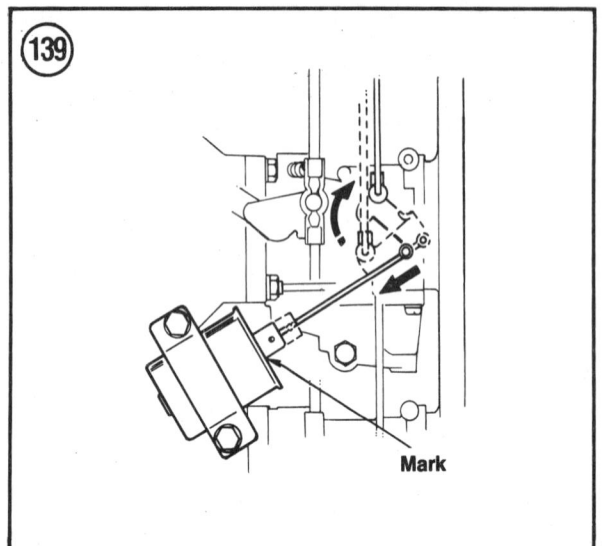

Mark

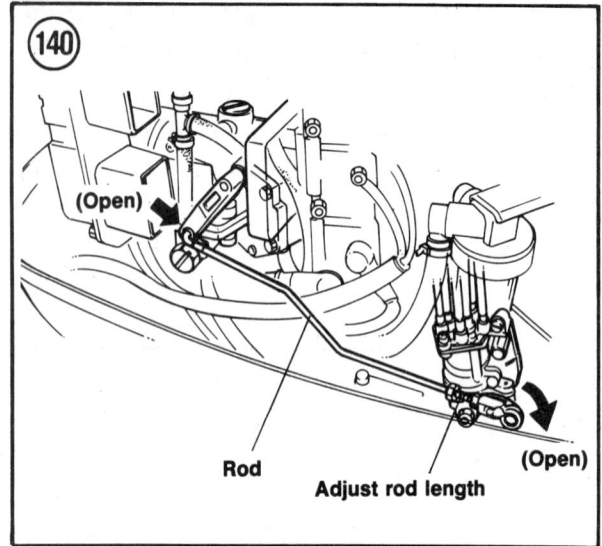

(Open)

Rod   Adjust rod length   (Open)

Table 1 TEST WHEEL RECOMMENDATIONS

Engine	Part No.
9.9-15 hp	90890-1619
25-30 hp	90890-1621
40-50 hp	90890-01611
60-90 hp	90890-01620
115-130 hp	90890-01624
150-225 hp	90890-01626

Table 2 TUNE-UP SPECIFICATIONS (YAMAHA 2)

Firing order	Single cylinder
Spark plug	
Type	NGK B5HS
Gap	0.020-0.024 in.
Torque	14 ft.-lb. (20 N•m)
Static WOT (wide open throttle)	0.044 ±0.0047 in. BTDC
Breaker point gap	0.014 in.
Full throttle rpm	4,000-5,000 rpm
Idle rpm	1,150-1,250 rpm

Table 3 TUNE-UP SPECIFICATIONS (YAMAHA 3)

Firing order	Single cylinder
Spark plug	
Type	NGK B6HS-10
Gap	0.035-0.039 in.
Torque	14 ft.-lb. (20 N•m)
Maximum timing	18° ±1° BTDC
Idle timing	18° ±1° BTDC
Full throttle rpm	4,500-5,500 rpm
Idle rpm	1,500-1,250 rpm

Table 4 TUNE-UP SPECIFICATIONS (YAMAHA 4 AND 5)

Firing order	Single cylinder
Spark plug	
Type	NGK B7HS
Gap	0.020-0.024 in.
Torque	18 ft.-lb. (25 N•m)
Maximum timing	28° ±3° BTDC
Idle timing	5° ±2° BTDC
Full throttle rpm	4,500-5,500 rpm
Idle rpm	1,100-1,200 rpm
Trolling rpm	950-1,050 rpm

Table 5 TUNE-UP SPECIFICATIONS (YAMAHA 6 AND 8)

Firing order	Alternate
Spark plug	
Type	NGK B7HS-10
Gap	0.035-0.039 in.
Torque	14 ft.-lb. (20 N•m)
Maximum timing	35° ±1 BTDC
Idle timing	Automatically set when maximum timing is correct
Full throttle rpm	4,500-5,500 rpm
Idle rpm	850-950 rpm
Trolling rpm	750-850 rpm

Table 6 TUNE-UP SPECIFICATIONS (YAMAHA 9.9 AND 15)

Firing order	Alternate
Spark plug	
Type	NGK B7HS-10
Gap	0.035-0.039 in.
Torque	14 ft.-lb. (20 N•m)
Static WOT (wide open throttle)	0.166 ±0.011 in.
Dynamic WOT (wide open throttle)	30° ±1 BTDC
Static idle timing	0.005 ±0.002 in.
Dynamic idle timing	5° ±1° BTDC
Full throttle rpm	4,500-5,500 rpm
Idle rpm	850-950 rpm
Trolling rpm	670-770 rpm

Table 7 TUNE-UP SPECIFICATIONS (YAMAHA 25)

Firing order	Alternate
Spark plug	
Type	
1984	NGK B7HS
1985-on	NGK B7HS-10
Gap	
1984	0.020-0.024 in.
1985-on	0.035-0.039 in.
Torque	
1984-1987	14 ft.-lb. (20 N•m)
1988-1989	18 ft.-lb. (25 N•m)
Dynamic WOT (wide open throttle)	25° BTDC
Full throttle rpm	
1984-1987	4,500-5,500 rpm
1988-1989	5,000-6,000 rpm
Idle rpm	
1984-1987	850-950 rpm
1988-1989	700-800 rpm
Trolling rpm	
1984-1987	750-850 rpm
1988-1989	600-700 rpm

Table 8 TUNE-UP SPECIFICATIONS (YAMAHA 30)

Firing order	
1984-1986	Alternate
1987-on	1-2-3
Spark plug	
Type	NGK B7HS-10
Gap	0.035-0.039 in.
Torque	14 ft.-lb. (20 N•m)
Dynamic WOT (wide open throttle)	25° BTDC
Full throttle rpm	4,500-5,500 rpm
Idle rpm	
1984-1986	850-950 rpm
1987	700-800 rpm
Trolling rpm	
1984-1986	750-850 rpm
1987	600-700 rpm

5

Table 9 TUNE-UP SPECIFICATIONS (YAMAHA 40 AND 50 [includes Pro 50])

Firing order	1-2-3
Spark plug	
Type	
40 hp	NGK B7HS-10
50 hp	NGK B8HS-10
Gap	0.035-0.039 in.
Torque	14 ft.-lb. (20 N•m)
Dynamic WOT (wide open throttle)	25° ±1° BTDC
Dynamic idle timing	5° ±1° ATDC
Pickup timing	1° ±1° ATDC
Full throttle rpm	4,500-5,500 rpm
Idle rpm	750-850 rpm
Trolling rpm	550-650 rpm

Table 10 TUNE-UP SPECIFICATIONS (YAMAHA 70)

Firing order	1-2-3
Spark plug	
Type	NGK B8HS-10
Gap	0.035-0.039 in.
Torque	18 ft.-lb. (25 N•m)
Dynamic WOT (wide open throttle)	20° ±1° BTDC
Dynamic idle timing	
1984	4° ±1° ATDC
1985-on	7° ±1° ATDC
Pickup timing	
1984	NA
1985-on	7° ATDC
Full throttle rpm	4,500-5,500 rpm
Idle rpm	750-850 rpm
Trolling rpm	550-650 rpm

Table 11 TUNE-UP SPECIFICATIONS (YAMAHA 90)

Firing order	1-2-3
Spark plug	
Type	NGK B8HS-10
Gap	0.035-0.039 in.
Torque	18 ft.-lb. (25 N•m)
Dynamic WOT (wide open throttle)	22° ±1° BTDC
Dynamic idle timing	
1984	5° ±1° ATDC
1985-on	10° ±1° ATDC
Pickup timing	
1984	5° ATDC
1985-on	10° ±1° ATDC
Full throttle rpm	4,500-5,500 rpm
Idle rpm	750-850 rpm
Trolling rpm	550-650 rpm

Table 12 TUNE-UP SPECIFICATIONS (YAMAHA 115 AND 130)

Firing order	1-2-3-4
Spark plug	
Type	
115 hp	NGK B8HS-10
130 hp	NGK B9HS-10
Gap	0.035-0.039 in.
Torque	14 ft.-lb. (20 N•m)
Dynamic WOT (wide open throttle)	
115 hp	25° ±1° BTDC
130 hp	23° ±1° BTDC
Dynamic idle timing	5° ±1° ATDC
Pickup timing	4° ±1° ATDC
Full throttle rpm	4,500-5,500 rpm
Idle rpm	700-800 rpm
Trolling rpm	600-700 rpm

Table 13 TUNE-UP SPECIFICATIONS (YAMAHA V6)

Firing order	1-2-3-4-5-6
Spark plug	
Type	
150-200 hp	NGK B8HS-10
220 hp V6 Special	NGK BR8HS-10
225 hp Excel	NGK BR9HS-10
Gap	0.035-0.039 in.
Torque	14 ft.-lb. (20 N•m)
Dynamic WOT (wide open throttle)	
150 hp	24° ±1° BTDC
150 hp Pro V	28° ±1° BTDC
175-200 hp	22° ±1° BTDC
220 hp[1]	26° ±2° BTDC
225 hp Excel[1]	22° ±1° BTDC
	(continued)

Table 13 TUNE-UP SPECIFICATIONS (YAMAHA V6) (continued)

Dynamic idle timing	
1984 150-200 hp	5° ±1° ATDC
1985-on 150-200 hp	7° ±1° ATDC
220 hp[1]	5° ±2° ATDC
225 hp Excel[1]	6° ±1° ATDC
Pickup timing	
1984 150-200 hp	4° ±1° ATDC
1985-on 150 and 200 hp	6° ±1° ATDC
175 hp	7° ±1° ATDC
220 hp[1]	5° ±2° ATDC
225 hp Excel[1]	6° ±1° ATDC
Full throttle rpm	
150-200 hp	4,500-5,500 rpm
220 hp and 225 hp Excel	4,800-5,800 rpm
Idle rpm	
1984 150 hp	795-845 rpm
1984 175-200 hp	775-825 rpm
1984 220 hp	755-805 rpm
1985-on 150-175 hp	775-825 rpm
1985-on 200 hp	675-725 rpm
1985-on 220 hp and 225 hp Excel	725-775 rpm
Trolling rpm	
1984 (all)	600-650 rpm
1985-on 150-200 hp	550-600 rpm
1985-on 220 hp and 225 hp Excel	575-625 rpm
Initial idle screw	
adjustment[2]	
150 hp w/carburetor	
#6G400	2 1/8-2 5/8 turns
#6G401	1-1 1/2 turns
#6G402	7/8-1 3/8 turns
#6G403	1-1 1/2 turns
150 hp Pro V w/carburetor	
#6J900	
Starboard	3/4-1 1/4 turns
Port	1 1/4-1 3/4 turns
175 hp w/carburetor	
#6G500	1 7/8-2 3/8 turns
#6G501	1-1 1/2 turns
#6G502	
Starboard	7/8-1 3/8 turns
Port	1 1/2-2 turns
#6G503	1 1/8-1 5/8 turns
200 hp w/carburetor	
#6G600	1 7/8-2 3/8 turns
#6G601	1 1/8-1 5/8 turns
#6G602	
Starboard	7/8-1 3/8 turns
Port	1 1/4-1 3/4 turns
#6G603	
Starboard	5/8-1 1/8 turns
Port	1 3/8-1 7/8 turns

(continued)

5

Table 13 TUNE-UP SPECIFICATIONS (YAMAHA V6) (continued)

220 hp w/carburetor
  #6G700   1-1 1/2 turns
  #6G701   1 1/4-1 3/4 turns
  #6G702
    Starboard   1/2-1 turn
    Port   1 3/8-1 7/8 turns
225 hp Excel w/carburetor
  #6K700   3/4-1 1/4 turns

1. Computer controlled.
2. Lightly seat the idle mixture screw, then back it out the specified number of turns as an initial adjustment point.

Chapter Six

Fuel System

This chapter contains removal, overhaul, installation and adjustment procedures for fuel pumps, carburetors, portable fuel tanks and connecting lines used with the Yamaha outboards covered in this book. Carburetor (**Table 1**) and reed stop specifications (**Table 2**) are at the end of the chapter.

FUEL PUMP

Yamaha outboards equipped with an integral fuel tank use a gravity flow fuel system (**Figure 1**) and require no fuel pump. The diaphragm-type fuel pump used on Yamaha outboards is operated by crankcase pressure. Since this type of fuel pump cannot create sufficient pressure to draw fuel from the tank during cranking, fuel is transferred to the carburetor for starting by operating the primer bulb installed in the fuel line.

Pressure pulsations created by movement of the pistons reach the fuel pump through a passageway between the crankcase and the pump.

Upward piston motion creates a low pressure on the pump diaphragm. This low pressure opens the inlet check valve in the pump, drawing fuel from the line into the pump. At the same time, the low pressure draws the air-fuel mixture from the carburetor into the crankcase.

Downward piston motion creates a high pressure on the pump diaphragm. This pressure closes the inlet check valve and opens the outlet check valve, forcing the fuel into the carburetor and drawing the air-fuel mixture from the crankcase into the cylinder for combustion. **Figure 2** shows the operational sequence of a typical Yamaha outboard fuel pump (high output models use two pumps).

Outboard fuel pumps are extremely simple in design and reliable in operation. Diaphragm failures are the most common problem, although the use of dirty or improper fuel-oil mixtures can cause check valve problems.

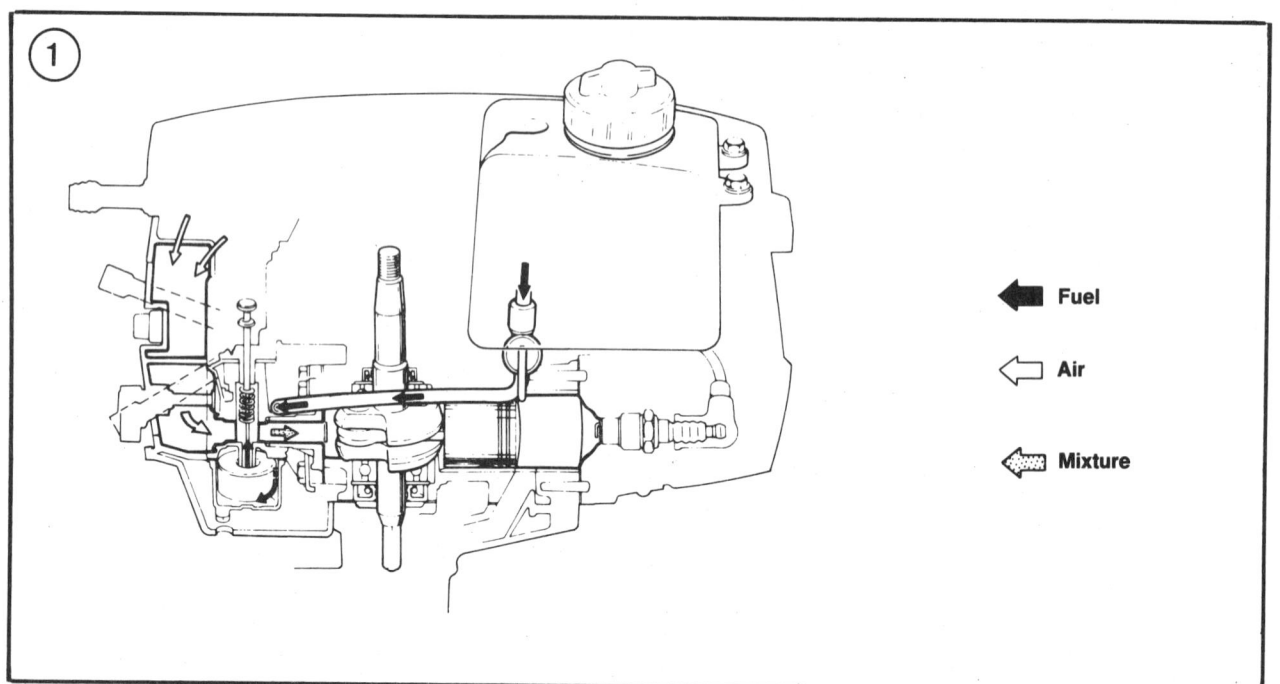

① Fuel

◁ Air

◁ Mixture

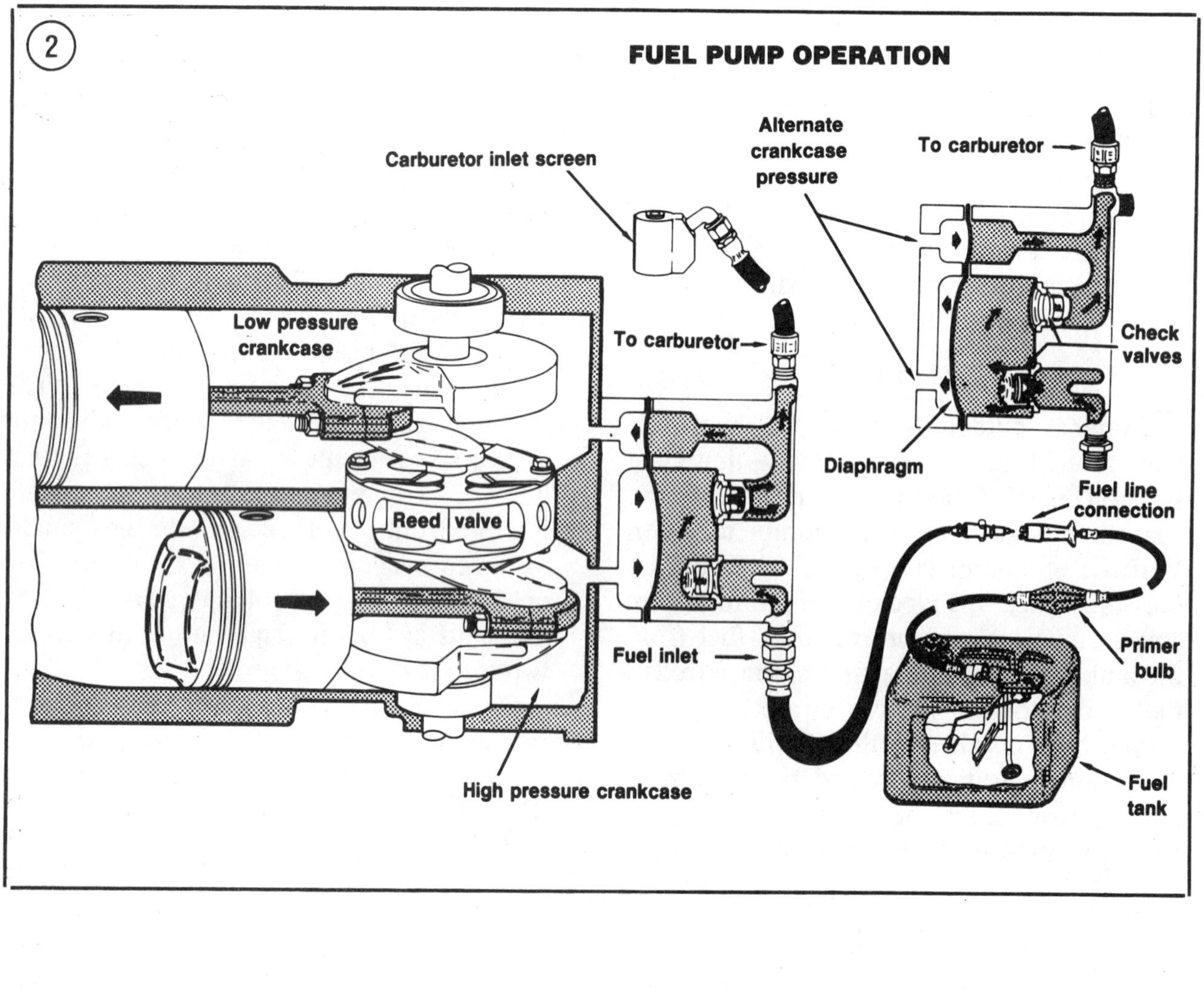

② **FUEL PUMP OPERATION**

Carburetor inlet screen

Alternate crankcase pressure

To carburetor

Low pressure crankcase

To carburetor

Check valves

Diaphragm

Reed valve

Fuel line connection

Fuel inlet

Primer bulb

High pressure crankcase

Fuel tank

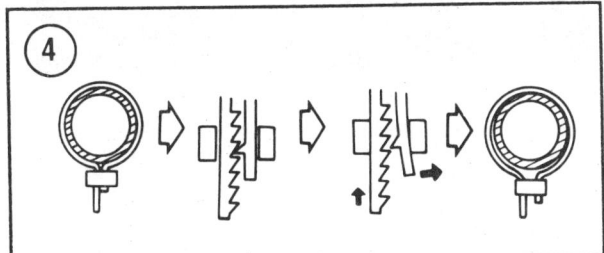

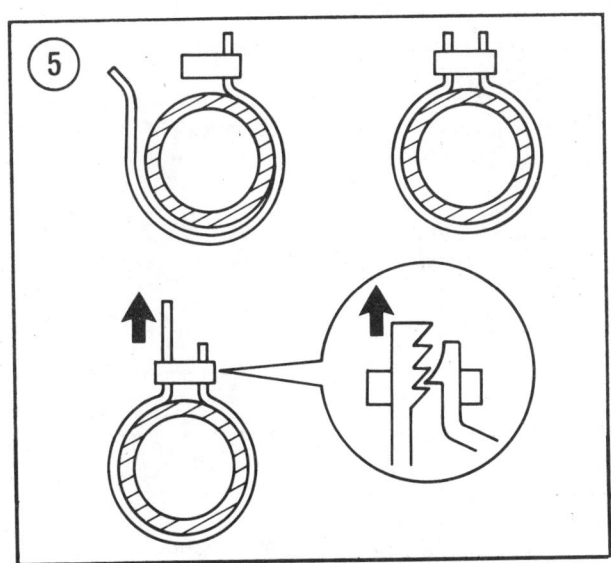

Fuel System Connections

Connections between fuel system components may be secured with wire spring clips or plastic band clamps. To remove the wire clip type, squeeze the ends together with pliers and slide the clip off the fuel fitting (**Figure 3**). Installation is the reverse of removal. Wire clips should be replaced when they have lost tension.

Plastic band clamps are reuseable if removed properly. This type of clamp depends upon tension maintained by serrated teeth when the clamp is pulled tight. See **Figure 4**. To remove the clamp, insert the tip of a small flat blade screwdriver into the ends of the band. Pry the ends apart, pushing the longer end of the band to loosen it sufficiently to slide the clamp off the fuel fitting. To reinstall the clamp, position it on the fuel fitting and insert the loose end into the band, then pull it tight with a pair of pliers. The teeth will engage and hold the tension applied. See **Figure 5**.

Remote Fuel Pump
Removal/Installation

1. Remove and discard any straps holding the carburetor fuel line to the fuel pump. Disconnect the lines at the pump (**Figure 6**). Plug the lines to prevent leakage.
2. Remove the screws holding the pump assembly to the engine. Remove the pump from the engine and discard the gasket.
3. Clean all gasket residue from the engine mounting pad. Work carefully to avoid gouging or damaging the mounting surface.
4. Installation is the reverse of removal. Use a new gasket. Install new straps.

Integral Fuel Pump
Removal/Installation

1. Disconnect the fuel line at the inlet cover (A, **Figure 7**).

2. Remove 4 screws holding the fuel pump components to the side of the carburetor (B, **Figure 7**).

3. Remove the components.

4. Installation is the reverse of removal.

Remote Fuel Pump
Disassembly/Assembly

Refer to **Figure 8** or **Figure 9** as appropriate for this procedure.

1. Test the pump outlet check valve by sucking and blowing through the outlet

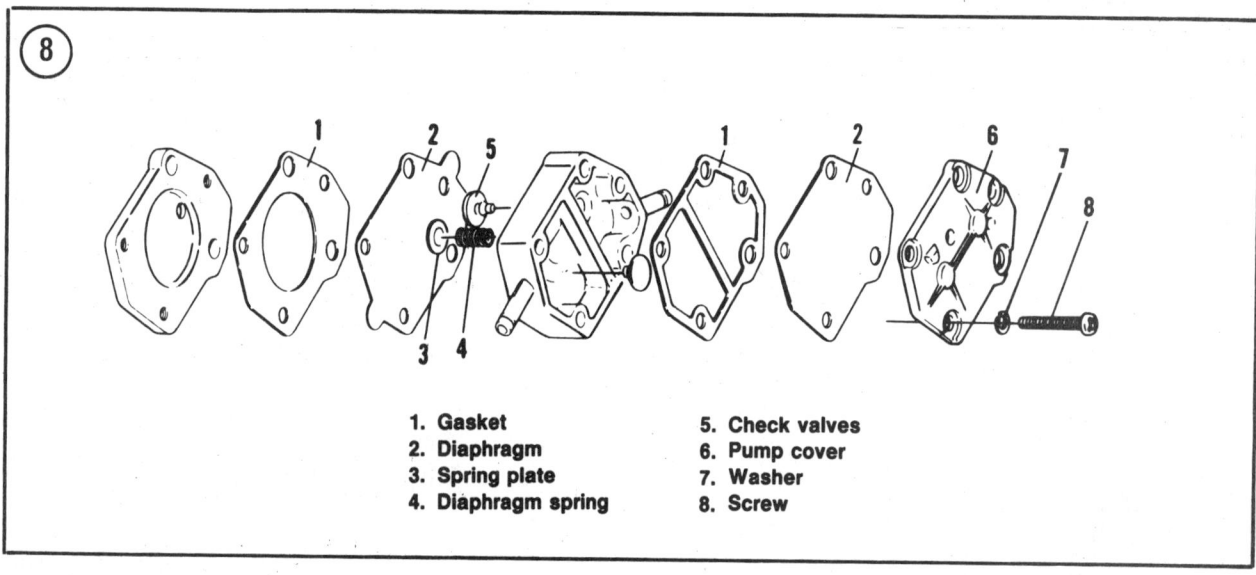

1. Gasket
2. Diaphragm
3. Spring plate
4. Diaphragm spring
5. Check valves
6. Pump cover
7. Washer
8. Screw

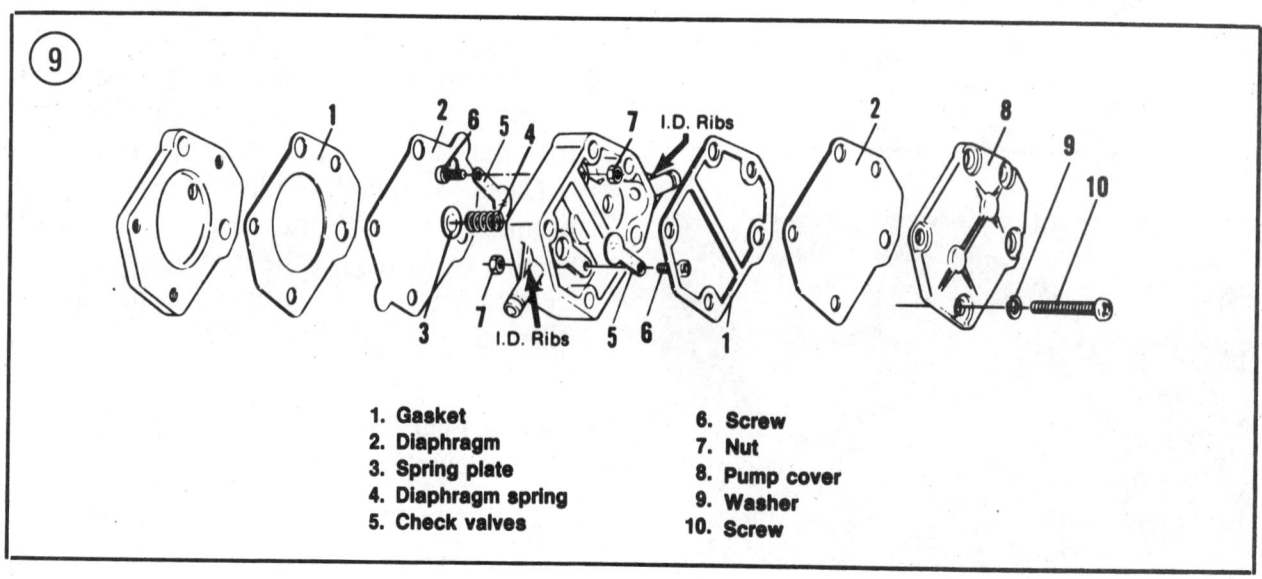

1. Gasket
2. Diaphragm
3. Spring plate
4. Diaphragm spring
5. Check valves
6. Screw
7. Nut
8. Pump cover
9. Washer
10. Screw

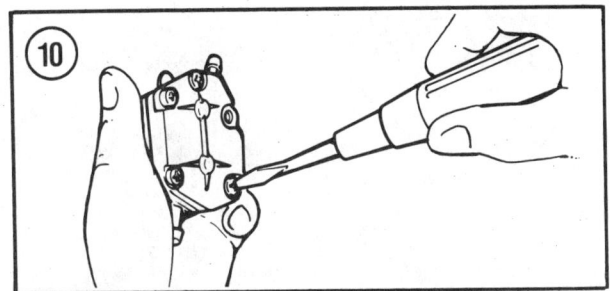

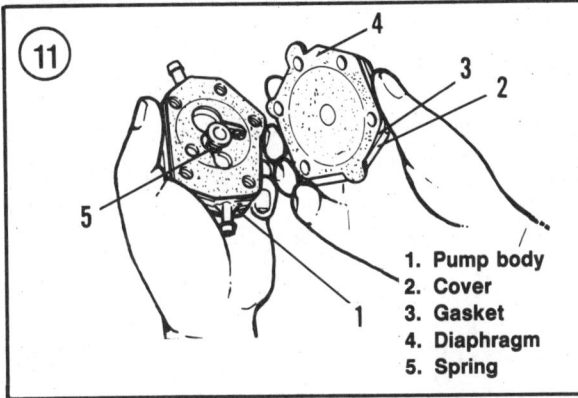

1. Pump body
2. Cover
3. Gasket
4. Diaphragm
5. Spring

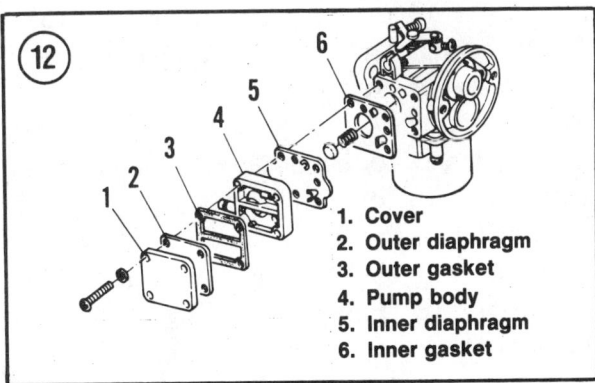

1. Cover
2. Outer diaphragm
3. Outer gasket
4. Pump body
5. Inner diaphragm
6. Inner gasket

opening. You should be able to draw air through the valve but not blow air through it.

2. Test the pump inlet check valve by sucking and blowing through the inlet opening. You should be able to blow through the valve, but not draw air through it.

3. If the check valves do not operate as specified in Step 1 or Step 2, replace them as described in this chapter.

4. Remove the 3 screws holding the covers on each side of the pump body (**Figure 10**). Separate the covers from the body.

5. Remove the diaphragm(s) and gaskets. Separate the diaphragm(s) from the gaskets. Discard the gaskets. See **Figure 11**.

6. Note the location and positioning of the spring (5, **Figure 11**). Remove the spring and check valves. Some pumps use a flap-type check valve diaphragm; others use metal reed-type valves secured by a small bolt and nut.

7. Clean and inspect the pump components as described in this chapter.

NOTE
Do not use any form of gasket sealer with fuel pump gaskets.

8. Assembly is the reverse of disassembly. Repeat Step 1 and Step 2 to test the check valves after installation. Use new gaskets and diaphragm(s).

9. Install the fuel pump to the engine as described in this chapter.

Integral Fuel Pump
Disassembly/Assembly

Refer to **Figure 12** (typical) for this procedure.

1. Separate the pump cover, diaphragms, valve and strainer assembly and fuel pump body. Peel gaskets from cover and body and discard.

2. Clean and inspect the pump components as described in this chapter.

3. Assemble components with new gaskets in reverse of order given in Step 1.

Cleaning and Inspection
(Except Integral Fuel Pump)

1. Clean pump body, covers and check valves in solvent. Dry housing and covers with compressed air. Let check valves air dry.

2. Inspect check valves for warpage and spring tension. Replace any valve that is slightly warped, has weak tension or broken springs.

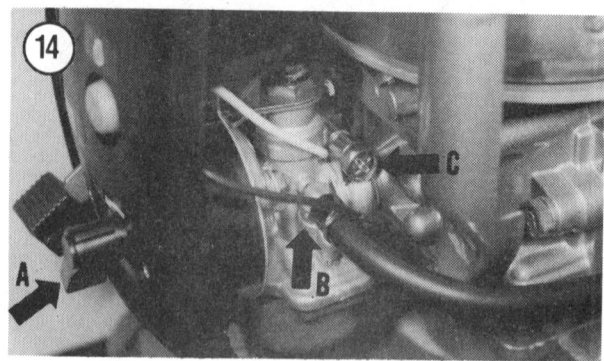

3. Check body condition. Make sure the valve seats provide a flat contact area for the valve disc. Replace the body if cracks or rough gasket mating surfaces are found.

4. Check the body fitting. If loose or damaged, tighten or replace as required.

Cleaning and Inspection (Integral Fuel Pump)

1. Clean strainer and fuel pump bodies in solvent. Blow dry with compressed air.

2. Inspect check valve diaphragm for damage. Replace as required.

CARBURETORS

All carburetors used on Yamaha outboards use a fixed main jet and require no high-speed adjustment. A main jet with a smaller number should be substituted when the engine is used primarily at low speeds or at high elevations. A larger number jet should be substituted for use in extremely cold areas.

When removing and installing a carburetor, make sure that the mounting fasteners are tightened securely. A carburetor that is loose will cause a lean-running condition.

THROTTLE PLUNGER CARBURETOR

This carburetor design is used on Yamaha 2 hp models.

Removal/Installation

1. Remove the screws holding the 2 halves of the engine cover. Remove the engine cover.

2. Loosen the screw holding the throttle lever knob (A, **Figure 13**). Remove the knob from the lever.

3. Loosen the screw holding the choke lever knob (A, **Figure 14**). Remove the knob from the lever.

4. Disconnect the white stop button lead at the magneto base connector (B, **Figure 13**).

5. Disconnect the black ground lead at the rewind starter bracket (C, **Figure 13**).

6. Remove the 2 screws holding the air intake silencer to the carburetor. Remove the air intake silencer.

7. Make sure the fuel petcock is in the OFF position. Disconnect the fuel line at the carburetor (B, **Figure 14**). Plug the line to prevent leakage.

8. Loosen the clamp screw holding the carburetor to the reed valve housing (C, **Figure 14**). Remove the carburetor.

9. Remove and discard the O-ring inside the carburetor throat.

10. Installation is the reverse of removal. Install a new O-ring in the carburetor throat. Tighten the clamp screw securely to prevent an air or fuel leak.

Disassembly

Refer to **Figure 15** for this procedure.

1. Remove the throttle lever tensioner screw, spring and washer.

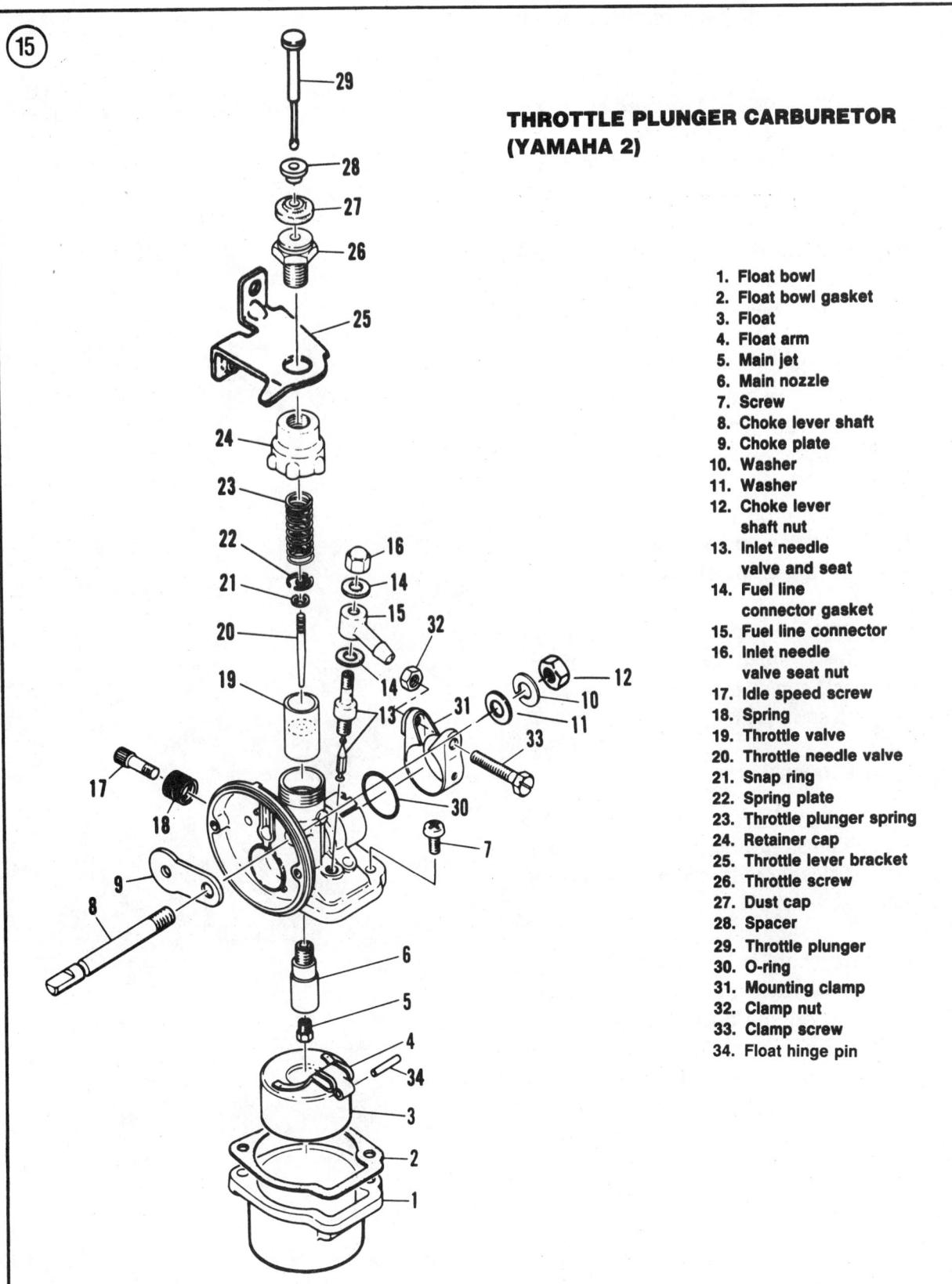

THROTTLE PLUNGER CARBURETOR (YAMAHA 2)

1. Float bowl
2. Float bowl gasket
3. Float
4. Float arm
5. Main jet
6. Main nozzle
7. Screw
8. Choke lever shaft
9. Choke plate
10. Washer
11. Washer
12. Choke lever shaft nut
13. Inlet needle valve and seat
14. Fuel line connector gasket
15. Fuel line connector
16. Inlet needle valve seat nut
17. Idle speed screw
18. Spring
19. Throttle valve
20. Throttle needle valve
21. Snap ring
22. Spring plate
23. Throttle plunger spring
24. Retainer cap
25. Throttle lever bracket
26. Throttle screw
27. Dust cap
28. Spacer
29. Throttle plunger
30. O-ring
31. Mounting clamp
32. Clamp nut
33. Clamp screw
34. Float hinge pin

6

2. Remove the snap ring and pin holding the throttle lever to the bracket. Remove the throttle lever.

3. Remove float bowl and gasket from carburetor body (**Figure 16**). Discard gasket.

4. Remove the float. Invert carburetor and slide float hinge pin from float arm. Remove float arm and needle valve (**Figure 17**).

5. Remove main nozzle (A) and main jet (B) assembly (**Figure 18**).

6. Unscrew inlet valve seat nut from valve seat. Remove fuel line connector and gaskets from valve seat. See **Figure 19**.

7. Remove the valve seat (**Figure 20**).

8. Loosen throttle screw and unscrew retainer cap. Remove cap with bracket and valve assembly (**Figure 21**).

9. Disconnect throttle plunger from needle valve. Remove valve and spring plate. Make sure tiny snap ring is intact on needle valve. See **Figure 22**.

10. Lightly seat idle adjust screw, counting number of turns required for reassembly reference, then back screw out and remove from carburetor with spring (**Figure 23**).

Cleaning and Inspection

A sealing compound is used to eliminate porosity problems. Submersion in a hot tank or carburetor cleaner will remove this sealing compound.

1. Wipe the carburetor casting and linkage with a cloth moistened in solvent to remove any contamination and operating film.

2. Clean the carburetor castings and metal parts with aerosol solvent and a brush. Spray the aerosol solvent on the casting and scrub off any gum or varnish with a small bristle brush.

CAUTION
Do not clean carburetor passages with a wire or drill bit. This can enlarge the passages and change the carburetor calibration.

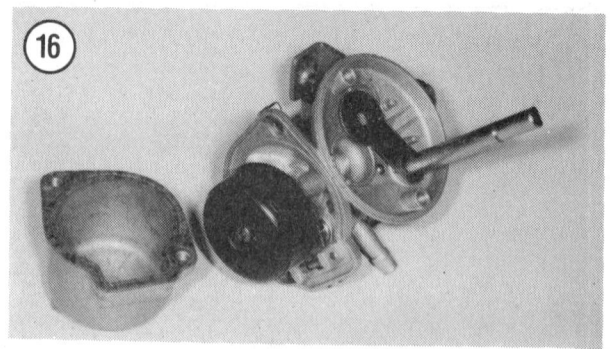

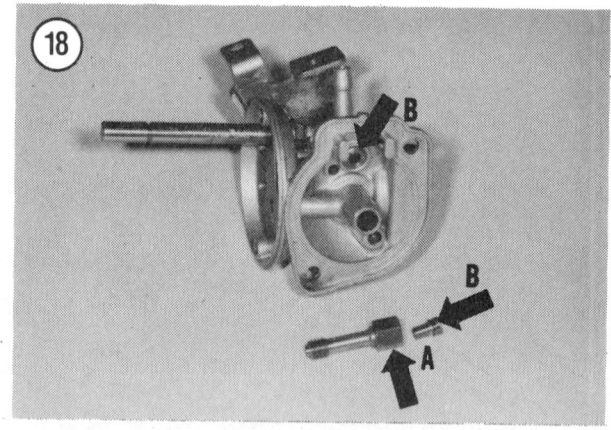

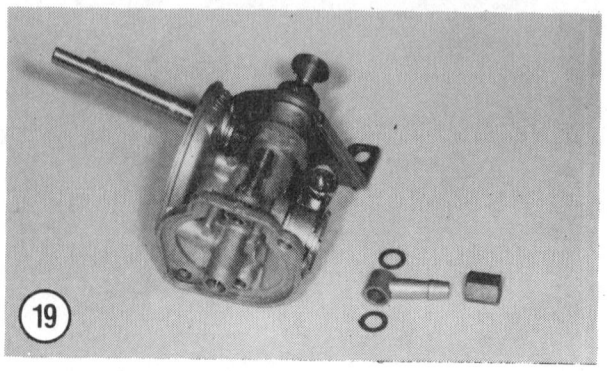

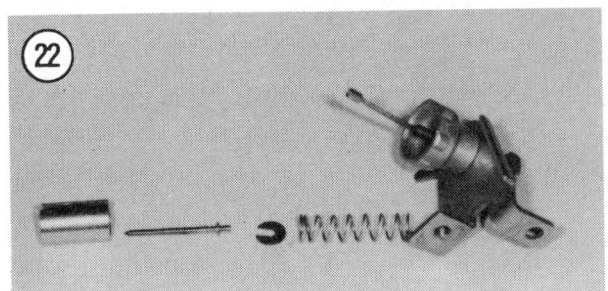

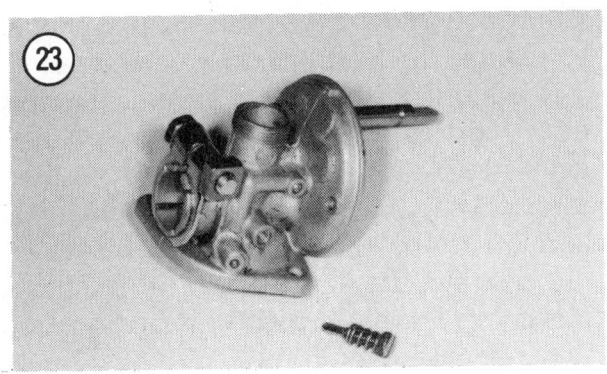

3. Spray cleaner through the metering passages in the casting.

4. Rinse the carburetor components thoroughly in clean solvent after removing them from the cleaning solution.

5. Blow the carburetor casting and passages dry with low-pressure (25 psi or less) compressed air. The use of higher pressures can damage the sealing compound.

6. Check carburetor body and fuel bowl for stripped threads, cracks or other defects. Replace as required.

7. Check the float for fuel saturation, deterioration or excessive wear where it contacts the float arm. Replace as required.

8. Check float arm for excessive wear where it contacts the needle valve. Replace as required.

9. Check inlet needle valve tip and seat contact areas for grooving, nicks, scratches or excessive wear. Replace as required.

10. Check the throttle needle tip for grooving, nicks or scratches. Replace the throttle needle if tip is damaged.

11. Check the throttle and choke shafts for excessive wear or play. The throttle and choke valves must move freely without binding. Replace carburetor if any of these defects is noted.

12. Clean all gasket residue from mating surfaces and remove any nicks, scratches or slight distortion with a surface plate and emery cloth.

Reassembly

Refer to **Figure 15** for this procedure.

1. Slide the spring onto the idle mixture screw. Install the screw in the carburetor body until it lightly seats, then back out the number of turns noted during disassembly.

2. Make sure circlip is installed in the second notch from the top of the throttle needle valve. Fit needle valve inside throttle valve, then insert the spring plate.

3. Place spring over plunger cable. Compress spring and connect plunger to throttle valve by inserting plunger cable through slot in side of valve.

4. Install throttle valve assembly to carburetor, positioning slot in valve over the idle mixture screw. Thread retainer cap onto the carburetor body and tighten securely.

5. Install the inlet needle valve seat.

6. Install a gasket over the needle valve seat assembly, then thread the fuel line connector on the seat. Place other gasket over fuel line connector and install needle valve nut.

7. Thread main nozzle into carburetor body. Throttle needle valve must enter nozzle.

8. Install main jet in main nozzle and tighten securely with a suitable screwdriver blade.

9. Insert inlet needle valve in float arm slot. Install float arm and needle valve assembly in carburetor. Secure with float hinge pin.

CAUTION
*Bend the float adjustment arm or tang carefully when adjustment is required—do **not** press down on the float. Downward pressure on the float will press the inlet needle tip into its seat and can damage the tip surface.*

10. On 1984-1988 models, invert carburetor body and adjust float arm height to 0.157 ±0.020 in. above the float chamber gasket surface. See **Figure 24**. On 1989 models, a one-piece float arm and float assembly is used. With the inlet needle and float assembly installed, the base of the float should be parallel with the float bowl gasket surface on the carburetor body. Bend the float arm to adjust.

11. Install a new float bowl gasket on the carburetor.

12. Place float over float arm and install float bowl. Tighten bowl screws to 1.1 ft.-lb. (1.5 N•m).

13. Insert mounting pin through throttle lever and bracket. Install snap ring.

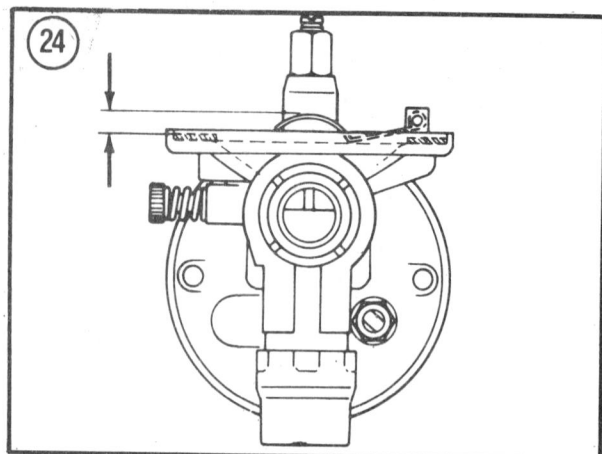

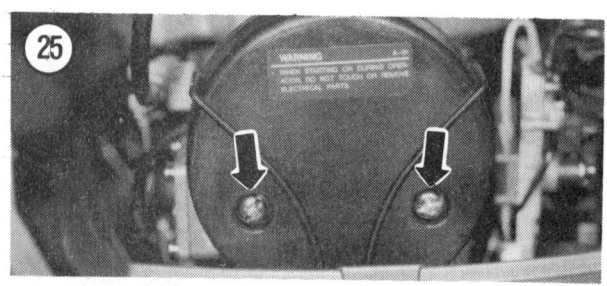

14. Install spring and washer over throttle lever tensioner screw. Install screw and tighten until throttle lever can be moved with light force. This establishes the throttle lever bracket position.

15. Tighten retainer cap throttle screw.

Mid-range Throttle Valve Needle Adjustment

The position of the circlip on the throttle valve needle determines the proper low- and medium-speed range mixture. The standard setting (the 2nd groove from the top) should be suitable for most operating conditions.

If a too-rich or too-lean condition results from extreme changes in elevation, temperature or humidity, the circlip position can be changed. The closer the circlip is installed to the top of the needle, the leaner the mixture. Remember that it is always preferable to run a slightly richer mixture than one that is too lean.

INTEGRAL FUEL PUMP CARBURETOR

An integral carburetor and fuel pump assembly is used on Yamaha 3, 4, 5, 6, 8, 9.9, 15, 25 (1988-on) and and 30 (1987-on) hp models. Only the bottom carburetor is equipped with a fuel pump on 25 hp models and only the center carburetor is equipped with a fuel pump on 30 hp models. The carburetor on Yamaha 4 and 5 hp models has the main jet attached to the main nozzle. The carburetor on Yamaha 9.9 and 15 hp models has the main jet installed in the bottom of the float bowl. The carburetor on Yamaha 3, 6, 8, 25 and 30 hp models has the main jet screwed into the carburetor body and is used to retain the main nozzle. The fuel pump on 3, 6, 8, 25 and 30 hp models is attached to the float bowl. The fuel pump on all other models is attached to the carburetor body.

Removal/Installation

1. 6-8 hp—Remove the screws holding the front panel. Remove the panel.

2. Remove the air intake cover. **Figure 25** shows the 4-5 hp cover. The 6-8 hp, 9.9-15 hp and 25-30 hp covers are shaped differently but are removed in a similar manner.

3. Disconnect the choke lever and bushing from carburetor linkage. See **Figure 26** (typical).

4. Disconnect the throttle cable or linkage at the carburetor. **Figure 27** shows the 4 and 5 hp cable arrangement; 3 hp is similar. The 6 and 8 hp linkage is connected to the throttle lever on the port side of the carburetor. The 9.9 and 15 hp linkage is similar. Disconnect linkage from carburetor being serviced on 25 and 30 hp models or remove carburetors as an assembly.

5. Disconnect the fuel line at the fuel pump inlet. Plug the line to prevent leakage.

6. Disconnect the crankcase recirculation lines at the carburetor, if so equipped.

7. Remove the port side mounting nut and washer. Partially loosen the starboard nut, pull the carburetor forward and slide it to the port side. Remove the carburetor and gasket. Discard the gasket.

8. Installation is the reverse of removal. Use a new gasket. Tighten mounting nuts securely to prevent carburetor warpage or an air leak.

Disassembly

Refer to **Figure 28** (4 and 5 hp), **Figure 29A** (3, 6 and 8 hp), **Figure 29B** (9.9 and 15 hp), **Figure 30A** (25 hp) and **Figure 30B** (30 hp) for this procedure.

1. Disassemble the fuel pump components (**Figure 31**) as described in this chapter.

6

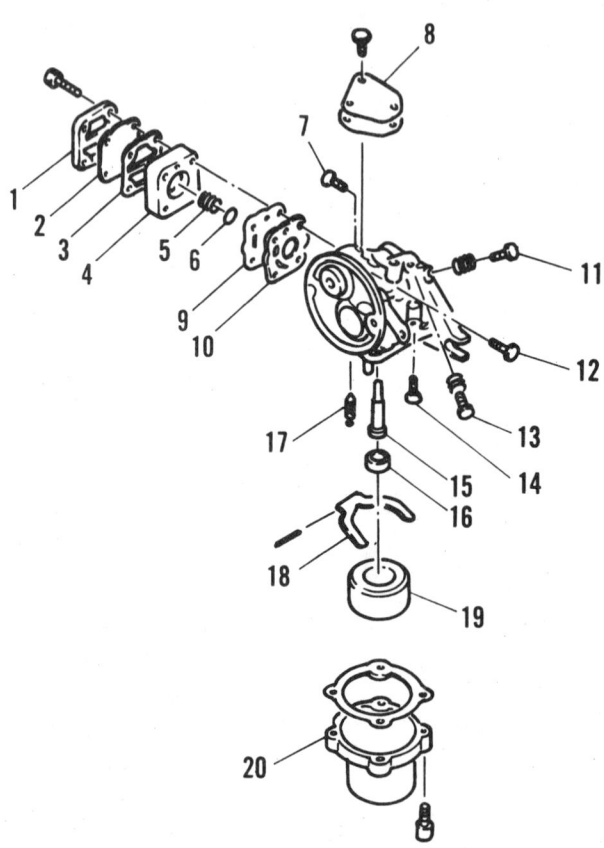

INTEGRAL FUEL PUMP CARBURETOR (YAMAHA 4 AND 5)

1. Fuel pump cover
2. Outer diaphragm
3. Pump cover gasket
4. Pump body
5. Spring
6. Plate
7. Guide screw
8. Cover
9. Inner diaphragm
10. Pump body gasket
11. Idle mixture screw
12. Screw
13. Idle speed screw
14. Idle jet
15. Main nozzle
16. Main jet
17. Inlet needle valve
18. Float arm
19. Float
20. Float chamber

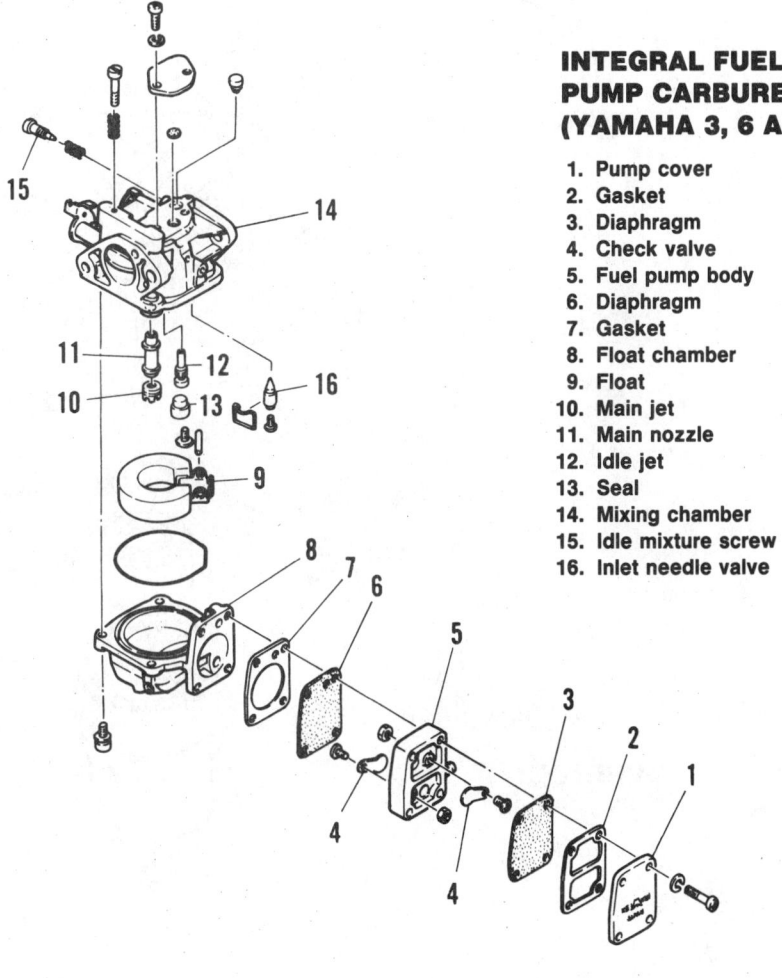

INTEGRAL FUEL PUMP CARBURETOR (YAMAHA 3, 6 AND 8)

1. Pump cover
2. Gasket
3. Diaphragm
4. Check valve
5. Fuel pump body
6. Diaphragm
7. Gasket
8. Float chamber
9. Float
10. Main jet
11. Main nozzle
12. Idle jet
13. Seal
14. Mixing chamber
15. Idle mixture screw
16. Inlet needle valve

6

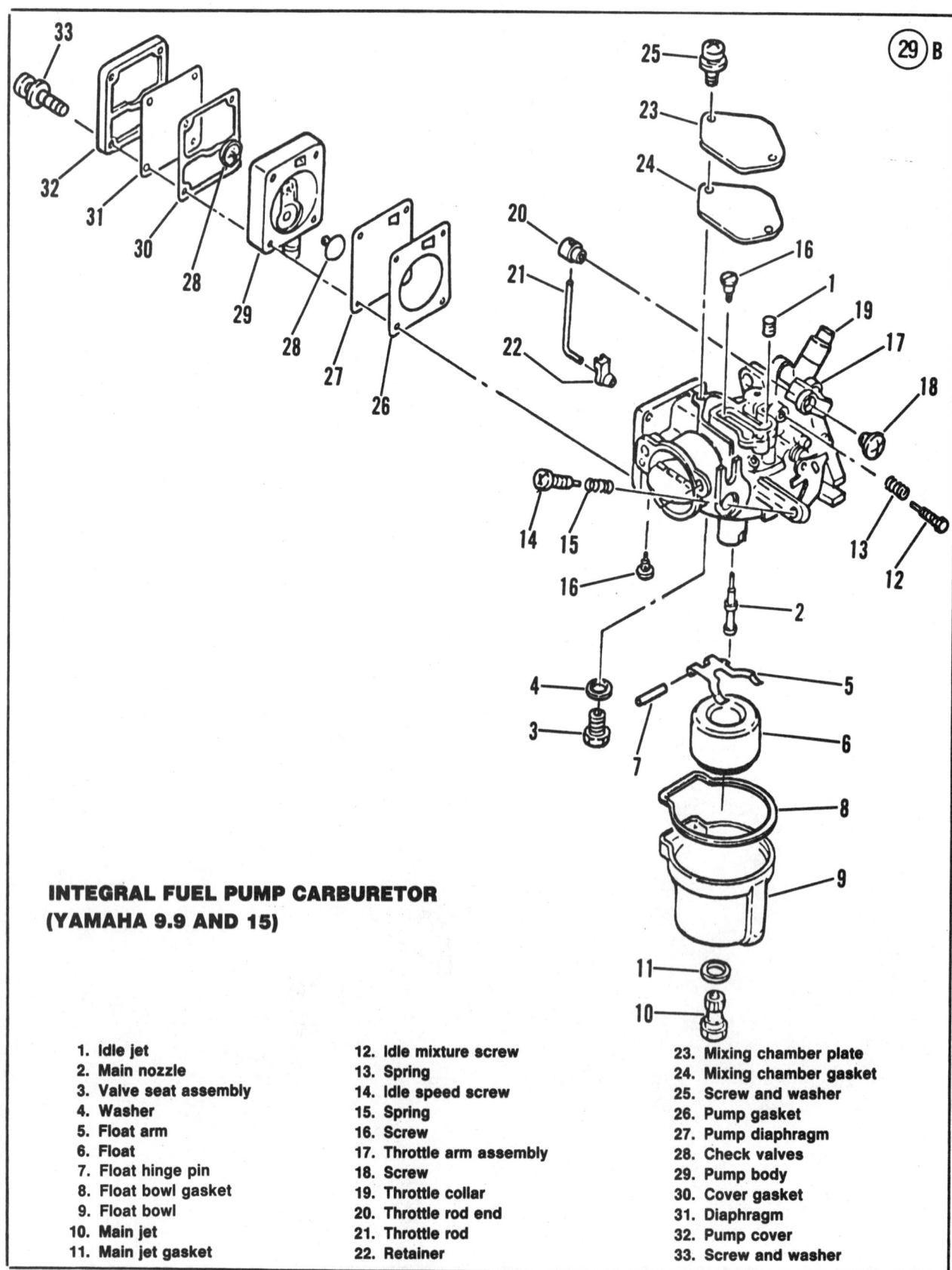

**INTEGRAL FUEL PUMP CARBURETOR
(YAMAHA 9.9 AND 15)**

1. Idle jet	12. Idle mixture screw	23. Mixing chamber plate
2. Main nozzle	13. Spring	24. Mixing chamber gasket
3. Valve seat assembly	14. Idle speed screw	25. Screw and washer
4. Washer	15. Spring	26. Pump gasket
5. Float arm	16. Screw	27. Pump diaphragm
6. Float	17. Throttle arm assembly	28. Check valves
7. Float hinge pin	18. Screw	29. Pump body
8. Float bowl gasket	19. Throttle collar	30. Cover gasket
9. Float bowl	20. Throttle rod end	31. Diaphragm
10. Main jet	21. Throttle rod	32. Pump cover
11. Main jet gasket	22. Retainer	33. Screw and washer

30 A

INTEGRAL FUEL PUMP
CARBURETOR (YAMAHA 25)

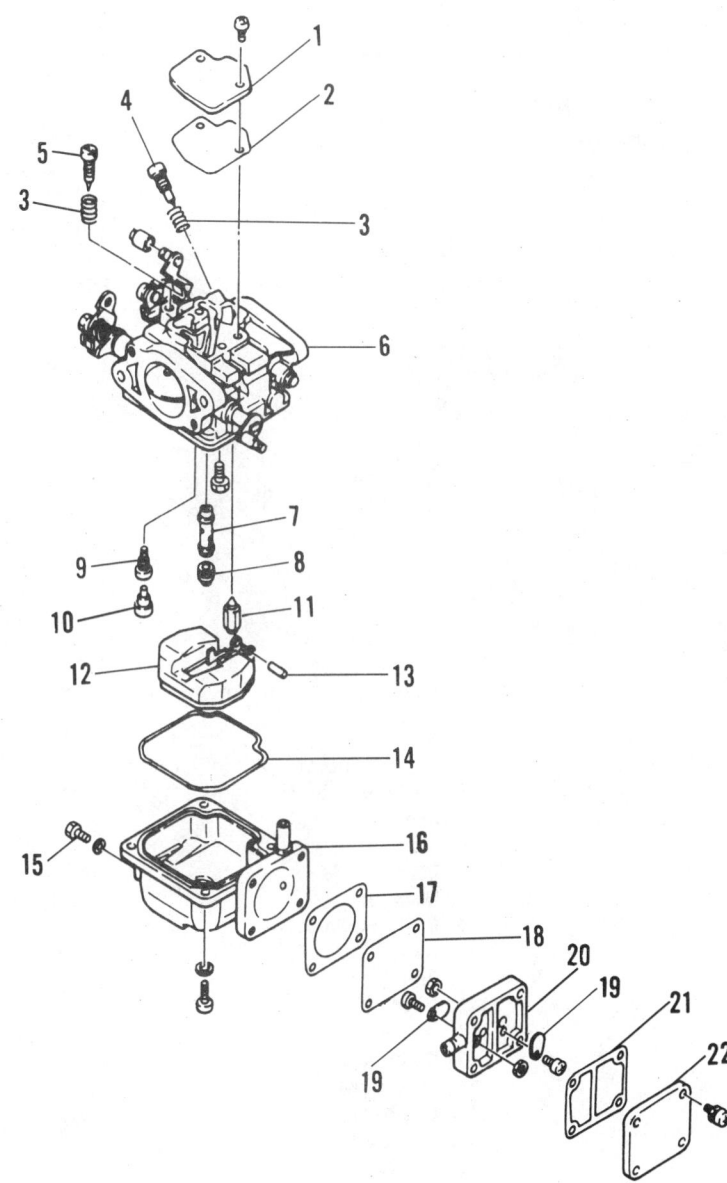

1. Cover
2. Gasket
3. Spring
4. Idle mixture screw
5. Idle speed screw
6. Carburetor body
7. Main nozzle
8. Main jet
9. Idle jet
10. Plug
11. Inlet needle valve
12. Float
13. Pin
14. Seal
15. Drain plug
16. Float bowl
17. Gasket
18. Diaphragm
19. Check valve
20. Fuel pump body
21. Gasket
22. Pump cover

6

2. Remove the mixing chamber plate and gasket (**Figure 32**). Discard the gasket.

NOTE
The main jet on all models except Yamaha 9.9 and 15 carburetors is installed in the main nozzle or carburetor body instead of in the float bowl. Remove the jet on these models after Step 4.

3. Remove the main jet and gasket from the float bowl, then separate the float bowl from the carburetor (**Figure 33**). Remove and discard the float bowl gasket.

4. Remove the float. Slide the float hinge pin to one side and remove the float arm, hinge pin and needle valve. See **Figure 34**.

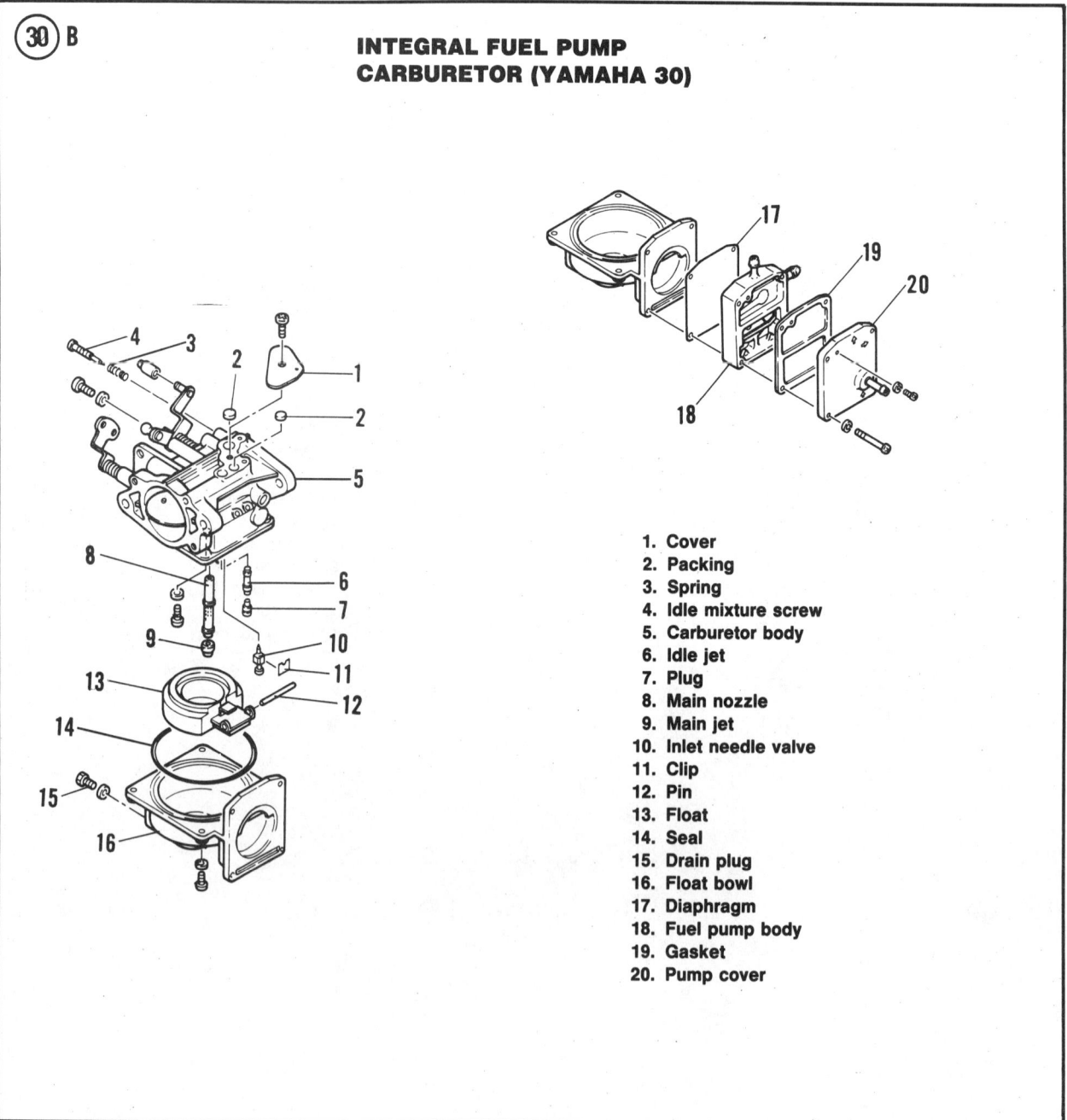

**INTEGRAL FUEL PUMP
CARBURETOR (YAMAHA 30)**

1. Cover
2. Packing
3. Spring
4. Idle mixture screw
5. Carburetor body
6. Idle jet
7. Plug
8. Main nozzle
9. Main jet
10. Inlet needle valve
11. Clip
12. Pin
13. Float
14. Seal
15. Drain plug
16. Float bowl
17. Diaphragm
18. Fuel pump body
19. Gasket
20. Pump cover

5. Remove the main nozzle (A) with a suitable screwdriver. Remove needle valve seat (B) with a suitable socket or open-end wrench. See **Figure 35**.

6. Remove the idle (pilot) jet (A, **Figure 36**).

7. Lightly seat the idle mixture screw (B, **Figure 36**), counting the number of turns required for reassembly reference, then back

screw out and remove from carburetor with spring.

Cleaning and Inspection

A sealing compound is used to eliminate porosity problems. Submersion in a hot tank

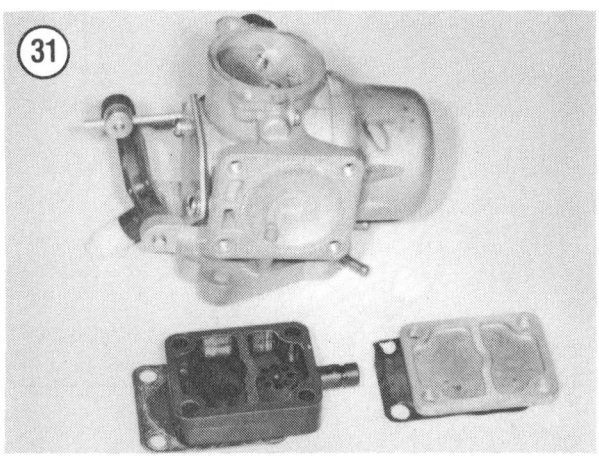

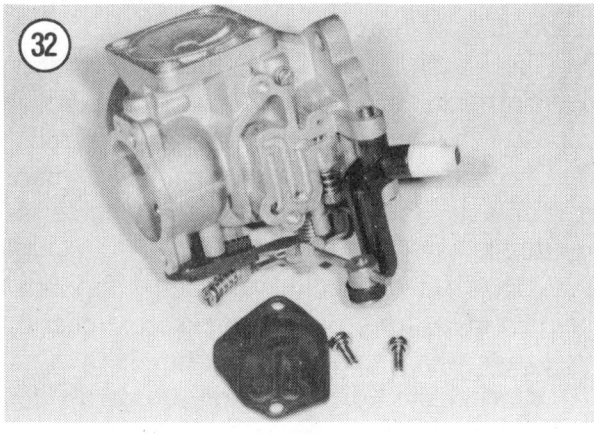

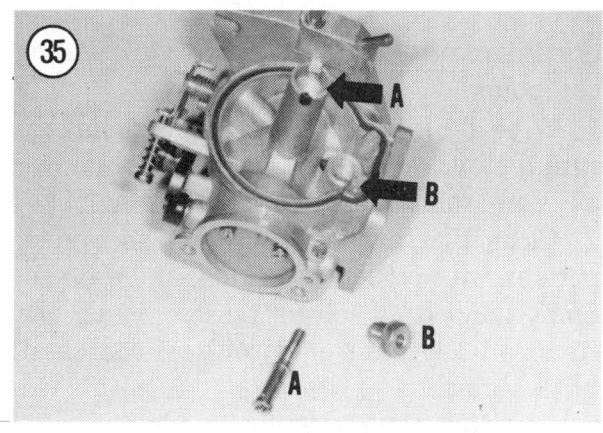

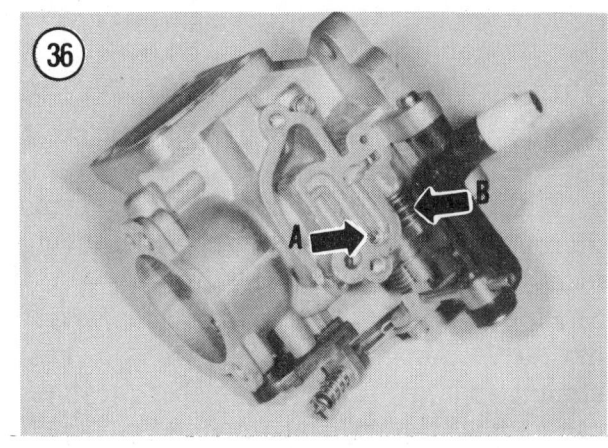

6

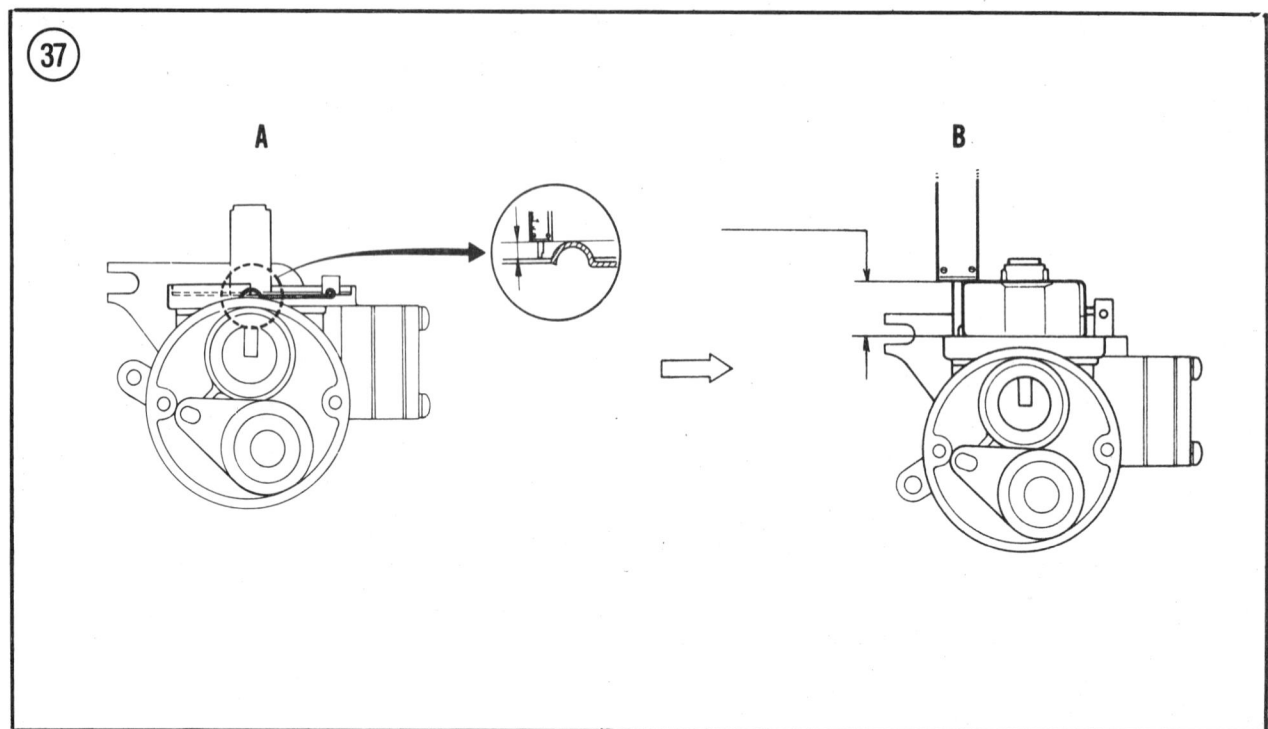

or carburetor cleaner will remove this sealing compound.

1. Wipe the carburetor casting and linkage with a cloth moistened in solvent to remove any contamination and operating film.

2. Clean the carburetor castings and metal parts in with an aerosol solvent and a brush. Spray the aerosol solvent on the casting and scrub off any gum or varnish with a small bristle brush.

> *CAUTION*
> *Do not clean carburetor passages with a wire or drill bit. This can enlarge the passages and change the carburetor calibration.*

3. Spray cleaner through the metering passages in the casting.

4. Rinse the carburetor components thoroughly in clean gasoline after removing them from the cleaning solution.

5. Blow the carburetor casting and passages dry with low-pressure (25 psi or less) compressed air. The use of higher pressures can damage the sealing compound.

6. Check carburetor body and fuel bowl for stripped threads, cracks or other defects. Replace as required.

7. Check the float for fuel saturation, deterioration or excessive wear where it contacts the float arm. Replace as required.

8. Check float arm for excessive wear where it contacts the needle valve. Replace as required.

9. Check inlet needle valve tip and seat contact areas for grooving, nicks, scratches or excessive wear. Replace as required.

10. Check the idle mixture screw tip for grooving, nicks or scratches. Replace the screw if tip is damaged.

11. Check the throttle and choke shafts for excessive wear or play. The throttle and choke valves must move freely without binding. Replace carburetor if any of these defects is noted.

12. Clean all gasket residue from mating surfaces and remove any nicks, scratches or slight distortion with a surface plate and emery cloth.

Reassembly

Refer to **Figure 28** (4 and 5 hp), **Figure 29A** (3, 6 and 8 hp), **Figure 29B** (9.9 and 15 hp), **Figure 30A** (25 hp) and **Figure 30B** (30 hp) procedure.

1. Install the idle (pilot) jet (A, **Figure 36**).
2. Slide spring over idle mixture screw. Install screw until it seats lightly, then back out the number of turns noted during disassembly. See B, **Figure 36**.
3. Install mixing chamber plate with a new gasket (**Figure 32**). Tighten screws securely.
4. Install a new float bowl gasket.
5. Install needle valve seat with a new gasket (B, **Figure 35**). Tighten snugly and drop needle valve into seat.
6. Install main nozzle (A, **Figure 35**). On all models except Yamaha 9.9 and 15, install the main jet.
7. Install float arm and pin in float bowl. Install float.
8. Invert carburetor and check float level as follows:

   a. 1984-1985 Yamaha 4 and 5—The float arm hump should be 0.55 ±0.08 in. above the float bowl casting surface. See A, **Figure 37**.
   b. 1986-on Yamaha 4—Distance between float bowl casting and top of float should be 0.87 ±0.02 in. See B, **Figure 37**.
   c. Yamaha 3, 6 and 8—Distance between float bowl casting and top of float should be 0.55 ± 0.08 in. See **Figure 38**.
   d. 1984-1985 Yamaha 9.9-15—The float arm hump should be 0.078 ±0.02 in. above the float bowl casting surface. See A, **Figure 39**.
   e. 1986-on Yamaha 9.9 and 15—Distance between float bowl casting and top of float should be 0.79 ±0.02 in. See B, **Figure 39**.
   f. Yamaha 25—Distance between float bowl casting and top of float should be 0.57 in. See **Figure 38**.
   g. Yamaha 30—Distance between float bowl casting and top of float should be 0.69 ±0.02 in. See **Figure 38**.

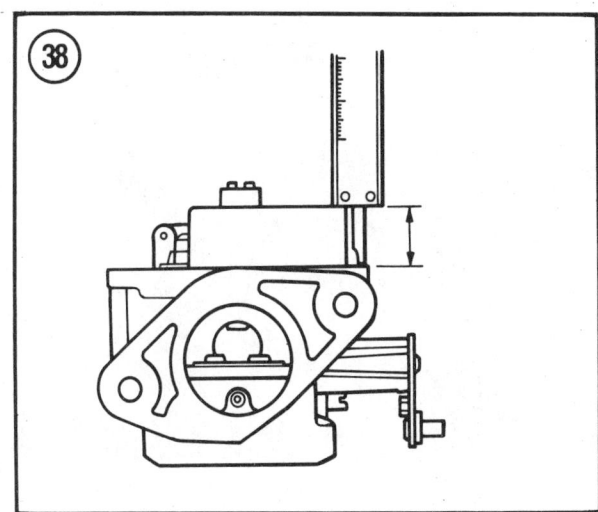

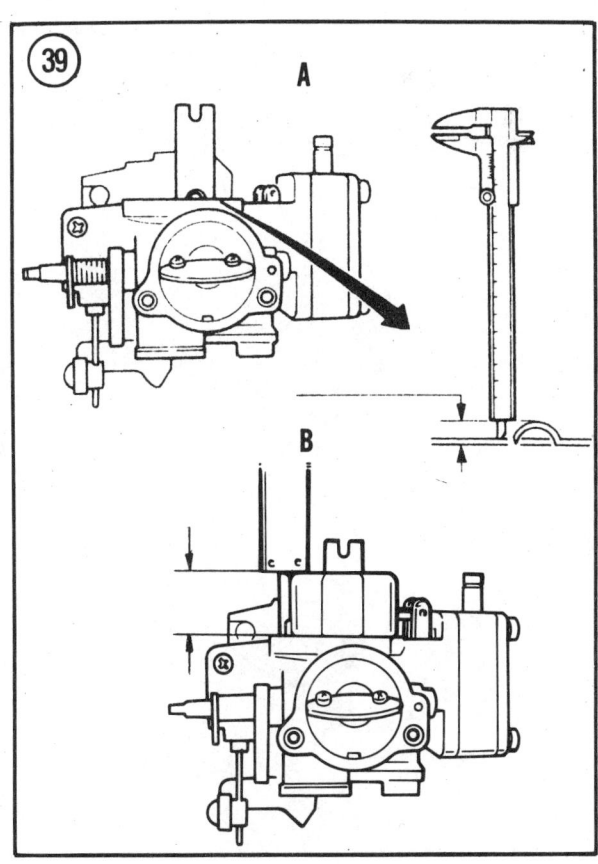

CAUTION
*Bend the float adjustment arm or tang carefully when adjustment is required—do **not** press down on the float. Downward pressure on the float will press the inlet needle tip into its seat and can damage the tip surface.*

9. If float level is incorrect, bend float arm behind hump as required to bring it within specifications.

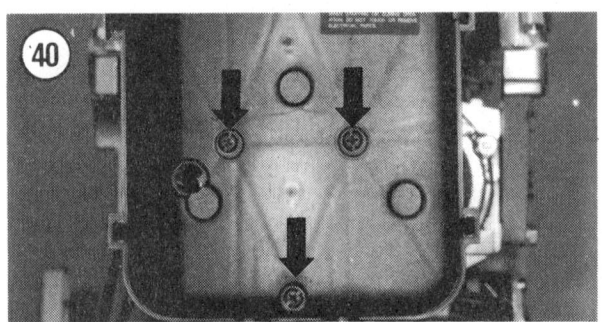

10. Install float bowl to main body. Install attaching screws on Yamaha 3, 4, 5, 6, 8, 25 and 30 carburetors. Install main jet with a new gasket and tighten snugly on 9.9 and 15 carburetors.

CENTER BOWL CARBURETOR (YAMAHA 25 [1984-1987] AND 30 [1984-1986])

This carburetor design is used on Yamaha 25 and 30 hp models.

Removal/Installation

1. Disconnect the choke link rod from the choke lever.
2. Remove 3 screws holding the air silencer cover and choke lever to the carburetor (**Figure 40**). Disconnect silencer cover from choke lever and remove from carburetor.

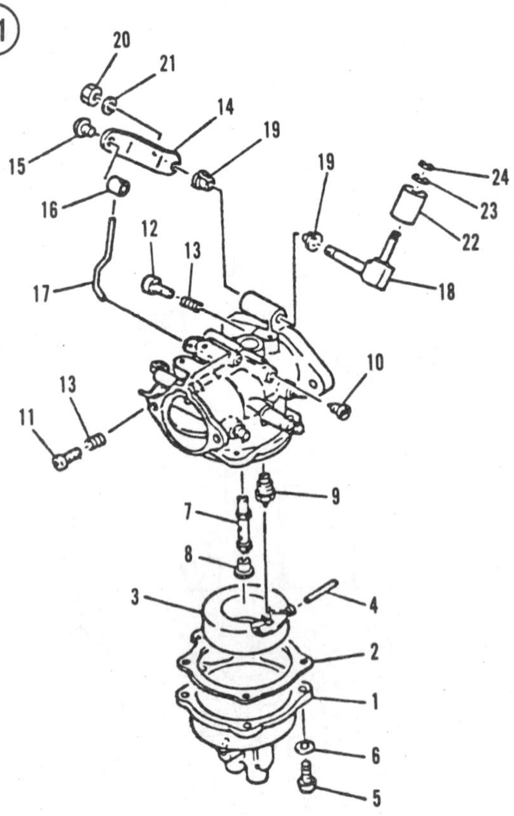

CENTER BOWL CARBURETOR (YAMAHA 25 [1984-1987] AND 30 [1984-1986])

1. Float bowl
2. Gasket
3. Float
4. Float pin
5. Screw
6. Washer
7. Main nozzle
8. Main jet
9. Inlet needle valve assembly
10. Idle jet
11. Idle speed screw
12. Idle mixture screw
13. Spring
14. Throttle rod lever
15. Screw
16. Throttle rod retainer
17. Throttle rod
18. Throttle arm
19. Bushing
20. Nut
21. Washer
22. Throttle arm collar
23. Washer
24. Snap ring

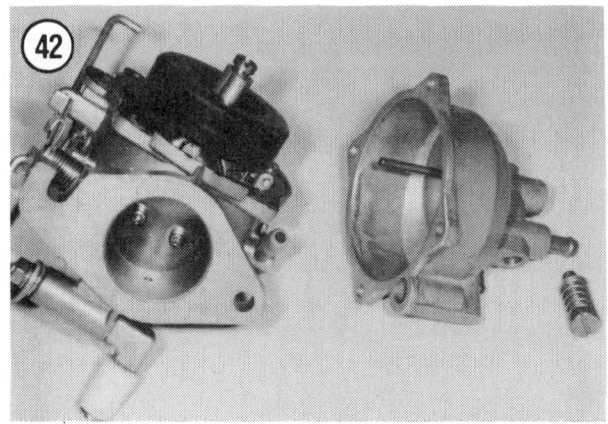

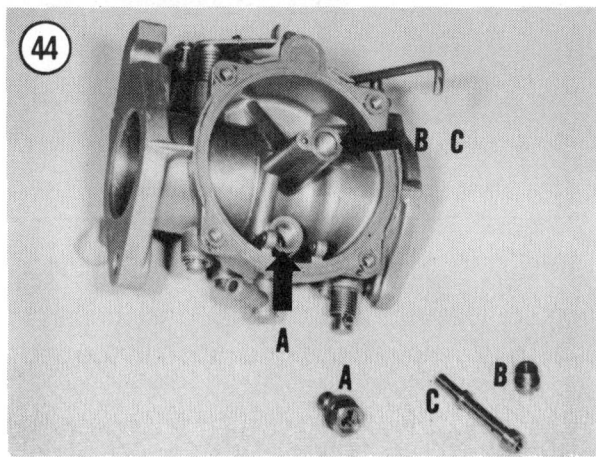

3. 30 hp—Disconnect the choke solenoid lead connector. Remove the solenoid link O-ring. Remove the link. Remove the 2 solenoid screws. Remove the solenoid.

4. Disconnect the fuel line at the carburetor. Plug the line to prevent leakage.

5. Remove the mounting nuts and washers. Remove carburetor and gasket. Discard the gasket.

6. Installation is the reverse of removal. Use a new gasket and tighten mounting nuts evenly and securely to prevent carburetor warpage or an air leak.

Disassembly

Refer to **Figure 41** for this procedure.

1. Remove the float bowl from the carburetor. See **Figure 42**.

2. Remove and discard the float bowl gasket (A, **Figure 43**).

3. Slide float hinge pin in direction of arrow cast on float hinge. Remove float and hinge pin assembly from carburetor (B, **Figure 43**).

CAUTION
Inlet needle is permanently installed in valve seat. Do not try to remove needle individually in Step 4. If replacement is required, the entire assembly is replaced.

4. Remove inlet needle valve assembly (A, **Figure 44**). Discard the gasket.

CAUTION
Use a jet remover or an appropriate size screwdriver to remove the components in Step 5. If the screwdriver used to remove the main nozzle is too large, it may damage the housing threads and prevent proper installation of the main jet.

5. Remove main jet (B) and main nozzle (C). See **Figure 44**.

6. Remove the idle (pilot) jet (**Figure 45**).

6

7. Lightly seat idle mixture screw, counting number of turns required for reassembly reference, then back screw out and remove from carburetor with spring (**Figure 46**).

Cleaning and Inspection

A sealing compound is used to eliminate porosity problems. Submersion in a hot tank or carburetor cleaner will remove this sealing compound. Clean with an aerosol cleaner.

1. Wipe the carburetor casting and linkage with a cloth moistened in solvent to remove any contamination and operating film.
2. Clean the carburetor castings and metal parts with an aerosol solvent and a brush. Spray the aerosol solvent on the casting and scrub off any gum or varnish with a small bristle brush.

> *CAUTION*
> *Do not clean carburetor passages with a wire or drill bit. This can enlarge the passages and change the carburetor calibration.*

3. Spray cleaner through the metering passages in the casting.
4. Rinse the carburetor components thoroughly in clean solvent after cleaning them.
5. Blow the carburetor casting and passages dry with low-pressure (25 psi or less) compressed air. The use of higher pressures can damage the sealing compound.
6. Check carburetor body and fuel bowl for stripped threads, cracks or other defects. Replace as required.
7. Check the float for fuel saturation, deterioration or excessive wear where it contacts the float arm. Replace as required.
8. Check float arm for excessive wear where it contacts the needle valve. Replace as required.

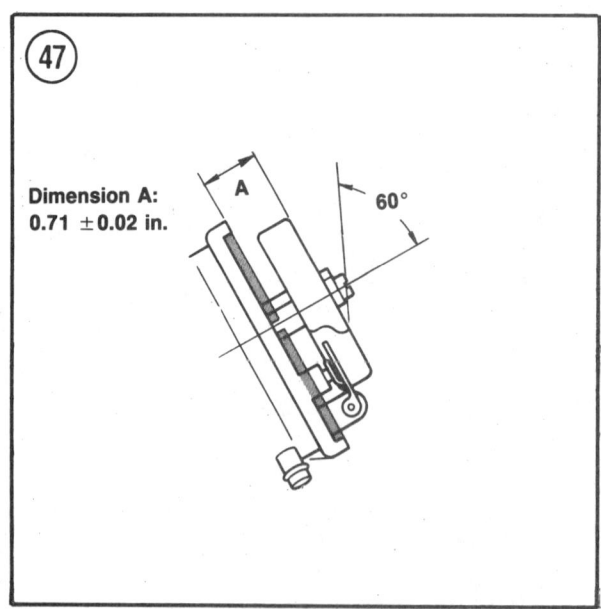

Dimension A:
0.71 ±0.02 in.

60°

A

9. Check inlet needle valve tip and seat contact areas for grooving, nicks, scratches or excessive wear. Replace the carburetor body as required.
10. Check the idle mixture screw tip for grooving, nicks or scratches. Replace the screw if tip is damaged.
11. Check the throttle and choke shafts for excessive wear or play. The throttle and choke valves must move freely without binding. Replace carburetor if any of these defects is noted.
12. Clean all gasket residue from mating surfaces and remove any nicks, scratches or slight distortion with a surface plate and emery cloth.

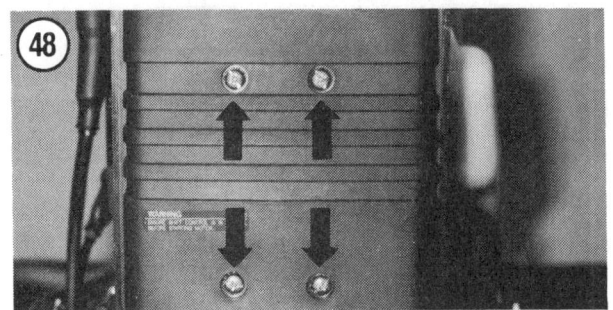

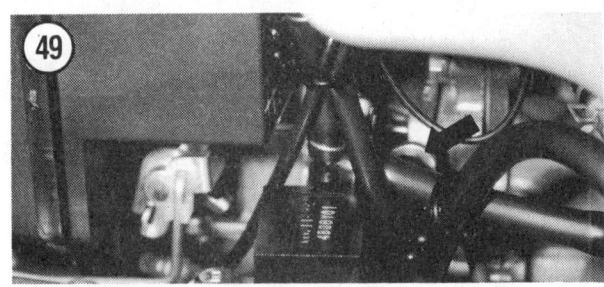

opposite side of the hinge. It should be 0.71 ±0.02 in. See **Figure 47**. If it is not, remove the float and bend the hinged tang on the float as required to position the float properly.

5. Press a new gasket into the float bowl gasket groove and make sure it is properly seated, then install the float bowl to the carburetor body. Tighten the screws evenly and securely to prevent warpage.

6. Slide the spring onto the idle mixture screw. Install the screw in the carburetor body until it lightly seats, then back it out the number of turns noted during disassembly.

7. Install the idle (pilot) jet.

CENTER BOWL CARBURETOR (40-90 HP)

Yamaha 40-90 hp models use one carburetor of this design for each cylinder.

Removal/Installation (40-50 hp)

1. Remove the 4 bolts holding the air silencer cover to the carburetor assembly (**Figure 48**). Remove the cover and O-rings.

2. Disconnect the choke lever, choke link and throttle rod at the carburetor.

3. Disconnect the fuel lines at the carburetors. Plug lines to prevent leakage.

4. Disconnect the choke solenoid lead connector, if so equipped. Remove the choke solenoid link O-ring and disconnect the solenoid link from the choke lever.

5. Disconnect the oil pump operating rod at the carburetor (**Figure 49**), if so equipped.

NOTE
A single gasket is used with the bank of 3 carburetors.

6. Remove the mounting bolts and washers. Remove the carburetors and gasket. Discard the gasket.

7. Disconnect the connecting throttle linkage from each carburetor.

Reassembly

Refer to **Figure 41** for this procedure.

CAUTION
Use a jet remover or an appropriate size screwdriver to install the components in Step 1. If the screwdriver used to install the main nozzle is too large, it may damage the housing threads and prevent proper installation of the main jet.

1. Install the main nozzle, then install the main jet.

2. Install inlet needle valve assembly with a new gasket.

3. Position float in fuel chamber, then install float hinge pin from the side of the cast hinge marked with an arrow.

CAUTION
Bend the float adjustment arm or tang carefully when adjustment is required—do not press down on the float. Downward pressure on the float will press the inlet needle tip into its seat and can damage the tip surface.

4. Invert the carburetor and incline at a 60° angle. Measure the float height on the

6

8. Installation is the reverse of removal. Use a new gasket and tighten mounting bolts evenly and securely to prevent carburetor warpage or an air leak.

Removal/Installation (70 hp)

1. Disconnect the oil line between the oil tank and injection pump at the tank. Plug the line with an M6 bolt or suitable equivalent to prevent oil leakage.

2. Remove the oil tank mounting bolts. Disconnect the green/red oil level sensor lead (**Figure 50**) and remove the oil tank.

3. Remove the 4 bolts holding the air silencer cover to the carburetor assembly. Remove the cover and O-rings.

4. Disconnect the choke link and accelerator link rods at the carburetors.

5. Remove the choke solenoid link O-ring and disconnect the solenoid link from the choke lever.

6. Disconnect the oil pump operating rod at the carburetor.

7. Disconnect the fuel lines at the carburetors. Plug lines to prevent leakage.

> *NOTE*
> *A single gasket is used with the bank of 3 carburetors.*

8. Remove the mounting bolts and washers. Remove the carburetors, bracket assembly and gasket. Discard the gasket.

> *CAUTION*
> *The oil pump must be bled whenever the oil tank hose is disconnected or the engine may seize when started.*

9. Installation is the reverse of removal. Use a new gasket and tighten mounting bolts securely to prevent carburetor warpage or an air leak. Bleed the oil injection pump. See Chapter Twelve.

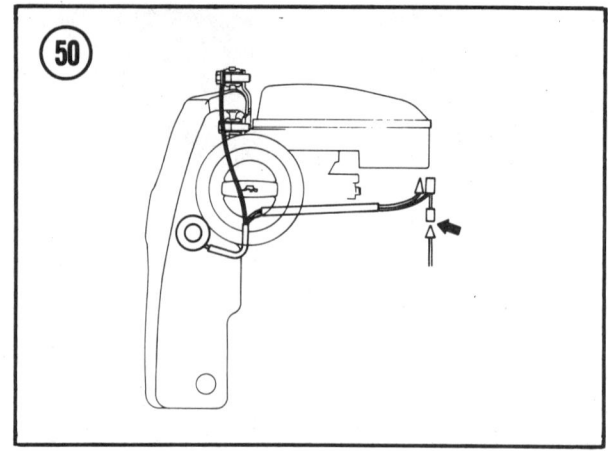

Removal/Installation (90 hp)

1. Disconnect the oil line between the oil tank and injection pump at the tank. Plug the line with an M6 bolt or suitable equivalent to prevent oil leakage.

2. Remove the oil tank mounting bolts. Disconnect the green/red oil level sensor lead and the ground lead (**Figure 51**), then remove the oil tank.

3. Remove the 4 bolts holding the air silencer cover to the carburetor assembly. Remove the cover and O-rings.

4. Remove the bolts holding the air silencer bracket to the carburetor assembly. Remove the bracket.

5. Disconnect the choke link and accelerator link rods at the carburetors.

6. Remove the choke solenoid link O-ring and disconnect the solenoid link from the choke lever.

7. Disconnect the oil pump operating rod at the carburetor.

8. Disconnect the fuel lines at the carburetors. Plug lines to prevent leakage.

> *NOTE*
> *A single gasket is used with the bank of 3 carburetors.*

9. Remove the mounting nuts and washers. Remove the carburetors, bracket assembly and gasket. Discard the gasket.

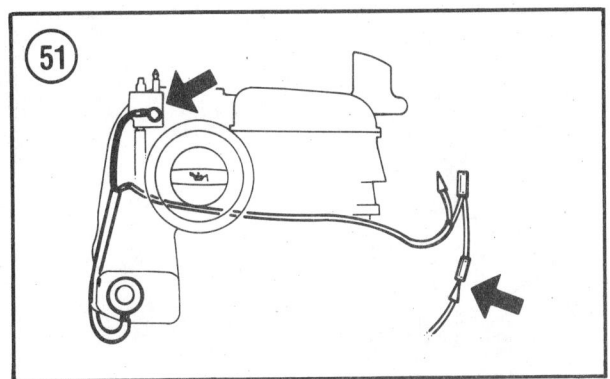

CAUTION
The oil pump must be bled whenever the oil tank hose is disconnected or the engine may seize when started.

10. Installation is the reverse of removal. Use a new gasket and tighten mounting nuts securely to prevent carburetor warpage or an air leak. Bleed the oil injection system. See Chapter Twelve.

Disassembly

Refer to **Figure 52** (typical) for this procedure.
1. Remove the drain screw from the carburetor float bowl. Drain the fuel into a suitable container and dispose of it safely.
2. Remove the float bowl from the carburetor.
3. Remove and discard the float bowl gasket.
4. Remove the setscrew holding the float hinge pin. Remove float and hinge pin assembly from carburetor with the needle valve.
5. Remove needle valve seat with a suitable socket wrench.

CAUTION
Use a jet remover or an appropriate size screwdriver to remove the components in Step 6. If the screwdriver used to remove the main nozzle is too large, it may damage the housing threads and prevent proper installation of the main jet.

6. Remove main jet (A) and main nozzle (B). See **Figure 53**.
7. Remove the cap over the idle (pilot) jet, if so equipped. Remove the jet.
8. Lightly seat idle mixture screw, counting number of turns required for reassembly reference, then back screw out and remove from carburetor with spring.
9. Remove the cover plate screws. Remove the cover plate. Remove the packings and circular plates housed in the carburetor casting.

Cleaning and Inspection

A sealing compound is used to eliminate porosity problems. Submersion in a hot tank or carburetor cleaner will remove this sealing compound. Clean the carburetor with aerosol solvent.
1. Wipe the carburetor casting and linkage with a cloth moistened in solvent to remove any contamination and operating film.
2. Clean the carburetor castings and metal parts with an aerosol solvent and a brush. Spray or aerosol solvent on the casting and scrub off any gum or varnish with a small bristle brush.

CAUTION
Do not clean carburetor passages with a wire or drill bit. This can enlarge the passages and change the carburetor calibration.

3. Spray cleaner through the metering passages in the casting.
4. Rinse the carburetor components thoroughly in clean solvent after cleaning them.
5. Blow the carburetor casting and passages dry with low-pressure (25 psi or less) compressed air. The use of higher pressures can damage the sealing compound.
6. Check carburetor body and fuel bowl for stripped threads, cracks or other defects. Replace as required.

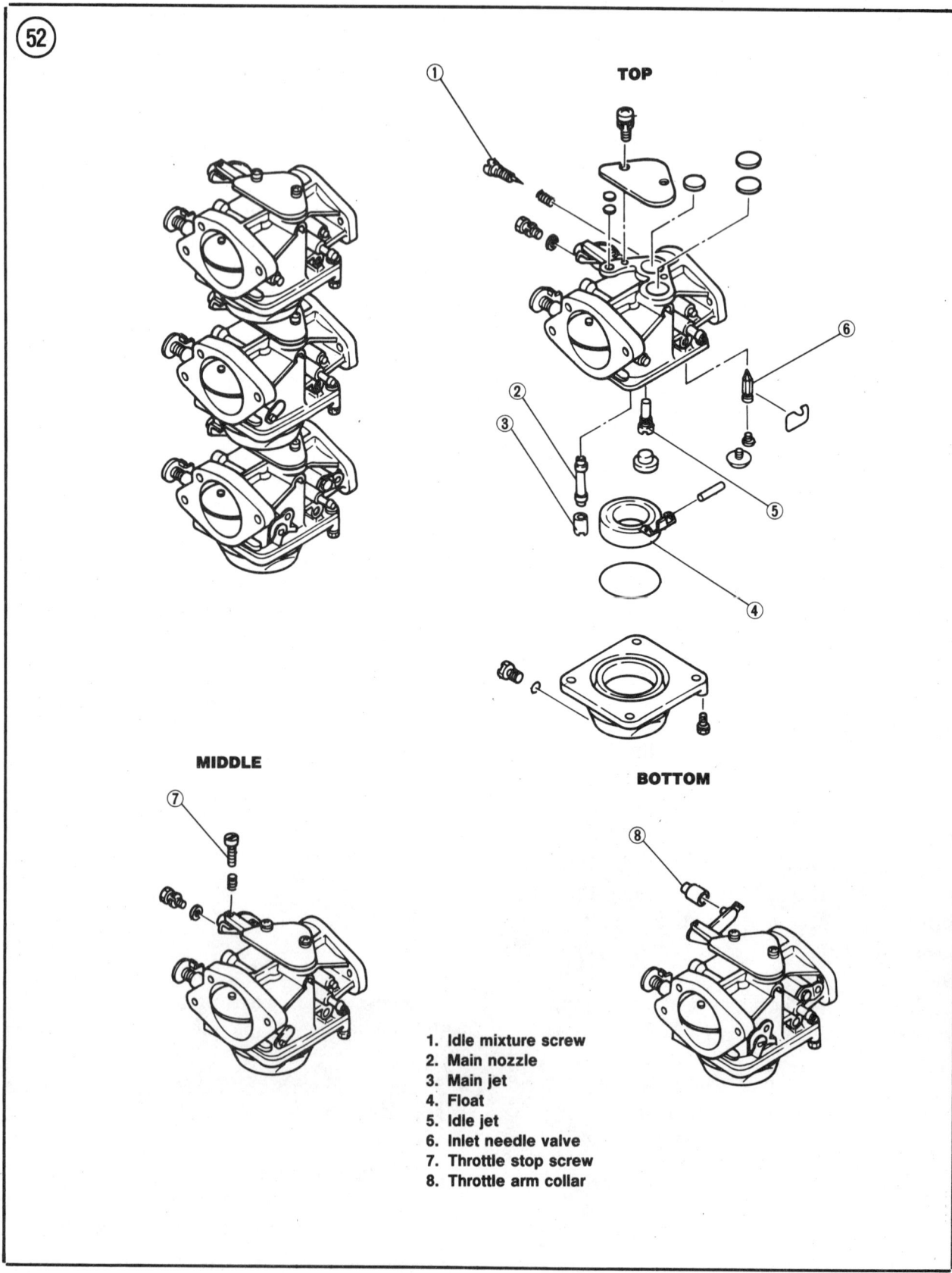

TOP

MIDDLE

BOTTOM

1. Idle mixture screw
2. Main nozzle
3. Main jet
4. Float
5. Idle jet
6. Inlet needle valve
7. Throttle stop screw
8. Throttle arm collar

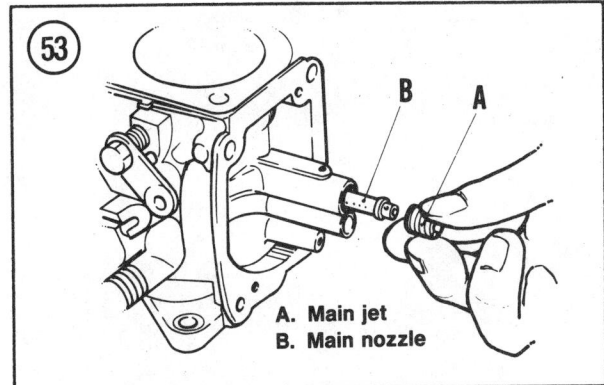

A. Main jet
B. Main nozzle

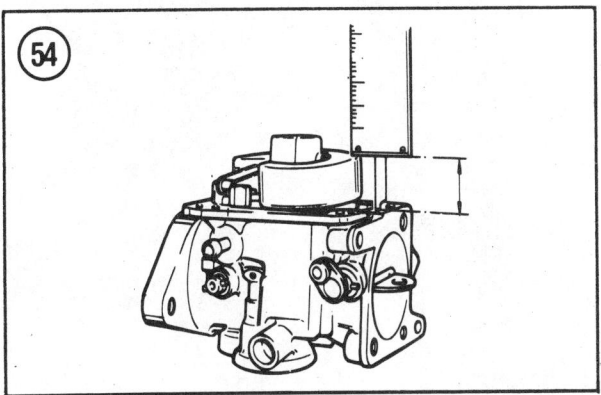

slight distortion with a surface plate and emery cloth.

Reassembly

Refer to **Figure 52** for this procedure.

> *CAUTION*
> *Use a jet remover or an appropriate size screwdriver to install the components in Step 1. If the screwdriver used to install the main nozzle is too large, it may damage the housing threads and prevent proper installation of the main jet.*

1. Install the main nozzle, then install the main jet. See **Figure 53**.
2. Install the idle (pilot) jet, then install the cap over the jet, if used.
3. Install the needle valve seat with a new gasket.
4. Attach needle valve to float hinge, then install float assembly in the float bowl with the hinge pin and tighten the setscrew securely.

> *CAUTION*
> *Bend the float adjustment arm or tang carefully when adjustment is required—do* ***not*** *press down on the float. Downward pressure on the float will press the inlet needle tip into its seat and can damage the tip surface.*

5. Invert the carburetor and measure the float height from the carburetor casting as shown in **Figure 54**. It should be 0.55 ±0.08 in. (40-70 hp) or 0.77 ±0.12 in. (90 hp). If it is not, remove the float and check the needle valve seat for proper installation. If the seat is properly installed, bend the hinged tang on the float as required to position the float properly.
6. Press a new O-ring gasket into the float bowl groove and make sure it is properly seated, then install the float bowl to the carburetor body. Tighten the screws evenly and securely to prevent warpage.

7. Check the float for fuel saturation, deterioration or excessive wear where it contacts the float arm. Replace as required.
8. Check float arm for excessive wear where it contacts the needle valve. Replace as required.
9. Check inlet needle valve tip and seat contact areas for grooving, nicks, scratches or excessive wear. Replace as required.
10. Check the idle mixture screw tip for grooving, nicks or scratches. Replace the screw if tip is damaged.
11. Check the throttle and choke shafts for excessive wear or play. The throttle and choke valves must move freely without binding. Replace carburetor if any of these defects is noted.
12. Check the throttle arm collar for excessive wear. Replace as required.
13. Clean all gasket residue from mating surfaces and remove any nicks, scratches or

6

7. Slide the spring onto the idle mixture screw. Install the screw in the carburetor body until it lightly seats, then back it out the number of turns noted during disassembly.

8. Install the circular plate(s) and new packings in the cover plate recess. Install the cover plate and tighten the screws securely.

V4/V6 CARBURETOR

A twin barrel carburetor is used with Yamaha V-model engines. Each carburetor supplies fuel to one right and one left cylinder. V4 engines use 2 carburetors; V6 engines have 3 carburetors.

Although a single float supplies both carburetor barrels, each barrel contains a full line of fuel circuits and has its own idle mixture screw.

Removal/Installation (V4)

1. Remove the 6 bolts holding the outer air silencer cover to the carburetor assembly. Remove the cover and gasket.

2. Disconnect the choke solenoid lead connector. Remove the choke solenoid link O-ring and disconnect the solenoid link from the choke lever.

3. Remove the 4 bolts holding the inner air silencer cover.

4. Disconnect the fuel recirculation line and main oil tank breather hose at the inner air silencer cover.

5. Remove the inner air silencer cover and gasket.

6. Disconnect the choke lever, choke link rods and throttle link at the carburetor.

7. Disconnect the fuel lines at the carburetors. Plug lines to prevent leakage.

8. Disconnect the oil pump operating rod at the oil pump.

9. Remove the mounting nuts and washers (each carburetor is retained by 4 nuts and washers). Remove the carburetor(s) and gasket(s). Discard the gasket(s).

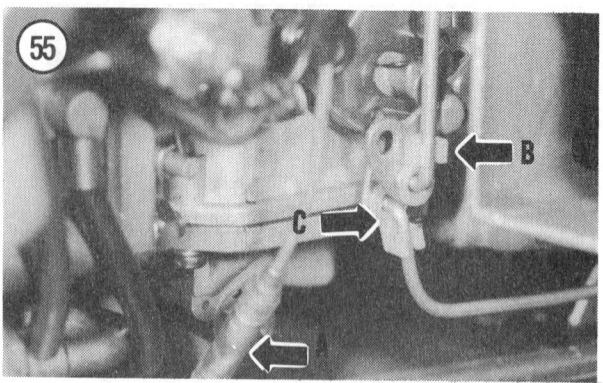

10. Installation is the reverse of removal. Use a new gasket and tighten mounting nuts securely to prevent carburetor warpage or an air leak.

Removal/Installation (V6)

1. Remove the 10 bolts holding the outer air silencer cover to the carburetor assembly. Remove the cover and gasket.

2. Remove the 6 bolts holding the inner air silencer cover.

3. Disconnect the fuel recirculation line and main oil tank breather hose at the inner air silencer cover.

4. Remove the inner air silencer cover and gasket.

5. Disconnect the choke solenoid lead connector (A, **Figure 55**). Remove the choke solenoid link O-ring and disconnect the solenoid link from the choke lever.

6. Disconnect the choke lever and choke link rods (B and C, **Figure 55**) and the throttle link at the carburetor.

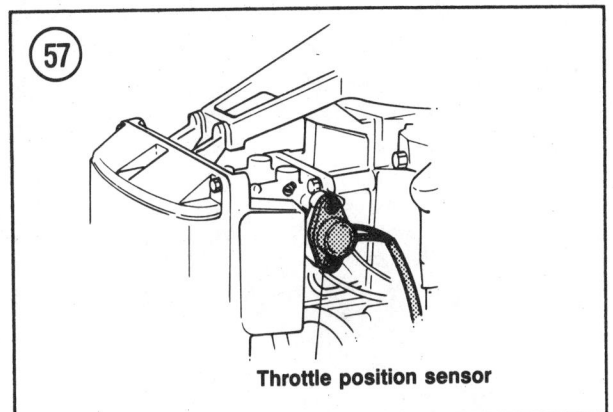

Throttle position sensor

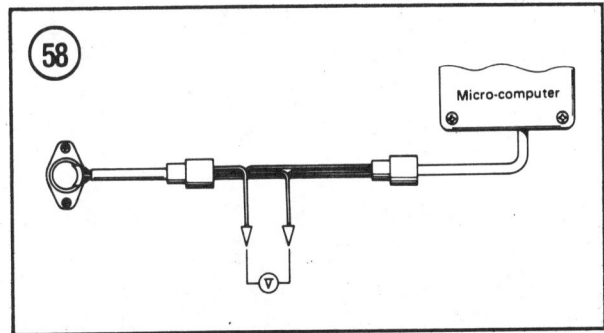

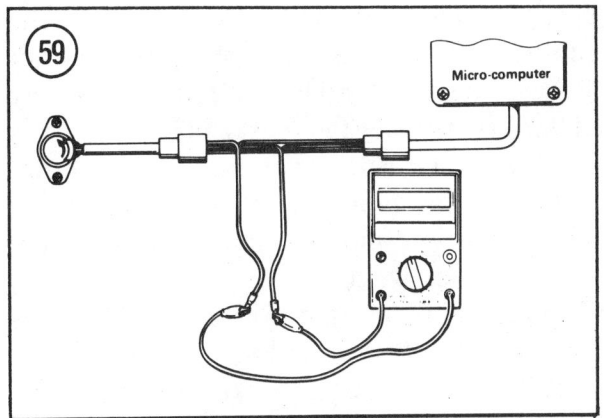

7. Disconnect the fuel lines at the carburetors. Plug lines to prevent leakage.

8. Disconnect the oil pump operating rod at the oil pump (**Figure 56**).

9. 220-225 hp—If the upper carburetor is to be overhauled, disconnect the throttle position sensor lead from the YMIS computer and remove the throttle position sensor with bracket (**Figure 57**). If carburetor is removed for other service and will not be disassembled, disconnect the sensor lead and remove the sensor with the carburetor.

10. Remove the mounting nuts and washers (each carburetor is retained by 4 nuts and washers). Remove the carburetor(s) and gasket(s). Discard the gasket(s).

11. Installation is the reverse of removal. Use a new gasket and tighten mounting nuts securely to prevent carburetor warpage or an air leak. On upper carburetors of 220-225 hp models, adjust the throttle position sensor as described in this chapter.

Throttle Position Sensor Adjustment (220-225 hp V6)

V6 Special and Excel models equipped with YMIS ignition use a throttle position sensor connected to the throttle shaft to detect the throttle valve angle. Throttle valve movement changes the sensor resistance. Sensor resistance is converted to a voltage signal which is transmitted to the YMIS computer.

Whenever the throttle position sensor is removed, it must be adjusted with the throttle valve closed, then the linkage must be adjusted (see Chapter Five for linkage adjustment).

A digital multi-meter (part No. YU-33263) and test wire harness (part No. YB-6283) are required for this procedure.

1. Coat the sensor mounting screws with NEJILOCK TB1041 or Thread Lock 1342 and temporarily fasten the sensor to the upper carburetor bracket. Leave the screws loose enough to allow the sensor to be rotated.

2. Loosen the throttle valve screw and close the throttle valve completely.

3. Disconnect the sensor from the YMIS computer and install the test wire harness between the sensor and computer. See **Figure 58**.

4. Connect the digital multi-meter to the test harness terminals. See **Figure 59**.

6

5. Turn on the main switch, read the multi-meter and adjust the sensor output voltage to 0.41 ±0.01 volt (throttle fully closed) by turning the sensor body.

6. When sensor voltage is within specifications, hold the body of the sensor from moving and alternately tighten the mounting screws tightly.

7. Open and close the throttle several times to make sure that its output voltage changes and returns to the specified level when the throttle is completely closed.

8. If the sensor does not perform as specified in Step 7, repeat Steps 5-7 as required.

Disassembly

Refer to **Figure 60** (typical) for this procedure.

1. Remove the main jet on each side of the carburetor float chamber. See A, **Figure 61**.

2. Remove the screw over each of the 2 idle jets in the float chamber (B, **Figure 61**) with a wide-blade screwdriver.

3. Remove the idle jets with a suitable jet remover or wide-blade screwdriver. See **Figure 62**.

4. Drain any fuel remaining in the float chamber into a suitable container and dispose of it safely.

5. Lightly seat each idle mixture screw, counting the number of turns required for reassembly reference, then back the screws out and remove from carburetor with springs. See A, **Figure 63**.

6. Unscrew and remove the bypass covers from the mixing chamber with a wide-blade screwdriver. See B, **Figure 63**.

7. Remove the main and idle air jets from the mixing chamber with a suitable jet remover or wide-blade screwdriver. See **Figure 64**.

8. Remove the float chamber from the carburetor. Remove and discard the float chamber gasket.

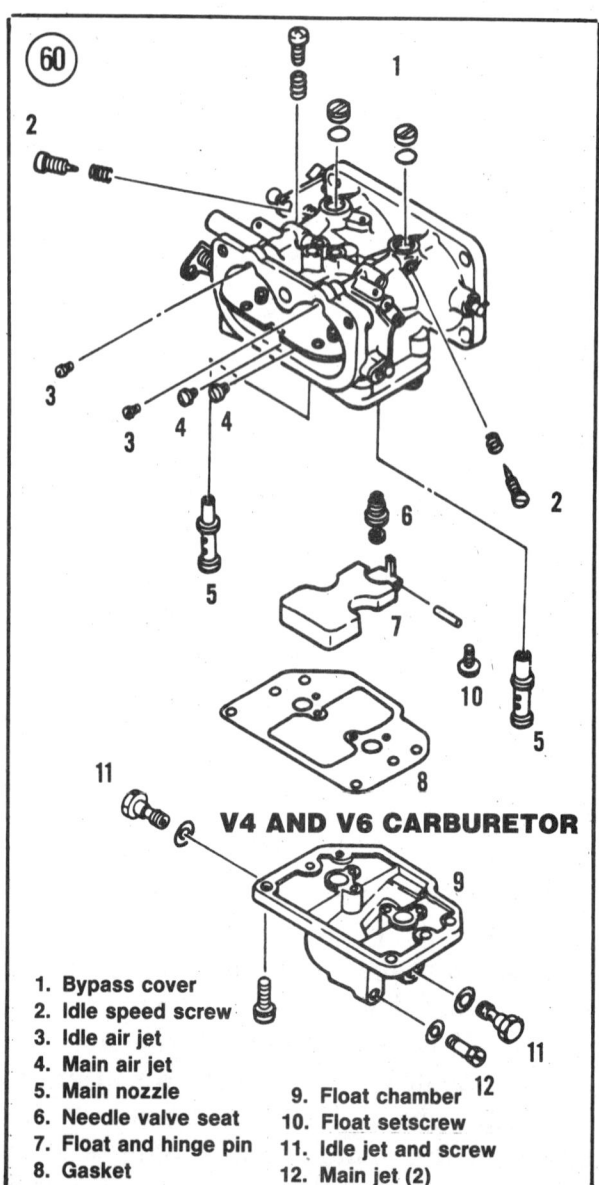

V4 AND V6 CARBURETOR

1. Bypass cover
2. Idle speed screw
3. Idle air jet
4. Main air jet
5. Main nozzle
6. Needle valve seat
7. Float and hinge pin
8. Gasket
9. Float chamber
10. Float setscrew
11. Idle jet and screw
12. Main jet (2)

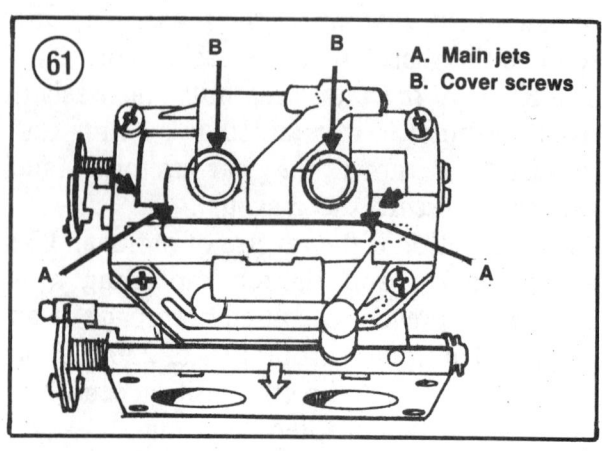

A. Main jets
B. Cover screws

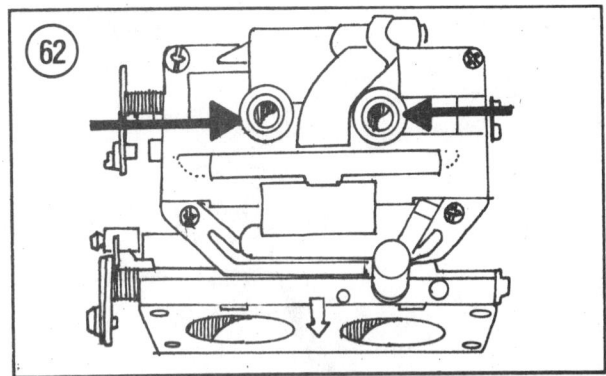

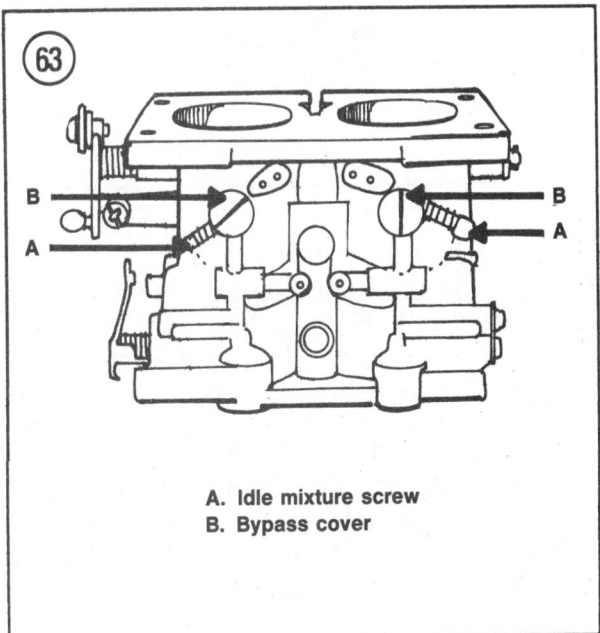

A. Idle mixture screw
B. Bypass cover

9. Remove the setscrew holding the float hinge pin. Remove float and hinge pin assembly from carburetor with the needle valve.

10. Remove the needle valve seat with a suitable socket wrench. Discard the gasket.

11. Remove the 2 main nozzles from the mixing chamber. See **Figure 65**.

Cleaning and Inspection

A sealing compound is used to eliminate porosity problems. Submersion in a hot tank or carburetor cleaner will remove this sealing compound. Clean the carburetor with an aerosol solvent.

1. Wipe the carburetor casting and linkage with a cloth moistened in solvent to remove any contamination and operating film.

2. Clean the carburetor castings and metal parts with an aerosol solvent and a brush. Spray the aerosol solvent on the casting and scrub off any gum or varnish with a small bristle brush.

CAUTION
Do not clean carburetor passages with a wire or drill bit. This can enlarge the passages and change the carburetor calibration.

6

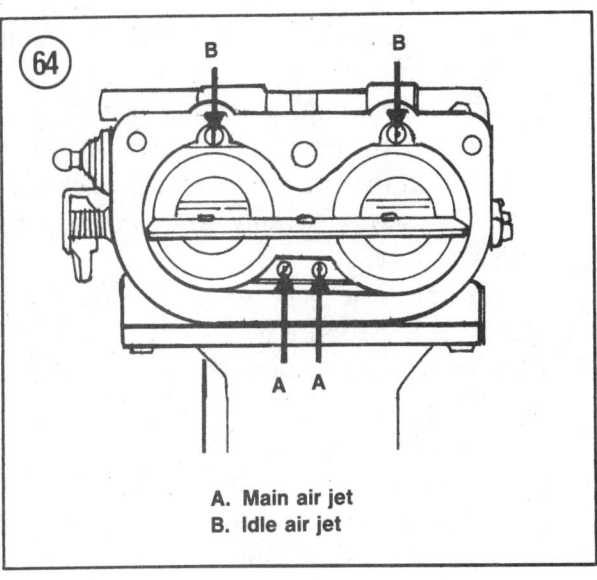

A. Main air jet
B. Idle air jet

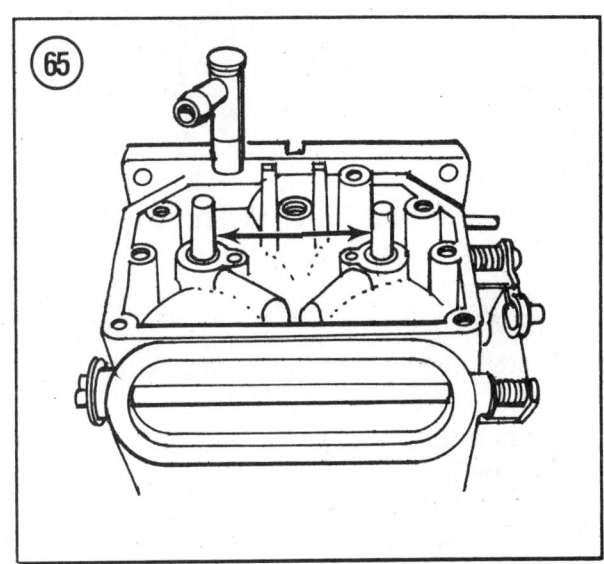

3. Spray cleaner through the metering passages in the casting.

4. Rinse the carburetor components thoroughly in clean solvent after cleaning them.

5. Blow the carburetor casting and passages dry with low-pressure (25 psi or less) compressed air. The use of higher pressures can damage the sealing compound.

6. Check carburetor body and fuel bowl for stripped threads, cracks or other defects. Replace as required.

7. Check the float for fuel saturation, deterioration or excessive wear where it contacts the float arm. Replace as required.

8. Check float arm for excessive wear where it contacts the needle valve. Replace as required.

9. Check inlet needle valve tip and seat contact areas for grooving, nicks, scratches or excessive wear. Replace as required.

10. Check each idle mixture screw tip for grooving, nicks or scratches. Replace the screw if tip is damaged.

11. Check the throttle and choke shafts for excessive wear or play. The throttle and choke valves must move freely without binding. Replace carburetor if any of these defects is noted.

12. Clean all gasket residue from mating surfaces and remove any nicks, scratches or slight distortion with a surface plate and emery cloth.

Reassembly

Refer to **Figure 60** for this procedure.

1. Thread each main nozzle into the carburetor float chamber until it stops. See **Figure 65**.

2. Install the needle valve seat with a new gasket.

3. Attach needle valve to float hinge (**Figure 66**), then install float assembly in the float bowl with the hinge pin and tighten the setscrew securely.

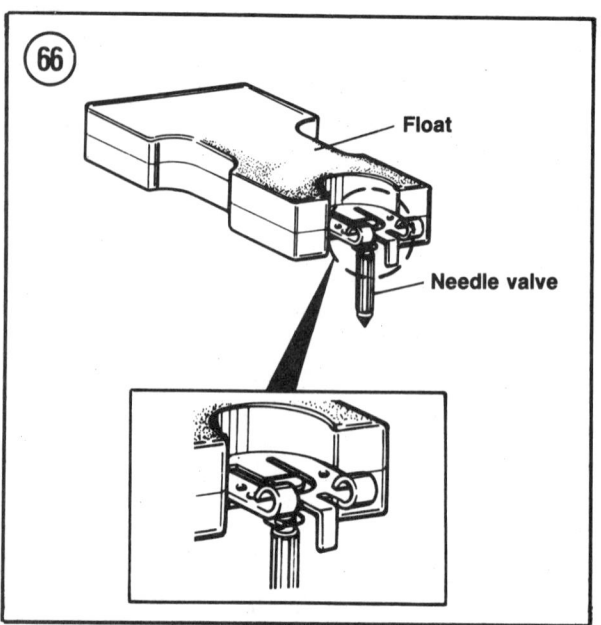

Float

Needle valve

CAUTION
Be sure to remove the inlet needle from the float before bending the float adjustment tang when adjustment is required.

4. Invert the carburetor and measure the float height from the carburetor casting as shown in **Figure 67**. It should be 0.50 in. If it is not, remove the float and check the needle valve seat for proper installation. If the seat is properly installed, remove the needle valve and valve clip, then bend the hinged tang on the float as required to position the float properly.

5. Install a new float chamber gasket and make sure it is properly seated, then install the float chamber to the carburetor body. Tighten the screws evenly and securely to prevent warpage.

6. Install the idle jets with a wide-blade screwdriver. See **Figure 62**.

7. Install the screws over the idle jets and each main jet. See **Figure 61**.

8. Slide the spring onto the idle mixture screw. Install the screw in the carburetor body until it lightly seats, then back it out the number of turns noted during disassembly.

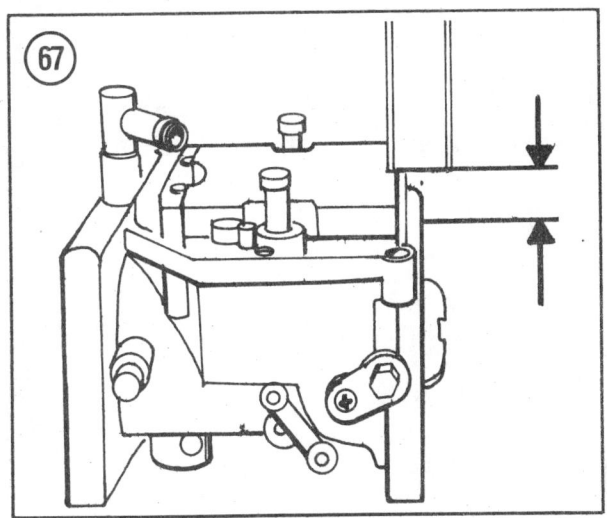

9. Install the bypass covers in the mixing chamber body.

NOTE
The main and idle jets apppear identical. If there is any confusion in identifying them for installation in Step 10, remember that the main air jet calibration number is larger than that of the idle air jet.

10. Install the main and idle air jets in the mixing chamber.

FUEL FILTER

Yamaha 2, 3 and 4 hp engines are equipped with an integral fuel tank containing a fuel filter screen in the petcock. Yamaha 5, 6 and 8 hp engines use a small disposable inline filter installed in the fuel line between the fuel line connector and carburetor. Service these filters at the intervals specified in Chapter Four.

All other Yamaha engines are equipped with a water-separating filter canister positioned between the fuel line connector and fuel pump. See **Figure 68** (typical). This canister traps moisture and prevents it from entering the carburetor when the engine is in its normal vertical position. The filter should be serviced whenever water can be seen in the

bowl or at the intervals specified in Chapter Four.

REED VALVE ASSEMBLY

The reed valve assembly is mounted between the intake manifold and crankcase on Yamaha 6-225 hp engines. See **Figure 69** and **Figure 70** for typical reed valve installations. On 2-5 hp engines, the reed valve assembly is attached directly to the intake manifold. See **Figure 71** (2 hp engine) and **Figure 73** (3-5 hp engine).

Reed valves control the passage of air-fuel mixture into the crankcase by opening and closing as crankcase pressure changes. When crankcase pressure is high, the reeds maintain contact with the reed plate to which they are attached. As crankcase pressure drops on the compression stroke, the reeds move away from the plate and allow air-fuel mixture to pass. Reed travel is limited by the reed stop. As crankcase pressure increases, the reeds return to the reed plate.

Removal/Installation
(All Models)

1. Remove the carburetor(s) as described in this chapter.
2. Label and disconnect any hoses connected to the intake manifold.

NOTE
*On some models, the reed valve assembly may be secured to the crankcase by one or more separate fasteners. With this design, the intake manifold must be removed first, then the reed valve assembly fastener(s) removed before the assembly and gasket can be removed. See **Figure 72**.*

3. Remove the screws holding the intake manifold to the crankcase cover. Remove the intake manifold, gasket, reed valve assembly and gasket from the crankcase. Discard the gaskets.
4. Clean all mating surfaces of gasket or sealant residue.

6

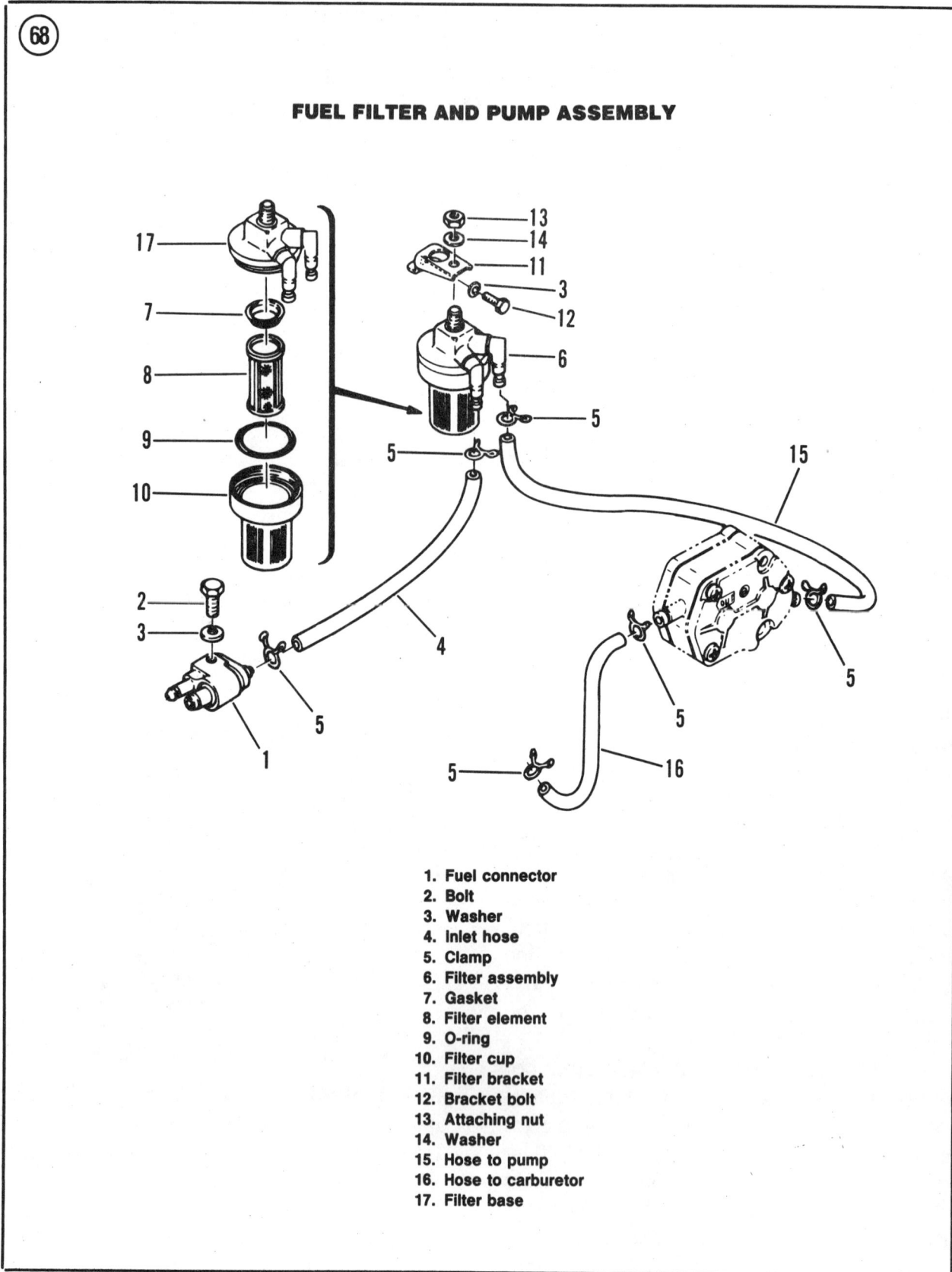

FUEL FILTER AND PUMP ASSEMBLY

1. Fuel connector
2. Bolt
3. Washer
4. Inlet hose
5. Clamp
6. Filter assembly
7. Gasket
8. Filter element
9. O-ring
10. Filter cup
11. Filter bracket
12. Bracket bolt
13. Attaching nut
14. Washer
15. Hose to pump
16. Hose to carburetor
17. Filter base

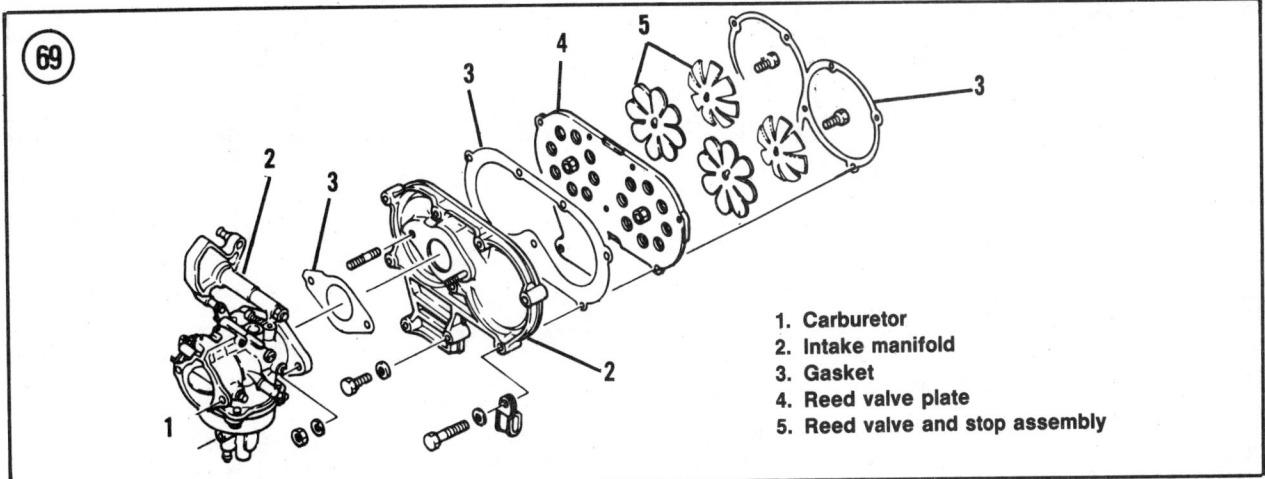

1. Carburetor
2. Intake manifold
3. Gasket
4. Reed valve plate
5. Reed valve and stop assembly

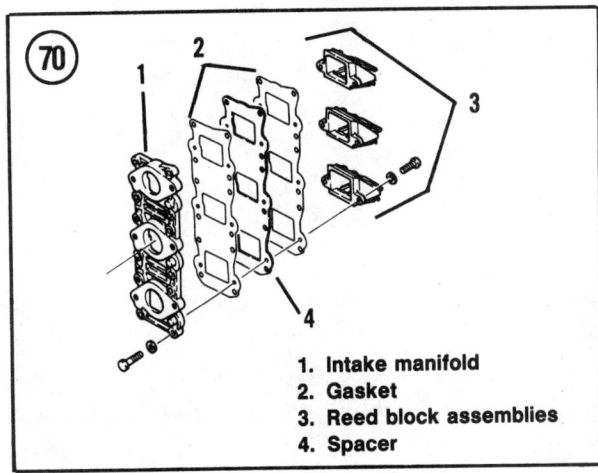

1. Intake manifold
2. Gasket
3. Reed block assemblies
4. Spacer

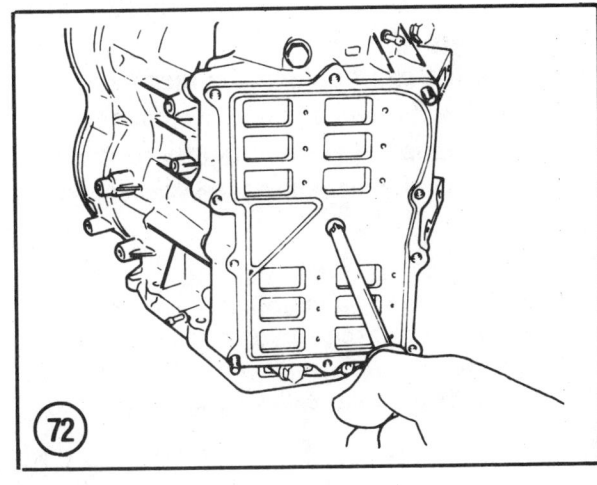

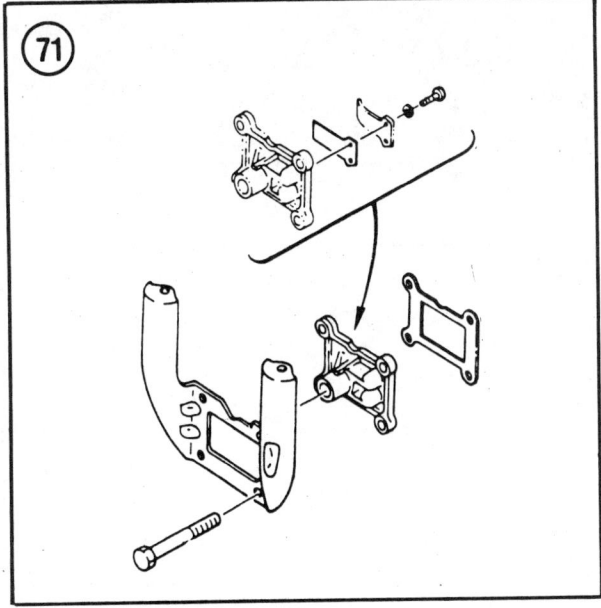

5. Clean and inspect the reed block assembly as described in this chapter.

6. Installation is the reverse of removal. Use new gaskets and make sure that the reed assembly faces the crankcase.

Cleaning and Inspection

1. Clean reed block assembly with solvent and blow dry with low-pressure compressed air, taking care not to direct the air stream directly on or through the reed valves.

2. Remove all gasket or sealant residue from the reed block assembly.

3. Check the intake side of the assembly to make sure that the reeds are not sticking tightly to the valve face.

4. Check the crankcase side of the assembly to make sure that the reeds are lying flat on the valve face with no preload. To check flatness, gently push each reed petal out. Constant resistance should be felt with no audible noise.

CAUTION
Always replace reeds in sets. Never turn a reed over for reuse.

5. Check for chipped, cracked or broken reeds. Replace if any defects are noted. See **Figure 73** (3, 4 and 5 hp), **Figure 74** (9.9 and 15 hp) and **Figure 75** (25 [1984-1987] and 30 [1984-1986] hp). Other models are similar to those shown.

6. Check the gap between the reeds and reed block with a flat feeler gauge. Replace reeds if they are preloaded (adhere tightly to the reed block) or if the gap is excessive. Refer to **Table 2**.

NOTE
Reed stop adjustment is not recommended for Yamaha 2, 25 or 30 hp models. If reed stop is defective, replace the entire reed valve assembly.

7. Check each reed stop opening by measuring from inside of reed stop to top of closed reed. See **Figures 76-78** for typical reed block designs. If reed stop is not within specifications (**Table 2**), carefully bend reed stop to obtain the specified opening.

Reed and Reed Stop Replacement

NOTE
When installing new reeds on 25 (1984-1987) and 30 (1984-1986) hp reed blocks, the side of the reed assembly stamped "HEAD" must face the crankcase and the double dot on the reed block must be centered between every other reed.

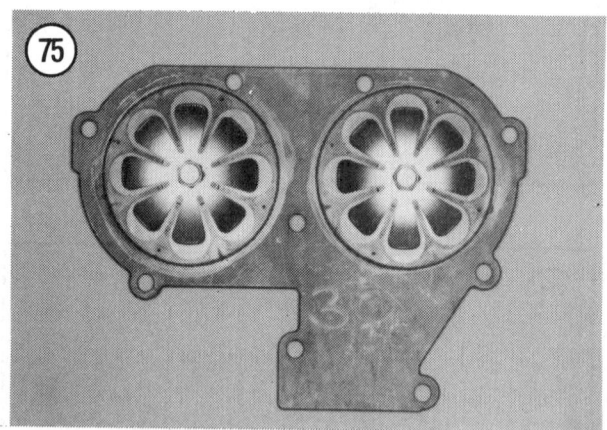

1. Remove the screws holding the reed stop and reeds to the valve seat. See **Figure 79** (typical).

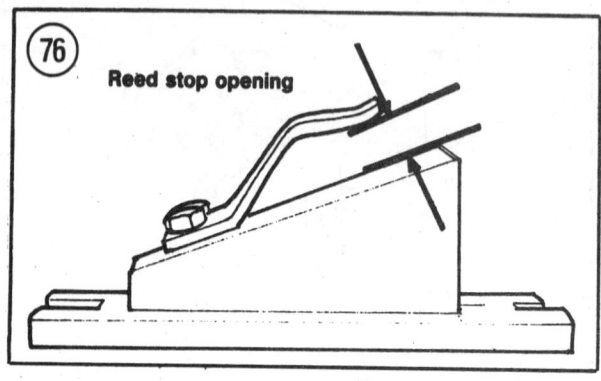

Reed stop opening

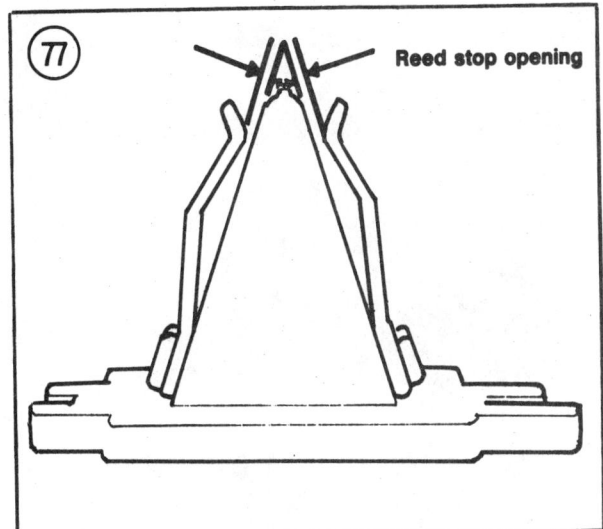

Reed stop opening

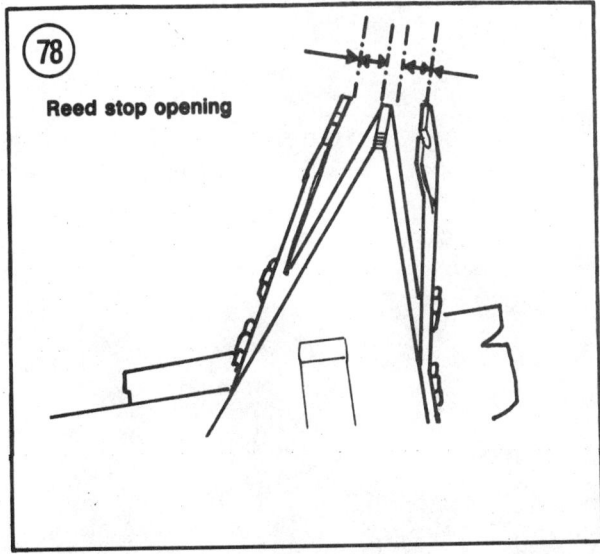

Reed stop opening

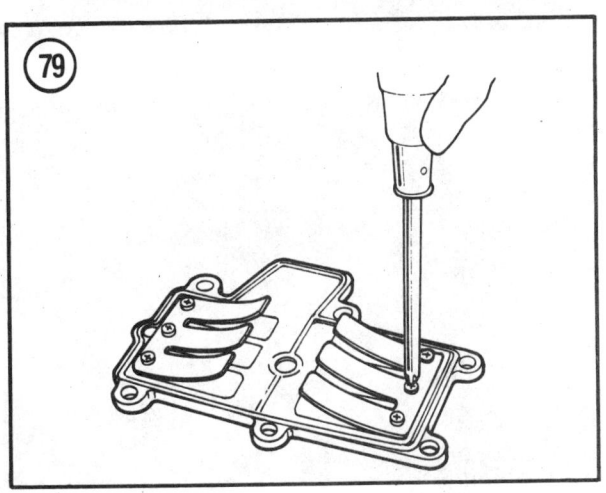

2. Remove the reed stop and reeds.

3. Place a new reed on the valve seat and check for flatness.

4. Center the reed over the valve seat openings.

5. Wipe reed stop screw threads with Loctite Type A or Thread Lock 1342. Install reed stop and gradually tighten screws to 0.7 ft.-lb. (1 N•m) to prevent warping.

6. Check reed tension and opening. See *Cleaning and Inspection* in this chapter.

INTEGRAL FUEL TANK

The 2 and 4 hp models have an integral fuel tank. Fuel is transported from the tank to the carburetor by gravity feed. See **Figure 1** (typical). A non-serviceable petcock controls fuel flow. If the petcock leaks, it must be replaced as an assembly.

Removal/Installation (2 hp)

1. Remove the right and left engine cowling.

2. If the fuel petcock is in the ON position (**Figure 80**), turn it to OFF and disconnect the fuel line at the fuel petcock.

3. Remove the 2 nuts (**Figure 81**) and 1 bolt (**Figure 82**) holding the fuel tank to the engine. Remove the fuel tank.

4. If there is fuel remaining in the tank, open the petcock and drain the fuel into a suitable container.

5. If petcock removal is required for filter cleaning, loosen the screw holding the band

and unscrew the petcock from the tank. See **Figure 83** (typical).

6. Installation is the reverse of removal.

Removal/Installation (3 hp)

1. Remove the engine cover.

2. Make sure the fuel lever is turned to the OFF position.

3. Disconnect the fuel line at the petcock outlet or the fuel pump inlet.

4. Remove the fasteners retaining the fuel petcock and the fuel tank, then remove the components.

> *NOTE*
> *Dampers are installed at each mounting position on the fuel tank to absorb vibration. They are removable and must be reinstalled with the tank.*

5. Installation is the reverse of removal.

Removal/Installation (4 hp)

1. Remove the engine cover.

2. Make sure the fuel lever is turned to the OFF position.

3. Disconnect the fuel line at the fuel pump (**Figure 84**). Plug the line to prevent leakage when the tank is removed.

4. Remove the 2 bolts holding the fuel tank to the bottom cowling.

> *NOTE*
> *Dampers are installed on each side of the fuel tank to absorb vibration. They are removable and must be reinstalled with the tank.*

5. Carefully unsnap the fuel lever from the petcock connection with a flat blade screwdriver. See arrow, **Figure 85**. Remove the fuel tank.

6. If petcock removal is required for filter cleaning:

  a. Remove the tank cap and drain the fuel into a suitable container.

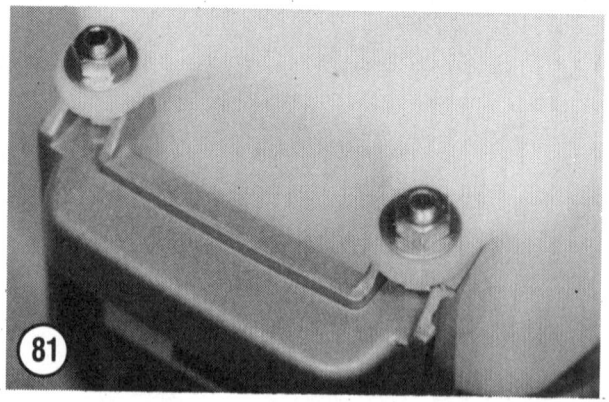

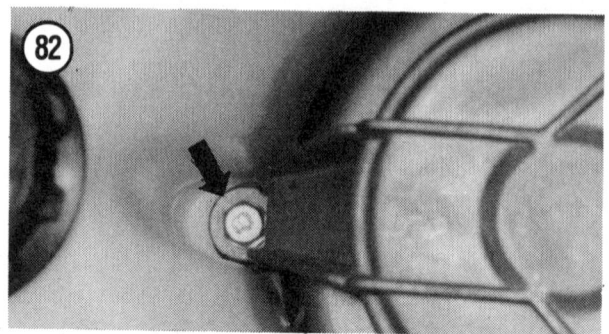

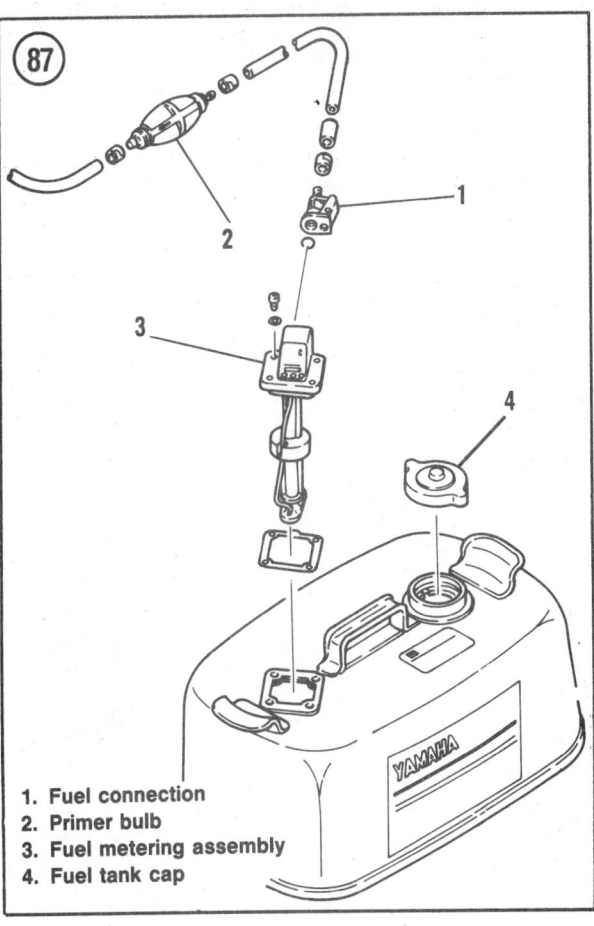

1. Fuel connection
2. Primer bulb
3. Fuel metering assembly
4. Fuel tank cap

b. Unplug the fuel hose and drain it into the container.

c. Loosen the petcock nut with an open end wrench (**Figure 86**).

d. Unscrew the petcock from the tank. See **Figure 83** (typical).

7. Installation is the reverse of removal. Make sure the petcock faces in the same direction as before removal. When properly installed, the petcock should connect easily to the fuel lever. Tighten the petcock nut to 3.6 ft.-lb. (5 N•m).

Cleaning and Inspection

1. Remove and drain the fuel tank as described in this chapter.

2. Pour several ounces of solvent into the tank. Reinstall the tank cap and slosh the solvent around, then drain it into a suitable container.

3. Check the fuel tank cap for a worn gasket or packing. Make sure the air vent is not clogged.

4. Clean the petcock filter in solvent and blow dry with low-pressure compressed air.

PORTABLE FUEL TANK

Figure 87 shows the components of the portable tank including the fuel gauge sender and in-tank filter housed in the fuel metering assembly and the primer bulb.

When some oils are mixed with gasoline and stored in a warm place, a bacterial substance will form. This colorless substance covers the fuel pickup, restricting flow through the fuel system. Bacterial formation can be prevented by using Yamalube Fuel Conditioner on a regular basis. If present, bacteria can be removed with a good marine engine cleaner.

To remove any dirt or water that may have entered the tank during refilling and to prevent the build-up of gum and varnish,

6

clean the inside of the tank once each season by flushing with clean lead-free gasoline or kerosene.

Check the inside and outside of the tank for signs of rust, leakage or corrosion. Replace as required. Do not attempt to patch the tank with automotive fuel tank repair materials. Portable marine fuel tanks are subject to much greater pressure and vacuum than automotive tanks.

To check the in-tank filter screen for possible restrictions, remove the fuel meter housing and inspect the screen on the end of the suction pipe. Clean the screen with solvent, then blow low-pressure compressed air through the screen to remove any particles. See **Figure 88** (typical).

FUEL LINE AND PRIMER BULB

Figure 87 shows the primer bulb installation in the fuel line. **Figure 89** shows the inlet and outlet check valves inside the primer bulb. When priming the engine, the primer bulb should gradually become firm. If it does not become firm or if it stays firm even when disconnected, a check valve is malfunctioning.

The line should be checked periodically for cracks, breaks, restrictions and chafing. The bulb should be checked periodically for proper operation. Make sure all fuel line connections are tight and securely clamped.

ANTI-SIPHON DEVICES

In accordance with industry safety standards, late-model boats equipped with a built-in fuel tank will have some form of anti-siphon device installed between the fuel tank outlet and outboard fuel inlet. This device is designed to shut the fuel supply off in case the boat capsizes or is involved in an accident. Quite often, the malfunction of such devices leads the owner to replace a fuel pump in the belief that it is defective.

Anti-siphon devices can malfunction in one of the following ways:
a. Anti-siphon valve: orifice in valve is too small or clogs easily; valve sticks in closed or partially closed position; valve fluctuates between open and closed position; thread sealer, metal filings or dirt/debris clogs orifice or lodges in the relief spring.
b. Solenoid-operated fuel shut-off valve: solenoid fails with valve in closed position; solenoid malfunctions, leaving valve in partially closed position.
c. Manually-operated fuel shut-off valve: valve is left in completely closed position; valve is not fully opened.

The easiest way to determine if an anti-siphon valve is defective is to bypass it by operating the engine with a remote fuel supply. If a fuel system problem is suspected, check the fuel filter first. See Chapter Four. If the filter is not clogged or dirty, bypass the anti-siphon device. If the engine runs properly with the anti-siphon device bypassed, contact the boat manufacturer for replacement of the anti-siphon device.

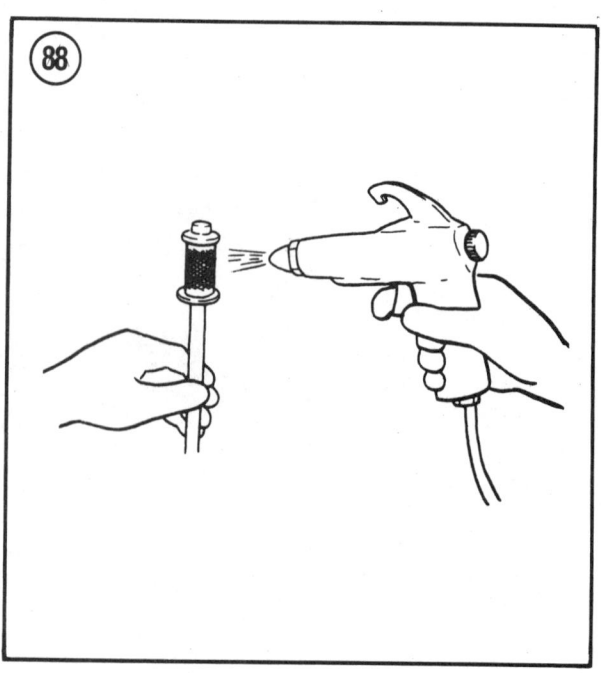

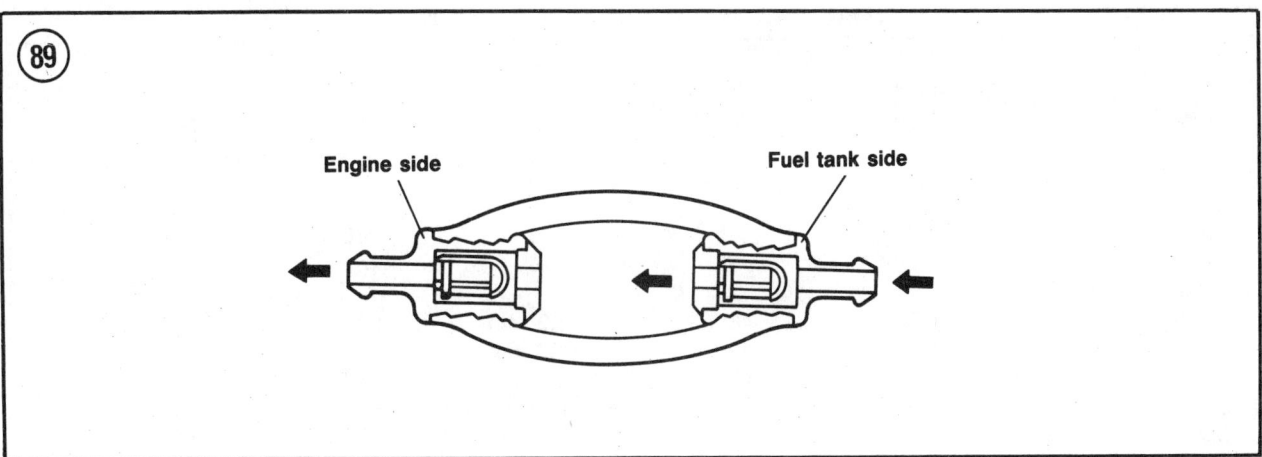

Table 1 CARBURETOR SPECIFICATIONS

Model	Carburetor stamped mark	Main jet	Main air jet	Idle jet	Idle air jet
2 hp	K	#96	1 mm	—	—
3 hp					
1988	6L500	#65	#85	#40	#70
1989	6L501	#68	#85	#40	#70
4 hp					
1984-1985	6E071	#80	1.2 mm	#46	1.4 mm
1986-on	6E002	#80	1.2 mm	#46	1.4 mm
5 hp	6E471	#82	1.5 mm	#46	1.2 mm
6-8 hp	6G110	#98	#110	#45	#80
9.9 hp					
1984-1985	6E770	#104	0.8 mm	#52	0.8 mm
1986-on	6E772	#104	0.8 mm	#52	0.8 mm
15 hp					
1984-1985	6E870	#135	0.8 mm	#56	0.8 mm
1986-on	6E802	#135	0.8 mm	#56	0.8 mm
25 hp					
1984	6H770	#133	1.2 mm	#60	1.1 mm
1985-1987	6H771	#133	1.2 mm	#52	1.1 mm
1988-on	6L200	#125	2.3 mm	#60	1.3 mm
30 hp					
2-cyl.	6H972	#135	#100	#54	#120
3-cyl.	6J800	#102	N/A	#50	See Note 1
40 hp					
1984	6H401	See Note 2	#175	#75	#80
1985-1986	6H402	See Note 3	#175	#70	#80
1987	6H405	See Note 4	#155	#68	#85
1988	6H406	See Note 4	#155	#68	#85
1989	6H407	See Note 4	#155	#62	#85
1989	6H410	See Note 5	#175	#62	#85
50 hp					
1984	6H501	#150	#175	#70	#75
1985-1986	6H502	#135	#175	#75	#75
1987	6H505	#130	#160	#70	#80
1988	6H506	#130	#160	#70	#80
1989	6H510	#135	#170	#62	#80
70 hp					
1984	6H301	#150	#170	#78	#80
1985-1987	6H302	#150	#185	#78	#70
1988	6H306	#150	#185	#78	#70
1989	6H307	#145	#170	#72	#70
90 hp					
1984	6H101	#165	#195	#80	#90
1985-1987	6H110	#165	#185	#80	#70
1988	6H113	#165	#185	#80	#70
1989	6H114	#165	#185	#78	#70
115 hp					
1984	6H401	#175	#160	#94	#50
1985	6H402	#175	#250	#84	#60
1986	6E510	#170	2.5 mm	#88	0.5 mm
1986-on	6E511	#180	2.5 mm	#78	#60

(continued)

Table 1 CARBURETOR SPECIFICATIONS (continued)

Model	Carburetor stamped mark	Main jet	Main air jet	Idle jet	Idle air jet
130 hp	6L1-00	#180	#220	#82	#60
150 hp					
1984	6G401[6]	#125	#130	#94	#50
1985-1986	6G402	#130	#190	#90	#50
1987	6G403	#124	#210	#74	#60
1988-on	6G403	#124	#210	#80	#60
150 hp Pro V	6J902	#124	#210	#82	#60
175 hp					
1984	6G501[7]	#125	0.9 mm	#94	#50
1985-1986	6G502	#130	#180	#90	#50
1987	6G503	#124	#200	#72	#60
1988-on	6G503	#124	#200	#76	#60
200 hp					
1984	6G601[8]	#155	#160	#94	#60
1985-1986	6G602	#160	#210	#88	#50
1987	6G603	#146	#210	#82	#60
1988-on	6G603	#124	#270	#84	#60
220-225 hp					
1984	6G700	#165	#170	#100	#60
1985-1986	6G701	#170	#230	#90	#50
1987-on	6G702	See Note 9	#270	#76	#60
	6SK700	#160	#270	#76	#60

1. No. 1 and 2 cylinders, #100; No. 3 cylinder, #120.
2. Top and 3rd carburetors, #130; center carburetor, #135.
3. Top and 3rd carburetors, #125; center carburetor, #130.
4. Top and 3rd carburetors, #115; center carburetor, #120.
5. Top and 3rd carburetors, #120; center carburetor, #125.
6. May be equipped with 6G400.
7. May be equipped with 6G500.
8. May be equipped with 6G600.
9. Top carburetor, #160; center and bottom carburetors, #156.

Table 2 REED STOP SPECIFICATIONS

Model	Reed stop clearance (in.)	Bending limit (in.)
2 hp	0.24	0.012
3 hp	0.15 ±0.008	0.060
4 and 5 hp	0.28	0.008
6 and 8 hp	0.177 ±0.008	0.024
9.9 hp	0.05	0.008
15 hp	0.16	0.008
25 hp		
1984	0.11 ±0.012	0.008
1985-1987	0.07 ±0.012	0.008
1988-on	0.24	0.060
(continued)		

6

Table 2 REED STOP SPECIFICATIONS (continued)

Model	Reed stop clearance (in.)	Bending limit (in.)
30 hp		
2-cylinder (1984-1986)	0.21	0.016
3-cylinder (1987-on)	0.087 ±0.006	0.035
40 and 50 hp		
1984-1986	0.06-0.07	0.035
1987-on	0.24	0.012
70 and 90 hp	0.39 ±0.004	0.008
115 and 130 hp	0.26 ±0.012	0.008
150, 175 and 200 hp	0.26 ±0.008	0.035
220 and 225 hp		
1985-1986	0.26 ±0.012	0.035
1987-on	0.24-0.26	0.035

Chapter Seven

Ignition and Electrical Systems

This chapter provides service procedures for the battery, starter motor (electric start models) and each ignition system used on Yamaha outboard motors during the years covered by this manual. Wiring diagrams are included at the end of the book. **Tables 1-4** are at the end of the chapter.

BATTERY

Since batteries used in marine applications endure far more rigorous treatment than those used in an automotive charging system, they are constructed differently. Marine batteries have a thicker exterior case to cushion the plates inside during tight turns and rough weather. Thicker plates are also used, with each one individually fastened within the case to prevent premature failure. Spill-proof caps on the battery cells prevent electrolyte from spilling into the bilges.

Automotive batteries are not designed to be run down and recharged repeatedly. For this reason, they should *only* be used in an emergency situation when a suitable marine battery is not available.

Yamaha recommends that any battery used with its outboard motors have a minimum 70 amp-hour rating.

> *CAUTION*
> *Sealed or maintenance-free batteries are **not** recommended for use with the unregulated charging systems used on Yamaha outboards. Excessive charging during continued high-speed operation will cause the electrolyte to boil, resulting in its loss. Since water cannot be added to such batteries, such overcharging will ruin the battery.*

Separate batteries may be used to provide power for any accessories such as lighting, fish finders, depth finder, etc. To determine the required capacity of such batteries, calculate the average discharge rate of the accessories and refer to **Table 1**.

Batteries may be wired in parallel to double the ampere hour capacity while maintaining a 12-volt system. See **Figure 1**. For accessories which require 24 volts, batteries may be wired in series (**Figure 2**) but only accessories specifically requiring 24 volts should be connected into the system. Whether wired in parallel or in series, charge the batteries individually.

Battery Installation in Aluminum Boats

If a battery is not properly secured and grounded when installed in an aluminum boat, it may contact the hull and short to ground. This will burn out remote control cables, tiller handle cables or wiring harnesses.

The following preventive steps should be taken when installing a battery in a metal boat:

The following preventive steps should be taken when installing a battery in a metal boat:

1. Choose a location as far as practical from the fuel tank.

2. Install the battery in a plastic battery box with cover and tie-down strap (**Figure 3**).

3. If a covered container is not used, cover the positive battery terminal with a non-conductive shield or boot (**Figure 4**).

4. Make sure the battery is secured inside the battery box and that the box is fastened in position with the tie-down strap.

Care and Inspection

1. Remove the battery container cover (**Figure 3**) or hold-down (**Figure 4**).

2. Disconnect the negative battery cable. Disconnect the positive battery cable.

NOTE
*Some batteries have a built-in carry strap (**Figure 5**) for use in Step 3.*

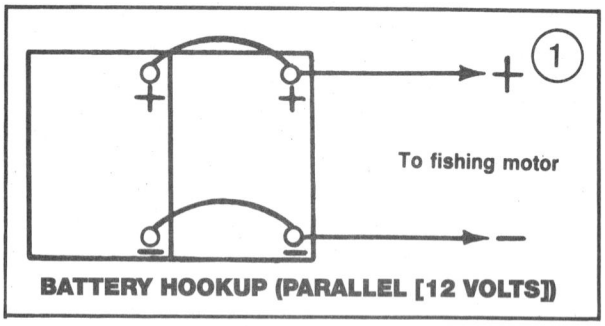

BATTERY HOOKUP (PARALLEL [12 VOLTS])
To fishing motor

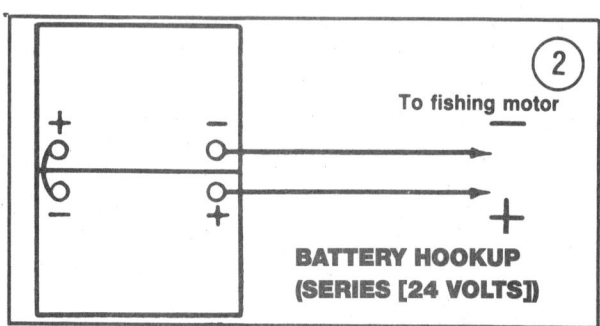

To fishing motor
BATTERY HOOKUP (SERIES [24 VOLTS])

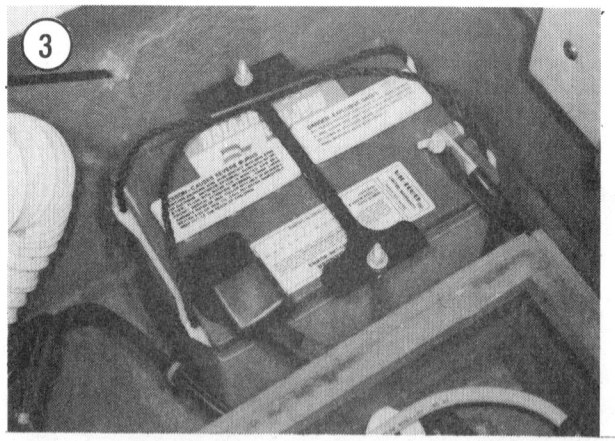

3. Attach a battery carry strap to the terminal posts. Remove the battery from the battery tray or container.

4. Check the entire battery case for cracks.

5. Inspect the battery tray or container for corrosion and clean if necessary with a solution of baking soda and water.

NOTE
Keep cleaning solution out of the battery cells in Step 6 or the electrolyte will be seriously weakened.

6. Clean the top of the battery with a stiff bristle brush using the baking soda and water solution (**Figure 6**). Rinse the battery case with clear water and wipe dry with a clean cloth or paper towel.

7. Position the battery in the battery tray or container.

8. Clean the battery cable clamps with a stiff wire brush or one of the many tools made for this purpose (**Figure 7**). The same tool is used for cleaning the battery posts. See **Figure 8**.

9. Reconnect the positive battery cable, then the negative cable.

CAUTION
Be sure the battery cables are connected to their proper terminals. Connecting the battery backwards will reverse the polarity and damage the rectifier.

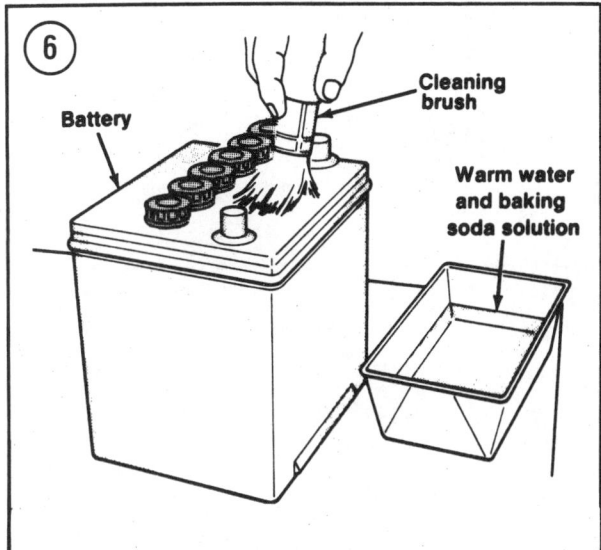

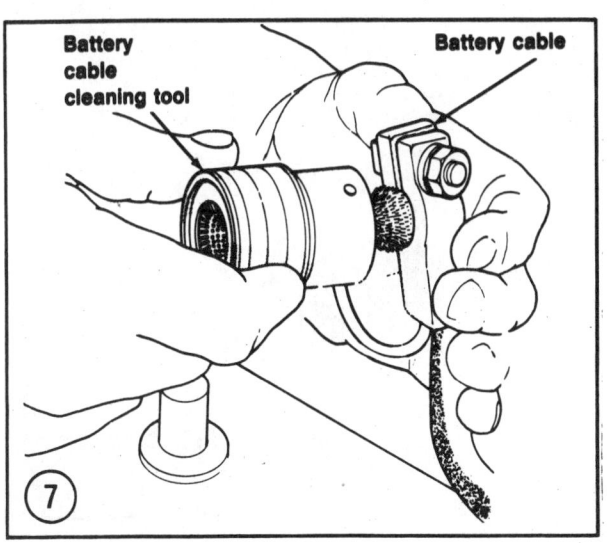

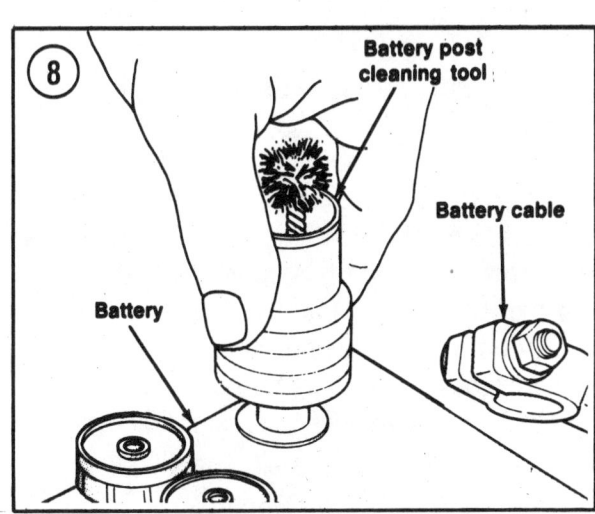

10. Tighten the battery connections and coat with a petroleum jelly such as Vaseline or a light mineral grease.

NOTE
Do not overfill the battery cells in Step 11. The electrolyte expands due to heat from charging and will overflow if the level is more than 3/16 in. above the battery plates.

11. Remove the filler caps and check the electrolyte level. Add distilled water, if necessary, to bring the level up to 3/16 in. above the plates in the battery case. See **Figure 9**.

Testing

Hydrometer testing is the best way to check battery condition. Use a hydrometer with numbered graduations from 1.100-1.300 rather than one with just color-coded bands. To use the hydrometer, squeeze the rubber ball, insert the tip in a cell and release the ball (**Figure 10**).

NOTE
Do not attempt to test a battery with a hydrometer immediately after adding water to the cells. Charge the battery for 15-20 minutes at a rate high enough to cause vigorous gassing and allow the water and electrolyte to mix thoroughly.

Draw enough electrolyte to float the weighted float inside the hydrometer. When using a temperature-compensated hydrometer, release the electrolyte and repeat this process several times to make sure the thermometer has adjusted to the electrolyte temperature before taking the reading.

Hold the hydrometer vertically and note the number in line with the surface of the electrolyte (**Figure 11**). This is the specific gravity for the cell. Return the electrolyte to the cell from which it came.

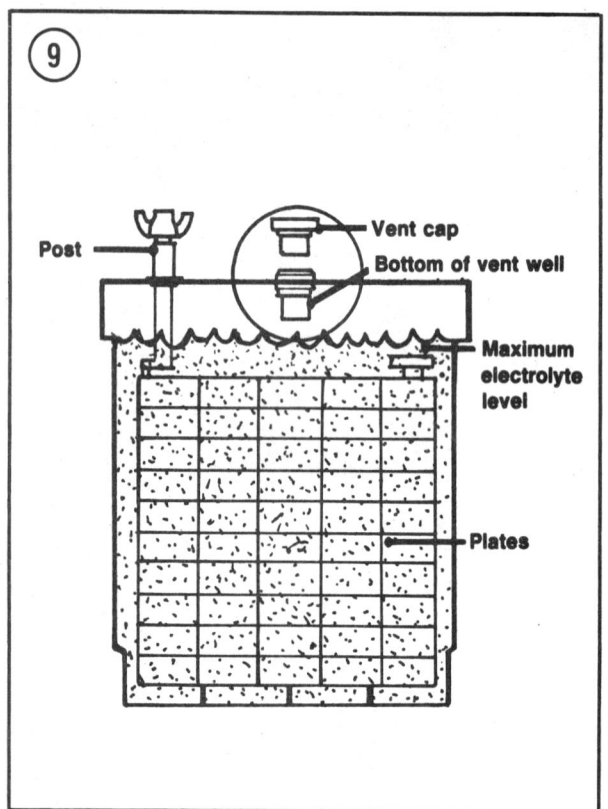

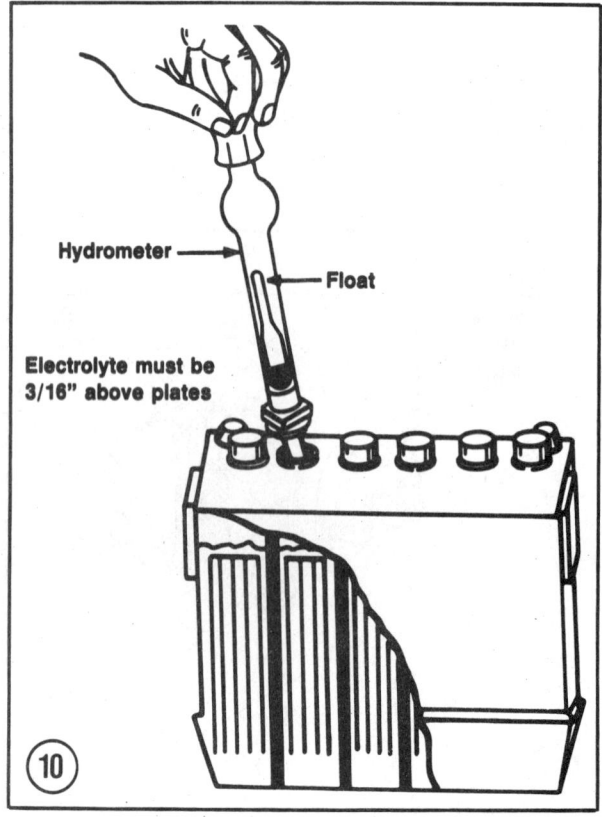

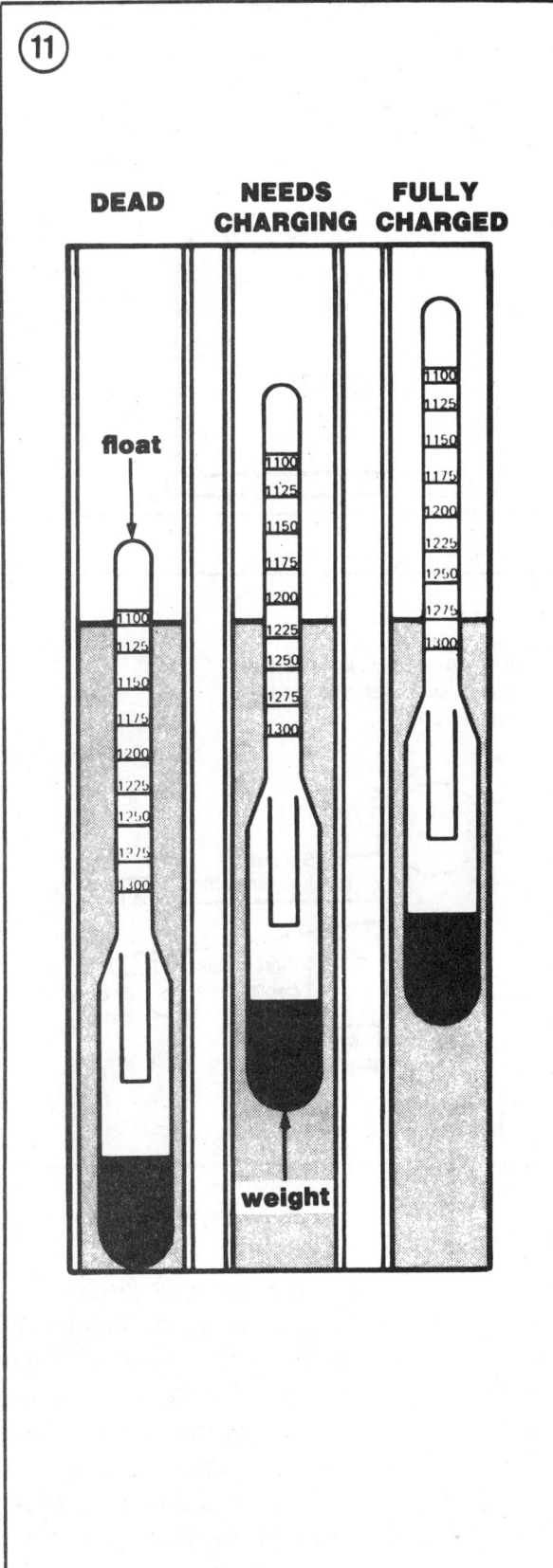

The specific gravity of the electrolyte in each battery cell is an excellent indicator of that cell's condition. A fully charged cell will read 1.260 or more at 68° F (20° C). A cell that is 75 percent charged will read from 1.220-1.230 while one with a 50 percent charge reads from 1.170-1.180. If the cell tests below 1.120, the battery must be recharged and one that reads 1.100 or below is dead. Charging is also necessary if the specific gravity varies more than 0.050 from cell to cell.

NOTE
If a temperature-compensated hydro-meter is not used, add 0.004 to the specific gravity reading for every 10° above 80° F (25° C). For every 10° below 80° F (25° C), subtract 0.004.

Battery Storage

Wet cell batteries slowly discharge when stored. They discharge faster when warm than when cold. See **Table 2**. Before storing a battery for the season, clean the case with a solution of baking soda and water. Rinse with clear water and wipe dry. The battery should be fully charged (no change in specific gravity when 3 readings are taken 1 hour apart) and then stored in as cool and dry a place as possible.

Charging

A good state of charge should be maintained in batteries used for starting. Check the battery with a voltmeter as shown in **Figure 12**. Any battery that cannot deliver at least 9.6 volts under a starting load should be recharged. If recharging does not bring it up to strength or if it does not hold the charge, replace the battery.

The battery does not have to be removed from the boat for charging, but it is a recommended safety procedure since a charging battery gives off highly explosive

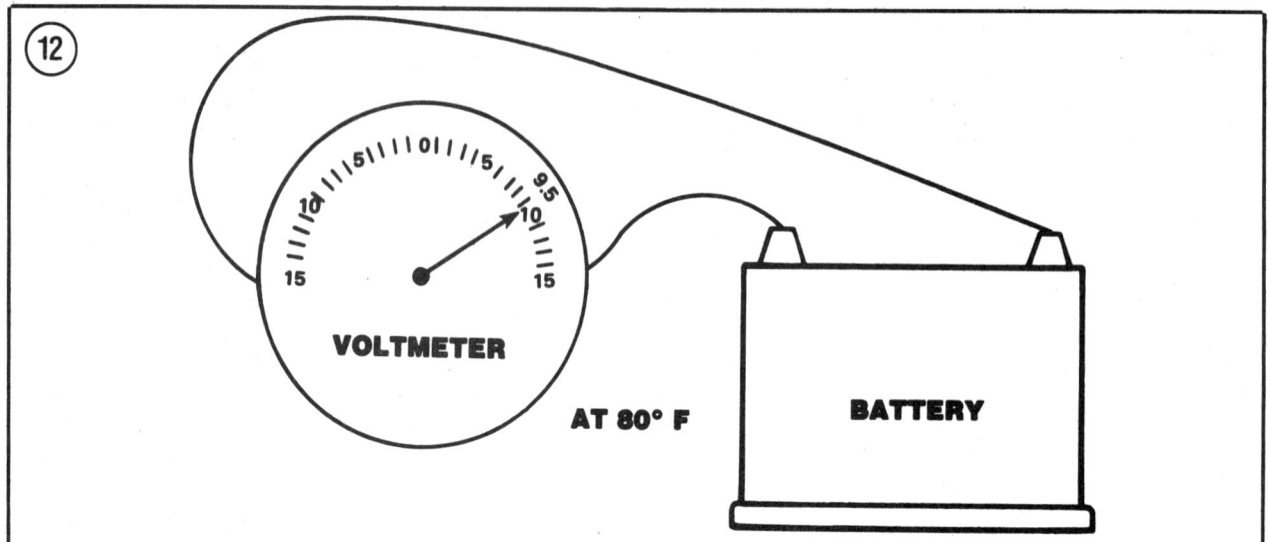

hydrogen gas. In many boats, the area around the battery is not well ventilated and the gas may remain in the area for hours after the charging process has been completed. Sparks or flames occurring near the battery can cause it to explode, spraying battery acid over a wide area.

For this reason, it is important that you observe the following precautions:

a. Do not smoke around batteries that are charging or have been recently charged.

b. Do not break a live circuit at the battery terminals and cause an electrical arc that can ignite the hydrogen gas.

Disconnect the negative battery cable first, then the positive cable. Make sure the electrolyte is fully topped up.

Connect the charger to the battery—negative to negative, positive to positive. If the charger output is variable, select a 4 amp setting. Set the voltage regulator to 12 volts and plug the charger in. If the battery is severely discharged, allow it to charge for at least 8 hours. Batteries that are not as badly discharged require less charging time. **Table 3** gives approximate state of charge for batteries used primarily for cranking. Check the charging progress with the hydrometer.

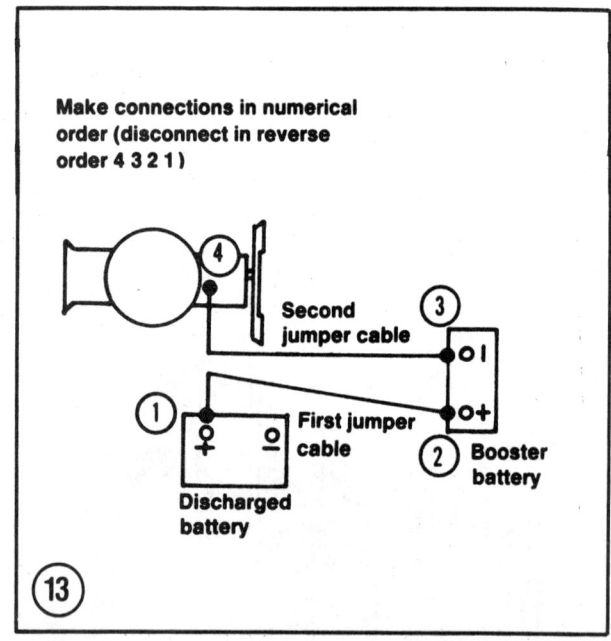

Jump Starting

If the battery becomes severely discharged, it is possible to start and run an engine by jump starting it from another battery. If the proper procedure is not followed, however, jump starting can be dangerous. Check the electrolyte level before jump starting any battery. If it is not visible or if it appears to be frozen, do not attempt to jump start the battery.

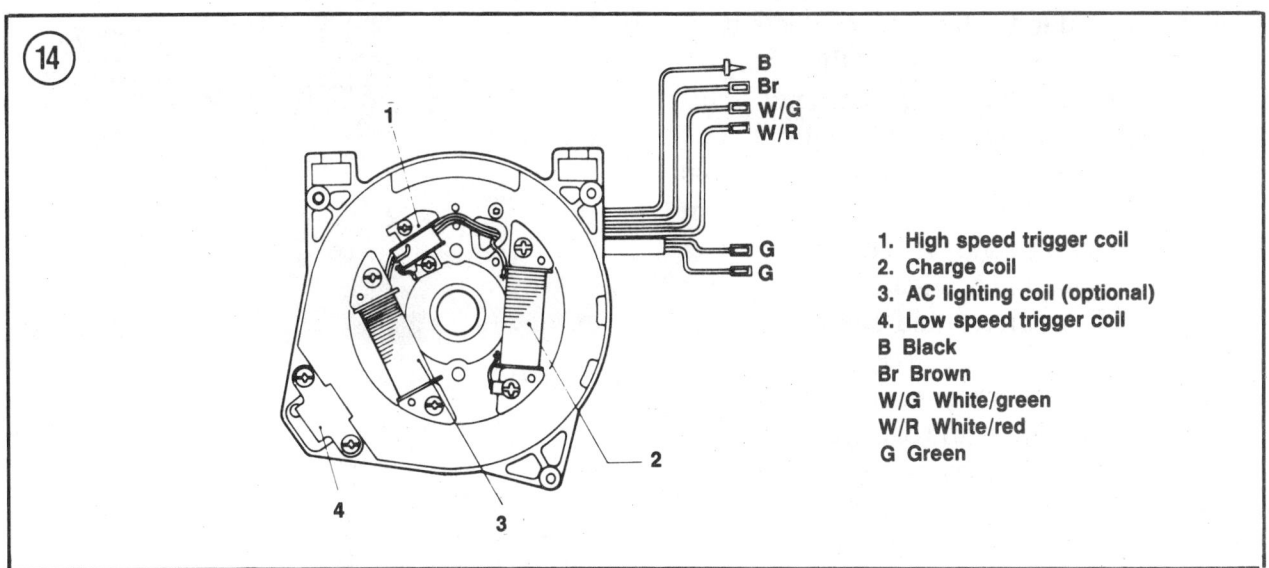

(14)

1. High speed trigger coil
2. Charge coil
3. AC lighting coil (optional)
4. Low speed trigger coil
B Black
Br Brown
W/G White/green
W/R White/red
G Green

(15)

WARNING
Use extreme caution when connecting a booster battery to one that is discharged to avoid personal injury or damage to the system.

1. Connect the jumper cables in the order and sequence shown in **Figure 13**.

WARNING
An electrical arc may occur when the final connection is made. This could cause an explosion if it occurs near the battery. For this reason, the final connection should be made to a good ground away from the battery and not to the battery itself.

2. Check that all jumper cables are out of the way of moving engine parts.
3. Start the engine. Once it starts, run it at a moderate speed.

CAUTION
Running the engine at wide-open throttle may cause damage to the electrical system.

4. Remove the jumper cables in the exact reverse order shown in **Figure 13**. Remove the cables at point 4, then 3, 2 and 1.

BATTERY CHARGING SYSTEM

A battery charging system is standard on electric start models, with an AC lighting coil system used on 6-30 hp manual start models to operate lights and other electrical accessories. All 70 hp and smaller engines use a 6.7 amp system. A 10 amp system is used on all 1984-1985 90-220 hp and 1986 and later 90-130 hp engines. The 1986-on 150-225 hp engines have a 15 amp system.

The battery charging system consists of a battery charging coil on the magneto base of 6-90 hp models (**Figure 14**) or a series of coils on the stator assembly of 115-225 hp models (**Figure 15**) and a series of permanent magnets located in the flywheel rim. Rotation of the flywheel magnets past the charging coil(s) creates alternating current. This current is sent from the AC lighting coil directly to the lights or accessories on manual models. On electric start models, current from the charging coil(s) is sent to a rectifier or rectifier/regulator (A, **Figure 16**) where it is changed to DC current and supplied to the battery or other electrical accessories through the engine wiring harness connector.

Figure 17 is a schematic of a typical battery charging system as used on 9.9 and 15 hp models (the components within the dotted block are found only on electric start models).

System Inspection

A malfunction in the battery charging system will result in an undercharged battery. Perform the following visual inspection to determine the cause of the problem. If the visual inspection proves satisfactory, test the charging coil(s) and rectifier or rectifier/regulator. See Chapter Three.

1. Check the fuse in the line between the rectifier and battery, if so equipped. See B, **Figure 16** (typical).

2. Make sure that the battery cables are connected properly. The red cable must be connected to the positive battery terminal. If polarity is reversed, check for a damaged rectifier.

3. Inspect the battery terminals for loose or corroded connections. Tighten or clean as required.

4. Inspect the physical condition of the battery. Look for bulges or cracks in the case, leaking electrolyte or corrosion build-up.

5. Carefully check the condition of the wiring between the magneto base or stator assembly and the battery for signs of chafing, deterioration or other damage.

6. Check the circuit wiring for corroded,

loose or disconnected connections. Clean, tighten or connect as required.

7. Determine if the electrical load on the battery from accessories is greater than the battery capacity.

Battery charging (lighting) coil replacement is described in this chapter. Rectifier or rectifier/regulator replacement involves disconnecting the electrical leads and removing the attaching bolts. See **Figure 16**. After installing the new rectifier or rectifier/regulator, reconnect the electrical leads.

Battery Charging/Lighting Coil Replacement (6-50 hp)

Refer to **Figure 14** (typical) for this procedure.

1. Remove the flywheel. See Chapter Eight.

2A. Manual model—Disconnect the 2 green lighting coil leads at their quick-disconnect terminals.

2B. Electric start model—Disconnect the 2 green charging coil leads at the rectifier (on 40 and 50 hp models, the rectifier is located under the CDI unit cover).

3. Remove the lighting/charging coil ground lead screw. Remove the 2 coil mounting screws. Remove the coil from the magneto base.

4. Installation is the reverse of removal. Be sure to reconnect the ground lead and route all wires so they do not contact or interfere with any moving components.

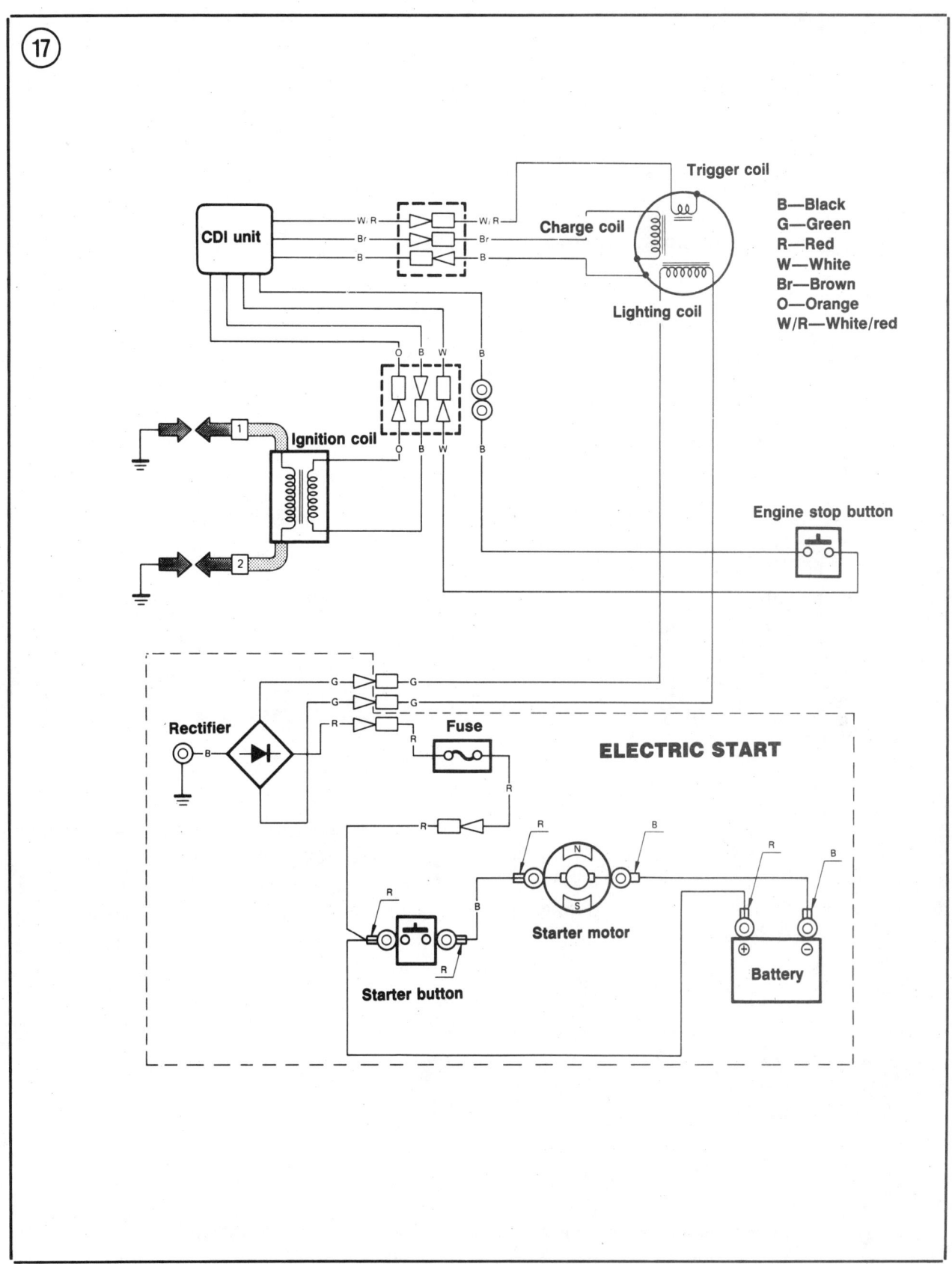

Charging Coil Replacement
(70-90 hp)

1. Remove the flywheel. See Chapter Eight.
2. Remove the CDI unit cover and disconnect the 2 green charging coil leads at the rectifier or rectifier/regulator.
3. Remove the charging coil ground lead screw. Remove the 2 coil mounting screws. Remove the coil from the magneto base.
4. Installation is the reverse of removal. Be sure to reconnect the stator leads correctly and route all wires so they do not contact or interfere with any moving components.

Charging Coil Replacement
(V4 and V6)

The battery charging coils are combined with the ignition system charge coils in an alternator stator on V4 and V6 models (**Figure 15**). The entire stator assembly is replaced if either the battery charging or charge coils are defective.
1. Remove the flywheel. See Chapter Eight.
2. Remove the CDI unit cover and disconnect the stator leads at the rectifier or rectifier/regulator and CDI unit.
3. Remove the bolts holding the stator to the powerhead (**Figure 18**). Remove the stator assembly.
4. Installation is the reverse of removal. Be sure to reconnect the stator leads correctly and route all wires so they do not contact or interfere with any moving components.

ELECTRIC STARTING SYSTEM

Outboards covered in this manual may use a rope-operated mechanical (rewind) starting system or an electric (starter motor) starting system. The electric starting circuit consists of the battery, an ignition switch, neutral start switch, the starter motor, starter relay and connecting wiring.

The plunger-operated neutral start switch is located in the remote control box. A stop button

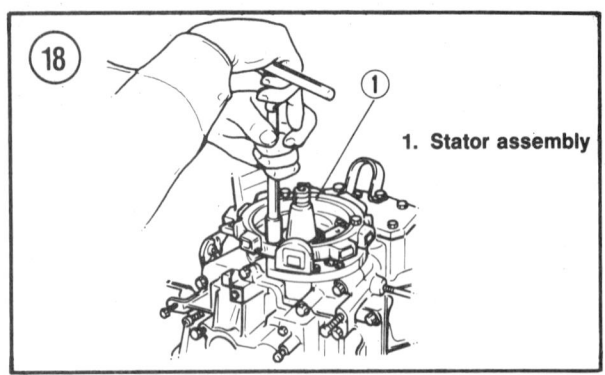

1. Stator assembly

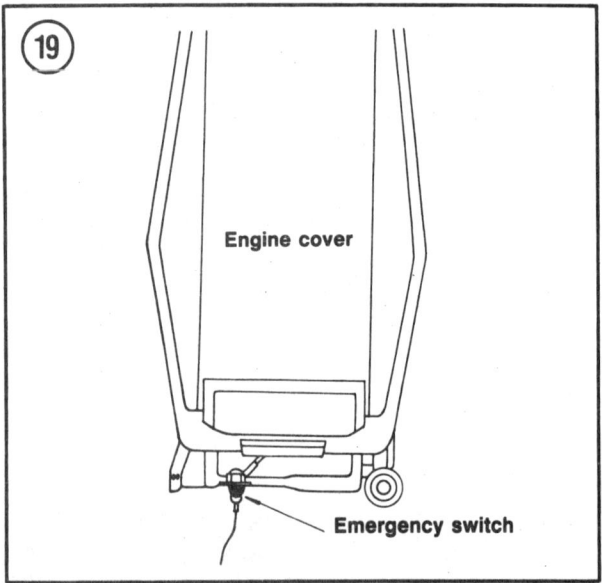

Engine cover

Emergency switch

(kill switch) connected into the ignition system can be used to shut the engine off in case of an emergency. Outboards equipped with a remote control box have this emergency switch located on the remote box. On outboards using a steering handle (tiller or twist grip), the switch is generally installed on the front of the lower cowling. **Figure 19** shows a typical switch installation as seen from the top of the engine.

Starting system operation and troubleshooting are described in Chapter Three.

Starter Motor

The Hitachi marine starter motors used on Yamaha outboards are very similar in design

and operation to those found on automotive engines. They use an inertia-type drive in which external spiral splines on the armature shaft mate with internal splines on the drive assembly.

The starter motor produces a very high torque but only for a brief period of time, due to heat buildup. Never operate the starter motor continuously for more than 10 seconds. Let the motor cool for at least 2 minutes before operating it again.

If the starter motor does not turn over, check the battery and all connecting wiring for loose or corroded connections. If this does not solve the problem, refer to Chapter Three. Except for brush replacement, service to the starter motor is limited to replacement with a new or rebuilt unit.

Removal/Installation

1. Disconnect the negative battery cable.
2. Remove the engine cover.
3. Disconnect the electrical cables from the starter motor terminals. See **Figure 20**.
4. Remove the mounting bolts. If equipped with a plastic pinion cover, remove the cover with the bolts. See **Figure 20**.
5. Remove the starter motor.
6. Installation is the reverse of removal. Tighten mounting bolts to specifications (**Table 4**).

Brush Replacement

Engines may use either a 2- or 3-brush starter design. See **Figure 21** for typical starter

components. Always replace brushes in complete sets.

1. Remove the starter as described in this chapter.
2. Remove the 2 through-bolts from the starter.
3. Remove the commutator end cap cover, circlip and thrust washer, if so equipped.
4. Lightly tap on end of starter drive with a rubber mallet until the lower end cap breaks free of starter housing. Remove end cap, taking care not to lose the brush springs.
5. Check brush spring tension by pulling spring back and releasing it. Replace the spring if it does not snap the brush firmly into position.

NOTE
If corrosion causes the brushes to stick during Step 6, replace the brush holder plate.

6. Remove the brushes and springs from the brush holder plate. See **Figure 22** or **Figure 23** (typical).
7. Inspect the brushes. Replace all brushes if any are oil-soaked or pitted. Replace brushes if worn to the following dimension or less. See **Figure 24** (typical).
   a. 9.9-15 hp—0.18 in.
   b. 25 hp—0.39 in. (1984-1987) or 0.18 in. (1988-on).
   c. 30 hp—0.39 in. (2-cylinder) or 0.35 in. (3-cylinder).
   d. 40-50 hp—0.35 in.
   e. 70-225 hp—0.47 in.
8A. If brush holder is installed in the end cap:
   a. Remove the positive terminal nut, insulators and O-ring. See **Figure 25** (typical).
   b. Remove the screws holding the brush holder to the end cap. See **Figure 26** (typical).

7

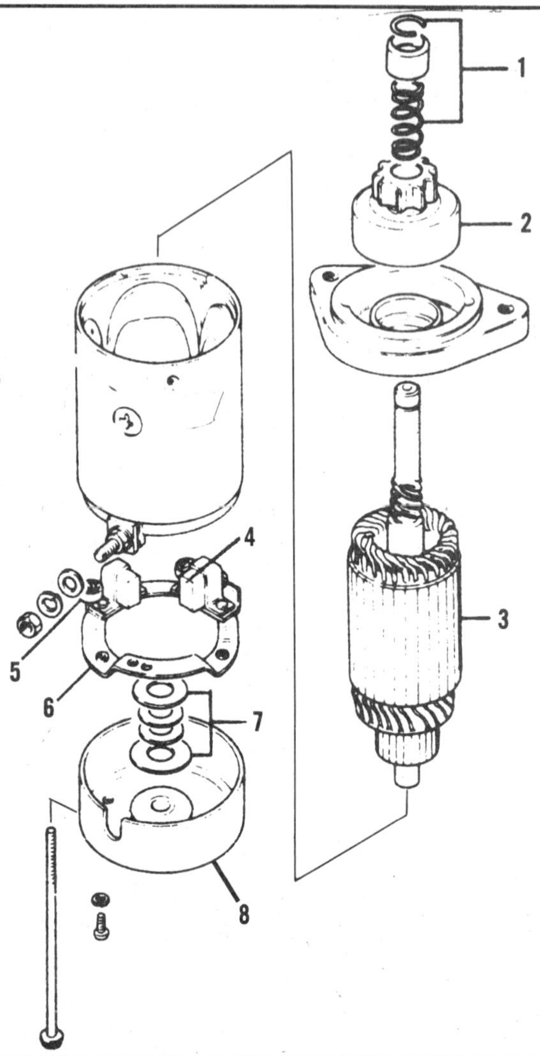

STARTER BRUSHES

1. Pinion stop assembly
2. Pinion gear
3. Armature
4. Brush
5. Brush spring
6. Brush holder
7. Thrust washers
8. End cap

c. Connect an ohmmeter between the insulated brush holder and ground (**Figure 27**). If continuity is shown, the brush holder is grounded and must be replaced.

d. Determine whether brushes are fitted with leads and terminals or not. Make sure you have the correct type of replacement brushes.

e. Brushes with leads and terminals— Remove screw holding old brush. Remove brush from end cap holder. Insert new brush in end cap holder and attach terminal to cap with screw.

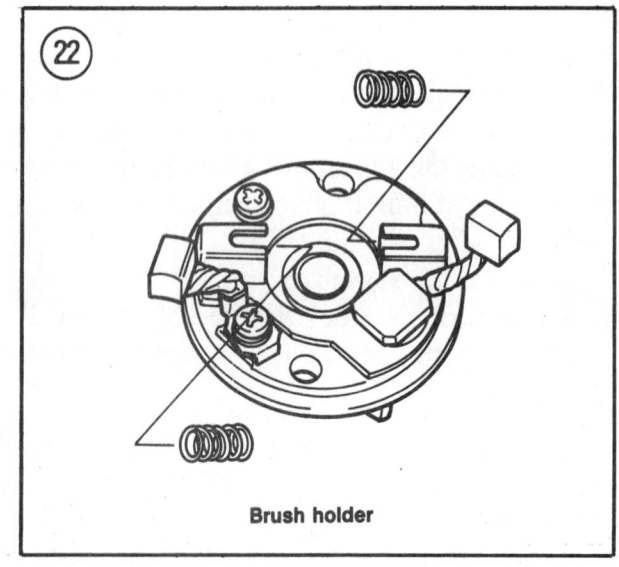

Brush holder

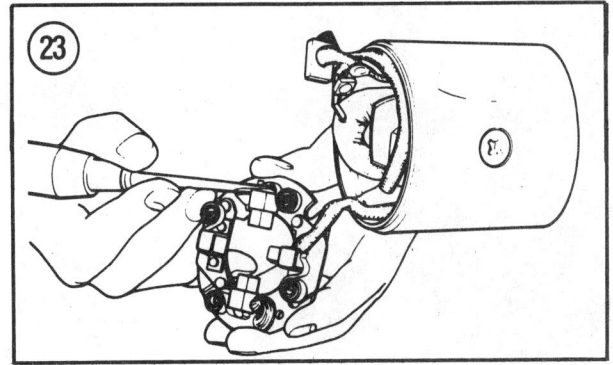

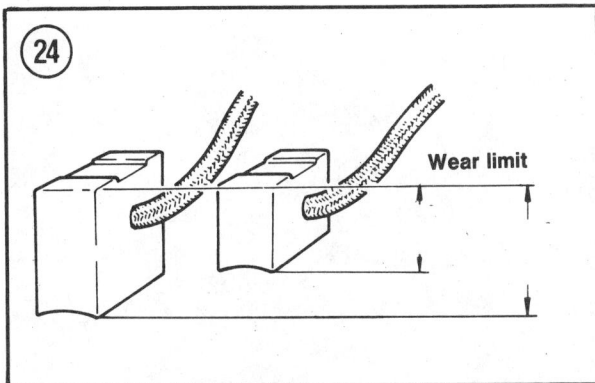

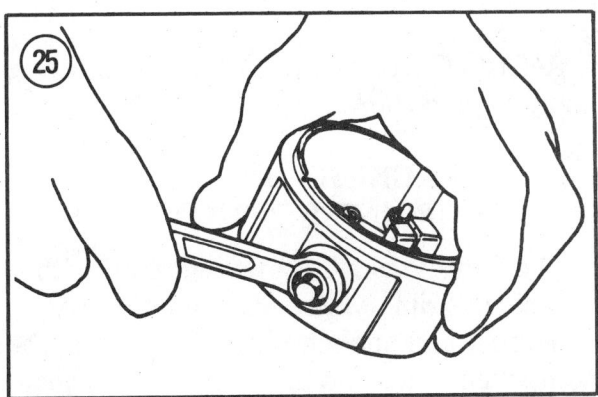

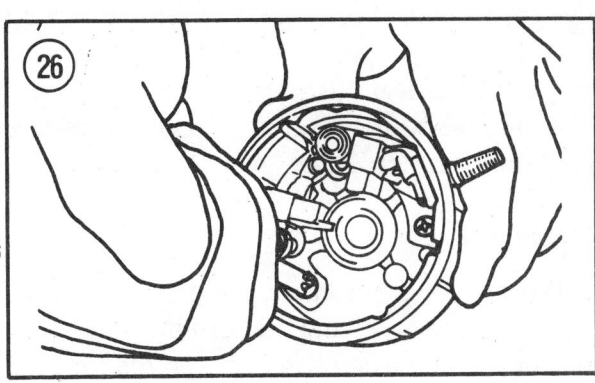

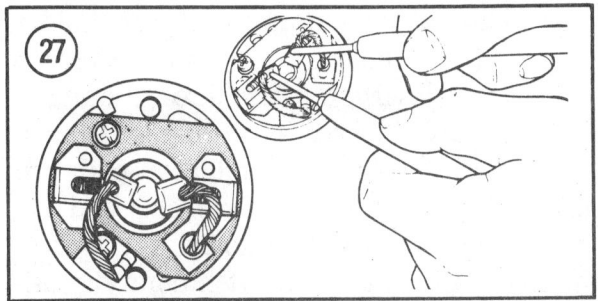

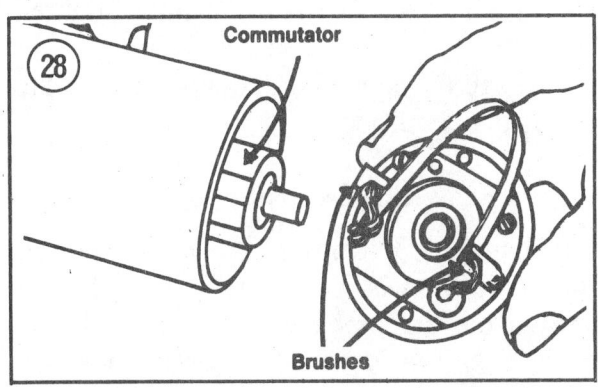

f. Brushes without leads—Remove old brush and spring from brush holder. Insert new brush with spring in holder. Reattach holder to end cap.

g. Press the brushes into the holders and use a narrow strip of flexible metal or plastic as shown in **Figure 28** to keep them in place.

8B. If brush holder is separate from end cap and remains on commutator end of the armature:

a. Check brush holder straightness. If brushes do not show full-face contact with commutator, holder is probably bent.

b. Connect an ohmmeter between the insulated brush holder and ground. If continuity is shown, the brush holder is grounded and must be replaced.

c. Install new brushes with springs in brush holder.

d. Keep brushes recessed in holder and install holder on commutator end of armature shaft.

e. Align the cutout in the brush holder with the hole in the front cover to permit insertion of the through-bolt.

9. Make sure the end cap O-ring is installed (if used). Align the commutator end cap notch with the frame tab and install end cap to starter frame. If brush holder is a part of the end cap, remove the temporary brush retainer as brushes slip over the commutator.

10. Install through-bolts and tighten securely.

11. Install thrust washer, circlip and end cap cover, if so equipped.

Starter Relay Replacement

A bank of 3 identical relays is used on 70-225 hp models. The starter relay is the top one on the bracket (**Figure 29**); the other 2 control the power trim/tilt system. A single relay mounted on the bottom cowling is used on other electric start models.

1. Disconnect the negative battery cable.

2. Remove the engine cover.

3. Disconnect the brown and red or brown and black relay leads.

4. Disconnect the battery and starter cables from the relay terminals.

5. Remove the bolts holding the relay to the cowling or relay bracket. On some models, a ground lead is installed under one of the mounting bolts.

6. Installation is the reverse of removal.

IGNITION SYSTEMS

The outboards covered in this manual use one of the following ignition systems:

a. Magneto breaker point.
b. Simultaneous CDI (capacitor discharge ignition).
c. Independent CDI (capacitor discharge ignition).
d. Yamaha micro-computer ignition (YMIS).

Refer to Chapter Three for troubleshooting and test procedures.

MAGNETO BREAKER POINT IGNITION

The 2 hp uses a magneto ignition with a combined primary/secondary ignition coil, a condenser and one set of breaker points. The components are mounted on the magneto base under the flywheel which revolves with the crankshaft. The system is self-energizing and does not require the use of a battery to provide electrical current.

Troubleshooting and test procedures are given in Chapter Three.

Operation

As the flywheel rotates, magnets around its outer diameter create a current that flows through the closed breaker points on the

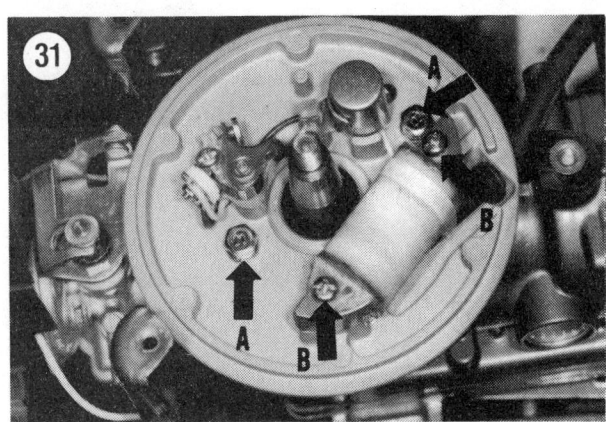

magneto base. This flow of current through the coil primary winding builds a strong magnetic field. When the cam opens the point set, the magnetic field collapses, inducing a high voltage (approximately 18,000 volts) in the secondary winding of the coil. This voltage is sent to the spark plug, where it jumps the plug gap and fires the cylinder. The condenser absorbs any residual current remaining in the primary windings. This eliminates arcing at the points and produces a stronger spark at the plug. The breaker points close and the flywheel continues to rotate as the process starts again.

Magneto Base
Removal/Installation

Magneto base removal is not required to replace a component on the base. Remove the base only if damaged or if power head is being disassembled.
1. Remove the engine cowling.
2. Remove the fuel tank. See Chapter Six.
3. Remove the rewind starter. See Chapter Ten.
4. Remove the flywheel. See Chapter Eight.
5. Disconnect the spark plug lead from the spark plug.
6. Disconnect the magneto base lead wire at the bullet connector. Disconnnect the ground wire attached to the cylinder block. See **Figure 30**.

7. Remove the 2 screws holding the magneto base to the power head (A, **Figure 31**). Remove the magneto base.
8. Installation is the reverse of removal.

Breaker Point and
Condenser Replacement

See *Tune-up*, Chapter Four.

Ignition Coil
Removal/Installation

1. Remove the flywheel. See Chapter Eight.
2. Disconnect the coil lead wires at the breaker point set and ground.
3. Disconnect the spark plug lead. Remove the plug boot from the lead.
4. Remove the screws holding the ignition coil to the magneto base. See B, **Figure 31**.
5. Pull the spark plug lead through the magneto base grommet and remove the coil.
6. Installation is the reverse of removal. Insert the spark plug lead through the magneto base cutout before installing coil mounting screws.

YAMAHA 3-70 HP CDI
(CAPACITOR DISCHARGE IGNITION)

Yamaha 3, 4 and 5 hp models have a CDI ignition using a single ignition coil, a charge coil and 2 trigger coils with different output voltages. The CDI unit combines the 2 different signal waves to electronically advance spark timing when firing the spark plug.

Yamaha 6-25 (1984-1987) hp and 2-cylinder 30 hp models have a simultaneous ignition which uses a single charge coil to supply current to the CDI unit. The CDI unit in turn fires the single ignition coil (with 2 spark plug leads).

Yamaha 2-cylinder 25 hp (1988-on) models have an independent ignition using a single charge coil and 2 trigger coils to supply current to the CDI unit. The CDI unit in turn

7

triggers the 2 ignition coils to fire the spark plugs as required.

Yamaha 3-cylinder 30 hp and 40-70 hp models have an independent ignition using a single charge coil and 3 trigger coils to supply current to the CDI unit. The CDI unit in turn triggers the 3 ignition coils to fire the spark plugs as required.

Spark timing on 6-70 hp engines is changed by rotating the trigger coil position relative to the flywheel hub magnets.

Operation

The major components of the CDI system on all 4-70 hp models are the flywheel, charge coil, trigger coil assembly, CDI unit, ignition coil, spark plugs and connecting wiring. **Figure 32** shows a typical 2-cylinder CDI system.

The charge coil is located under the flywheel on the magneto base plate. The flywheel is fitted with permanent magnets inside its outer rim (**Figure 33**). As the crankshaft and flywheel rotate, the flywheel magnets pass the stationary charge coil, creating an AC voltage in the charge coil. The AC voltage is sent to the CDI unit (switchbox) where it is rectified and stored in a capacitor for release.

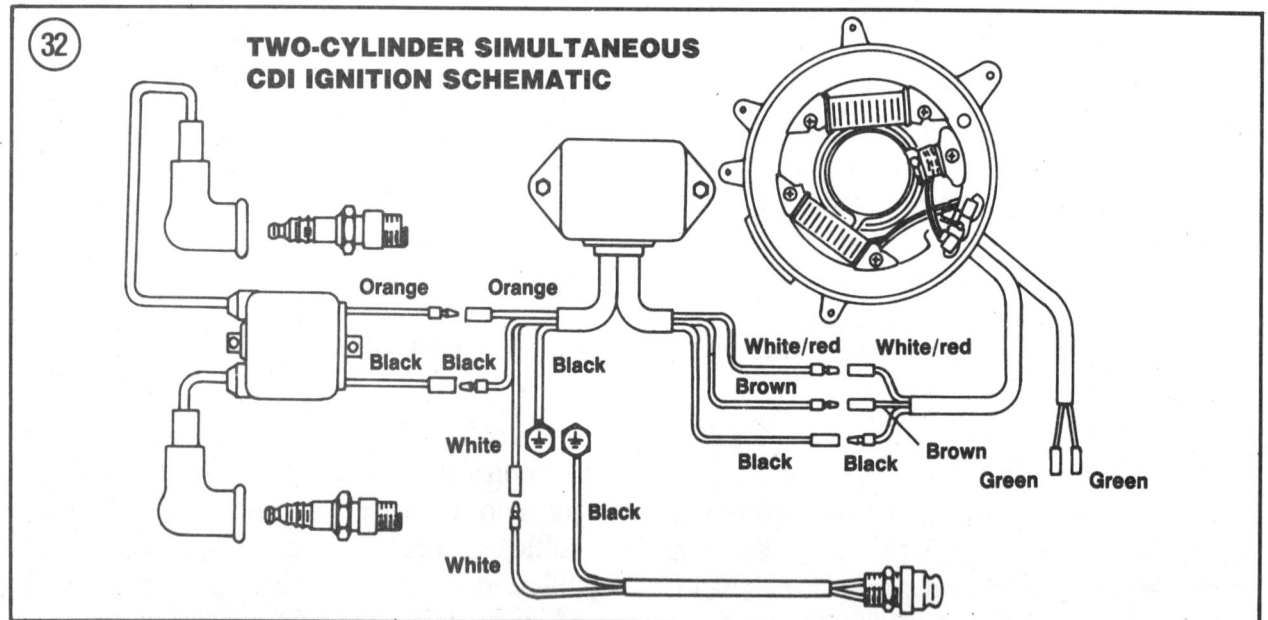

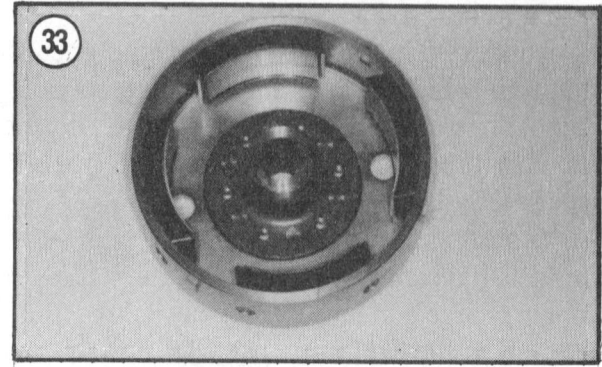

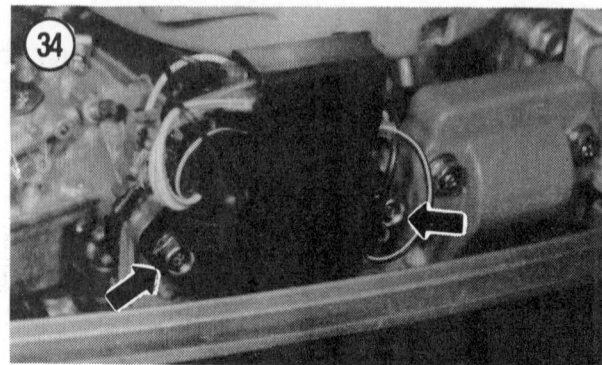

A trigger coil assembly is also mounted under the flywheel (3, 4 and 5 hp models use a low- and high-speed coil). A second set of magnets located around the flywheel hub pass the stationary trigger coil assembly, creating an AC voltage in the trigger coil(s). The AC voltage is routed to an electronic switch (SCR) in the CDI unit. The SCR discharges the capacitor into the ignition coil at the proper time and in the correct firing order sequence for multi-cylinder engines.

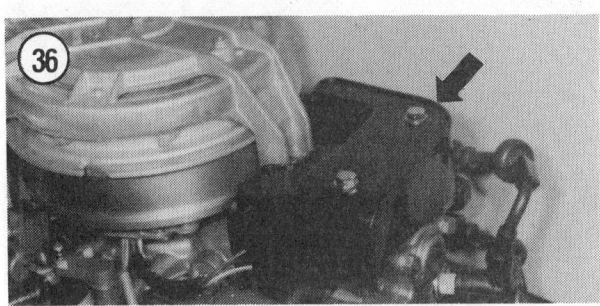

The simultaneous ignition system used on 6-30 hp engines is also called a "waste spark" system. When the piston in one cylinder is at TDC, the other is at BTDC. If both spark plugs fire at the same time, the spark in the cylinder with the piston at TDC is "used" while the spark in the other cylinder is "wasted."

Ignition advance on 3, 4 and 5 hp models is controlled by the CDI unit, which combines the 2 different trigger coil signal waves electronically. Ignition advance on 6-70 hp models is mechanical, with the magneto base and throttle interlocked.

Ignition timing should be adjusted (Chapter Five) whenever a component is replaced.

Charge Coil, Battery Charging Coil, Lighting Coil or Trigger Coil Replacement (Except 3, 4 and 5 hp)

See **Figures 34-36** for various types of CDI unit connector housings. The leads can be pulled out of the housing shown in **Figure 34** and **Figure 35**. The cover screws and cover must be removed from the types shown in **Figure 36** to gain access to the leads. **Figure 36** shows the cover installed; **Figure 37** shows access to the CDI unit and connectors with the cover removed.

1. Disconnect the negative battery cable, if so equipped.
2. Remove the engine cover.
3. Remove the flywheel. See Chapter Eight.
4A. Charge coil—Disconnect the brown lead at the CDI unit connector.
4B. Trigger coil—Disconnect the red-white lead at the CDI unit connector.
4C. Lighting coil—Disconnect the green leads at their quick-disconnect points.
4D. Battery charging coil—Disconnect the green leads at the rectifier.
5. Disconnect the magneto base link from the magneto control lever.

6. Remove the screws holding the magneto base to the retainer plate or power head. Remove the magneto base (**Figure 38**).

7. Loosen the clamp screw holding the defective coil lead to the magneto base.

8. Remove the 2 screws and washers holding the defective coil to the magneto base.

9. Remove the defective coil from the magneto base, pulling the disconnected lead(s) through the clamp and protective sleeve (if used).

10. Installation is the reverse of removal. Apply a light coat of Yamalube All-purpose marine grease to the retainer ring and all other points where the base will contact other components. Wipe attaching screw threads with Loctite Type A or Threadlock 1342 and tighten securely.

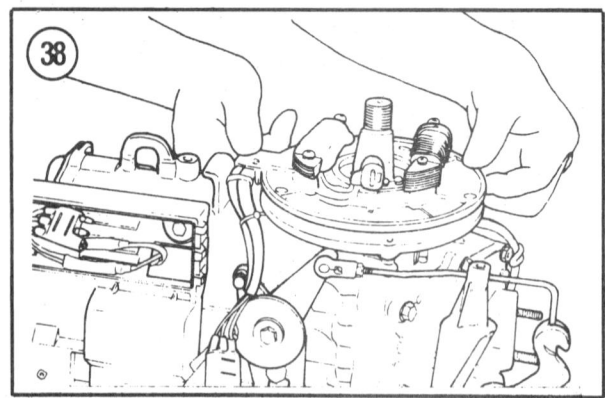

Charge Coil or Trigger
Coil Replacement
(3, 4 and 5 hp)

1. Remove the flywheel. See Chapter Eight.

2. Disconnect the lead(s) from the defective coil at the connector holder above the CDI unit. See **Figure 34** (typical).

3A. 3 hp—Remove the 2 bolts holding the magneto base to the crankcase halves. Remove the base. See **Figure 39** (typical).

3B. 4 and 5 hp—Remove the 2 bolts holding the magneto base to the retainer ring. Remove the base. See **Figure 39**.

4. Remove the screws holding the defective coil to the magneto base. **Figure 40** shows the location of the charge coil (A), low-speed trigger coil (B) and high-speed trigger coil (C).

5. Loosen the wire harness clamp (A, **Figure 41**) and remove the coil from the magneto base, pulling the lead wire(s) from the harness sleeve.

6. Installation is the reverse of removal. Check O-ring on magneto base (B, **Figure 41**) and replace if damaged. Lubricate O-ring with Yamalube All-purpose Marine Grease. Apply a light coat of Yamalube All-purpose

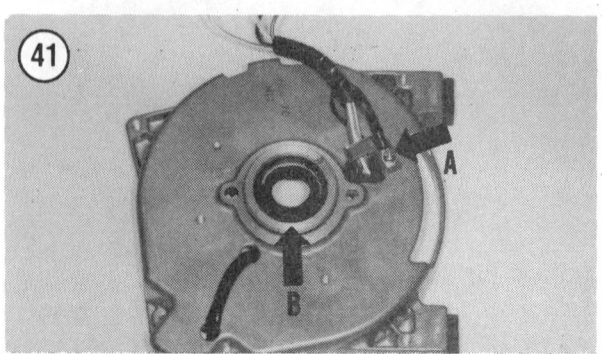

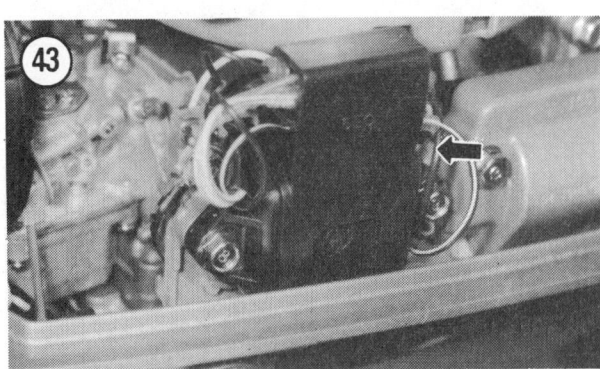

Marine Grease to the retainer ring and all other points where the base will contact other components. Wipe attaching screw threads with Loctite Type A or Threadlock 1342 and tighten securely.

CDI Unit Replacement

1. Disconnect the negative battery cable, if so equipped.

2. Remove the engine cover.

3. Remove CDI housing cover (if so equipped) and disconnect all CDI unit leads at their connectors.

4. Disconnect the black CDI unit ground lead.

5. Remove the bolts holding the CDI unit to the mounting bracket or power head. See **Figure 42** or **Figure 43** (typical).

6. Remove the CDI unit. On Yamaha 4 and 5 hp models, the CDI unit is contained within the connector housing. Slide the connector housing off the CDI unit.

7. Installation is the reverse of removal. Tighten attaching bolts to 30-40 in.-lb.

Secondary Ignition Coil Replacement

1. Disconnect the negative battery cable, if so equipped.

2. Remove the engine cover.

3. Disconnect the spark plug lead(s) at the spark plug(s).

4A. Simultaneous ignition—Disconnect the magneto base lead at the secondary coil.

4B. Independent ignition—Disconnect the coil primary leads at their bullet connectors.

5. Remove the attaching bolts and washers. See **Figure 44** (1-cylinder) or **Figure 45** (2-cylinder) for typical installations. Remove coil from mounting bracket.

6. Installation is the reverse of removal. If coil has a black ground lead attached, be sure to reinstall it under one of the attaching bolts.

7

YAMAHA 90-200 HP CDI (CAPACITOR DISCHARGE IGNITION)

Yamaha 90-200 hp models use an alternator stator containing 2 charge coils and 8 battery charging coils. See **Figure 46** (typical). A timer base underneath the stator contains 2 trigger coils (90-130 hp) or 3 trigger coils (150-200 hp) enclosed in iron to help build up the magnetic field and prevent interference from external sources.

With the 90 hp engine, one trigger coil controls the No. 1 and No. 3 cylinders (**Figure 47**). The other trigger coil controls the No. 2 cylinder. On V4 engines, one trigger coil controls the No. 1 and No. 2 cylinders; the second trigger coil controls the No. 3 and No. 4 cylinders (**Figure 48**). The third trigger coil on V6 engines controls the No. 5 and No. 6 cylinders.

Operation

The outer rim of the flywheel contains a series of magnets which create a magnetic field during rotation. This magnetic field cuts through the charge coil windings and produces an alternating current of positive and negative waveforms. This current is sent to the CDI unit where it is changed into direct current by an internal rectifier and stored in a capacitor.

The rotation of a timing magnet in the flywheel hub past the trigger coils on the timer base also creates a magnetic field. See **Figure 49**. As the flywheel continues to rotate, this magnetic field collapses, inducing a small voltage pulse in the trigger coil. This pulse causes an electronic switch in the CDI unit to close, discharging the stored voltage into the appropriate ignition coil where it is stepped up to a higher voltage and sent to the spark plug.

Ignition advance is mechanical, with the timer base and throttle interlocked. Ignition

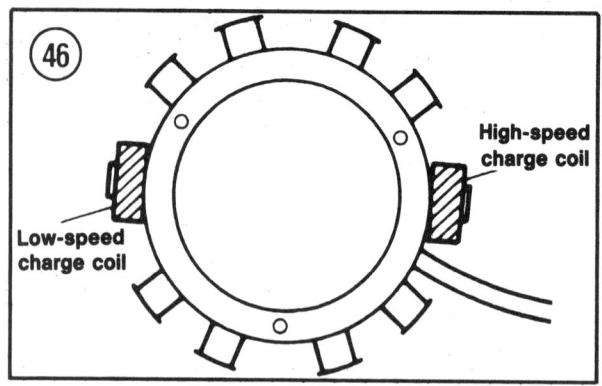

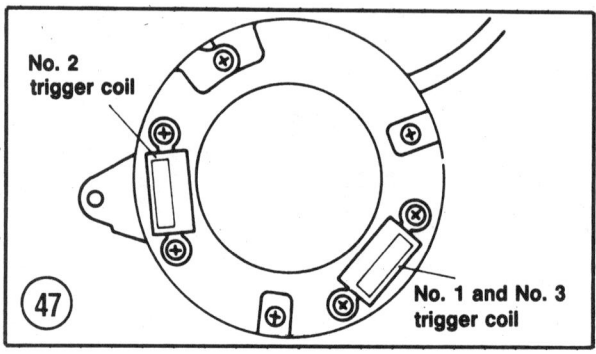

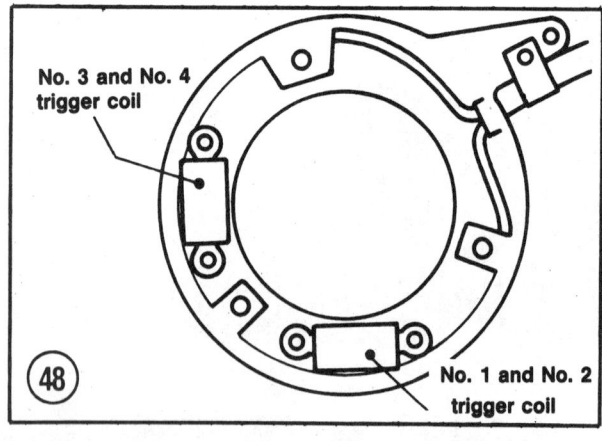

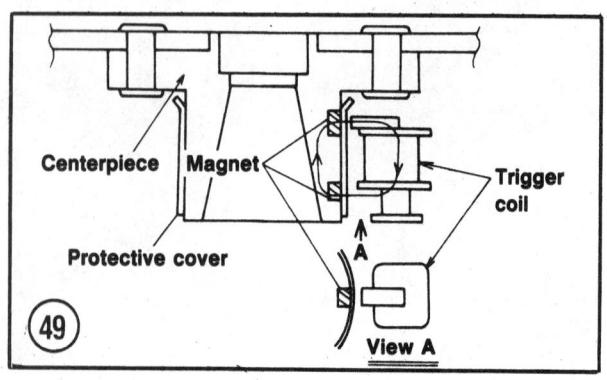

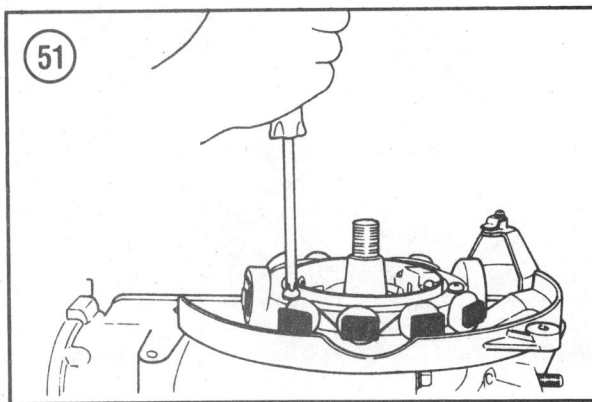

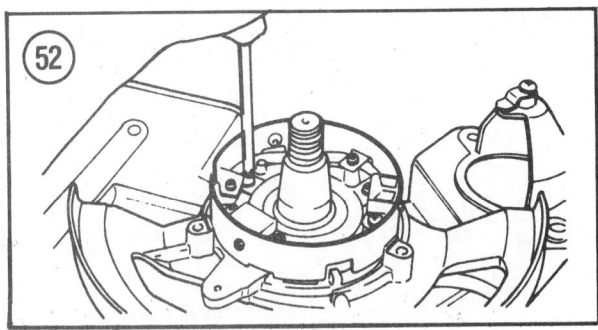

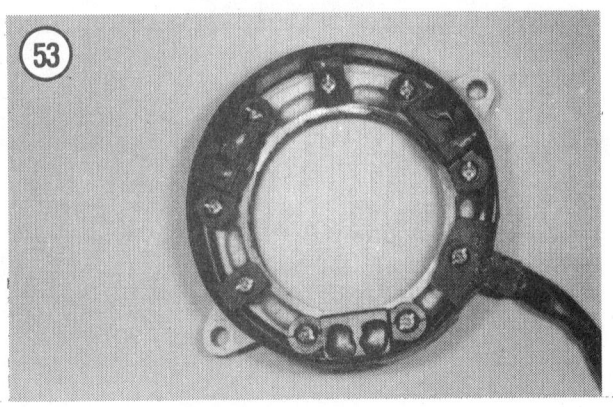

timing should be adjusted (Chapter Five) whenever a component is replaced.

Alternator Stator
Removal/Installation

1. Disconnect the negative battery cable, if so equipped.
2. Remove the engine cover.
3. Remove the flywheel. See Chapter Eight.
4. Remove the CDI unit cover. Disconnect all stator leads at the CDI unit. See **Figure 50** (typical).
5. Remove the screws holding the alternator stator to the timer base (**Figure 51**). Remove the alternator stator.
6. Installation is the reverse of removal.

Timer Base Removal/Installation

1. Remove the alternator stator as described in this chapter.
2. Disconnect the timer base link from the control lever.
3. Remove the CDI unit cover. Disconnect the trigger coil leads at the CDI unit. See **Figure 50** (typical).
4. Remove the timer base retaining screws. Remove the timer base. See **Figure 52** (typical).

NOTE
The 1984 timer base has a sheet metal enclosure (Figure 53) not used on 1985 and later models.

5. Installation is the reverse of removal.

Charge Coil or Battery
Charging Coil Replacement

The alternator stator is replaced as an assembly if any coil is defective.

Trigger Coil Replacement

The timer base is replaced as an assembly if any coil is defective.

**Secondary Ignition
Coil Replacement**

1. Disconnect the negative battery cable.
2. Remove the engine cover.
3. Disconnect the coil primary leads at their bullet connectors.
4. Disconnect the coil secondary lead at the spark plug.
5. Remove the attaching bolts and washers. See **Figure 54** (typical). Remove coil from mounting bracket.
6. Installation is the reverse of removal. If coil has a black ground lead attached, be sure to reinstall it under one of the attaching bolts.

CDI Unit Removal/Installation

1. Disconnect the negative battery cable.
2. Remove the engine cover.
3. Remove the CDI unit cover.

> *NOTE*
> *The CDI unit leads are grouped in sets of 3 or 4 and installed in grommets on some engines to assist in proper reconnection. Do not remove the leads from the grommets unless replacing a component.*

4. Disconnect all leads at the CDI unit terminals (**Figure 50**).
5. Remove the mounting screws at the top and bottom of the CDI unit. Remove the CDI unit from its mounting bracket.
6. Installation is the reverse of removal.

**Ignition Control Unit
Removal/Installation
(115-220 hp Only)**

1. Disconnect the negative battery cable.
2. Remove the engine cover.
3. Remove the CDI unit cover.
4. Remove the ignition control unit mounting screws (**Figure 55**). Remove the ignition control unit from the CDI unit mounting bracket.

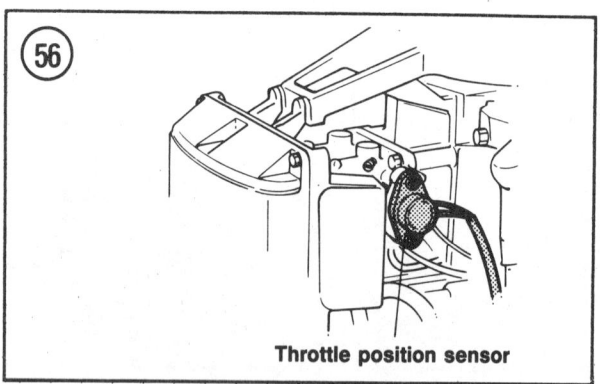

Throttle position sensor

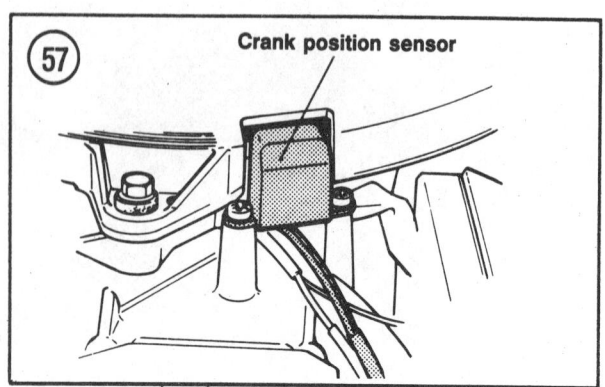

Crank position sensor

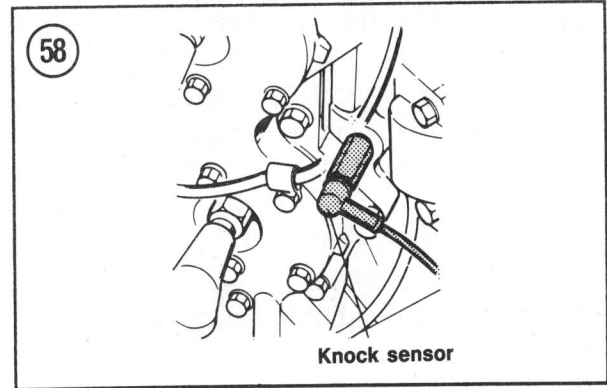

Knock sensor

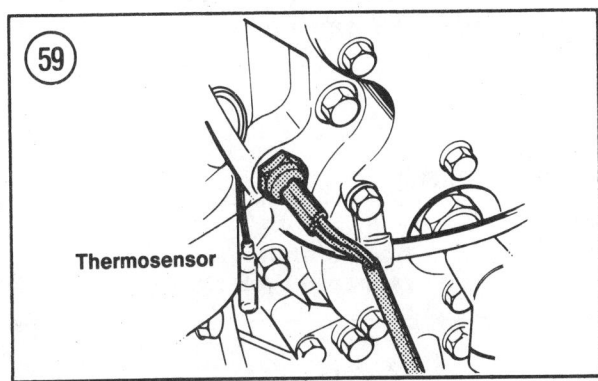

Thermosensor

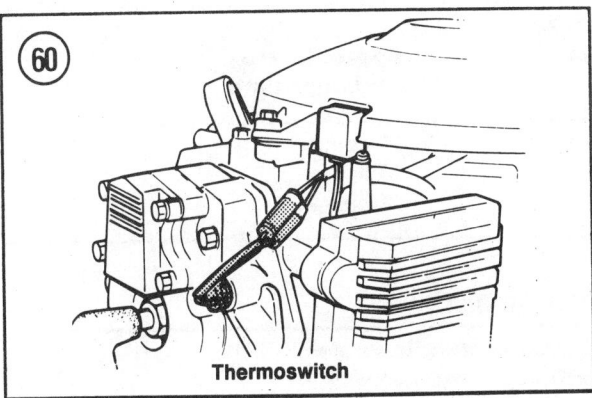

Thermoswitch

5. Disconnect the ignition control unit leads from the CDI unit.

6. Installation is the reverse of removal.

YAMAHA MICRO-COMPUTER IGNITION (YMIS)

The 220 V6 Special and 225 hp Excel ignition system is similar to that used on the other V6 models, but incorporates a

micro-computer to electronically control engine operation and ignition timing according to data received from a variety of sensors. The micro-computer central processing unit (CPU) receives voltage signals from the throttle position sensor (**Figure 56**), crank position sensor (**Figure 57**), cylinder head knock sensor (**Figure 58**), a thermo sensor (**Figure 59**) and a thermo switch in each cylinder head (**Figure 60**). The micro-computer compares the data provided by these sensors to the operating strategies program in its memory and determines the correct ignition timing for a given engine operating condition. The micro-computer then signals the CDI unit to advance the timing as desired.

The micro-computer also:

a. Provides a fixed amount of advance during engine cranking.

b. Retards timing whenever engine detonation is detected by the knock sensor.

c. Alternately misfires the cylinders if engine speed exceeds 5,900-6,300 rpm until the engine speed returns to a safe level.

d. Misfires all 6 cylinders to stop the engine if it is shifted into REVERSE while cruising. At low speeds, it prevents stalling and allows REVERSE to be used as a brake when docking.

e. Misfires the cylinders in a specified order whenever engine temperature exceeds a predetermined level until temperature returns to a safe level.

Proper operation of an outboard equipped with YMIS depends upon the use of the correct fuel (see Chapter Four), correct linkage adjustment (see Chapter Five) and a fully charged battery. If battery voltage drops below 7 volts, ignition timing is automatically set at 7° BTDC, engine speed will vary by 1,000-1,500 rpm and maximum engine speed will not exceed 4,300-4,800 rpm. Yamaha

7

recommends the installation of a voltmeter to quickly check battery condition before operating the engine.

The YMIS system is quite complex, sophisticated and expensive. *Never* remove the cover from the micro-computer unit and always handle it carefully when removal from the engine is required.

Component Replacement

The YMIS micro-computer is attached to the ignition coil bracket on the aft of the engine by 4 bolts. Remove the bolts and disconnect the wiring to remove the unit.

To remove/install all other electrical components (alternator stator, timer base, ignition coil and CDI unit), see *Yamaha 90-200 HP CDI (Capacitor Discharge Ignition)* in this chapter.

Throttle position sensor removal, installation and adjustment procedures are given in Chapter Five. Replacement procedures for all other sensors and thermo switches are provided in Chapter Eight under *Disassembly, V4 and V6*.

Table 1 BATTERY CAPACITY (HOURS)

Accessory draw	80 amp-hour battery provides continuous power for:	Approximate recharge time
5 amps	13.5 hours	16 hours
15 amps	3.5 hours	13 hours
25 amps	1.8 hours	12 hours
Accessory draw	105 amp-hour battery provides continuous power for:	Approximate recharge time
5 amps	15.8 hours	16 hours
15 amps	4.2 hours	13 hours
25 amps	2.4 hours	12 hours

Table 2 SELF-DISCHARGE RATE

Temperature	Approximate allowable self-discharge per day for first 10 days (specific gravity)
100° F (37.8° C)	0.0025 points
80° F (26.7° C)	0.0010 points
50° F (10.0° C)	0.0003 points

Table 3 STATE OF BATTERY CHARGE

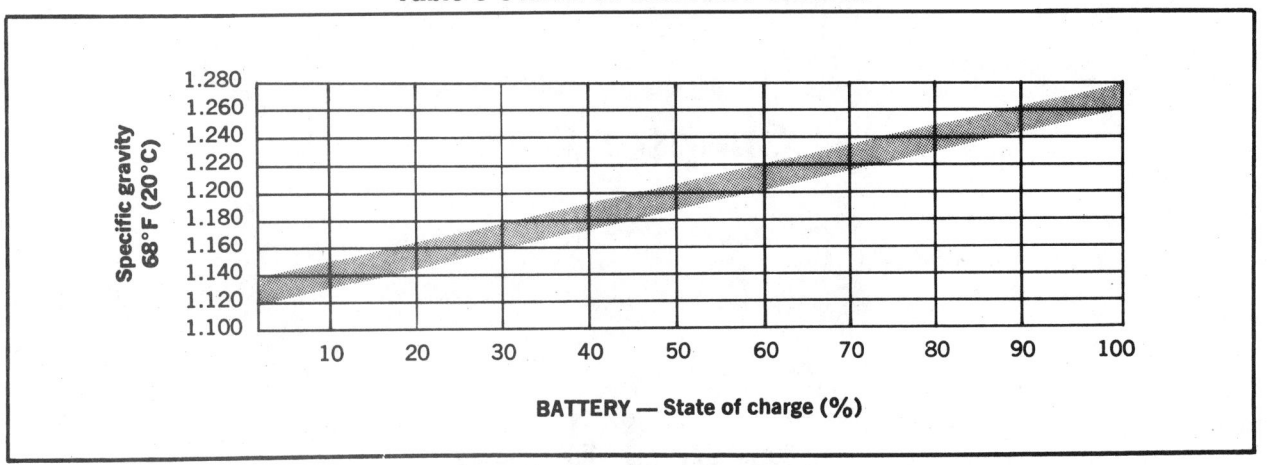

Table 4 TIGHTENING TORQUES

Fastener	ft.-lb.	N•m
M5 bolt or 8 mm nut	3.6	5
M6 bolt or 10 mm nut	5.8	8
M8 bolt or 12 mm nut	13	18
M10 bolt or 14 mm nut	25	36
M12 bolt or 17 mm nut	30	42

7

Chapter Eight

Power Head

This chapter covers the basic repair of Yamaha outboard power heads. The procedures involved are similar from model to model, with minor differences. Some procedures require the use of special tools. These can be purchased from a dealer or directly from Kent-Moore, SPX Corporation, 29784 Little Mack, Roseville, MI 48006-2298. Certain tools may also be fabricated by a machinist, often at substantial savings. Power head stands are available from specialty shops such as Bob Kerr's Marine Tool Co. (P.O. Box 1135, Winter Garden, FL 32787).

Work on the power head requires considerable mechanical ability. You should carefully consider your own capabilities before attempting any operation involving major disassembly of the engine.

Much of the labor charge for dealer repairs involves the removal and disassembly of other parts to reach the defective component.

Even if you decide not to tackle the entire power head overhaul after studying the text and illustrations in this chapter, it can be less expensive to perform the preliminary operations yourself and then take the power head to your dealer. Since many marine dealers have lengthy waiting lists for service (especially during the spring and summer season), this practice can reduce the time your unit is in the shop. If you have done much of the preliminary work, your repairs can be scheduled and performed much quicker.

Repairs go much faster and easier if your motor is clean before you begin work. There are special cleaners for washing the motor and related parts. Just spray or brush on the cleaning solution, let it stand, then rinse it away with a garden hose. Clean all oily or greasy parts with fresh solvent as you remove them. Place the parts with their fasteners in trays or cupcake tins in the order of removal.

This will speed assembly while helping to ensure that all parts are properly reinstalled.

WARNING
Never use gasoline as a cleaning agent. It presents an extreme fire hazard. Be sure to work in a well-ventilated area when using cleaning solvents. Keep a fire extinguisher rated for gasoline and oil fires nearby in case of emergency.

Once you have decided to do the job yourself, read this chapter thoroughly until you have a good idea of what is involved in completing the overhaul satisfactorily. Make arrangements to buy or rent any special tools necessary and obtain replacement parts before you start. It is frustrating and time-consuming to start an overhaul and be unable to complete it because the necessary tools or parts are not at hand.

Before beginning the job, re-read Chapter Two of this manual. You will do a better job with this information fresh in your mind.

Remember that new engine break-in procedures should be followed after an engine has been overhauled. Refer to your owner's manual for specific instructions.

Since this chapter covers a large range of models over a lengthy time period, the procedures are sometimes generalized to accommodate all models. Where individual differences occur, they are specifically pointed out. The power heads shown in the accompanying pictures are current designs. While it is possible that the components shown in the pictures may not be identical with those being serviced, the step-by-step procedures may be used with all models covered in this manual. The serial and model numbers should be used when ordering any replacement parts for all engine components.

Tables 1-3 are at the end of the chapter.

FASTENERS AND TORQUE

Always replace a worn or damaged fastener with one of the same size, type and torque require-ment. Unless a thread locking agent (e.g. Loctite) is specified, bolt threads should be lubricated with Yamalube Two-Cycle Outboard oil before torque is applied.

Unless otherwise specified, power head fasteners should be tightened in 2 steps. Tighten to 50 percent of the torque value in the first step, then to 100 percent in the second step. Power head tightening torques are listed in **Table 1**.

Power head torque sequences are provided in this chapter. They are also embossed on the power head components. The embossed sequence should be followed if it differs from that given in this chapter, as it reflects a product change during the model run.

Retighten the power head bolts after the engine has been run for 15 minutes and allowed to cool. It is a good idea to retorque them again after 10 hours of operation. To retighten the power head mounting fasteners properly, back them out one turn and then tighten to the torque specifications.

When spark plug(s) are reinstalled after an overhaul, tighten to the specified torque. Warm the engine to normal operating temperature, let it cool down and retorque the plug(s).

Where a specification is not provided for a given bolt or nut, use the standard bolt and nut torque according to fastener size as listed at the end of **Table 1**.

ENGINE SERIAL NUMBER

Yamaha engines are identified by engine serial number and model number. The engine serial numbers are stamped on a welch plug installed on the power head (**Figure 1**). Exact location of the welch plug varies according to model. The model and serial number is stamped on a plate riveted to the port size clamp bracket.

This information identifies the outboard engine if there are unique parts or if internal changes have been made during the model run. The serial and model numbers should be used when ordering any replacement parts for your outboard.

GASKETS AND SEALANT

Yamaha Gasket Maker is applied in a thin, even coat covering the entire mating surface when sealing the crankcase cover and cylinder block. Both mating surfaces must be free of all sealant residue, dirt and oil. Lacquer thinner, acetone or similar solvents work well when used with a broad, flat scraper or a somewhat dull putty knife. Solvents with an oil, wax or petroleum base should not be used. Clean the aluminum surfaces carefully to avoid nicking them with the scraper or putty knife.

Once the gasket surfaces are cleaned, place the mating surface of each component on a large pane of glass. Apply uniform downward pressure on the component and check for warpage (**Figure 2**). Replace any component that shows more than a 0.004 in. (0.1 mm) warpage. In cases where warpage is slight, it can often be eliminated by placing the mating surface of each component on a large sheet of 400-600 grit wet sandpaper. Apply a slight amount of pressure and move the component in a figure-8 pattern. See **Figure 3**. Remove the component and sandpaper and recheck

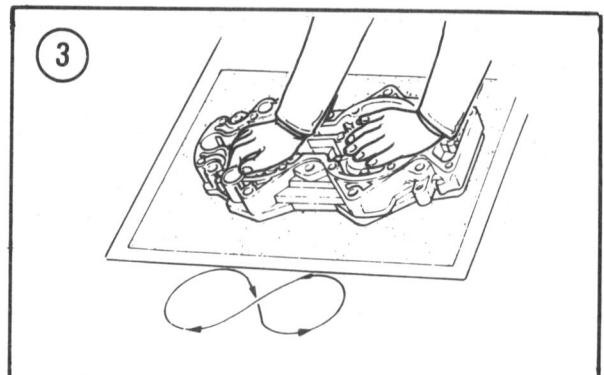

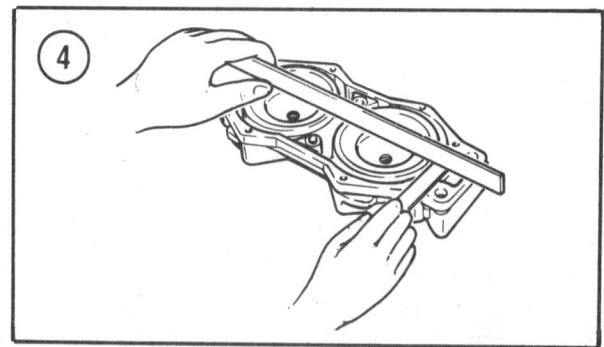

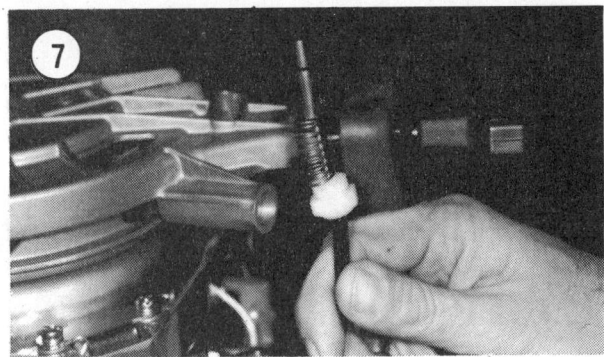

the surface flatness on the pane of glass. It may be necessary to repeat this procedure 2-3 times to remove small pieces of gasket material, eliminate warpage and produce a uniform mating surface. When finished, check for warpage with a straightedge and feeler gauge as shown in **Figure 4**.

Unless otherwise specified, gaskets used with other components contain a heat-activated sealant and require no additional sealant. When newly installed,

heat-activated gaskets may leak slightly until the engine has warmed up sufficiently to trigger the necessary chemical reaction in the gasket. For this reason, some service technicians feel that the application of Gasket Maker on a power head mounting gasket (**Figure 5**) provides additional protection against water leakage.

FLYWHEEL

Removal/Installation

1. Remove the engine cowling or cover.
2. Disconnect the spark plug lead(s) to prevent accidental starting of the engine.
3. If equipped with a rewind starter:
   a. Unscrew and disconnect the starter lockout cable from the starter housing. See **Figure 6** (typical).
   b. Remove the lockout plunger and spring from the cable end and place in a small container for reinstallation. See **Figure 7**.
   c. Remove the attaching bolts and any related hardware from the starter housing. Remove the starter housing.
4. Remove the plastic flywheel protector, if so equipped.
5. Remove the bolt(s) holding the starter rewind cup, if so equipped. Remove the starter cup. See **Figure 8** (typical).
6. Install flywheel holder (part No. YB-06139 or equivalent) to hold the flywheel while loosening the flywheel nut.
7. Loosen the flywheel nut with an appropriate socket wrench. See **Figure 9** (typical). Unthread the nut until it is flush with the top of the crankshaft—do not remove at this time.

> *CAUTION*
> *Do not thread the puller bolts into the flywheel more than 1/2 in. in Step 8 or damage may result to the charge coil or trigger assemblies located beneath the flywheel.*

8. Install flywheel puller part No. YU-33270 (2-30 hp) or part No. YB-6117 (40-225 hp) to the flywheel with the puller bolts.

CAUTION
Do not strike puller screw with excessive force in Step 9 or crankshaft and/or bearing damage may result. Heat should not be used, as it may cause the flywheel to seize on the crankshaft.

9. Hold the puller body with the puller handle and tighten the center screw (**Figure 10**). If flywheel does not pop from the crankshaft taper, lightly tap the puller center screw with a brass hammer (**Figure 11**).

10. Remove puller from flywheel. Remove flywheel nut and washer. Remove flywheel from crankshaft.

NOTE
*The Woodruff key may fall out of the crankshaft keyway as the flywheel is removed in Step 10. **Figure 12** shows the key lodged betwen the crankshaft and condenser. Be sure to locate and retrieve the key.*

11. Remove Woodruff key (**Figure 13**) from crankshaft slot if it does not come off with the flywheel.

12. Inspect flywheel carefully as described in this chapter.

13. Inspect the crankshaft and flywheel tapers. They must be perfectly dry and free of oil. Swab the tapered surfaces with solvent and blow dry with compressed air.

14. To install, place Woodruff key in crankshaft slot (**Figure 13**). Position the flywheel over the crankshaft with the flywheel keyway aligned with the crankshaft key. Install the flywheel.

15. Install flywheel washer and nut. Thread washer as far as it will go by hand, then install flywheel holder (part No. YB-06139 or equivalent) and tighten flywheel nut to specifications (**Table 1**) with an appropriate socket wrench.

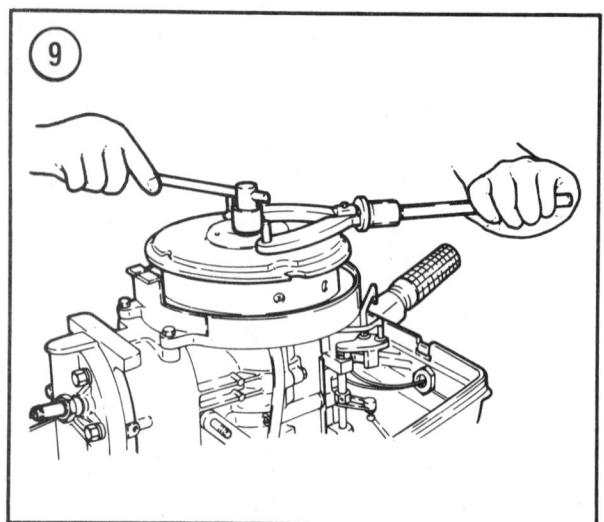

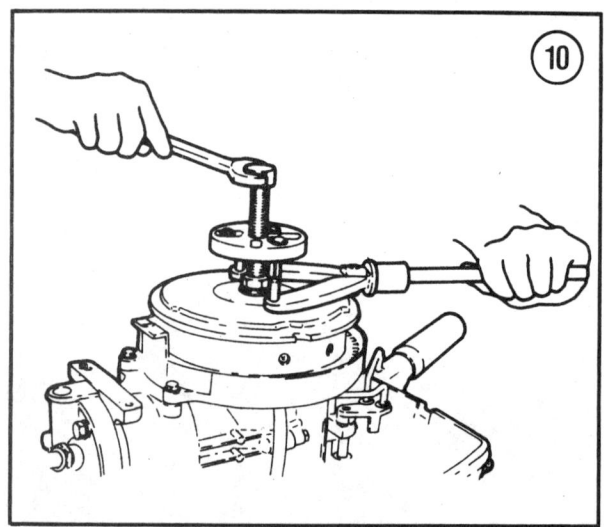

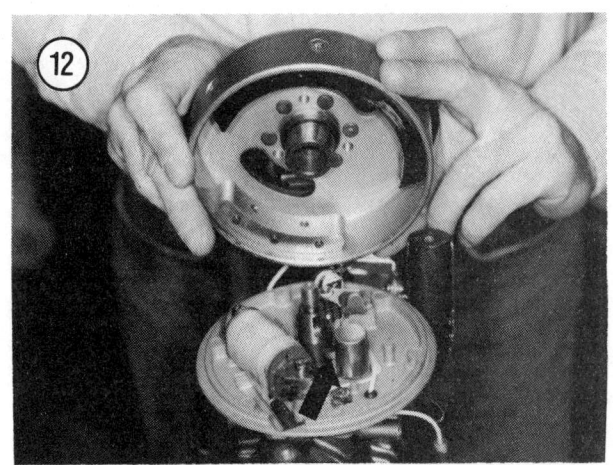

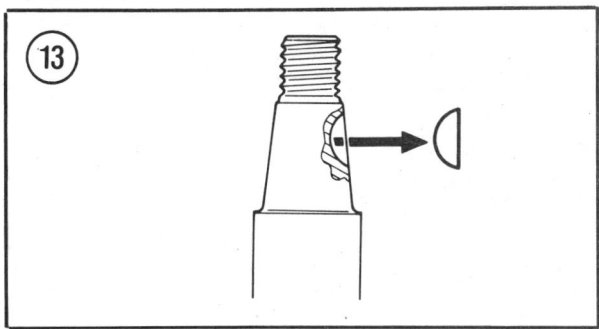

16. Reverse Steps 1-5 to complete installation.

Inspection

1. Check the flywheel carefully for cracks or breaks.

WARNING
A cracked or chipped flywheel must be replaced. A damaged flywheel may fly apart at high rpm, throwing metal fragments over a large area. Do not attempt to repair a damaged flywheel.

2. Check the tapered bore of the flywheel and the crankshaft taper for signs of fretting or working.

3. On electric start models, check the flywheel teeth for signs of excessive wear or damage.

4. Check the crankshaft and flywheel nut threads for wear or damage.

5. Replace flywheel, crankshaft and/or flywheel nut as required.

**POWER HEAD
REMOVAL/INSTALLATION**

When removing any power head, it is a good idea to make a sketch or take an instant picture of the location, routing and positioning of electrical wiring, brackets and J-clamps for reassembly reference. This is especially important with 70-225 hp engines, as the wiring and hose arrangement becomes progressively more complex as engine size increases.

Take note as you remove wires, washers and engine grounds so they may be reinstalled in their correct position. Unless otherwise specified, install lockwashers on the engine side of the electrical lead to assure a good ground.

Professional mechanics find that the removal/installation of a power head is much faster and easier when they disconnect or remove only those components necessary to allow the power head to be pulled. This is especially important with the larger displacement outboards.

For example, there is no point in disconnecting all the lines to the fuel pumps on a V-block engine unless the pumps require service. Simply unbolt the pumps from the block and allow them to rest against the bottom cowling.

Components such as the CDI unit must be disconnected to remove the power head, but need not be taken off the assembly until it is out of the cowling and on a clean work bench.

Other components may not require removal at all, depending upon the purpose of power head removal. For example, if the power head is being removed to replace a base gasket, there is no need to pull the flywheel.

8

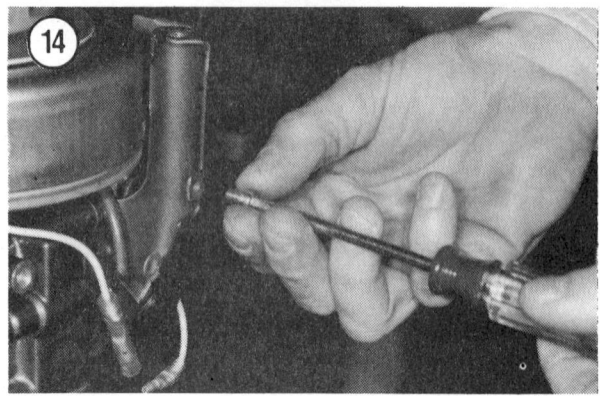

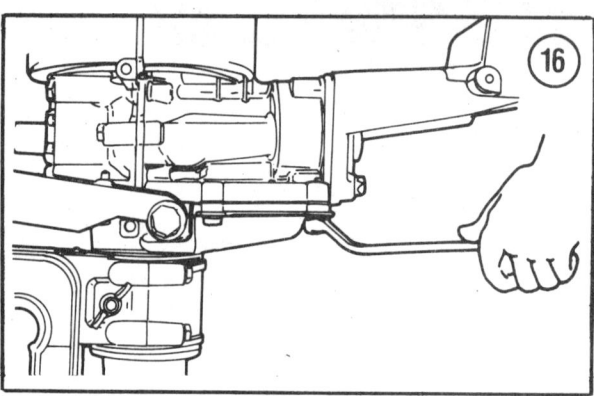

The procedures provided below assume that the power head is being removed for disassembly and overhaul. If you are removing it for some other reason, apply logic, common sense and good judgement—perform only those steps necessary to remove the power head with the least effort.

CAUTION
*After overhauling an oil-injected engine, it should be run with a 50:1 (25:1 for 40 hp) fuel-oil mixture **in addition** to the lubricant supplied by the injection pump. See **Break-in Procedure**, Chapter Thirteen.*

Yamaha 2 hp

1. Remove the engine cowling.
2. Remove the fuel tank. See Chapter Six.
3. Disconnect the spark plug lead and remove the spark plug.
4. Disconnect the white magneto lead at the bullet connector. Remove the screw holding the black ground lead on the rewind starter bracket. See **Figure 14**.
5. Remove the flywheel as described in this chapter.
6. Remove the 2 magneto base fasteners. Remove the magneto base (**Figure 15**).
7. Remove the 6 bolts and washers underneath the drive shaft housing which hold the power head and exhaust plate in place. See **Figure 16**.

8. Carefully pry the power head and exhaust plate assembly from the drive shaft housing.
9. Separate the exhaust plate from the power head. Place power head on a clean work bench and set exhaust plate to one side out of the way. Remove and discard gaskets.
10. Clean gasket residue from drive shaft housing mating surface and make sure that the 2 locating pins are installed as shown in **Figure 17**.
11. Installation is the reverse of removal, plus the following:
   a. Coat new gaskets with Yamabond No. 4 and install between the power head, exhaust plate and drive shaft housing.
   b. Lightly coat the drive shaft splines with Yamalube All-purpose Marine grease or equivalent.
   c. Coat power head attaching bolt threads with Loctite Type A.
   d. Tighten all fasteners to specifications in 2 steps (**Table 1**).

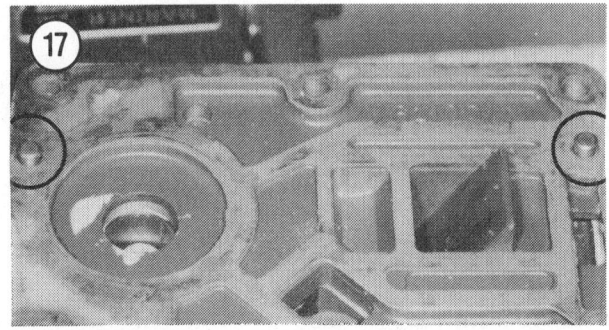

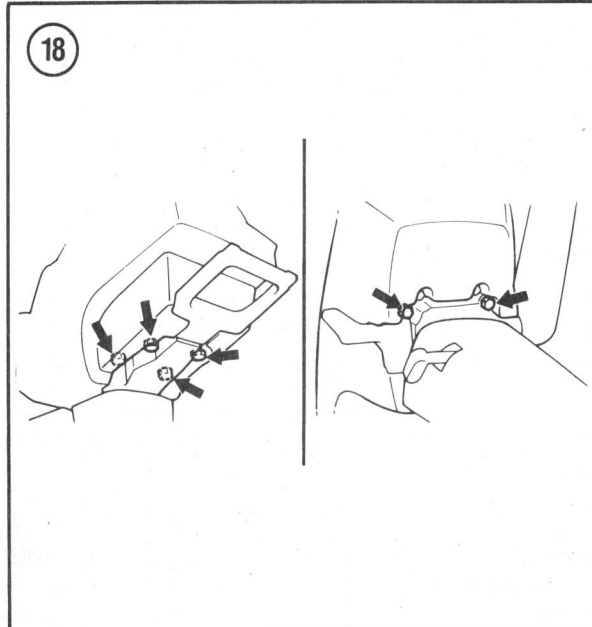

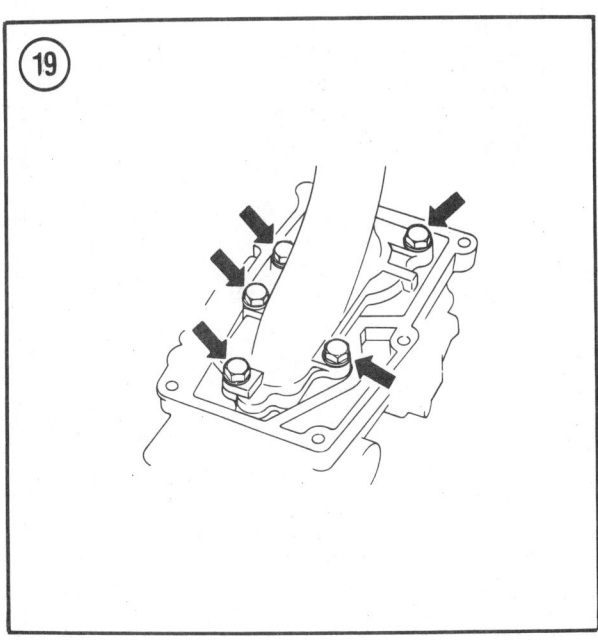

e. Open the fuel petcock and check for leakage.

f. Perform engine timing, synchronizing and adjustment sequence. See Chapter Five.

Yamaha 3 hp

1. Remove the engine cover.

2. Remove the fuel tank. See Chapter Six.

3. Disconnect the spark plug lead and remove the spark plug.

4. Remove the bolts holding the rewind starter. Remove the starter assembly.

5. Remove the flywheel as described in this chapter.

6. Disconnect the 4 magneto base-to-CDI unit leads at the bullet connectors.

7. Remove the 2 magneto base fasteners. Remove the magneto base.

8. Remove the air silencer cover, disconnect the throttle wire and choke linkage. If power head is to be disassembled, remove the carburetor. See Chapter Six.

9. Remove the 6 bolts underneath the drive shaft housing which hold the power head to the drive shaft housing. See **Figure 18**.

NOTE
At this point, there should be no hoses, wires or linkage connecting the power head to the bottom cowling. Recheck this to make sure nothing will hamper power head removal.

10. If necessary, tap power head lightly with a rubber mallet to break the gasket seal. Lift power head up and off the drive shaft housing.

11. Place power head on a clean work bench. If power head is to be disassembled, remove the exhaust manifold (**Figure 19**).

12. Installation is the reverse of removal, plus the following:

a. Install new seals in top end of exhaust manifold. Open side of seals must face toward bottom of exhaust manifold.

8

Lubricate the seal lips with Yamalube All-purpose Marine grease before installing the manifold on the power head.

b. Make sure locating pins (A, **Figure 20**) are correctly positioned in drive shaft housing.

c. Install a new gasket on drive shaft housing making sure locating pin holes are properly aligned.

d. Lightly lubricate drive shaft splines (B, **Figure 20**) with Yamalube All-purpose Marine grease.

e. Lower power head in place. Rotate propeller if necessary to mesh drive shaft and crankshaft splines.

f. Connect the CDI-to-magneto base leads as shown in **Figure 21**.

g. Tighten all fasteners to specifications (**Table 1**).

h. Complete timing, synchronizing and adjustment sequence. See Chapter Five.

Yamaha 4 and 5 hp

1. Remove the engine cover.

2. 4 hp—Remove the fuel tank. See Chapter Six.

3. Disconnect the spark plug lead and remove the spark plug.

4. Remove the tool kit from the top of the power head, if so equipped.

5. Disconnect the 4 magneto base-to-CDI unit leads at the bullet connectors, then remove the leads from the bracket clamp on the magneto base side of the power head. See **Figure 22**.

6. Remove the bolts (located underneath the drive shaft housing) which hold the CDI unit and coil mounting bracket. Move the bracket, CDI unit and coil assembly (**Figure 23**) out of the way (these components can be left in the bottom cowl).

7. Remove the flywheel as described in this chapter.

8. Remove the air silencer cover, disconnect the throttle wire and choke linkage and disconnect/plug the fuel line. If power head is to be disassembled, remove the carburetor. See Chapter Six.

9. Disconnect the magneto base linkage. Remove the 2 magneto base fasteners. Tap the underside of the magneto base with a plastic hammer to free it from the oil seal housing. Remove the magneto base.

10. Loosen the starter stop cable adjustment nut. Disconnect the cable from the stay, then remove the cable end from the linkage arm. See **Figure 24**.

11. Remove the hairpin retaining clip from the starter lockout arm and disconnect the link from the arm. See **Figure 24**.

12. Tilt the drive shaft housing up and remove the 3 bolts and washers shown in **Figure 25**. Remove the 3 bolts and washers on the opposite side. Loosen but do not remove the 7th bolt located at the front of the drive shaft housing.

13. Bring drive shaft housing back to a horizontal position and remove the bolt loosened in Step 12.

NOTE
At this point, there should be no hoses, wires or linkage connecting the power head to the bottom cowling. Recheck this to make sure nothing will hamper power head removal.

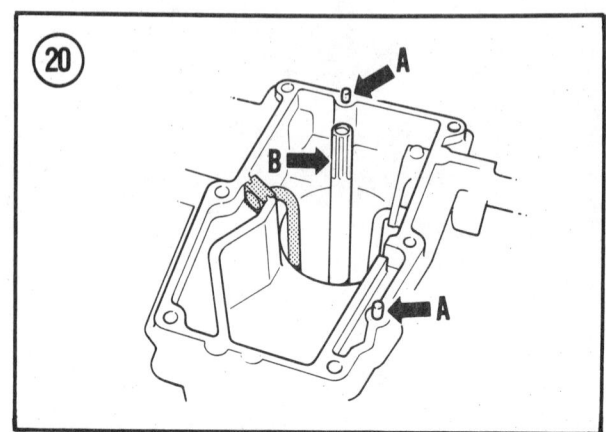

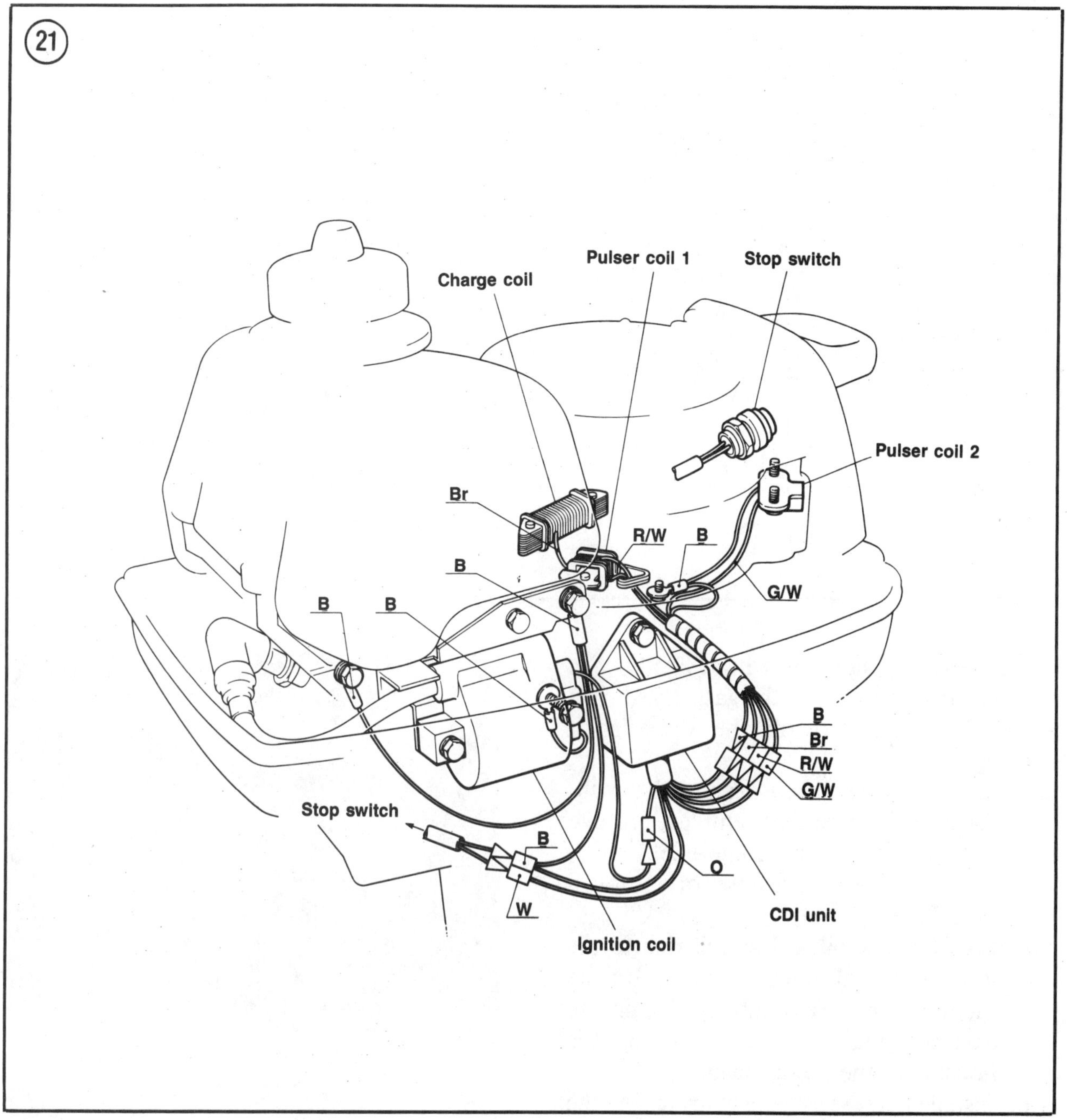

(21)

Charge coil

Pulser coil 1

Stop switch

Pulser coil 2

Br

R/W B

G/W

B

B B

B
Br
R/W
G/W

Stop switch

B

O

W

CDI unit

Ignition coil

8

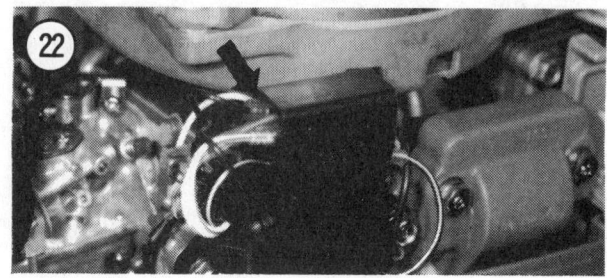

(22)

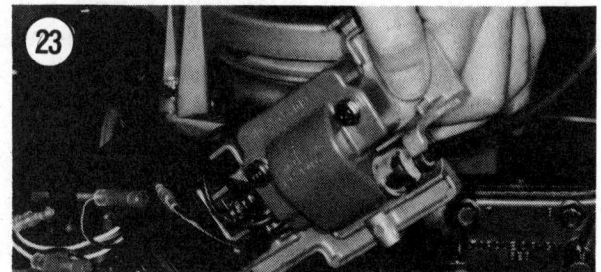

(23)

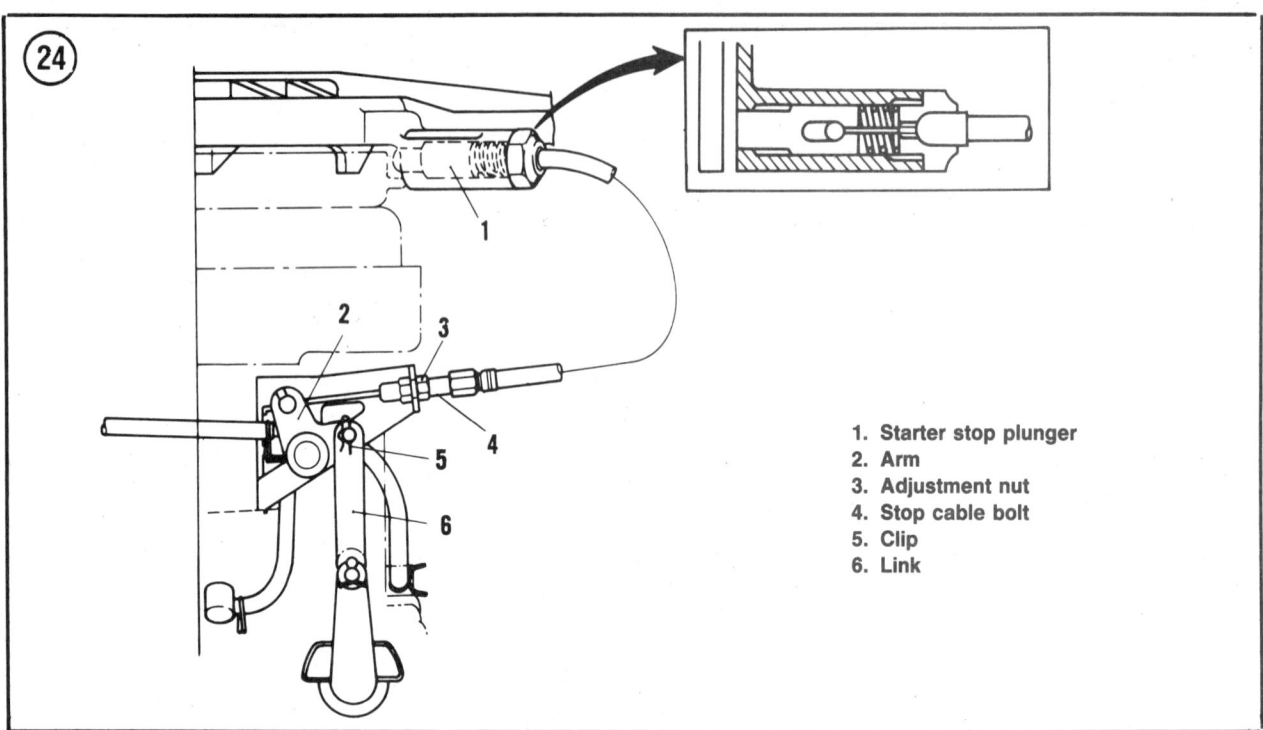

1. Starter stop plunger
2. Arm
3. Adjustment nut
4. Stop cable bolt
5. Clip
6. Link

14. If necessary, tap power head lightly with a rubber mallet to break the gasket seal. Lift power head up and off the drive shaft housing.

15. Place power head on a clean work bench. If power head is to be disassembled, remove the lower oil seal housing (**Figure 26**).

16. Installation is the reverse of removal, plus the following:

   a. Install a new oil seal and O-ring to the lower oil seal housing and lubricate them with Yamalube All-purpose Marine grease before installing the housing to the power head.

   b. Install new seals and O-ring in the magneto base.

   c. Lightly lubricate drive shaft splines (A, **Figure 27**) with Yamalube All-purpose Marine grease and install a new gasket on the 2 housing locating pins (B, **Figure 27**).

   d. Lower power head in place. Rotate propeller if necessary to mesh drive shaft and crankshaft splines.

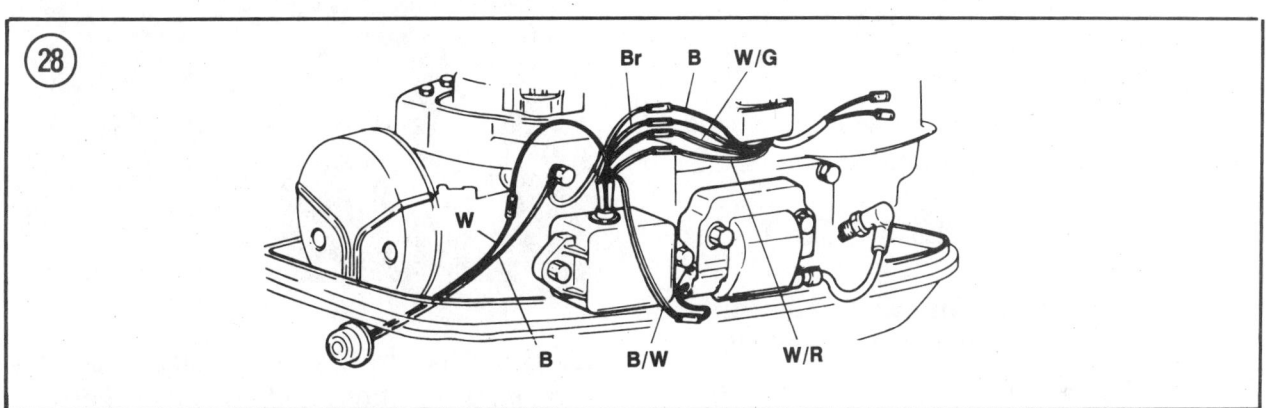

8

e. Connect the CDI leads as shown in **Figure 28**.

f. Tighten all fasteners to specifications (**Table 1**).

g. Complete timing, synchronizing and adjustment sequence. See Chapter Five.

Yamaha 6, 8, 9.9 and 15 hp

1. Disconnect the negative battery cable, if so equipped.

2. Remove the engine cover.

3. Disconnect the spark plug leads and remove the spark plugs.

4A. Manual start models—Disconnect the starter lockout cable at the rewind starter (**Figure 6**). Be careful not to lose the cable plunger or spring (**Figure 7**). Remove the bolts holding the rewind starter. Remove the starter assembly.

4B. Electric start models—Disconnect the lead at the starter motor positive terminal. Remove the mounting bolts. Remove the starter.

5A. 6 and 8 hp:

a. Remove the collar holding the CDI leads (**Figure 29**).

b. Disconnect the white stop switch lead.

c. Remove the bolt holding the 2 black leads. Reinstall bolt in power head to prevent its loss.

5B. 9.9 and 15 hp:

a. Remove the CDI unit cover.

b. Disconnect and unclamp the white stop switch lead and the 2 black ground leads.

c. Remove the bullet connectors from their holders on each side of the CDI unit (**Figure 30**). Disconnect all leads.

d. Remove the ignition coil from the CDI bracket.

e. Remove the CDI bracket.

6. Remove the flywheel as described in this chapter.

7. Disconnect the fuel line at the inlet fitting. Plug the line to prevent leakage.

8. Disconnect the magneto base linkage at the control lever. On 9.9 and 15 hp models, disconnect the linkage on the opposite side of the magneto base which connects to the shift handle.

9. Remove the screws holding the magneto base to the retaining plate. Remove the magneto base with an upward rotating motion.

10. Remove the retaining plate bolts and washers. Remove the retaining plate.

11A. 6 and 8 hp:

a. Disconnect the choke link at the carburetor.

b. Loosen the throttle wire adjustment bolts. Remove the adjustment bolts from the throttle wire stay. Disconnect the throttle wire ends from the throttle control lever.

11B. 9.9 and 15 hp:

a. Remove the neutral start arm and link, then disconnect the stop wire from the arm.

b. Loosen the stop wire adjustment bolt locknut and remove the wire from the bracket.

c. Remove the bolt holding the magneto base control lever. Remove the control lever.

d. Disconnect the choke and throttle linkage at the carburetor. See **Figure 31**.

e. Remove the inner and outer air silencer covers.

f. Disconnect the water pump tell-tale hose at the outer exhaust cover.

12. Tilt the drive shaft housing up and remove 5 of the 6 retaining bolts and washers. Loosen but do not remove the 6th bolt.

13. Bring the drive shaft housing back to a horizontal position and remove the bolt loosened in Step 13.

NOTE
At this point, there should be no hoses, wires or linkage connecting the power head to the bottom cowling. Recheck this to make sure nothing will hamper power head removal.

14. If necessary, tap the power head with a rubber mallet to break the gasket seal. Carefully lift the power head from the drive shaft housing and place it on a clean work bench.

15. Unbolt and remove the exhaust manifold (6 and 8 hp only) and the oil seal housing from the bottom of the power head. Install power head on an appropriate power head stand secured in a vise.

16. Remove old power head gasket and O-rings (if used) from drive shaft housing (A, **Figure 32**).

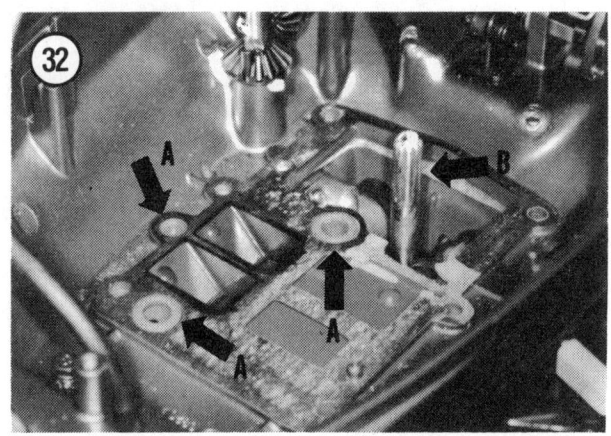

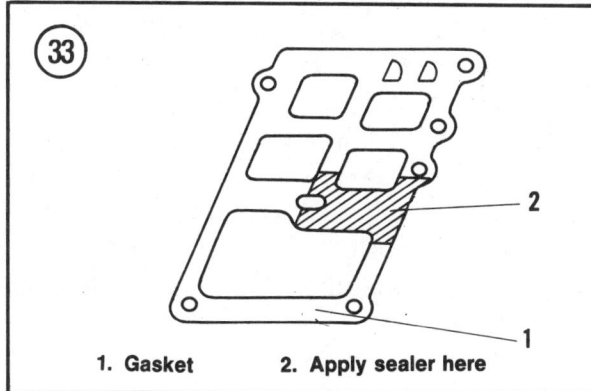

1. Gasket   2. Apply sealer here

17. Installation is the reverse of removal, plus the following:

  a. 6 and 8 hp—Install exhaust manifold with a new gasket.

  b. Install a new oil seal and O-ring to the lower oil seal housing and lubricate them with Yamalube All-purpose Marine grease before installing the housing to the power head.

  c. Install a new power head gasket on the drive shaft housing locating pins. On 9.9 and 15 hp models, coat the cross-hatched area of the gasket as shown in **Figure 33** with Yamabond No. 7 or Permatex No. 27 sealant and install new O-rings on the bottom cowling (A, **Figure 32**).

  d. Lightly lubricate the drive shaft splines (B, **Figure 32**) with Yamalube All-purpose Marine grease.

  e. Lower power head in place. Rotate propeller if necessary to mesh drive shaft and crankshaft splines.

  f. Lubricate magneto base retainer ring with Yamalube All-purpose Marine grease. On 6 and 8 hp models, apply a light coat of the same lubricant to the inner and outer surfaces of the magneto base bushing.

  g. Tighten all fasteners to specifications (**Table 1**).

  h. Complete timing, synchronizing and adjustment sequence. See Chapter Five.

Yamaha 25 and 30 hp

1. Disconnect the negative battery cable, if so equipped.

2. Remove the engine cover.

3. Disconnect the spark plug leads and remove the spark plugs.

4A. Manual start models:

  a. Disconnect the starter lockout cable at the rewind starter (**Figure 6**). Be careful not to lose the cable plunger or spring (**Figure 7**).

  b. Remove the bolts holding the rewind starter. Remove the starter assembly.

  c. Disconnect the stop switch leads at the CDI unit bullet connectors.

4B. Electric start models:

  a. Disconnect the leads at the starter motor terminals. Remove the mounting bolts. Remove the starter.

  b. Disconnect the leads at the starter relay.

  c. Disconnect the neutral switch leads.

  d. Remove the starter mounting bracket screw holding the 2 ground leads. Reinstall screw in bracket to prevent its loss.

  e. Disconnect the rectifier leads.

  f. Disconnect the CDI unit lead(s), at the wiring harness, that will interfere with power head removal. Remove the nut on the CDI unit ground lead bolt. Remove the lead from the bolt.

8

5. Remove the cotter pin, washer and wave washer connecting the shift lever arm to the linkage. Separate the linkage from the arm.

6. Disconnect all linkage at the magneto control arm. See **Figure 34**.

7. Disconnect the fuel line at the inlet side of the filter. Plug the line to prevent leakage.

8. Disconnect the choke and throttle linkage at the carburetor.

9. Remove the flywheel as described in this chapter.

10. Disconnect the magneto base linkage. Remove the 4 screws holding the magneto base assembly to the retainer plate. Remove the magneto base assembly.

11. Remove the 4 retainer plate bolts. Remove the retainer plate and retainer ring.

12. Tilt the drive shaft housing up and remove the 3 bolts and washers on one side. Remove 2 of the 3 bolts and washers on the opposite side. Loosen but do not remove the 6th bolt.

13. Bring drive shaft housing back to a horizontal position and remove the bolt loosened in Step 12.

NOTE
At this point, there should be no hoses, wires or linkage connecting the power head to the bottom cowling. Recheck this to make sure nothing will hamper power head removal.

14. If necessary, tap power head lightly with a rubber mallet to break the gasket seal. Lift power head up and off the drive shaft housing.

15. Place power head on a clean work bench. If power head is to be disassembled, remove the lower oil seal housing, and all remaining fuel and electrical components.

16. Installation is the reverse of removal, plus the following:

  a. Make sure the 2 locating pins are securely installed in the base of the power head.

  b. Install a new oil seal and O-ring to the lower oil seal housing and lubricate them with Yamalube All-purpose Marine grease before installing the housing to the power head.

  c. Install a new power head gasket on the power head locating pins.

  d. Lightly lubricate the drive shaft splines with Yamalube All-purpose Marine grease.

  e. Lower power head in place. Rotate propeller if necessary to mesh drive shaft and crankshaft splines.

  f. Lubricate magneto base retainer ring with Yamalube All-purpose Marine grease.

  g. Tighten all fasteners to specifications (**Table 1**).

  h. Complete timing, synchronizing and adjustment sequence. See Chapter Five.

Yamaha 40 and 50 hp

1. Disconnect the negative battery cable, if so equipped.

2. Remove the engine cover.

3. Disconnect the spark plug leads and remove the spark plugs.

4A. Manual start models:

  a. Disconnect the starter lockout cable at the rewind starter (**Figure 6**). Be careful not to lose the cable plunger or spring (**Figure 7**).

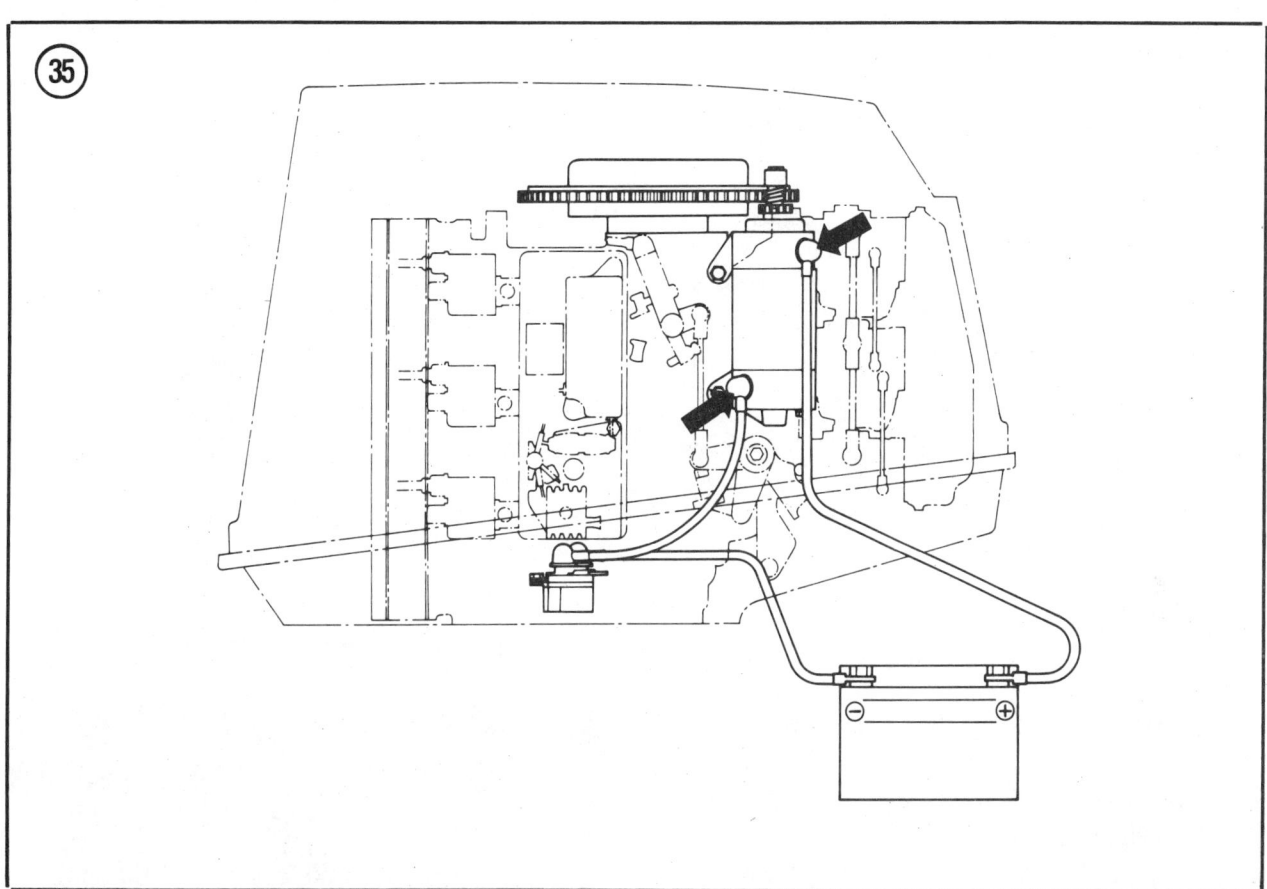

b. Remove the bolts holding the rewind starter. Remove the starter assembly.

c. Disconnect the white and black stop switch leads at the CDI unit bullet connectors.

d. Remove the cotter pin, washer and wave washer connecting the shift lever arm to the linkage. Separate the linkage from the arm.

4B. Electric start models:

a. Disconnect the leads at the starter motor terminals (**Figure 35**). Remove the mounting bolts. Remove the starter.

b. Disconnect the leads at the starter relay (**Figure 35**).

c. Disconnect the neutral switch leads.

d. Remove the hairpin clips and disconnect the remote control cables (**Figure 36**).

5. Unsnap the wiring harness connector from the CDI unit bracket cover (A, **Figure 37**) and separate the connectors. Remove the CDI unit cover (B, **Figure 37**).

6. Disconnect the CDI unit leads. Disconnect the rectifier leads. Remove the CDI unit and rectifier.

7. Remove the ignition coils. See Chapter Seven.

8. Remove the oil tank. Disconnect the injection pump and remove the control unit. See Chapter Twelve.

9. Remove the air silencer cover (**Figure 38**).

10. Disconnect the choke solenoid leads. Disconnect the choke linkage at the lower carburetor.

12. Disconnect the linkage at the magneto control arm.

13. Disconnect the fuel line at the inlet side of the filter. Plug the line to prevent leakage.

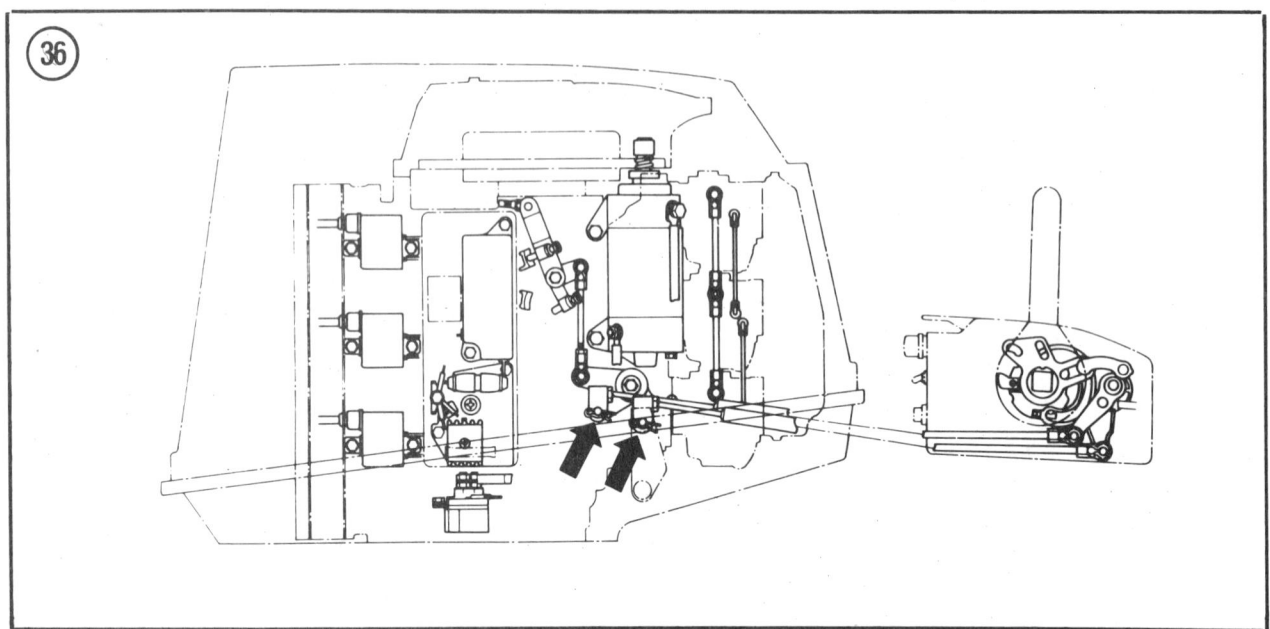

14. Remove the flywheel as described in this chapter.

15. Disconnect the magneto base linkage. Remove the screws holding the magneto base assembly to the retainer plate. Remove the magneto base assembly.

16. Remove the retainer plate bolts. Remove the retainer plate and retainer ring.

17. Remove the bolts holding the apron to the bottom cowling (**Figure 39**). Remove the apron.

> *NOTE*
> *At this point, there should be no hoses, wires or linkage connecting the power head to the bottom cowling. Recheck this to make sure nothing will hamper power head removal.*

18. Remove the 8 bolts and washers holding the power head to the bottom cowling.

19. If necessary, tap power head lightly with a rubber mallet to break the gasket seal. Lift power head up and off the bottom cowling.

20. Place power head on a clean work bench. If power head is to be disassembled, remove the lower oil seal housing and all remaining fuel and electrical components.

21. Installation is the reverse of removal, plus the following:

   a. Make sure the 2 locating pins are securely installed in the bottom cowling.

   b. Install a new oil seal and O-ring to the lower oil seal housing and lubricate them with Yamalube All-purpose Marine grease before installing the housing to the power head.

   c. Install a new power head gasket on the bottom cowling locating pins.

   d. Lightly lubricate the drive shaft splines with Yamalube All-purpose Marine grease.

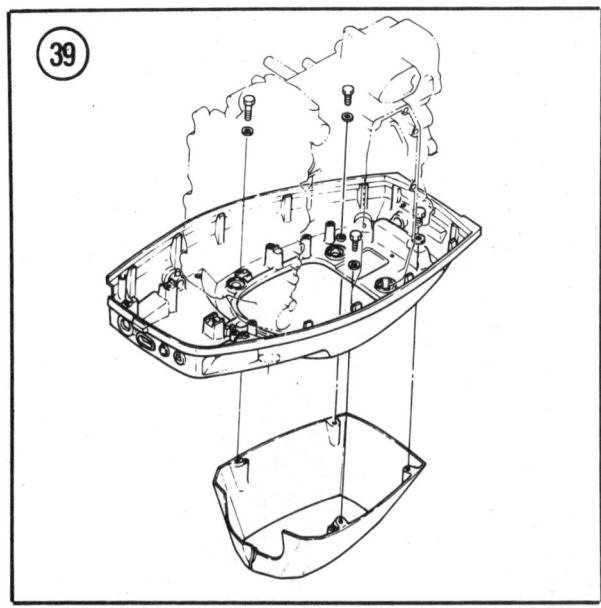

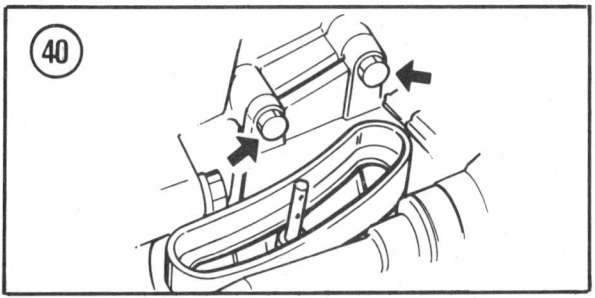

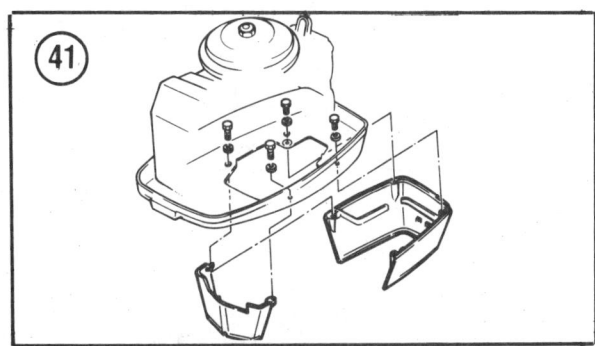

e. Lower power head in place. Rotate propeller if necessary to mesh drive shaft and crankshaft splines.

f. Lubricate magneto base retainer ring with Yamalube All-purpose Marine grease.

g. Tighten all fasteners to specifications (**Table 1**).

h. Complete timing, synchronizing and adjustment sequence. See Chapter Five.

Yamaha 70 and 90 hp

1. Disconnect the negative battery cable, if so equipped.

2. Remove the engine cover.

3. Disconnect the spark plug leads and remove the spark plugs.

4. Disconnect the power trim/tilt motor and relay leads. Remove the relay(s).

5. Remove the oil injection tank. See Chapter Twelve.

6. Remove the CDI unit cover. Disconnect the brown starter relay and red fused lead. Remove the starter relay from the bottom cowling.

7. Disconnect the fuel line at the inlet side of the fuel filter. Plug the line to prevent leakage.

8. Disconnect the water pump tell-tale hose at the outer exhaust cover.

9. Disconnect the ground lead at the cylinder head cover.

10. Remove the shift rod bracket bolts. See **Figure 40**. Remove the hairpin clip holding the shift cable. Remove the shift cable and bushing.

11. Disconnect the choke linkage at the carburetor.

12. Remove the 4 bolts and washers holding the front and rear aprons (**Figure 41**). Remove the aprons.

8

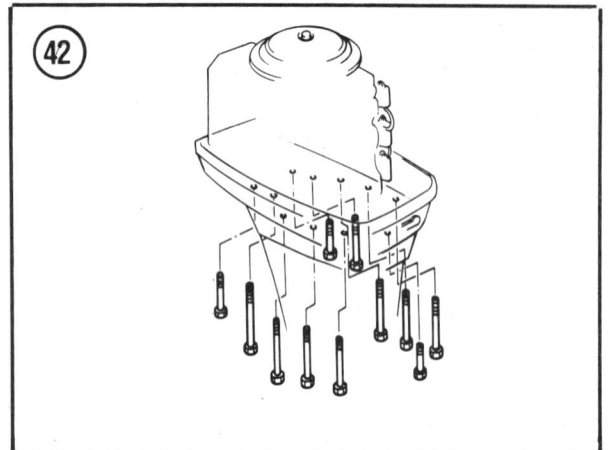

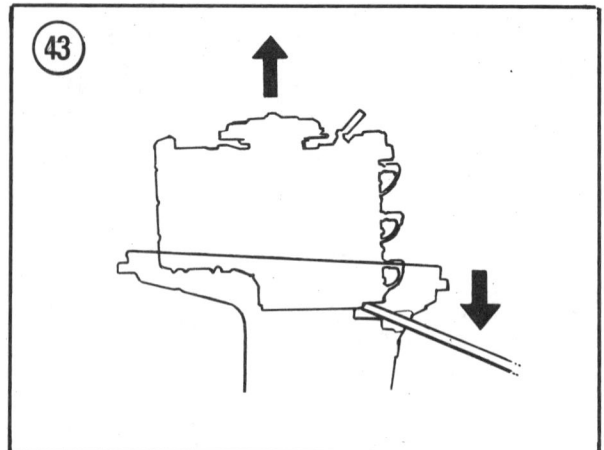

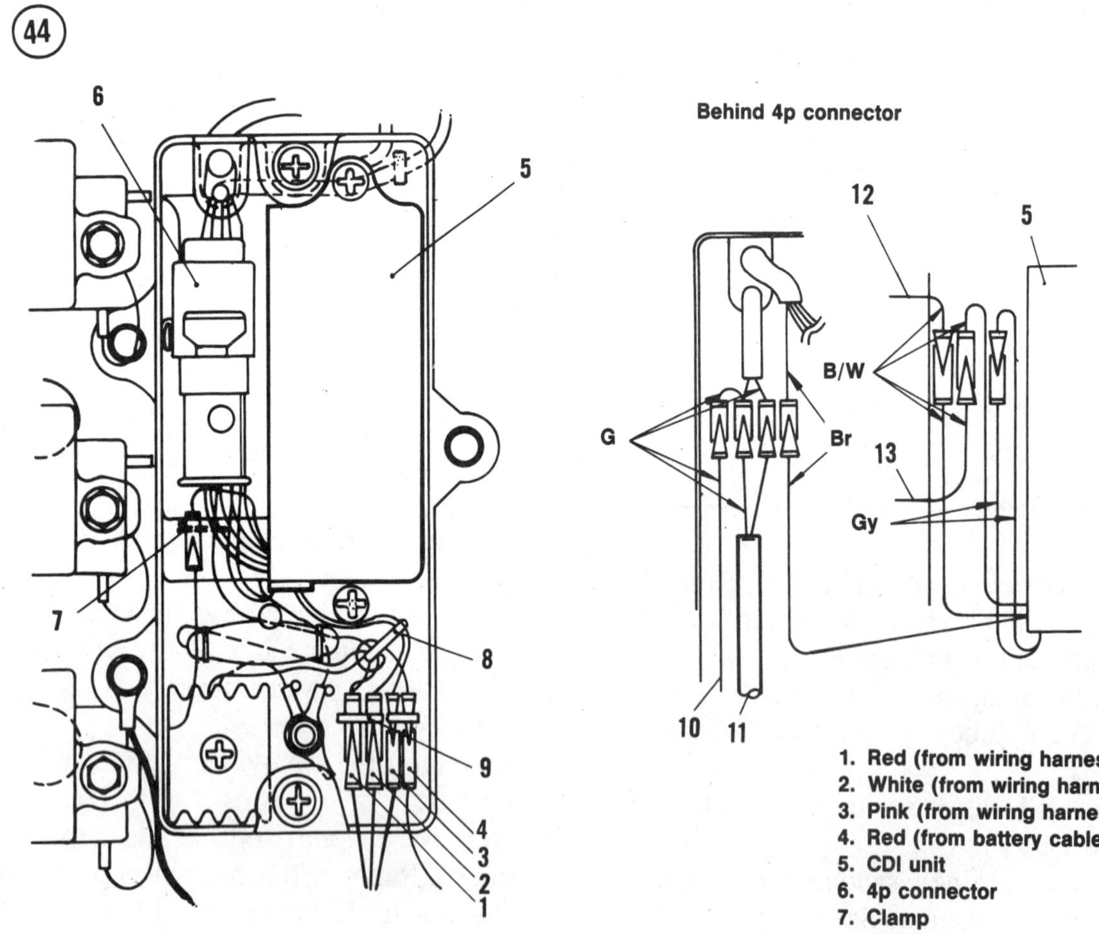

Behind 4p connector

1. Red (from wiring harness)
2. White (from wiring harness)
3. Pink (from wiring harness)
4. Red (from battery cable)
5. CDI unit
6. 4p connector
7. Clamp
8. Clamp
9. Clamp
10. From wiring harness
11. From rectifier
12. From No. 1 ignition coil
13. From No. 2 ignition coil

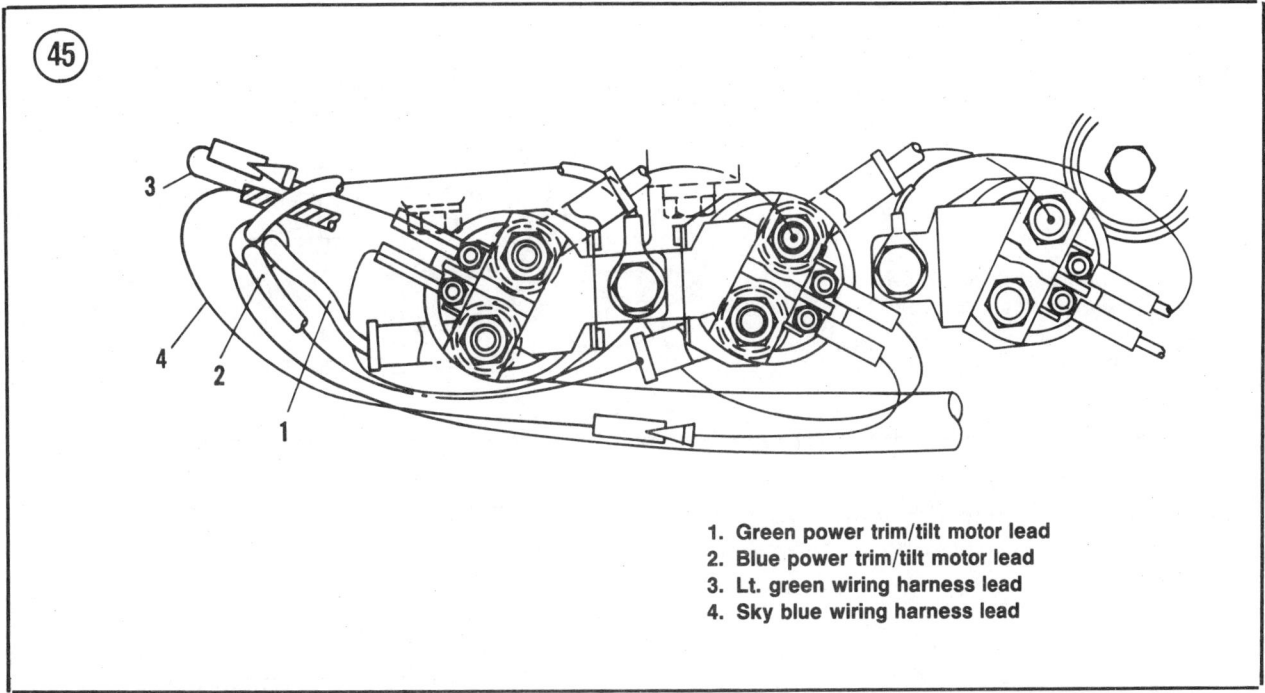

(45)

1. Green power trim/tilt motor lead
2. Blue power trim/tilt motor lead
3. Lt. green wiring harness lead
4. Sky blue wiring harness lead

NOTE
At this point, there should be no hoses, wires or linkage connecting the power head to the bottom cowling. Recheck this to make sure nothing will hamper power head removal.

13. Remove the bolts and washers holding the power head to the bottom cowling. The 70 hp uses 8 fasteners; the 90 hp uses 11. See **Figure 42** (90 hp).

14. If necessary, insert a wooden pry bar between the power head and bottom cowling as shown in **Figure 43** to break the gasket seal. Lift power head up and off the bottom cowling.

15. Place power head on a clean work bench. If power head is to be disassembled, remove the flywheel and magneto base (70 hp) or alternator stator/timer base (90 hp), the oil injection pump and driven gear and all remaining fuel and electrical components.

16. Installation is the reverse of removal, plus the following:

   a. Make sure the 2 locating pins are securely installed in the bottom cowling.

   b. Install new oil seals and O-ring to the lower oil seal housing and lubricate them with Yamalube All-purpose Marine grease before installing the housing to the power head.

   c. Install a new power head gasket on the bottom cowling locating pins.

   d. Lightly lubricate the drive shaft splines with Yamalube All-purpose Marine grease.

   e. Lower power head in place. Rotate propeller if necessary to mesh drive shaft and crankshaft splines.

   f. 70 hp—Lubricate magneto base retainer ring with Yamalube All-purpose Marine grease.

   g. 70 hp—Install the connectors in the CDI unit box as shown in **Figure 44**.

   h. Install the power trim/tilt relays and connect the wiring as shown in **Figure 45** (70 hp) or **Figure 46** (90 hp).

   i. 90 hp—Connect the battery and ground leads as shown in **Figure 47**.

   j. Tighten all fasteners to specifications **(Table 1)**.

8

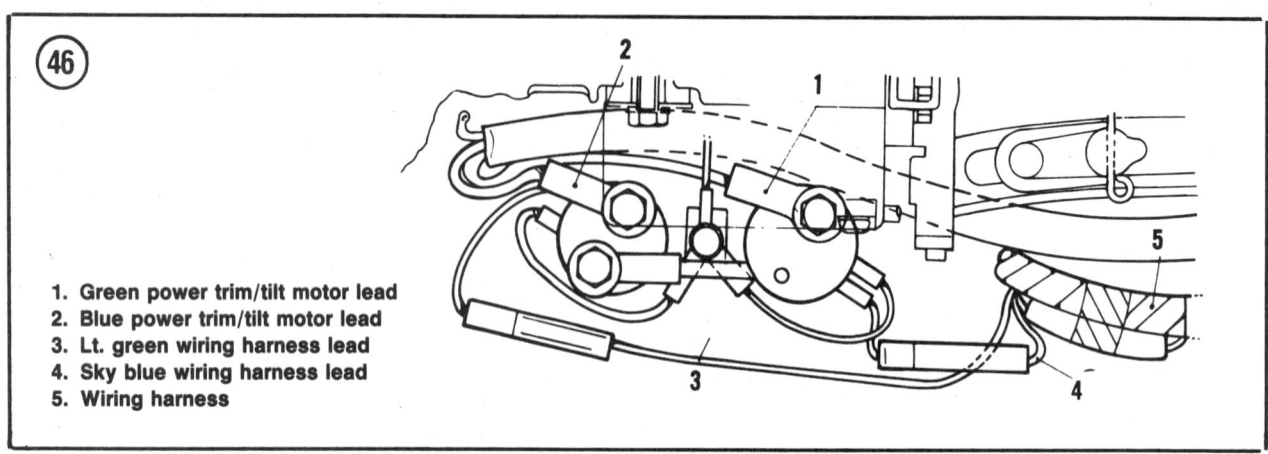

1. Green power trim/tilt motor lead
2. Blue power trim/tilt motor lead
3. Lt. green wiring harness lead
4. Sky blue wiring harness lead
5. Wiring harness

k. Complete timing, synchronizing and adjustment sequence. See Chapter Five.

Yamaha 115-130 hp

1. Disconnect the negative battery cable, if so equipped.
2. Remove the engine cover.
3. Disconnect the spark plug leads and remove the spark plugs.
4. Disconnect the electrical leads at the starter motor and relay.
5. Disconnect the red lead between the fuse and starter relay.
6. Disconnect the power trim/tilt motor and relay leads.
7. Remove the plate covering the wiring harness from the bottom cowling. Remove the wiring harness.
8. Disconnect the fuel line at the inlet side of the fuel filter. Plug the line to prevent leakage.
9. Disconnect the water pump tell-tale hose at the outer exhaust cover.
10. Disconnect the ground leads at the outer exhaust cover.
11. Remove the hairpin clip holding the shift cable. Remove the shift cable and bushing.
12. Remove the bolts holding the shift rod bracket to the crankcase. Remove the bracket.
13. Disconnect the choke link rod at the carburetor.

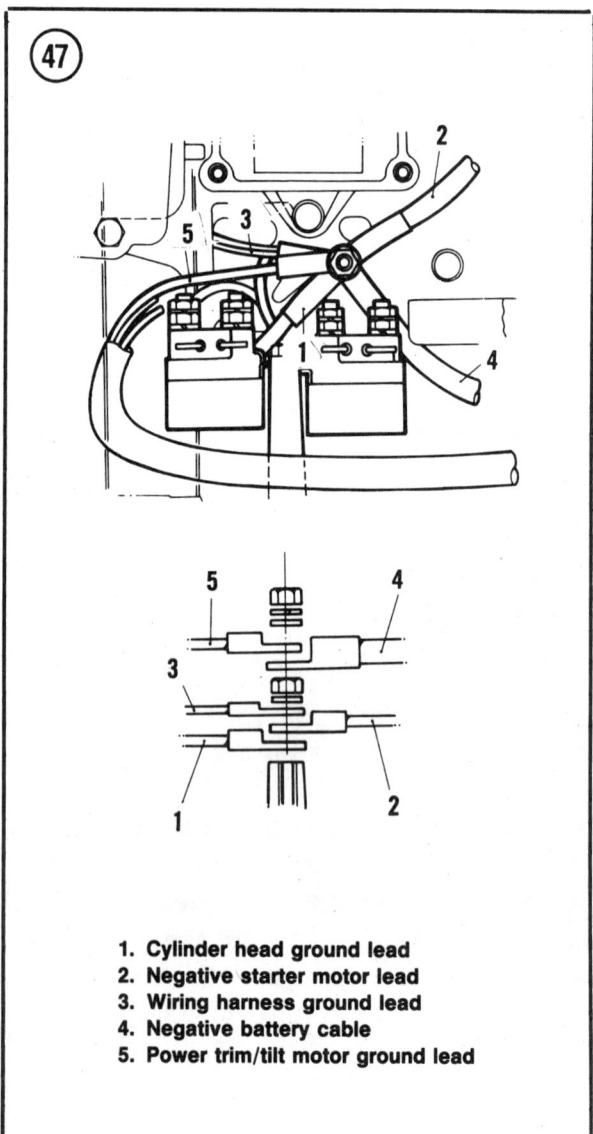

1. Cylinder head ground lead
2. Negative starter motor lead
3. Wiring harness ground lead
4. Negative battery cable
5. Power trim/tilt motor ground lead

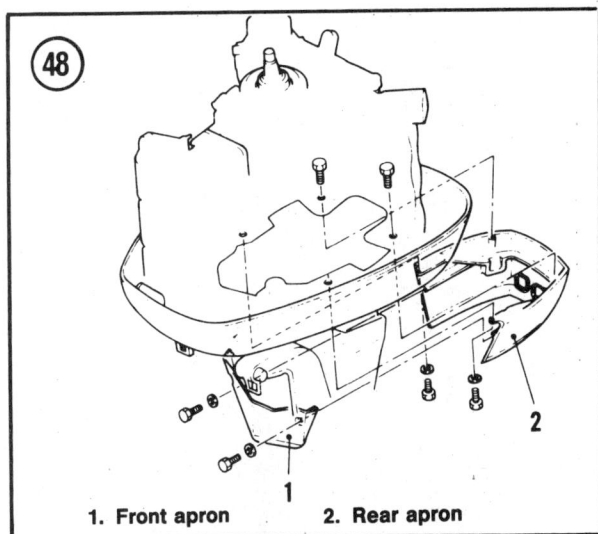

1. Front apron   2. Rear apron

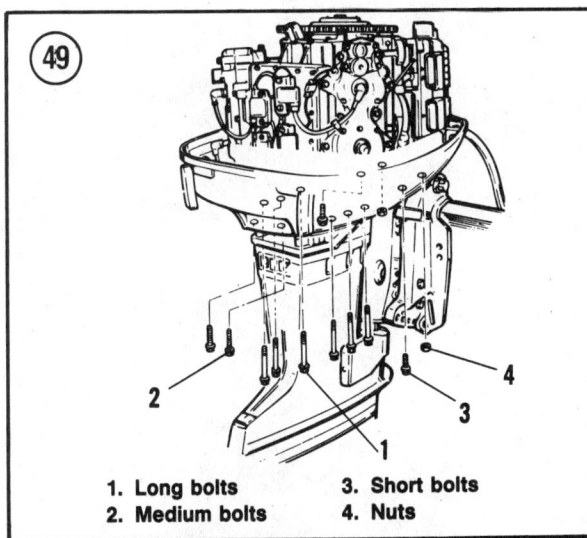

1. Long bolts   3. Short bolts
2. Medium bolts   4. Nuts

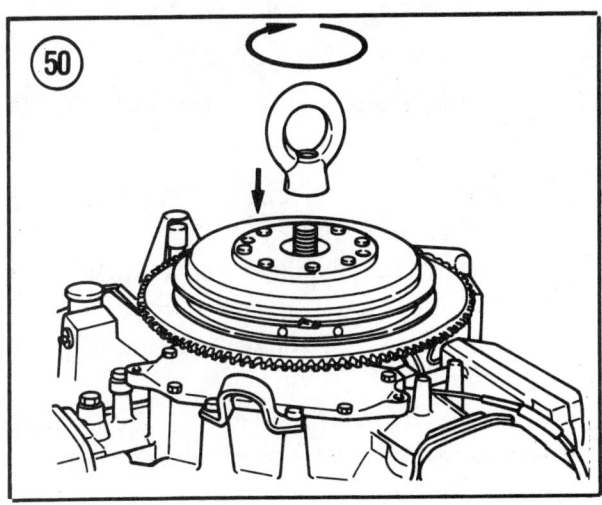

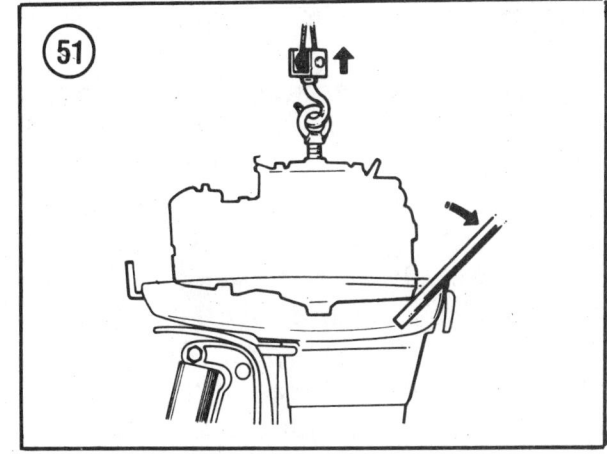

14. Remove the flywheel cover and nut as described in this chapter.

15. Remove the 2 bolts and washers holding the front apron. Remove the 2 bolts at the bottom and the 2 bolts at the inside of the bottom cowling. Remove the rear apron. See **Figure 48**.

NOTE
At this point, there should be no hoses, wires or linkage connecting the power head to the bottom cowling. Recheck this to make sure nothing will hamper power head removal.

16. Remove the 10 bolts and 2 nuts holding the power head to the bottom cowling. See **Figure 49**.

17. Thread lifting hook part No. YB-6202 onto the crankshaft. See **Figure 50**.

CAUTION
The power head must be lifted straight up in Step 18. Since most of its weight is on the cylinder side, the power head will tend to lift at an angle unless the pry bar is used as specified. Lifting at an angle may damage the crankshaft and drive shaft splines.

18. Attach a hoist to the lifting hook. Insert a wooden pry bar between the power head and bottom cowling as shown in **Figure 51** to help break the gasket seal and allow the power

8

head to be lifted straight up and off the bottom cowling.

19. Check the power head base for locating pins. If any came off with the power head, remove and reinstall in the bottom cowling.

20. Bolt the power head to a suitable engine stand (**Figure 52**). Remove the hoist and lifting eye. If power head is to be disassembled, remove the flywheel as described in this chapter. Remove the alternator stator/timer base (Chapter Eight), the oil injection pump and driven gear (Chapter Twelve) and all remaining fuel (Chapter Six) and electrical (Chapter Seven) components.

21. Installation is the reverse of removal, plus the following:

   a. Make sure the 2 locating pins are securely installed in the bottom cowling.

   b. Install new oil seals and O-ring to the lower oil seal housing and lubricate them with Yamalube All-purpose Marine grease before installing the housing to the power head.

   c. Install a new power head gasket on the bottom cowling locating pins.

   d. Lightly lubricate the drive shaft splines with Yamalube All-purpose Marine grease.

   e. Lower power head in place. Rotate propeller if necessary to mesh drive shaft and crankshaft splines.

   f. Tighten all fasteners to specifications (**Table 1**).

   g. Complete timing, synchronizing and adjustment sequence. See Chapter Five.

Yamaha 150-225 hp

1. Disconnect the negative battery cable, if so equipped.
2. Remove the engine cover.
3. Disconnect the spark plug leads and remove the spark plugs.

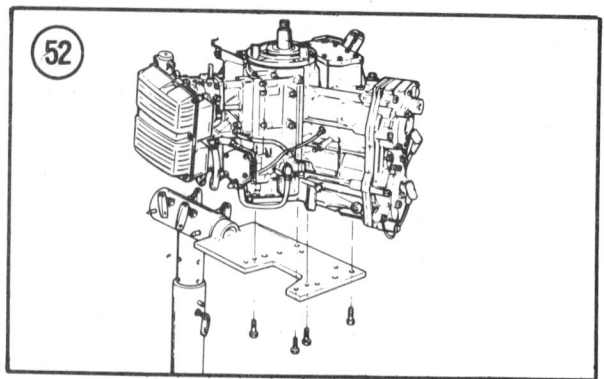

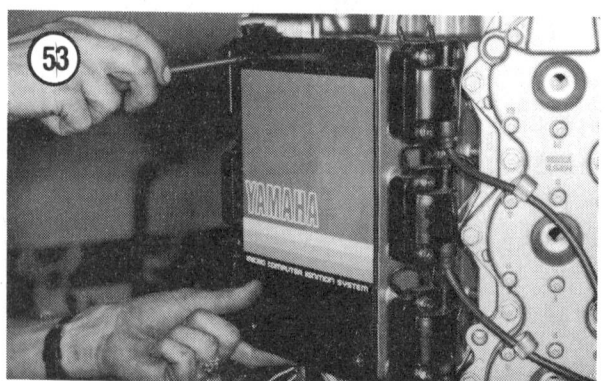

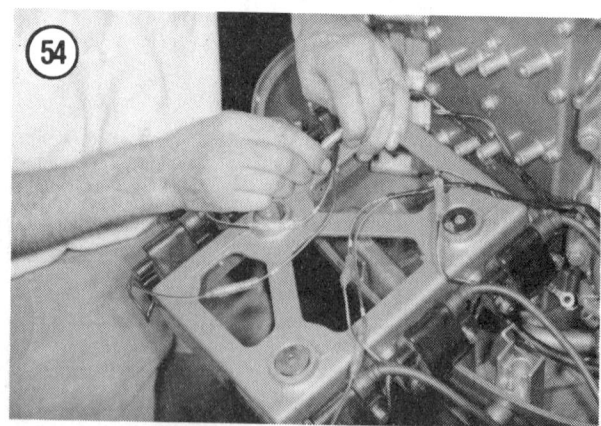

4. Disconnect the electrical leads at the starter motor and relay.
5. Disconnect the red lead between the fuse and starter relay.
6. Disconnect the power trim/tilt motor and relay leads.
7. Remove the plate covering the wiring harness from the bottom cowling. Remove the wiring harness.

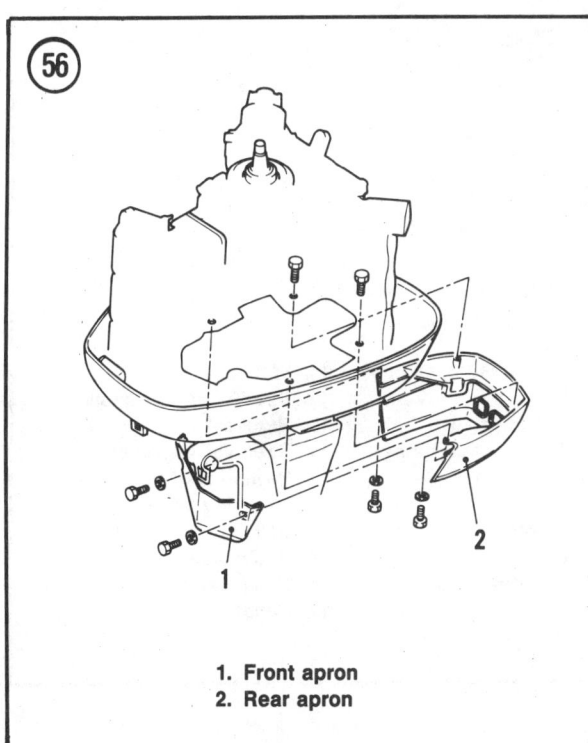

1. Front apron
2. Rear apron

8. Disconnect the fuel line at the inlet side of the fuel filter. Plug the line to prevent leakage.

9. Disconnect the water pump tell-tale hose and the bypass hose at the outer exhaust cover.

10. Disconnect the ground leads on each side of the power head at the outer exhaust cover.

11. Remove the hairpin clip holding the shift cable. Remove the shift cable and bushing.

12. Remove the bolts holding the shift rod bracket to the crankcase. Remove the bracket.

13. Disconnect the choke link rod at the carburetor.

14. Remove the flywheel cover and nut as described in this chapter.

15. 220-225 hp:
   a. Disconnect the YMIS control unit electrical connectors.
   b. Remove the 4 screws holding the control unit to its bracket. See **Figure 53**.
   c. Remove the bolts holding the control unit bracket and ignition coil assembly to the power head. Disconnect the electrical leads (**Figure 54**) and remove the bracket and coil assembly.

16. Remove the CDI unit cover. Remove the 3 bolts holding the CDI unit bracket to the power head (**Figure 55**). Remove the bracket and let it hang over the bottom cowling.

17. Remove the 2 bolts and washers holding the front apron. Remove the 2 bolts at the bottom and the 2 bolts at the inside of the bottom cowling. Remove the rear apron. See **Figure 56**.

NOTE
At this point, there should be no hoses, wires or linkage connecting the power head to the bottom cowling. Recheck this to make sure nothing will hamper power head removal.

18. Remove the 10 bolts and 2 nuts holding the power head to the bottom cowling. See **Figure 57** (typical).

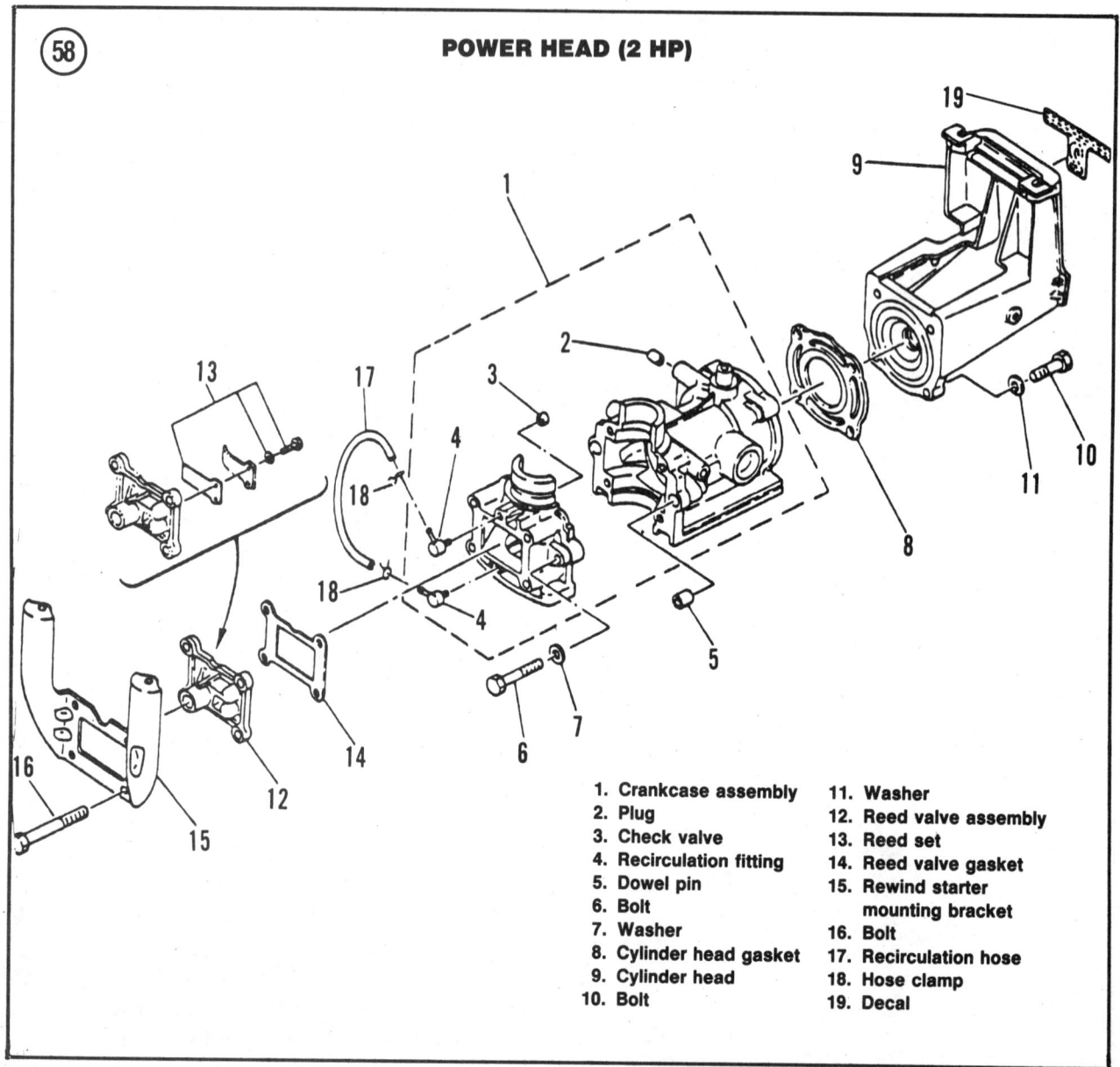

POWER HEAD (2 HP)

58

1. Crankcase assembly	11. Washer
2. Plug	12. Reed valve assembly
3. Check valve	13. Reed set
4. Recirculation fitting	14. Reed valve gasket
5. Dowel pin	15. Rewind starter
6. Bolt	mounting bracket
7. Washer	16. Bolt
8. Cylinder head gasket	17. Recirculation hose
9. Cylinder head	18. Hose clamp
10. Bolt	19. Decal

19. Thread lifting hook part No. YB-6202 onto the crankshaft. See **Figure 50**.

CAUTION
The power head must be lifted straight up in Step 20. Since most of its weight is on the cylinder side, the power head will tend to lift at an angle unless the pry bar is used as specified. Lifting at an angle may damage the crankshaft and drive shaft splines.

20. Attach a hoist to the lifting hook. Insert a wooden pry between the power head and bottom cowling as shown in **Figure 51** to help break the gasket seal and allow the power head to be lifted straight up and off the bottom cowling.

21. Check the power head base for locating pins. If any came off with the power head, remove and reinstall in the bottom cowling.

22. Bolt the power head to a suitable engine stand (**Figure 52**). Remove the hoist and

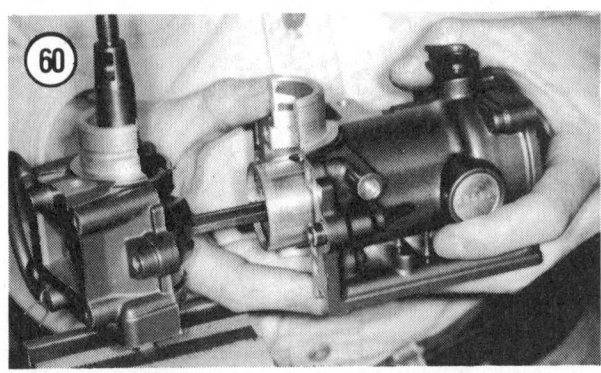

lifting eye. If power head is to be disassembled, remove the flywheel as described in this chapter. Remove the alternator stator/timer base (Chapter Eight), the oil injection pump and driven gear (Chapter Twelve) and all remaining fuel (Chapter Six) and electrical (Chapter Seven) components.

23. Installation is the reverse of removal, plus the following:

a. Make sure the 2 locating pins are securely installed in the bottom cowling.

b. Install new oil seals and O-ring to the lower oil seal housing and lubricate them with Yamalube All-purpose Marine grease before installing the housing to the power head.

c. Install a new power head gasket on the bottom cowling locating pins.

d. Lightly lubricate the drive shaft splines with Yamalube All-purpose Marine grease.

e. Lower power head in place. Rotate propeller if necessary to mesh drive shaft and crankshaft splines.

f. Tighten all fasteners to specifications (**Table 1**).

g. Complete timing, synchronizing and adjustment sequence. See Chapter Five.

POWER HEAD DISASSEMBLY

A large number of bolts and screws of different lengths are used to secure the various covers and components. It is a good idea to use a cupcake tin or similar compartmented container to hold the various fasteners removed from each cover or component. This will make reassembly easier and faster.

Remove all electrical, fuel and mechanical components attached to the power head, noting electrical connections and use of J-clamps or other lead routing devices. If necessary, make a sketch of wire routing and J-clamp location to assist you in reassembly. This is especially important in properly reconnecting crankcase recirculation lines that will have to be removed.

Some power heads have pry points in the casting for easier separation of the components. If no pry points are provided, break the gasket seal with a wide-blade putty knife and mallet, then carefully separate the components.

Yamaha 2 hp

Refer to **Figure 58** for this procedure.

1. Secure the power head in a vise with protective jaws.

2. Remove the carburetor.

3. Remove the front bracket, reed valve assembly and gasket. Discard the gasket.

4. Remove the cylinder head (**Figure 59**).

5. Remove the crankcase bolts. Carefully pry the cylinder block from the crankcase and remove (**Figure 60**).

8

6. Remove the crankcase from the crankshaft assembly (**Figure 61**).

7. Remove the drive shaft O-ring from the exhaust plate.

8. Disconnect the upper recirculation line fitting on cylinder block. Blow into hose end and air should pass. Draw suction and air should not pass. If air flow is not as specified, replace the check valve in the block.

Yamaha 3 hp

Refer to **Figure 62** for this procedure.

1. Secure the power head in a vise with protective jaws.

2. Remove the reed valve assembly and gasket. Discard the gasket.

3. Remove the cylinder head (**Figure 59**, typical).

4. Remove the thermostat cover, gasket and thermostat.

5. Remove the crankcase bolts. Carefully pry the cylinder block from the crankcase and remove (**Figure 60**, typical).

6. Remove the crankcase from the crankshaft assembly (**Figure 61**, typical).

Yamaha 4 and 5 hp

Refer to **Figure 63** for this procedure.

1. Note location of the crankcase recirculation lines. Use pliers to slide each line clamp away from the fitting enough to disconnect the line. Pull the lines off the fittings at the cylinder block. Air should pass when you blow into the line. Air should not pass when you draw a suction. If air flow is not as specified, replace the check valve in the block.

2. Remove the 2 nuts holding the carburetor to the reed valve assembly (**Figure 64**). Remove the carburetor and gasket. Discard the gasket.

3. Remove the 2 bolts holding the reed valve assembly to the crankcase (**Figure 64**). Remove the reed valve assembly.

4. If oil seal housing has not been removed, remove the bolt holding it to the bottom of the power head and remove the housing.

5. Unbolt the exhaust cover assembly. Remove the outer and inner covers with gaskets (**Figure 65**). Discard the gaskets.

6. On 5 hp models, remove the thermostat from the inner exhaust cover (**Figure 66**).

7. Remove the cylinder head cover bolts. Remove the cover (**Figure 67**).

NOTE
On 5 hp models, the starter lockout linkage and bracket are removed with the crankcase bolts. Place assembly to one side out of the way.

8. Remove the 6 crankcase bolts. Carefully pry crankcase cover from cylinder block and remove (**Figure 68**).

9. Tap on underneath of crankshaft with a rubber mallet to loosen piston and bearings, then lift crankshaft assembly from cylinder block (**Figure 69**).

10. Install crankshaft assembly on a suitable power head stand or clamp in a vise with protective jaws.

All 2-cylinder Models

Refer to **Figure 70** (9.9 and 15 hp), **Figure 71** (25 hp [1984-1987]), **Figure 72** (25 hp [1988-on]) and **Figure 73** (30 hp) for this procedure. Other models are similar to those shown.

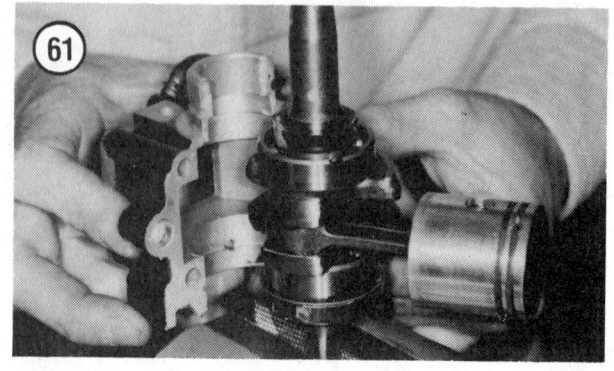

POWER HEAD (3 HP)

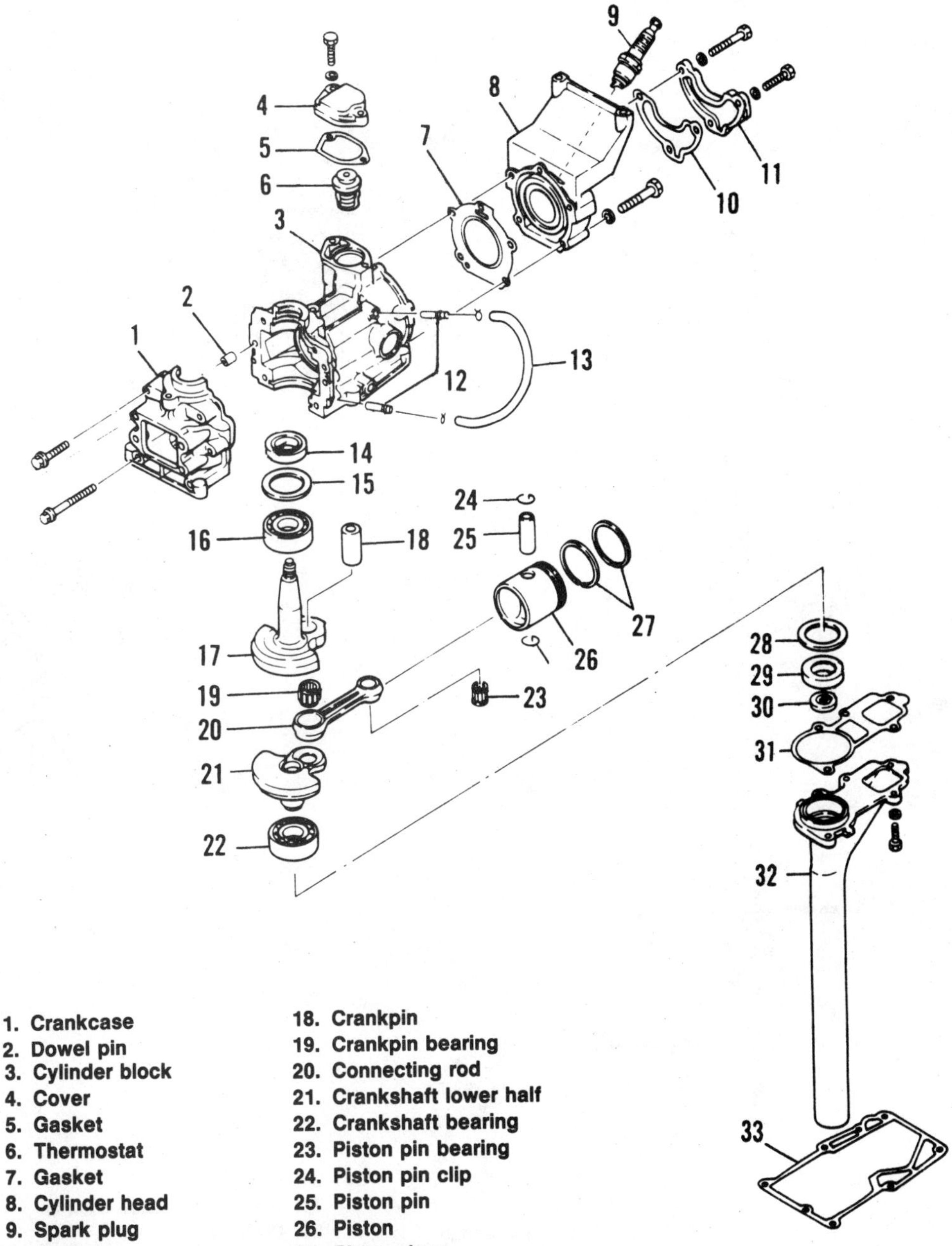

1. Crankcase
2. Dowel pin
3. Cylinder block
4. Cover
5. Gasket
6. Thermostat
7. Gasket
8. Cylinder head
9. Spark plug
10. Gasket
11. Cylinder head cover
12. Hose fittings
13. Recirculation hose
14. Oil seal
15. Washer
16. Crankshaft bearing
17. Crankshaft upper half

18. Crankpin
19. Crankpin bearing
20. Connecting rod
21. Crankshaft lower half
22. Crankshaft bearing
23. Piston pin bearing
24. Piston pin clip
25. Piston pin
26. Piston
27. Piston rings
28. Lower crankcase washer
29. Oil seal
30. Oil seal
31. Gasket
32. Exhaust manifold
33. Gasket

8

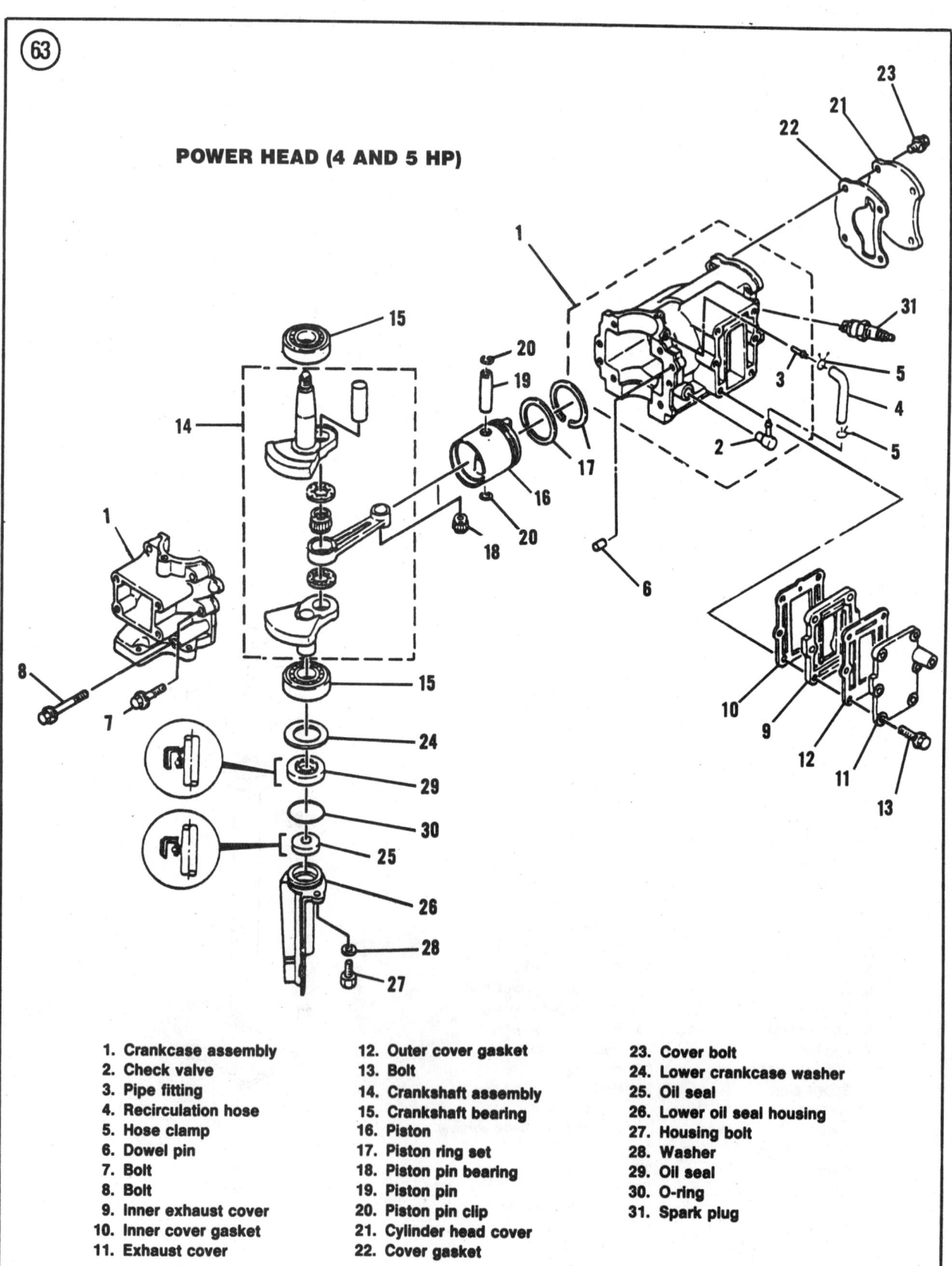

POWER HEAD (4 AND 5 HP)

1. Crankcase assembly
2. Check valve
3. Pipe fitting
4. Recirculation hose
5. Hose clamp
6. Dowel pin
7. Bolt
8. Bolt
9. Inner exhaust cover
10. Inner cover gasket
11. Exhaust cover
12. Outer cover gasket
13. Bolt
14. Crankshaft assembly
15. Crankshaft bearing
16. Piston
17. Piston ring set
18. Piston pin bearing
19. Piston pin
20. Piston pin clip
21. Cylinder head cover
22. Cover gasket
23. Cover bolt
24. Lower crankcase washer
25. Oil seal
26. Lower oil seal housing
27. Housing bolt
28. Washer
29. Oil seal
30. O-ring
31. Spark plug

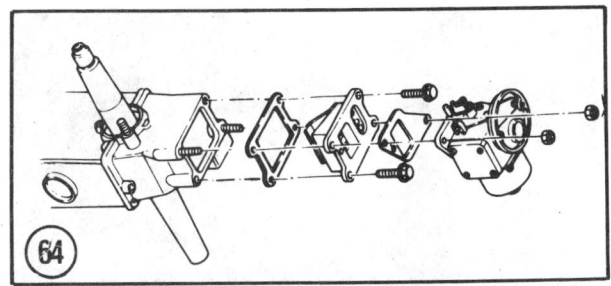

1. Note location of the crankcase recirculation lines. Use pliers to slide each line clamp away from the fitting enough to disconnect the line. Pull the lines off the fittings at the cylinder block. Air should pass when you blow into the line. Air should not pass when you draw suction. If air flow is not as specified, replace the check valve in the block.

2. Remove the intake manifold with reed valve assembly. See **Figure 74** (typical).

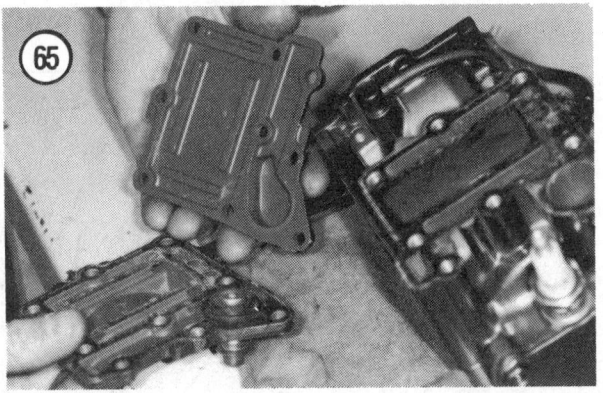

3. 9.9-30 hp—Remove the thermostat housing. See **Figure 75** (typical). Remove the thermostat and rubber washer.

4. Unbolt and remove the cylinder head. See **Figure 76** (typical).

5. 6 and 8 hp—Remove the thermostat and rubber washer from the cylinder block.

6. If anodes are installed in the cylinder head water passages, remove and check their condition. Replace any anode that is less than 60 percent of its original size.

8

7. Unbolt and remove the outer and inner exhaust covers. See **Figure 77** (typical). Remove and discard the inner and outer exhaust cover gaskets.

8. As equipped, carefully pry the upper oil seal housing free, then remove it from the power head.

NOTE
With larger displacement engines, it may be more convenient at this time to remove the assembly from the power head stand and place it on a clean work bench before performing Step 9 and Step 10.

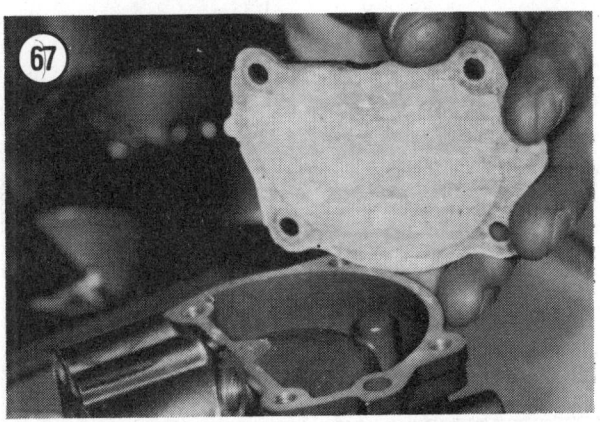

9. Unbolt the crankcase halves. Carefully pry the cover from the cylinder block and remove. See **Figure 78** (typical).

**POWER HEAD
(9.9 AND 15 HP)**

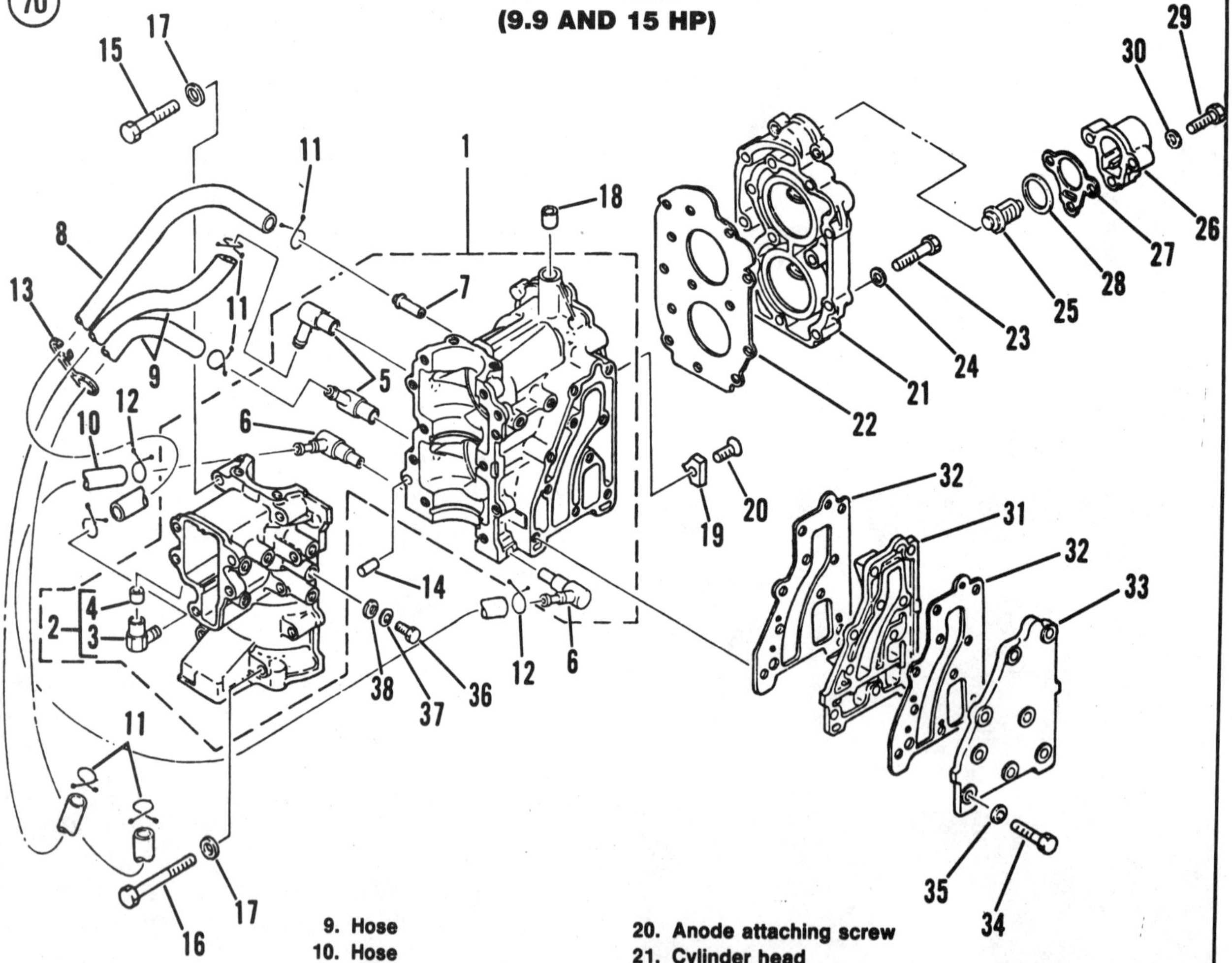

1. Crankcase assembly
2. Check valve assembly
3. Drain fitting
4. Check valve
5. Check valve assembly
6. Drain fitting
7. Drain fitting
8. Hose
9. Hose
10. Hose
11. Clamp
12. Clamp
13. Clamp
14. Dowel pin
15. Crankcase bolt (short)
16. Crankcase bolt (long)
17. Washer
18. Dowel
19. Anode
20. Anode attaching screw
21. Cylinder head
22. Cylinder head gasket
23. Head bolt
24. Washer
25. Thermostat
26. Thermostat cover
27. Thermostat cover gasket
28. Washer
29. Thermostat cover bolt
30. Washer
31. Inner exhaust cover
32. Exhaust cover gasket
33. Outer exhaust cover
34. Exhaust cover bolt
35. Washer
36. Bolt
37. Washer
38. Gasket

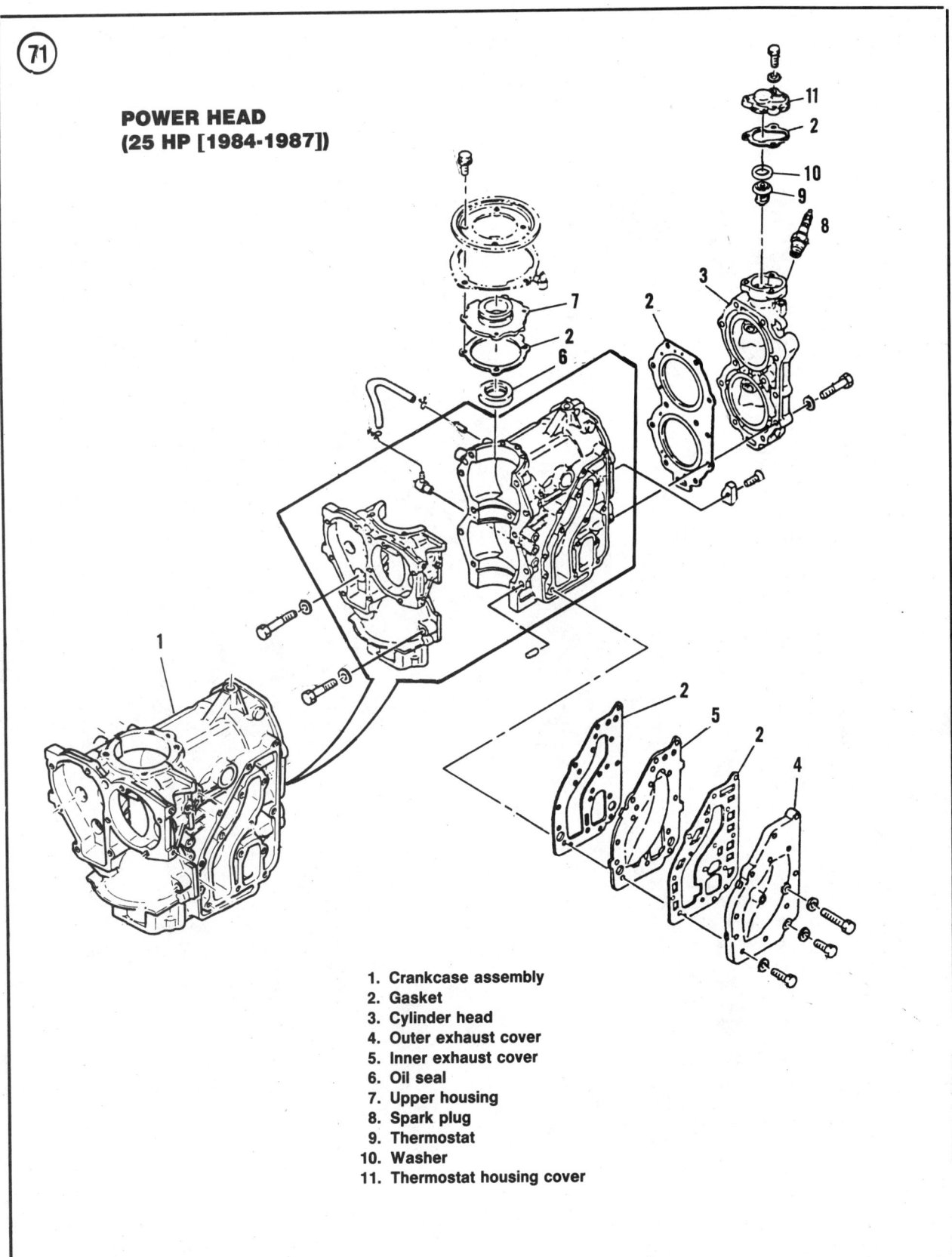

71

**POWER HEAD
(25 HP [1984-1987])**

1. Crankcase assembly
2. Gasket
3. Cylinder head
4. Outer exhaust cover
5. Inner exhaust cover
6. Oil seal
7. Upper housing
8. Spark plug
9. Thermostat
10. Washer
11. Thermostat housing cover

8

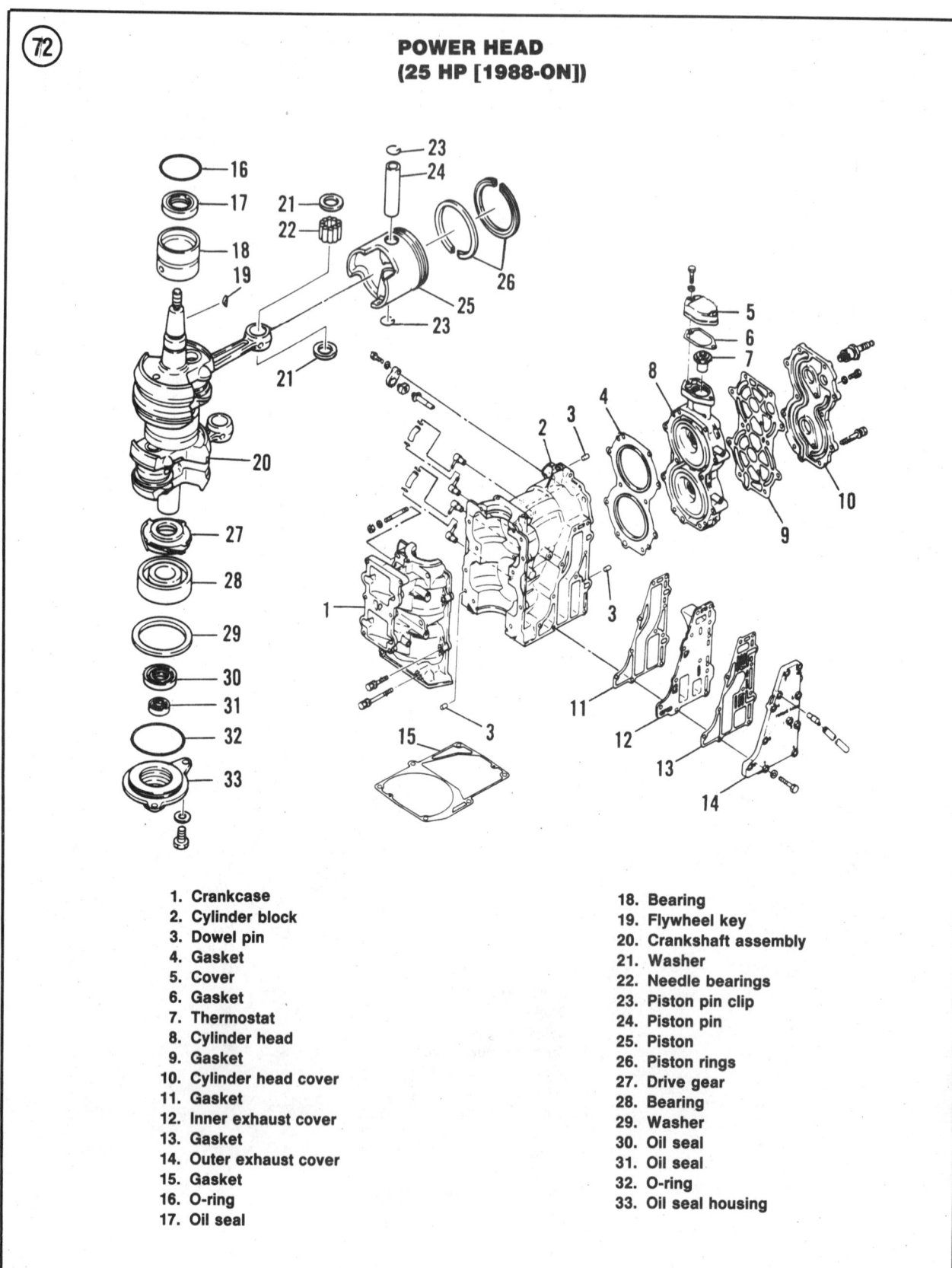

**POWER HEAD
(25 HP [1988-ON])**

1. Crankcase
2. Cylinder block
3. Dowel pin
4. Gasket
5. Cover
6. Gasket
7. Thermostat
8. Cylinder head
9. Gasket
10. Cylinder head cover
11. Gasket
12. Inner exhaust cover
13. Gasket
14. Outer exhaust cover
15. Gasket
16. O-ring
17. Oil seal
18. Bearing
19. Flywheel key
20. Crankshaft assembly
21. Washer
22. Needle bearings
23. Piston pin clip
24. Piston pin
25. Piston
26. Piston rings
27. Drive gear
28. Bearing
29. Washer
30. Oil seal
31. Oil seal
32. O-ring
33. Oil seal housing

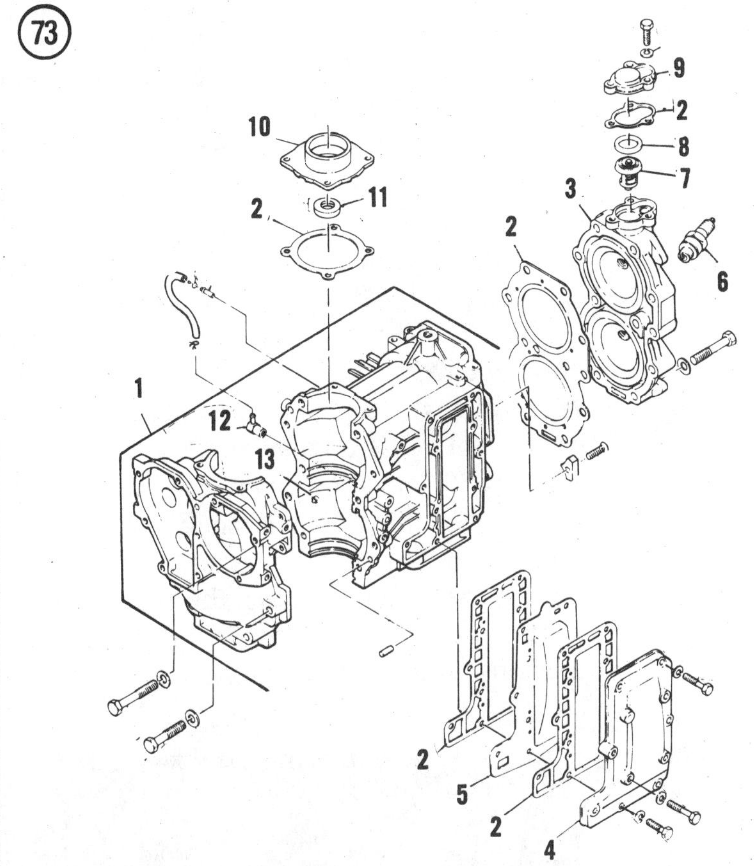

**POWER HEAD
(30 HP [1984-1986])**

1. Crankcase assembly
2. Gasket
3. Cylinder head
4. Outer exhaust cover
5. Inner exhaust cover
6. Spark plug
7. Thermostat
8. Washer
9. Thermostat housing cover
10. Upper housing
11. Oil seal
12. Fitting
13. Dowel

8

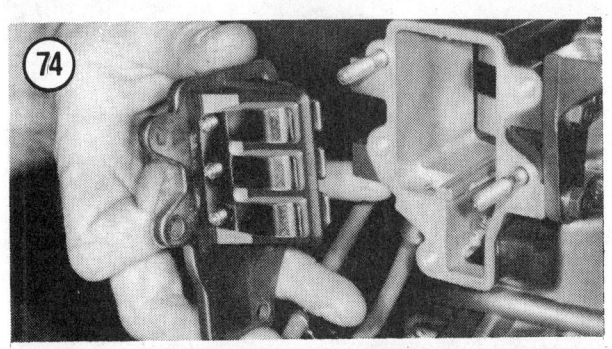

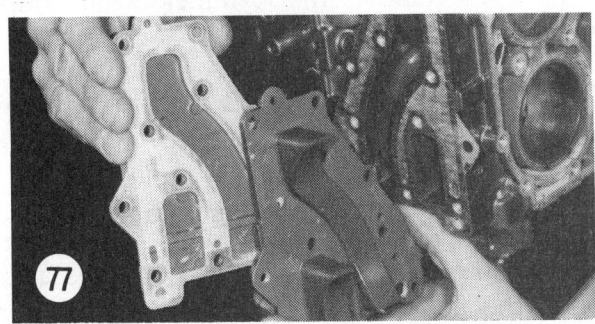

10. If installed on the power head stand, remove the cylinder block from the crankshaft assembly (**Figure 79**). If block is on the work bench, tap the underside of the crankshaft and remove assembly from the block, then reinstall it on the power head stand.

All 3-cylinder Models

Refer to **Figure 80** (30 hp), **Figure 81** (40 and 50 hp), **Figure 82** (70 hp) and **Figure 83** (90 hp) for this procedure.

1. Note location of the crankcase recirculation lines. Use pliers to slide each line clamp away from the fitting enough to disconnect the line. Pull the lines off the fittings at the cylinder block. Air should pass when you blow into the line. Air should not pass when you draw suction. If air flow is not as specified, replace the check valve in the block.

2. Remove the intake manifold with reed valve assembly.

3. 90 hp—Remove the thermostat cover from the cylinder head. Remove the thermostat and pressure valve.

4. Remove the thermoswitch from the cylinder head.

5A. 30-70 hp:

a. Remove the cylinder head bolts.

b. Tap the edge of the cylinder head with a plastic hammer to break the gasket seal, then remove the cylinder head and cover assembly.

c. Place the assembly on a solid surface and carefully pry the cover and head apart using a suitable pry tool at the pry points provided. See **Figure 84**.

d. Separate the cover and head. Discard the gaskets.

5B. 90 hp:

a. Remove the cylinder head cover bolts.

b. Remove the cylinder head cover and gasket. Discard the gasket.

c. Remove the cylinder head bolts.

d. Tap the edge of the cylinder head with a plastic hammer to break the gasket seal, then remove the head and gasket assembly. Discard the gasket.

6. 70 hp—Remove the upper and lower anode covers. Remove and check their condition. Replace any anode that is less than 60 percent of its original size.

7A. 30-50 hp—Remove the thermostat cover from the cylinder block. Remove the thermostat and rubber washer.

7B. 70 hp—Remove the thermostat cover from the exhaust cover. Remove the thermostat and rubber washer.

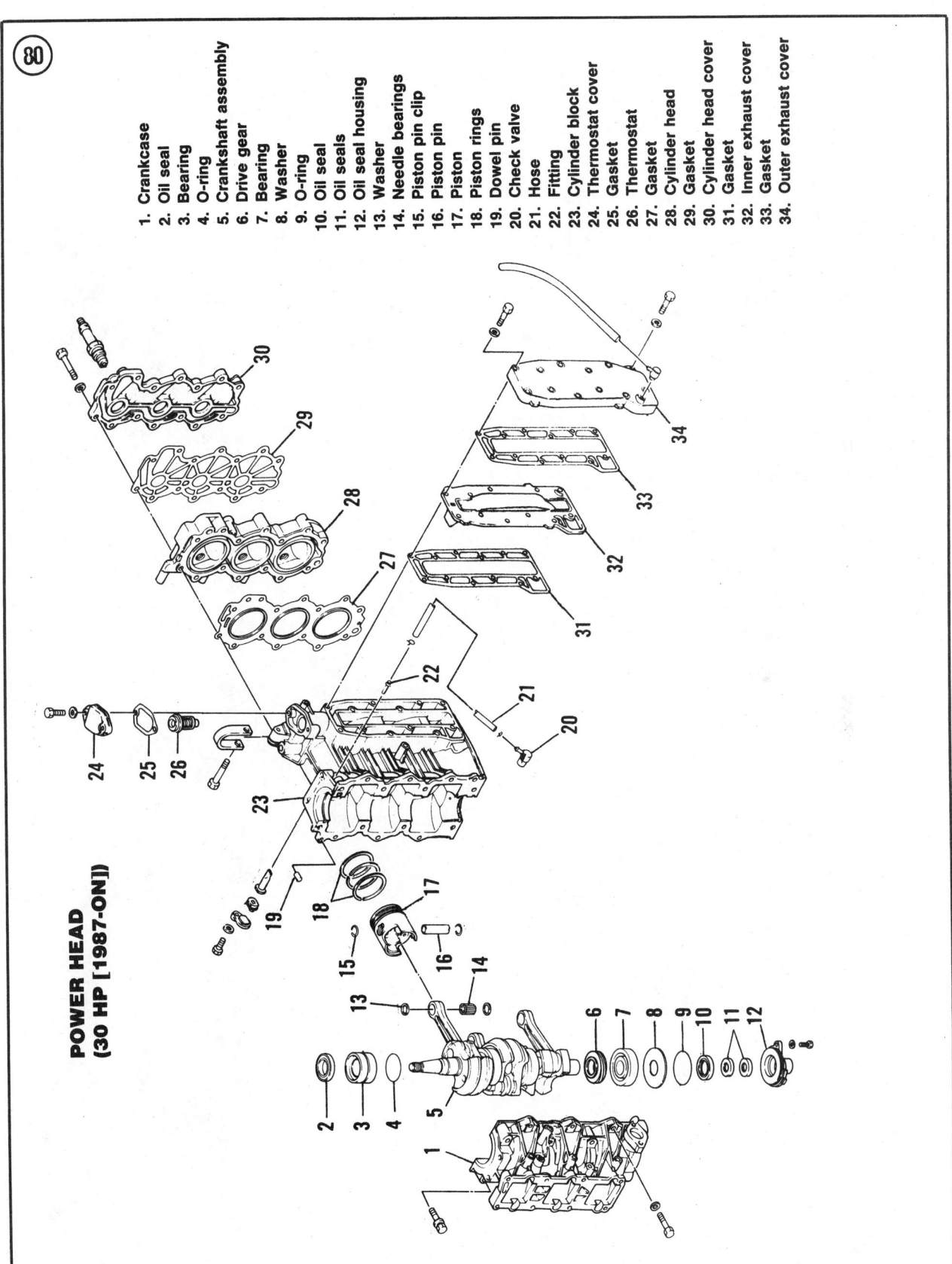

**POWER HEAD
(30 HP [1987-ON])**

80

1. Crankcase
2. Oil seal
3. Bearing
4. O-ring
5. Crankshaft assembly
6. Drive gear
7. Bearing
8. Washer
9. O-ring
10. Oil seal
11. Oil seals
12. Oil seal housing
13. Washer
14. Needle bearings
15. Piston pin clip
16. Piston pin
17. Piston
18. Piston rings
19. Dowel pin
20. Check valve
21. Hose
22. Fitting
23. Cylinder block
24. Thermostat cover
25. Gasket
26. Thermostat
27. Gasket
28. Cylinder head
29. Gasket
30. Cylinder head cover
31. Gasket
32. Inner exhaust cover
33. Gasket
34. Outer exhaust cover

8

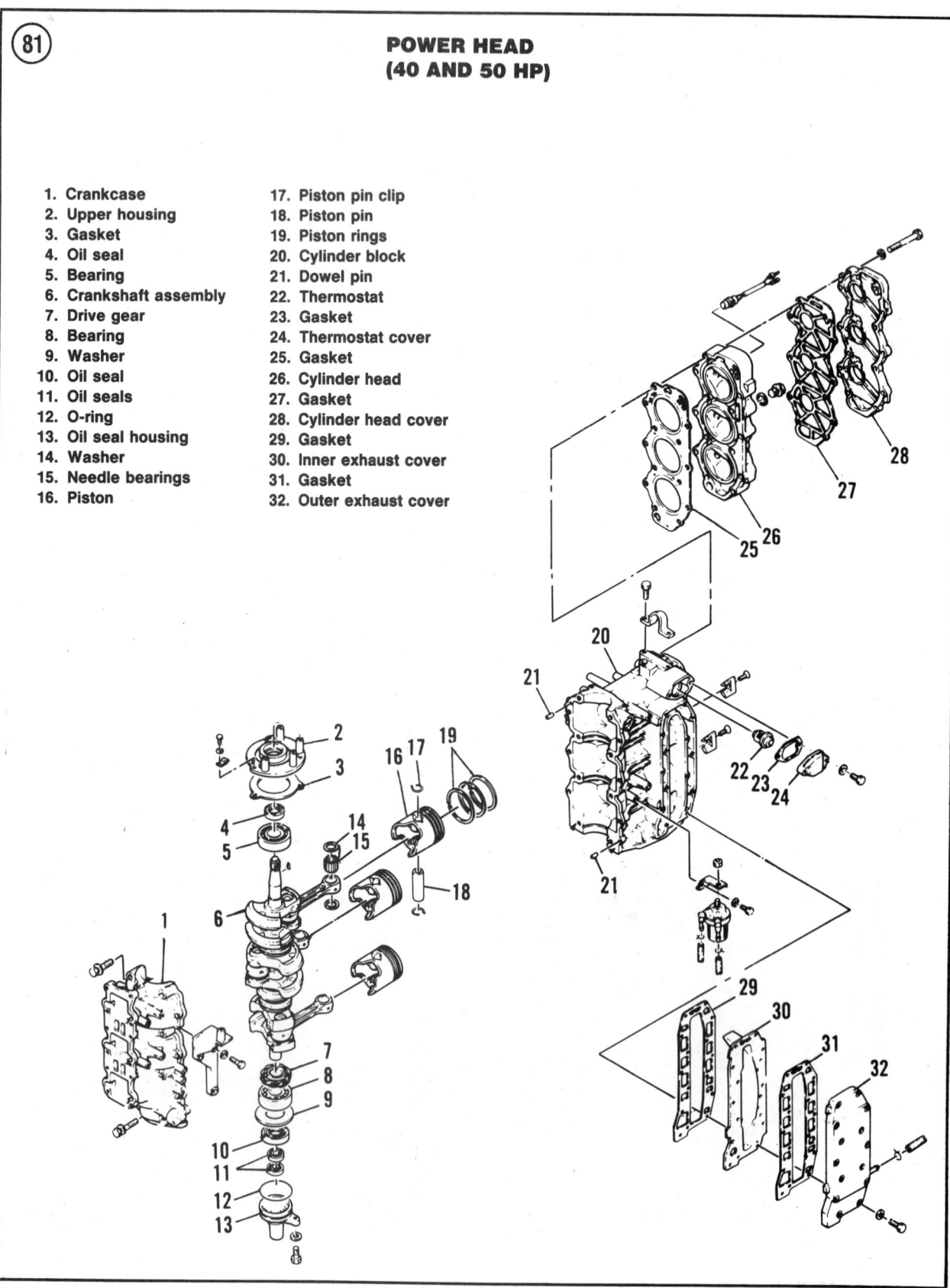

⑧1

**POWER HEAD
(40 AND 50 HP)**

1. Crankcase
2. Upper housing
3. Gasket
4. Oil seal
5. Bearing
6. Crankshaft assembly
7. Drive gear
8. Bearing
9. Washer
10. Oil seal
11. Oil seals
12. O-ring
13. Oil seal housing
14. Washer
15. Needle bearings
16. Piston

17. Piston pin clip
18. Piston pin
19. Piston rings
20. Cylinder block
21. Dowel pin
22. Thermostat
23. Gasket
24. Thermostat cover
25. Gasket
26. Cylinder head
27. Gasket
28. Cylinder head cover
29. Gasket
30. Inner exhaust cover
31. Gasket
32. Outer exhaust cover

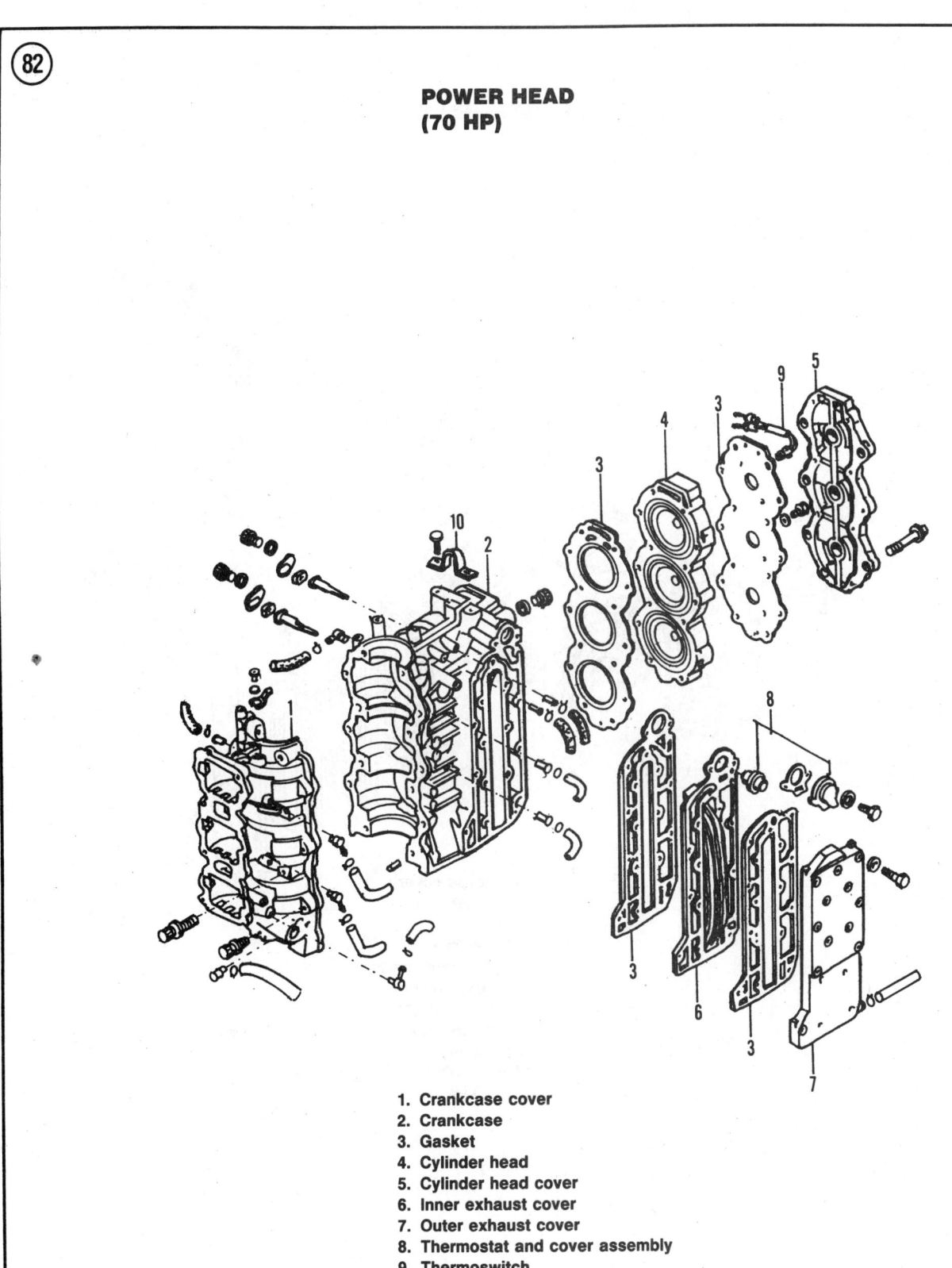

⑧²

**POWER HEAD
(70 HP)**

8

1. Crankcase cover
2. Crankcase
3. Gasket
4. Cylinder head
5. Cylinder head cover
6. Inner exhaust cover
7. Outer exhaust cover
8. Thermostat and cover assembly
9. Thermoswitch
10. Lifting eye

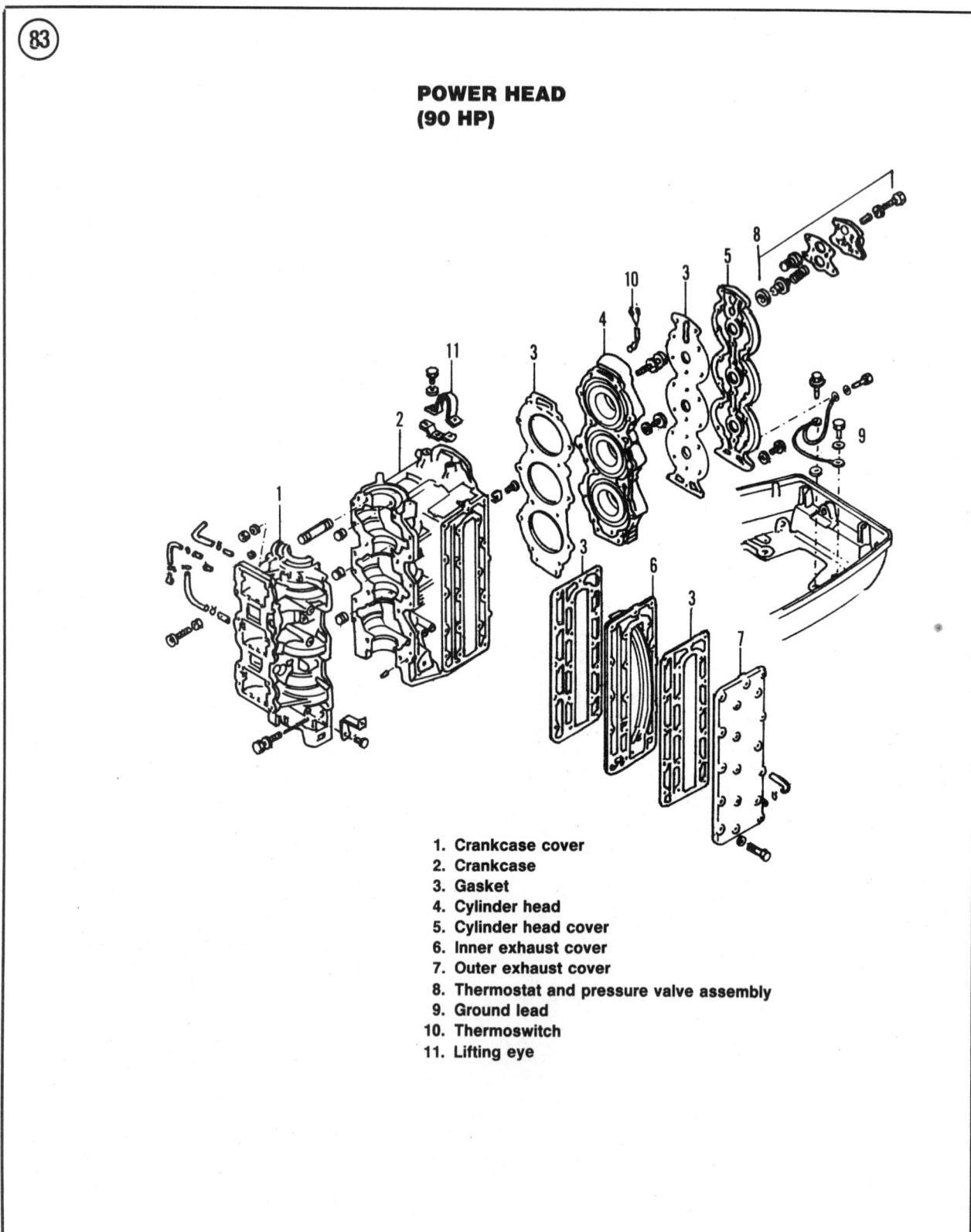

83

**POWER HEAD
(90 HP)**

1. Crankcase cover
2. Crankcase
3. Gasket
4. Cylinder head
5. Cylinder head cover
6. Inner exhaust cover
7. Outer exhaust cover
8. Thermostat and pressure valve assembly
9. Ground lead
10. Thermoswitch
11. Lifting eye

8. Remove the exhaust cover bolts. Insert a screwdriver blade between the cover lugs and exhaust manifold and pry the cover assembly free. Separate the covers and discard the gaskets.

9. Unbolt the lower oil seal housing. Remove lower oil seal housing on 30-50 hp models.

NOTE
It may be more convenient at this time to remove the assembly from the power head stand and place it in a horizontal (30-70 hp) or vertical (90 hp) position on a clean work bench before performing Step 10.

10. Remove the crankcase bolts. Insert a screwdriver blade between the tabs on each side of the crankcase and cylinder and pry the two apart.

11. 70 hp—Remove the lower oil seal housing.

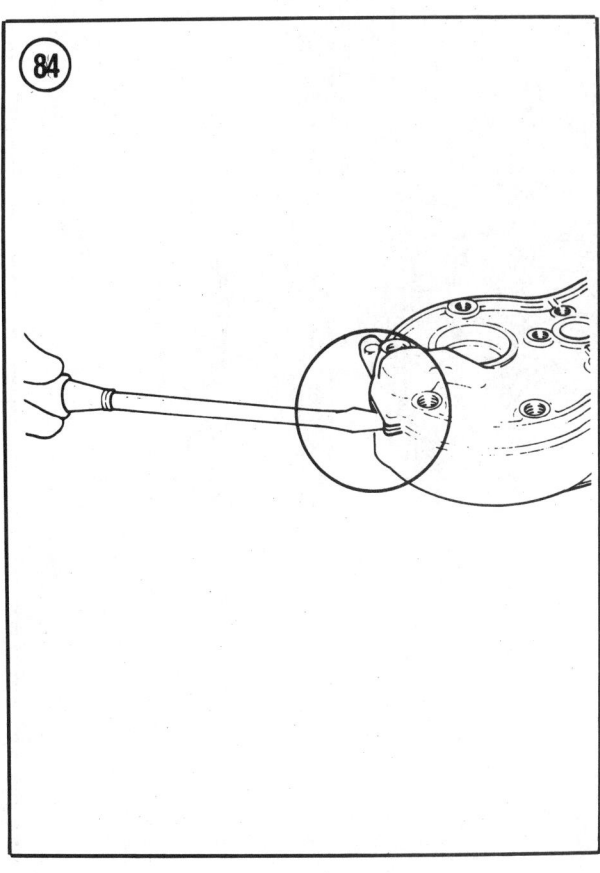

12A. 30-70 hp—If installed on the power head stand, remove the cylinder block from the crankshaft assembly. If block is in a horizontal position on the work bench, tap the underside of the crankshaft and remove assembly from the block, then reinstall it on the power head stand.

12B. 90 hp—With power head in a vertical position:

  a. Mark the connecting rod and caps. Remove each connecting rod cap, roller bearings and bearing cage.

  b. Place bearings and cage from each rod in separate containers.

  c. Remove the lower oil seal housing.

  d. Remove the crankshaft from the cylinder block. If necessary, lightly tap on crankshaft taper with a rubber mallet to break the seal.

  e. Remove the remaining connecting rod bearings and cages. Place in their respective containers.

  f. Reinstall each rod cap to its respective connecting rod.

  g. Carefully push each connecting rod and piston assembly from its cylinder and remove from the top of the cylinder block. Place on a clean workbench.

  h. Mark the cylinder number on the top of the piston with a felt-tipped pen.

All V-block Models

Refer to **Figure 85** (typical) for this procedure.

1. Note location of the crankcase recirculation lines. Use pliers to slide each line clamp away from the fitting enough to disconnect the line. Pull the lines off the fittings at the cylinder block. Air should pass when you blow into the line. Air should not pass when you draw a suction. If air flow is not as specified, replace the check valve in the block.

2. Unbolt and remove the shift slide bracket from the power head.

8

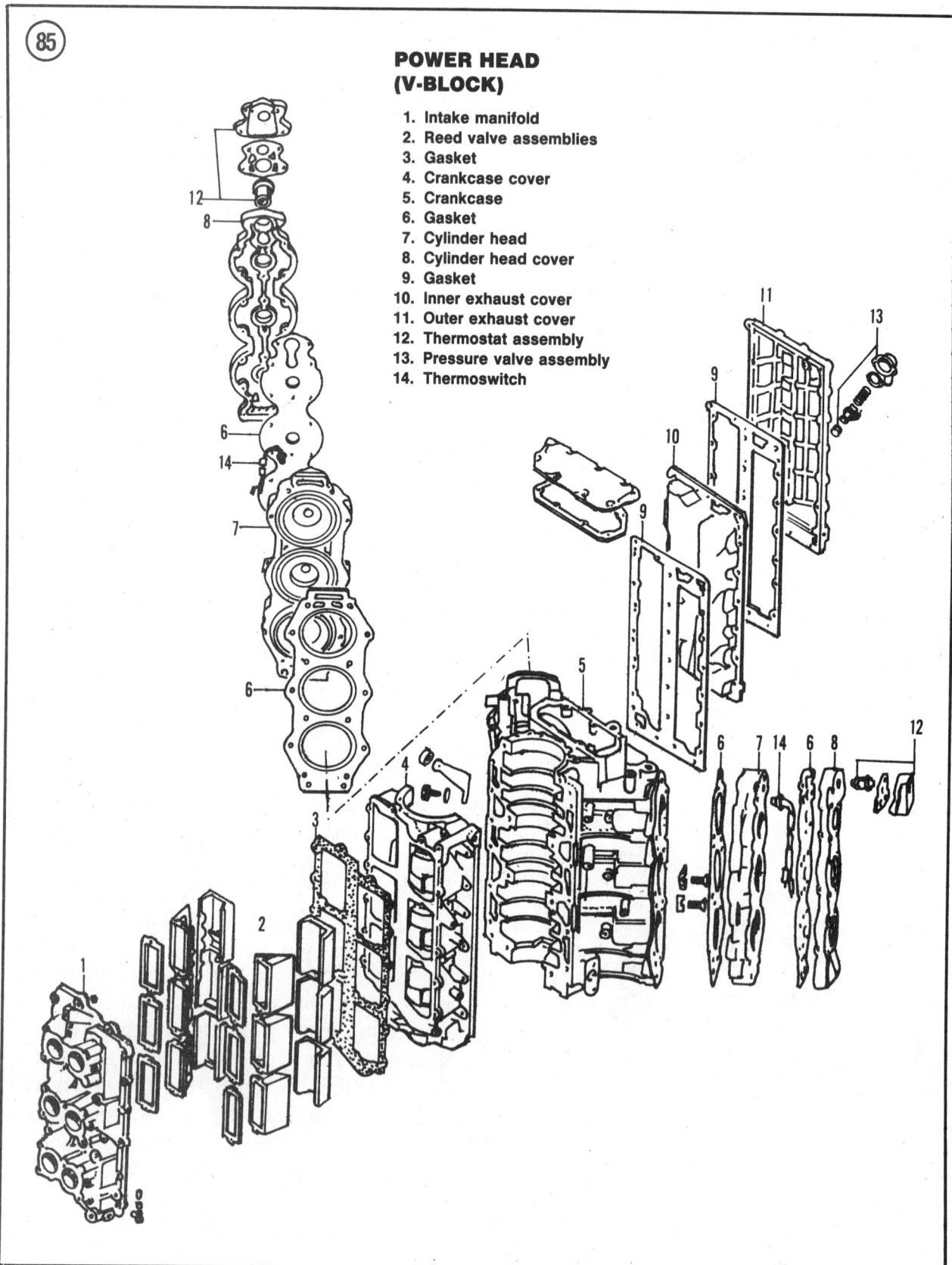

85

POWER HEAD (V-BLOCK)

1. Intake manifold
2. Reed valve assemblies
3. Gasket
4. Crankcase cover
5. Crankcase
6. Gasket
7. Cylinder head
8. Cylinder head cover
9. Gasket
10. Inner exhaust cover
11. Outer exhaust cover
12. Thermostat assembly
13. Pressure valve assembly
14. Thermoswitch

3. Unbolt and remove the engine stop on each side of the cylinder block. See **Figure 86**.

4. Remove the intake manifold with reed valve assembly.

5. Remove the bolts holding the upper and lower crankshaft oil seal housings.

6. Remove the crankcase bolts. Insert a screwdriver blade between the tabs on each side of the crankcase and cylinder and pry the two apart.

7. Mark the connecting rod and caps. Remove each connecting rod cap, roller bearings and bearing cage. Place bearings and cage from each rod in separate containers.

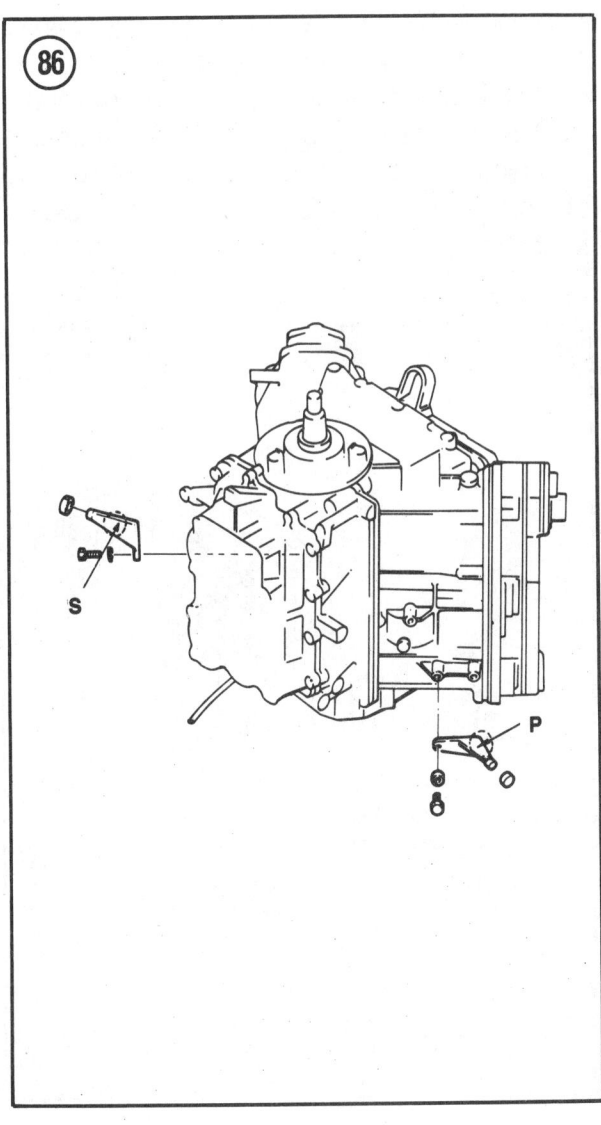

8. Grasp the lower and upper bearing oil seal housings and use them as handles to remove the crankshaft from the crankcase. If necessary, tap lightly on crankshaft taper with a rubber mallet to break the seal. Place the crankshaft on a clean work bench and remove the housings at each end.

9. Remove the remaining connecting rod bearings and cages. Place in their respective container.

10. V-4—With the power head in a vertical position, remove the thermostat cover from each cylinder head cover. Remove the thermostat, spring and pressure valve from each cylinder head.

11. 220 and 225 hp—Remove the knock sensor and crankshaft position sensor from the cylinder head and crankcase.

12. Remove the cylinder head cover bolts. Insert a screwdriver blade between the cover lugs and cylinder head and pry the cover assembly free. Discard the gasket.

13. Grasp the cylinder head thermoswitch and pull it straight out from the cylinder head. Repeat this step to remove the other cylinder head thermoswitch.

14. Loosen all cylinder head bolts on one cylinder head. Remove all bolts and insert a screwdriver blade between the cylinder head and crankcase lugs. Pry the head free and discard the gasket. Repeat this step to remove the other cylinder head.

15. Reinstall each rod cap to its respective connecting rod. Carefully push each connecting rod and piston assembly from its cylinder and remove from the top of the cylinder block. Mark the cylinder number on the top of the piston with a felt-tipped pen and place on a clean workbench.

16. V6—Remove the pressure valve cover from the exhaust cover. Remove the valve and spring.

17. Remove the exhaust cover bolts. Insert a screwdriver blade between the cover lugs and exhaust manifold and pry the cover assembly

8

free. Separate the covers and discard the gaskets. See **Figure 87**.

18. Unbolt and remove the cylinder cover from the top of the power head. See **Figure 87**.

CRANKSHAFT ASSEMBLY AND CYLINDER BLOCK

Crankshaft Disassembly
(2-70 hp)

Connecting rods cannot be disassembled from the crankshaft. If either the rod(s) or crankshaft is defective, replace the entire assembly. Refer to **Figure 88** (typical 1-cylinder), **Figure 89** (typical 2-cylinder) or **Figure 90** (typical 3-cylinder) for this procedure.

1. Mark the upper and lower piston on 2- and 3-cylinder crankshafts with a felt-tipped pen for reassembly to the same connecting rod. The No. 1 piston is at the top when the crankshaft is installed on a power head stand.

NOTE
It is a good idea to install new piston rings whenever the crankshaft is disassembled.

NOTE
Some 1989 30, 40, 50 and 70 hp models are equipped with two ring pistons. Both piston rings are semi-keystone shaped. This was a running change for the 1989 model year so some early 1989 models will be equipped with the three ring pistons used in 1988 and prior engines.

2. Remove the rings from each piston with a suitable ring expander tool. If rings are to be reused, tie a parts tag around each set and identify the tag with the piston number.

WARNING
Wear protective eyeglasses while performing Step 3.

3. Remove the piston pin clips with needlenose pliers (**Figure 91**). Discard the clips.

NOTE
If piston pin cannot be pushed out with a finger in Step 4, insert an appropriate size punch in the hollow end of the piston pin. Support bottom of piston with one hand and tap end of punch with a hammer to push pin from piston.

4. Holding the piston in position, push the piston pin up through the piston and connecting rod. Remove the piston from the rod and catch the bearings. One-cylinder connecting rods use a caged bearing (**Figure 92**). Two- and three-cylinder rods use uncaged needle bearings with retaining washers which will fall into the piston as it is removed. See **Figure 93**. Be sure not to drop or lose any uncaged bearings if reuse is intended.

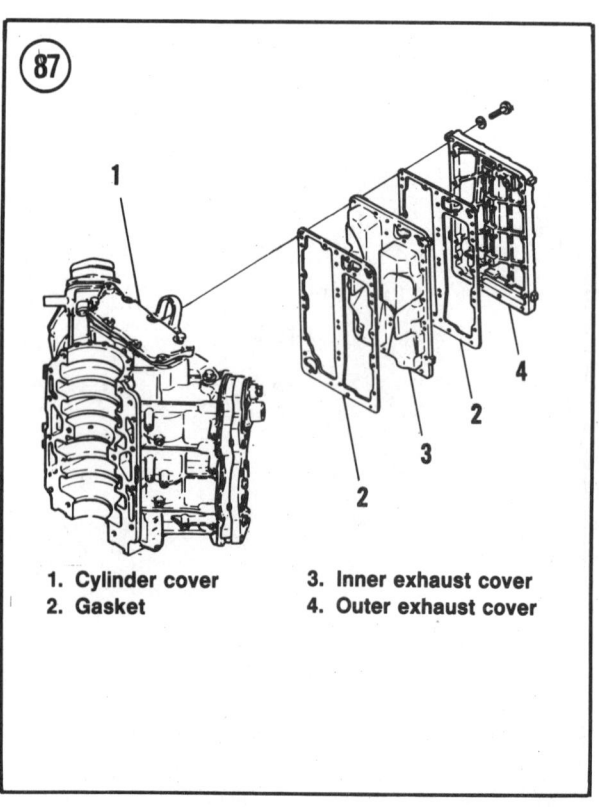

1. Cylinder cover
2. Gasket
3. Inner exhaust cover
4. Outer exhaust cover

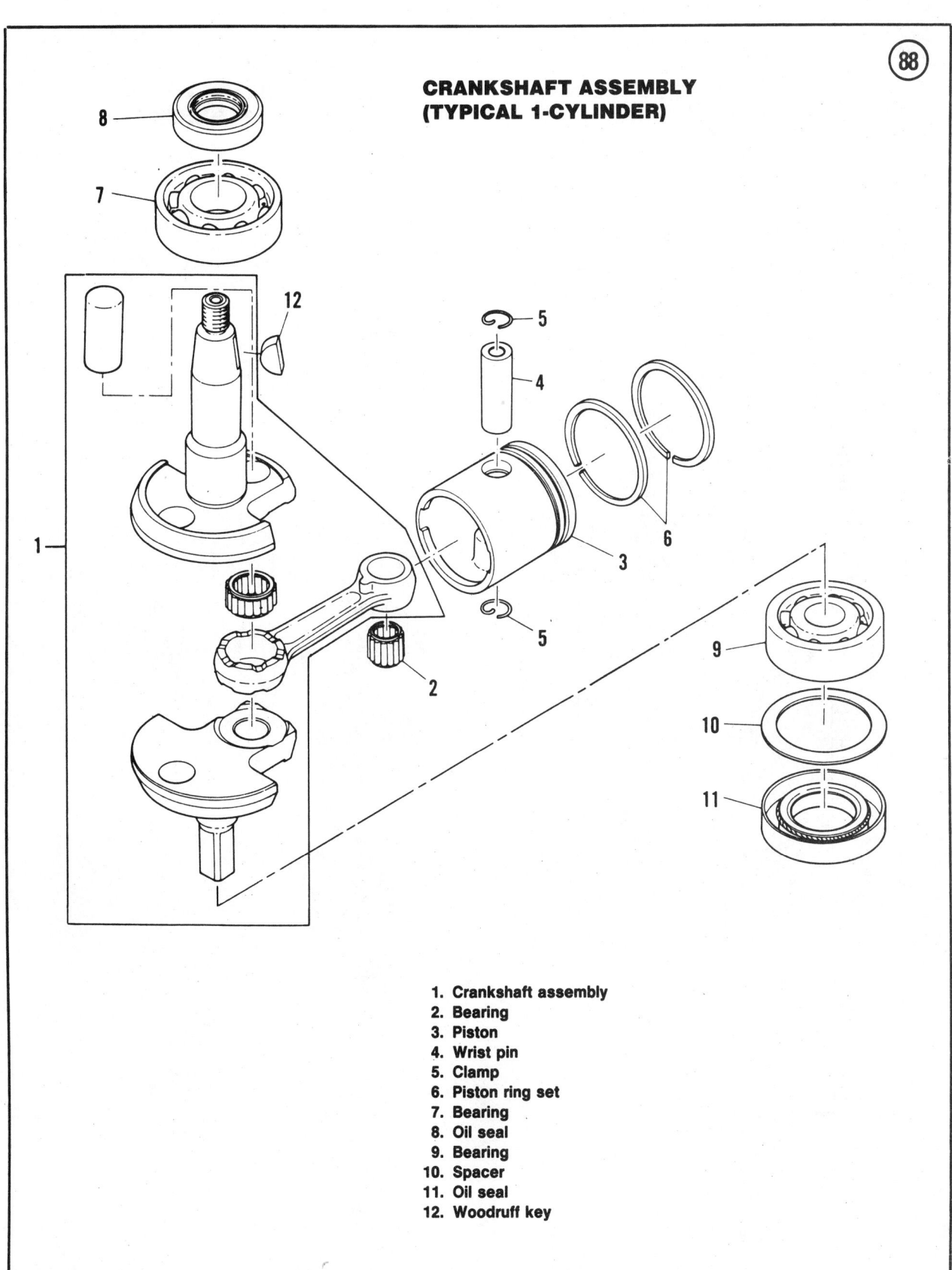

**CRANKSHAFT ASSEMBLY
(TYPICAL 1-CYLINDER)**

1. Crankshaft assembly
2. Bearing
3. Piston
4. Wrist pin
5. Clamp
6. Piston ring set
7. Bearing
8. Oil seal
9. Bearing
10. Spacer
11. Oil seal
12. Woodruff key

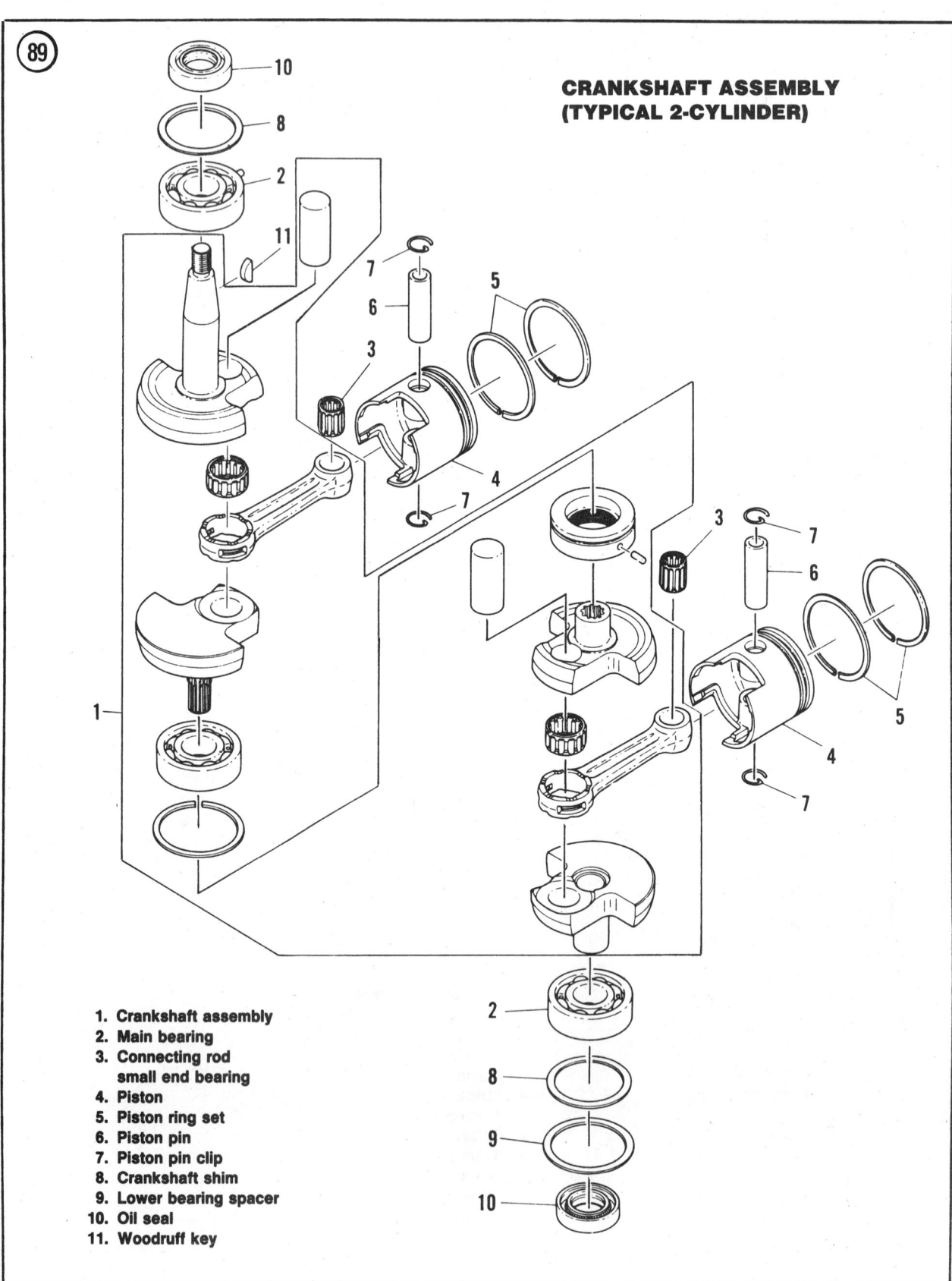

89

**CRANKSHAFT ASSEMBLY
(TYPICAL 2-CYLINDER)**

1. Crankshaft assembly
2. Main bearing
3. Connecting rod
   small end bearing
4. Piston
5. Piston ring set
6. Piston pin
7. Piston pin clip
8. Crankshaft shim
9. Lower bearing spacer
10. Oil seal
11. Woodruff key

**CRANKSHAFT ASSEMBLY
(TYPICAL 3-CYLINDER)**

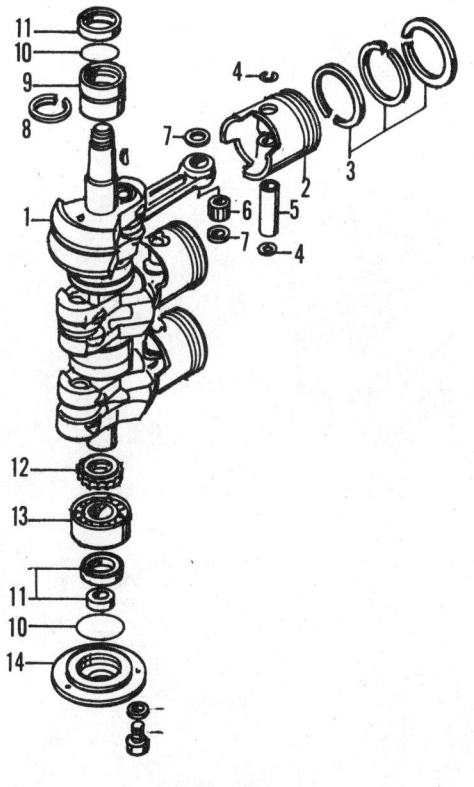

1. Crankshaft
2. Piston
3. Piston rings
4. Piston lock pin
5. Piston pin
6. Needle bearings
7. Washer
8. Bearing sleeve retaining clip
9. Bearing sleeves
10. O-ring
11. Oil seal
12. Oil pump drive gear
13. Ball bearing
14. Lower oil seal housing

5. Repeat Step 3 and Step 4 to remove each remaining piston from its connecting rod.

NOTE
It is a good idea to install new needle bearings whenever the crankshaft is disassembled. If old bearings must be reused, store them in clean numbered containers corresponding to the piston and rod number for reinstallation.

6. Remove any oil seals or spacers from the crankshaft.

7. If upper or lower crankshaft ball bearing removal is necessary, press bearing from the crankshaft with a universal puller plate and arbor press.

**Crankshaft Disassembly
(90-225 hp)**

Refer to **Figure 90** (3-cylinder) for this procedure.

1. Remove the flywheel Woodruff key, if not removed when the flywheel was removed.

2. Slide the upper bearing housing and lower oil seal housing off the crankshaft.

3. 90 hp—Remove the upper crankshaft bearing.

4. Pry the main bearing sleeve retaining ring from the bearing sleeve groove. Note that the retaining ring groove in the bearing sleeve faces toward the top of the crankshaft, then separate the bearing sleeve halves and remove from the crankshaft.

5. Carefully remove bearings from crankshaft. Store bearings and sleeve halves in a clean container for reinstallation on the same journal.

6. Carefully slip the retaining ring off the crankshaft journal to prevent excessive stretching.

8

7. Repeat Steps 4-6 to remove the other main bearing.

NOTE
If seal rings require replacement, remove and install with a piston ring expander using the same method as described for piston ring removal/installation.

8. On V-blocks check but do not remove sealing rings from crankshaft unless excessively worn or damaged. Insert a flat feeler gauge between the sealing ring and ring groove in the crankshaft. See **Figure 94**. The wear limit on either side of the ring should not exceed 0.002 in. (0.1 mm).

9. If the lower main bearing or oil pump drive gear requires removal, remove the bearing snap ring and then remove the bearing with an appropriate puller. Remove the pump drive gear.

NOTE
It is a good idea to install new piston rings whenever the crankshaft is disassembled.

NOTE
Some 1989 90 hp models are equipped with two ring pistons. Both piston rings are semi-keystone shaped. This was a running change for the 1989 model year so some early 1989 models will be equipped with the three ring pistons used in 1988 and prior engines. Some 1989 V4 and V6 models are equipped with two semi-keystone shaped piston rings. This was a running change for the 1989 model year so some early 1989 models will be equipped with the one semi-keystone (upper ring) and one square (lower ring) piston ring used in 1988 and prior engines.

10. Remove the piston rings with a suitable ring expander tool. If rings are to be reused, tie a parts tag around each set and identify the tag with the piston number.

WARNING
Wear protective eyeglasses while performing Step 11.

11. If piston is to be removed from the connecting rod, remove the piston pin clips with needlenose pliers (**Figure 91**). Discard the clips.

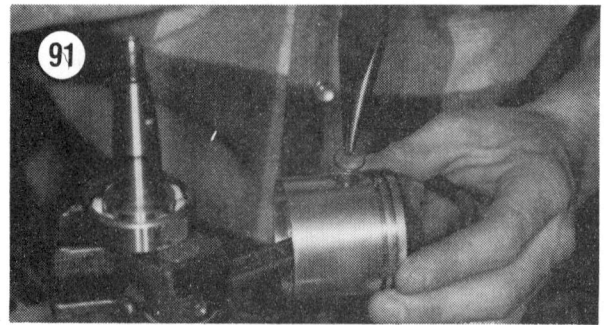

NOTE
If piston pin cannot be pushed out with a finger in Step 12, use a suitable driver and carefully drive it out.

12. Holding the piston in position, push the piston pin up through the piston and connecting rod. Remove the piston from the rod and catch the bearings. All connecting rods use uncaged needle bearings with retaining washers which will fall into the piston as it is removed. See **Figure 93**. Be sure not to drop or lose any uncaged bearings if reuse is intended.

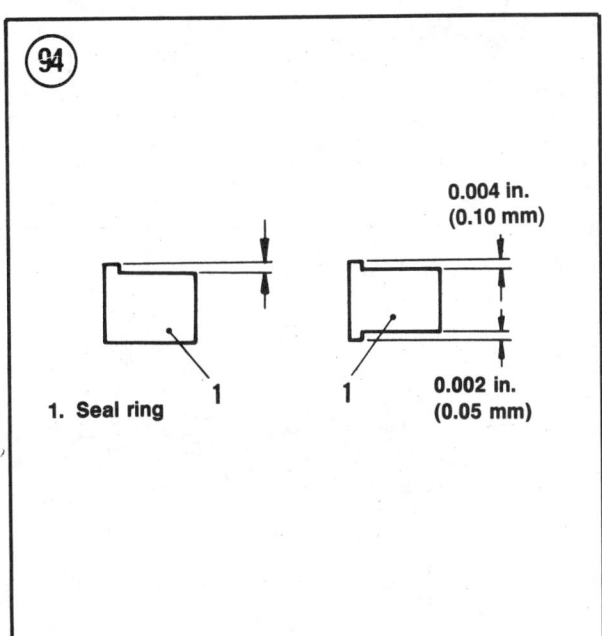

1. Seal ring

0.004 in. (0.10 mm)

0.002 in. (0.05 mm)

NOTE
It is a good idea to install new needle bearings whenever the crankshaft is disassembled. If old bearings must be reused, store them in clean numbered containers corresponding to the piston and rod number for reinstallation.

Cylinder Block Cleaning and Inspection (All Engines)

Yamaha outboard cylinder blocks and crankcase covers are matched and line-bored assemblies. For this reason, you should not attempt to assemble an engine with parts salvaged from other blocks. If the following inspection procedure indicates that either the block or cover requires replacement, replace both as an assembly.

Carefully remove all gasket and sealant residue from the cylinder block and crankcase cover mating surfaces with lacquer thinner. Clean the aluminum surfaces carefully to avoid nicking them. A dull putty knife can be used, but a piece of Lucite with one edge ground to a 45 degree angle is more efficient and will also reduce the possibility of damage to the surfaces. When reassembling the crankcase cover and cylinder block, both mating surfaces must be free of all sealant residue, dirt and oil or leaks will develop.

1. Remove all carbon deposits from the combustion chambers, exhaust ports and cylinder head. See **Figure 95** (typical). Use a hardwood or Lucite scraper and solvent. Be careful not to scratch or gouge the areas while cleaning.

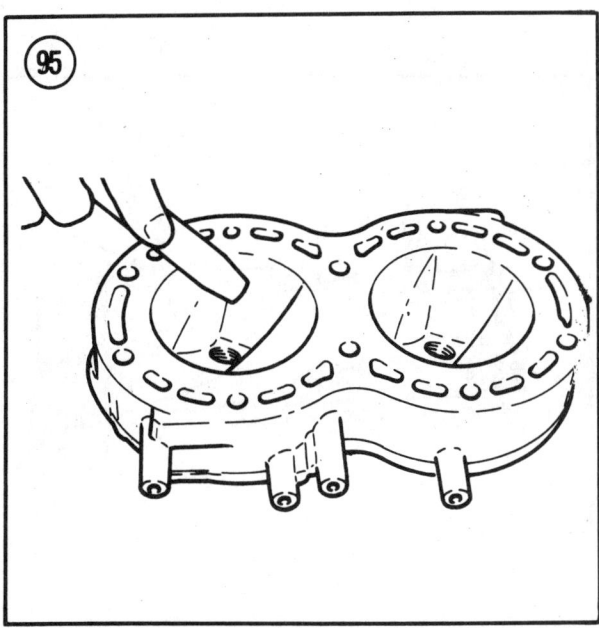

8

2. Once all carbon is removed, clean the cylinder block and cylinder head(s) thoroughly with solvent and a brush.

3. Clean all salt and dirt from the cylinder block and cylinder head water jackets and cooling passages. Check the condition of the anti-corrosion paint applied to the water jacket area and touch up as required.

4. Check the cylinder head and cylinder block gasket mating surfaces for warpage in the directions shown in **Figure 96-98** and correct as required. See *Gaskets and Sealants* in this chapter.

5. Check the exhaust port surfaces for warpage as described in Step 4. Follow the directions shown in **Figure 99** or **Figure 100** and correct as required.

6. Check the block, cylinder head(s) and cover for cracks, fractures, stripped bolt or spark plug threads or other defects.

7. Check all gasket mating surfaces for nicks, grooves, cracks or excessive distortion. Any of these defects will cause compression leakage. Replace as required.

8. Check all oil and water passages in the block and cover for obstructions. Make sure any plugs installed are properly tightened.

9. On 1-cylinder models, check cylinder wall for scoring. If lightly scored, clean up with No. 400 grit paper. Scrub bore with hot water and detergent, then rinse with hot water.

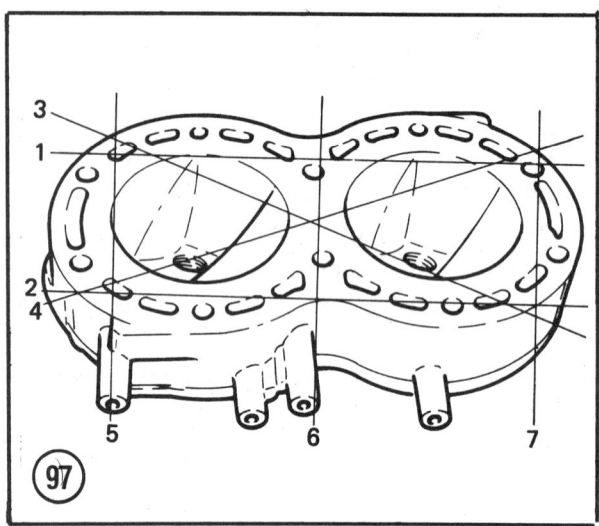

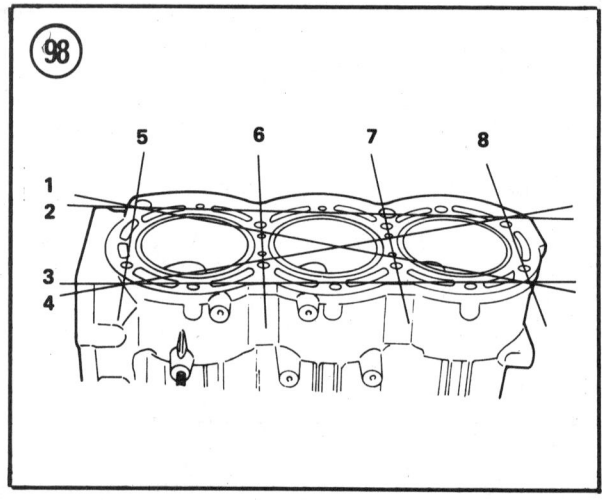

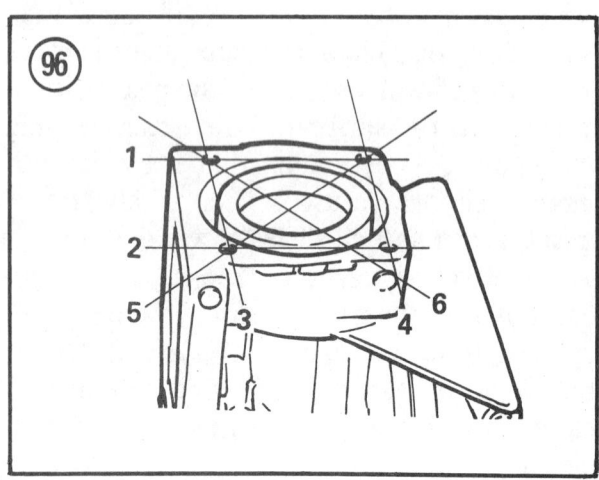

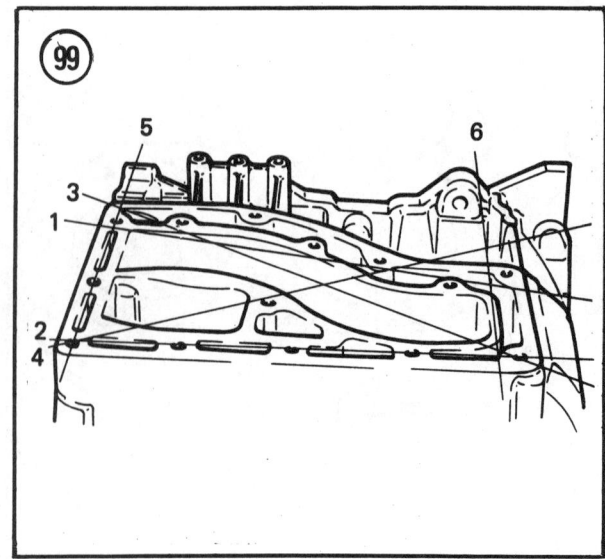

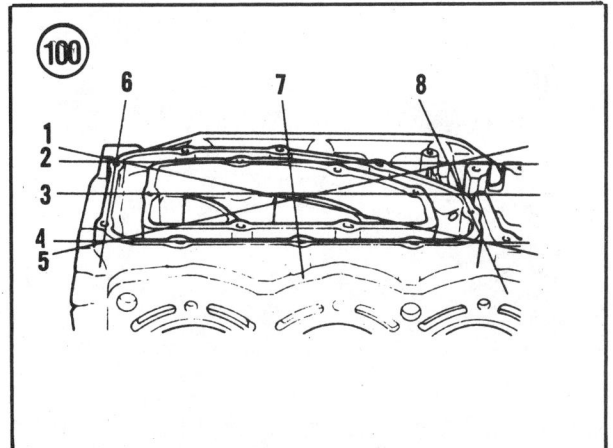

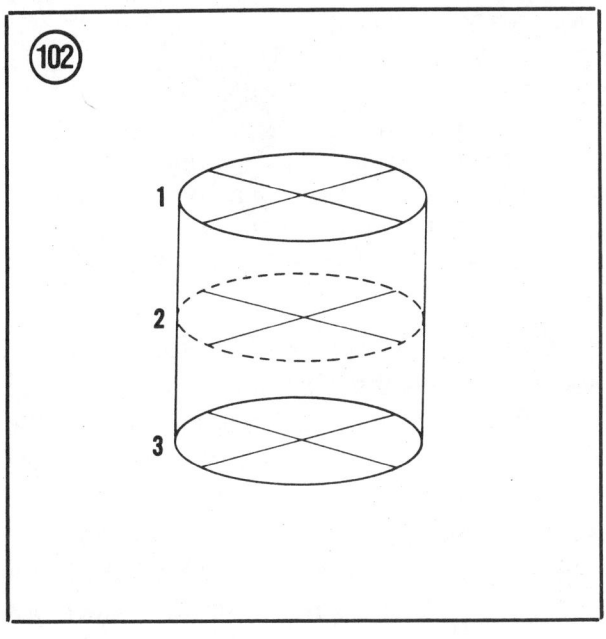

Wipe bore with a cloth moistened in Yamalube Two-cycle lubricant (for outboards) and wipe again with a clean dry cloth. If scoring is excessive or cylinder is out-of-round, replace the block.

NOTE
With older engines, it is a good idea to have the cylinder walls lightly honed with a medium stone even if they are in good condition. This will break up any glaze that might reduce compression.

10. Check each bore on multi-cylinder engines for any signs of aluminum transfer from the pistons to the cylinder walls. If scoring is present but not excessive, have the cylinders honed by a dealer or qualified machine shop.

11. Measure each cylinder bore front-to-rear with a bore gauge or inside micrometer (**Figure 101**) at the top, center and bottom just above the exhaust port (**Figure 102**). Record the readings.

12. Repeat Step 11 to measure the cylinder bore from side-to-side and record the measurements. Subtract the smallest from the largest reading. If the difference between the 2 measurements exceeds the maximum allowable taper or out-of-round (**Table 2**) have the cylinders rebored by a dealer or qualified machine shop and install oversize pistons.

NOTE
*Obtain the new pistons and measure them before having the cylinders bored. This allows for variations in piston size due to manufacturing tolerances. Remember to allow for piston-to-cylinder wall clearance (**Table 2**) when reboring.*

13. If check valve replacement is required, remove the defective valve with pliers and a rotating motion. Clean all sealing compound from the mounting hole, coat a new check

8

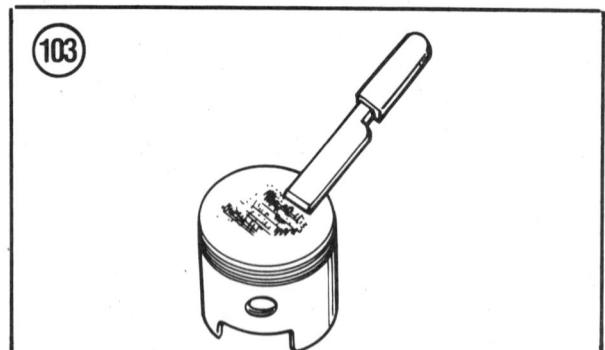

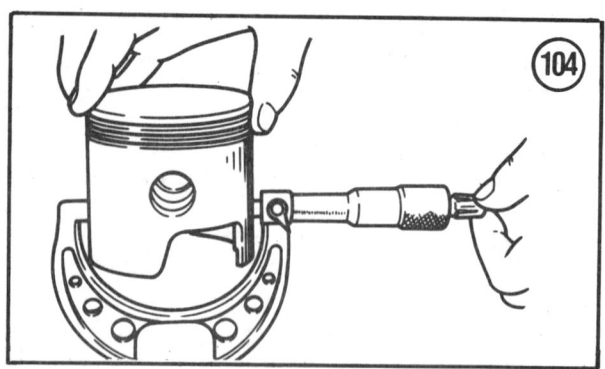

valve stem with Gasket Maker or equivalent and install with a suitable driver.

14. If oil pump driven gear shaft bushing requires replacement, be sure to install the new bushing with its slit facing the lower oil seal housing.

Crankshaft and Connecting Rod Bearing Cleaning and Inspection (All Engines)

Bearings can be reused if they are in good condition. To be on the safe side, however, it is a good idea to discard all removable bearings and install new ones whenever the engine is disassembled. New bearings are inexpensive compared to the cost of another overhaul caused by the use of marginal bearings.

1. Remove any sealer from outer edge of ball bearings with a scraper, then clean the bearing surface with kerosene.

2. Place ball bearings in a wire basket and submerge in a suitable container of fresh kerosene. The bottom of the basket should not touch the bottom of the container.

3. Agitate basket containing bearings to loosen all grease, sludge and other contamination.

4. Dry ball bearings with dry filtered compressed air. Be careful not to spin the bearings.

5. Lubricate the dry bearings with a light coat of Yamalube Two-cycle lubricant (for

outboards) and inspect for rust, wear, scuffed surfaces, heat discoloration or other defects. Replace as required.

6. If caged or loose needle bearings are to be reused, repeat Steps 2-5, cleaning one set at a time to prevent any possible mixup. Check bearings for flat spots. If one needle bearing is defective, replace the entire set.

Piston Cleaning and Inspection (All Engines)

1. Check the piston(s) for signs of scoring, cracking, cracked or worn piston pin bosses or metal damage. Replace piston and pin as an assembly if any of these defects are noted.

2. Remove any carbon deposits on the piston crown with a hardwood or plastic scraper. See **Figure 103**.

> *NOTE*
> *To remove stubborn carbon deposits in Step 3, carefully scrape the ring groove with the recessed end of a broken ring. Do not use an automotive ring groove cleaning tool, as it can damage the piston ring locating pin.*

3. Check piston ring grooves for distortion, loose ring locating pins or excessive wear. If the flexing action of the rings has not kept the lower surface of the ring grooves free of carbon, clean with a bristle brush and solvent.

4. Immerse pistons in a carbon removal solution to remove any carbon deposits not removed in Step 3. If the solution does not

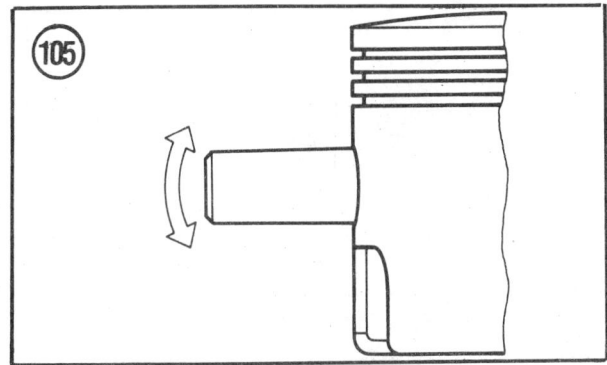

remove all of the carbon, carefully use a fine wire brush and avoid burring or rounding of the machined edges.

5. Remove all score marks and/or lacquer deposits from the piston side with No. 600 or No. 800 grit wet sandpaper, sanding in a crisscross pattern.

6. Measure the piston skirt diameter at right angles to the piston pin with a micrometer at a point 0.40 in. (10 mm) above the bottom edge of the piston (**Figure 104**).

7. Measure the cylinder bore with a bore gauge or inside micrometer (**Figure 101**) just above the exhaust port.

8. Subtract the measurement obtained in Step 6 from the Step 7 measurement to determine the cylinder-to-piston clearance. Compare clearance to specifications (**Table 2**). If clearance exceeds specifications, have the cylinder block rebored by a dealer or qualified machine shop and install oversize pistons or replace the block and piston assemblies.

NOTE
Obtain the new pistons and measure them before having the cylinders bored. This allows for variations in piston size due to manufacturing tolerances.

9. 1-cylinder piston—Lightly lubricate the piston pin with Yamalube Two-cycle lubricant (for outboards). Install bearing in connecting rod small end and insert pin through bearing to check for wear. There should be no noticeable vertical play. If there is, check the connecting rod small end for excessive wear and replace the pin, connecting rod and/or bearing as required.

10. Lightly lubricate the piston pin with Yamalube Two-cycle lubricant (for outboards) and insert its end in the piston pin boss. Check for free play by moving it in the directions shown in **Figure 105**. The pin should be a hand press-fit with no noticeable vertical play. If play exists or pin is excessively loose, replace the pin and/or piston as required.

11. Check the piston pin for signs of heat discoloration, fretting, pitching or other defects. This is especially important if defective small end needle bearings have been found, as they have probably damaged the piston pin surface. If the pin shows signs of damage but the bearings do not, replace both pin and bearings.

Crankshaft Cleaning and Inspection (All Engines)

Only 90-225 hp engine connecting rods can be disassembled from the crankshaft. If either the rod(s) or crankshaft is defective on 2-70 hp engines, replace the entire assembly.

1. Clean the crankshaft thoroughly with kerosene and a brush. Blow dry with dry filtered compressed air and lubricate with a light coat of Yamalube Two-cycle lubricant (for outboards).

2. Check the crankshaft journals and crankpins for scratches, heat discoloration or other defects.

3. Check drive shaft splines, flywheel taper threads, keyway and oil injection pump drive gear (if so equipped) for wear or damage. Replace crankshaft as required.

4. Check crankshaft oil seal surfaces for grooving, pitting or scratches.

5. Check crankshaft bearing surfaces for rust, water marks, chatter marks and excessive or uneven wear. Minor cases of rust and water

8

or chatter marks may be cleaned up with 320 grit carborundum cloth.

6. If 320 grit cloth is used, clean crankshaft in solvent and check surfaces. If they did not clean up properly, replace the crankshaft.

7. If lower crankshaft ball bearing has not been removed, grasp inner race and try to work it back and forth. Replace bearing if excessive axial play is noted.

8. Lubricate ball bearing with Yamalube Two-cycle lubricant (for outboards) and rotate outer race. Replace bearing if it sounds or feels rough or if it does not rotate smoothly.

9. Check crankshaft runout with a dial indicator and V-blocks. Check runout at the points marked A in **Figure 106** (1-cylinder), **Figure 107** (2-cylinder), **Figure 108** (3-cylinder) or **Figure 109** (V4-V6). Replace crankshaft if runout exceeds 0.0012 in. (0.3 mm) for 4-30 hp engines, 0.0020 in. (0.5 mm) for 40-70 hp engines or 0.0008 in. (0.2 mm) for all others.

10. 2-70 hp—Check connecting rod deflection at the piston end of the rod. See **Figure 110** (typical). Maximum permissible deflection is 0.08 in. (2 mm). If deflection exceeds the maximum, replace the crankshaft and rod assembly.

11. 1- and 2-cylinder crankshafts—Check crankshaft dimension marked B in **Figure 106** (1-cylinder) or **Figure 107** (2-cylinder). Compare to specification in **Table 3**.

12. Slide the connecting rod to one side of the crankshaft and measure the clearance between the connecting rod and crankshaft. This is dimension C in **Figures 106-108**. Compare to specification in **Table 3**.

13. 115-225 hp—Check but do not remove sealing rings from crankshaft unless excessively worn or damaged. Insert a flat feeler gauge between the sealing ring and ring groove at the end of the ring. See **Figure 94**. The wear limit on either side of the ring should not exceed 0.002 in. (0.1 mm).

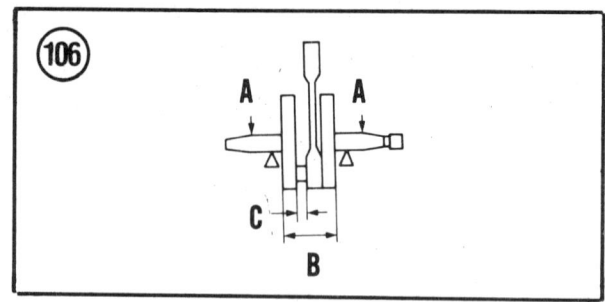

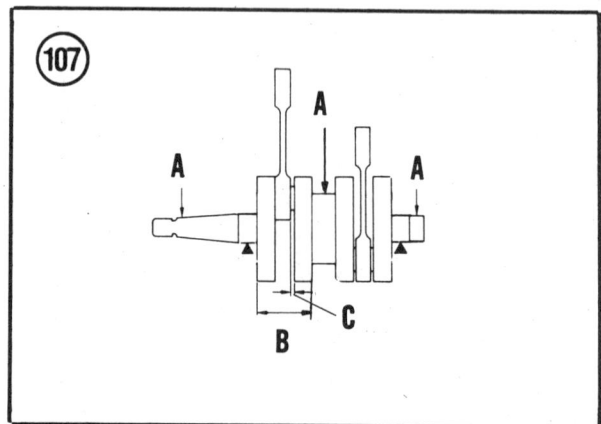

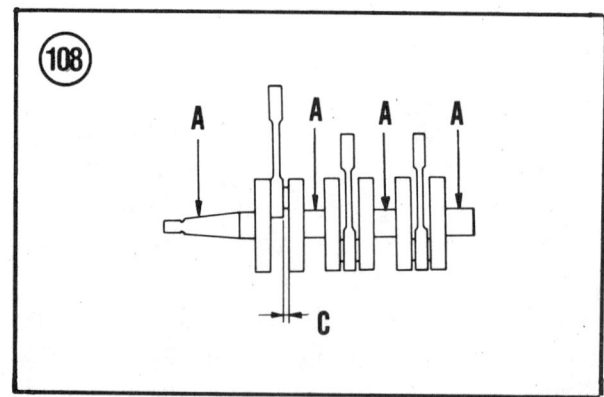

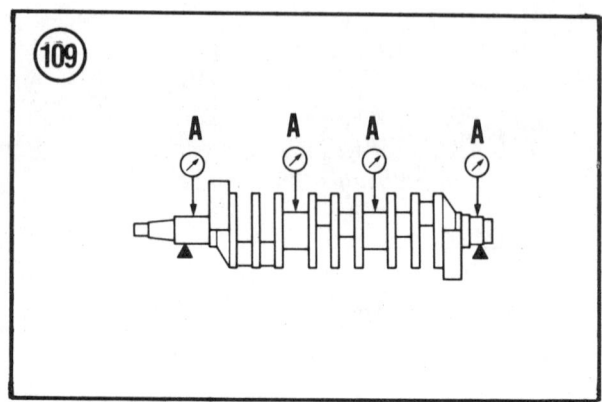

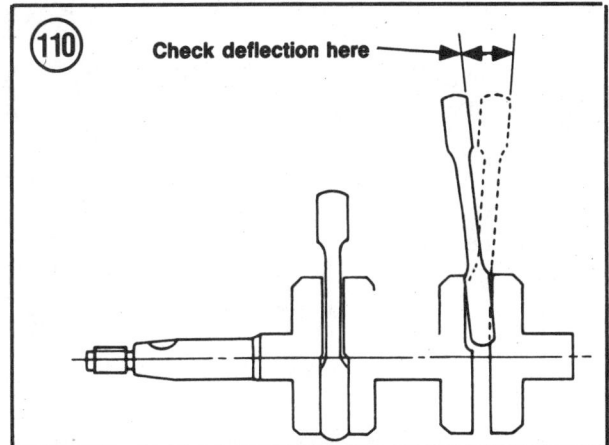

Check deflection here →

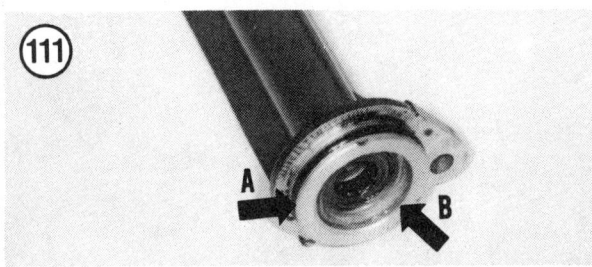

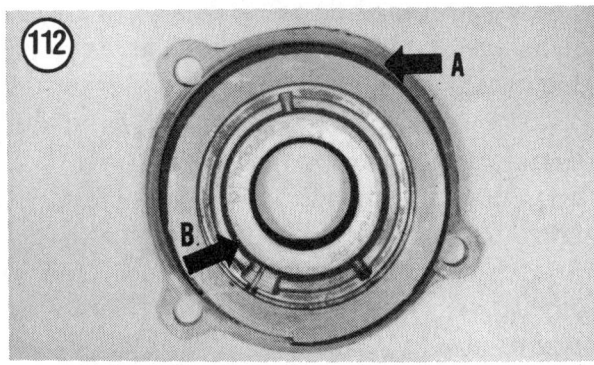

14. Replace entire assembly if any dimension in Steps 9-12 is not within specifications. Replace sealing rings as required.

Connecting Rod Cleaning and Inspection (90-225 hp)

1. Check connecting rod straightness. Place each rod and cap assembly on a smooth, flat surface and press downward on the rod—it should not wobble under pressure. Try inserting a 0.002 in. flat feeler gauge between

the machined portion of the rod and the flat surface. If it fits, replace the rod.

2. Check connecting rod bearings for rust or bearing failure.

3. Check the connecting rod big and small end bearing surfaces for rust, water marks, spalling, chatter marks, heat discoloration and excessive or uneven wear.

4. Slight defects noted in Step 3 may be cleaned up as follows:

   a. Reassemble connecting rod cap to rod end and make sure markings made before disassembly are aligned. Tighten cap fasteners securely.

   b. Clean bearing surfaces with crocus cloth.

   c. Clean piston pin bearing surfaces with 320 grit carborundum cloth.

   d. Wash connecting rod thoroughly in solvent to remove any abrasive grit, then recheck the bearing surface condition.

   e. Replace any rod and cap assembly that does not clean up properly.

   f. Lightly oil bearing surfaces of rod and cap assemblies that will be reused with Yamalube Two-cycle Lubricant (for outboards).

Upper and Lower Oil Seal Housing (All Engines)

Figure 111 shows a lower housing used on the smaller displacement engines. **Figure 112** shows the lower housing used on larger displacement engines; the upper housing, used on some engines, is similar and may contain a roller bearing.

1. Remove O-ring(s) (A, **Figure 111** and A, **Figure 112**). Some models use one O-ring; others use two.

2. Note positioning of oil seal lips and remove the oil seal(s) (B, **Figure 111** and B, **Figure 112**) from lower housing. Depending

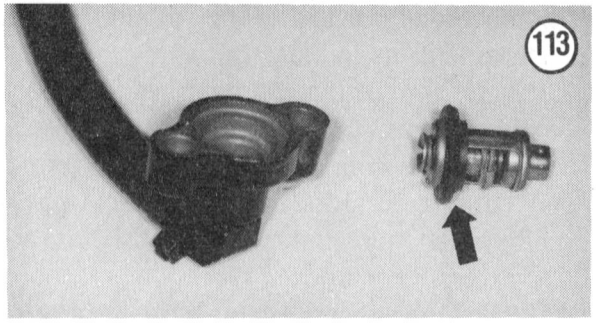

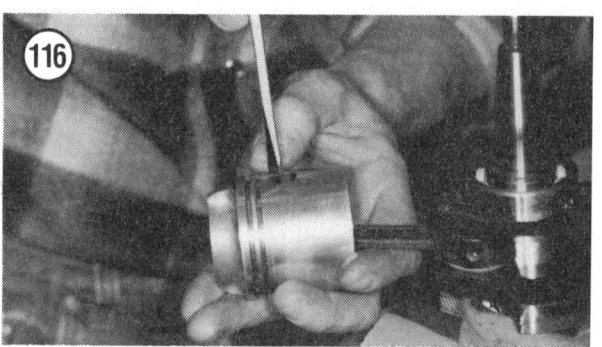

upon the model, the lower housing may contain 1-3 seals.

3. If upper housing contains a roller bearing that must be replaced, remove old bearing and install a new one with a hydraulic press.

4. Clean housing in solvent and blow dry with compressed air.

5. Lubricate the outer diameter of seal(s) with Gasket Maker or equivalent.

6. Install new oil seal(s) with a suitable mandrel. Seal lip(s) should face in the same direction as noted in Step 2.

7. Remove any excess sealer and lubricate the seal lips with Yamalube All-purpose Marine grease.

8. Install new O-rings and lubricate with Yamalube All-purpose Marine grease.

Thermostat
(All Engines So Equipped)

Yamaha engines use a 118° F (48° C) or a 126° F (52° C) thermostat. The temperature

rating is stamped on the thermostat flange. If defective, replace the thermostat with one that has the same temperature rating.

1. Clean all gasket residue from the thermostat and cover (**Figure 113**).

2. Check thermostat cover for cracks or corrosion damage. Replace as required.

3. Remove rubber washer or collar (arrow, **Figure 113**) and wash thermostat with clean water.

4. Suspend the thermostat and a thermometer in a container of water that can be heated. Support the thermostat with wire so it does not touch the sides or bottom of the container. See **Figure 114**.

5. Heat the water and note the temperature at which the thermostat starts to open. It should be approximately the same temperature as that stamped on the thermostat flange. If not, replace the thermostat.

6. Measure the maximum lift of the thermostat valve. To do this, mark a small

screwdriver at a point 0.12 in. (3 mm) from the tip. Use the screwdriver tip to measure the valve lift when the valve is fully open. If the valve lift is less than 0.12 in. (3 mm) at stamped temperature rating or more, replace the thermostat.

7. Remove the thermostat from the water and let it cool. Reinstall the rubber washer or collar.

Piston and Connecting Rod Assembly (2-70 hp Engines)

If the pistons were removed from the connecting rods, they must be correctly oriented when reassembling. The UP mark on the piston crown must face toward the flywheel end of the crankshaft.

Use Yamalube Two-cycle lubricant (for outboards) whenever the procedure specifies lubrication with oil. Use Yamalube Marine grease where grease is specified.

1. Coat the piston pin, piston pin bore, and connecting rod small end bore with oil.

2. To assemble 1-cylinder piston:

a. Partially insert piston pin in piston.

b. Install piston pin bearing in connecting rod. See **Figure 115**.

c. Hold piston with stamped UP mark facing flywheel end of crankshaft and fit over connecting rod.

d. Push piston pin through connecting rod.

e. Install new piston pin clips with needlenose pliers. Use a screwdriver to make sure they are properly seated in the piston bore groove (**Figure 116**).

3. To assemble 2- and 3-cylinder pistons:

a. Check the piston dome number made during disassembly and match piston with its correct connecting rod.

b. Partially insert piston pin in piston.

c. Wipe the inside of the conneting rod piston bore with grease.

d. Install a suitable bushing to act as a spacer and insert needle bearings individually (**Figure 117**).

e. When all bearings are in place, fit washers at top and bottom of bearing assembly (**Figure 118**) with their convex sides facing the piston.

f. Hold piston with stamped UP mark facing flywheel end of crankshaft and fit carefully over connecting rod to prevent disturbing the bearings.

g. When piston pin bore and connecting rod bore are aligned, push piston pin through connecting rod and catch bearing installer spacer as it falls out.

h. Install new piston pin clips with needlenose pliers. Use a screwdriver to make sure they are properly seated in piston bore groove (**Figure 116**.)

i. Repeat Steps a-h to install remaining piston(s).

Piston Ring Installation (All Engines)

NOTE
Some 1989 30-90 hp models are equipped with two ring pistons. Both

8

piston rings are semi-keystone shaped. This was a running change for the 1989 model year so some early 1989 models will be equipped with the three ring pistons used in 1988 and prior engines. Some 1989 V4 and V6 models are equipped with two semi-keystone shaped piston rings. This was a running change for the 1989 model year so some early 1989 models will be equipped with the one semi-keystone (upper ring) and one square (lower ring) piston ring used in 1988 and prior engines.

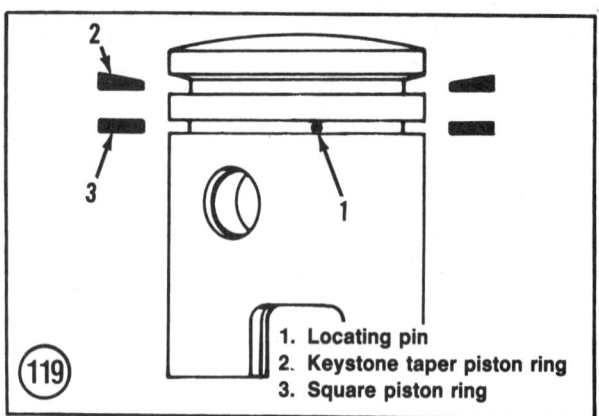

1. Locating pin
2. Keystone taper piston ring
3. Square piston ring

Pistons use a keystone design top ring. On some 1989 models, a keystone design is used on the second ring. On all other models, the second, and if so equipped, the third ring is a flat design. See **Figure 119**. Check new rings carefully before installation and install them in their proper grooves.

1. Check end gap of new rings before installing on piston. Place ring in cylinder bore just above the intake and exhaust ports, then square it up by inserting the bottom of an old piston. Measure the gap with a feeler gauge (**Figure 120**) and compare to specifications (**Table 2**).

2. If ring gap is excessive in Step 1, repeat the step with the ring in another cylinder. If gap is also excessive in that cylinder, discard and replace with another new ring.

3. If ring gap is insufficient in Step 1, the ends of the ring can be filed slightly. Clean ring thoroughly and recheck gap as in Step 1.

NOTE
*Piston rings must be installed in Step 4 with the mark on the end of the ring facing the piston crown. See **Figure 121**.*

4. Once the ring gaps are correctly established, spread the bottom ring just enough to fit it over the piston head and into position. Repeat this step to install the remaining ring(s).

5. Position each ring so that the piston groove locating pin fits into the ring gap (**Figure 122**). Proper ring positioning is

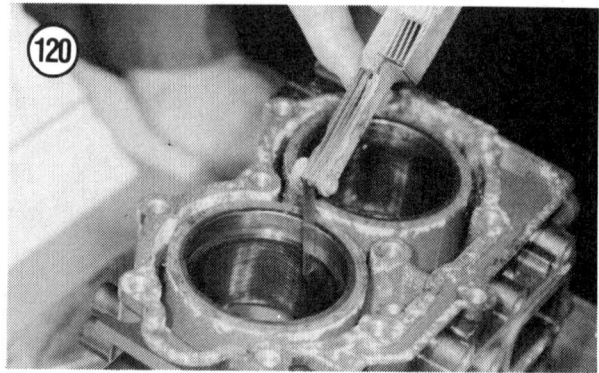

necessary to minimize compression loss and prevent the ring ends from catching on the cylinder ports.

Crankshaft Assembly (All Engines)

Use Yamalube Two-cycle Lubricant (for outboards) whenever the procedure specifies lubrication with oil. Use Yamalube All-purpose Marine grease where grease is specified.

1. 2-30 hp and 40 hp manual start—If upper (as equipped) or lower crankshaft ball bearing was removed, install with a suitable mandrel, crankshaft support and hydraulic press. Apply pressure only on inner race of bearing. Lubricate the bearings with oil.

2. Secure the 2 and 3 hp crankshaft in a vise with protective jaws. Install other crankshafts on a suitable power head stand secured in a vise with protective jaws.

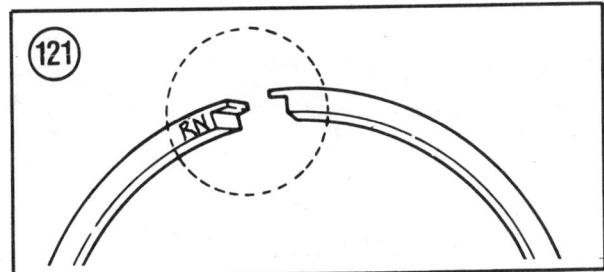

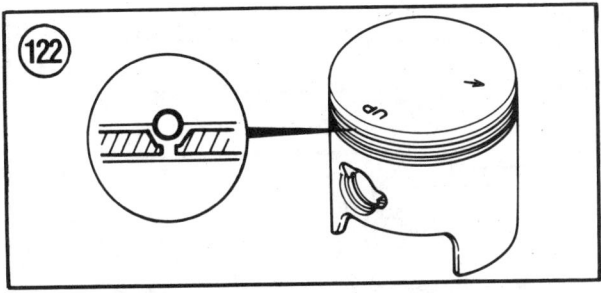

3. 2-70 hp—Install piston(s) to connecting rod(s) as described in this chapter. Remove crankshaft from vise or power stand and place on a clean work bench.

4. 2-3 hp—Lubricate the lip of 2 new oil seals with grease. Install one seal on the upper crankshaft. Install the washer and the other seal on the lower end of the crankshaft.

5. 4-5 hp—Lubricate the lip of a new oil seal with grease. Install the spacer and the oil seal on lower end of the crankshaft.

6. 9.9-15 hp—Lubricate the lip of a new oil seal with grease. Install spacer and oil seal on the upper end of the crankshaft. If one side of spacer is stamped INSIDE, that side should face the flywheel end of the crankshaft.

7. If crankshaft uses split sleeve center main bearings:

    a. Slip the retaining ring around the journal and let it rest on the crankshaft web.

    b. Lightly lubricate the bearing journal and bearings with grease.

    c. Install the bearings around the bearing journal.

    d. Install the split sleeves around the bearings. Each set of sleeves has a

locating hole which must be positioned to engage a pin in the crankcase. Generally speaking, the sleeves should be installed with the hole facing away from the flywheel end of the crankshaft. It is, however, a good idea to recheck the location of the pin in the crankcase and match it up with the hole in the bearing sleeves before installing the sleeves.

    e. Carefully install the retaining ring around the bearing sleeves. Do not stretch ring excessively during installation—it must fit snugly in the sleeve groove.

8. 115-225 hp—If crankshaft seal rings were removed, reinstall rings with a suitable piston ring expander.

NOTE
On some models, the oil pump drive gear is installed on the crankshaft before the bearing. On other models, the bearing must be installed first, then the drive gear. On all models, the bearing must be installed with its stamped side facing away from the flywheel end.

9. 25-225 hp—If lower crankshaft ball bearing or oil injection pump drive gear was removed, reinstall in the same order as removed. The drive gear teeth must face in the same direction as removed. Install the bearing or drive gear snap ring, as required.

10. 115-225 hp—Make sure the bearing in the upper bearing housing has been inspected and/or replaced and that new O-rings have been installed on the housing. Install the upper bearing housing on the crankshaft.

Crankshaft Installation
(2-70 hp Engines)

1. Lubricate the crankcase cylinder walls and the entire crankshaft, connecting rod and piston assembly with Yamalube Two-cycle lubricant (for outboards).

8

2. Place the cylinder block on a clean, lever work bench surface. Install the lower bearing washer, if used.

CAUTION
Be sure that piston ring end gaps remain aligned with ring groove locating pin as each ring enters the cylinder in Step 3 and Step 4.

3. Grasp the cylinder block assembly with one hand and position over crankcase. Slowly lower the assembly until the No. 1 piston enters the cylinder block, then compress each piston ring by hand and push the piston downward until all rings have entered the cylinder.

4. Rotate the crankshaft until the No. 2 piston is about to enter its cylinder, then compress each piston ring by hand and push the piston downward until all rings have entered the cylinder. Repeat this step to install the No. 3 piston on 3-cylinder engines.

5. When all piston-ring assemblies have entered the cylinder bores, apply sufficient downward pressure to nearly seat the pistons in their bores and the crankshaft in the cylinder block.

6. Rotate the crankshaft ball bearing races to align their locating pin with the pin recess in the cylinder block. See A, **Figure 123**. With some crankshafts, it is also necessary to align the lower bearing washer and the labyrinth seal and center bearing circlips in their respec-

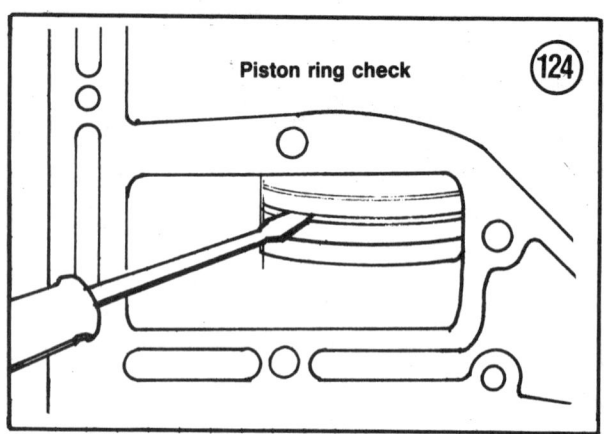

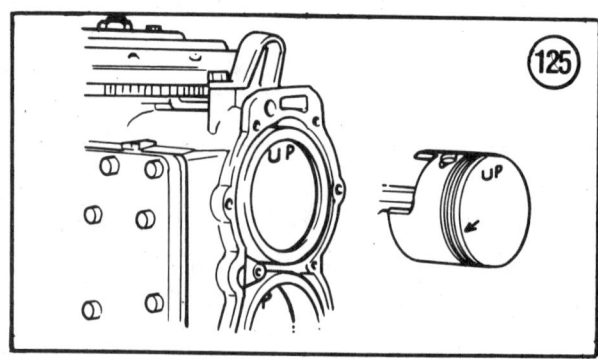

tive crankcase cutouts. See B, **Figure 123** (typical).

7. Insert a screwdriver blade or pencil point through the exhaust ports and depress each piston ring lightly (**Figure 124**). The ring should snap back when pressure is released. If it does not, the ring was broken during piston installation. Remove the crankshaft assembly from the cylinder block and replace the broken ring.

Piston and Connecting Rod Installation (90-225 hp Engines)

1. Lubricate pistons, rings and cylinder bores with Yamalube Two-cycle lubricant (for outboards).

2. Place crankcase on a clean, level work bench surface.

3. Check the piston dome number made during disassembly and match the piston with it correct cylinder bore.

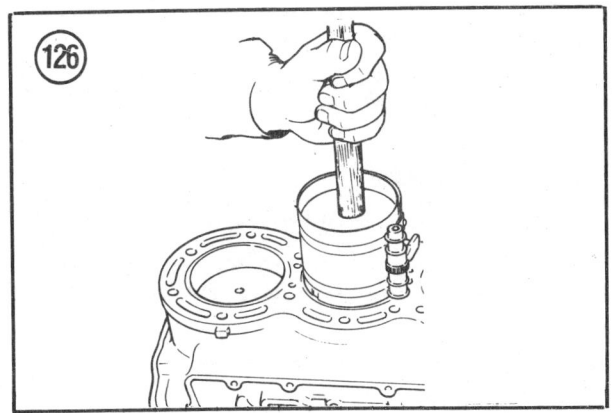

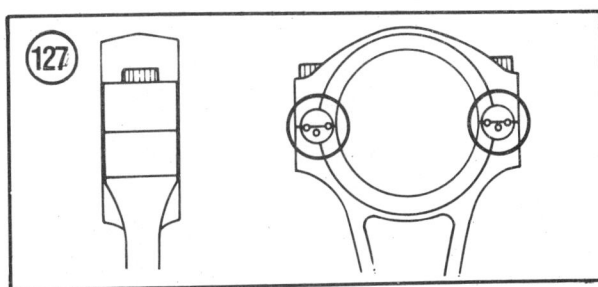

CAUTION
The piston crown on V4 and V6 pistons is marked P or S to indicate port or starboard side. Be sure to install the pistons in their proper location in Step 4 or the engine will not start.

4. Insert the piston into its cylinder bore with the UP mark on its crown facing the flywheel end of the crankcase. See **Figure 125**.

5. Make sure the rings are properly positioned in their grooves and that the locating pins are positioned in the ring end gaps.

6. Install a suitable ring compressor over the piston dome and rings. With the compressor resting on the crankcase, tighten it until the rings are compressed sufficiently to enter the piston bore. See **Figure 126**.

7. Hold the connecting rod end with one hand to prevent it from scraping or scratching the cylinder bore and slowly push the piston into its cylinder.

8. Remove the ring compressor and repeat Steps 3-7 to install the remaining pistons.

9. Temporarily install the cylinder heads to prevent the pistons from falling out.

10. Insert a screwdriver blade or pencil point through the exhaust ports and depress each piston ring lightly (**Figure 124**). The ring should snap back when pressure is released. If it does not, the ring was broken during piston installation and will have to be replaced.

**Crankshaft Installation
(90-225 hp Engines)**

Use Yamalube Two-cycle Lubricant (for outboards) whenever the procedure specifies lubrication with oil.

1. Remove the connecting rod caps. Coat the connecting rod bearing cages with oil. Install a bearing cage in each rod.

2. Lubricate the crankshaft with oil. Rotate the bearings to position their locating pin holes at the bottom of the crankshaft and install it in the cylinder block.

3. Lift up slightly on the crankshaft and align the locating pin holes in the bearings with the pins in the cylinder block until they engage.

4. V4 and V6—Rotate the seal rings so that their end gaps face upward.

5. Coat the oil injection pump driven gear with oil and insert it in the cylinder block. The drive gear boss should fit into the cylinder block bushing as the driven gear is meshed with the drive gear. Install the washer and spacer.

6. Position each connecting rod under the crankshaft and pull it up to the crankpin.

7. Lightly coat the exposed part of each crankpin with oil and install the remaining bearing cages.

8. Install the rod caps with their match marks aligned with those on the rods (**Figure 127**) and install new cap bolts.

9. Screw bolts in by hand to make sure the cap and rod ends align, then tighten the bolts to the Step 1 torque value (**Table 1**).

8

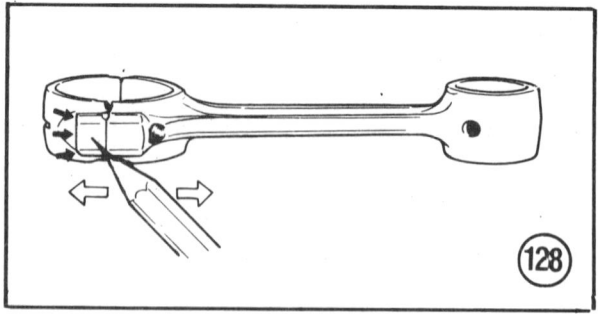

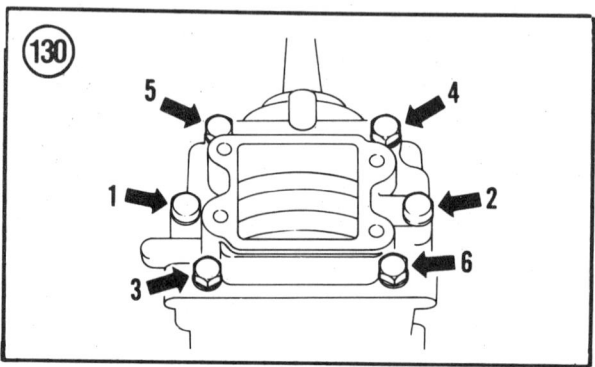

CAUTION
The procedure detailed in Step 10 is very important to proper engine operation, as it affects bearing action. If not done properly, major engine damage can result. It can also be a time-consuming and frustrating process. Work slowly and with patience. If alignment cannot be achieved, replace the connecting rod and cap assembly.

10. Tighten the cap bolts to the Step 2 torque value (**Table 1**) and run a dental pick or pencil point along the cap match marks to check cap offset. See **Figure 128**. Rod and cap must be aligned so that the dental pick or pencil point will pass smoothly across the fracture line at each of the 3 faces indicated by the arrows in **Figure 128**. If alignment is not correct, return to Step 8.

11. Once rod and cap alignment is correct, loosen the cap bolts 1/2 turn and retighten to Step 1, then Step 2 values (**Table 1**).

12. Remove the cylinder heads which were temporarily installed in *Piston and Connecting Rod Installation, 90-225 hp Engines* to keep the pistons from falling out.

POWER HEAD ASSEMBLY

Yamaha 2 hp

Refer to **Figure 58** as required for this procedure.

1. Apply a thin coat of Gasket Maker or equivalent on the crankcase and cylinder block mating surfaces. See **Figure 129**.

2. Position crankcase on block and install the 2 attaching bolts. Be sure to install the J-clamp under the bolt on the starboard side of the engine.

3. Connect recirculation line between the crankcase fittings.

4. Install the cylinder head with a new gasket. Wipe bolt threads with Loctite Type A and tighten bolts to specifications (**Table 1**).

5. Install a new O-ring in the exhaust plate groove and lubricate with Yamalube All-purpose Marine grease.

6. Install the reed valve assembly with a new gasket. Position front bracket against reed valve assembly and install fasteners, tightening to specifications (**Table 1**).

7. Install the carburetor. See Chapter Six.

8. Install power head as described in this chapter.

Yamaha 3 hp

Refer to **Figure 62** for this procedure.

1. Apply a thin coat of Gasket Maker or equivalent on the crankcase and cylinder block mating surfaces. See **Figure 129**.

2. Install dowel pins into cylinder block.

3. Position crankcase on cylinder block and install the six attaching bolts. Tighten bolts to specifications (**Table 1**) in the sequence shown in **Figure 130**.

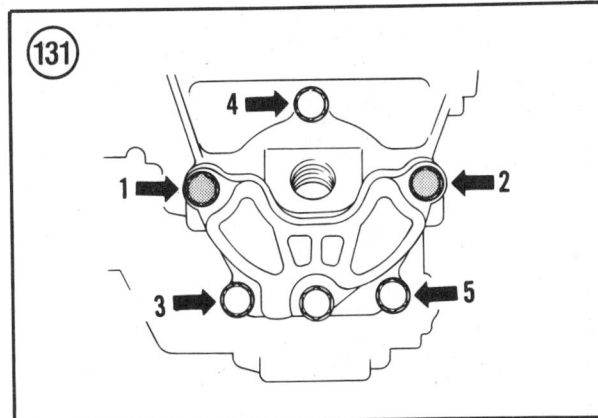

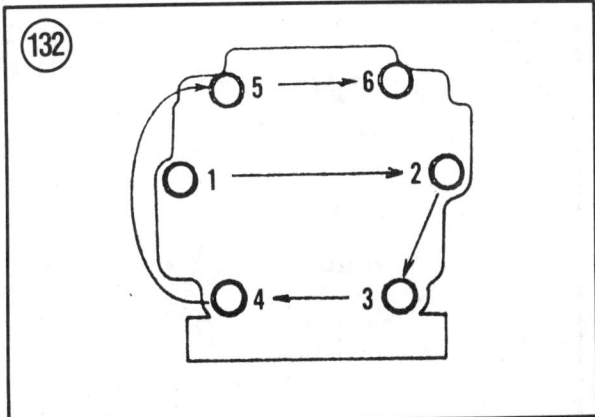

4. Connect recirculation line between the crankcase fittings.

5. Install the cylinder head with a new gasket. Tighten bolts to specifications (**Table 1**) in the sequence shown in **Figure 131**.

6. Install the thermostat with a new gasket and securely tighten thermostat cover retaining bolts.

7. Install the reed valve and intake manifold assembly with a new gasket. Tighten bolts to specifications (**Table 1**).

8. Install the carburetor.

9. Install the power head as described in this chapter.

Yamaha 4 and 5 hp

Yamaha engines have the correct bolt torque sequence cast in each housing component. Follow the numbered sequence when tightening bolts to specifications. Refer to **Figure 63** as required for this procedure.

1. Apply a thin coat of Gasket Maker or equivalent on the crankcase and cylinder block mating surfaces. See **Figure 129**.

2. Position crankcase on block and install attaching bolts. Be sure to install starter lockout linkage and bracket under bolts 1 and 5 (**Figure 132**). Tighten bolts to specifications (**Table 1**) in the sequence shown in **Figure 132**.

3. Wipe the outer diameter of a new crankcase oil seal with Loctite Type A and install it in the crankcase opening around the drive shaft end of the crankshaft. The seal lip should face away from the cylinder block.

4. Remove any excess Loctite and lubricate the seal lip with Yamalube All-purpose Marine grease.

5. Install the lower oil seal housing on the crankcase and tighten the bolt securely.

6. Sandwich the inner exhaust cover between 2 new gaskets and install the thermostat. Add outer exhaust cover to sandwich and position assembly on crankcase. See **Figure 133**.

8

7. Install the long bolt at position A, **Figure 134**. Install the short bolt at position B, **Figure 134**. Install the remaining bolts at each position marked C, **Figure 134** and tighten to specifications (**Table 1**) following the numerical sequence shown in **Figure 134**.

8. Install the cylinder head cover with a new gasket. Tighten bolts to specifications (**Table 1**) in a clockwise direction.

9. Connect the recirculation line.

10. Install reed block assembly with a new gasket. Tighten bolts to specifications (**Table 1**).

11. Install the carburetor.

12. Install the power head as described in this chapter.

All 2-cylinder Models

Refer to **Figures 70-73** as required for this procedure.

1. Install the lower oil seal housing and align the bolt holes in the housing with those in the cylinder block.

2. Apply a thin coat of Gasket Maker or equivalent on the cylinder block and crankcase mating surfaces. Install crankcase to cylinder block and tighten bolts to specifications (**Table 1**) following the sequence shown in **Figure 135** (6-8 hp),

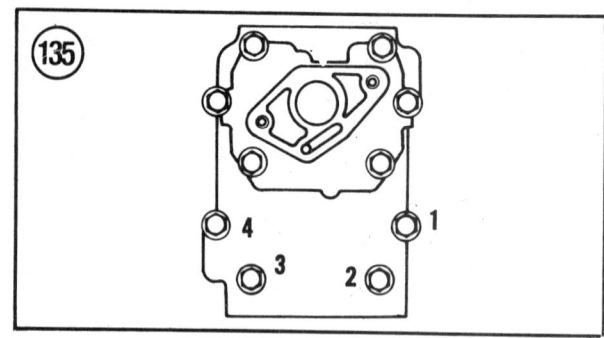

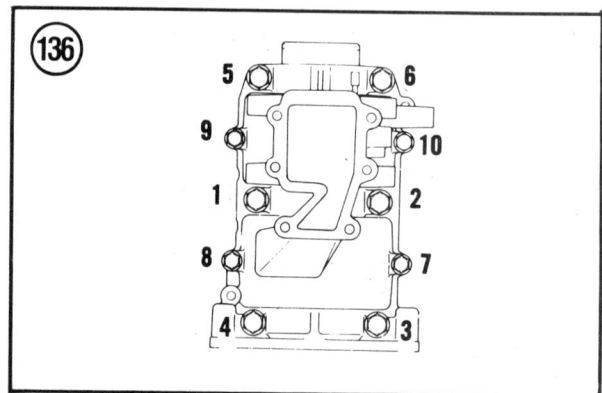

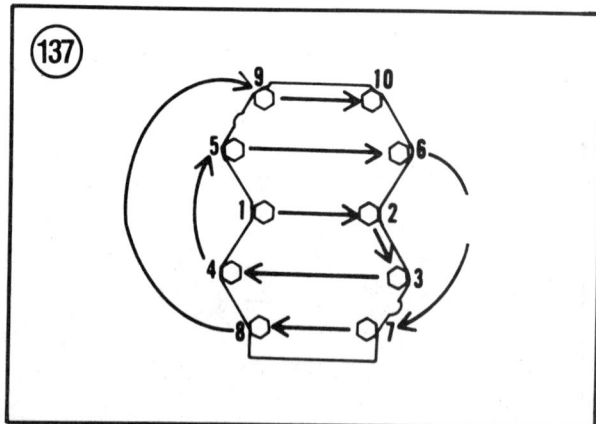

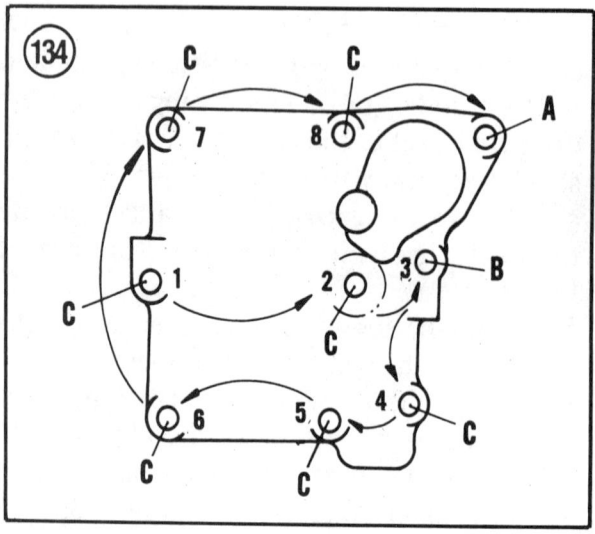

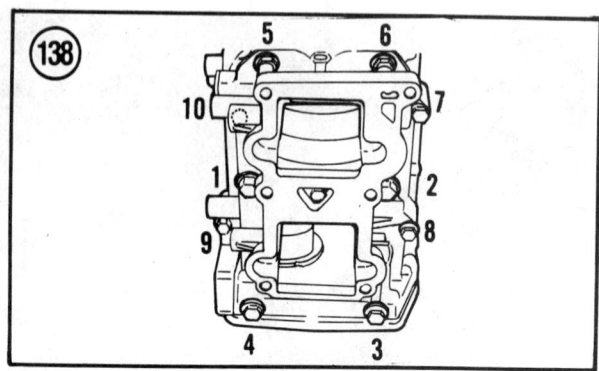

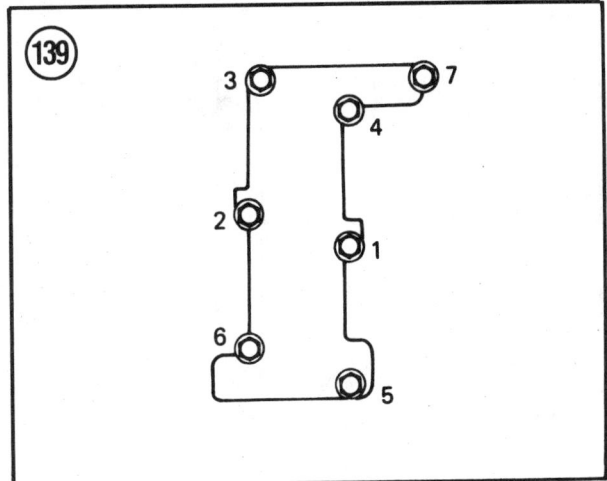

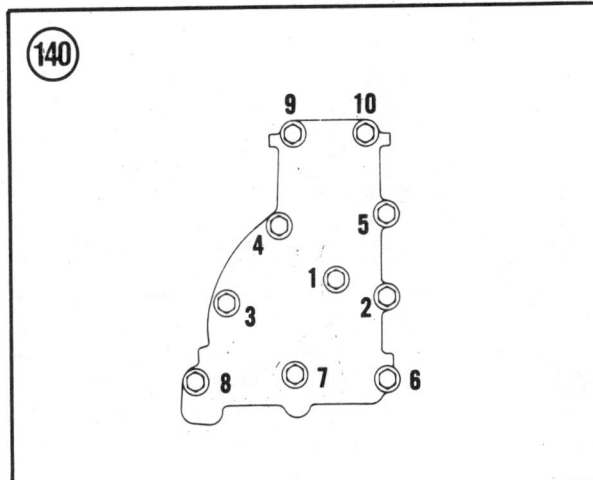

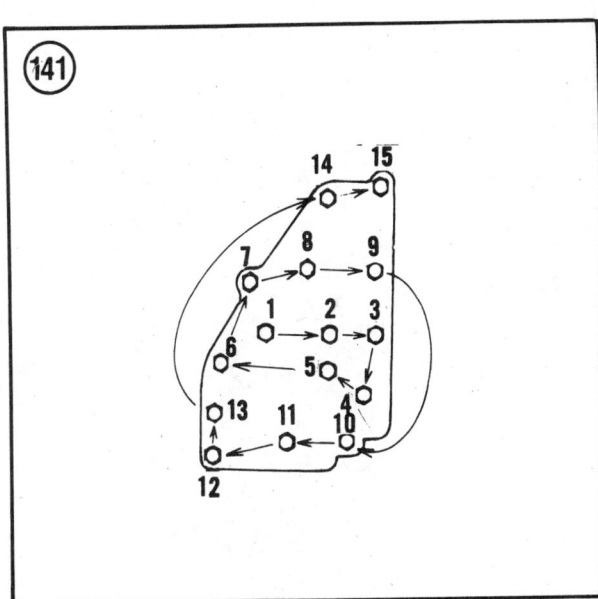

Figure 136 (9.9-15 hp), **Figure 137** (25 [1984-1987]-30 hp) or **Figure 138** (25 [1988-on] hp).

3. Install and tighten the lower oil seal housing bolt to specifications (**Table 1**).

NOTE
Do not install the upper housing bolts in Step 4 or Step 12 during this procedure. They are installed when the magneto base is reattached.

4. Except 1988-on 25 hp—install the crankshaft upper oil seal housing with a new gasket, aligning housing/gasket holes with those in the crankcase.

5. Sandwich the inner exhaust cover between 2 new gaskets. Position the assembly on the crankcase and install the outer exhaust cover on top. Install and tighten attaching bolts to specifications (**Table 1**) following the sequence shown in **Figure 139** (6-8 hp), **Figure 140** (9.9-15 hp), **Figure 141** (25 [1984-1987]-30 hp) or **Figure 142** (25 [1988-on] hp). Connect recirculation lines to cover, if so equipped.

6. If an anode is installed in the cylinder head water passages, make sure it is in good condition and properly installed. See **Figure 143** (typical).

7. 6-8 hp—Install thermostat and rubber collar or washer in cylinder block.

8

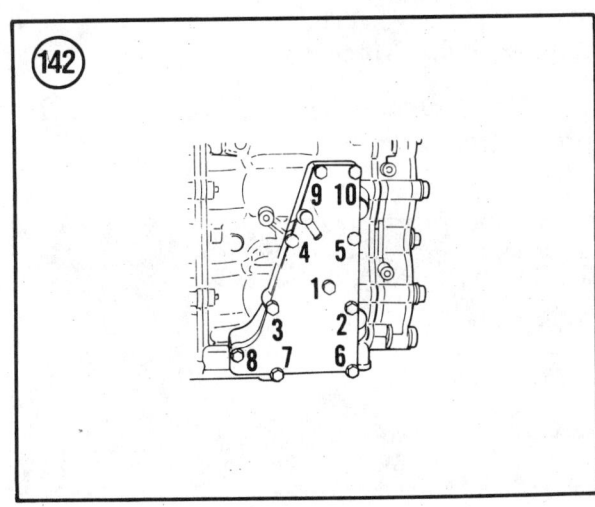

8. Install cylinder head with a new gasket and tighten fasteners to specifications (**Table 1**) following the sequence shown in **Figure 144** (6-8 hp), **Figure 145** (9.9-15 hp), **Figure 146** (25 [1984-1987]-30 hp) or **Figure 147** (25 [1988-on] hp). On 6-8 hp, be sure to reinstall the plug wire clamp under the correct bolt.

9. 9.9-30 hp—Install the thermostat and rubber collar or washer. See **Figure 148** (typical). Install the cover with a new gasket and tighten to specifications (**Table 1**).

10. On 25 (1984-1987)-30 hp, sandwich the reed valve assembly between 2 new gaskets.

11. Install the reed valve assembly and intake manifold to cylinder block with a new gasket. On 9.9-15 hp, coat gasket portion shown in **Figure 149** with Gasket Maker or equivalent. Coat bolt threads with Loctite Type A and tighten to specifications (**Table 1**). Follow the tightening sequence shown in **Figure 150** for 6-8 hp models and **Figure 151** for 25 (1988-on) hp models.

12. 6-8 hp—Install the upper oil seal housing, aligning the bolt holes in the housing with the crankcase. Install the starter stays to the crankcase, if removed.

13. Install the power head as described in this chapter.

All 3-cylinder Models

Refer to **Figures 80-83** as required for this procedure.

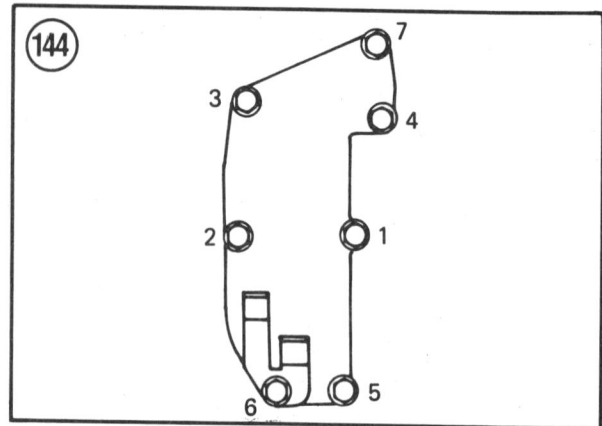

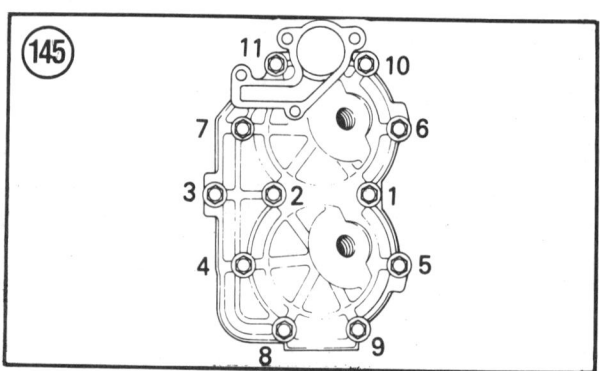

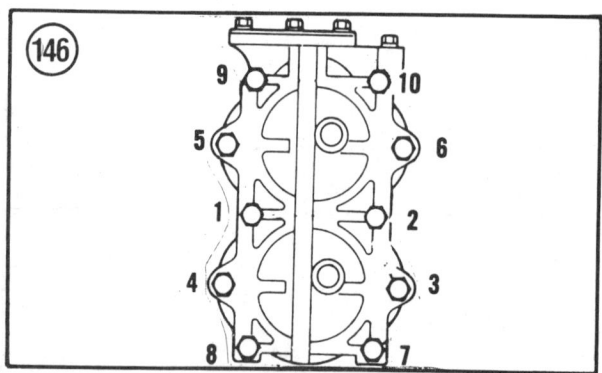

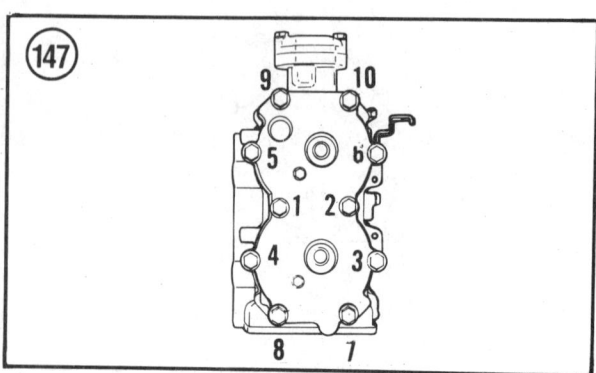

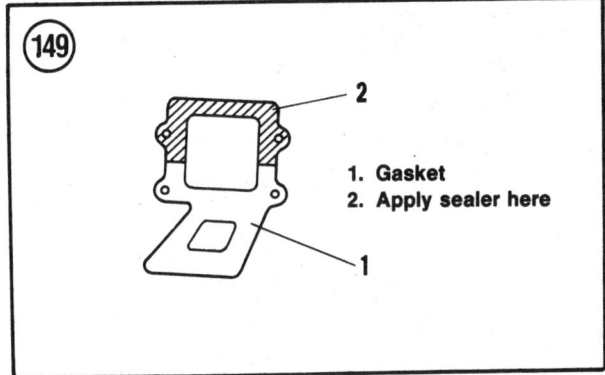

1. Gasket
2. Apply sealer here

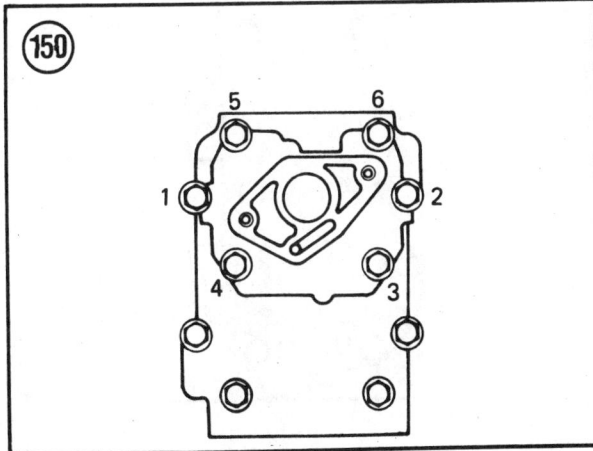

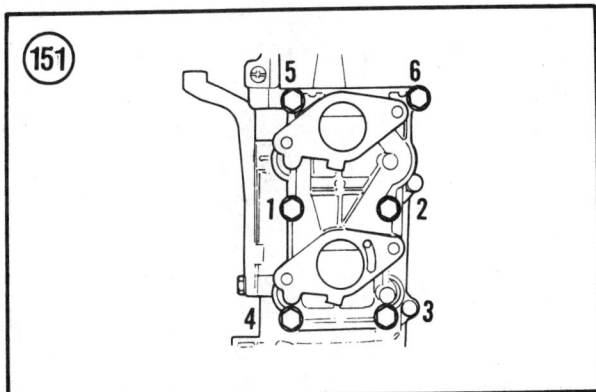

1. Make sure the lower oil seal housing (installed with the crankshaft) bolt holes align with those in the cylinder block.

2. Apply a thin coat of Gasket Maker or equivalent on the cylinder block and crankcase mating surfaces. Install crankcase to cylinder block and tighten bolts to specifications (**Table 1**) following the sequence shown in **Figure 152** (30 hp), **Figure 153** (40-70 hp) or **Figure 154** (90 hp).

3. Install and tighten the lower oil seal housing bolts to specifications (**Table 1**).

4. If anodes are installed in the cylinder head water passages, make sure they are in good condition and properly installed. See **Figure 143** (typical).

5. 30-70 hp—Install the cylinder head gasket, cylinder head, head cover gasket and head cover (**Figure 155**). Tighten head bolts to specifications (**Table 1**) following the sequence shown in **Figure 156** (30 hp) or **Figure 157** (40-70 hp). Install the thermoswitch in the cylinder head.

8

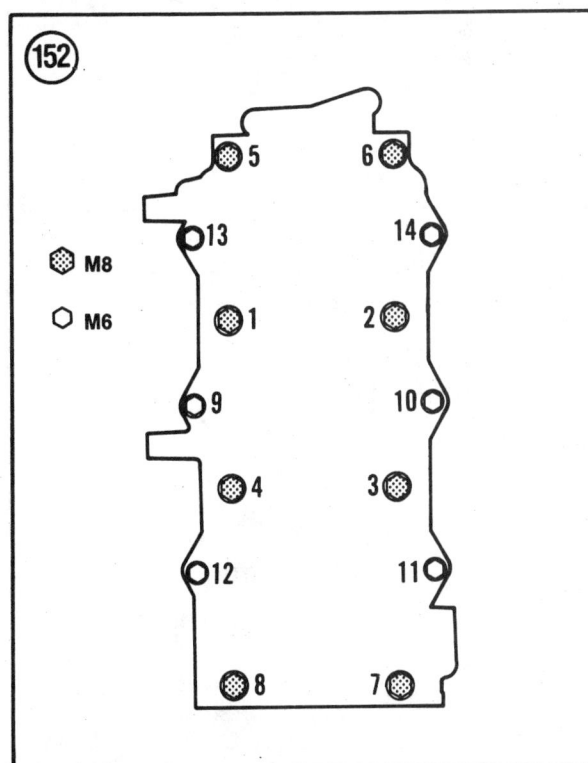

6. 90 hp:

   a. Install the cylinder head with a new gasket. Tighten head bolts to specifications (**Table 1**) following the sequence shown in **Figure 157**.

   b. Temporarily install the cylinder head cover with a new gasket.

   c. Install the thermoswitch on the cylinder head.

   d. Install the grommet, pressure valve, compression spring and thermostat.

   e. Install the thermostat cover with a new gasket.

   f. Tighten the cylinder head cover and thermostat covers to specifications (**Table 1**) following the sequence shown in **Figure**

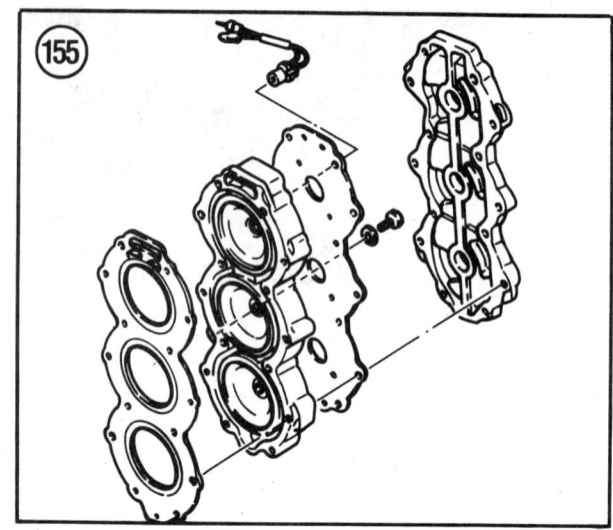

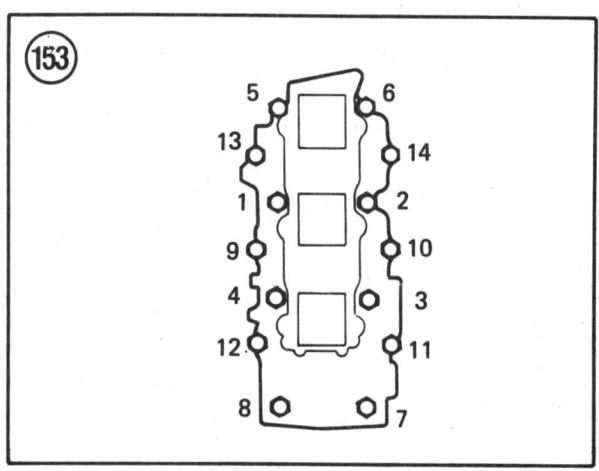

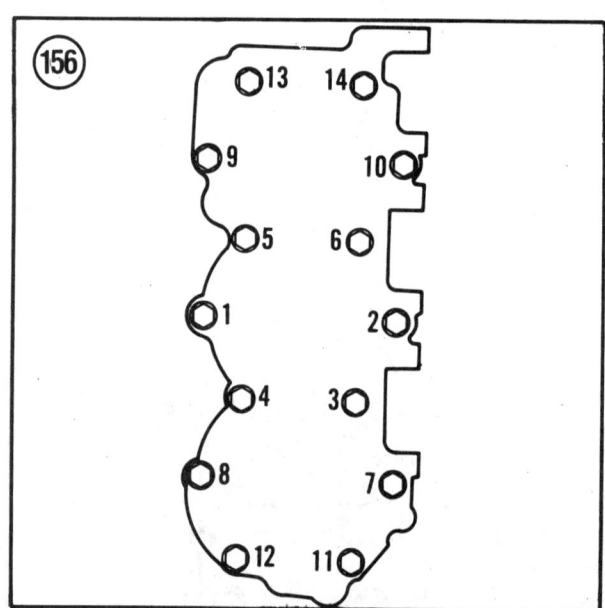

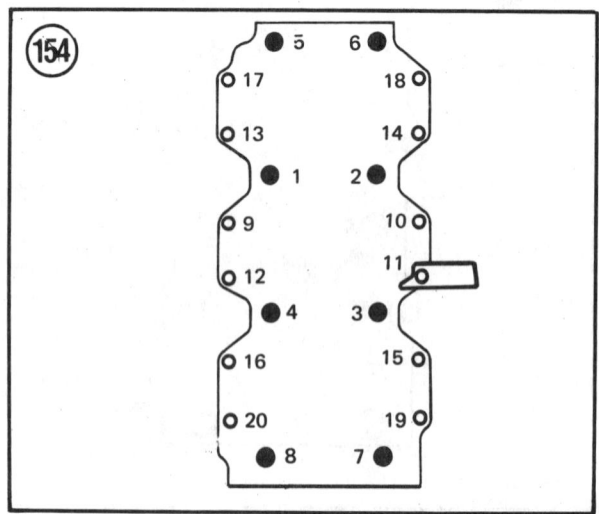

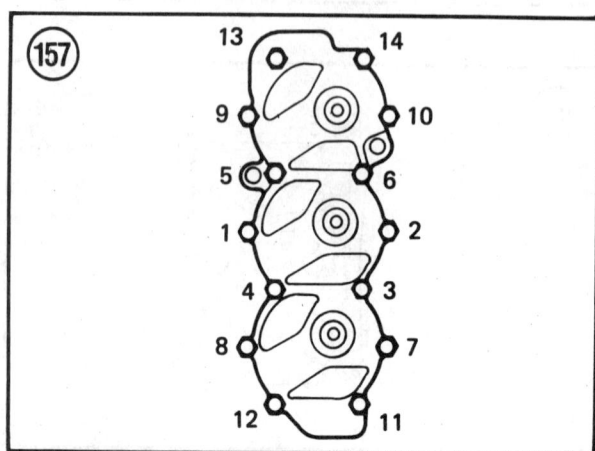

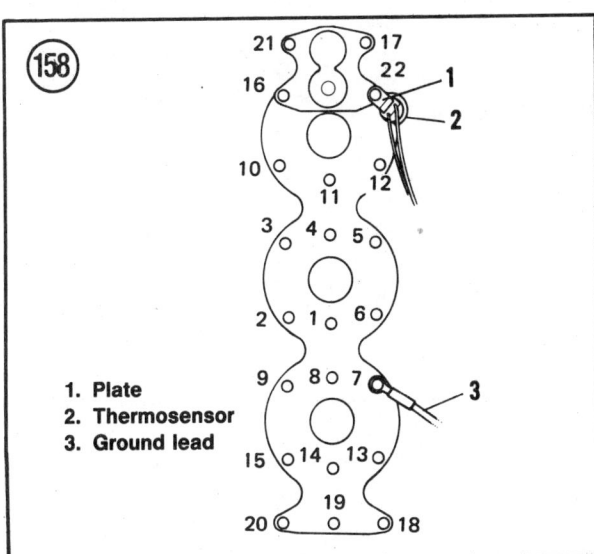

158.

1. Plate
2. Thermosensor
3. Ground lead

158. Be sure to install the thermoswitch plate under the thermostat cover bolts. See **Figure 158**.

7. Sandwich the inner exhaust cover between 2 new gaskets. Position the assembly on the crankcase and install the outer exhaust cover on top. Install and tighten attaching bolts to specifications (**Table 1**) following the sequence shown in **Figure 159** (30 hp), **Figure 160** (40-50 hp), **Figure 161** (70 hp) or **Figure 162** (90 hp). Connect recirculation lines to check valves.

8. 30-70 hp—Install thermostat and rubber collar or washer. Install thermostat cover

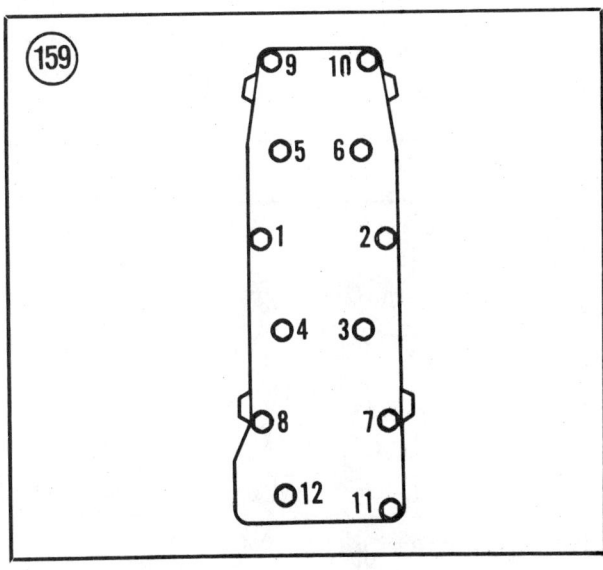

159.

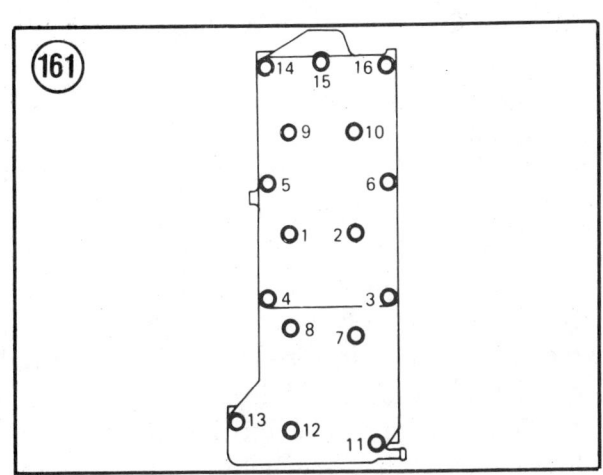

161.

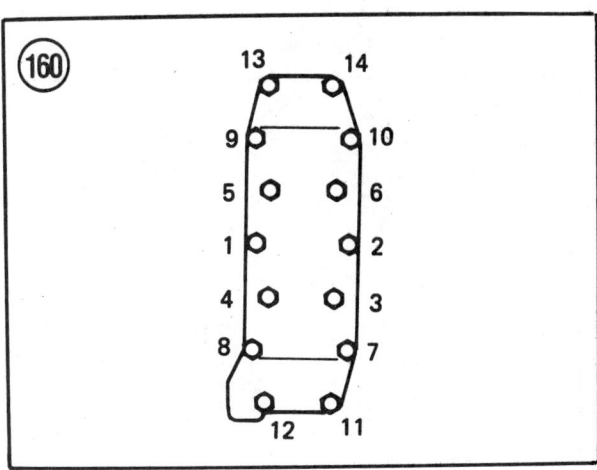

160.

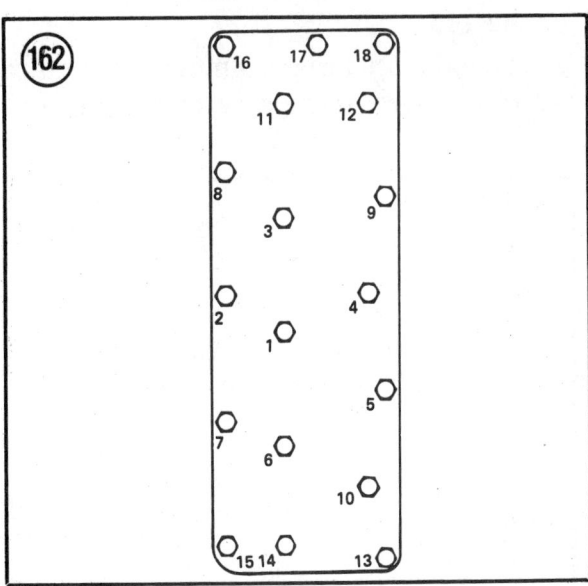

162.

8

with a new gasket and tighten cover bolts to specifications (**Table 1**).

9. Install the reed valve assembly and intake manifold to cylinder block with a new gasket. Coat bolt threads with Loctite and tighten to specifications (**Table 1**).

10. Install the power head as described in this chapter.

All V-block Models

Refer to **Figure 85** (typical) as required for this procedure.

1. Align the arrow stamped on the upper bearing housing with the cylinder cover opening. Install bolts finger-tight.

2. Temporarily bolt the lower oil seal housing in place finger-tight with its 2 tabs facing the exhaust cover.

3. Apply a thin coat of Gasket Maker or equivalent on the cylinder block and crankcase mating surfaces. Install crankcase to cylinder block and tighten bolts to specifications (**Table 1**) following the sequence shown in **Figure 163** (115-130 hp) or **Figure 164** (150-225 hp).

4. Tighten the upper bearing housing and lower oil seal housing bolts to specifications (**Table 1**) following the sequence shown in **Figure 165**.

5. If anodes are installed in the cylinder head water passages, make sure they are in good

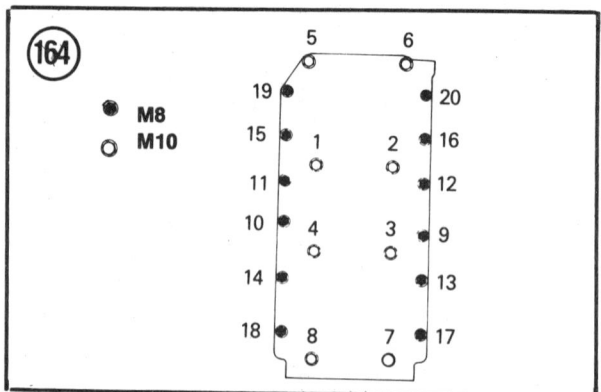

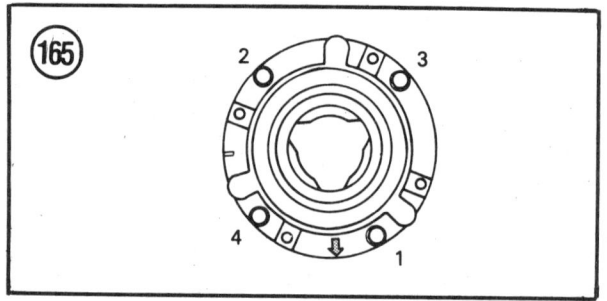

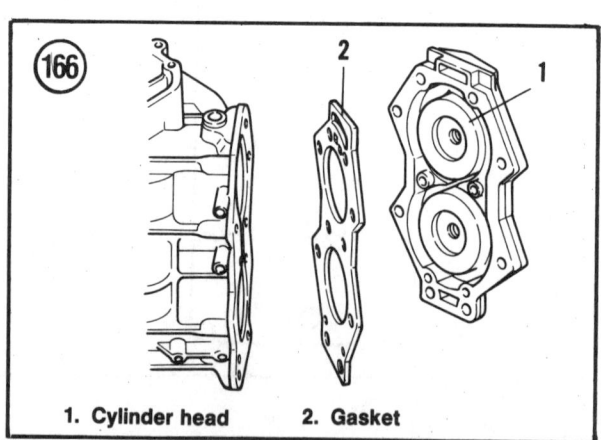

1. Cylinder head   2. Gasket

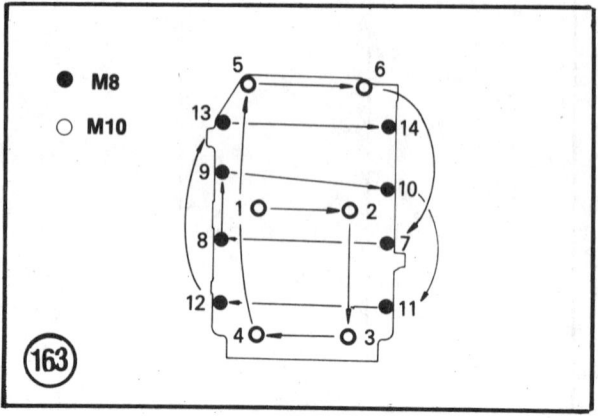

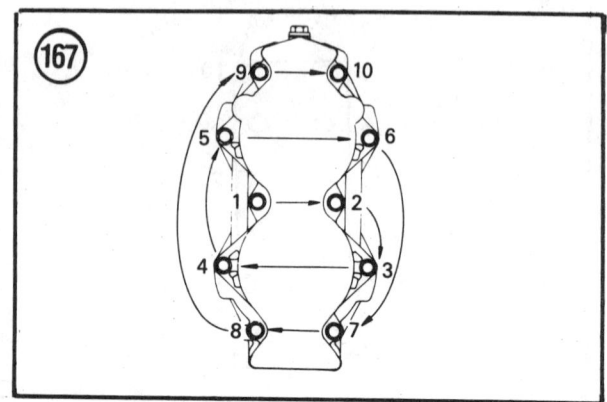

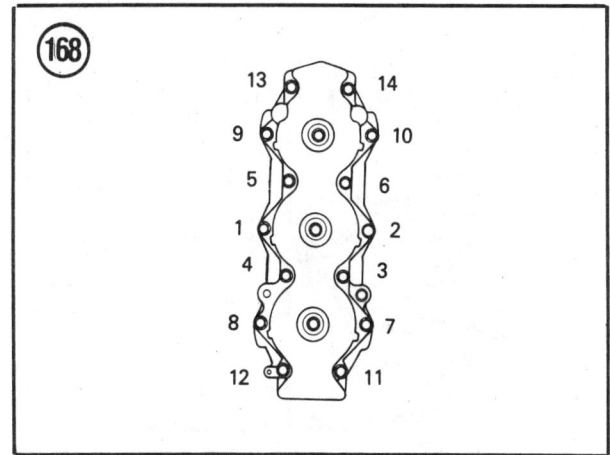

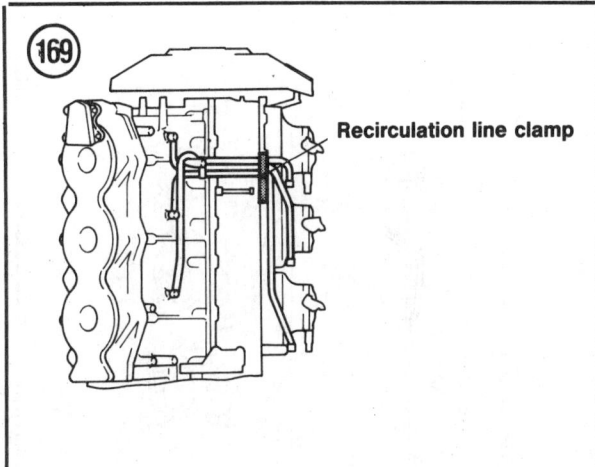

Recirculation line clamp

condition and properly installed. See **Figure 143** (typical).

6. Install the cylinder head with a new gasket (**Figure 166**). Tighten head bolts to specifications (**Table 1**) following the sequence shown in **Figure 167** (115-130 hp) or **Figure 168** (150-225 hp). Be sure to install the recirculation line clamp on the starboard side of V6 crankcases (**Figure 169**).

7. Install the reed valve assembly and intake manifold with a new gasket. Tighten the fasteners to specifications (**Table 1**) following the sequence shown in **Figure 170** (115-130 hp) or **Figure 171** (150-225 hp). On V6 models, be sure to install the recirculation clamps under the proper intake manifold bolts.

8. Install the exhaust cover with a new gasket. Tighten bolts to specifications (**Table 1**).

9. Connect the recirculation lines as shown in **Figure 172** (115-130 hp) or **Figure 173** (150-225 hp).

10. Install the thermoswitch in each cylinder head. High step on the thermoswitch grommet should face the cylinder head cover. See **Figure 174**.

8

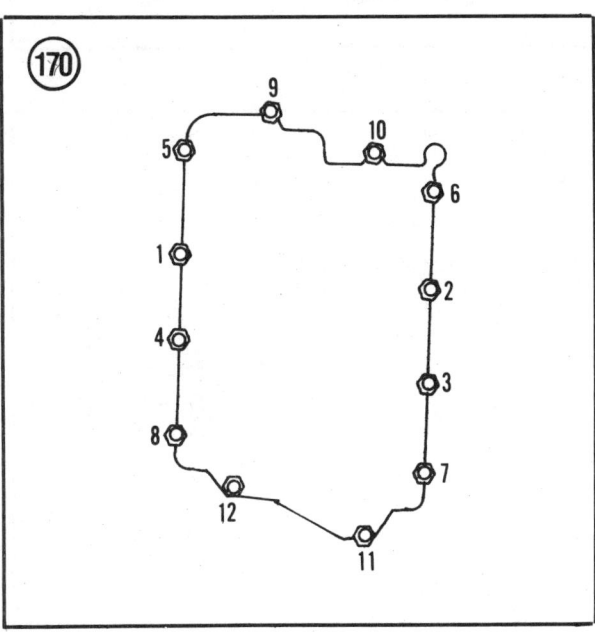

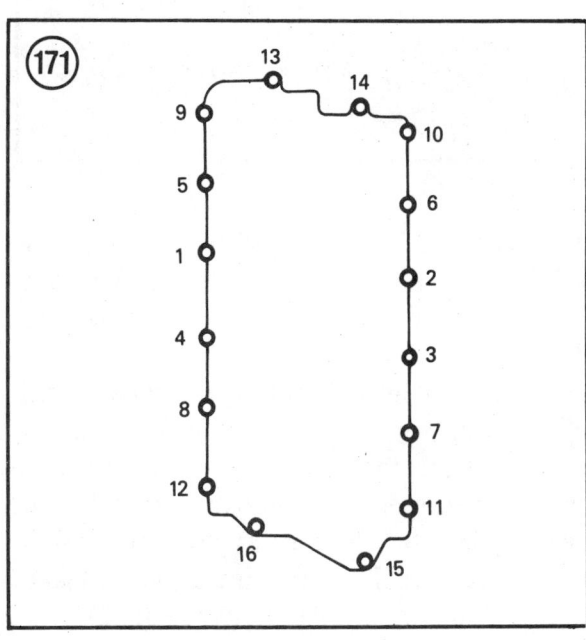

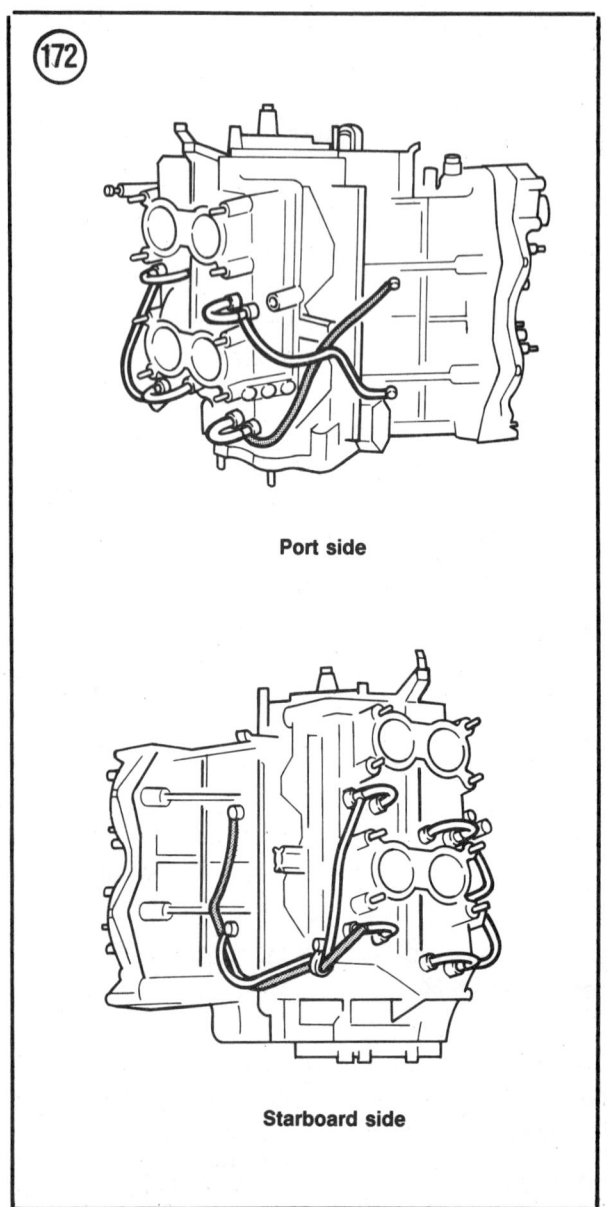

Port side

Starboard side

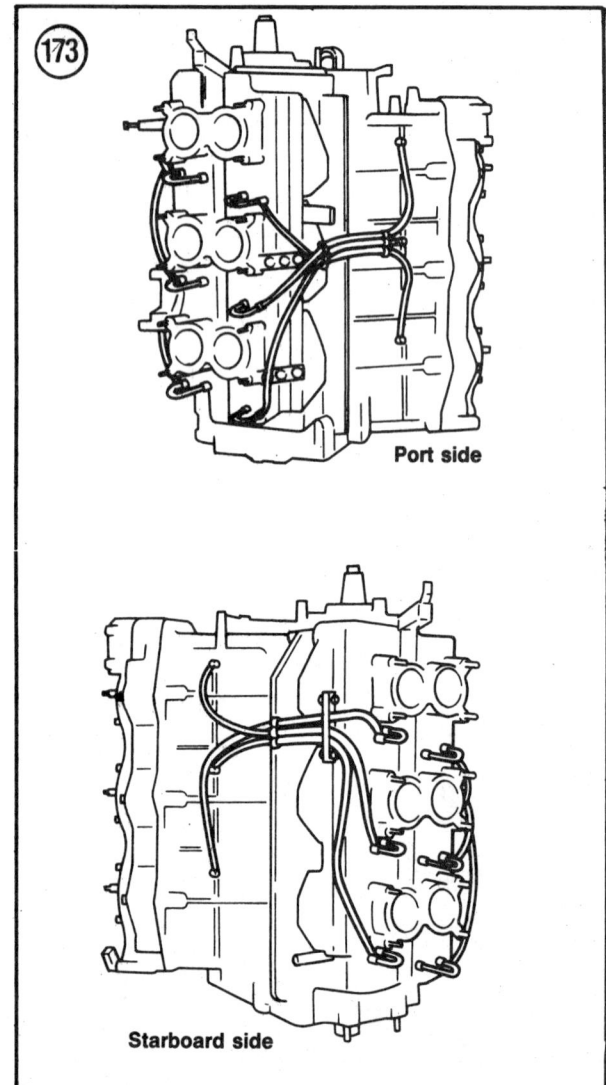

Port side

Starboard side

11. Install the cylinder head covers with new gaskets. Tighten cover bolts to specifications (**Table 1**) following the sequence shown in **Figure 174**.

12A. 115-130 hp—Install the thermostat and rubber collar or washer in the cylinder head cover. Install the pressure valve with its longer side facing the exhaust cover side. Fit the spring into the thermostat cover and install on cylinder head cover with a new gasket. Tighten bolts to specifications (**Table 1**).

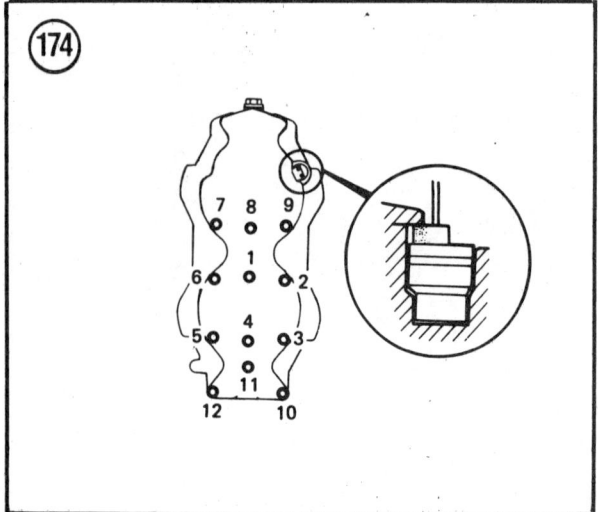

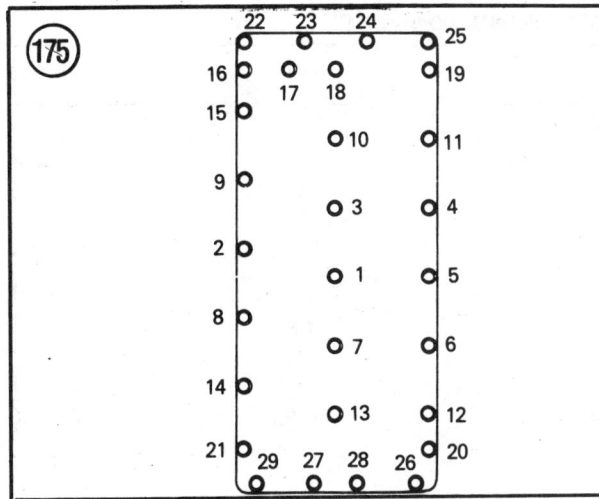

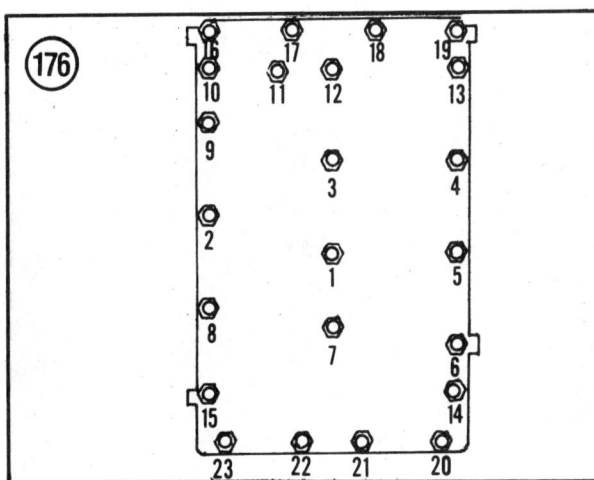

12B. 150-225 hp—Install the thermostat and rubber collar or washer in the cylinder head cover. Install the thermostat cover with a new gasket and tighten bolts to specifications (**Table 1**).

13. 220-225 hp—Install the knock sensor and crankshaft position sensors.

14. Sandwich the inner exhaust cover between 2 new gaskets. Position the assembly on the crankcase and install the outer exhaust cover on top. Install and tighten attaching bolts to specifications (**Table 1**) following the sequence shown in **Figure 175** (115-130 hp) or **Figure 176** (150-225 hp).

15. 150-225 hp—Install the pressure valve in the exhaust cover with its longer side facing outward. Fit the sping into the pressure valve cover and install on exhaust cover with a new gasket. Tighten bolts to specifications (**Table 1**).

16. Install the cylinder cover to the cylinder block. Be sure to install the J-clamp under the end bolts facing the crankshaft.

17. Install the engine stop on each side of the cylinder block.

18. Install the shift slide bracket.

19. Install the power head as described in this chapter.

8

Table 1 POWER HEAD TIGHTENING TORQUES

Fastener	ft.-lb.	N•m
Connecting rod bolts		
90-225 hp		
Step 1 [2]	12	17
Step 2	25	35
Crankcase-to-block		
2 hp		
Step 1 [1]	3.6	5
Step 2	7.2	10
3 hp		
Step 1	2.2	3
Step 2	4.3-6.9	6-9.5
4 and 5 hp		
Step 1 [1]	4.3	6
Step 2	8.7	12
6 and 8 hp		
Step 1 [2]	4.3	6
Step 2	8.7	12
9.9 and 15 hp		
M6 bolts		
Step 1 [2]	4.3	6
Step 2	8.7	12
M8 bolts		
Step 1 [2]	11	15
Step 2	22	30
25 (1984-1987) and 2-cyl. 30 hp		
Step 1	10	15
Step 2	19	27
25 (1988-on),		
3-cyl. 30, 40 and 50 hp		
M6 bolts		
Step 1	3.6	5
Step 2	8	11
M8 bolts		
Step 1	11	15
Step 2	20	28
70 hp		
M8 bolts		
Step 1	7.2	10
Step 2	14	20
M10 bolts		
Step 1	14	20
Step 2	29	40
90 hp		
M6 bolts		
Step 1	4.3	6
Step 2	8.7	12
M10 bolts		
Step 1	14	20
Step 2	29	40

(continued)

Table 1 POWER HEAD TIGHTENING TORQUES (continued)

Fastener	ft.-lb.	N•m
115-225 hp		
M8 bolts		
Step 1	7.2	10
Step 2	13	18
M10 bolts		
Step 1	14	20
Step 2	29	40
Cylinder head		
2 hp		
Step 1[3]	3.6	5
Step 2	7.2	10
3 hp		
Step 1	2.2	3
Step 2	4.3-6.9	6-9.5
4 and 5 hp		
Step 1[3]	2.9	4
Step 2	5.8	8
6 and 8 hp		
Step 1	5.8	8
Step 2	7.7	10
9.9 and 15 hp		
Step 1	5.8	8
Step 2	12	17
25 (1984-1987) and 2-cyl. 30 hp		
Step 1	10	15
Step 2	19	27
25 (1988-on) hp		
M6 bolts		
Step 1	2.9	4
Step 2	5.8	8
M8 bolts		
Step 1	11	15
Step 2	20	28
3-cyl. 30, 40 and 50 hp		
Step 1	11	20
Step 2	20	28
70 and 90 hp		
Step 1[2]	11	15
Step 2	22	30
115-130 hp		
Step 1	10	14
Step 2	22	30
150-225 hp		
Step 1	11	15
Step 2	22	30
Cylinder head cover		
90-225 hp	5.8	8
Exhaust cover		
4 and 5 hp		
Step 1[3]	2.9	4
Step 2	5.8	8

8

(continued)

Table 1 POWER HEAD TIGHTENING TORQUES (continued)

Fastener	ft.-lb.	N•m
6 and 8 hp		
Step 1	4.3	6
Step 2	7.1	10
9.9 and 15 hp		
Step 1	4.3	6
Step 2	8.7	12
25 (1984-1987) and 2-cyl. 30 hp		
Step 1	10	15
Step 2	19	27
25 (1988-on) hp		
Step 1	2.9	4
Step 2	5.8	8
3-cyl. 30, 40 and 50 hp		
Step 1	11	15
Step 2	20	28
40 and 50 hp		
Step 1[3]	11	15
Step 2	20	28
70 hp	5.8	8
90 hp	13	18
115-225 hp		
Step 1	2.9	4
Step 2	5.8	8
Intake manifold		
25 (1988-on) hp		
Step 1	2.9	4
Step 2	5.8	8
90 hp	8.7	12
70, 115-225 hp		
Step 1	2.9	4
Step 2	5.8	8
Lower oil seal housing		
4, 5, 6 and 8 hp	5.8	8
9.9 and 15 hp	4.3	6
All others	5.8	8
Flywheel nut		
2 and 3 hp	29-36	40-50
4 and 5 hp	32	45
6 and 8 hp	45	61
9.9 and 15 hp	75	102
25 (1988-on) and 3-cyl. 30 hp	72	100
40 and 50 hp	100	140
All others	115	160
Power head mounting fasteners		
3 hp	4.3-6.9	6-9.5
9.9, 15, 25 (1984-1987), 30 (2-cyl.) and 150-225 hp	13	18
25 (1988-on), 30 (3-cyl.) and 40-130 hp	15	21
Spark plug		
2, 25 (1984-1987) and 30 hp	14	20
6, 8, 9.9 and 15 hp	20	28
All others	18	25
		(continued)

Table 1 POWER HEAD TIGHTENING TORQUES (continued)

Fastener	ft.-lb.	N•m
Starter motor		
V4 and V6	22	30
Thermostat cover		
9.9 and 15 hp	4.3	6
All others	5.8	8
Upper bearing housing	5.8	8
Standard bolt/nut torque		
M5 bolt or 8 mm nut	3.6	5
M6 bolt or 10 mm nut	5.8	8
M8 bolt or 12 mm nut	13	18
M10 bolt or 14 mm nut	25	36
M12 bolt or 17 mm nut	30	42

1. Coat threads with Gasket Maker of equivalent.
2. Coat threads with Yamalube Two-cycle lubricant (for outboards).
3. Coat threads with Loctite Type A.

Table 2 POWER HEAD SPECIFICATIONS

8

Connecting rod	
Big end side clearance	
2 hp	0.012-0.024 in. (0.3-0.6 mm)
3 hp	0.011-0.037 in. (0.30-0.95 mm)
4 and 5 hp	0.008-0.028 in. (0.2-0.7 mm)
6-15 hp	0.008-0.030 in. (0.2-0.7 mm)
25 hp (1984-1987)	0.015-0.019 in. (0.38-0.48 mm)
25 hp (1988-on)	0.075-0.082 in. (1.90-2.10 mm)
2-cyl. 30 hp	0.008-0.012 in. (0.20-0.30 mm)
90 hp	0.008-0.013 in. (0.20-0.33 mm)
Small end free play limit	0.08 in. (2 mm)
Crankshaft runout limit	
3-25 hp, 2-cyl. 30 hp	0.0012 in. (0.03 mm)
3-cyl. 30 hp, 40 and 50 hp	0.0019 in. (0.05 mm)
All others	0.0008 in.(0.02 mm)
Cylinder head warp limit	0.0039 in. (0.1 mm)
Cylinder bore	
Size	
2 hp	1.535 in. (39 mm)
3 hp	1.811-1.812 in. (46.00-46.02 mm)
5 hp	2.13 in. (54 mm)
4, 6 and 8 hp	1.97 in. (50 mm)
9.9 and 15 hp	2.20 in. (56 mm)
25 (1984-1987), 2-cyl. 30 and 70 hp	2.83 in. (72 mm)
3-cyl. 30 hp	2.34 in. (59.5 mm)
25 (1988-on), 40 and 50 hp	2.638-2.639 in. (67.00-67.02 mm)
90 hp	3.23 in. (82 mm)
115-225 hp	3.54 in. (90 mm)

(continued)

Table 2 POWER HEAD SPECIFICATIONS (continued)

Taper limit	0.003 in. (0.08 mm)
Out-of-round limit	0.002 in. (0.05 mm)
Piston	
Size	
2 hp	1.535 in. (39 mm)
3 hp	1.8096-1.8106 in. (45.965-45.990 mm)
5 hp	2.13 in. (54 mm)
4, 6 and 8 hp	1.97 in. (50 mm)
9.9 and 15 hp	2.20 in. (56 mm)
25 (1984-1987), 2-cyl. 30 and 70 hp	2.83 in. (72 mm)
25 (1988-on) hp	2.637-2.638 in. (66.98-67.00 mm)
3-cyl. 30 hp	2.34 in. (59.5 mm)
40 and 50 hp	2.6376-2.6378 in. (66.994-67.000 mm)
90 hp	3.23 in. (82 mm)
115-225 hp	3.54 in. (90 mm)
Measuring point	
2 hp	0.197 in. (5 mm) above bottom of skirt
All others	0.39 in. (10 mm) above bottom of skirt
Clearance-to-bore	
2 hp	0.0012-0.0016 in. (0.03-0.04 mm)
3, 4 and 5 hp	0.0012-0.0014 in. (0.030-0.035 mm)
6, 8 and 25 (1988-on) hp	0.0016-0.0018 in. (0.040-0.045 mm)
9.9, 15 and 3-cyl. 30 hp	0.0014-0.0016 in. (0.035-0.040 mm)
25 (1984-1987), 2-cyl. 30, 40, 50 and 90 hp	0.0024-0.0026 in. (0.060-0.065 mm)
70 hp	0.0020-0.0022 in. (0.050-0.055 mm)
115-225 hp (1984-1988)	0.0033-0.0035 in. (0.085-0.090 mm)
115-225 hp (1989)	0.0031-0.0033 in. (0.080-0.085 mm)
Oversize	
2 hp	
1st	1.545 in. (39.25 mm)
2nd	1.555 in. (39.50 mm)
3 hp	
1st	1.821 in. (46.25 mm)
2nd	1.831 in. (46.50 mm)
9.9 and 15 hp	
1st	2.21 in. (56.35 mm)
2nd	2.22 in. (56.60 mm)
3-cyl. 30 hp	
1st	2.352 in. (59.75 mm)
2nd	2.362 in. (60.00 mm)
25 (1984-1987), 2-cyl. 30 and 70 hp	
1st	2.844 in. (72.25)
2nd	2.854 in. (72.50 mm)
25 (1988-on), 40 and 50 hp	
1st	2.648 in. (67.25 mm)
2nd	2.657 in. (67.50)
90 hp	
1st	3.238 in. (82.25 mm)
2nd	3.248 in. (82.50 mm)
115-225 hp	
1st	3.553 in. (90.250 mm)
2nd	3.563 in. (90.500 mm)

(continued)

Table 2 POWER HEAD SPECIFICATIONS (continued)

Offset	
4-8 hp	0.02 in. (0.5 mm)
70 hp	0.0197 in. (0.5 mm)
90 hp	0.039 in. (1 mm)
115-200 hp	
Starboard side	0.000 in. (0 mm)
Port (exhaust) side	0.059 in. (1.5 mm)
220-225 hp	
Starboard side	0.02 in. (0.5 mm)
Port (exhaust) side	0.08 in. (2 mm)
All others	No offset
Piston ring	
Width	
2 hp	0.071 ±0.0039 in. (1.8 ±0.1 mm)
3 hp	0.08 in. (2.0 mm)
4 hp	0.087 ±0.0004 in. (2.2 ±0.1 mm)
5 hp	0.079 ±0.0004 in. (2.0 ±0.1 mm)
6 and 8 hp	0.075-0.083 in. (2.0 ±0.1 mm)
9.9 and 15 hp	0.1 ±0.004 in. (2.5 ±1 mm)
25 (1984-1987) and 2-cyl. 30 hp	
Top	0.12 in. (3 mm)
Bottom	0.13 in. (3.2 mm)
25 (1988-on), 40 and 50 hp	0.102 in. (2.6 mm)
3-cyl. 30 hp	
Top ring	0.083 in. (2.1 mm)
No. 2 and No. 3	0.098 in. (2.5 mm)
70 hp	0.118 in. (3 mm)
90 hp	0.098 in. (2.5 mm)
115-225 hp	0.110 in. (2.8 mm)
Height	
25 (1988-on) and 70 hp	0.059 in. (1.5 mm)
3-cyl. 30 hp	
Top	0.059 in. (1.5 mm)
No. 2 and No. 3	0.079 in. (2.0 mm)
All others	0.079 in. (2.0 mm)
End gap	
2 and 3 hp	0.004-0.012 in. (0.1-0.3 mm)
4-15 hp	0.006-0.014 in. (0.15-0.35 mm)
25 hp (1984-1987) and 2-cyl. 30 hp	0.008-0.024 in. (0.2-0.6 mm)
3-cyl. 30 hp	0.006-0.012 in. (0.15-0.30 mm)
25 (1988-on), 40, 50 and 90 hp	0.016-0.024 in. (0.4-0.6 mm)
70, 115-225 hp	0.012-0.020 in. (0.3-0.5 mm)
Side clearance	
2, 4 and 5 hp	0.001-0.003 in. (0.03-0.07 mm)
3, 6, 8 and 25 (1988-on) hp	
Top	0.0008-0.0024 in. (0.02-0.06 mm)
Bottom	0.0012-0.0028 in. (0.03-0.07 mm)
9.9 and 15 hp	0.0016-0.0031 in. (0.04-0.08 mm)

8

(continued)

Table 2 POWER HEAD SPECIFICATIONS (continued)

25 (1984-1987) and 2-cyl. 30 hp	
Top	0.0008-0.0024 in. (0.02-0.06 mm)
Bottom	0.0016-0.0032 in. (0.04-0.08 mm)
3-cyl. 30 hp	
Top	0.0028-0.0043 in. (0.07-0.11 mm)
No. 2 and No. 3	0.0020-0.0035 in. (0.05-0.09 mm)
40 and 50 hp	
1st	0.001-0.002 in. (0.03-0.05 mm)
2nd and 3rd	0.001-0.003 in. (0.03-0.07 mm)
70 hp	
Top	0.0008-0.0024 in. (0.02-0.06 mm)
Bottom	0.0012-0.0028 in. (0.03-0.07 mm)
90 hp	
Top	0.0012-0.0026 in. (0.03-0.065 mm)
Bottom	0.0012-0.0028 in. (0.03-0.07 mm)
115-225 hp	
Top	0.0012-0.0026 in. (0.03-0.065 mm)
Bottom	0.0016-0.0029 in. (0.04-0.075 mm)
Seal ring wear limit	0.004 in. (1 mm)

Table 3 CRANKSHAFT SPECIFICATIONS

Model	Runout	Dimension B	Dimension C
2 hp	0.0008 in. (0.02 mm)	1.100-1.106 in. (27.95-28.10 mm)	0.012-0.024 in. (0.3-0.6 mm)
3 hp	0.0012 in. (0.03 mm)	1.41-1.42 in. (35.9-36.0 mm)	0.011-0.037 in. (0.30-0.95 mm)
4-8 hp	0.0012 in. (0.03 mm)	1.568-1.574 in. (39.95-40.10 mm)	0.008-0.028 in. (0.2-0.7 mm)
9.9-15 hp	0.0012 in. (0.03 mm)	1.85 in. (47 mm)	0.008-0.028 in. (0.2-0.7 mm)
25 (1984-1987) hp	0.0012 in. (0.03 mm)	2.244 in. (57 mm)	0.015-0.019 in. (0.38-0.48 mm)
25 (1988-on) hp	0.0012 in. (0.03 mm)	1.97 in. (50 mm)	0.075-0.082 in. (1.90-2.10 mm)
2-cyl. 30 hp	0.0012 in. (0.03 mm)	2.244 in. (57 mm)	0.000-0.000 in. (0.00-0.00 mm)
3-cyl. 30, 40 and 50 hp	0.0020 in. (0.05 mm)	2.1 in. (54 mm)	— —
70 hp	0.0020 in. (0.05 mm)	— —	— —
90 hp	0.0008 in. (0.02 mm)	— —	0.008-0.013 in. (0.20-0.33 mm)
115-225 hp	0.0008 in. (0.02 mm)	— —	— —

8

Chapter Nine

Gearcase and Drive Shaft Housing

Torque is transferred from the engine crankshaft to the gearcase by a drive shaft. A pinion gear on the drive shaft meshes with a drive gear in the gearcase to change the vertical power flow into a horizontal flow through the propeller shaft. The power head drive shaft rotates continuously in a clockwise direction when the engine is running, but propeller rotation is controlled by the gear train shifting mechanism.

On Yamaha outboards with a reverse gear, a sliding clutch engages the appropriate gear in the gearcase when the shift mechanism is placed in FORWARD or REVERSE. This creates a direct coupling that transfers the power flow from the pinion to the propeller shaft. **Figure 1** (typical) shows the operation of the gear train.

The gearcase can be removed without removing the entire outboard from the boat. This chapter contains removal, overhaul and installation procedures for the gearcase, water

pump and propeller. **Table 1** is at the end of the chapter.

The gearcases covered in this chapter differ somewhat in design and construction over the years covered and thus require slightly different service procedures. The chapter is arranged in a normal disassembly/assembly sequence. When only a partial repair is required, follow the procedure(s) for your gearcase to the point where the faulty parts can be replaced, then reassemble the unit.

Since this chapter covers a wide range of models, the gearcases shown in the accompanying illustrations are the most common ones. While it is possible that the components shown in the pictures may not be identical with those being serviced, the step-by-step procedures may be used with all models covered in this manual.

SERVICE PRECAUTIONS

Whenever you work on a Yamaha outboard, there are several good procedures

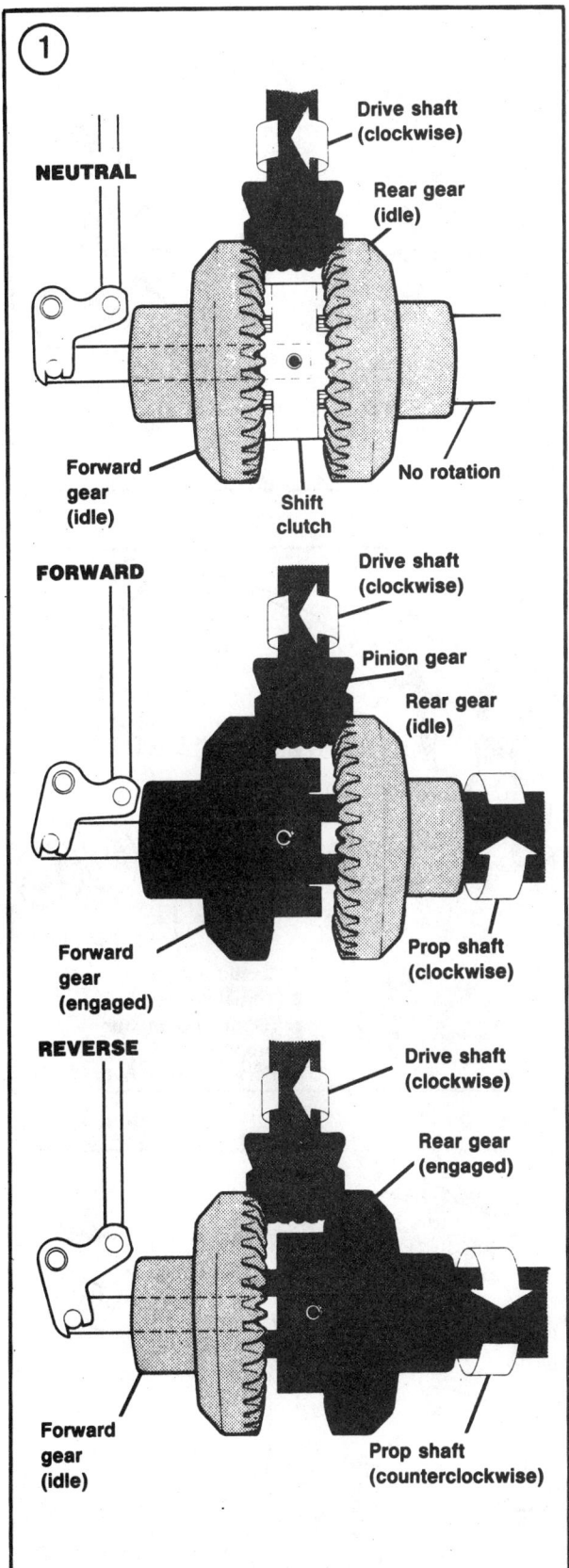

NEUTRAL

Drive shaft (clockwise)

Rear gear (idle)

Forward gear (idle)

Shift clutch

No rotation

FORWARD

Drive shaft (clockwise)

Pinion gear

Rear gear (idle)

Forward gear (engaged)

Prop shaft (clockwise)

REVERSE

Drive shaft (clockwise)

Rear gear (engaged)

Forward gear (idle)

Prop shaft (counterclockwise)

to keep in mind that will make your work easier, faster and more accurate.

1. Never use elastic stop nuts more than twice. It is a good idea to replace such nuts with new ones each time they are removed. Never use non-locking nuts or worn-out stop nuts.

2. Use special tools where noted. In some cases, it may be possible to perform the procedure with makeshift tools, but this procedure is not recommended. The use of makeshift tools can damage the components and may cause serious personal injury.

3. Use a vise with protective jaws to hold housings or parts. If protective jaws are not available, insert wooden blocks on either side of the part(s) before clamping them in the vise.

4. Remove and install pressed-on parts with an appropriate mandrel, support and hydraulic press. Do not try to pry, hammer or otherwise force them on or off.

5. Refer to **Table 1** at the end of the chapter for torque values, if not given in the text. Proper torque is essential to assure long life and satisfactory service from outboard components.

6. Apply Yamalube All-purpose Marine grease to the outer surfaces of all bearing carrier and retainer mating surfaces during reassembly.

7. Discard all O-rings and oil seals during disassembly. Apply Yamalube All-purpose Marine grease to new O-rings and seal lips to prevent damage when the motor is first started.

8. Keep a record of all shims and where they came from. As soon as the shims are removed, inspect them for damage and write down their thickness and location. Wire the shims together for reassembly and store them in a safe place. Follow shimming instructions closely. If gear backlash is not properly set, the unit will be noisy and suffer premature

9

gear failure. Incorrect bearing preload will result in premature bearing failure.

9. Work in an area where there is good lighting and sufficient space for component storage. Keep an ample number of clean containers available for storing small parts. Cover parts with clean shop cloths when you are not working with them.

PROPELLER

Yamaha outboards use 2 different methods of attaching the propeller. All 2-5 hp models use the shear pin design (**Figure 2**). In this design, a metal pin installed in the propeller shaft engages a recessed slot in the propeller hub. As the shaft rotates, the pin rotates the propeller. The shear pin is designed to break if the propeller hits an obstruction in the water. This system has 2 advantages. The pin absorbs the impact to prevent possible propeller damage. It also alerts the user to the fact that something is wrong, since the engine speed will increase if the pin breaks. Yamaha outboards are equipped with spare shear and cotter pins installed on the bottom cowl near the power head.

Propellers on Yamaha 6-225 hp models use some variation of the thrust hub design (**Figure 3**). The propeller rides on a ratchet-type bushing or a spline drive rubber hub and is retained by a castellated nut and cotter pin. Any underwater impact is absorbed by the propeller bushing/hub.

> *WARNING*
> *To prevent accidental engine starting when working on or around the propeller, disconnect all spark plug leads from the plugs on manual and electric start models. In addition, electric start models should be shifted into NEUTRAL and the ignition key removed (if so equipped).*

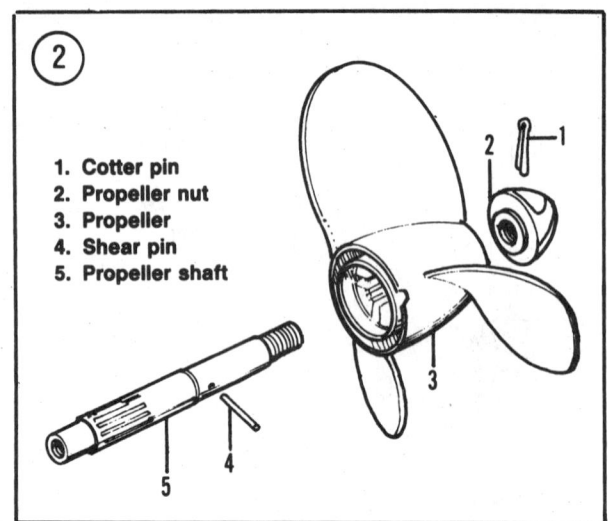

1. Cotter pin
2. Propeller nut
3. Propeller
4. Shear pin
5. Propeller shaft

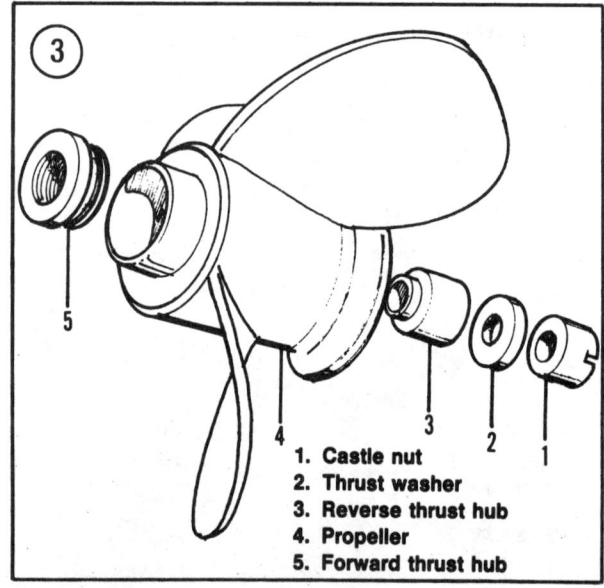

1. Castle nut
2. Thrust washer
3. Reverse thrust hub
4. Propeller
5. Forward thrust hub

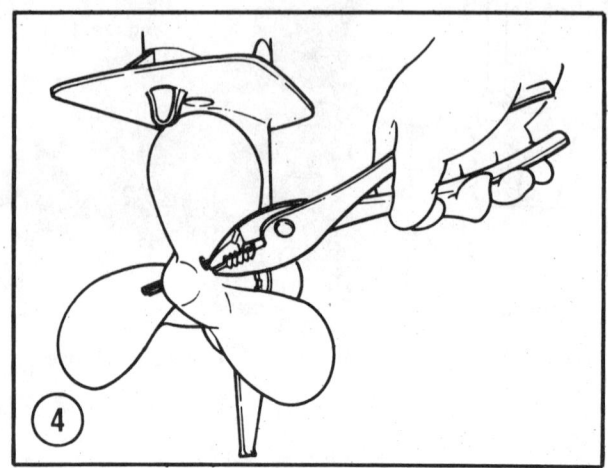

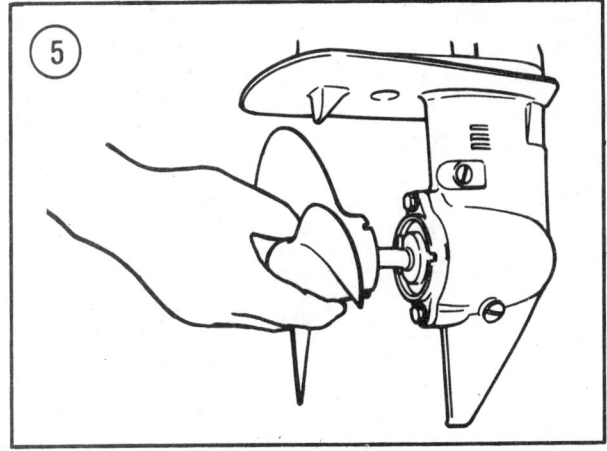

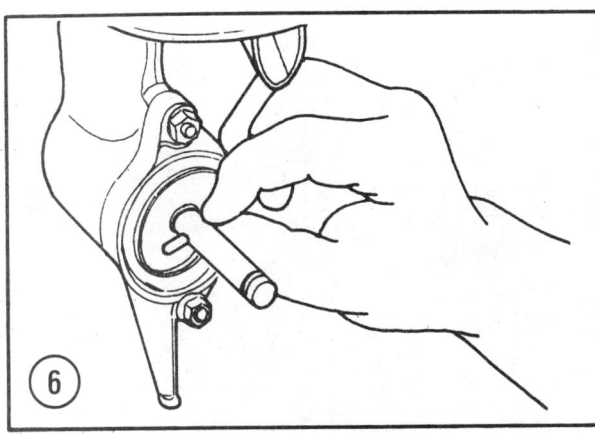

Propeller Removal/Installation (Thrust Hub Design)

1. Remove and discard the propeller nut cotter pin.

2. Install a block of wood between the propeller blades and anti-ventilation plate to prevent the propeller from rotating.

3. Remove the castellated nut and washer. Remove the reverse thrust hub.

4. Remove the propeller and forward thrust hub.

5. Clean the propeller shaft splines thoroughly. Check splines for excessive wear or damage.

6. Installation is the reverse of removal. Lubricate the propeller shaft splines with Yamalube All-purpose Marine grease. Tighten propeller nut to specifications (**Table 1**) while holding the propeller from moving with a block of wood positioned as in Step 2. Install a new cotter pin.

7. Recheck propeller nut torque after the first time the engine is operated.

9

Propeller Removal/Installation (Shear Pin Design)

1. Remove and discard the cotter pin (**Figure 4**).

2. 2 hp—Unscrew and remove the plastic propeller nut.

3. Remove the propeller (**Figure 5**).

4. If bearing housing or housing cap is to be removed, remove the shear pin from the propeller shaft with an appropriate punch (**Figure 6**).

5. Clean the propeller shaft thoroughly. Inspect the pin engagement slot in the propeller hub and shaft for wear or damage.

6. Installation is the reverse of removal. Lubricate the propeller shaft with Yamalube All-purpose Marine grease. Use a new cotter pin.

WATER PUMP

The water pump is mounted on top of the gearcase housing on all outboards. The 2-5 hp water pump impeller is fastened to the drive shaft by a key that fits into the drive shaft and a similar cutout in the impeller hub. The impeller on other models is secured by a key which engages a flat on the drive shaft or engages a slot in the drive shaft and a cutout in the impeller hub. As the drive shaft rotates, the impeller rotates with it. Water between the impeller blades and pump housing is pumped up to the power head through the water tube.

The offset center of the pump housing cause the impeller vanes to flex during rotation. At low speeds, the pump acts as a displacement type; at high speeds, water

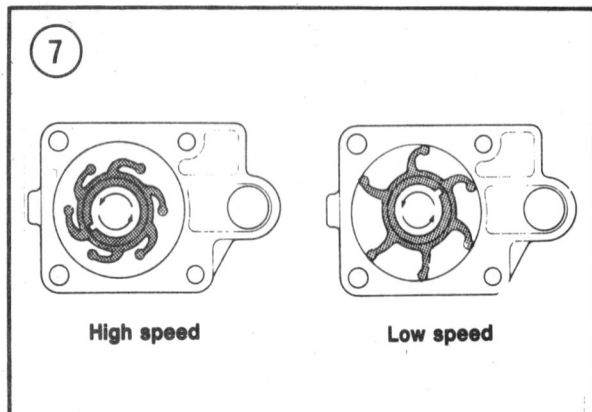

High speed   Low speed

resistance forces the vanes to flex inward and the pump becomes a centrifugal type. See **Figure 7**.

All seals and gaskets should be replaced whenever the water pump is removed. Since proper water pump operation is critical to outboard operation, it is also a good idea to install a new impeller at the same time.

Never turn a used impeller over and reuse it. The impeller rotates in a clockwise direction with the drive shaft and the vanes gradually take a set in one direction. Turning the impeller over will cause the vanes to move in a direction opposite to that which caused the set. This can result in premature impeller failure and cause extensive power head damage.

An optional chrome water pump kit is available for some models and is recommended for use in salt water or areas where the water contains a large amount of sand or silt.

Removal and Disassembly (All Models)

Refer to **Figure 8** (typical) for this procedure.

1. Remove the gearcase as described in this chapter.

2. Secure the gearcase in a suitable holding device or a vise with protective jaws. If protective jaws are not available for the vise,

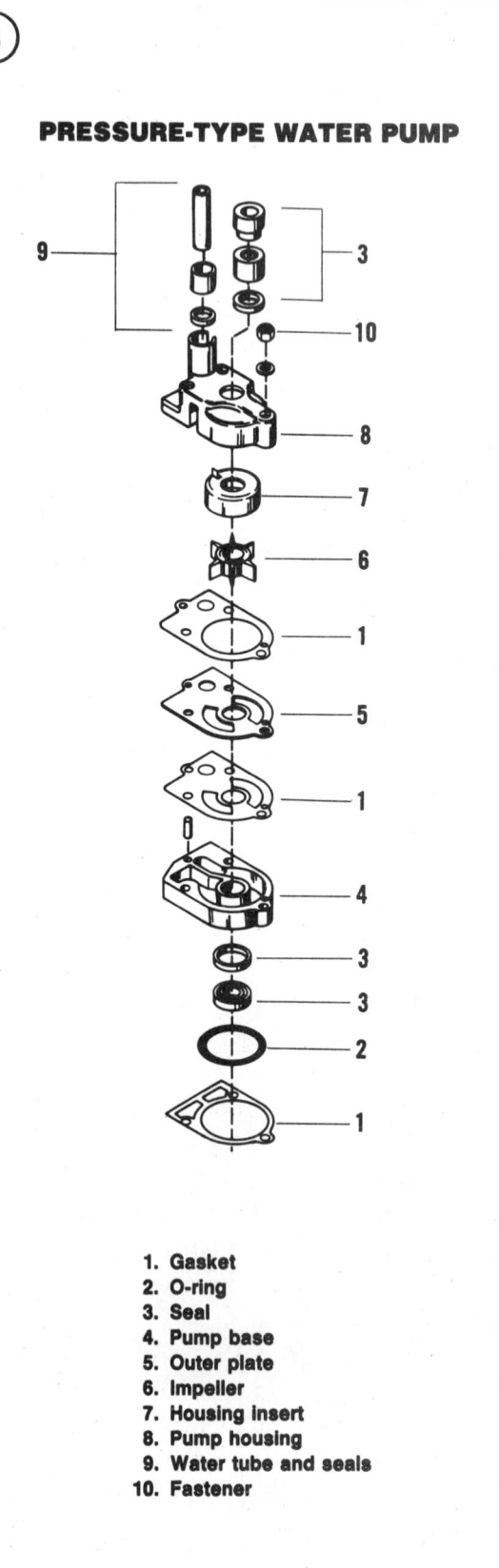

PRESSURE-TYPE WATER PUMP

1. Gasket
2. O-ring
3. Seal
4. Pump base
5. Outer plate
6. Impeller
7. Housing insert
8. Pump housing
9. Water tube and seals
10. Fastener

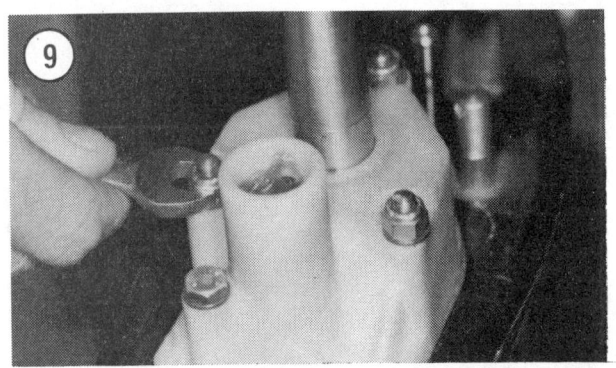

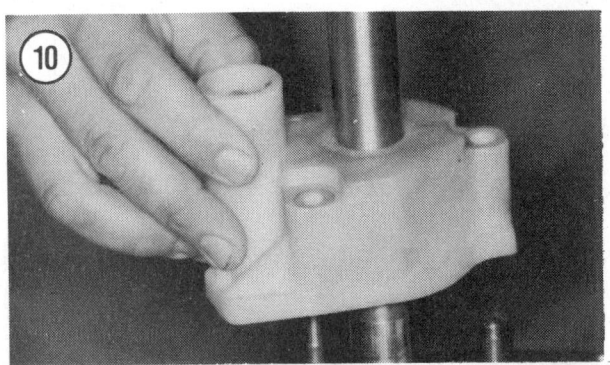

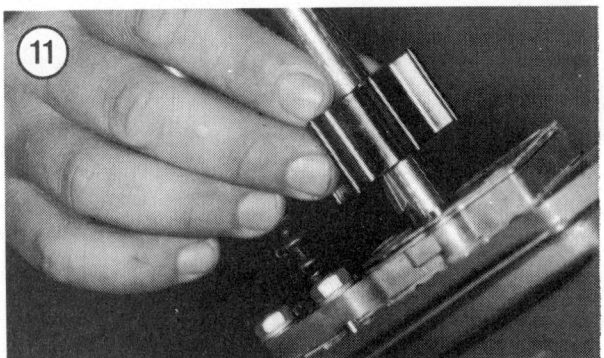

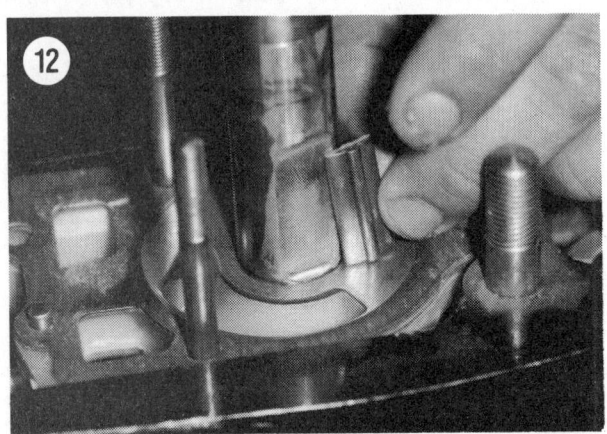

position the unit upright with the skeg between wooden blocks in the vise.

3. Remove the water tube from the pump housing. Slide rubber centrifugal slinger up and off drive shaft, if so equipped.

4. Remove the fasteners holding the pump cover to the gearcase. See **Figure 9** (typical). If support plates are used under the fasteners, do not bend, distort or damage the plates during removal.

5. Insert screwdrivers at the fore and aft ends of the pump cover and pry the cover up. Remove cover from drive shaft (**Figure 10**).

> *NOTE*
> *In some cases, the impeller may come off with the pump cover. In extreme cases, the impeller hub may have to be split with a hammer and chisel to remove it in Step 6.*

6. If impeller does not come off with pump cover, drive it upward on the drive shaft with a punch and hammer. Remove the impeller. See **Figure 11** (typical).

> *NOTE*
> *The impeller drive pin is installed in the side of the drive shaft on 2-5 hp models and need not be removed in Step 7.*

7. Remove the impeller drive key (**Figure 12**).

> *NOTE*
> *Smaller displacement units may have only one face plate gasket.*

8. Remove the water pump face plate with top and bottom gaskets. See **Figure 13** (typical). Separate face plate from gaskets. Discard the gaskets.

9. 4-5 hp—Remove bolt holding water pump base to gearcase. Remove shift rod boot from shift rod. Pry water pump base free of gearcase and slide off drive shaft and shift rod.

10A. 2 hp—Remove the pump base cartridge and oil seal protector.

NOTE
Water pump base on 6-8 hp models also serves as the oil seal housing and need not be removed during this procedure. If removal is required, see Step 9.

10B. 3 and 9.9-225 hp—Insert screwdrivers at the fore and aft ends of the pump base. Pad the pry areas under each screwdriver with clean shop cloths and pry pump base loose. Remove pump base. See **Figure 14** (typical).

Cleaning and Inspection

When removing seals from water pump cover and base, note and record the direction in which the lip of each seal faces for proper reinstallation.

1. Remove the water pump cartridge insert from the pump cover. See **Figure 15** (typical). Drive insert from oil seal side of base or through drive shaft opening as required.

2. Remove the water tube seal. See **Figure 16** (typical).

3. Remove the seals from the water pump base. See **Figure 17** (typical).

4. Turn the water pump base over and remove the O-ring, if so equipped. See **Figure 18** (typical).

5. Clean all metal parts in solvent and blow dry with compressed air, if available.

6. Clean gasket residue from all mating surfaces.

7. Check pump cover and base for cracks or distortion from overheating.

8. Check face plate and cartridge insert for grooves or rough surfaces. Replace if any defects are found.

9. If original impeller is to be reused, check bonding to hub. Check side seal surfaces and blade ends for cracks, tears, wear and a glazed or melted appearance. See **Figure 19** (typical).

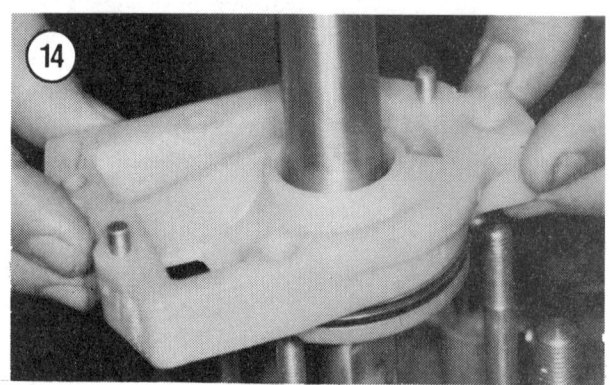

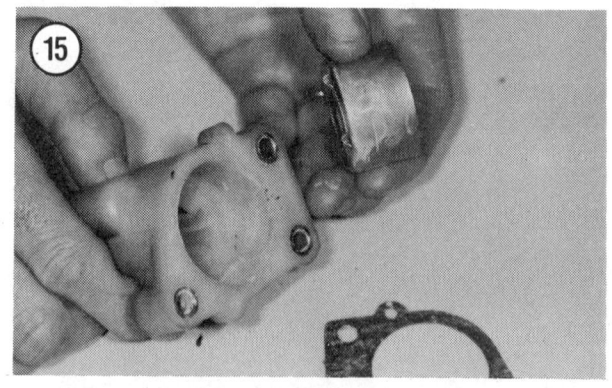

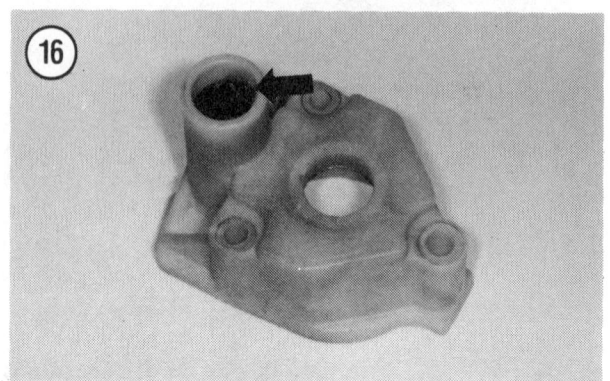

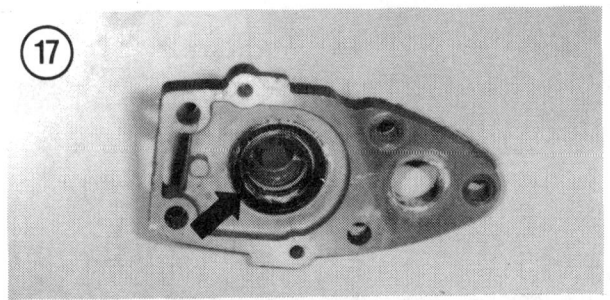

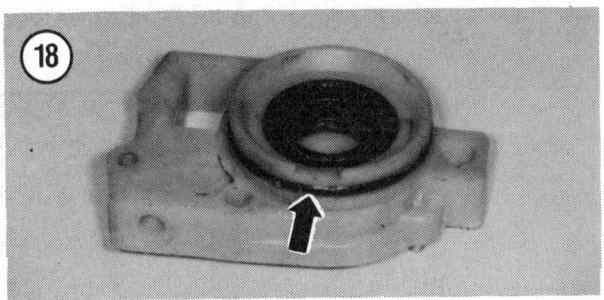

If any of these defects are noted, do *not* reuse the original impeller.

Assembly and Installation (All Models)

When new seals are installed in the pump cover and base, their lips should face in the direction recorded during disassembly. After installation, pack the seal lip cavity with Yamalube All-purpose Marine grease.

1. Secure the gearcase in a suitable holding device or a vise with protective jaws. If protective jaws are not available for the vise, position the unit upright with the skeg between wooden blocks in the vise.

2. If pump cover cartridge insert was removed, coat insert area in cover with Perfect Seal and install new insert with bent tab(s) engaging slot(s) in pump cover.

3. Wipe outer diameter of pump cover oil seal(s) with Loctite Type A. Install seal(s) in pump cover with an appropriate mandrel and wipe any excess Loctite from seal(s) and pump cover.

4. Wipe outer diameter of pump base oil seal(s) with Loctite Type A. Install seal(s) in pump base with an appropriate mandrel and wipe any excess Loctite from seal(s) and pump base.

5. Install a new O-ring in housing groove (2 hp) or pump base groove (all others), if so equipped. Wipe O-ring with Yamalube All-purpose Marine grease.

6A. 2 hp—Install oil seal protector in gearcase. Install pump cartridge in oil seal protector.

6B. All others—Coat outer surface of a new gasket with Perfect Seal and install on the gearcase. Install the pump base. Make sure base is fully seated.

7. 4-5 hp—Install and tighten base attaching bolt. Install shift rod boot. See **Figure 20** (typical).

8. Install pump face plate on the pump base with appropriate number of gaskets. See **Figure 21** (typical). Outer surface of gaskets should be coated with Perfect Seal.

9A. 2-5 hp—Install impeller drive pin in drive shaft, if removed.

9B. All others—Wipe key flat on drive shaft with Yamalube All-purpose Marine grease to hold drive key in place. Install key to drive shaft flat.

> *CAUTION*
> *If the original impeller is to be reused, install it in the same rotational direction as removed to avoid premature failure. The curl of the blades should be positioned in a counterclockwise direction, as seen from the top of the unit. See **Figure 22**.*

10. Lubricate impeller hub with Yamalube All-purpose Marine grease. Slide impeller on drive shaft. Align slot in impeller hub with drive key or pin and seat on water pump face plate gasket. See **Figure 23** (typical).

11. Check impeller installation by rotating drive shaft clockwise. Impeller should rotate with the shaft. If not, remove and reposition impeller to properly engage drive key or pin.

12. Slide pump cover over drive shaft and align cover holes with base holes or studs. Push downward on cover while rotating the drive shaft clockwise to assist the impeller in entering the cover without damage.

> *CAUTION*
> *Do not over-tighten fasteners in Step 13 or pump cover may crack during operation.*

13. Wipe pump housing bolt threads with Loctite Type A. Install the cover attaching fasteners. On 3-8 hp models, install the support plates with the fasteners. Tighten all fasteners to specifications (**Table 1**).

14. Install centrifugal slinger over drive shaft (if so equipped) and position against pump cover.

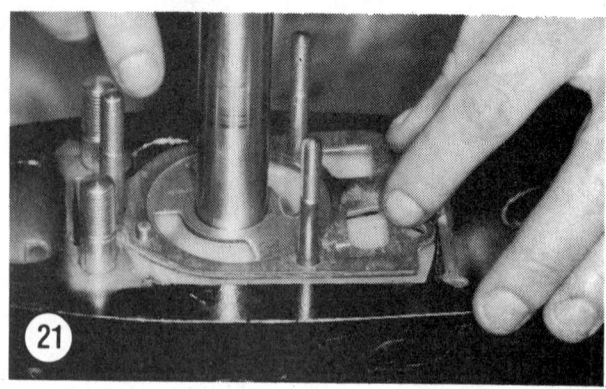

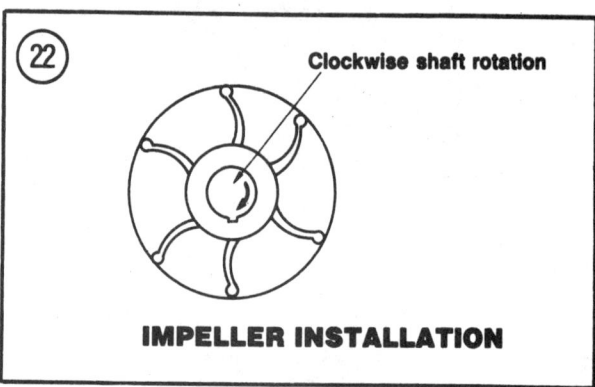

Clockwise shaft rotation

IMPELLER INSTALLATION

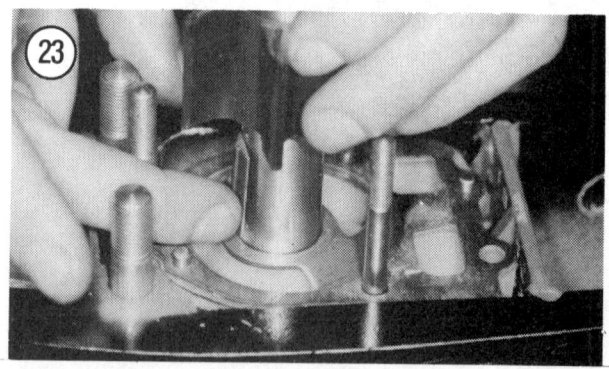

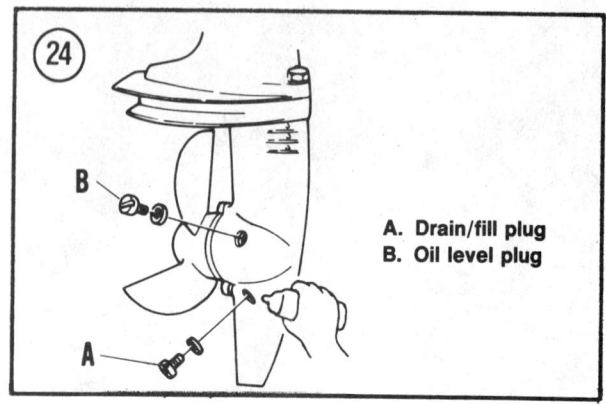

A. Drain/fill plug
B. Oil level plug

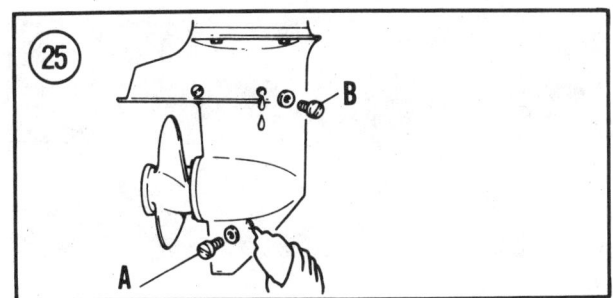

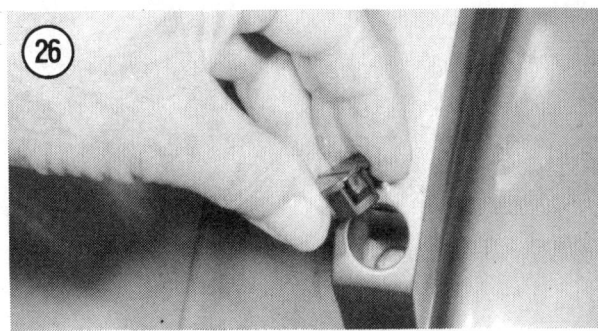

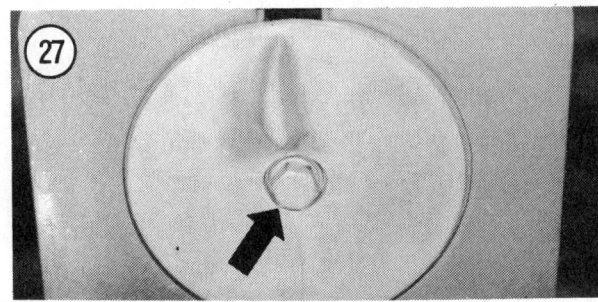

15. Install water tube in pump cover with a new tube seal.

GEARCASE

When removing gearcase mounting fasteners, it is not uncommon to find that they are corroded. Such fasteners should be discarded and new ones used when the gearcase is installed.

If water has entered the unit, it may have also corroded the water tubes and the drive shaft/crankshaft connection, making it impossible to remove the unit with hand pressure. If this happens, it will be necessary to pry the gearcase free from the drive shaft housing. Extreme cases may require the use

of heat to free the corroded components. Remember that excessive use of heat and/or prying can result in damage to the drive shaft, crankshaft and/or gearcase housing.

Removal

1. Disconnect the spark plug lead(s) as a safety precaution to prevent accidental starting of the engine during gearcase removal.
2. Shift engine into NEUTRAL.
3. Tilt engine to full out position and engage tilt lock lever.
4. Place a container under the drain/fill plug and remove it. See A, **Figure 24** (2 hp) or A, **Figure 25** (all others). Remove the oil level plug. See B, **Figure 24** (2 hp) or B, **Figure 25** (all others). Drain the lubricant from the unit.

> *NOTE*
> *If the lubricant is white or creamy in color or metallic particles are found in Step 5, remove and disassemble the gearcase to determine and correct the cause of the problem.*

5. Wipe a small amount of lubricant on your finger and rub finger and thumb together. Check for the presence of metallic particles in the lubricant. Note the color of the lubricant. A white or creamy color indicates water in the lubricant. Check the drain container for signs of water separation from the lubricant.
6. Remove the propeller as described in this chapter. On 2-5 hp models, if gearcase is to be disassembled, also remove the shear pin.
7. 4-5 hp—Shift the engine into REVERSE.
8. 25-225 hp—Scribe a mark on the gearcase and trim tab for reassembly reference.
9A. 25-50 hp—Unbolt and remove the trim tab.
9B. 70-225 hp—Pry the plastic plug from the rear of the drive shaft housing to expose the trim tab bolt (**Figure 26**). Unbolt and remove the trim tab (**Figure 27**).

9

10. 3-5 hp—Remove rubber cover on drive shaft housing for access to shift rod connector. See **Figure 28**.

11. 3-8 hp—Loosen the bolt holding the shift connector halves about 3 turns; do not remove it.

12. 9.9-50 hp—Hold connector on upper shift shaft with one wrench and loosen the lower shift shaft jam nut with a second wrench. See **Figure 29**. Record the number of turns required to unscrew the connector and disconnect the lower shift shaft from the upper shift shaft.

13. 70-225 hp—Unclip the pivot tube from the swivel bracket.

14. Remove the bolts and washers holding the gearcase to the drive shaft housing. See **Figure 30** (typical).

15A. 3-8 hp—Partially separate gearcase from drive shaft housing. Disconnect shift rods at connector.

15B. All others—Pull downward to separate the gearcase from the drive shaft housing. Remove the gearcase.

16. 2 hp—Remove the drive shaft sleeve and anti-ventilation plate.

17. Mount the gearcase in a suitable holding fixture.

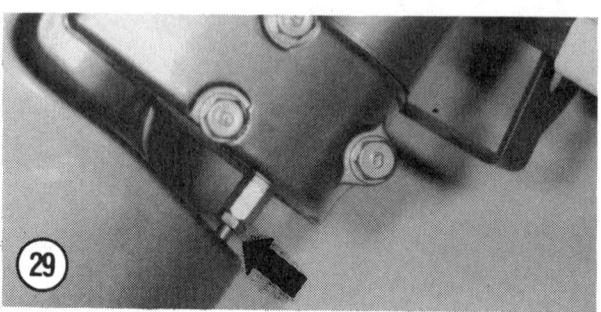

Installation (2 hp)

1. Lubricate the inside of drive shaft sleeve and water tube seals with Yamalube All-purpose Marine grease.

2. Fit drive shaft sleeve over drive shaft and position it in the water pump housing.

3. Make sure the locating dowel is positioned in the gearcase.

4. Install anti-ventilation plate with a new gasket.

5. Wipe threads of gearcase mounting bolts with Loctite Type A.

6. Install gearcase to drive shaft housing, guiding drive shaft and water tube into housing properly. Rotate the propeller shaft as required to align crankshaft and drive shaft splines.

7. Install and tighten mounting bolts to specifications (**Table 1**).

8. Install shear pin, if removed. Install the propeller as described in this chapter.

9. Reconnect the spark plug lead.

10. Fill gearcase with the correct amount and type of lubricant. See Chapter Four.

Installation (3 hp)

> *CAUTION*
> *Do not grease the top of the drive shaft in Step 1. This may excessively preload the drive shaft and crankshaft when the mounting bolts are tightened and cause a premature failure of the power head or gearcase.*

1. Lightly lubricate the drive shaft and splines with Yamalube All-purpose Marine grease. Wipe any excessive grease off the top of the drive shaft.

2. Fit the drive shaft into the drive shaft housing, aligning the lower shift rod with the shift rod connector and the water pump with the water tube. Keep housing and gearcase mating surfaces parallel.

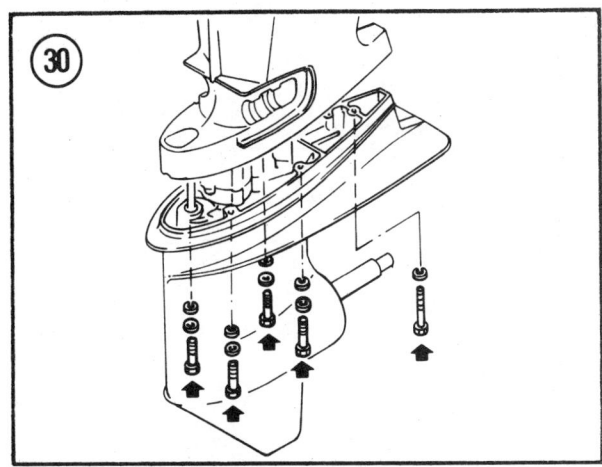

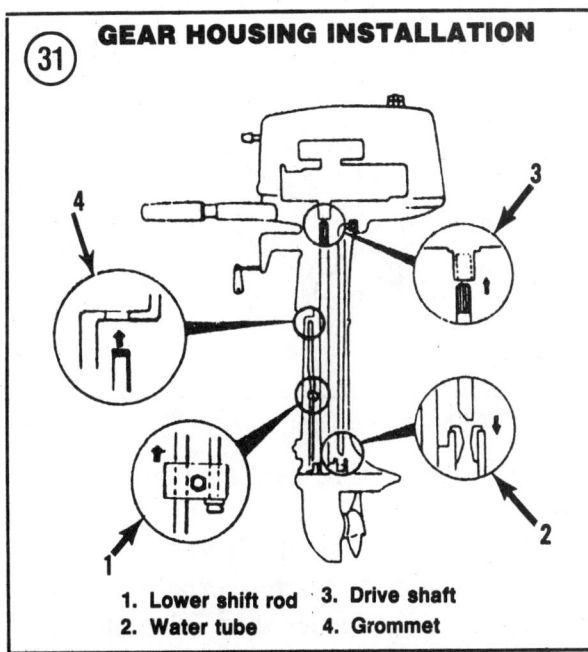

GEAR HOUSING INSTALLATION

1. Lower shift rod   3. Drive shaft
2. Water tube   4. Grommet

3. Rotate the flywheel as required to align the crankshaft and drive shaft splines as gearcase is pushed into place.

4. Wipe the gearcase mounting bolt threads with Loctite Type A. Install bolts with washers and tighten to specification (**Table 1**).

5. If gearcase was installed properly, the lower shift rod should be properly positioned in the upper shift rod connector. Tighten the connector bolt. See **Figure 28** (typical).

6. Check shift operation in FORWARD and NEUTRAL. If the unit does not shift

properly, the problem may either be in the connector (**Figure 28**) or in the gearcase.

7. If the unit shifts properly, install the access cover into intermediate housing. If it does not shift properly, remove the gearcase and locate the problem.

8. Install the propeller as described in this chapter.

9. Reconnect the spark plug lead.

10. Fill gearcase with the correct amount and type of lubricant. See Chapter Four.

Installation (4-8 hp)

1. Make sure the engine shift handle is in REVERSE.

2. Push the gearcase shift rod downward to shift unit into REVERSE.

> *CAUTION*
> *Do not grease the top of the drive shaft in Step 3. This may excessively preload the drive shaft and crankshaft when the mounting bolts are tightened and cause a premature failure of the power head or gearcase.*

3. Lightly lubricate the drive shaft and splines with Yamalube All-purpose Marine grease. Wipe any excessive grease off the top of the shaft.

> *NOTE*
> *The lower shift rod must pass through a connector and grommet located in the reverse lock plate in the drive shaft housing when installed in Step 4. See* ***Figure 31***. *If grommet came off during removal, reinstall after completing Step 4.*

4. Fit the drive shaft into the drive shaft housing, aligning the lower shift rod with the shift rod connector and the water pump with the water tube. Keep housing and gearcase mating surfaces parallel.

5. Rotate the flywheel as required to align the crankshaft and drive shaft splines as gearcase is pushed into place.

9

6. Wipe the gearcase mounting bolt threads with Loctite Type A. Install bolts with washers and tighten to specifications (**Table 1**).

7. If gearcase was installed properly, the lower shift rod should be properly positioned in the upper shift rod connector. Tighten the connector bolt. See **Figure 28** (typical).

8. Check shift operation in FORWARD, NEUTRAL and REVERSE. If the unit does not shift properly, the problem may either be in the connector (**Figure 28**) or in the gearcase.

9. If the unit shifts properly, install the rubber access cover over the connector on 4-5 hp models. If it does not shift properly, remove the gearcase and locate the problem.

10. Install the propeller as described in this chapter.

11. Reconnect the spark plug lead(s).

12. Fill gearcase with the correct amount and type of lubricant. See Chapter Four.

Installation (9.9-50 hp)

> *CAUTION*
> *Do not grease the top of the drive shaft in Step 1. This may excessively preload the drive shaft and crankshaft when the mounting bolts are tightened and cause a premature failure of the power head or gearcase.*

1. Lightly lubricate the drive shaft and splines with Yamalube All-purpose Marine grease. Wipe any excessive grease off the top of the shaft.

2. Fit drive shaft into drive shaft housing. Work slowly and carefully, making sure that the drive shaft, water tube and shift shaft align properly with their respective connections.

3. Wipe mounting bolt threads with Loctite Type A. Install bolts with washers and tighten to specifications (**Table 1**).

4. Shift upper and lower shift shafts into REVERSE. Engage connector on lower shift

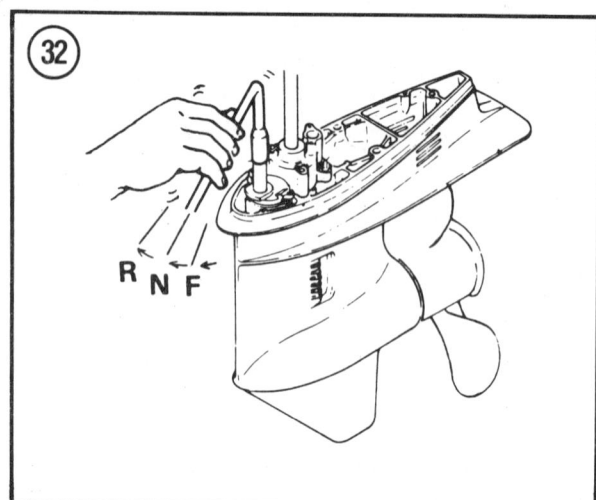

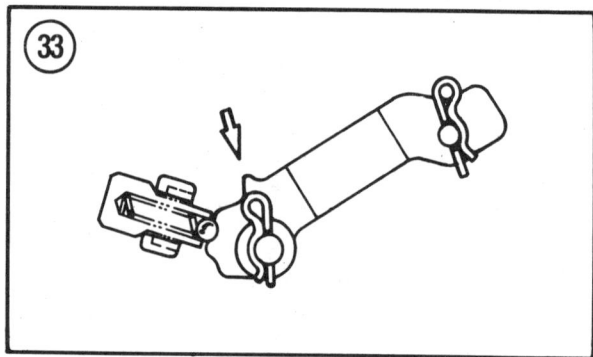

shaft threads (**Figure 29**). Tighten connector the number of turns noted during disassembly. This should give a reasonably close starting point for proper adjustment.

5. Rotate the propeller shaft while adjusting the connector. Adjustment is correct when you cannot feel or hear contact between the sliding clutch and forward or reverse gear with the shift handle in NEUTRAL.

6. Once adjustment is satisfactory, tighten the connector jam nut securely.

7. Install the propeller as described in this chapter.

8. Install the trim tab, if so equipped. Align marks scribed during disassembly and tighten bolt securely.

9. Reconnect the spark plug leads.

10. Fill gearcase with the correct amount and type of lubricant. See Chapter Four.

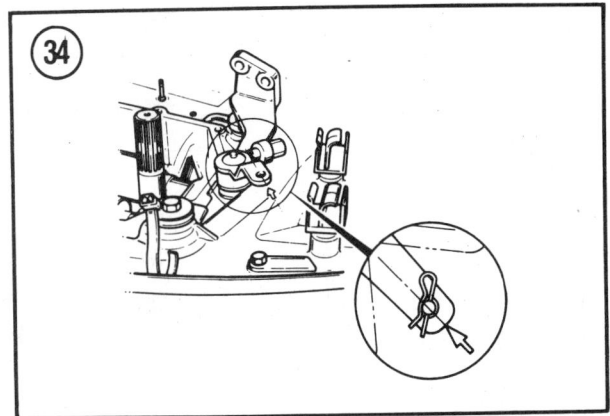

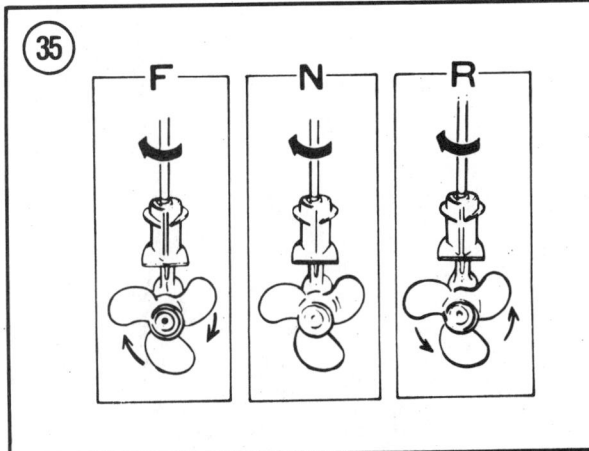

Installation (70-225 hp)

CAUTION
Do not grease the top of the drive shaft in Step 3. This may excessively preload the drive shaft and crankshaft when the mounting bolts are tightened and cause a premature failure of the power head or gearcase.

1. Lightly lubricate the drive shaft and splines with Yamalube All-purpose Marine grease. Wipe any excessive grease off the top of the drive shaft.

2. Install shift shaft handle part No. YB-6052 on shift shaft and check gearcase shifting action (**Figure 32**). Leave unit in NEUTRAL and align the shift lever mark with the arrow on the bottom cowling. See **Figure 33** (70-90 hp) or **Figure 34** (115-225 hp).

CAUTION
Work carefully and slowly in Step 3 to avoid damaging the crankshaft lower oil seal with the drive shaft.

3. Fit the drive shaft into the drive shaft housing. Make sure that the drive shaft, water tube and shift shaft align properly with their respective connections. The pivot tube should enter the hole in the drive shaft housing.

4. Join the gearcase to the drive shaft housing, rotating the flywheel clockwise to align the crankshaft and drive shaft splines while exerting upward pressure on the gearcase.

NOTE
If gearcase and drive shaft housing do not mate in Step 5, the 2 shift shafts are not aligned. Separate the units and repeat Steps 2-5.

5. Push upper shift shaft onto lower shift shaft with a punch or drift and seat the gearcase against the drive shaft housing.

NOTE
If the gearcase does not shift as described in Step 6, remove and repeat Steps 2-5.

6. Hold gearcase in place and check shift operation as follows. See **Figure 35**.
   a. FORWARD—Propeller shaft will rotate only clockwise.
   b. NEUTRAL—Propeller shaft turns freely in either direction.
   c. REVERSE—Propeller shaft will rotate only counterclockwise.

7. Wipe the gearcase mounting bolt threads with Loctite Type A. Install bolts with washers and snug them down evenly, then tighten to specifications (**Table 1**).

8. Install trim tab. Align marks scribed during disassembly and tighten bolt securely. Press plastic cap (if so equipped) into trim tab adjustment bolt hole.

9

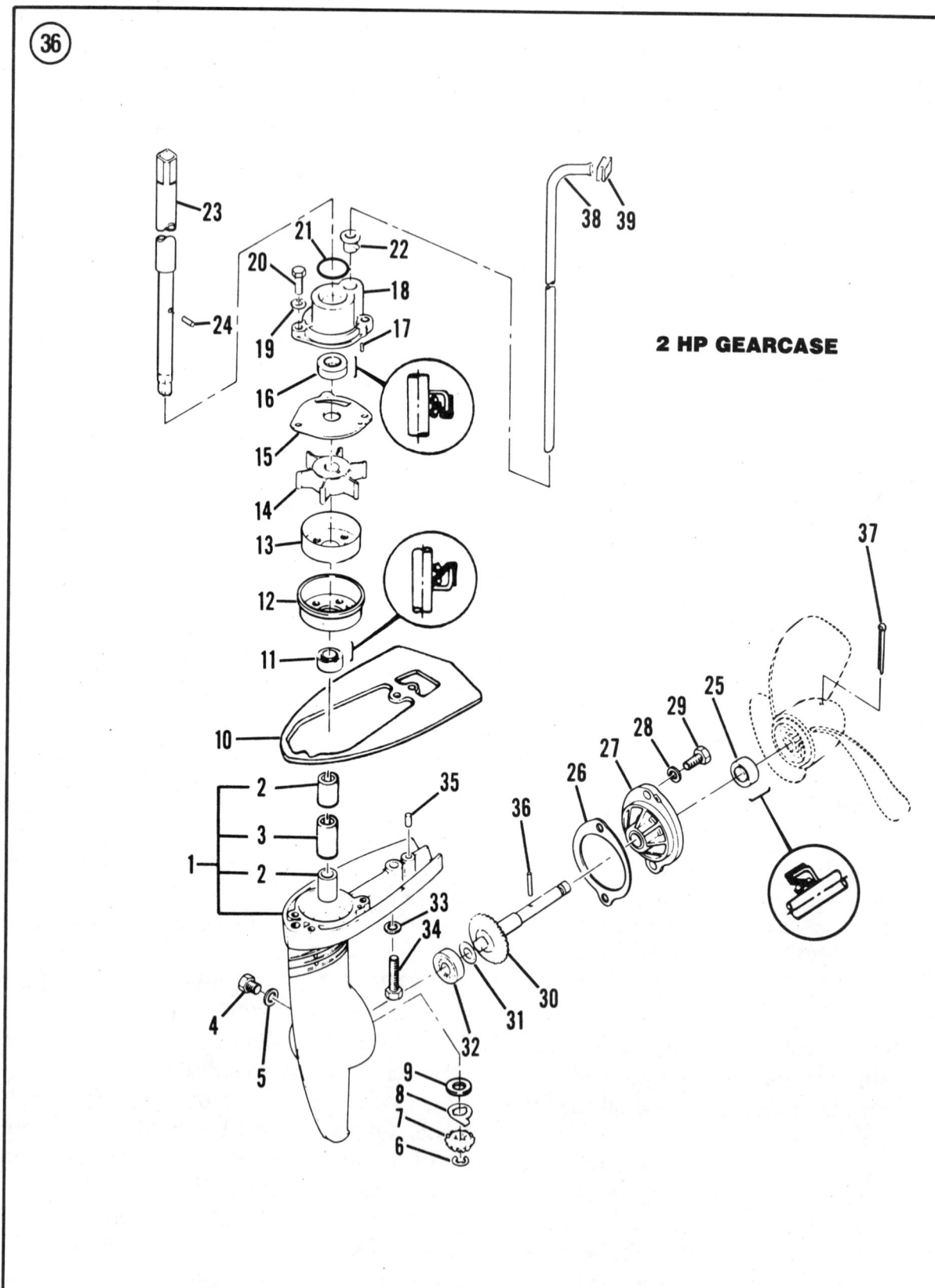

2 HP GEARCASE

1. Gear housing
2. Drive shaft bushing
3. Drive shaft bushing
4. Vent and drain screws
5. Gasket
6. Snap ring
7. Pinion gear
8. Thrust washer
9. Pinion gear depth shim
10. Anti-cavitation plate
11. Drive shaft oil seal
12. Oil seal protector
13. Water pump cartridge
14. Water pump impeller
15. Outer plate
16. Drive shaft oil seal
   (Two separate seals on 1989 models)
17. Dowel pin
18. Water pump housing
19. Washer
20. Water pump bolt
21. O-ring
22. Water tube seal
23. Drive shaft
24. Impeller drive pin
25. Propeller shaft oil seal
26. Bearing carrier gasket
27. Bearing carrier
28. Washer
29. Bearing carrier bolt
30. Forward gear and
   propeller shaft assembly
31. Forward gear shim
32. Forward gear bearing
33. Washer
34. Bolt
35. Dowel pin
36. Shear pin
37. Cotter pin
38. Water tube
39. Water tube seal

9. Install the propeller as described in this chapter.
10. Clip the pivot tube to the swivel bracket.
11. Reconnect the spark plug leads.
12. Fill gearcase with the correct amount and type of lubricant. See Chapter Four.

BEARING CARRIER AND PROPELLER SHAFT

Figures 36-46 are exploded views of the gearcases covered in this chapter. Refer to the appropriate drawing for your model for all service procedures.

Bearing Carrier and Propeller Shaft Removal (2-15 hp and 1988-on 25 hp)

1. Remove the propeller as described in this chapter.
2. Secure the gearcase in a suitable holding device or a vise with protective jaws. If protective jaws are not available for the vise, position the unit upright with the skeg between wooden blocks in the vise.
3. Remove the 2 gear housing cap bolts and washers (**Figure 47**).

NOTE
If assembly is corroded in prop shaft bore and cannot be removed easily by hand in Step 4, tap side of cap with a rubber mallet to rotate cap ears and then pry it off.

4. Pry the gear housing cap free. Remove cap and gasket (2-5 hp). Remove cap and bearing carrier (all others). See **Figure 48**.
5. Remove the propeller shaft from the prop shaft bore. See **Figure 49** (typical).
6. Remove and retain reverse gear thrust washer (4-25 hp).

9

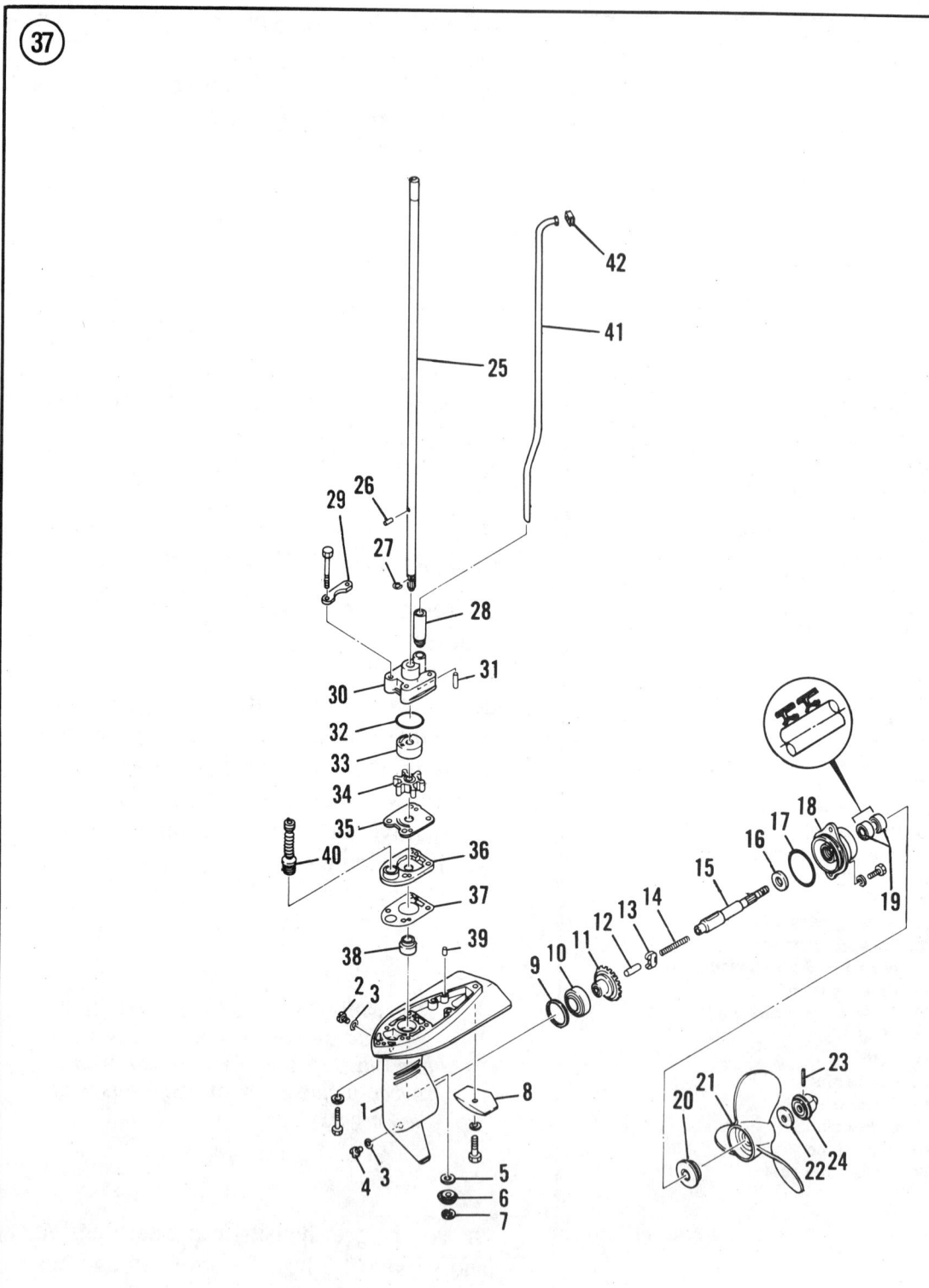

3 HP GEARCASE

1. Gear housing
2. Vent plug
3. Gasket
4. Fill plug
5. Shim
6. Pinion gear
7. Snap ring
8. Anode
9. Shim
10. Bearing
11. Forward gear
12. Cam follower
13. Sliding clutch
14. Spring
15. Propeller shaft
16. Thrust washer
17. O-ring
18. Bearing carrier
19. Oil seals
20. Spacer
21. Propeller
22. Washer
23. Shear pin
24. Propeller nut
25. Drive shaft
26. Pin
27. Clip
28. Lower water tube seal
29. Plate
30. Water pump housing
31. Dowel pin
32. O-ring
33. Insert
34. Impeller
35. Plate
36. Water pump base
37. Gasket
38. Oil seals
39. Dowel pin
40. Boot
41. Water tube
42. Upper water tube seal

9

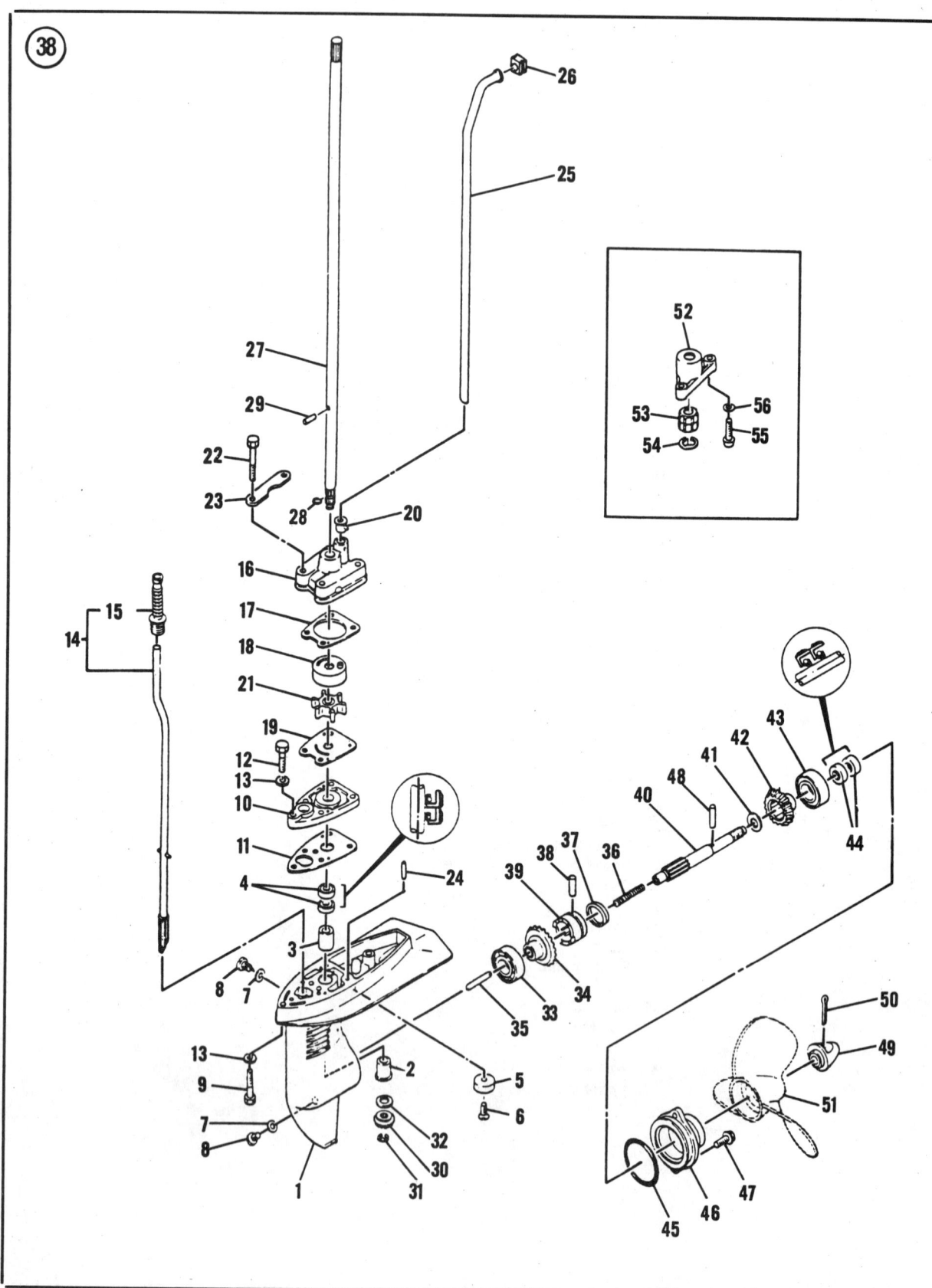

4 AND 5 HP GEARCASE

1. Gear housing
2. Lower drive shaft bushing
3. Upper drive shaft bushing
4. Upper drive shaft oil seal
5. Anode
6. Anode attaching screw
7. Drain/vent screw gaskets
8. Drain/vent/flush screws
9. Bolt
10. Water pump base
11. Water pump base gasket
12. Pump base attaching bolt
13. Washer
14. Lower shift rod
15. Boot
16. Water pump housing
17. Pump housing gasket
18. Insert
19. Outer plate
20. Lower water tube seal
21. Impeller
22. Bolt
23. Support plate
24. Dowel pin
25. Water tube
26. Upper water tube seal
27. Drive shaft
28. Clip
29. Dowel pin
30. Pinion gear
31. Pinion gear snap ring
32. Pinion gear thrust washer
33. Forward gear ball bearing
34. Forward gear
35. Cam follower
36. Compression spring
37. Cross pin retainer ring
38. Cross pin
39. Sliding clutch
40. Propeller shaft
41. Reverse gear thrust washer
42. Reverse gear
43. Reverse gear ball bearing
44. Propeller shaft oil seal
45. Gear housing cap O-ring
46. Gear housing cap
    (bearing carrier)
47. Bolt
48. Shear pin
49. Propeller nut
50. Cotter pin
51. Propeller
52. Bushing housing*
53. Bushing*
54. Snap ring*
55. Bushing housing bolt*
56. Washer*
* Long drive shaft model only

9

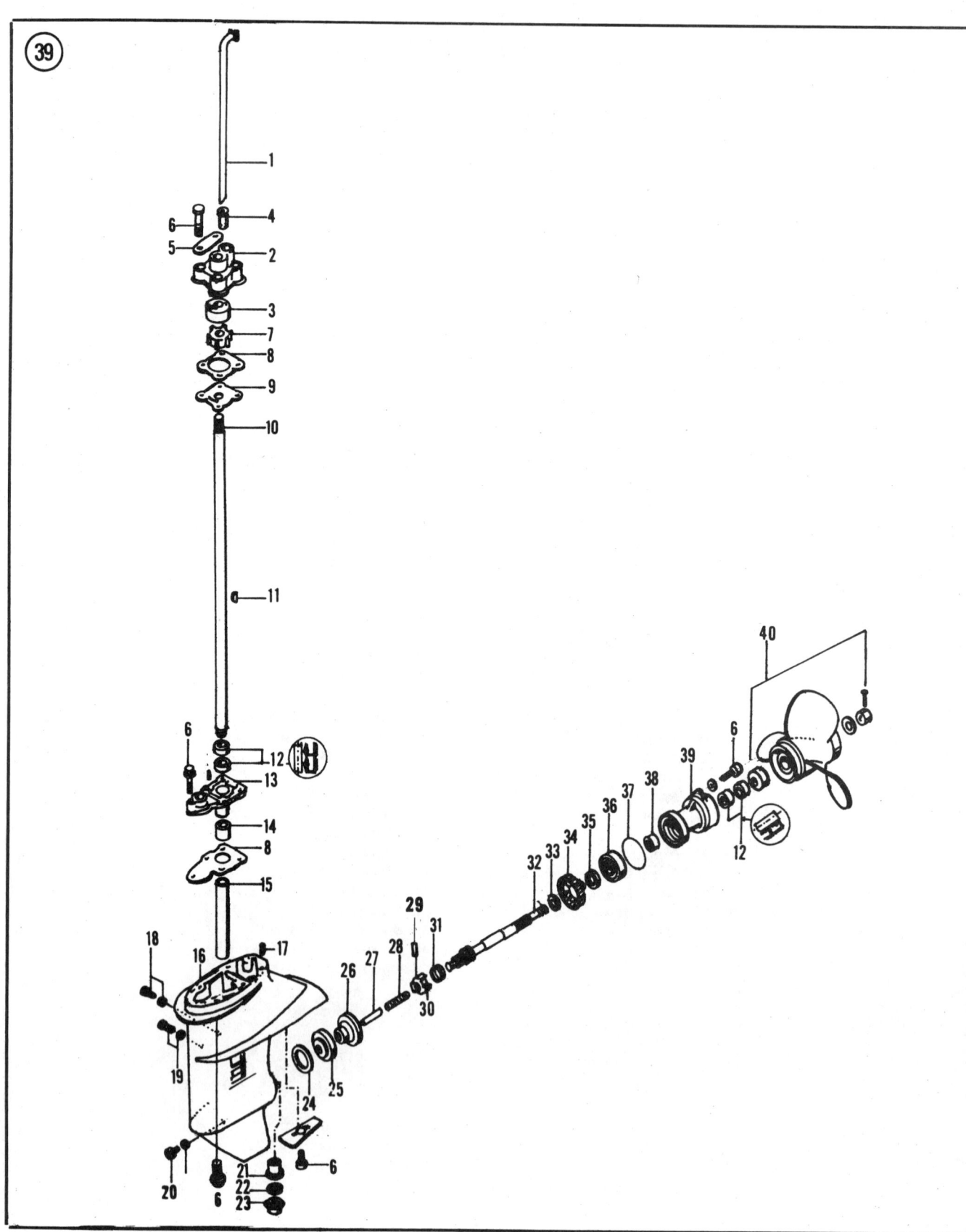

6 AND 8 HP GEARCASE

1. Water tube
2. Water pump cover
3. Cover insert
4. Water tube seal
5. Support plate
6. Bolt
7. Impeller
8. Gasket
9. Face plate
10. Drive shaft
11. Impeller drive key
12. Oil seals
13. Oil seal housing (water pump base)
14. Upper bushing
15. Drive shaft sleeve
16. Gearcase
17. Dowel pin
18. Vent plug/washer
19. Oil level plug/washer
20. Drain/fill plug/washer
21. Lower bushing
22. Thrust washer
23. Pinion gear
24. Shim
25. Forward gear bearing
26. Forward gear
27. Shift plunger
28. Spring
29. Cross pin
30. Clutch dog
31. Retaining spring
32. Propeller shaft
33. Thrust bearing
34. Reverse gear
35. Shim
36. Reverse gear bearing
37. O-ring
38. Needle bearing
39. Bearing housing
40. Propeller assembly

9

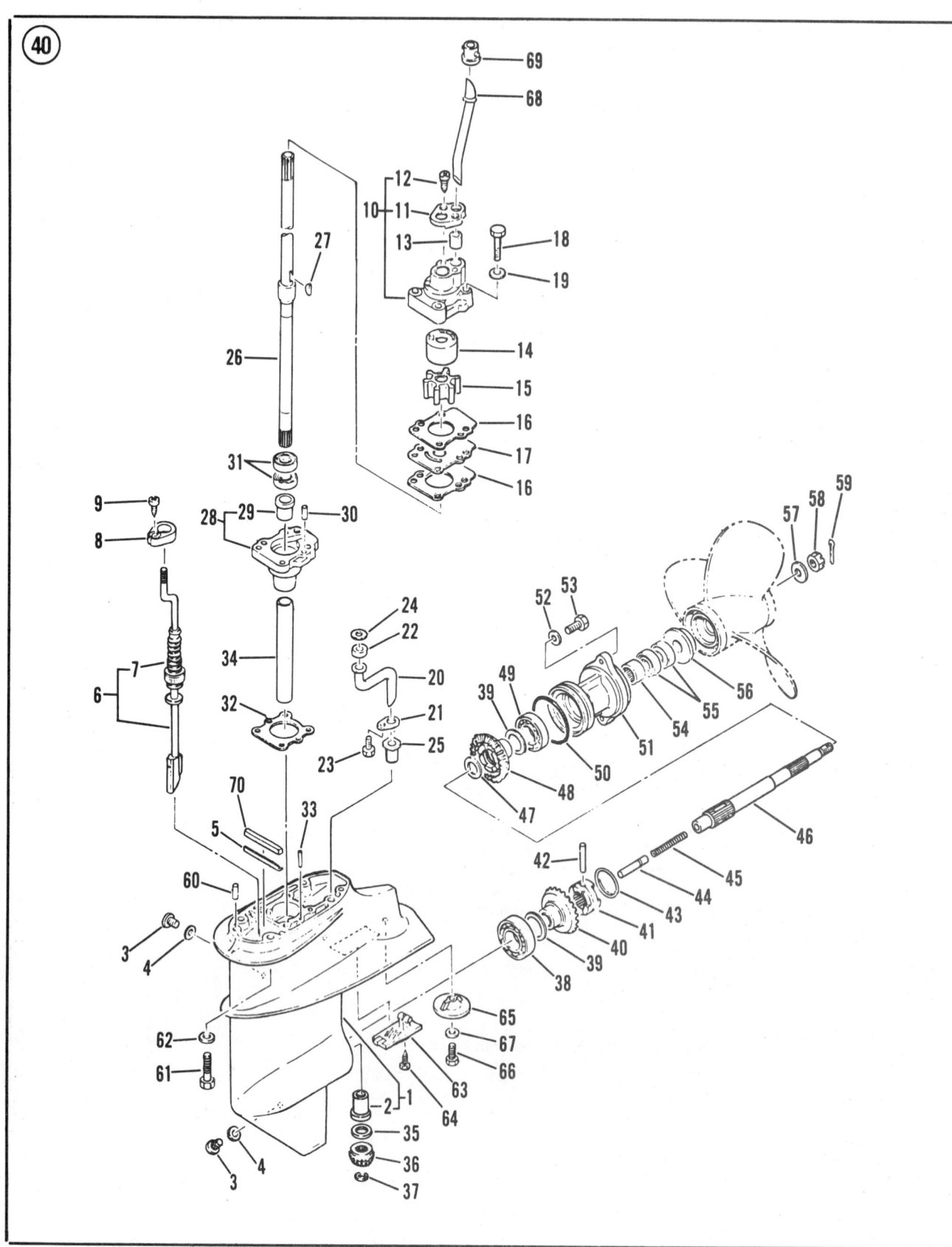

9.9 AND 15 HP GEARCASE

1. Gear housing
2. Lower drive shaft bushing
3. Drain, vent and flush screws
4. Screw gaskets
5. Seal guide
6. Shift cam assembly
7. Shift rod boot
8. Shift cam assembly bracket
9. Bracket separation screw
10. Water pump housing
11. Water pump cover
12. Screw
13. Seal
14. Water pump insert
15. Impeller
16. Gasket
17. Outer plate
18. Bolt
19. Washer
20. Water pickup tube
21. Pickup tube retainer plate
22. Seal
23. Screw
24. Washer
25. Seal
26. Drive shaft
27. Drive shaft key
28. Bearing housing
    (water pump base)
29. Upper drive shaft bushing
30. Dowel pin
31. Oil seal
32. Gasket
33. Dowel pin
34. Drive shaft tube
35. Pinion gear thrust washer
    (adjusting shim)
36. Pinion gear
37. Snap ring
38. Forward gear
    ball bearing
39. Shims
40. Forward gear
41. Sliding clutch
42. Cross pin
43. Cross pin
    retainer ring
44. Cam follower
45. Spring
46. Propeller shaft
47. Reverse gear
    thrust washer
48. Reverse gear
49. Reverse gear
    ball bearing
50. O-ring
51. Bearing carrier
52. Bolt
53. Washer
54. Propeller shaft
    needle bearing
55. Oil seal
56. Thrust hub
57. Washer
58. Propeller nut
59. Cotter pin
60. Dowel pin
61. Bolt
62. Washer
63. Inlet cover
64. Cover screw
65. Anode
66. Anode bolt
67. Washer
68. Water tube
69. Seal
70. Seal

9

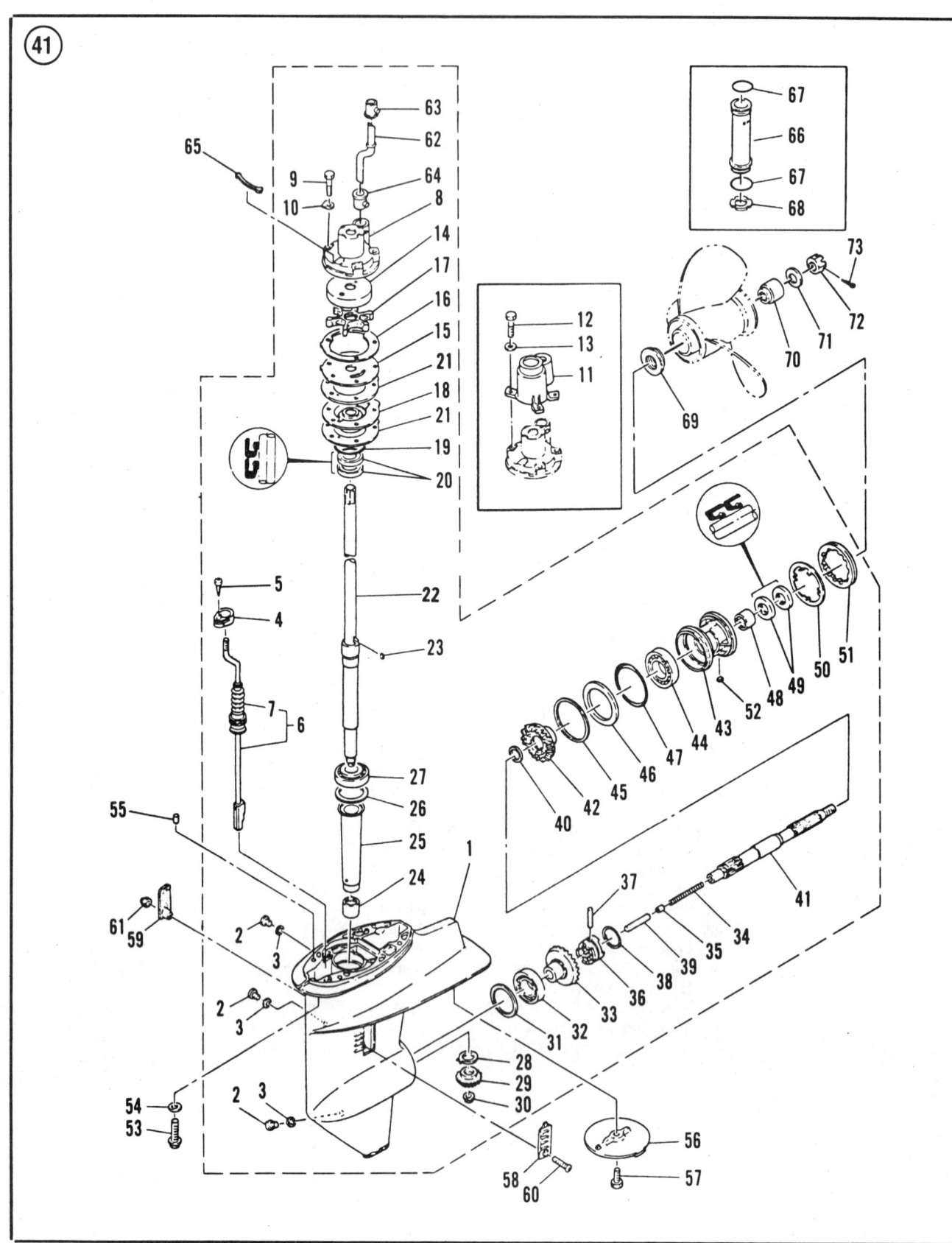

25 (1984-1987) AND
30 (1984-1986) HP GEARCASE

1. Gear housing
2. Drain, vent and
   fill screws
3. Screw gaskets
4. Lower shift shaft
   attaching bracket
5. Screw
6. Lower shift shaft assembly
7. Boot
8. Water pump housing
9. Bolt
10. Washer*
11. Housing extension
12. Housing and
    extension bolt**
13. Washer**
14. Cartridge insert
15. Outer plate
16. Gasket
17. Impeller
18. Water pump base
19. O-ring
20. Drive shaft oil seal
21. Base gasket
22. Drive shaft
23. Drive shaft key
24. Lower drive shaft
    needle bearing
25. Drive shaft sleeve
26. Pinion gear shim
27. Upper drive shaft
    tapered roller bearing
28. Pinion gear
    thrust washer
29. Pinion gear
30. Pinion nut
31. Forward gear shim
32. Forward gear
    tapered roller bearing
33. Forward gear
34. Compression spring
35. Shift slide
36. Sliding clutch

37. Cross pin
38. Cross pin retainer ring
39. Shift plunger
40. Propeller shaft
    thrust washer
41. Propeller shaft
42. Reverse gear
43. Bearing carrier
44. Reverse gear
    ball bearing
45. Reverse gear shim
46. Reverse gear
    thrust washer
47. O-ring
48. Propeller shaft
    needle bearing
49. Propeller shaft
    oil seal
50. Tab washer
51. Cover nut
52. Key
53. Bolt
54. Washer
55. Dowel pin
56. Trim tab
57. Trim tab bolt
58. Port water inlet cover
59. Starboard water
    inlet cover
60. Cover screw
61. Nut
62. Water tube
63. Upper water tube seal
64. Lower water tube seal
65. Spacer
66. Water tube extension***
67. O-ring
68. Plate
69. Forward thrust hub
70. Reverse thrust hub
71. Washer
72. Propeller nut
73. Cotter pin

*Short shaft only.
**Long shaft only.
***Long and Super Ultra Long shaft only.

9

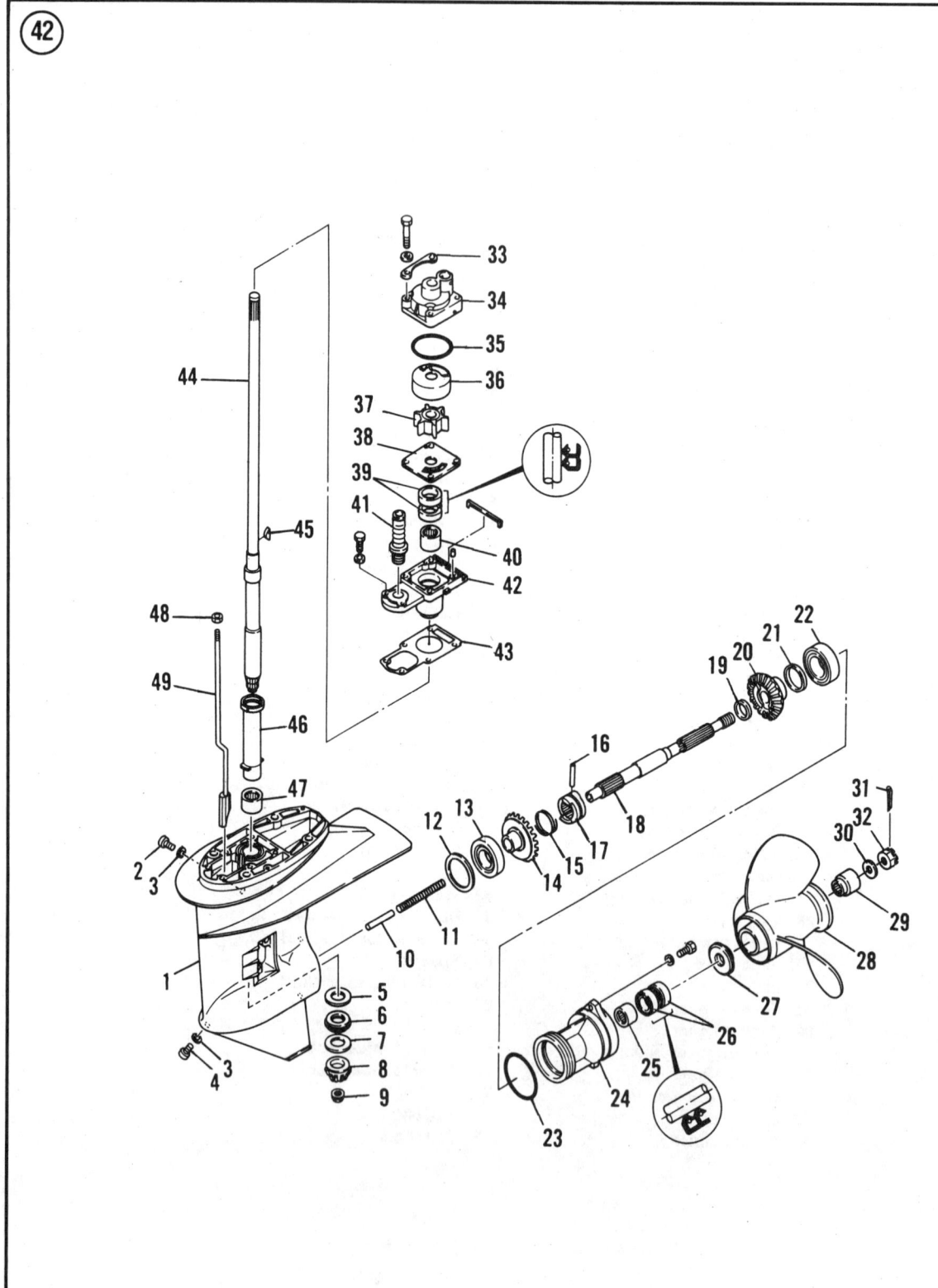

25 HP (1988-ON) GEARCASE

1. Gear housing
2. Vent plug
3. Gasket
4. Fill plug
5. Thrust washer
6. Bearing
7. Shim
8. Pinion gear
9. Nut
10. Cam follower
11. Spring
12. Shim
13. Bearing
14. Forward gear
15. Cross pin retainer ring
16. Cross pin
17. Sliding clutch
18. Propeller shaft
19. Plate washer
20. Reverse gear
21. Shim
22. Bearing
23. O-ring
24. Bearing carrier
25. Bearing
26. Oil seals
27. Spacer
28. Propeller
29. Spacer
30. Washer
31. Cotter pin
32. Nut
33. Plate
34. Water pump housing
35. O-ring
36. Insert
37. Impeller
38. Plate
39. Oil seals
40. Bearing
41. Boot
42. Oil seal housing
43. Gasket
44. Drive shaft
45. Woodruff key
46. Sleeve
47. Bearing
48. Nut
49. Shift rod assembly

9

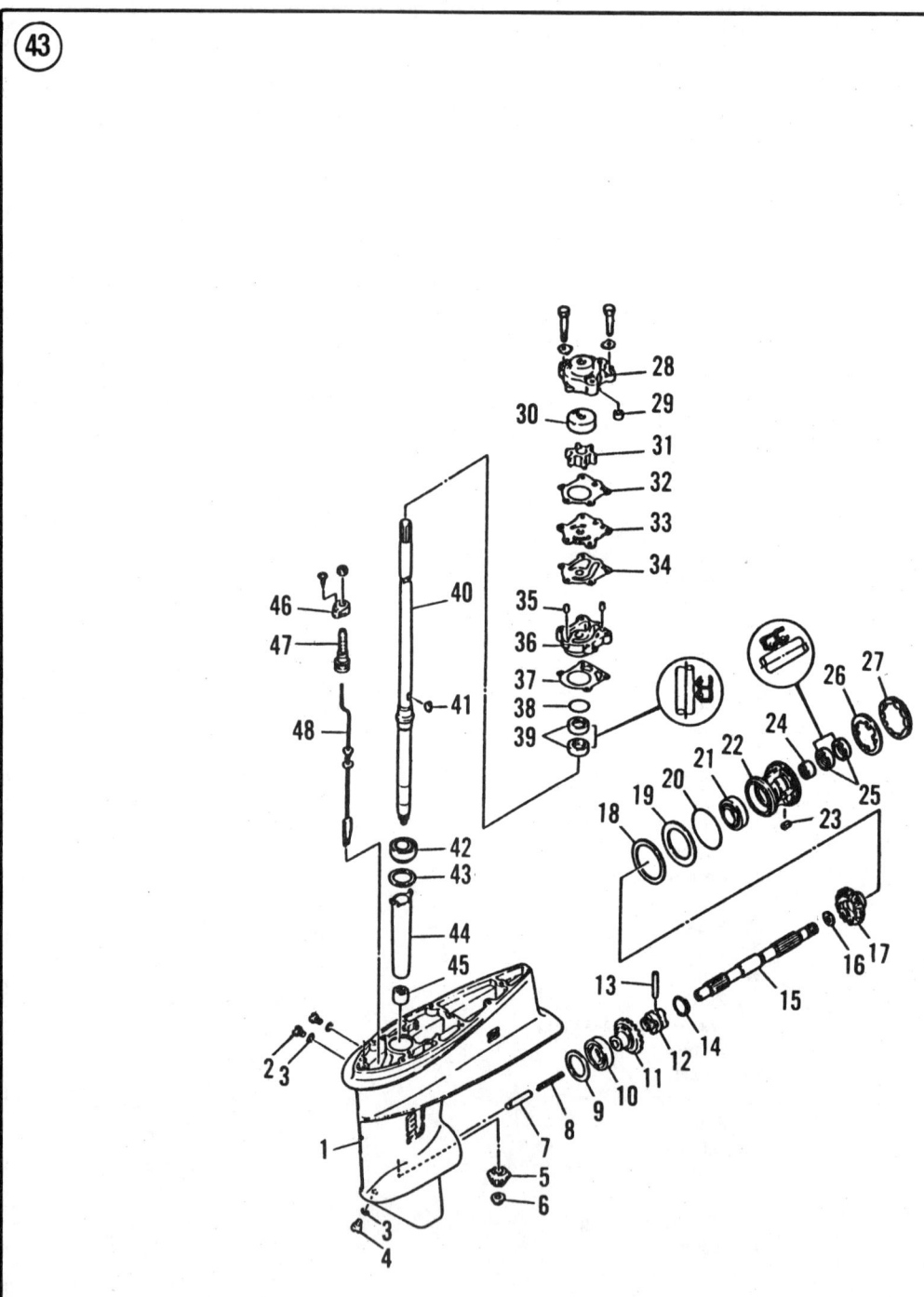

30 HP (1987-ON) GEARCASE

1. Gear housing
2. Vent plug
3. Gasket
4. Fill plug
5. Pinion gear
6. Nut
7. Cam follower
8. Spring
9. Shim
10. Bearing
11. Forward gear
12. Sliding clutch
13. Cross pin
14. Cross pin retainer ring
15. Propeller shaft
16. Plate washer
17. Reverse gear
18. Shim
19. Thrust washer
20. O-ring
21. Bearing
22. Bearing carrier
23. Key
24. Bearing
25. Seals
26. Washer
27. Nut
28. Water pump housing
29. Damper
30. Insert
31. Impeller
32. Gasket
33. Plate
34. Gasket
35. Dowel pin
36. Water pump base
37. Gasket
38. O-ring
39. Oil seals
40. Drive shaft
41. Woodruff key
42. Bearing
43. Shim
44. Sleeve
45. Bearing
46. Bracket
47. Boot
48. Shift rod assembly

9

1. Bolt
2. Washer
3. Water pump housing
4. Cartridge insert
5. Impeller
6. Water pump gasket
7. Outer plate
8. Outer plate gasket
9. Water pump housing
10. Water pump gasket
11. O-ring
12. Oil seal
13. Bolt
14. Washer
15. Water seal
16. Drive shaft
17. Tapered roller bearing
18. Shim
19. Drive shaft sleeve
20. Needle bearing
21. Bearing (not used on 1988-on)
22. Pinion gear
23. Nut
24. Woodruff key
25. Reverse rod guide
26. Reverse rod cam
27. Oil seal
28. Plate
29. O-ring
30. Shift rod
31. Shift cam
32. Plug
33. Dowel pin
34. Gasket
35. Nut
36. Water inlet cover
37. Washer
38. Bolt
39. Circlip
40. Plug
41. Gasket
42. Bolt
43. Lower casing
44. Spring washer
45. Bolt
46. Trim tab
47. Seal rubber
48. Seal rubber guide
49. Screw
50. Water inlet cover
51. Shim
52. Taper bearing
53. Forward gear
54. Cross pin
55. Sliding clutch
56. Cross pin
    retainer ring
57. Shift plunger
58. Shift slide
59. Compression spring
60. Propeller shaft
61. Plate washer

62. Reverse gear
63. Shim
64. Thrust washer
65. Cotter pin
66. Castle nut
67. Spacer
68. Thrust plate cap
69. Spacer
70. Cover nut
71. Tab washer
72. Oil seal
73. Needle bearing
74. Bearing carrier key
75. Bearing carrier
76. O-ring
77. Ball bearing
78. Propeller

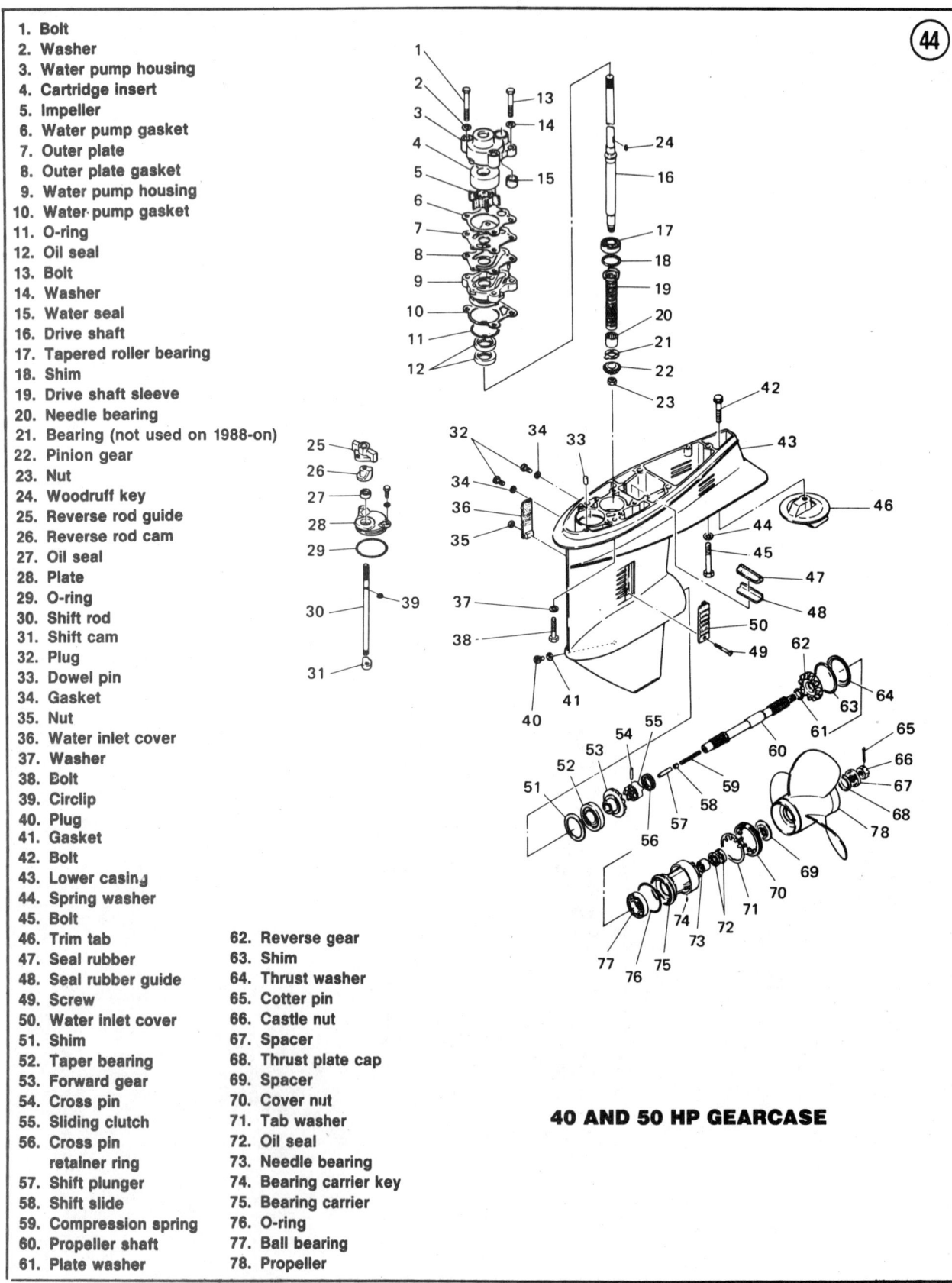

40 AND 50 HP GEARCASE

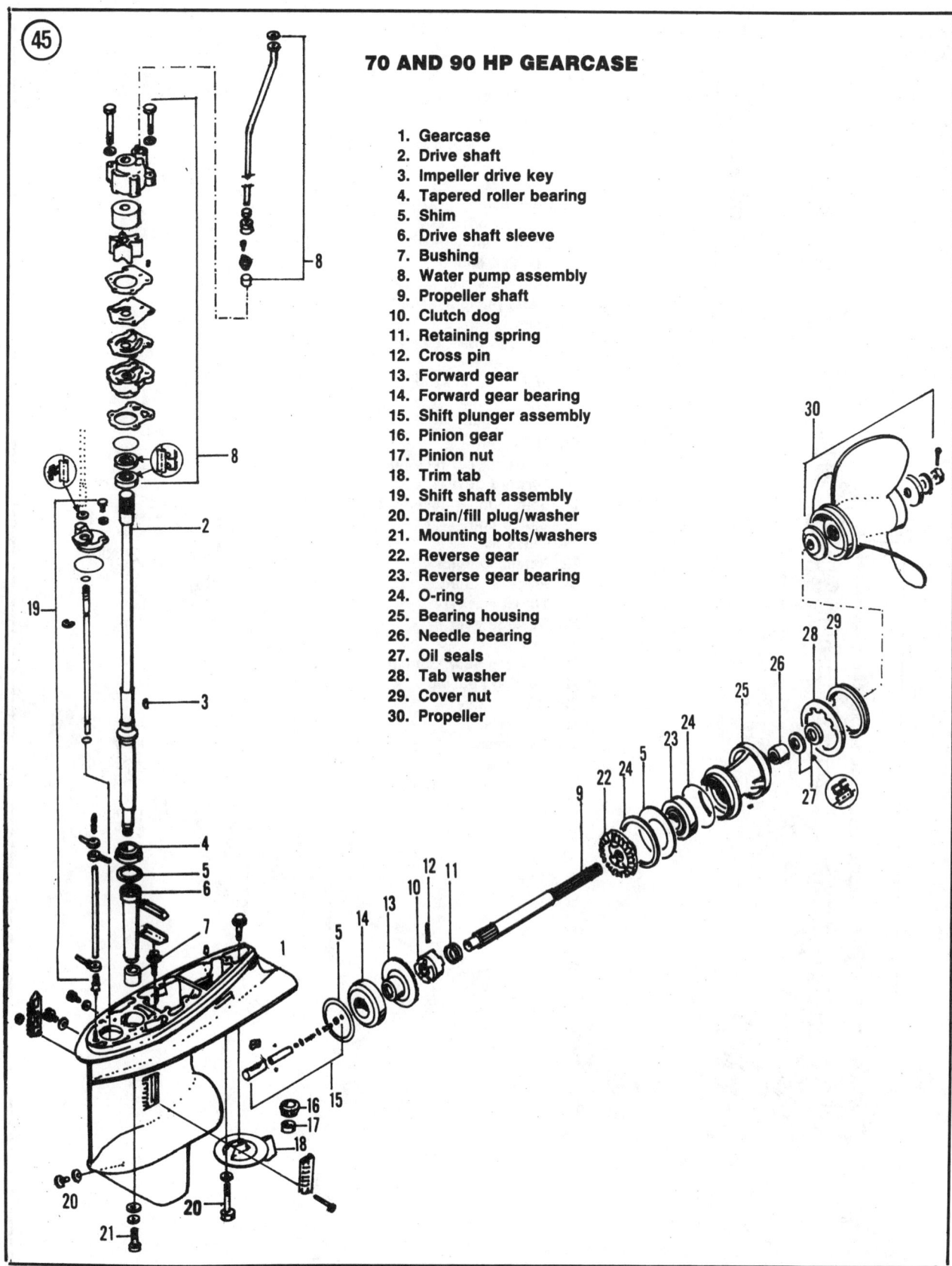

70 AND 90 HP GEARCASE

1. Gearcase
2. Drive shaft
3. Impeller drive key
4. Tapered roller bearing
5. Shim
6. Drive shaft sleeve
7. Bushing
8. Water pump assembly
9. Propeller shaft
10. Clutch dog
11. Retaining spring
12. Cross pin
13. Forward gear
14. Forward gear bearing
15. Shift plunger assembly
16. Pinion gear
17. Pinion nut
18. Trim tab
19. Shift shaft assembly
20. Drain/fill plug/washer
21. Mounting bolts/washers
22. Reverse gear
23. Reverse gear bearing
24. O-ring
25. Bearing housing
26. Needle bearing
27. Oil seals
28. Tab washer
29. Cover nut
30. Propeller

9

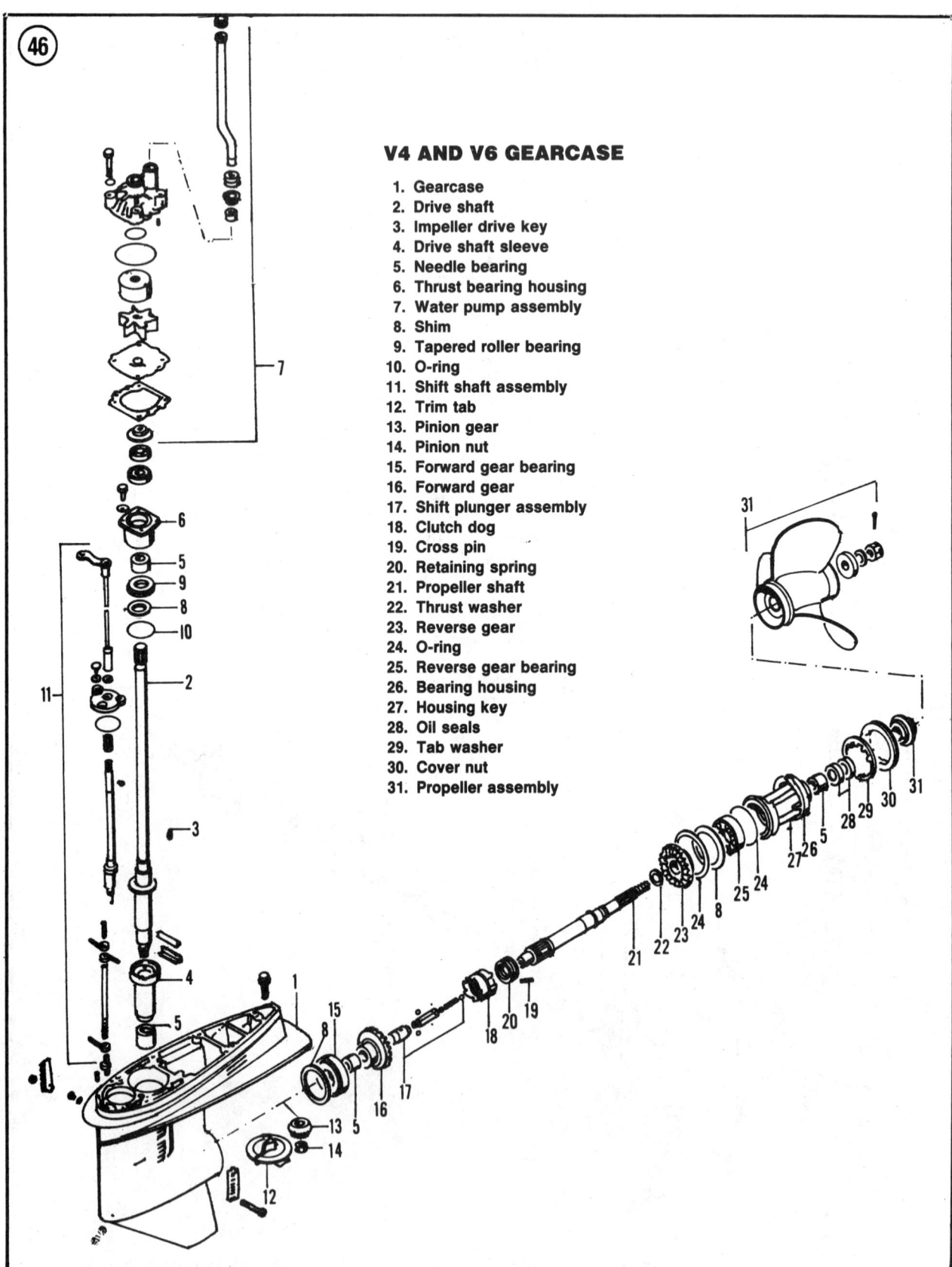

46

V4 AND V6 GEARCASE

1. Gearcase
2. Drive shaft
3. Impeller drive key
4. Drive shaft sleeve
5. Needle bearing
6. Thrust bearing housing
7. Water pump assembly
8. Shim
9. Tapered roller bearing
10. O-ring
11. Shift shaft assembly
12. Trim tab
13. Pinion gear
14. Pinion nut
15. Forward gear bearing
16. Forward gear
17. Shift plunger assembly
18. Clutch dog
19. Cross pin
20. Retaining spring
21. Propeller shaft
22. Thrust washer
23. Reverse gear
24. O-ring
25. Reverse gear bearing
26. Bearing housing
27. Housing key
28. Oil seals
29. Tab washer
30. Cover nut
31. Propeller assembly

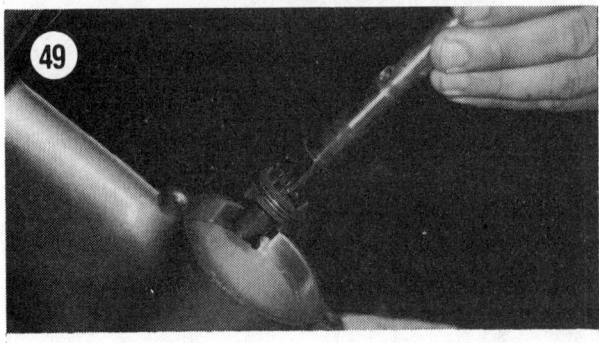

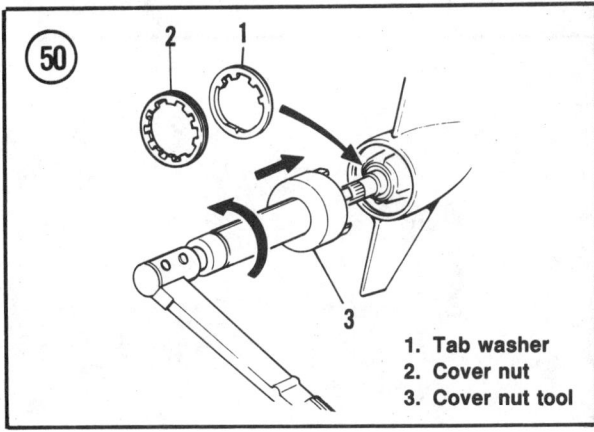

1. Tab washer
2. Cover nut
3. Cover nut tool

Bearing Carrier and Propeller Shaft Removal (25-225 hp [except 1988-on 25 hp])

1. Remove the propeller as described in this chapter.

2. Secure the gearcase in a suitable holding device or a vise with protective jaws. If protective jaws are not available for the vise, position the unit upright with the skeg between wooden blocks in the vise.

3. Straighten the bent tab(s) of the prop shaft housing tab washer with a punch and hammer.

NOTE
If the cover nut is frozen in place and cannot be moved in Step 4 even with the assistance of heat and tapping on the tool, use an electric drill to drill out one side of the nut. This will permit removal of the nut without damage to the gearcase housing.

4. Install cover nut tool part No. YB-6075 (25-30 hp), part No. YB-6048 (40-50 hp), part No. YB-34447 (all others) in cover nut. See **Figure 50**. Turn the tool counterclockwise and remove the cover nut and tab washer.

5. Install bearing housing puller jaws part No. YB-6234 (30-50 hp) or part No. YB-6207 (all others) to appropriate flywheel puller plate.

6. Slide puller jaw rings over prop shaft and insert puller jaws to grip horizontal legs in bearing carrier. Tighten the puller to break the carrier free of the prop shaft bore. See **Figure 51** (typical). Remove bearing housing puller plate and jaws.

NOTE
The small locating key installed in the bearing carrier may come out with the carrier or drop into the prop shaft bore when the carrier is removed. Be sure to locate and save this key for reassembly.

9

7A. 25-50 hp—Remove the bearing carrier and propeller shaft assembly from the prop shaft bore.

7B. All others—Remove the bearing carrier from the prop shaft bore.

NOTE
The shift rod and shift cam on the prop shaft are splined and must be removed as specified in Step 8.

8A. 70-90 hp—Unbolt the shift shaft plate. Remove the shift shaft and plate, then withdraw the prop shaft straight out of its bore without rotating it.

8B. 115-225 hp—Install shift shaft handle part No. YB-6052 on shift shaft and make sure gearcase is in NEUTRAL (**Figure 32**). Unbolt the shift shaft retainer and remove shift shaft, then withdraw the prop shaft straight out of its bore without rotating it.

Cleaning and Inspection (All Models)

1. Clean all parts in fresh solvent. Blow dry with compressed air, if available.
2. Check bearing carrier and bearing contact points on propeller shaft (**Figure 52**). Replace shaft and bearing carrier bearing if shaft shows signs of pitting, grooving, scoring, heat discoloration or embedded metallic particles.
3. Check propeller shaft surfaces where oil seal lips contact shaft (**Figure 52**). If grooved, replace the shaft and oil seals.
4. Install V-blocks under the propeller shaft bearing surface at each end of the shaft (**Figure 53**). Position a dial indicator at the center of the shaft and set the indicator gauge at zero. Slowly rotate the shaft 360° and note the indicator reading. Replace the shaft if runout exceeds 0.0008 in. (0.02 mm).
5. Check propeller shaft splines and threads for wear, rust or corrosion damage (**Figure 52**). Replace shaft as required.
6. Apply a light coat of Yamalube Gearcase Lubricant to reverse gear ball bearing and

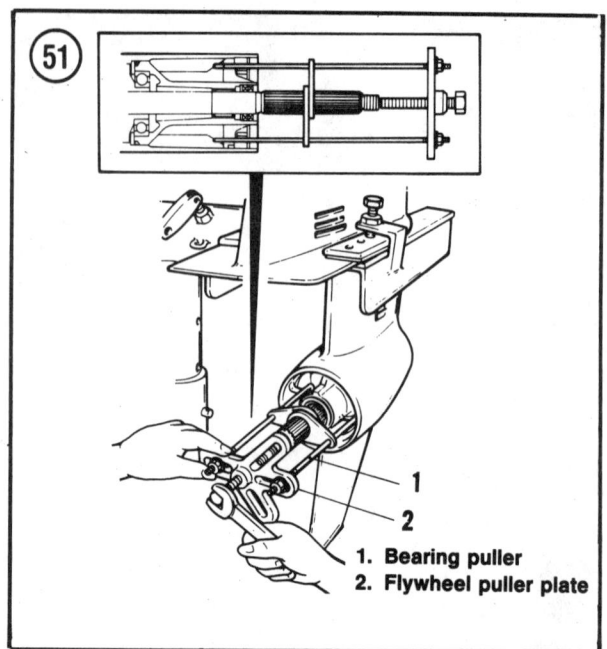

1. Bearing puller
2. Flywheel puller plate

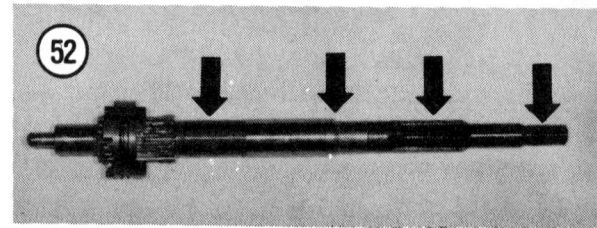

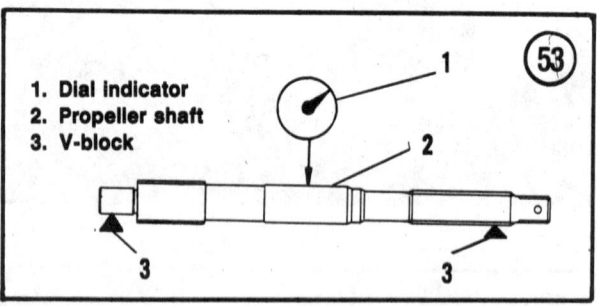

1. Dial indicator
2. Propeller shaft
3. V-block

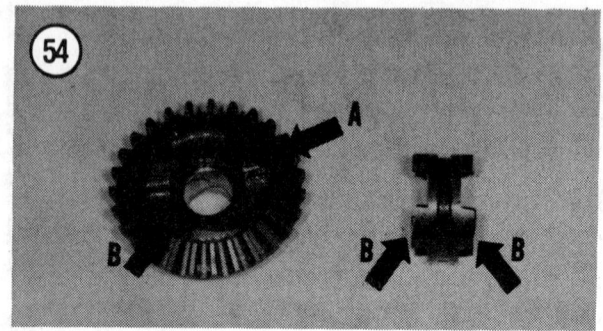

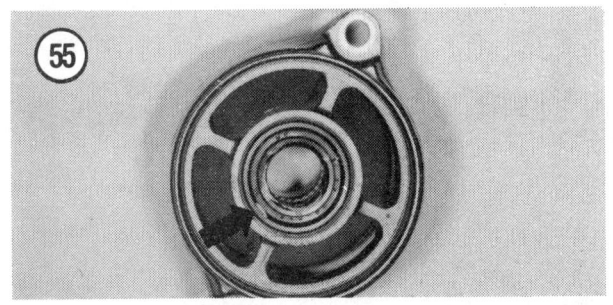

9. Inspect cover nut for cracks and broken or corroded threads. Replace as required.

10. Inspect tab washer. There must be sufficient tabs in good condition for reinstallation. If not, replace the washer.

Bearing Carrier Disassembly/ Assembly (2-5 hp)

1. Remove all gasket residue from the prop shaft housing cap.

2. Remove and discard the prop shaft oil seals from the housing cap. See **Figure 55**.

3. 4-5 hp—Remove the prop shaft bearing from the housing cap (A, **Figure 56**).

4. Remove and discard the housing cap O-ring (B, **Figure 56**).

5. Clean cap with fresh solvent and blow dry with compressed air, if available.

6. 4-5 hp—Install a new prop shaft bearing (lettered side facing up) with driver part No. YB-6014 or equivalent.

7. Press a new oil seal or seals into the cap (lip facing outward) with tool part No. YB-6025 or a suitable mandrel.

8. Lubricate seal lip with Yamalube All-purpose Marine grease.

9. Install a new O-ring on the housing cap. Lubricate O-ring with Yamalube All-purpose Marine grease.

Bearing Carrier Disassembly/ Assembly (6-15 hp and 1988-on 25 hp)

1. Remove and discard the bearing carrier O-ring (A, **Figure 57**).

2. Remove and discard the prop shaft oil seals from the housing cap. See B, **Figure 57**.

NOTE
If reverse gear cannot be removed by prying off in Step 3, secure bearing carrier assembly in a vise with protective jaws and remove the gear with a suitable puller.

rotate bearing to check for rough spots. Push and pull on reverse gear to check for side wear. If movement is excessive, replace bearing with a hydraulic press.

7. Check reverse gear teeth (A, **Figure 54**). If teeth are pitted, chipped, broken or excessively worn, replace the gear.

8. Check clutch dogs (B, **Figure 54**) on reverse gear and sliding clutch. If clutch dogs are chipped or rounded off, replace gear and sliding clutch. Such damage can be caused by improper shift adjustment, excessive rpm during idle or shifting too slowly from REVERSE to NEUTRAL.

9

3. Remove reverse gear from carrier by prying off with screwdrivers as shown in **Figure 58**. Save any shims installed behind reverse gear.

4. If reverse gear bearing remained in carrier when gear was removed in Step 2, remove bearing with a suitable puller. If bearing is attached to reverse gear, press off with a universal puller, support plate and arbor press.

5. Remove the bushing or needle bearing from the bearing carrier with a suitable driver.

6. Clean housing with fresh solvent and blow dry with compressed air, if available.

7. Install a new bushing or needle bearing (lettered side facing up) in bearing carrier with a suitable mandrel and driver.

8. Install new oil seals in the bearing carrier with a suitable mandrel and driver. Seal lips must face to the rear. Lubricate lips with Yamalube All-purpose Marine grease.

9. If housing cap ball bearing was removed, lubricate the bearing carrier bore with Yamalube Two-cycle Lubricant (for outboards). Install a new bearing (lettered side facing up) with a suitable mandrel and driver.

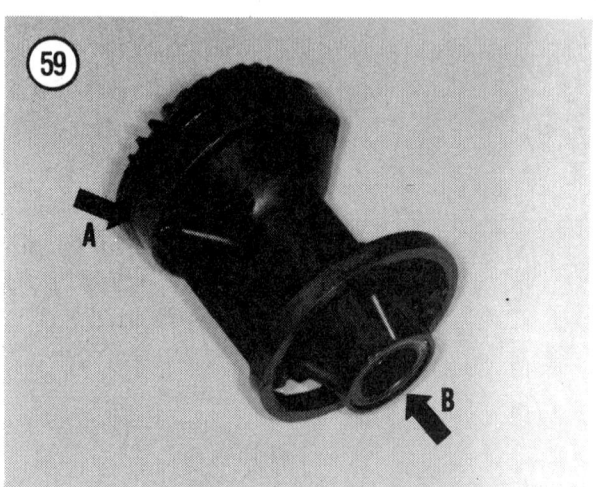

NOTE
If old shim(s) were damaged during disassembly or a new bearing carrier is being installed, install the same thickness shim(s) in Step 10 that were removed.

10. Install reverse gear (and shimming) on bearing carrier with an arbor press.

11. Wipe new O-ring with Yamalube All-purpose Marine grease and install on bearing carrier. See B, **Figure 57**.

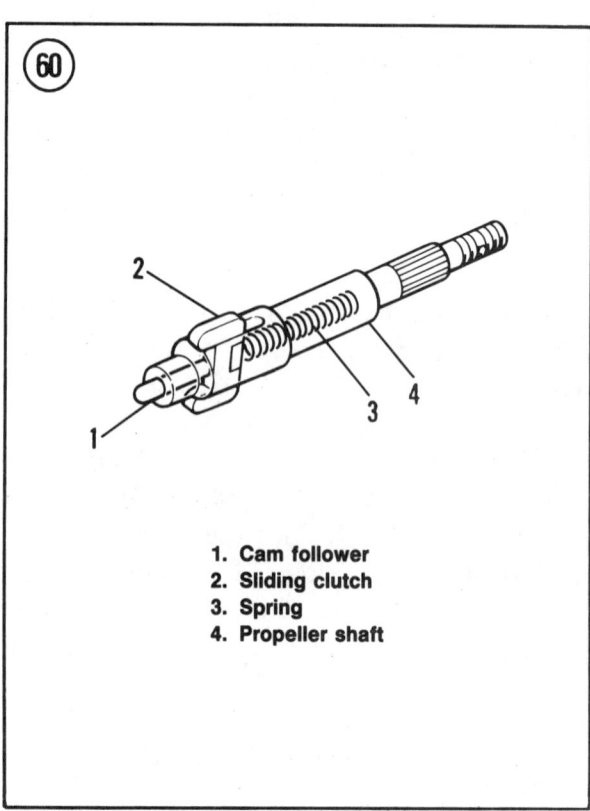

1. Cam follower
2. Sliding clutch
3. Spring
4. Propeller shaft

**Bearing Carrier Disassembly/
Assembly (25-225 hp [except 1988-on 25 hp])**

1. Remove and discard the bearing carrier O-ring. See A, **Figure 59**.

NOTE
The reverse gear ball bearing may come out with the gear in Step 2 or it may remain inside the carrier.

2. If reverse gear cannot be removed by hand, install a suitable puller and remove the gear. Remove and record any reverse gear shimming. Save shimming for reassembly.

3. If reverse gear ball bearing came out with the gear, press it off the gear with a universal puller, support plate and arbor press.

4. If reverse gear ball bearing remained in the carrier, remove it with a suitable puller.

5. Pry both oil seals from the bearing carrier. See B, **Figure 59**. Do not remove the needle bearing under the seals unless inspection indicates that it must be replaced.

6. If needle bearing must be removed, drive bearing from carrier with a suitable mandrel and driver.

7. Clean bearing carrier with fresh solvent and blow dry with compressed air, if available.

8. Position reverse gear on a suitable support. Install shimming removed during disassembly. Position a new ball bearing (lettered side facing down) on reverse gear and install with a suitable mandrel and an arbor press. Apply pressure only on the inner race.

9. If carrier needle bearing was removed, install a new one (lettered side facing to the rear) with a suitable mandrel and driver.

10. Wipe the outer diameter of 2 new oil seals with Loctite Type A. Install seals in the carrier (lips facing to the rear) with a suitable mandrel and driver.

11. Wipe bearing carrier bore with Yamalube Two-cycle Lubricant (for outboards). With reverse gear on a suitable

support, fit a new carrier O-ring on the gear. Press carrier assembly onto gear with an arbor press, making sure that the O-ring seats properly and is not pinched or damaged.

**Propeller Shaft Disassembly/
Assembly (2 hp)**

See *Forward Gear and Bearing Assembly Removal/Inspection/Installation, 2 hp* in this chapter.

Propeller Shaft Disassembly/Assembly (3 hp)

Refer to **Figure 60** for this procedure.

1. Withdraw cam follower from end of propeller shaft.

2. Push sliding clutch toward propeller end of propeller shaft to compress spring.

3. Withdraw sliding clutch from propeller shaft.

4. Rotate propeller shaft on end to remove spring.

5. To reassemble, reverse disassembly procedure.

6. Make sure eared side of sliding clutch faces toward the forward gear.

**Propeller Shaft Disassembly/
Assembly (4-50 hp)**

Refer to **Figure 61** for this procedure.

1. Slip the end of an awl or thin-blade screwdriver under the end of the sliding clutch cross pin retainer ring. See **Figure 62**. Lift end of ring up and rotate propeller shaft to unwind ring from clutch dog.

2. Place the propeller shaft cam follower against a solid object and apply pressure on the shaft. Push cross pin from clutch dog with the awl or screwdriver. See **Figure 63**.

3. Release pressure on the shaft. Remove the cam follower or shift plunger, the shift slide and compression spring. Slide the clutch dog off the shaft.

9

4. To assemble, install the clutch dog on the prop shaft clutch splines. The "F" (forward) stamped on the clutch must point to the front of the shaft.

5. Align the cross pin holes in the clutch dog and shaft.

6. Insert the compression spring, shift slide (if so equipped) and shift plunger or cam follower in the propeller shaft.

7. Apply pressure to the shift plunger or cam follower to compress the spring. Hold pressure and install cross pin through the clutch dog.

8. Insert a small punch through the opposite clutch dog and propeller shaft holes. Pry spring to rear of shaft until cross pin can be installed. Cross pin must pass between shift slide or cam follower and spring.

9. Release pressure on the shift plunger or cam follower and carefully install retainer ring over clutch dog. Do not stretch ring excessively during installation.

10. Remove shift plunger or cam follower. Place a small amount of Yamalube All-purpose Marine grease in the propeller shaft end and reinstall plunger or follower.

Propeller Shaft Disassembly/ Assembly (70-90 hp)

Refer to **Figure 64** for this procedure.

1. Remove the shift cam from the shifter.

2. Slip the end of an awl or thin-blade screwdriver under the end of the sliding clutch cross pin retainer ring. See **Figure 62** (typical). Lift end of ring up and rotate propeller shaft to unwind ring from clutch dog.

3. Place the propeller shaft shifter against a solid object and apply pressure on the shaft. Push cross pin from clutch dog with the awl or screwdriver. See **Figure 63** (typical). Release pressure on the shaft and remove the clutch dog from the shaft.

4. Remove the 2 shift slide balls from the neutral position with an awl or thin-blade

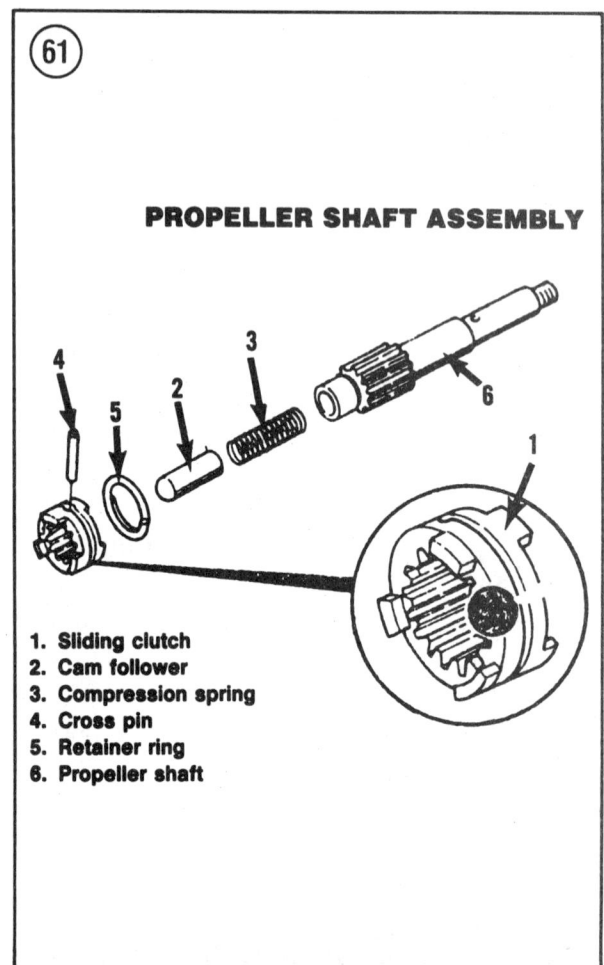

PROPELLER SHAFT ASSEMBLY

1. Sliding clutch
2. Cam follower
3. Compression spring
4. Cross pin
5. Retainer ring
6. Propeller shaft

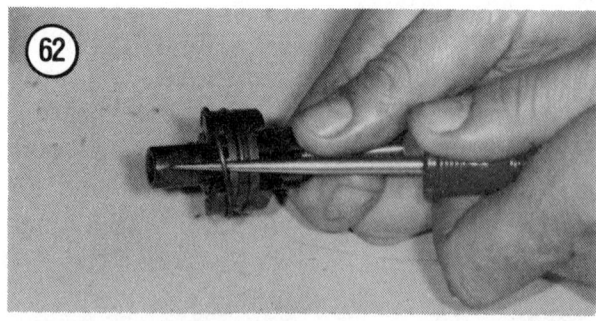

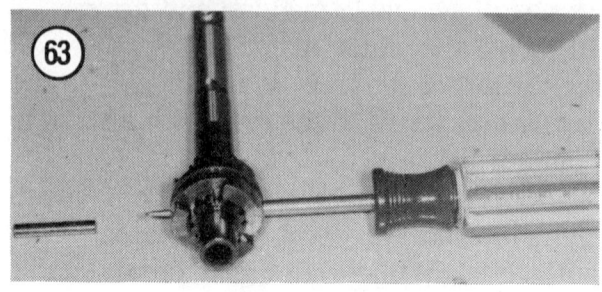

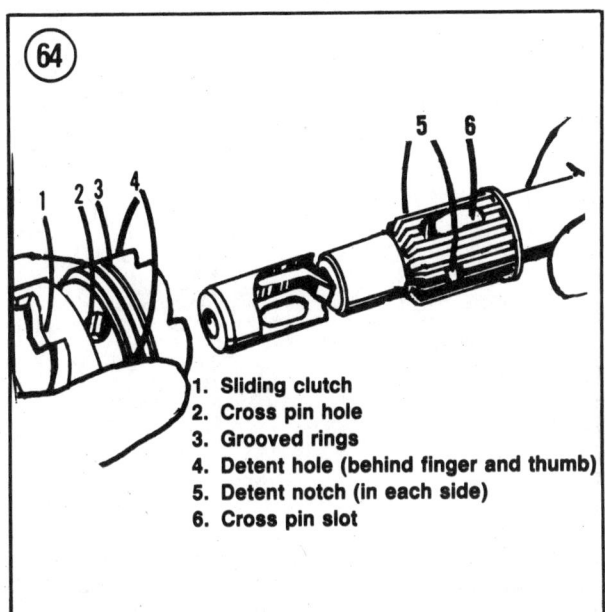

64

1. Sliding clutch
2. Cross pin hole
3. Grooved rings
4. Detent hole (behind finger and thumb)
5. Detent notch (in each side)
6. Cross pin slot

screwdriver, then pull the shift slide and shifter from the propeller shaft.

5. Separate the free shaft and spring from the shift slide.

6. To assemble, insert the free shaft and spring into the shift slide.

7. Liberally coat the 2 balls with Yamalube All-purpose Marine grease to hold them in place, then install the balls and shifter on the shift slide.

8. Align the cross pin hole in the shift slide with the cross pin slot in the propeller shaft. Insert the shift slide into the propeller shaft.

9. Install the shifter, pushing it inward until the 2 balls click into the neutral groove in the shift shaft.

10. Align the clutch dog cross pin hole with the propeller shaft slot and install the clutch dog on the prop shaft clutch splines. The "F" (forward) stamped on the clutch must point to the front of the shaft.

11. Align the cross pin holes in the clutch dog and shaft and install the cross pin through the clutch dog and propeller shaft.

12. Carefully install retainer ring over clutch dog. Do not stretch ring excessively during installation.

13. Place a small amount of Yamalube All-purpose Marine grease on the shifter and install the shift cam with the stamped "F" facing toward the front of the shaft.

Propeller Shaft Disassembly/Assembly (115-225 hp)

1. Slip the end of an awl or thin-blade screwdriver under the end of the sliding clutch cross pin retainer ring. See **Figure 62** (typical). Lift end of ring up and rotate propeller shaft to unwind ring from clutch dog.

2. Push cross pin from clutch dog with the awl or screwdriver. See **Figure 63** (typical). Remove the clutch dog from the shaft.

3. Slowly pull the shift slide and shifter from the propeller shaft, catching the large steel detent balls.

4. Separate the shifter from the shift slide, then remove the small steel balls and spring from the shift slide.

5. To assemble, insert a small steel ball, compression spring and the 2nd small steel ball in the shift slide. Push the 2 balls completely into the shift slide holes.

6. Insert the shift slide into the propeller shaft. Make sure the 2 balls do not fall out of place.

7. Liberally coat the large steel balls with Yamalube All-purpose Marine grease to hold them in place, then install the balls in the shift slide.

8. Align the cross pin hole in the shift slide with the cross pin slot in the propeller shaft. Insert the shift slide into the propeller shaft.

9. Install the shifter, pushing it inward until the 2 balls click into the neutral groove in the shift shaft.

10. Align the clutch dog cross pin hole with the propeller shaft slot and install the clutch dog on the prop shaft clutch splines.

11. Align the cross pin holes in the clutch dog and shaft and install the cross pin through the clutch dog and propeller shaft.

9

12. Carefully install retainer ring over clutch dog. Do not stretch ring excessively during installation.

13. Make sure the clutch dog moves correctly between each shift position, then place it in the NEUTRAL position.

Bearing Carrier and Propeller Shaft Installation (2-15 hp and 1988-on 25 hp)

1. Secure the gearcase in a suitable holding device or a vise with protective jaws. If protective haws are not available for the vise, position the unit upright in the vise with the skeg between wooden blocks.

2. As equipped, fit the reverse gear thrust washer over the threaded end of the propeller shaft.

3. Wipe the housing cap and bearing carrier O-rings with Yamalube All-purpose Marine grease.

NOTE
It may be necessary to rotate the drive shaft to properly align the pinion, forward and reverse gears in Step 5.

4. Install the gear housing cap with a new gasket (2 hp). Install cap and bearing carrier assembly (all others). Make sure cap is fully seated.

5. Install the gear housing cap bolts and washers. Tighten bolts to specifications (**Table 1**).

6. 6-25 hp—Check forward and reverse gear backlash as described in this chapter.

7. Install the propeller as described in this chapter.

Bearing Carrier and Propeller Shaft Installation (25-225 hp [except 1988-on 25 hp])

1. Secure the gearcase in a suitable holding device or a vise with protective jaws. If

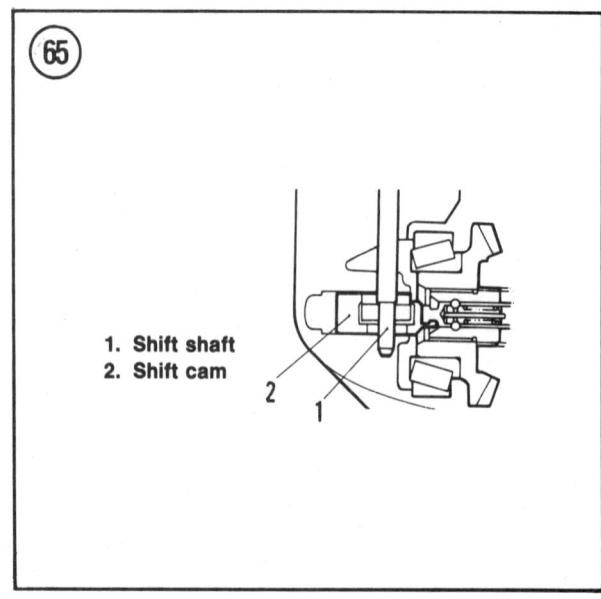

1. Shift shaft
2. Shift cam

protective jaws are not available for the vise, position the unit upright in the vise with the skeg between wooden blocks.

2. Fit the propeller shaft thrust washer over the threaded end of the propeller shaft.

3. Apply a light coat of Yamalube All-purpose Marine grease to the outer diameter of the bearing carrier at all points where the carrier contacts the prop shaft bore.

4. Install reverse gear shimming between the gear and thrust washers.

5. Install the propeller shaft in the housing bore. On 70-90 hp models, the shift cam must face upward. On 115-225 hp models, the flat side of the shifter must face upward.

6. 70-225 hp:

a. Rotate propeller shaft slightly as required to align splined hole in shift cam with the shift shaft hole in the gearcase and install the shift shaft. When properly installed, shift shaft will engage shift cam as shown in **Figure 65**.

b. Install a new O-ring on the shift shaft plate. Lubricate O-ring with Yamalube All-purpose Marine grease.

c. Install shift shaft plate over shift shaft and seat in gearcase bore. Install and tighten fastener securely.

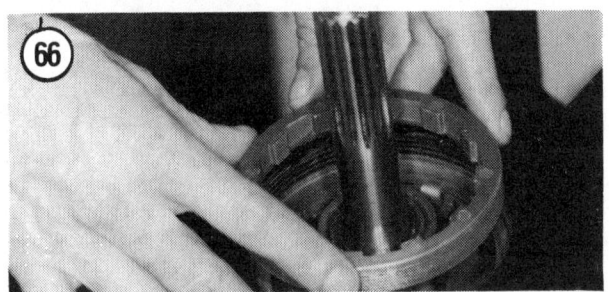

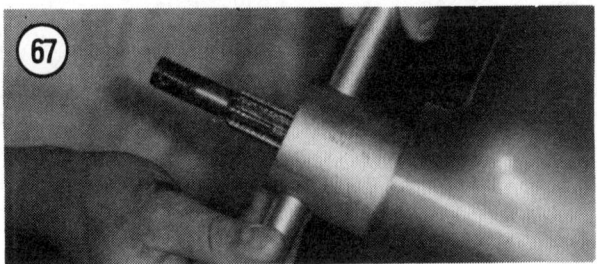

7. Install the bearing carrier. Align locating key slots in carrier and housing bore, then rotate the drive shaft clockwise to mesh with the pinion and reverse gears.

8. When carrier is fully seated, install the locating key.

9. Install a new cover nut tab washer. If a new washer is not available, the one removed may be reinstalled, but different tabs should be bent in Step 12.

10. Wipe the cover nut threads with a light coat of Yamalube All-purpose Marine grease. Install the cover nut and finger-tighten to make sure it does not cross-thread. See **Figure 66**.

11. Install the appropriate cover nut tool (**Figure 67**) and tighten as much as possible by hand, then torque to specifications (**Table 1**).

12. Check forward and reverse gear backlash as described in this chapter.

13. Bend one of the lock tabs on the washer into the cover nut slot. Bend the remaining tabs 90° in the opposite direction.

14. Install the propeller as described in this chapter.

PINION GEAR, DRIVE SHAFT AND SHIFT ROD

Removal (2-8 hp)

The pinion gear is retained on the drive shaft by a circlip.

1. Remove the water pump assembly as described in this chapter.

2. Remove the bearing carrier and propeller shaft as described in this chapter.

3. Invert the gearcase in the holding fixture or vise.

4. Reach into the prop shaft bore with a hooked awl or similar tool and remove the pinion gear circlip.

5. Remove the drive shaft.

6. Remove the pinion gear, thrust washers and shim(s) from the prop shaft bore.

7. Reinstall the gearcase upright in the holding fixture or vise.

8A. 3-5 hp—Remove the shift rod.

8B. 6-8 hp:
  a. Remove the shift rod boot.
  b. Unbolt the oil seal housing from the gearcase. Pry the housing free and remove from the gearcase.
  c. Remove the shift rod and cam.
  d. Remove the drive shaft sleeve from the gearcase.

9. 2, 4-8 hp—Remove lower drive shaft bushing from gearcase with a suitable driver.

10. Remove the upper drive shaft bushing and oil seals from the gearcase (3-5 hp) or oil seal housing (6-8 hp).

Removal (9.9-225 hp)

The pinion gear is retained on the drive shaft by a pinion gear nut.

1. Remove the water pump assembly as described in this chapter.

2. Remove the bearing carrier and propeller shaft as described in this chapter.

3. Install holding tool part No. YB-6079 (9.9-25 hp), part No. YB-6079 (40-50 hp),

9

part No. YB-6151 (70-90 hp) or part No. YB-6201 (115-225 hp) on drive shaft to protect crankshaft splines.

4. Install tool part No. YB-6078 or an appropriate size box wrench over the pinion nut and pad the sides of the prop shaft bore to prevent distortion or damage from contact with the wrench.

5. Holding pinion nut with the wrench installed in Step 4, install an appropriate size socket on the drive shaft holding tool and turn drive shaft to loosen the pinion nut. See **Figure 68**.

6. Remove the tools. Remove the pinion nut and pinion gear. Remove the thrust washer and any shim(s).

7A. 9.9-90 hp (except 1988-on 25 hp)—Withdraw the drive shaft from the gearcase.

7B. 25 hp (1988-on):

a. Remove the shift rod boot.
b. Unbolt the oil seal housing from the gearcase. Pry the housing free and remove from the gearcase.
c. Remove the shift rod and cam.
d. Remove the drive shaft.
e. Remove the drive shaft sleeve from the gearcase.

7C. 115-225 hp:

a. Unbolt the thrust bearing housing at the top of the gearcase.
b. Temporarily reinstall water pump drive key in drive shaft keyway.
c. Install drive shaft puller tool part No. YB-34449 over drive shaft to engage key.
d. Rotate drive shaft 180°, then attach a slide hammer to the drive shaft puller tool and remove the drive shaft.
e. Remove the thrust bearing housing and any shim(s) from the drive shaft.

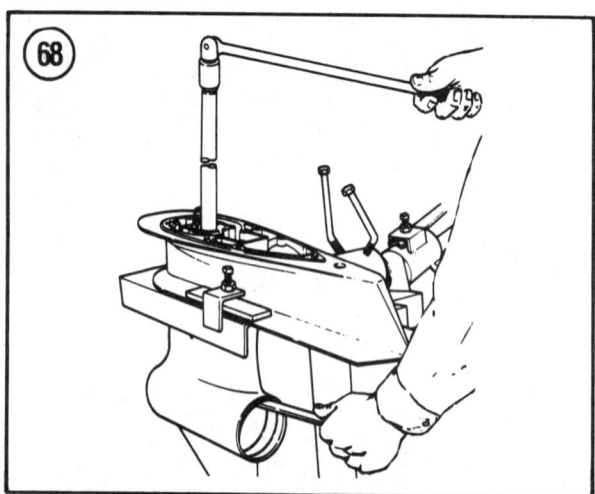

Cleaning and Inspection (All Models)

1. Clean drive shaft and pinion gear with solvent. Blow dry with compressed air, if available.

2. Check the pinion gear for pitting, grooving, scoring, uneven or excessive wear and heat discoloration. Replace the gear if any of these defects are noted.

3. Check the drive shaft bushing, needle bearing or roller bearing contact surfaces for the defects described in Step 2. Replace drive shaft and bushing or bearing if any defects are noted.

> *CAUTION*
> *Do not remove tapered roller bearing from the gearcase unless replacement is required. The bearing is destroyed during removal.*

4. Inspect the drive shaft tapered roller bearing, if so equipped, for pitting, cracks or missing rollers. If bearing or race is defective:

a. Remove bearing with a universal puller and arbor press.
b. Press a new bearing on the drive shaft.
c. Remove bearing race with a suitable puller and slide hammer. See **Figure 69** (typical). Remove shim(s).
d. Position shims under a new race and install with an appropriate driver.

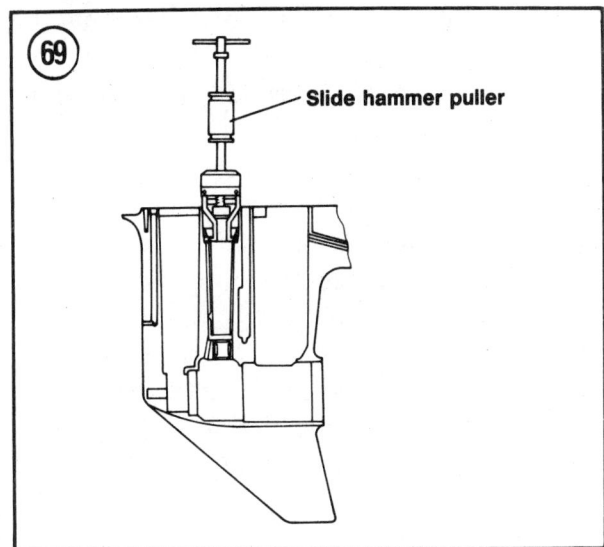

Slide hammer puller

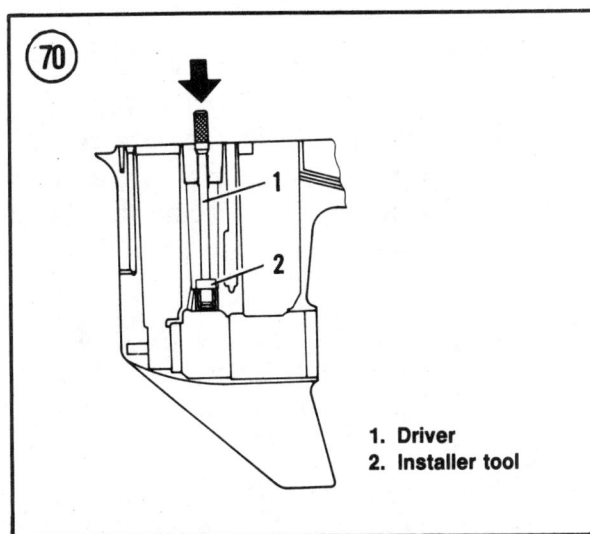

1. Driver
2. Installer tool

5. Check for excessive wear or damage to the drive shaft splines. Replace shaft as required.

6. Check water pump oil seal contact surfaces on the drive shaft. Replace the drive shaft if grooves are found.

7. Check shift plunger contact surface on shift cam. Replace cam and plunger if excessive wear is found.

8. Check shift rod boot for wear or damage. Replace as required.

9. If gearcase (lower drive shaft) needle bearing(s) require replacement:

   a. Remove the drive shaft sleeve.

   b. Drive needle bearing(s) out with a suitable driver. See **Figure 70** (typical).

   c. Install new needle bearing(s) (lettered side facing up) with a suitable driver.

   d. Install drive shaft sleeve, aligning locating rib with gearcase recess.

Installation (2-8 hp)

1. Secure the lower unit in a horizontal position in the holding fixture or vise.

2. 2 hp:

   a. Install lower drive shaft bushing on installer tool part No. YB-6169 and insert through propeller shaft bore. Insert bushing installer tool part No. YB-6029 in drive shaft bore to engage tool part No. YB-6169 and pull into place by tightening the installer center bolt. Remove installer tools.

   b. Install upper drive shaft bushing with installer part No. YB-6025 or equivalent.

   c. Install a new oil seal in the water pump cover and a new oil seal in the top of the gearcase drive shaft bore with a suitable installer. Seal lips must face power head.

3. 3 hp:

   a. Install 2 new oil seals in the top of the gearcase drive shaft bore with a suitable installer. Seal lips must face power head.

   b. Insert shift rod in shift rod boot attached to water pump base.

   c. Install shift rod and water pump base assembly with a new gasket.

4. 4-5 hp:

   a. Install lower drive shaft bushing on installer tool part No. YB-6022 and part No. YB-6169 and insert through propeller shaft bore. Insert bushing installer tool part No. YB-6029 in drive shaft bore to engage tool part No. YB-6169 and pull into place by tightening the installer center bolt. Remove installer tools.

9

b. Install upper drive shaft bushing with installer part No. YB-6025 or equivalent.

c. Install 2 new oil seals in the top of the gearcase drive shaft bore with a suitable installer. Seal lips must face power head.

d. Insert shift rod in shift rod boot attached to water pump base.

e. Install shift rod and water pump base assembly with a new gasket. Tighten attaching bolt.

5. 6-8 hp:

a. Install lower drive shaft bushing on installer tool part No. YB-6022 and part No. YB-6169 and insert through propeller shaft bore. Insert bushing installer tool part No. YB-6029 in drive shaft bore to engage tool part No. YB-6169 and pull into place by tightening the installer center bolt. Remove installer tools.

b. Install upper drive shaft bushing in oil seal housing with installer part No. YB-6071.

c. Install 2 new oil seals in the oil seal housing with installer part No. YB-6071.

d. Install shift rod boot on oil seal housing.

e. Insert shift rod through oil shift rod boot.

f. Install drive shaft sleeve in gearcase.

g. Install the shift rod and cam assembly in gearcase with cam lobe facing toward the gear.

h. Seat oil seal housing on gearcase and install attaching bolt.

6. Invert the gearcase in the holding fixture or vise.

7. Smear the top of the pinion gear with Yamalube All-purpose Marine grease and install the thrust washer.

8. 6-8 hp—Install the circlip on the pinion side of the drive shaft. Make sure it engages the groove properly.

9. Reach into the propeller bore and hold the pinion gear and thrust washer under the drive shaft bore, then insert drive shaft into gearcase, rotating it as required to align the shaft and gear splines.

10. When pinion gear is properly seated on drive shaft splines, install the circlip.

11. Reinstall the gearcase upright in the holding fixture or vise.

12. 6-8 hp—Check forward and reverse gear backlash as described in this chapter.

13. Install the bearing carrier and propeller shaft as described in this chapter.

14. Install the water pump assembly as described in this chapter.

Installation (9.9-225 hp)

1. Secure the gearcase in a horizontal position in the holding fixture or vise.

2. Install thrust washer and pinion gear in propeller shaft bore. Engage pinion gear teeth with the forward gear and locate the thrust washer tab in the housing slot provided.

NOTE
If any rollers fall off the drive shaft roller bearing on 115-225 hp models during Step 3, reinstall them in the outer race before installing the drive shaft.

3. Holding pinion gear and thrust washer in place, insert drive shaft in gearcase. Rotate shaft to align its splines with those of the pinion gear and seat gear on shaft.

4. Lightly coat pinion nut threads with Loctite Type A and position over drive shaft threads in prop shaft bore. Start nut by hand.

5. Install tool part No. YB-6078 or an appropriate size box wrench over the pinion nut and pad the sides of the prop shaft bore to prevent distortion or damage from contact with the wrench.

6. Install holding tool part No. YB-6079 (9.9-25 hp), part No. YB-6079 (40-50 hp), part No. YB-6151 (70-90 hp) or part No.

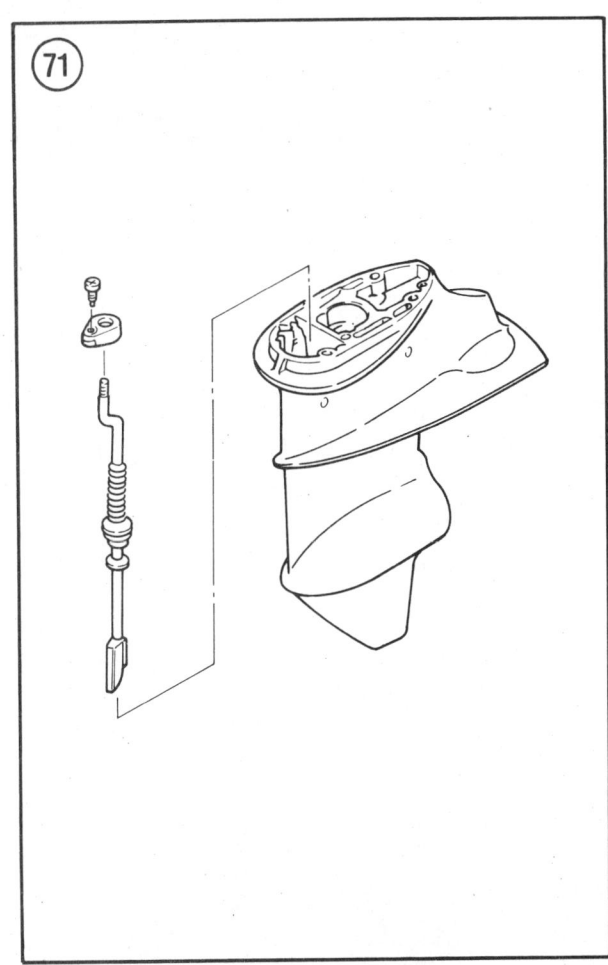

9. Check pinion gear depth and forward/reverse gear backlash as described in this chapter.

10. Install the bearing carrier and propeller shaft assembly as described in this chapter.

11. 25 hp (1988-on):
  a. Install upper drive shaft bushing in oil seal housing with installer part No. YB-6071.
  b. Install 2 new oil seals in the oil seal housing with installer part No. YB-6071.
  c. Install shift rod boot on oil seal housing.
  d. Insert shift rod through shift rod boot.
  e. Install drive shaft sleeve in gearcase.
  f. Install the shift rod and cam assembly in gearcase with cam lobe facing toward the gear.
  g. Seat oil seal housing on gearcase and install attaching bolts.

12. Install the water pump assembly as described in this chapter.

YB-6201 (115-225 hp) on drive shaft to protect crankshaft splines.

7. Holding pinion nut with the wrench installed in Step 5, install an appropriate size socket on the drive shaft holding tool and turn drive shaft to tighten the pinion nut. See **Figure 68**. Tighten nut to specifications (**Table 1**).

NOTE
If more than one shim is used in Step 8, place the thicker shim on top of the thinner one.

8. 115-225 hp—Install thrust bearing shim(s) and thrust bearing housing on drive shaft. Seat shim(s) and housing in gearcase, align housing and gearcase oil holes and install housing fasteners.

SHIFT ROD/SHAFT

Removal, Inspection and Installation (All Models)

1. 3-8 hp—Remove the water pump assembly as described in this chapter.

2A. 3-5 hp—Remove the shift rod.

2B. 6-8 hp and 1988-on 25 hp:
  a. Remove the shift rod boot.
  b. Unbolt the oil seal housing from the gearcase. Pry the housing free and remove from the gearcase.
  c. Remove the shift rod and cam.

2C. 9.9-50 hp (except 1988-on 25 hp)—Loosen screw in shift cam retainer bracket. See **Figure 71** (9.9-30 hp) or **Figure 72** (40-50 hp). Withdraw shift cam assembly from gearcase.

2D. 70-225 hp—Install shift rod handle (part No. YB-6052) and make sure gearcase is in NEUTRAL. Remove shift rod retaining plate and withdraw shift rod assembly from gearcase.

3. Clean shift shaft in solvent and blow dry with compressed air, if available.

4. Check shift cam ramps for wear or damage. Replace as required.

5. Check shift shaft boot, if so equipped, for cracks or deterioration. Replace if defective.

6. Check shift cam retainer, if so equipped, for cracks. Replace if defective.

7. To reinstall, wipe lower half of shift shaft with Yamalube All-purpose Marine grease.

8A. 3-5 hp:

   a. Insert shift rod in shift rod boot attached to water pump base.

   b. Install shift rod and water pump base assembly with a new gasket. Tighten attaching bolt on 4-5 hp models.

   c. Install water pump as described in this chapter.

8B. 6-8 hp:

   a. Install shift rod boot on oil seal housing.

   b. Insert shift rod through shift rod boot.

   c. Install the shift rod and cam assembly in gearcase with cam lobe facing toward the gear.

   d. Seat oil seal housing on gearcase and install attaching bolt.

   e. Install water pump as described in this chapter.

8C. 9.9-50 hp (except 1988-on 25 hp):

   a. Insert shift shaft in gearcase with ramped face of cam facing to the rear. See **Figure 71**.

   b. Install retainer over shift shaft and seat in gearcase bore.

   c. Align retainer hole with hole in gearcase and install fastener.

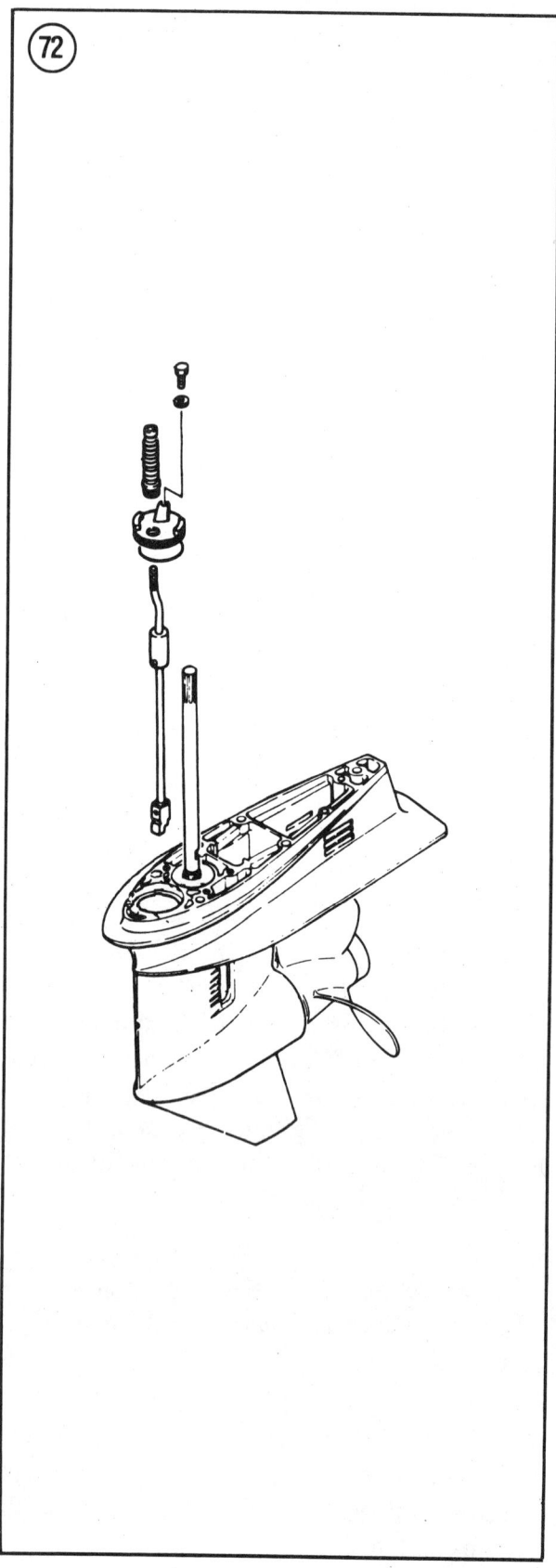

(72)

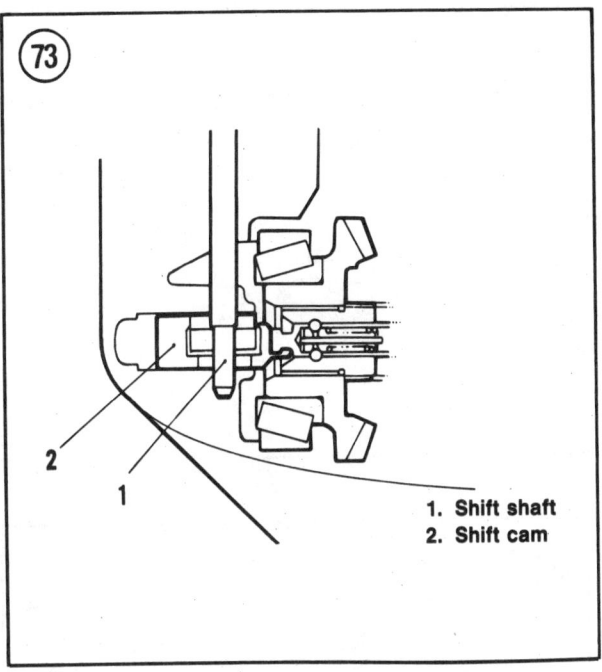

1. Shift shaft
2. Shift cam

8D. 25 hp (1988-on):

a. Install shift rod boot on oil seal housing.

b. Insert shift rod through shift rod boot.

c. Install the shift rod and cam assembly in gearcase with cam lobe facing toward the gear.

d. Seat oil seal housing with a new gasket on gearcase.

e. Install water pump as described in this chapter.

8E. 70-225 hp:

a. Rotate propeller shaft slightly as required to align splined hole in shift cam with the shift shaft hole in the gearcase and install the shift shaft. When properly installed, shift shaft will engage shift cam as shown in **Figure 73**.

b. Install a new O-ring on the shift shaft plate. Lubricate O-ring with Yamalube All-purpose Marine grease.

c. Install spring on shift shaft, if so equipped.

d. Install shift shaft plate over shift shaft and seat in gearcase bore. Install and tighten fastener securely.

FORWARD GEAR AND BEARING ASSEMBLY

Removal/Inspection/Installation (2hp)

1. Remove the water pump as described in this chapter.

2. Remove the pinion gear and drive shaft as described in this chapter.

3. Withdraw forward gear and propeller shaft assembly from the gearcase bore.

4. Reach into prop shaft bore and remove any shim(s) from the forward gear bearing.

5. Remove the forward gear bearing with special tool part No. YB-6096 or an equivalent slide hammer fitted with appropriate puller jaws.

6. Clean the forward gear and propeller shaft assembly in fresh solvent and blow dry with compressed air.

NOTE
The forward gear and propeller shaft are a press-fit and must be replaced as an assembly.

7. Install propeller shaft on V-blocks and measure shaft runout with a dial indicator. If maximum runout exceeds 0.0008 in. (0.02 mm), replace the propeller shaft and forward gear assembly.

8. Check the forward and pinion gear teeth for pitting, grooving, scoring, uneven wear and heat discoloration. Replace the pinion gear and the forward gear/propeller shaft assembly if any defects are found.

9. Reposition gearcase in holding fixture with the prop shaft bore facing upward.

10. Lubricate forward gear bearing bore in gearcase with Yamalube Two-cylinder lubricant (for outboards).

11. Position the new bearing over the bearing bore in the gearcase with its lettered side visible.

12. Install bearing with mandrel part No. YB-6014 and a suitable driver. Seat bearing in bore.

9

13. Position forward gear shim(s) removed during Step 4 on forward gear end of propeller shaft.

14. Install propeller shaft assembly in prop shaft bore until forward gear engages the bearing.

15. Install the pinion gear and drive shaft as described in this chapter.

16. Check gear backlash as described in this chapter.

17. Install the water pump as described in this chapter.

Removal/Installation (3-225 hp)

1. Remove the propeller shaft as described in this chapter.

2. Remove the drive shaft as described in this chapter.

3. Remove the forward gear from the prop shaft bore.

NOTE
Discard any shim(s) installed behind forward gear bearing race on 6-8 hp models. New shim(s) must be installed upon reassembly.

4. Remove the forward gear bearing (3-5 hp) or bearing race (all others) with special tool part No. YB-6096 or an equivalent slide hammer fitted with appropriate puller jaws. See **Figure 74**. Remove any shim(s) from the prop shaft bore.

5. Reposition gearcase in holding fixture with prop shaft bore facing upward.

6. Lubricate bearing bore with Yamalube Two-cycle lubricant (for outboards).

7. Install shim(s) removed in Step 4 in bearing bore. If more than one shim is used, position the thickest one on top (facing the bearing).

8A. 3 hp—Install bearing with a suitable bearing installer and driver. See **Figure 75**.

8B. 4-5 hp—Install bearing with bearing installer part No. YB-6016 and a suitable driver. See **Figure 75**.

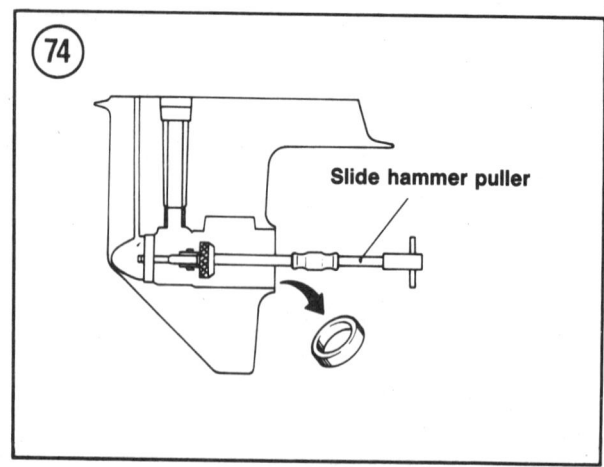

Slide hammer puller

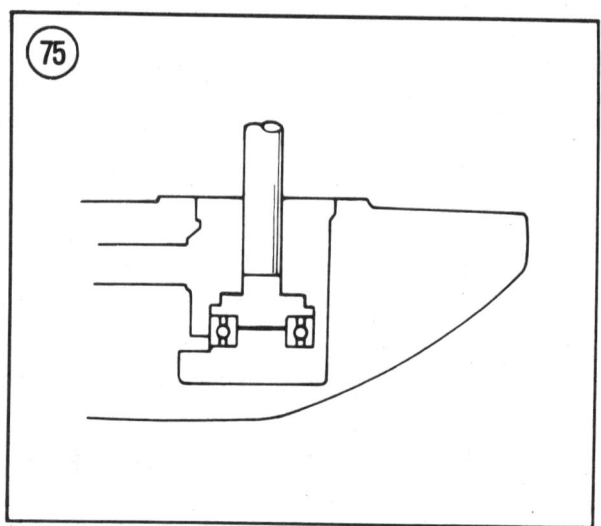

8C. 9.9-15 hp—Install bearing race with installer part No. YB-6085, a suitable driver and attachment part No. YB-6015. See **Figure 76**.

8D. 25 hp (1988-on)—Install bearing race with installer part No. YB-6085 and driver part No. YB-6229. See **Figure 77**.

8E. 30 hp (1987-on)—Install bearing race with installer part No. YB-6085 and driver part No. YB-6071. See **Figure 77**.

8F. All others—Install bearing race with installer part No. YB-6167 (6, 8, 25 and 30 hp), part No. YB-6277 (40-50 hp), part No. YB-6157 (70-90 hp) or part No. YB-6199 (115-225 hp) and a suitable driver. See **Figure 77**.

9. Install the forward gear.

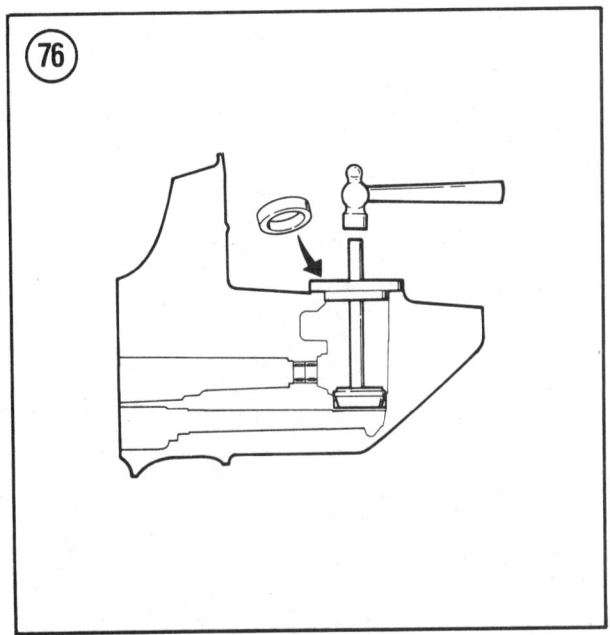

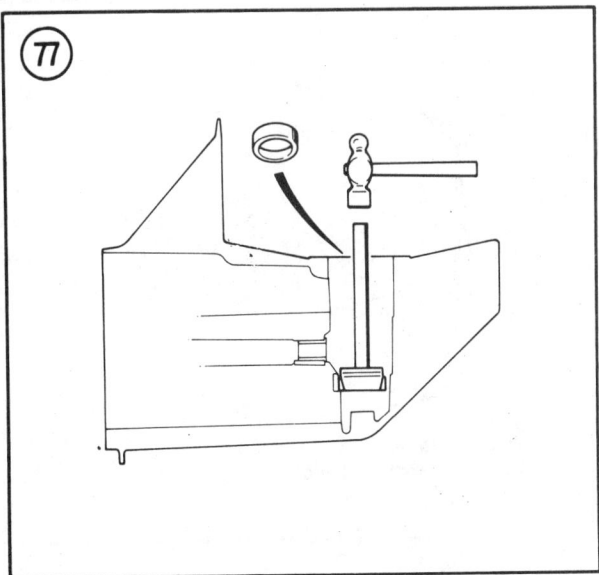

10. 6-225 hp—Check pinion depth and forward gear backlash as described in this chapter.

11. Install the drive shaft as described in this chapter.

12. Install the propeller shaft as described in this chapter.

Cleaning and Inspection (Except 2 hp)

1. Clean the forward gear and bearing with solvent. Blow dry with compressed air, if available. Do *not* spin bearing with compressed air.

2. Check forward gear teeth (A, **Figure 78**) and clutch dogs (B, **Figure 78**). If teeth are pitted, chipped, broken or excessively worn, replace gear. If clutch dogs are chipped or rounded off, replace gear. If gear or teeth show signs of heat discoloration, replace the gear.

3. Apply a light coat of oil to the forward gear bearing (**Figure 79**) and rotate bearing to check for rough spots. Push and pull on forward gear to check for side wear. If movement is excessive, replace bearing with a universal puller, support plate and arbor press.

GEARCASE HOUSING

Cleaning and Inspection

1. Clean gearcase with solvent and blow dry with compressed air, if available.

2. Check water inlet cover, screw hole threads and cover nut or bearing cap screw threads for corrosion damage or stripped threads. Replace gearcase as required.

3. Check gearcase housing for cracks or other impact damage. Replace gearcase as required.

4. Check the water inlet cover and all water passages for clogging or corrosion. Correct as required.

5. Check the gearcase paint for signs of flaking or corrosion. If found, touch up with Yamaha zinc primer (part No. LUB-84PNT-PR-MR) and Yamaha silver paint (part No. LUB-84PNT-SL-VR).

6. Check anode or trim tab as described in this chapter. Replace as required.

PINION GEAR DEPTH AND GEAR BACKLASH

Proper pinion gear engagement and forward and reverse gear backlash are important for smooth operation and long service life. When components are replaced, a shim adjustment is required to achieve the proper tolerances. Three shimming operations must be performed to properly set up the gearcase components:

    a. If the gearcase, drive shaft, bushing, pinion gear or thrust bearing is replaced, the pinion gear must be shimmed to a correct depth.

    b. If the gearcase, forward gear or forward gear bearing is replaced, the forward gear must be shimmed to the pinion gear for proper backlash.

    c. If the gearcase, reverse gear, reverse gear ball bearing or bearing housing/cap is replaced, the reverse gear must be shimmed to the pinion gear for proper backlash.

Not all models will require all 3 shimming procedures.

The gearcase on larger displacement outboards requires the use of precision measuring equipment. It is recommended

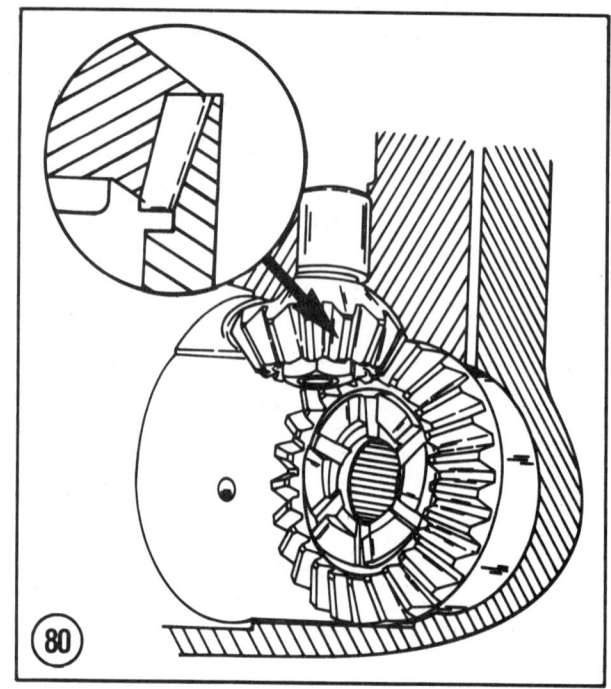

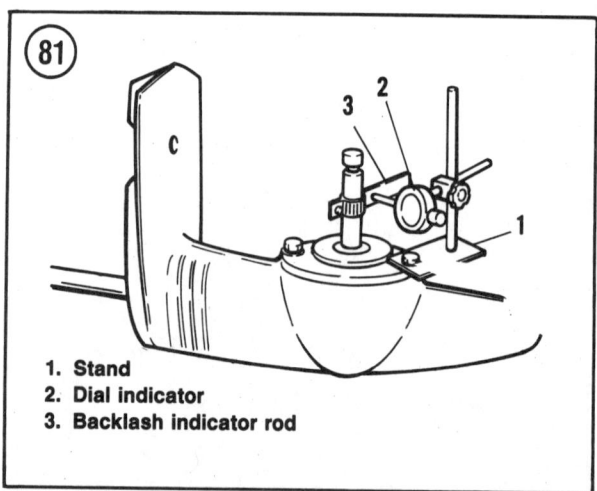

1. Stand
2. Dial indicator
3. Backlash indicator rod

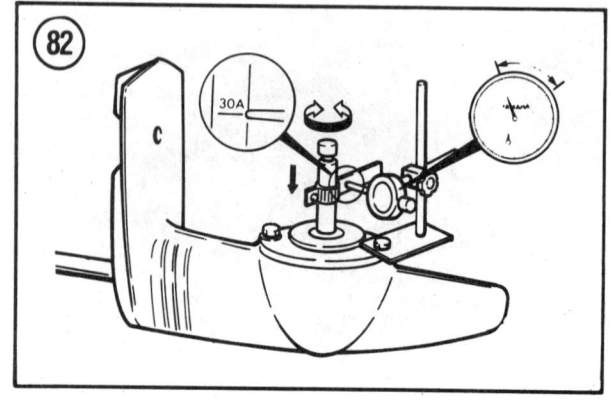

that measuring equipment also be used with smaller displacement models, but if it is not available, they may be checked by touch or sight, although this method requires some degree of experience to tell if the amount of backlash is within specifications. Yamaha 4 and 5 hp models require no shimming, provided the gearcase components have been properly assembled.

Pinion Gear Depth and Forward Gear Backlash (2 hp)

1. Depress and hold the drive shaft.
2. Reach into the prop shaft bore and lightly push the pinion gear upward.
3. Check pinion and forward gear tooth engagement by feel. The entire length of the gear teeth should be in contact. See **Figure 80**. There should be a minimum of up and down play in the drive shaft.
4. If the pinion gear is too high or too low, add or remove shimming to correct the gear height. The shimming is located between the pinion gear thrust washer and drive shaft bushing.

> *NOTE*
> *If the proper measuring equipment is not available for use in Steps 5-8, depress and hold the drive shaft. Push the propeller shaft forward, then rock it back and forth. The free play felt is forward gear backlash and should be 0.003-0.010 in. (0.08-0.3 mm). Proceed to Step 9.*

5. When proper pinion gear depth has been established, temporarily install the gearcase cap. Install the backlash indicator rod (part No. YB-6265) on the propeller shaft and a dial indicator. The dial indicator plunger must contact the mark on the backlash indicator rod. See **Figure 81**. Set the indicator gauge to zero.
6. Hold the drive shaft from moving and slowly rotate the propeller shaft back and forth while applying downward pressure until a click is felt. See **Figure 82**.
7. When the click is felt, stop rotating the propeller shaft and read the indicator gauge. See **Figure 82**.
8. Divide the measurement obtained from the gauge by 1.57 to obtain the backlash. It should be 0.003-0.010 in. (0.08-0.3 mm).
9. If backlash is incorrect, remove shims to increase or add shims to decrease it. Shimming is located between the forward gear and forward gear bearing.
10. If forward gear backlash requires adjustment, recheck pinion gear depth after changing the forward gear shimming.
11. Remove the measuring equipment and continue assembly.

Pinion Gear Depth and Forward Gear Backlash (3 hp)

1. Depress and hold the drive shaft.
2. Reach into the prop shaft bore and lightly push the pinion gear upward.
3. Check pinion and forward gear tooth engagement by feel. The entire length of the gear teeth should be in contact. See **Figure 80**. There should be a minimum of up and down play in the drive shaft.
4. If the pinion gear is too high or too low, add or remove shimming to correct the gear height. The shimming is located behind the pinion gear.

> *NOTE*
> *Do not install water pump assembly and position shift rod in forward position.*

5. Invert the gearcase in the holding fixture and install the backlash indicator rod (part No. YB-6265) on the drive shaft and a dial indicator. The dial indicator plunger must contact the mark on the backlash indicator rod. See **Figure 83**.

9

6. Set the indicator gauge to zero.

7. To check forward gear backlash:

   a. Use hand pressure to keep propeller shaft from rotating.

   b. Slowly rotate the drive shaft back and forth, noting the reading when the indicator reverses direction. See **Figure 83**.

   c. Correct backlash is 0.006-0.047 in. (0.15-1.2 mm). If backlash is incorrect, remove shim located behind the pinion gear and increase or decrease shim thickness as required.

   d. Shims are available in four thicknesses. They range from 0.079 in. (2.0 mm) to 0.091 in. (2.3 mm) in 0.004 in. (0.1 mm) increments.

NOTE
If forward gear backlash can not be brought within specification by altering pinion gear shim, then remove or add shim located behind forward gear. Forward gear shim is only available in one size.

Pinion Gear Depth and Forward/Reverse Gear Backlash (6-8 hp)

1. Depress and hold the drive shaft.

2. Reach into the prop shaft bore and lightly push the pinion gear upward.

3. Check pinion and foward gear tooth engagement by feel. The entire length of the gear teeth should be in contact. See **Figure 80**. There should be a minimum of up and down play in the drive shaft.

4. If the pinion gear is too high or too low, add or remove shimming to correct the gear height. The shimming is located between the pinion gear and drive shaft bushing.

NOTE
If the proper measuring equipment is not available for use in Steps 5-7, depress and hold the drive shaft. Push the propeller shaft forward, then rock it back and forth. The free play felt is

forward gear backlash and should be 0.010-0.030 in. (0.25-0.75 mm).

5. When proper pinion gear depth has been established:

   a. Temporarily install the propeller shaft and bearing housing (without O-ring).

   b. Install the backlash indicator rod (part No. YB-6265) on the drive shaft and a

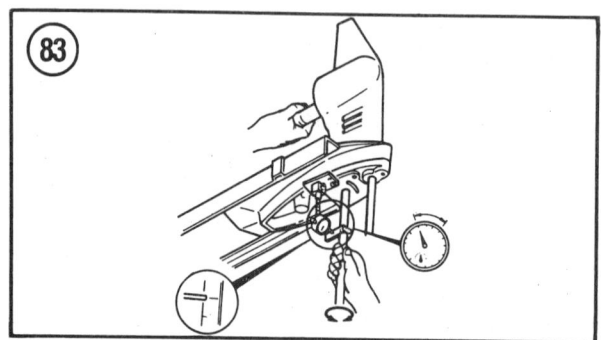

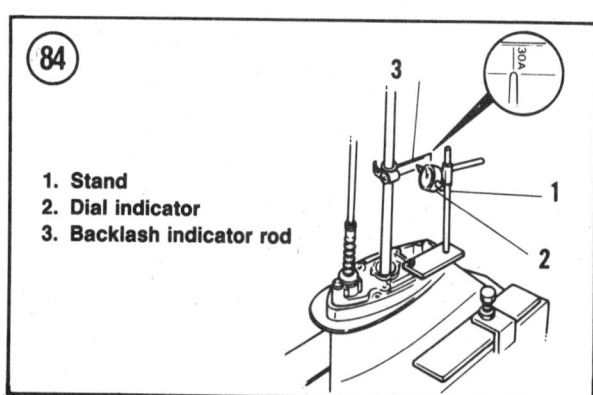

1. Stand
2. Dial indicator
3. Backlash indicator rod

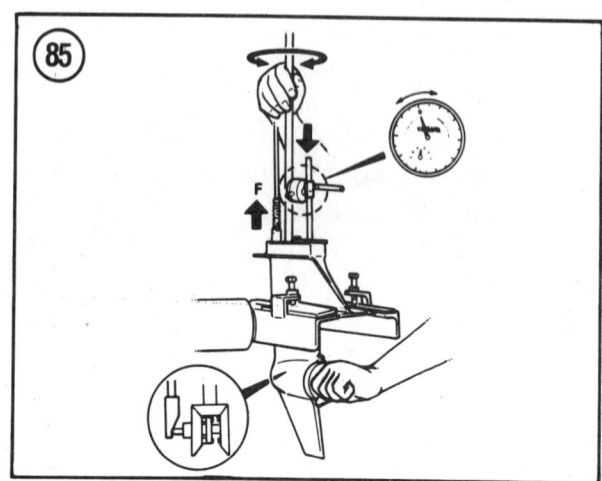

dial indicator. The dial indicator plunger must contact the mark on the backlash indicator rod. See **Figure 84**.

   c. Set the indicator gauge to zero.

6. To check forward gear backlash:

   a. Pull the shift rod upward to shift the unit into forward gear.

   b. Hold the prop shaft from moving and slowly rotate the drive shaft while applying downward pressure. See **Figure 85**.

   c. When a click is felt, stop rotating the drive shaft and read the indicator gauge. See **Figure 85**.

   d. Correct backlash is 0.01-0.03 in. (0.25-0.75 mm). If backlash is incorrect, remove shims to increase or add shims to decrease it. Shimming is located behind the forward gear bearing.

7. To check reverse gear backlash:

   a. Push the shift rod downward to shift the unit into reverse gear.

   b. Rotate the drive shaft clockwise and make sure the propeller shaft turns counterclockwise.

   c. Bring the shift rod back to a neutral position, hold propeller shaft from moving and rotate the drive shaft 30-45° clockwise to preload the reverse gear.

   d. Hold the propeller shaft from moving, push the shift rod downward into

reverse gear and slowly rotate the drive shaft while applying downward pressure.

   e. When a click is felt, stop rotating the drive shaft and read the indicator gauge. See **Figure 86**.

   f. Correct backlash is 0.01-0.03 in. (0.25-0.75 mm). If backlash is incorrect, add shims to decrease or remove shims to increase it. Shimming is located between the reverse gear and reverse gear bearing.

8. If forward or reverse gear backlash requires adjustment, recheck pinion gear depth after changing the shimming.

9. Remove the measuring equipment and continue assembly.

Pinion Gear Depth and Forward/Reverse Gear Backlash (9.9-15 hp)

1. Depress and hold the drive shaft.

2. Reach into the prop shaft bore and lightly push the pinion gear upward.

3. Check pinion and forward gear tooth engagement by feel. The entire length of the gear teeth should be in contact. See **Figure 80**. There should be a minimum of up and down play in the drive shaft.

4. If the pinion gear is too high or too low, add or remove shimming to correct the gear height. The shimming is located between the pinion gear thrust washer and drive shaft bushing.

5. When proper pinion gear depth has been established:

   a. Temporarily install the propeller shaft and bearing housing (without O-ring).

   b. Install the bearing housing puller to lock the propeller shaft and forward gear.

   c. Invert the gearcase in the holding fixture and install the backlash indicator rod (part No. YB-6265) on the drive shaft and a dial indicator. The dial indicator

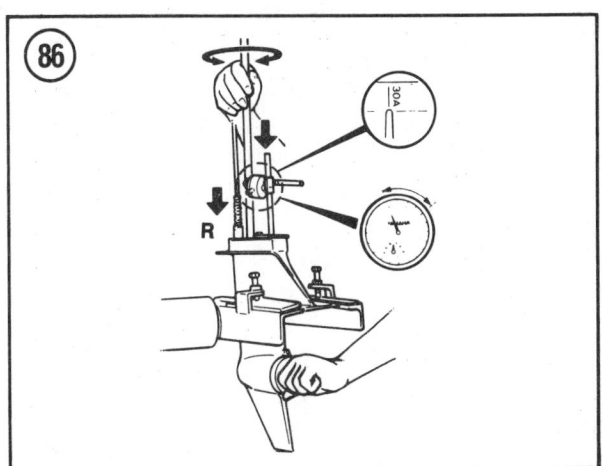

plunger must contact the mark on the backlash indicator rod. See **Figure 87**.

   d. Set the indicator gauge to zero.

6. To check forward gear backlash:

   a. Slowly rotate the drive shaft back and forth, noting the reading when the indicator reverses direction. See **Figure 87**.

   b. Correct backlash is 0.009-0.027 in. (0.23-0.69 mm). If backlash is incorrect, remove shims to increase or add shims to decrease it. Shimming is located behind the forward gear bearing.

7. To check reverse gear backlash:

   a. Remove the bearing housing puller and install reverse gear holder part No. YB-34232 on the propeller shaft. Tighten the holder screw and pull the propeller shaft outward to lock the reverse gear.

   b. Slowly rotate the drive shaft back and forth, noting the reading when the indicator reverses direction. See **Figure 88**.

   c. Correct backlash is 0.031-0.045 in. (0.80-1.15 mm). If backlash is incorrect, add shims to decrease or remove shims to increase it. Shimming is located between the reverse gear and reverse gear bearing.

8. If forward or reverse gear backlash requires adjustment, recheck pinion gear depth after changing the shimming.

9. Remove measuring equipment and continue assembly.

Pinion Gear Depth and Forward/Reverse Gear Backlash (25 [1984-1987] -30 [1984-1986] hp)

1. Depress and hold the drive shaft.

2. Reach into the prop shaft bore and lightly push the pinion gear upward.

3. Check pinion and forward gear tooth engagement by feel. The entire length of the gear teeth should be in contact. See **Figure 80**.

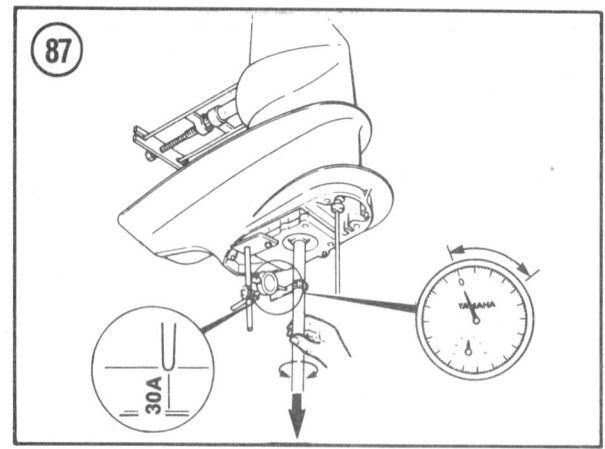

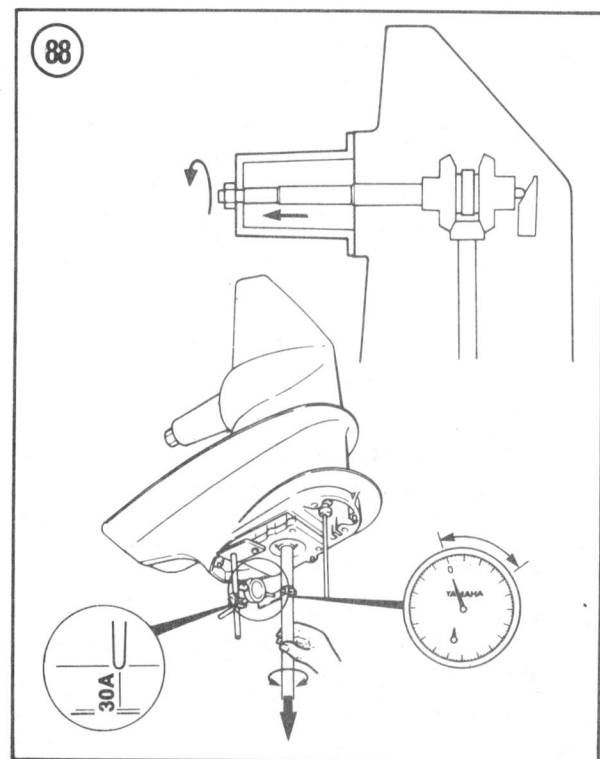

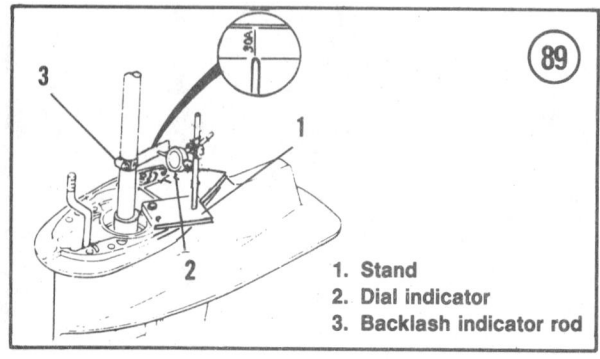

1. Stand
2. Dial indicator
3. Backlash indicator rod

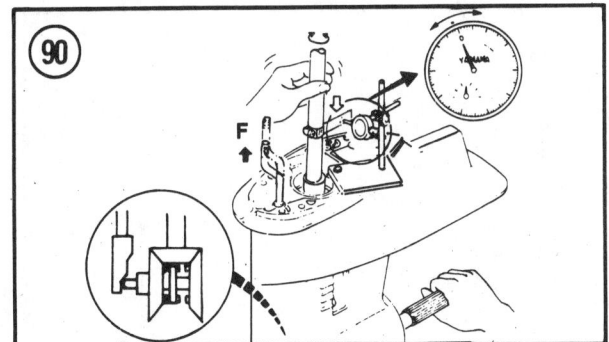

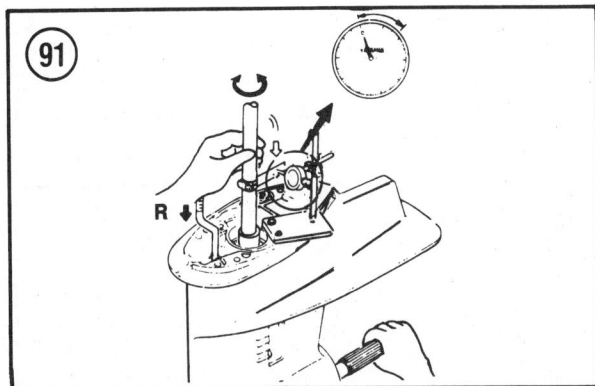

There should be a minimum of up and down play in the drive shaft.

4. If the pinion gear is too high or too low, add or remove shimming to correct the gear height. The shimming is located between the pinion gear and drive shaft bushing.

5. When proper pinion gear depth has been established:

   a. Temporarily install the propeller shaft and bearing housing (without O-ring).

   b. Align the bearing housing keyway and install the key, then install and tighten the cover nut to specifications (**Table 1**).

   c. Install the backlash indicator rod (part No. YB-6265) on the drive shaft and a dial indicator. The dial indicator plunger must contact the mark on the backlash indicator rod. See **Figure 89**.

   d. Set the indicator gauge to zero.

6. To check forward gear backlash:

   a. Pull the shift rod upward to shift the unit into forward gear. See **Figure 90**.

   b. Hold the prop shaft from moving and slowly rotate the drive shaft while applying downward pressure.

   c. When a click is felt, stop rotating the drive shaft and read the indicator gauge. See **Figure 90**.

   d. Correct backlash is 0.008-0.020 in. (0.20-0.50 mm). If backlash is incorrect, remove shims to increase or add shims to decrease it. Shimming is located behind the forward gear bearing.

7. To check reverse gear backlash:

   a. Push the shift rod downward to shift the unit into reverse gear.

   b. Rotate the drive shaft clockwise and make sure the propeller shaft turns counterclockwise.

   c. Bring the shift rod back to a neutral position, hold propeller shaft from moving and rotate the drive shaft 20-30° clockwise to preload the reverse gear.

   d. Hold the propeller shaft from moving, push the shift rod downward into reverse gear and slowly rotate the drive shaft while applying downward pressure.

   e. When a click is felt, stop rotating the drive shaft and read the indicator gauge. See **Figure 91**.

   f. Correct backlash is 0.03-0.04 in. (0.70-1.00 mm). If backlash is incorrect, add shims to decrease or remove shims to increase it. Shimming is located between the reverse gear and reverse gear bearing.

8. If forward or reverse gear backlash requires adjustment, recheck pinion gear depth after changing the shimming.

9. Remove measuring equipment and continue assembly.

Pinion Gear Depth and Forward/Reverse Gear Backlash (25 hp [1988-on])

1. Depress and hold the drive shaft.

9

2. Reach into the prop shaft bore and lightly push the pinion gear upward.

3. Check pinion and forward gear tooth engagement by feel. The entire length of the gear teeth should be in contact. See **Figure 80**. There should be a minimum of up and down play in the drive shaft.

4. If the pinion gear is too high or too low, add or remove shimming to correct the gear height. The shimming is located between the pinion gear and the drive shaft bearing.

NOTE
Do not install water pump assembly.

5. When proper pinion gear depth has been established:

   a. Move the shift shaft to the neutral position.

   b. Temporarily install the propeller shaft and bearing housing (without O-ring).

   c. Install the bearing housing puller (part No. YB-6234) and tighten the puller center screw to lock the propeller shaft. See **Figure 92**.

   d. Install the backlash indicator rod (part No. YB-6265) on the drive shaft and a dial indicator. The dial indicator plunger must contact the backlash indicator rod 0.35 in. (9 mm) from drive shaft. See **Figure 93**.

   e. Set the indicator gauge to zero.

6. To check forward gear backlash:

   a. Slowly rotate the drive shaft back and forth, noting the reading when the indicator reverses direction. See **Figure 93**.

   b. Correct backlash is 0.006-0.010 in. (0.15-0.25 mm). If backlash is incorrect, remove shims to increase or add shims to decrease it. Shimming is located behind the forward gear bearing.

7. To check reverse gear backlash:

   a. Remove the bearing housing puller and install special reverse gear holder tool on propeller shaft with the front side

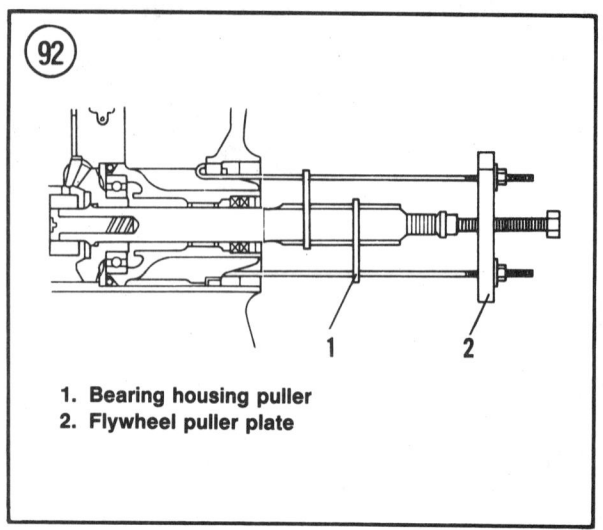

92

1. **Bearing housing puller**
2. **Flywheel puller plate**

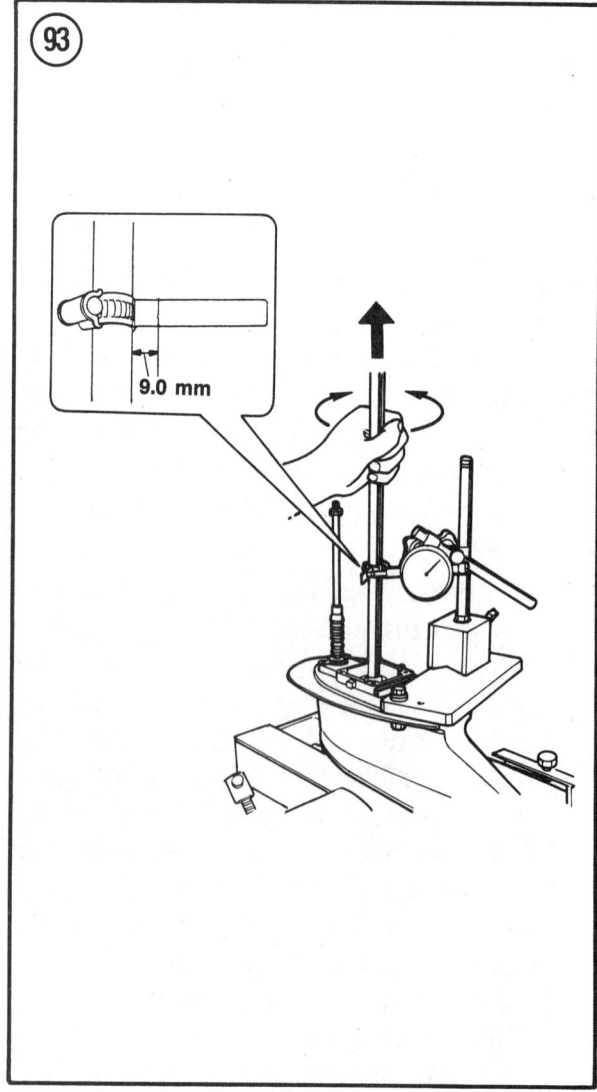

93

9.0 mm

facing aft and retain by securely tightening propeller nut. See **Figure 94**.

b. Slowly rotate the drive shaft back and forth, noting the reading when the indicator reverses direction. See **Figure 93**.

c. Correct backlash is 0.016-0.022 in. (0.40-0.55 mm). If backlash is incorrect, add shims to decrease or remove shims

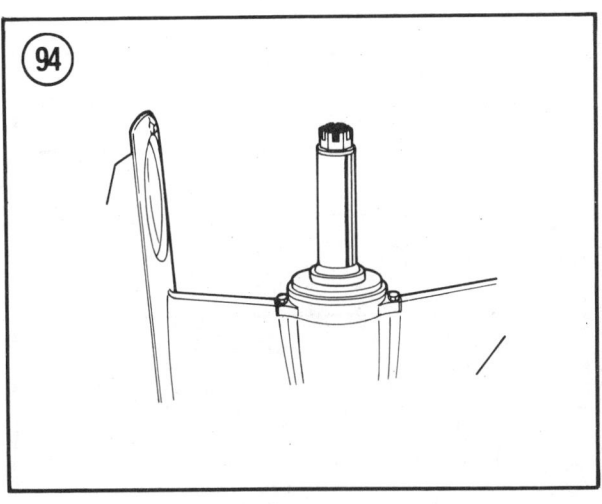

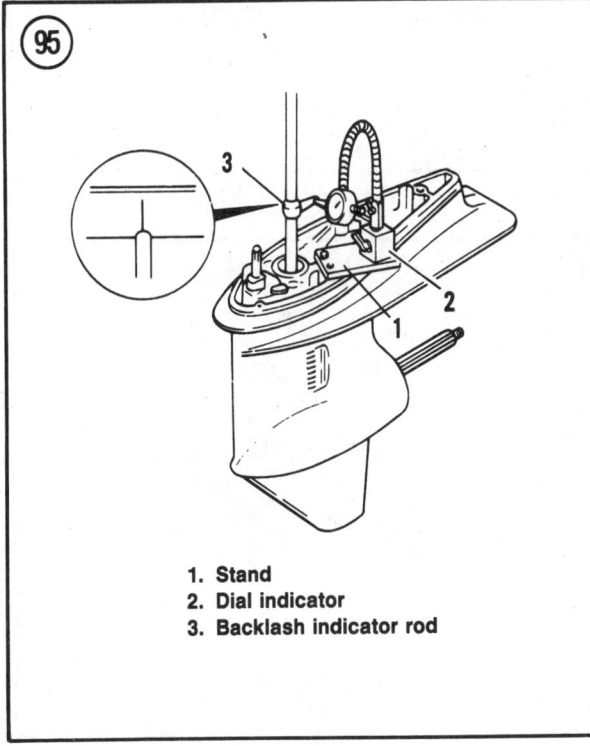

1. Stand
2. Dial indicator
3. Backlash indicator rod

to increase it. Shimming is located between the reverse gear and reverse gear bearing.

8. If forward or reverse gear backlash requires adjustment, recheck pinion gear depth after changing the shimming.

9. Remove measuring equipment and continue assembly.

Pinion Gear Depth and Forward/Reverse Gear Backlash (30 [1987-on] and 40-90 hp)

1. Depress and hold the drive shaft.

2. Reach into the prop shaft bore and lightly push the pinion gear upward.

3. Check pinion and forward gear tooth engagement by feel. The entire length of the gear teeth should be in contact. See **Figure 80**. There should be a minimum of up and down play in the drive shaft.

4. If the pinion gear is too high or too low, add or remove shimming to correct the gear height. The shimming is located between the drive shaft tapered roller bearing race and the drive shaft sleeve in the gearcase.

5. When proper pinion gear depth has been established:

a. Move the shift shaft to the neutral position.

b. Temporarily install the propeller shaft and bearing housing (without O-ring).

c. Install the bearing housing puller (part No. YB-6234) and tighten the puller center screw to lock the propeller shaft. See **Figure 92**.

d. Install the backlash indicator rod (part No. YB-6265) on the drive shaft and a dial indicator. The dial indicator plunger must contact the mark on the backlash indicator rod. See **Figure 95**.

e. Set the indicator gauge to zero.

9

6. To check forward gear backlash:
   a. Slowly rotate the drive shaft back and forth, noting the reading when the indicator reverses direction. See **Figure 96**.
   b. Correct backlash is 0.008-0.020 in. (0.2-0.5 mm) for 30 hp, 0.004-0.0102 in. (0.09-0.26 mm) for 40-50 hp or 0.004-0.011 in. (0.09-0.28 mm) for 70-90 hp. If backlash is incorrect, remove shims to increase or add shims to decrease it. Shimming is located behind the forward gear bearing.

7. To check reverse gear backlash:
   a. Remove the bearing housing puller and install the propeller (without thrust washer) backwards on the propeller shaft (**Figure 97**). Install and tighten propeller nut to specifications (**Table 1**).
   b. Slowly rotate the drive shaft back and forth, noting the reading when the indicator reverses direction. See **Figure 98**.
   c. Correct backlash is 0.028-0.039 in. (0.7-1.0 mm) for 30 hp, 0.030-0.035 in. (0.75-0.88 mm) for 40-50 hp or 0.030-0.044 in. (0.75-1.13 mm) for 70-90 hp. If backlash is incorrect, add shims to decrease or remove shims to increase it. Shimming is located between the reverse gear and reverse gear bearing.

8. If forward or reverse gear backlash requires adjustment, recheck pinion gear depth after changing the shimming.

9. Remove measuring equipment and continue assembly.

**Pinion Gear Depth and
Forward/Reverse Gear Backlash
(115-220 hp)**

1. Depress and hold the drive shaft.
2. Reach into the prop shaft bore and lightly push the pinion gear upward.

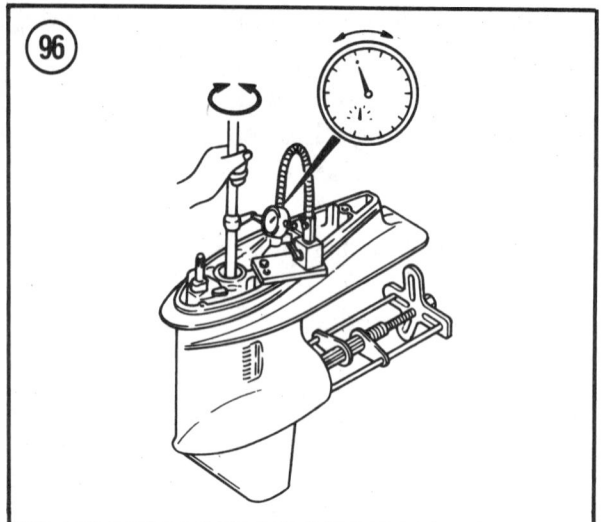

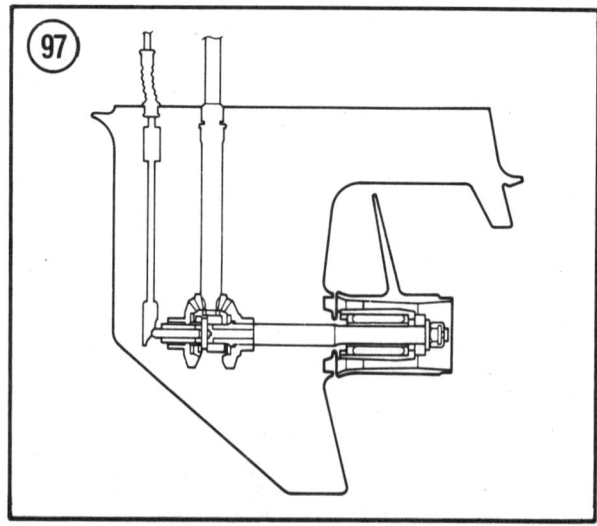

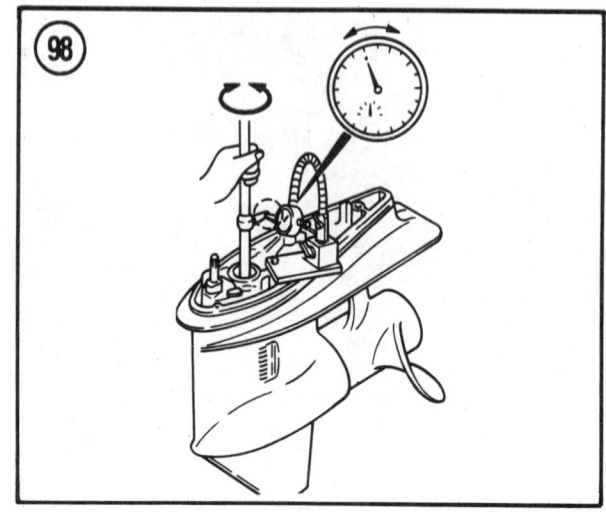

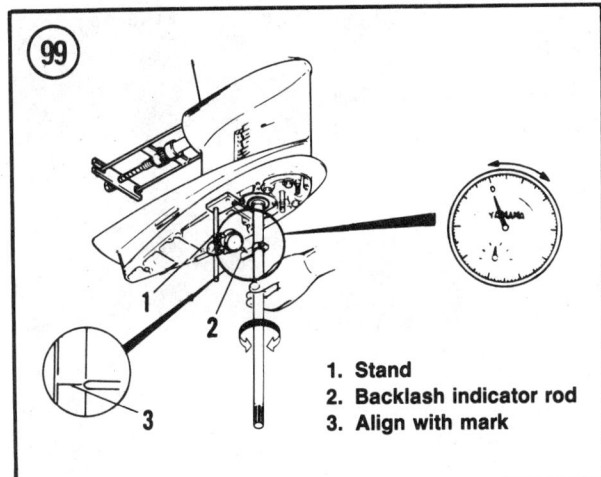

1. Stand
2. Backlash indicator rod
3. Align with mark

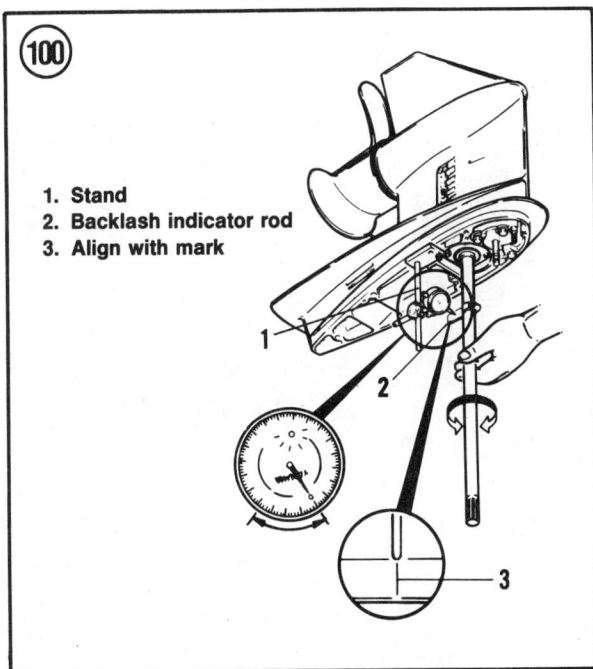

1. Stand
2. Backlash indicator rod
3. Align with mark

3. Check pinion and forward gear tooth engagement by feel. The entire length of the gear teeth should be in contact. See **Figure 80**. There should be a minimum of up and down play in the drive shaft.

4. If pinion depth is incorrect, remove the drive shaft as described in this chapter and add or remove shim(s) as required to bring the pinion gear teeth into full engagement with the forward gear teeth. The shimming is located between between the thrust bearing

housing and thrust bearing. Reinstall drive shaft as described in this chapter and recheck pinion gear depth.

5. When proper pinion gear depth has been established:

   a. Invert the gearcase in the holding fixture.

   b. Temporarily install the propeller shaft and bearing housing (without O-ring).

   c. Install the bearing housing puller (part No. YB-6207) and tighten the puller center screw to lock the propeller shaft. See **Figure 92** (typical).

   d. Install the backlash indicator rod (part No. YB-6265) on the larger diameter side of the drive shaft. Install a dial indicator. The dial indicator plunger must contact the mark on the backlash indicator rod. See **Figure 99**.

   e. Set the indicator gauge to zero.

6. To check forward gear backlash:

   a. Slowly rotate the drive shaft back and forth, noting the reading when the indicator reverses direction. See **Figure 99**.

   b. Correct backlash is 0.013-0.018 in. (0.32-0.45 mm) for 115-130 hp models, 0.011-0.015 in. (0.28-0.39 mm) for 150-200 hp models and 0.010-0.014 in. (0.26-0.36 mm) for 220-225 hp models. If backlash is incorrect, remove shims to increase or add shims to decrease it. Shimming is located behind the forward gear bearing.

7. To check reverse gear backlash:

   a. Remove the bearing housing puller and install the propeller (without thrust washer) backwards on the propeller shaft. See **Figure 97** (typical). Install and tighten propeller nut to specifications (**Table 1**).

   b. Slowly rotate the drive shaft back and forth, noting the reading when the indicator reverses direction. See **Figure 100**.

9

c. Correct backlash is 0.031-0.044 in. (0.80-1.12 mm) for 115-130 hp models, 0.038-0.049 in. (0.97-1.25 mm) for 150-200 hp models and 0.036-0.046 in. (0.91-1.17 mm) for 220-225 hp models. If backlash is incorrect, add shims to decrease or remove shims to increase it. Shimming is located between the reverse gear and reverse gear bearing.

8. If forward or reverse gear backlash requires adjustment, recheck pinion gear depth after changing the shimming.

9. Remove measuring equipment and continue assembly.

TRIM TAB AND ANODE

All Yamaha outboard gearcases use a sacrificial zinc anode as protection against galvanic corrosion. On models equipped with a trim tab, the trim tab is the zinc anode. Smaller models without a trim tab will have a zinc anode attached under or near the anti-ventilation plate. The 70-225 hp models also have an anode on the transom bracket for protection when the gearcase is tilted out of the water. The anode is attached with a single bolt and can be easily removed for inspection. See **Figure 101**. Trim tabs and zinc anodes should be replaced when they have been reduced to approximately 60 percent of their original size.

PRESSURE AND VACUUM TEST

Whenever a gearcase is disassembled and reassembled, it should be pressure and vacuum tested to check for leakage before refilling it with gear lubricant. If the gearcase fails either the pressure or vacuum test, it must be disassembled and repaired before returning the unit to service. Failure to perform a pressure and vacuum test after gearcase service will

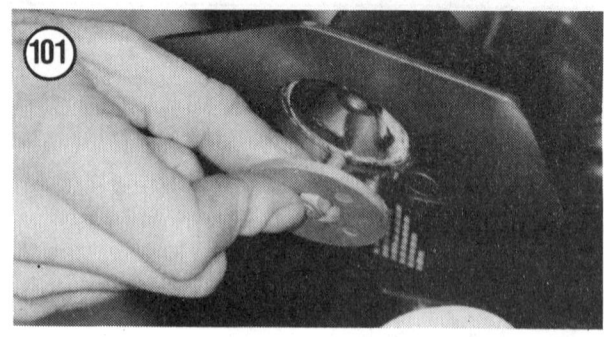

result in either lubricant leakage from the gearcase, or water leakage into the gearcase.

1. Thread a pressure test gauge into the drain/fill plug hole.

2. Pump the pressure tester to 3-6 psi (21-41 kPa). If the pressure holds steady, increase the pressure to 16-18 psi (110-124 kPa). Rotate, push, pull and wiggle all shafts while observing the pressure gauge.

3. If the pressure drops, submerge the gearcase in water and check for air bubbles to indicate the source of the leak.

4. If the pressure holds steady at 16-18 psi (110-124 kPa), release pressure and remove the pressure tester.

5. Install a vacuum tester into the drain/fill plug hole.

6. Draw a 3-5 in.-Hg vacuum and note the vacuum gauge. If the vacuum holds steady, increase to 15 in.-Hg. If the vacuum does not hold at this level, apply oil around the suspected area to determine if the leak stops.

7. If the vacuum holds steady at 15 in.-Hg, release the vacuum and remove the tester.

8. If the source of a leak can not be determined visually, disassemble the gearcase to locate the leak.

9. If no leakage is noted, fill the gearcase with the recommended lubricant.

Table 1 GEARCASE TIGHTENING TORQUES

Fastener	ft.-lb.	N•m
Bearing housing/gearcase cap	5.8	8.0
Cover nut		
25-30 hp	65	90
40-50 hp	94	130
70-130 hp	105	145
150-225 hp	140	190
Gearcase mounting fasteners		
2-8 hp	5.8	8.0
9.9-15, 25 (1984-1987), 30 (1984-1986) hp	13	18
25 (1988-on), 40, 50, 115-225 hp	29	40
30 (1987-on), 70-90 hp		
M8	15	21
M10	29	40
Pinion nut		
9.9-15 hp	19	26
25 (1984-1987), 30 (1984-1986) hp	25	35
25 (1988-on), 30 (1987-on) hp	36	50
40-50 hp	54	75
70-225 hp	68	95
Propeller nut		
3 hp	6-9	8-12
6-15 hp	12	17
25-90 hp	25	35
115-225 hp	40	55
Shift rod connector	7.1	10
Trim tab bolt	27	37
Water inlet screen	5.8	8.0
Water pump cover and base		
2 hp	3.2	4.4
All others	5.8	8.0
Standard torque values		
M5 bolt or 8 mm nut	3.6	5
M6 bolt or 10 mm nut	5.8	8
M8 bolt or 12 mm nut	13	18
M10 bolt or 14 mm nut	25	36
M12 bolt or 17 mm nut	30	42

9

Chapter Ten

Jet Drives

Yamaha first introduced jet drives as a factory option in 1987. Five models have been offered. They are: 40, 50, 90, 115 and 200. The jet drive model numbers reflect the equivalent prop-driven engines's horsepower. But because the jet drive is less efficient than a propeller, actual jet drive horsepower is lower. A Model 40 jet drive actual provides 28 horsepower of jet power. A Model 50 jet drive actual provides 35 horsepower of jet power. A Model 90 jet drive actual provides 65 horsepower of jet power. A Model 115 jet drive actual provides 90 horsepower of jet power. And a Model 200 jet drive actual provides 140 horsepower of jet power. Service to the engine and its related components and the power tilt and trim assembly is the same as prop-driven models. Refer to the appropriate chapter and service section for the engine model or component number your are servicing. Only service on the jet drive assembly will be covered in this chapter.

The jet drive serial number is stamped into the starboard side of the pump housing above the thrust gate pivot and the jet drive model number is stamped into the port side of the pump housing above the thrust gate pivot. See **Figure 1**.

MAINTENANCE

Outboard Mounting Height

A jet driven outboard motor must be mounted higher on the transom plate than an equivalent stock propeller driven outboard motor. If the jet drive is mounted too high, air will be allowed to enter the jet drive resulting in cavitation and power loss. If the jet drive is mounted too low, excessive drag, water spray and loss in speed will result.

To set initial height of outboard motor, proceed as follows:

1. Place a straightedge against boat bottom (*not* keel) and abutt end of straightedge with jet drive intake.

2. The fore edge of the water intake housing should align with the top edge of the straightedge (**Figure 2**).

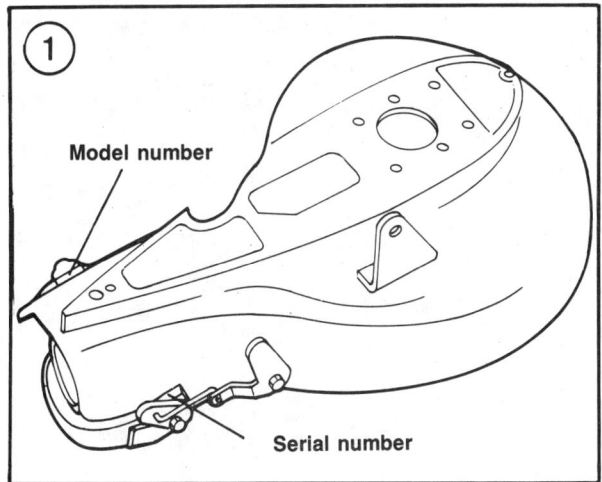

Model number

Serial number

3. Secure the outboard motor at the setting, then test run the boat.

4. If cavitation occurs (overrevving; loss of thrust), the outboard motor must be lowered 1/4 in. (6.35 mm) at a time until uniform operation is noted.

5. If uniform operation is noted with initial setting, the outboard motor should be raised at 1/4 in. (6.35 mm) intervals until cavitation is noted. Then lowered to last uniform setting.

NOTE
The outboard motor should be in a vertical position when the boat is on plane. Adjust motor trim setting as needed. If the outboard motor trim setting is altered, the outboard motor height must be checked and adjusted, if needed, as previously outlined.

Steering Torque

A minor adjustment to the trailing edge of the jet drive outlet nozzle may be made if the boat tends to pull in one direction when the boat and outboard motor are pointed in a straight-ahead direction. Should the boat tend to pull to the starboard side, bend the top and bottom trailing edge on the jet drive outlet nozzle 1/16 in. (1.6 mm) toward the starboard side of the jet drive. See **Figure 3**.

10

Bearing Lubrication

The jet drive bearing(s) should be lubricated after *each* operating period, after every 10 hours of operation and prior to storage. In addition, after every 50 hours of operation additional grease should, be pumped into the bearing(s) to purge any moisture. The bearing(s) is/are lubricated by first removing the cap on the end of the excess grease hose from the grease fitting on the side of the jet drive pump housing. See **Figure 4**. Use a grease gun and inject grease into the fitting until grease exits from the capped end

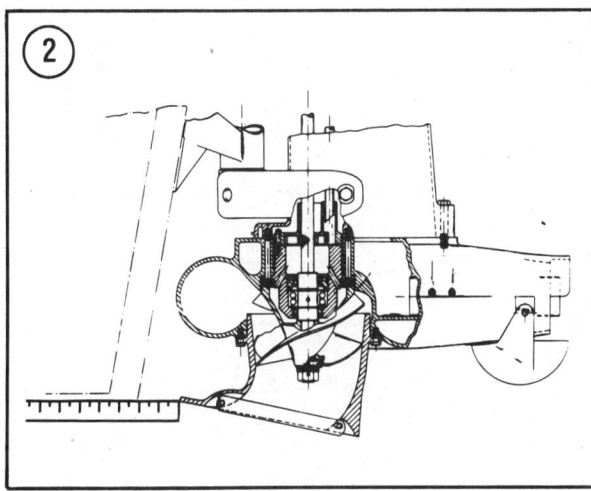

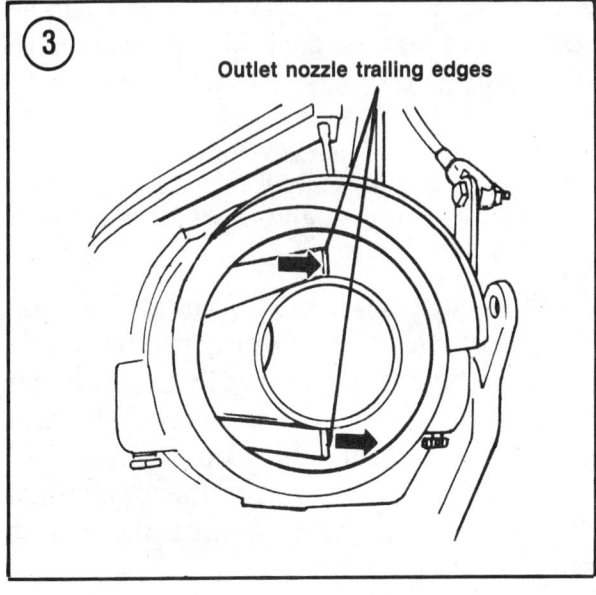

Outlet nozzle trailing edges

of the excess grease hose. Use only Yamalube All-purpose Marine grease or a NLGI No.1 rated grease to lubricate bearing(s).

Note the color of the grease being expelled from the excess grease hose. During the break-in period some discoloration of the grease is normal. After the break-in period, if the grease starts to turn a dark or dirty grey, then the jet drive assembly should be disassembled as outlined under *JET DRIVE* and the seals and bearing(s) inspected and replaced as needed. If moisture is noted being expelled from excess grease hose, then the jet drive should be disassembled and the seals replaced and the bearing(s) inspected and replaced if needed.

Directional Control

The boat's operational direction is controlled by a thrust gate via a cable and lever. When the directional control lever is placed in the full forward position, the thrust gate should completely uncover the jet drive housing's outlet nozzle opening and seat securely against the rubber pad on the jet drive pump housing. When the directional control lever is placed in the full reverse position, the thrust gate should completely close-off the pump housing's outlet nozzle opening. Neutral position is located midway between complete forward and complete reverse position.

> *NOTE*
> *On remote control models, the control box shift cable must be approximately 12 in. (30.5 cm) longer for a jet drive model than a propeller driven model as the control cable connects directly to the jet drive thrust gate linkage.*

The directional control cable is properly adjusted if after placing the directional control lever in the full forward position, the link between the thrust gate and the lower arm of the control cable pivot bracket are in

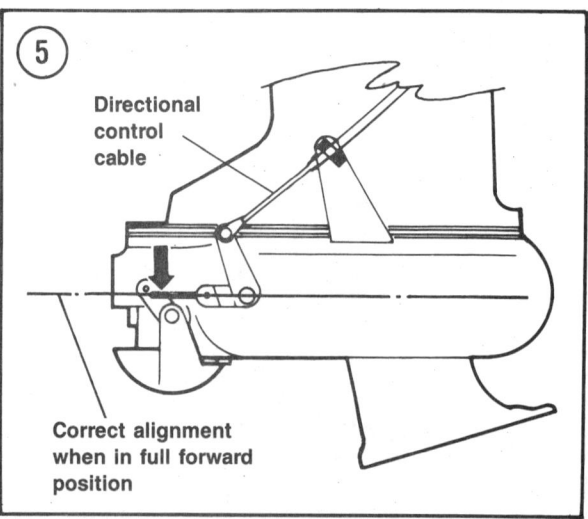

alignment (**Figure 5**). The thrust gate should seat securely against the rubber pad on the jet drive pump housing.

> *WARNING*
> *Always use the lower hole of the thrust gate lever to attach the control linkage. See arrow, **Figure 5**.*

On manual models, an adjustable neutral stop is provided. To adjust, first find true neutral with the engine at idle position and retain this setting. Then loosen the nut (**Figure 6**) securing the stop position and move the stop to contact the directional control lever. Securely tighten the nut to retain the adjustment.

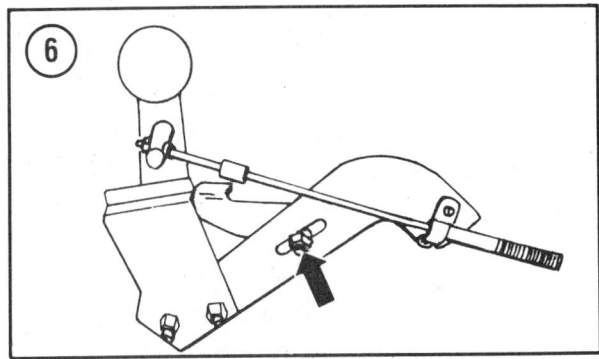

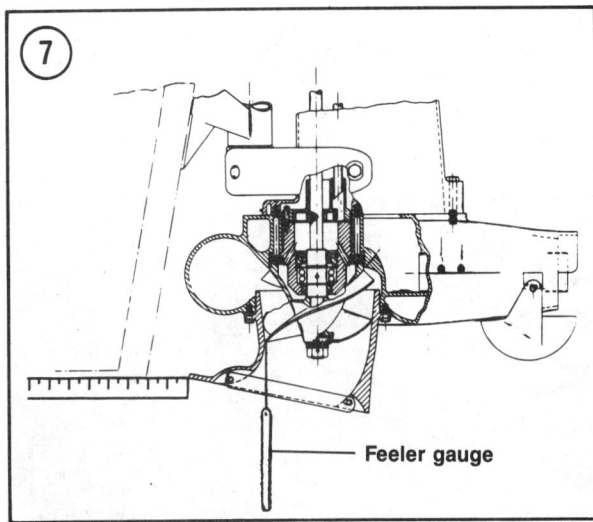

Feeler gauge

Impeller Clearance

1. Disconnect the spark plug leads as a safety precaution to prevent accidental starting of the engine.

2. Insert a selection of feeler gauge thicknesses through the intake housing opening and measure the clearance between the impeller blades and the intake housing. See **Figure 7**.

3. The impeller-to-intake housing clearance should be appproximately 1/32 in. (0.8 mm).

4. If the clearance is incorrect, remove the six intake housing mounting bolts (**Figure 12** or **Figure 13**). Remove intake housing.

5. Bend the ears on the washer retaining the jet drive impeller nut to allow a suitable tool to be installed on the impeller nut. Remove the nut, eared washer, lower shims, impeller and upper shims. Note the number of lower and upper shims.

NOTE
Eight 1/32 in. (0.8 mm) shims are used to adjust the impeller-to-intake housing clearance on 40 and 50 models. Nine 1/32 in. (0.8 mm) shims are used to adjust the impeller-to-intake housing clearance on 90, 115 and 200 models.

6. If clearance is excessive, remove shims as needed from below the impeller (lower shims) and position above the impeller.

7. Install the impeller with the selected number of shims below the impeller.

8. Install eared washer and impeller retaining nut onto drive shaft. Tighten the nut with hand pressure while ensuring that the eared washer does not lodge in the drive shaft threads and jam the nut. Tighten the nut to the specification shown in **Table 1**.

9. Grease the threads of the six intake housing mounting bolts with Yamalube All-purpose Marine grease or a NLGI No.1-rated grease.

10. Install the intake housing on the jet drive pump housing with the lowest part of the intake grill facing aft. Tighten the six intake housing bolts to the specification shown in **Table 1**.

11. Repeat Steps 2 and 3 to recheck impeller-to-intake housing clearance and Steps 4 through 10 if recommended clearance is not measured.

12. After correct clearance is measured, remove the six intake housing bolts and remove the intake housing.

13. Make sure the jet drive impeller retaining nut is correctly tightened to specification shown in **Table 1**.

NOTE
*If the ears on the washer located behind the jet drive impeller retaining nut do not align with flats on nut, remove the nut and turn over the washer then retighten nut to specification shown in **Table 1**.*

10

Make sure the selected number of lower shims do not fall free.

14. Bend the ears on the washer located behind the jet drive impeller retaining nut against the nut flats so to retain the torque setting on the nut.

15. Grease the threads of the six intake housing mounting bolts with Yamalube All-purpose Marine grease or a NLGI No.1-rated grease.

16. Install the intake housing on the jet drive pump housing with the lowest part of the intake grill facing aft. Tighten the six intake housing bolts to the specification shown in **Table 1**.

17. Install directional control cable in jet drive support bracket and reconnect cable end to lower arm of the control cable pivot bracket.

18. Refer to *Directional Control* in this chapter and adjust cable to provide correct operation of thrust gate.

19. Complete reassembly in the reverse of disassembly.

Cooling System Cleaning

The cooling system can become blocked by sand and salt deposits if it is not flushed occasionally. Clean the cooling system after each use in salt water.

All factory equipped jet drive models produced after January 1987 are equipped with a plug installed on the port side of the jet drive. See **Figure 8** for plug location. All factory equipped jet drive models produced prior to January 1987 must be modified as follows to permit cooling system flushing.

1. Remove the jet drive assembly and disassemble the jet drive pump housing as described in this chapter.

2. Remove the cap on the end of the excess grease hose from the grease fitting on the side of the jet drive pump housing and rotate toward aft end of housing.

3. Refer to **Figure 9** and note location of drilled hole.

4. Use a 15/64 in. drill bit and drill hole in location noted in Step 3.

5. Thread hole with a 8 mm×1.25 tap.

6. Clean jet drive housing of all metal chips.

7. Install gasket (part No. 90430-08020-00) on plug (part No. 90340-08014-00) and screw into newly tapped hole. Securely tighten.

8. Reassemble jet drive pump housing and reinstall jet drive assembly as outlined in this chapter.

9. Reconnect the excess grease hose cap onto the grease fitting in the side of the jet drive pump housing.

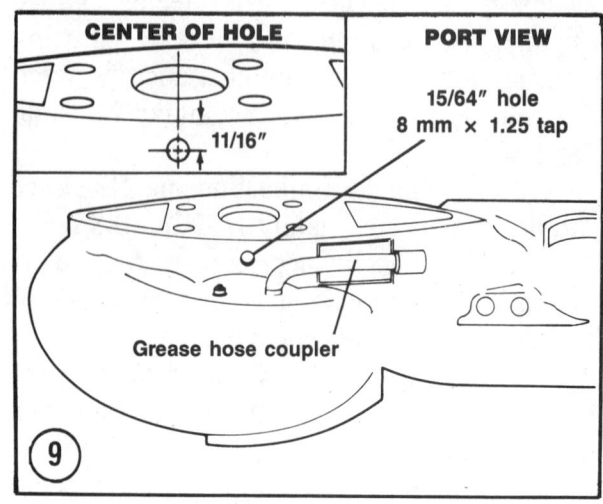

Flushing Cooling System

1. Remove plug and gasket from port side of jet drive pump housing to gain access to flush passage. See **Figure 8**.

2. Install adapter (part No. 6EO-28193-00-94) into flush passage.

3. Connect a suitable fresh water supply to adapter and turn on to full pressure (maximum output).

CAUTION
When the outboard motor is running, make sure a stream of water is noted being discharged from the motor's tell-tale outlet. If not, stop the motor immediately and diagnose the problem.

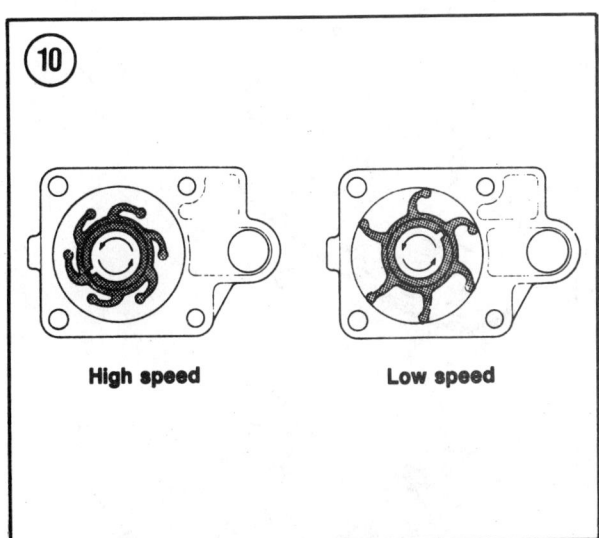

High speed   **Low speed**

CAUTION
Do not operate the outboard motor, when connected to the flush adapter, at a high rpm.

4. Start the motor and allow the fresh water to circulate for approximately 15 minutes.

5. Stop the motor, turn off the auxillary water supply and disconnect the water supply from the adapter.

6. Remove the adapter.

7. Replace gasket if damage is noted, then install on plug.

8. Install plug and gasket into jet drive pump housing flush passage and securely tighten.

NOTE
To flush jet drive impeller and intake housing, direct a fresh water supply into the intake housing area.

WATER PUMP

Refer to **Figure 12** or **Figure 13** for this procedure.

The water pump is mounted on top of the aluminum spacer on all models. The impeller is driven by a key which engages a groove in the drive shaft and a cutout in the impeller hub. As the drive shaft rotates, the impeller rotates with it. Water between the impeller blades and pump housing is pumped up to the power head through the water tube.

The offset center of the pump housing causes the impeller vanes to flex during rotation. At low speeds, the pump acts as a positive displacement type; at high speeds, water resistance forces the vanes to flex inward and the pump becomes a centrifugal type. See **Figure 10**.

All gaskets should be replaced whenever the water pump is removed. Since proper water pump operation is critical to outboard operation, it is also a good idea to install a new impeller at the same time. Note that the impeller will only slide over the Woodruff key in one direction.

10

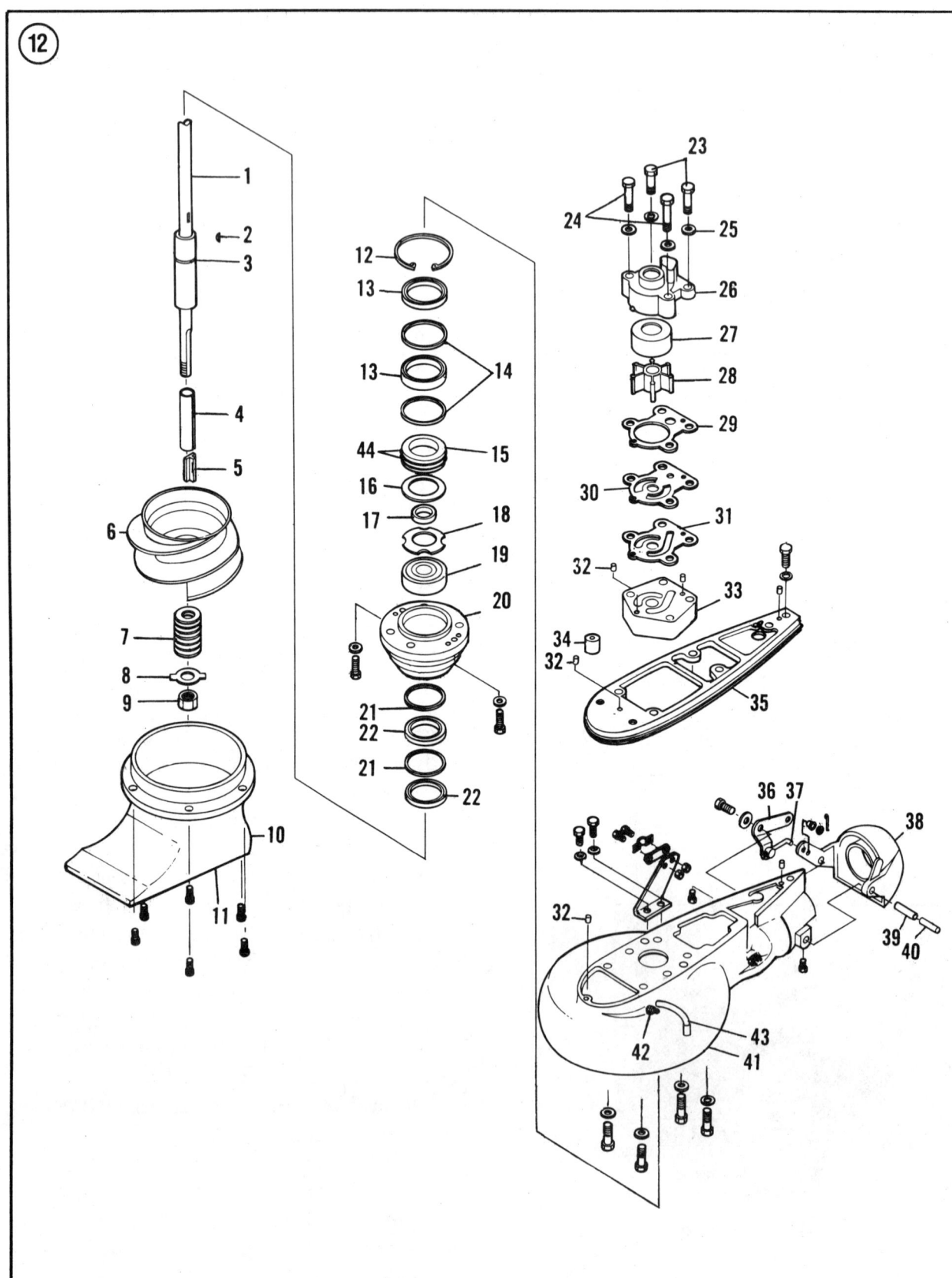

JET DRIVE (MODELS 40 [28 HP] AND 50 [35 HP])

1. Drive shaft
2. Woodruff key
3. Thrust ring
4. Nylon sleeve
5. Impeller drive key
6. Impeller
7. Shims
8. Tab washer
9. Nut
10. Intake housing
11. Intake grille
12. Snap ring
13. Grease seal
14. Retaining ring
15. Upper seal carrier
16. Spacer washer
17. Collar
18. Thrust washer
19. Bearing
20. Bearing housing
21. Retaining ring
22. Grease seal
23. Bolts (short)
24. Bolts (long)
25. Washers
26. Water pump housing
27. Cartridge insert
28. Impeller
29. Gasket
30. Plate
31. Gasket
32. Dowel pin
33. Spacer plate
34. Rubber sleeve
35. Adapter plate
36. Pivot bracket
37. Linkage rod
38. Thrust gate
39. Nylon sleeve
40. Pivot pin
41. Jet drive pump housing
42. Grease fitting
43. Hose
44. O-rings

10

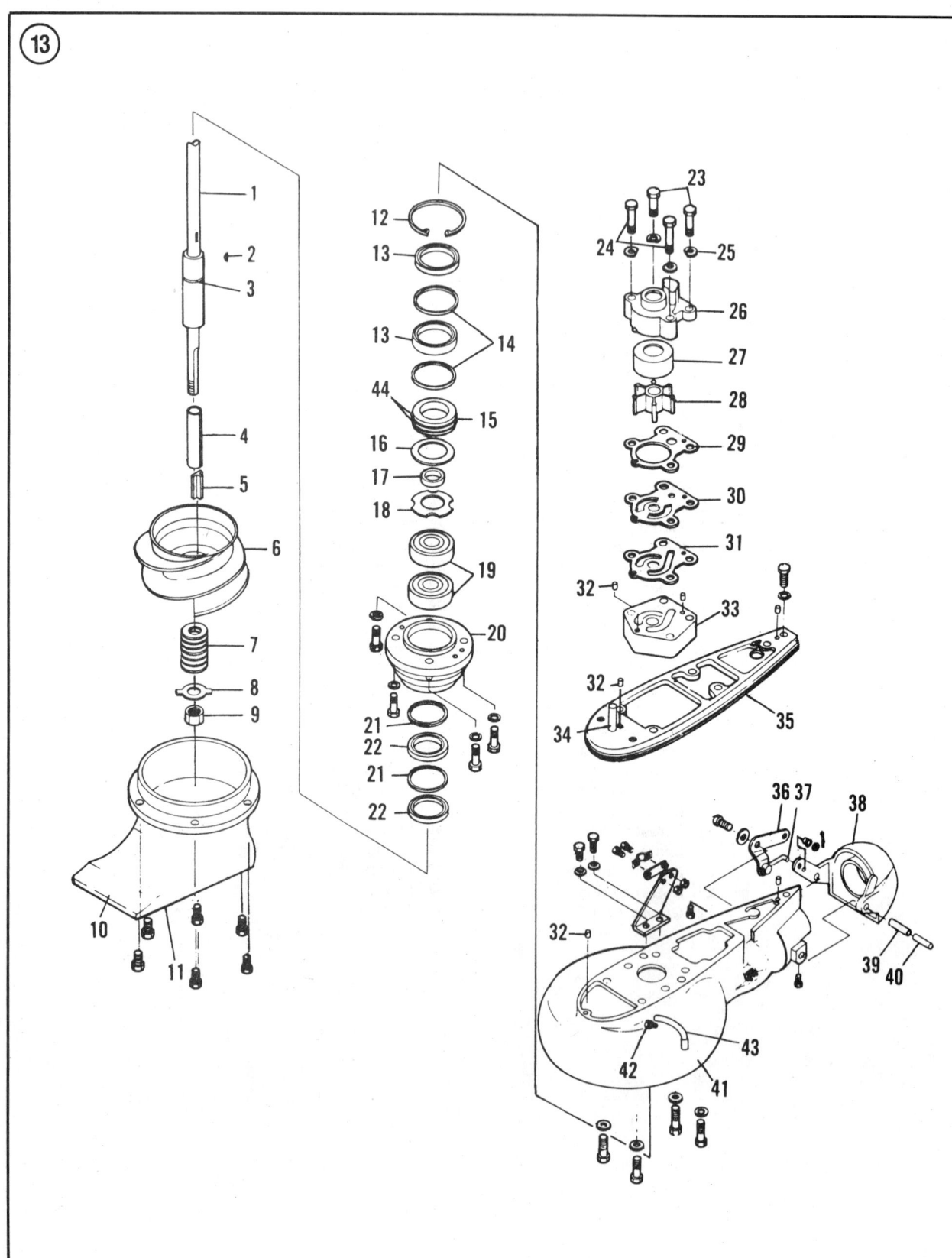

JET DRIVE (MODELS 90 [65 HP], 115 [90 HP] AND 200 [140 HP])

1. Drive shaft
2. Woodruff key
3. Thrust ring
4. Nylon sleeve
5. Impeller drive key
6. Impeller
7. Shims
8. Tab washer
9. Nut
10. Intake housing
11. Intake grille
12. Snap ring
13. Grease seal
14. Retaining ring
15. Upper seal carrier
16. Cupped washer
17. Collar
18. Thrust washer
19. Bearings
20. Bearing housing
21. Retaining ring
22. Grease seal
23. Bolts (short)
24. Bolts (long)
25. D-washers
26. Water pump housing
27. Cartridge insert
28. Impeller
29. Gasket
30. Plate
31. Gasket
32. Dowel pin
33. Spacer plate
34. Dowel pin
35. Adapter plate
36. Pivot bracket
37. Linkage rod
38. Thrust gate
39. Nylon sleeve
40. Pivot pin
41. Jet drive pump housing
42. Grease fitting
43. Hose
44. O-rings

Removal and Disassembly

Refer to **Figure 12** or **Figure 13** for this procedure.

1. Remove the jet drive assembly as described in this chapter.

2. Remove the four bolts and washers (D washers on 90, 115 and 200 models) retaining the pump housing to the jet drive pump housing.

3. Withdraw the pump cover off of the impeller. Rotate the pump cover counterclockwise if needed to assist in cover removal.

CAUTION
Do not rotate the pump cover clockwise to assist in its removal or damage to the impeller will result and necessitate it being replaced.

NOTE
In some cases, the impeller may come off with the pump cover. In extreme cases, the impeller hub may have to be split with a hammer and chisel to remove it in Step 4.

4. If impeller does not come off with pump cover, slide it upward on the drive shaft. Remove the impeller.

5. Remove the impeller drive key (Woodruff key).

6. Remove the water pump face plate with top and bottom gaskets. Separate face plate from gaskets. Discard the gaskets.

Cleaning and Inspection

Refer to **Figure 12** or **Figure 13** for this procedure.

1. Remove the water pump cartridge insert from the pump cover.

2. Remove the water tube seal from the pump cover.

3. Clean all metal parts in solvent and blow dry with compressed air, if available.

4. Clean gasket residue from all mating surfaces.

5. Check pump cover for damage or distortion from overheating.

6. Check face plate and cartridge insert for grooves or rough surfaces. Replace if any defects are found.

7. If original impeller is to be reused, check bonding to hub. Check side seal surfaces and blade ends for cracks, tears, wear and a glazed or melted appearance. See **Figure 11** (typical). If any of these defects are noted, do *not* reuse the original impeller.

Assembly and Installation

Refer to **Figure 12** or **Figure 13** for this procedure.

1. If pump cover cartridge insert was removed, reinstall cartridge with tab on top engaging slot in pump cover.

2. If the water tube seal was removed, replace the seal or reinstall original seal, if no damage was noted during inspection, into pump cover.

3. Install lower gasket, pump face plate and upper gasket.

4. Install the impeller drive key (Woodruff key).

5. Lightly lubricate impeller with Yamalube All-purpose Marine grease or a NLGI No.1-rated grease. Slide impeller onto drive shaft. Align slot in impeller hub with drive key and seat impeller on water pump face plate.

6. Check impeller installation by rotating drive shaft clockwise. Impeller should rotate with the shaft. If not, remove and reposition impeller to properly engage drive key.

7. Slide pump cover over drive shaft. Push downward on cover while rotating the drive shaft clockwise to assist the impeller in entering the cover without damage.

8. Grease the threads of the four bolts retaining the pump housing to the jet drive

pump housing with Yamalube All-purpose Marine grease or a NLGI No.1-rated grease. Install the bolts with washers (D washers on 90, 115 and 200 models) and tighten bolts in small increments following a crisscross pattern until torque specified in **Table 1** is obtained.

9. Install the jet drive assembly as described in this chapter.

JET DRIVE

When removing jet drive mounting fasteners, it is not uncommon to find that they are corroded. Such fasteners should be discarded and new ones used when the jet drive is installed. The threads on all mounting bolts should be greased with Yamalube All-purpose Marine grease or a NLGI No.1-rated grease.

Removal

1. Disconnect the spark plug leads as a safety precaution to prevent accidental starting of the engine during jet drive removal.
2. Tilt outboard motor to full out position and engage tilt lock lever.
3. Remove directional control cable from lower arm of the control cable pivot bracket. Remove cable from jet drive support bracket.
4. Remove the six intake housing mounting bolts. Remove intake housing. See **Figure 12** or **Figure 13**.
5. Bend the ears on the washer retaining the jet drive impeller nut to allow a suitable tool to be installed on the impeller nut. Remove the nut and eared washer.

NOTE
Eight 1/32 in. (0.8 mm) shims are used to adjust the impeller-to-intake housing clearance on 40 and 50 models. Nine 1/32 in. (0.8 mm) shims are used to adjust the impeller-to-intake housing clearance on 90, 115 and 200 models.

6. Remove the shims located below the impeller and note the number. Remove the impeller and the shims located above the impeller and note the number.
7. Slide impeller sleeve and drive key off drive shaft.
8. Remove the four bolts located on the inside of the jet drive pump housing and adjacent to the bearing housing and the one bolt at the external bottom aft end of the intermediate housing which secure the jet drive pump housing to the adapter plate, then withdraw the jet drive assembly.

Installation

Refer to **Figure 12** or **Figure 13** for this procedure.
1. Install the fore and aft dowel pin in the jet drive housing.
2. Lightly lubricate the drive shaft splines and the water tube seal with Yamalube All-purpose Marine grease or a NLGI No.1-rated grease. Wipe any excessive grease off the top of the shaft.
3. Grease the threads on all mounting bolts with Yamalube All-purpose Marine grease or a NLGI No.1-rated grease.
4. Install the jet drive assembly and tighten the securing bolts to the specification shown in **Table 1**.
5. Grease the drive shaft threads, drive key and jet drive impeller bore with Yamalube All-purpose Marine grease or a NLGI No.1-rated grease.
6. If the original bearing(s), bearing housing, jet drive pump housing, impeller and intake housing are being reinstalled, install the shims located above and below the impeller in the same number and location as removed. If any of the previously listed components were replaced, then install all of the adjustment shims below the impeller (next to retaining nut) and proceed to *Impeller Clearance* in this chapter after completion of installation.

10

NOTE
Eight 1/32 in. (0.8 mm) shims are used to adjust the impeller-to-intake housing clearance on 40 and 50 models. Nine 1/32 in. (0.8 mm) shims are used to adjust the impeller-to-intake housing clearance on 90, 115 and 200 models.

7. Install impeller sleeve into jet drive impeller bore.

8. Install shims above impeller as noted in Step 6.

9. Slide drive key and impeller onto drive shaft.

10. Install shims below impeller as noted in Step 6.

11. Install eared washer and impeller retaining nut onto drive shaft. Tighten the nut with hand pressure while ensuring that the eared washer does not lodge in the drive shaft threads and jam the nut. Tighten the nut to the specification shown in **Table 1**.

12. Grease the threads of the six intake housing mounting bolts with Yamalube All-purpose Marine grease or a NLGI No.1-rated grease.

13. Install the intake housing on the jet drive pump housing with the lowest part of the intake grill facing aft. Tighten the six intake housing bolts to the specification shown in **Table 1**.

14. Refer to *Impeller Clearance* in this chapter.

15. Remove the six intake housing bolts and remove the intake housing.

16. Make sure the jet drive impeller retaining nut is correctly tightened to specification shown in **Table 1**.

NOTE
*If the ears on the washer located behind the jet drive impeller retaining nut do not align with flats on nut, remove the nut and turn over the washer then retighten nut to specification shown in **Table 1**. Make sure the selected number of lower shims do not fall free.*

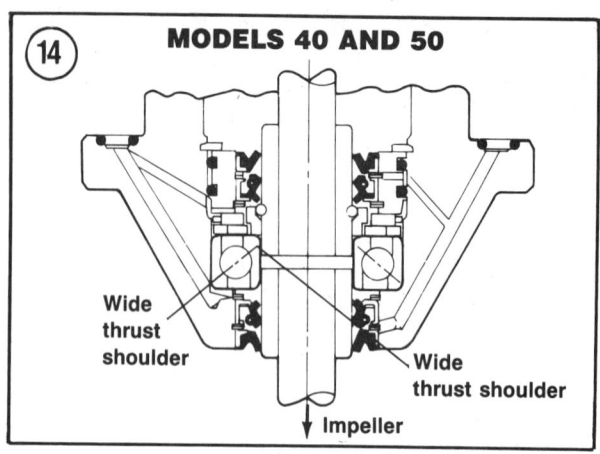

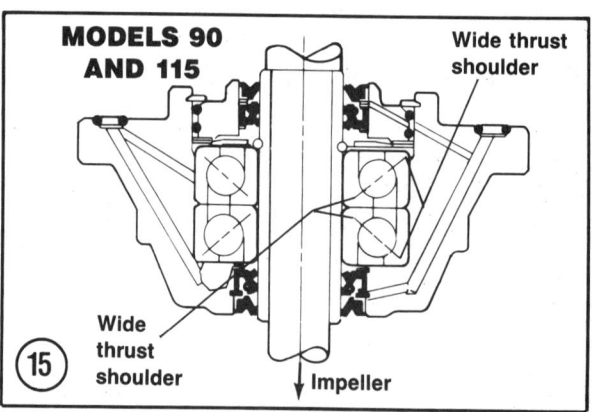

17. Bend the ears on the washer located behind the jet drive impeller retaining nut against the nut flats so to retain the torque setting on the nut.

18. Reinstall the intake housing as previously outlined under Steps 12 and 13.

19. Install directional control cable in jet drive support bracket and reconnect cable end to lower arm of the control cable pivot bracket.

20. Refer to *Directional Control* in this chapter and adjust cable to provide correct operation of thrust gate.

21. Complete reassembly in the reverse of disassembly.

BEARING HOUSING

It is recommended that service to the bearing housing assembly be performed by

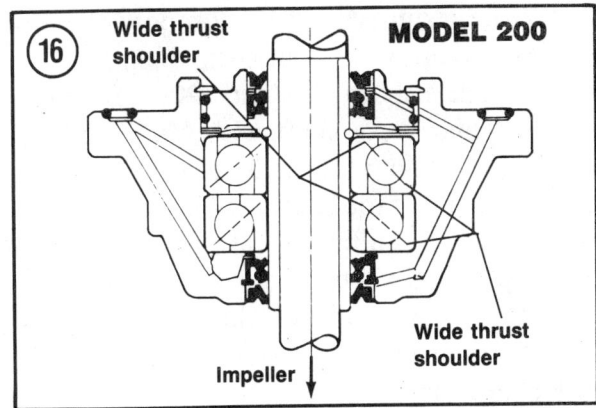

only an authorized Yamaha Outboard Motor dealer. You can save some labor rate cost by removing and installing the jet drive assembly as outlined in this chapter. Should you elect to service the bearing housing assembly yourself, read the following procedure completely before proceeding.

Refer to **Figure 12** or **Figure 13** for this procedure.

1. Remove the jet drive assembly as outlined under *Removal* in the *JET DRIVE* section.
2. Remove the water pump assembly and Woodruff key as outlined under *WATER PUMP* in this chapter.
3. Withdraw the aluminum spacer.
4. Remove the bolts and washers securing the bearing housing to the jet drive housing. Withdraw the bearing housing assembly and drive shaft from the jet drive housing and place on a clean work bench.
5. Remove snap ring from bore in top of bearing housing.

CAUTION
Do not apply excessive heat to bearing housing as grease seals may be damaged.

WARNING
*Components will be **hot**. Take necessary safety precautions to prevent personal injury.*

6. Apply heat to bearing housing in increments. After each application of heat, strike the impeller end of the drive shaft against a wooden block. If the bearing housing has been heated to a sufficient temperature, the housing should slide down off the bearing(s). If not, apply additional heat to bearing housing and reattempt removal of bearing housing.
7. Withdraw upper seal carrier assembly and spacer washer (Models 40 and 50) or cupped washer (Models 90, 115 and 200) from drive shaft.
8. Press bearing(s) off of drive shaft.
9. Slide thrust washer off of drive shaft.
10. Remove collar and thrust ring from drive shaft if needed.
11. Remove grease seals and retaining rings from bearing housing.
12. Remove grease seals and retaining rings from upper seal carrier.
13. Clean and inspect bearing housing, upper seal carrier and drive shaft.

NOTE
Whenever bearing(s) are pressed from drive shaft, it is recommended to replace bearing(s).

14A. Models 40 and 50—Install the thrust washer onto the drive shaft with the gray teflon coated side facing toward the jet drive impeller end. Press the new bearing onto the drive shaft. Press *only* against the inner race and position the bearing so that the bearing's thrust shoulders are positioned as shown in **Figure 14**.
14B. Models 90, 115 and 200—Install the thrust washer onto the drive shaft. Use Yamaha tool YB-06075 and press the new bearings onto the drive shaft. Press *only* against the inner races and position the bearings so that the bearings' thrust shoulders are positioned as shown in **Figure 15** for Models 90 and 115 and **Figure 16** for Model 200.
15. Install new grease seals and retaining rings into the bearing housing and upper seal carrier while noting the following:

10

a. Wipe a light film of Yamalube All-purpose Marine grease or a NLGI No.1-rated grease onto the surfaces of each retaining ring and grease seal prior to installing.

b. The grease seal open side or side with the lip tension spring faces outward. See **Figure 14** for Models 40 and 50, **Figure 15** for Models 90 and 115 **Figure 16** for Model 200.

c. Position the outer retaining ring so that its notched ends align with the small vent hole in the ring groove.

d. Fill the open side of each grease seal with Yamalube All-purpose Marine grease or a NLGI No.1-rated grease after installation.

e. Wipe a light film of Yamalube All-purpose Marine grease or a NLGI No.1-rated grease onto the inner suface of the bearing housing to ease installation of the bearing(s) and upper seal carrier.

16. Slide bearing housing onto drive shaft.

CAUTION
Do not apply excessive heat to bearing housing as grease seals may be damaged.

WARNING
*Components will be **hot**. Take necessary safety precautions to prevent personal injury.*

17A. Models 40 and 50—Apply heat to bearing housing in increments. If the bearing housing has been heated to a sufficient temperature, the housing should begin to slide onto the bearing. If not, apply additional heat to bearing housing and reattempt installation procedure. Make sure the bearing housing slides squarely onto the bearing. Use a hand press to fully seat bearing in bearing housing. Only press against the outer race of the bearing. If bearing and bearing housing are properly aligned, only light pressure should be required to seat components. *Do not* use a hammer to seat components.

17B. Models 90, 115 and 200—Support bearings and drive shaft with Yamaha tool YB-06205-2. Apply heat to bearing housing in increments. If the bearing housing has been heated to a sufficient temperature, the housing should begin to slide onto the bearings. If not, apply additional heat to bearing housing and reattempt installation procedure. Make sure the bearing housing slides squarely onto the bearings. Use a hand press to fully seat bearings in bearing housing. Only press against the outer race of the bearing. If bearings and bearing housing are properly aligned, only light pressure should be required to seat components. *Do not* use a hammer to seat components.

NOTE
Yamaha tool YB-06205-2 will prevent the bearing housing and bearings from completely seating. After initial pressing, remove tool YB-06205 and install upper seal carrier as a spacer then reinstall tool YB-06205 and completely seat components.

18A. Models 40 and 50—Install spacer washer.

18B. Models 90, 115 and 200—Install cupped washer with dished side facing upward.

19. Install two O-rings on outside of upper seal carrier.

20. Lightly grease O-rings with Yamalube All-purpose Marine grease or a NLGI No.1-rated grease.

21. Install upper seal carrier. Only light hand pressure should be required.

22. Install the snap ring with the beveled side facing upward.

23. Install the bearing housing assembly and drive shaft into the jet drive pump housing.

24. Apply Yamalube All-purpose Marine grease or a NGLI No.1-rated grease on the bearing housing retaining bolts.

25. Install bearing housing retaining bolts and washers and tighten to specification shown in **Table 1**.

26. Install the aluminum spacer on top of the jet drive housing.

27. Install the water pump assembly as outlined under *WATER PUMP* in this chapter.

28. Install the jet drive assembly as outlined under *Installation* in the *JET DRIVE* section.

29. Grease bearings as outlined under *Bearing Lubrication* in the *MAINTENANCE* section.

INTAKE HOUSING LINER

Replacement

1. Remove the six intake housing mounting bolts. Remove intake housing. See **Figure 12** or **Figure 13**.

2. Identify the liner bolts for reassembly in the same location, then remove the bolts.

3. Tap the liner loose by inserting a long drift punch through the intake housing grille and placing the punch on the edge of the liner and tapping with a suitable hammer.

4. Withdraw the liner from the intake housing.

5. Install a new liner into the intake housing.

6. Align liner bolt holes with respective intake housing holes. Tap liner into place with a suitable hammer if needed.

7. Apply Yamalube All-purpose Marine grease or a NGLI No.1-rated grease on liner retaining bolts prior to installation.

8. Install liner retaining bolts and tighten to specification shown in **Table 1**.

9. Remove any burrs from the liner retaining bolt area.

10. Grease the threads of the six intake housing mounting bolts with Yamalube All-purpose Marine grease or a NLGI No.1-rated grease.

11. Install the intake housing on the jet drive pump housing with the lowest part of the intake grill facing aft. Tighten the six intake housing bolts to the specification shown in **Table 1**.

12. Refer to *Impeller Clearance* in this chapter.

10

Table 1 JET DRIVE TIGHTENING TORQUES

Fastener	ft.-lb.	N·m
Adapter plate-to- intermediate housing		
8 mm	11*	15
10 mm	22*	30
Bearing housing-to- jet drive housing		
Models 40 and 50	Hand tighten*	
Models 90, 115 and 200	5*	7
Impeller nut	17*	23
Intake housing	5*	7
Intake housing liner	11*	15
Jet drive-to-adapter plate	11*	15
Water pump housing	11*	15

*Grease fastener threads with Yamalube All-purpose Marine grease or a NLGI No.1-rated grease prior to installation.

Chapter Eleven

Automatic Rewind Starter

All manual start (and some electric start) models are equipped with a rope-operated rewind starter. Electric start models not equipped with a rewind starter have a flywheel drive cup and starter rope for emergency starts.

The starter assembly is mounted above the flywheel on all models. See **Figure 1** (typical). Pulling the rope handle causes the starter sheave shaft to rotate against spring tension, moving the drive pawl to engage the flywheel and turn the engine over. When the rope handle is released, the spring inside the assembly reverses direction of the sheave shaft and winds the rope around the sheave.

All 4 hp and larger outboards are equipped with a starter lockout cable (**Figure 2**). When properly adjusted, this assembly prevents operation of the rewind starter unless the shift lever is positioned in NEUTRAL.

Automatic rewind starters are relatively trouble-free; a broken or frayed rope is the most common malfunction. This chapter covers removal and installation of the rewind starter assembly, rope and spring replacement and starter lockout cable adjustment. **Figure 3** shows the components of a typical Yamaha rewind starter assembly (5 hp and larger).

Rope and Spring Replacement (2 and 3 hp)

The spring should only be removed if broken. When replacing the rope, omit the

steps dealing with spring removal and installation. Refer to **Figure 4** for this procedure.

1. Remove the engine cowling.

2. Remove the 3 bolts and any related hardware holding the starter assembly to the engine mounting brackets. Remove the starter assembly (**Figure 5**).

3. Invert the rewind assembly on a clean work bench. Untie the knot in the starter rope and remove the rope from the handle.

4. Fit the rope in the sheave groove and let the sheave turn slowly to relax the starter spring.

5. Remove the starter shaft bolt and shaft.

6. Remove the drive pawl spring from the starter shaft.

7. Remove the drive pawl and return spring from the sheave.

WARNING
Removing the starter sheave without holding the spring in place with a screwdriver in Step 8 will allow the spring to unwind violently and may result in serious personal injury. Wear safety glasses while removing and installing the spring.

8. Insert a screwdriver in the sheave hole as shown in **Figure 6** to hold the starter spring in place.

9. Remove the sheave and rope.

10. Untie the rope knot and remove the rope from the sheave.

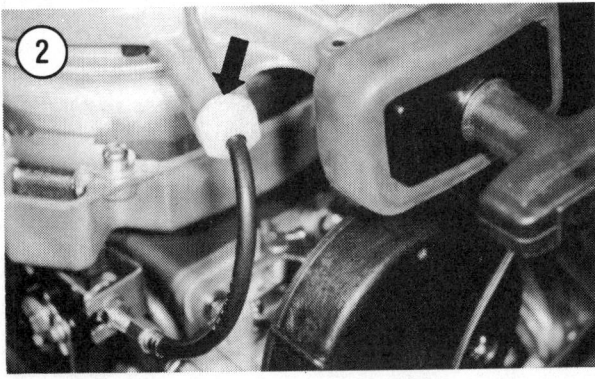

11. If the spring requires replacement, place the rewind assembly housing on the floor right side up. Tap lightly on the top of the housing while holding it tightly against the floor. The spring will fall out of the housing and unwind inside the housing mounting flanges. When the spring has unwound, pick the housing up off the floor and discard the spring.

12. Secure one housing leg in a vise with protective jaws and wipe the inside surface with Yamalube All-purpose Marine grease.

NOTE
Replacement springs are properly coiled at the factory. Do not uncoil the spring before installation.

13. Install the new spring coil in the starter housing. Fit the outer spring loop over the housing catch. With the spring fully seated and connected in the housing, carefully remove the retaining tie straps. The spring will uncoil slightly to fill the housing cavity.

14. Tie a knot in one end of the new starter rope and insert the rope through the sheave hole.

15. With the sheave facing flywheel end up, wind the rope 3 1/2 turns counterclockwise around it, then route the rope through the sheave groove.

16. Wipe the sheave with a light coat of Yamalube All-purpose Marine grease and install it in the starter housing. Make sure the inner spring hook is positioned over the sheave catch.

17. Install the drive pawl and return spring in the sheave.

18. Make sure the drive pawl spring is installed on the starter shaft (**Figure 7**).

19. Install starter shaft on sheave with open end of pawl spring engaging the round tab on the pawl.

20. Install the starter shaft bolt and tighten securely.

11

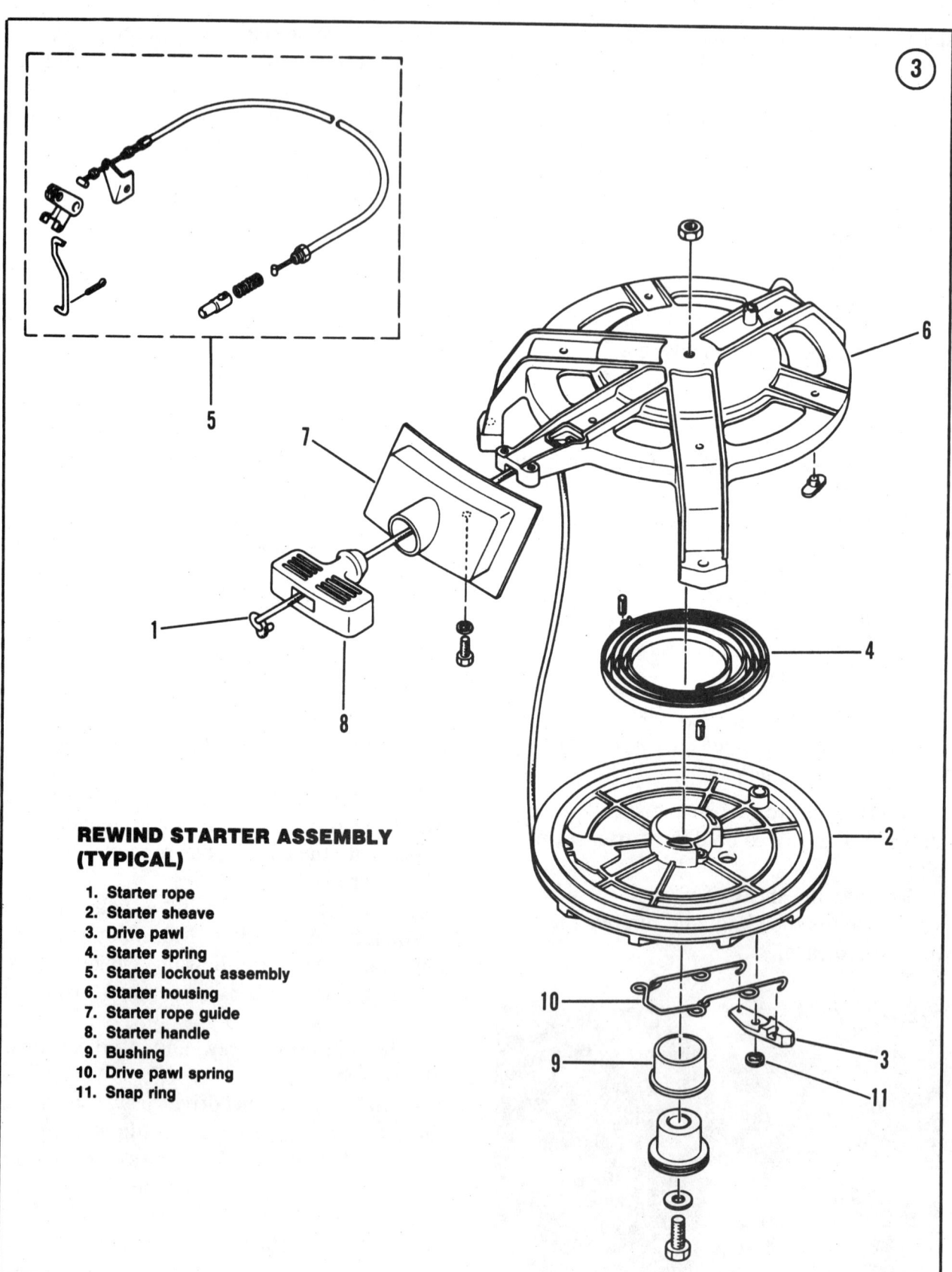

REWIND STARTER ASSEMBLY (TYPICAL)

1. Starter rope
2. Starter sheave
3. Drive pawl
4. Starter spring
5. Starter lockout assembly
6. Starter housing
7. Starter rope guide
8. Starter handle
9. Bushing
10. Drive pawl spring
11. Snap ring

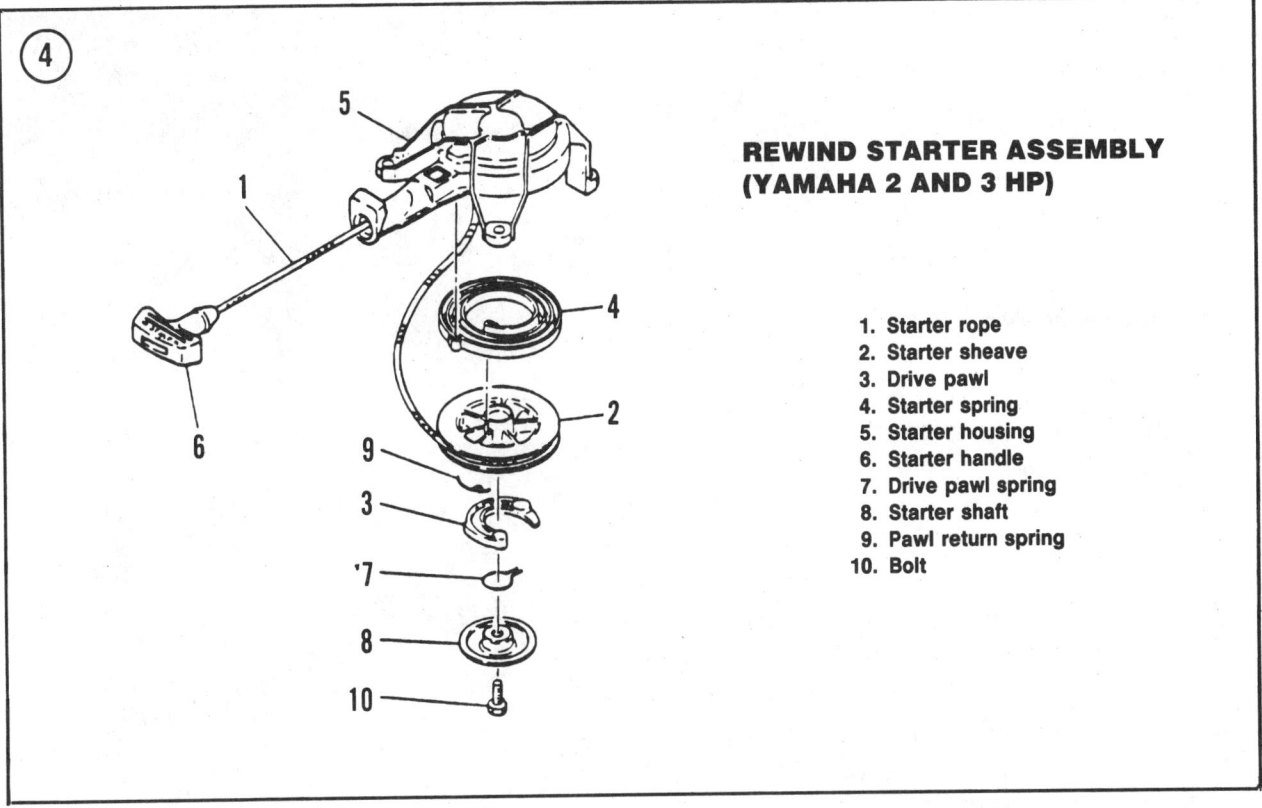

④

⑤

REWIND STARTER ASSEMBLY
(YAMAHA 2 AND 3 HP)

1. Starter rope
2. Starter sheave
3. Drive pawl
4. Starter spring
5. Starter housing
6. Starter handle
7. Drive pawl spring
8. Starter shaft
9. Pawl return spring
10. Bolt

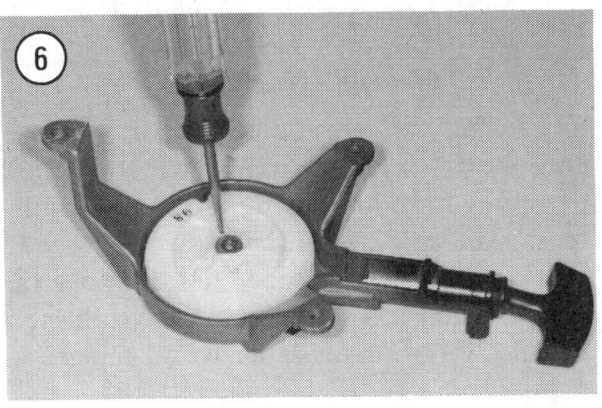

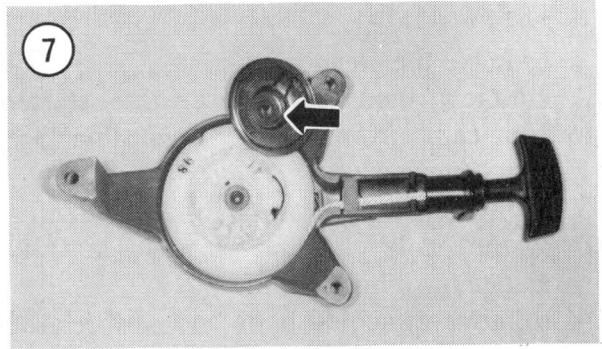

11

21. Depress the sheave and rotate it 3 turns counterclockwise. Maintain pressure on the sheave and route the rope through the rope guide in the starter housing and handle. Tie a knot in the end of the rope as shown in **Figure 8**. Release the sheave slowly and allow it to take up any slack remaining in the rope.

22. Install rewind starter assembly and any related hardware to the engine mounting brackets. Tighten bolts securely.

23. Install the engine cowling.

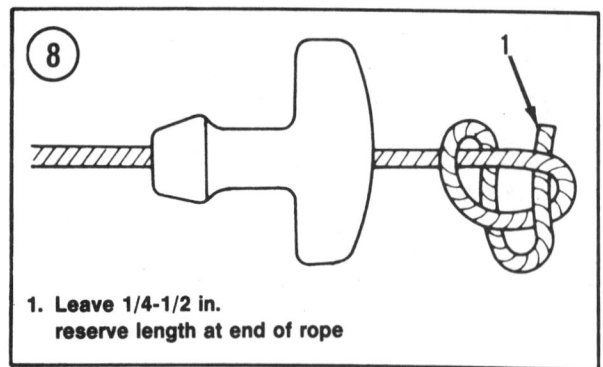

1. Leave 1/4-1/2 in.
reserve length at end of rope

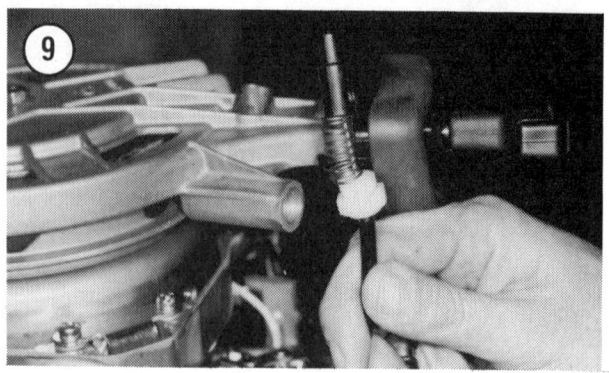

Rope and Spring Replacement (All Other Models)

The starter drive components used in some 9.9 and 15 hp models may be made of plastic. If such parts fail under continued hard use, replace them with the aluminum parts used in the 8 hp model. On-going production changes in starter component design require that you obtain the exact replacement parts to assure proper operation.

Refer to **Figure 3** (typical) for this procedure.

1. Remove the engine cover.
2. Unscrew and disconnect the starter lockout cable from the starter housing. See **Figure 9** (typical).
3. Remove the lockout plunger and spring from the cable end and place in a small container for reinstallation.
4. Remove attaching bolts and any related hardware from the starter housing. Remove the starter housing and place it flywheel side up on a clean work bench (**Figure 10**).
5. Remove the snap ring or circlip holding the drive pawl in place. Carefully remove the drive pawl and spring assembly (**Figure 11**).

> *WARNING*
> *Apply sufficient pressure on the sheave in Step 6 to prevent it from unwinding rapidly or serious personal injury may result.*

6. Hook the starter rope in the sheave groove. Depress the sheave and let it rotate slowly until all spring tension is removed.
7. Untie the knot in the starter rope. Remove handle from rope.
8. Remove the starter shaft bolt and nut. Remove the shaft (**Figure 12**).

> *WARNING*
> *Perform Step 9 slowly and carefully to prevent the spring from snapping out of the housing and unwinding. This can cause serious personal injury. Wear safety glasses while removing and installing the spring.*

9. Carefully lift one end of the starter sheave and insert a screwdriver under it to prevent the spring from flying up as the sheave is fully removed. See **Figure 13**.
10. Untie the starter rope knot. Remove the rope and bushing from the sheave. Inspect bushing for wear or damage and replace as required.

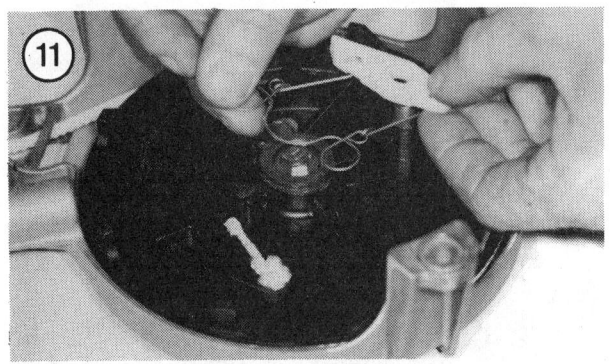

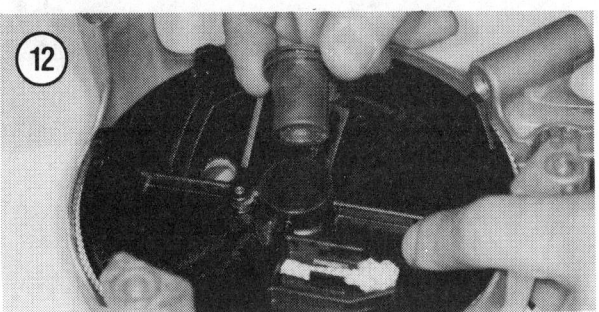

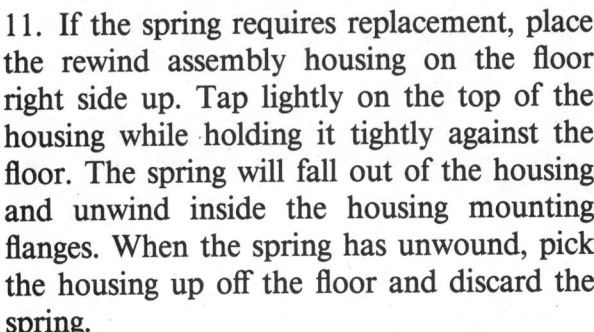

11. If the spring requires replacement, place the rewind assembly housing on the floor right side up. Tap lightly on the top of the housing while holding it tightly against the floor. The spring will fall out of the housing and unwind inside the housing mounting flanges. When the spring has unwound, pick the housing up off the floor and discard the spring.

12. Secure one housing leg in a vise with protective jaws and wipe the inside surface with Yamalube All-purpose Marine grease.

NOTE
Replacement springs are properly coiled at the factory. Do not uncoil the spring before installation.

13. Install the new spring coil in the starter housing. Fit the outer spring loop over the housing pin. With the spring fully seated and connected in the housing, carefully remove the retaining tie straps. The spring will uncoil slightly to fill the housing cavity. See **Figure 14.**

NOTE
If a pre-cut length of rope is not available and the dealer cuts one from a roll, make sure that it is the same length as the old rope.

14. Tie a knot in the end of a new rope. Insert the other end of the rope through the sheave hole.

15. Wind the rope counterclockwise 2 1/2 turns around the sheave, then route it through the sheave groove.

16. Lightly coat the sheave with Yamalube All-purpose Marine grease at contact points and install in the housing. Make sure the inner spring hook engages the sheave pin.

17. Wipe the starter shaft bushing and shaft with the same lubricant. Install the bushing and shaft in the housing.

NOTE
Be sure to tighten the nut securely on the starter shaft bolt in Step 18. If the nut works loose and comes off, the spring will unwind inside the housing and the starter will not function.

11

18. Wipe the shaft bolt threads with Loctite Type A and install with the washer and nut. Tighten the nut securely.

19. Depress the starter sheave and rotate it counterclockwise 5 full turns to tension the spring. Maintain the pressure to keep the sheave from unwinding and route the rope through the hole in the housing and handle. Tie a figure-8 knot in the rope and gradually release the pressure. The sheave will take up any slack in the rope.

20. Install the drive pawl and spring assembly. Install the snap ring or circlip to hold the assembly in place, then bend the spring ends (if necessary) to make sure they will not slip out of the drive pawl.

21. Install the starter assembly and any related hardware to the engine.

22. Fit the lockout cable spring and plunger on the cable. Install the lockout cable assembly to the starter housing and tighten the attaching nut securely.

23. Adjust the lockout cable as described in this chapter.

24. Install the engine cover.

Starter Lockout Cable Adjustment

1. Loosen the lockout cable adjusting locknuts. See **Figure 15** (typical).

2. Adjust the nuts as required to permit the starter rope to be pulled out with the shift lever in NEUTRAL but not when the lever is in FORWARD or REVERSE.

3. Tighten adjusting locknuts securely.

Chapter Twelve

Power Trim and Tilt Systems

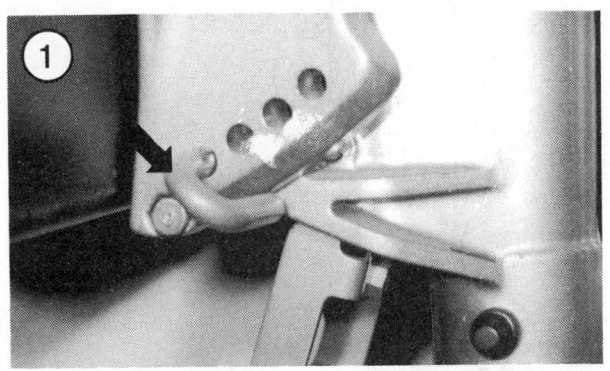

The usual method of raising and lowering the outboard gearcase is a mechanical one, consisting of a series of holes in the transom mounting bracket. To trim the engine, an adjustment stud or trim pin (**Figure 1**) is removed from the bracket, the outboard is repositioned and the stud or pin is reinserted in the proper set of holes to retain the unit in place.

A power tilt system is standard on all 1986-on 40 hp (40ETL only) models. See **Figure 2**. A power trim and tilt system is standard on Pro 50 models on 70-225 hp models. See **Figure 3**.

This chapter includes maintenance, troubleshooting procedures and removal/installation of the power tilt and power trim/tilt systems.

POWER TILT SYSTEM

Components

This system consists of a hydraulic pump (containing an electric motor, oil reservoir, oil pump and valve body) and a hydraulic tilt cylinder. See **Figure 4**. A tilt switch mounted in the remote control box handle or on the motor on models so equipped sends current to the up/down relays mounted on the bottom cowling at the rear of the starter relay. See **Figure 5** and **Figure 6**. The relays connect to a terminal assembly or junction box on the power head (**Figure 6**), sending current to the power tilt motor.

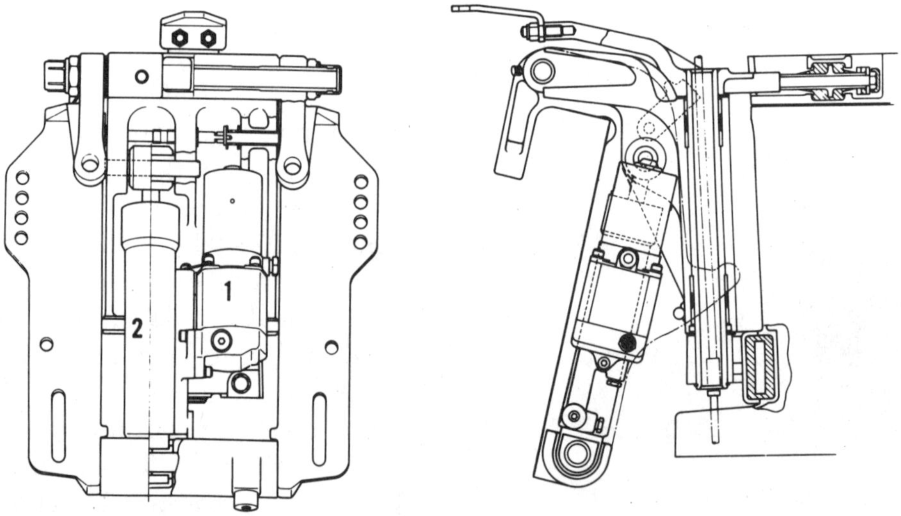

1. Hydraulic pump assembly
2. Tilt cylinder

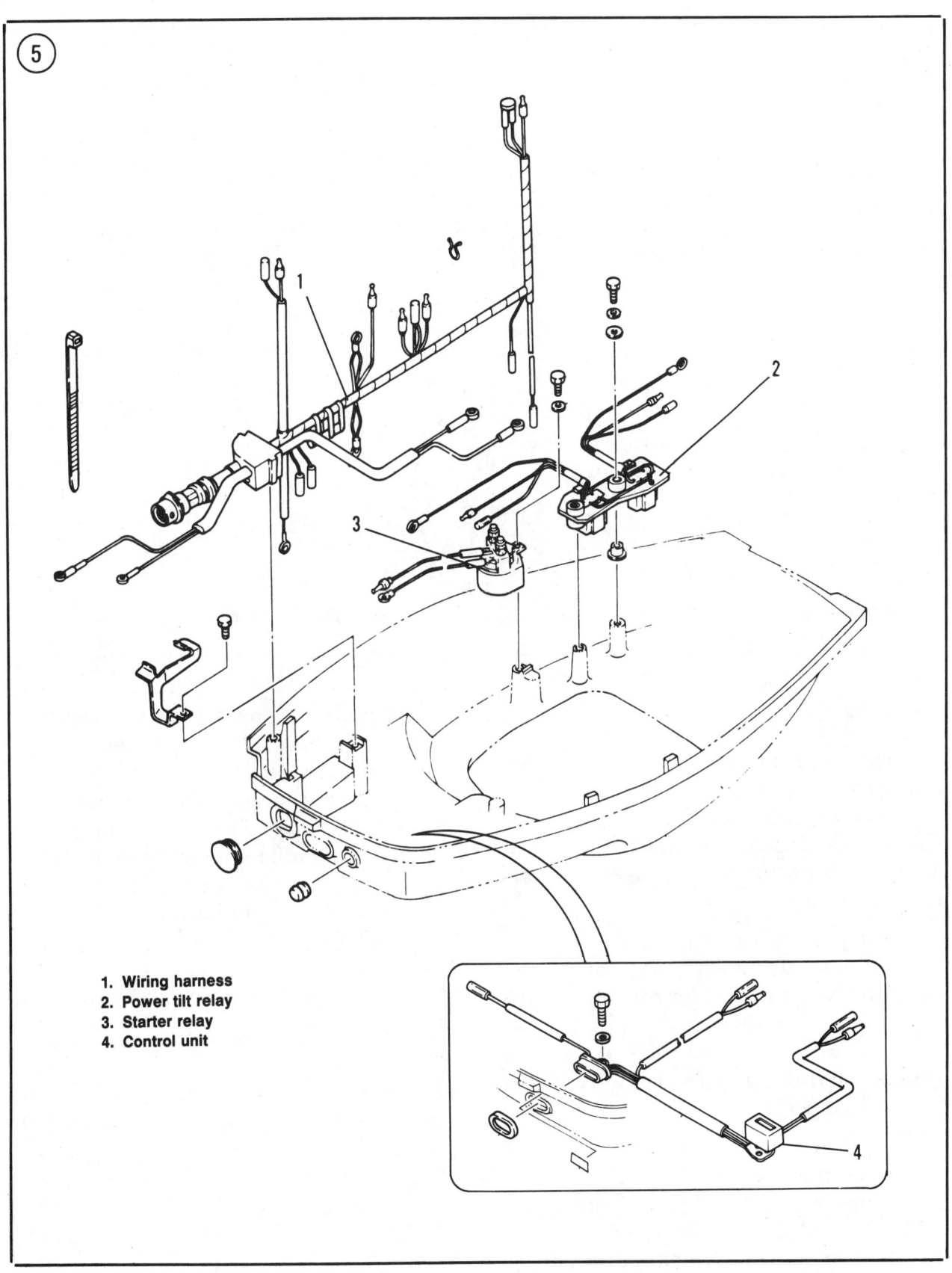

1. Wiring harness
2. Power tilt relay
3. Starter relay
4. Control unit

12

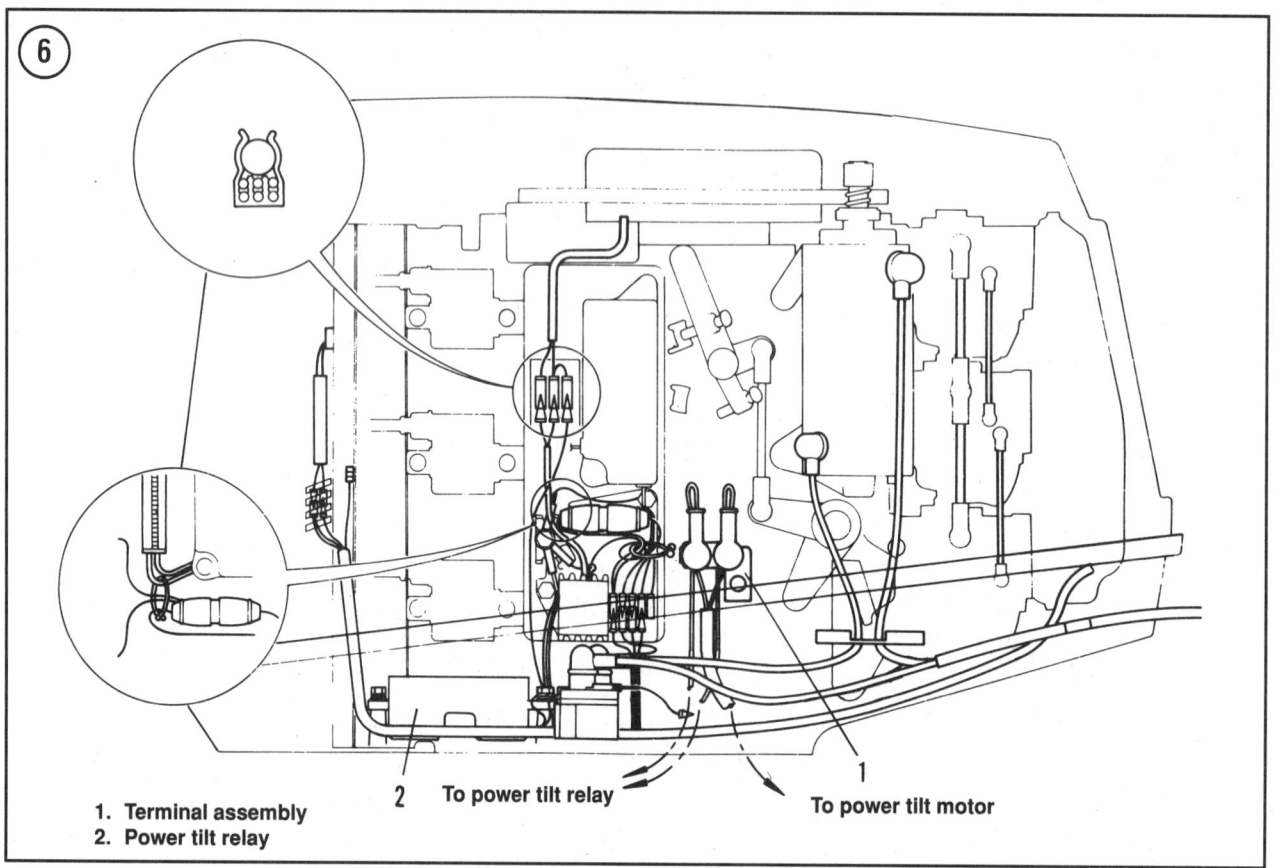

1. Terminal assembly
2. Power tilt relay

Operation

Moving the tilt switch to the UP position closes the pump motor circuit. The motor drives the oil pump, forcing oil into the up side of the tilt cylinder. The engine moves upward until it reaches its maximum position.

Moving the tilt switch to the DOWN position also closes the pump motor circuit. The reversible motor runs in the opposite direction, driving the oil pump to force oil into the down side of the tilt cylinder and bringing the engine back to the desired position.

To prevent damage if the outboard strikes an underwater object, the tilt piston moves up and hydraulic fluid flows through a valve in the piston. As the piston moves upward, a partial vacuum

forms under it and allows the piston to move downward, absorbing the shock.

A manual release valve is provided on the hydraulic pump. Turning the valve screw (A, **Figure 7**) opens the upper and lower tilt cylinder chambers to the reservoir. This permits manual raising and lowering of the engine if the electrical system fails.

Hydraulic Pump Fluid Check

1. Tilt the outboard to its fully UP position.
2. Clean area around pump fill plug. Remove the plug (B, **Figure 7**) and visually check the fluid level in the pump reservoir. It should be at the bottom of the fill hole threads (**Figure 8**).
3. Top up if necessary with Yamalube Power Trim/Tilt Fluid.

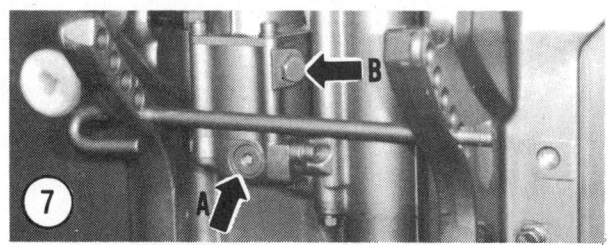

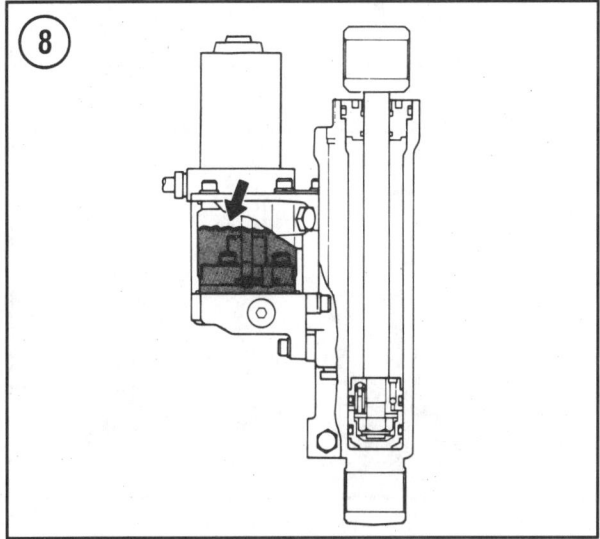

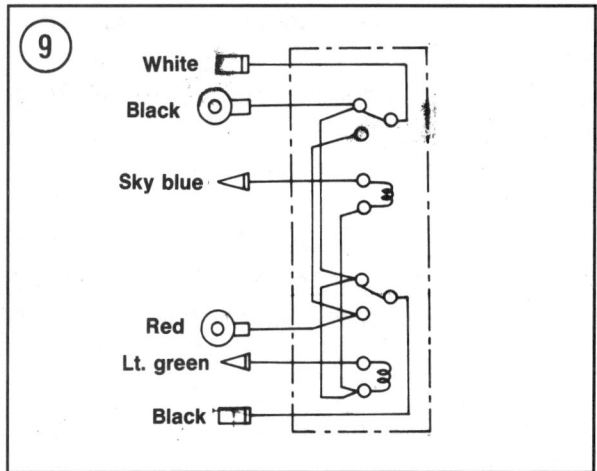

Hydraulic Pump Fluid
Bleed and Refill

Follow this procedure when a large amount of fluid has been lost due to an overhaul of the system or a leak that has been corrected.

1. Tilt the outboard to the fully UP position. Remove the fill plug (B, **Figure 7**) and top off the reservoir with Yamalube Power Trim/Tilt Fluid. Install the fill plug.

2. Turn the manual valve screw (A, **Figure 7**) clockwise until it stops.

3. Slowly pull upward on the tilt rod to allow the fluid to enter the lower tilt cylinder chamber.

4. Repeat Step 1. Turn the manual valve screw counterclockwise until it stops.

5. Slowly push down on the tilt rod to allow the fluid to enter the upper tilt cylinder chamber.

6. Repeat Steps 1-5 as required to completely bleed and fill the tilt cylinder with fluid.

Troubleshooting

Whenever a problem develops in the power tilt system, the initial step is to determine whether the problem is in the electrical or hydraulic system. Electrical and hydraulic tests are given in this chapter. If the problem appears to be in the hydraulic system, refer it to a dealer or qualified specialist for necessary service.

1. Make sure the plug-in connectors are properly engaged and that all terminals and wires are free of corrosion. Clean and tighten as required.

2. Make sure the battery is fully charged. Charge or replace as required.

3. Check the fluid level as described in this chapter. Top off if necessary.

4. Make sure the manual release valve (A, **Figure 7**) is fully closed.

Tilt Relay Test

1. Disconnect the negative battery lead, then the positive battery lead.

2. Disconnect the six tilt relay leads (**Figure 9**).

12

3. Connect an ohmmeter or test lamp between the disconnected red and black relay leads. The meter should show continuity (test lamp should light).

4. Connect a voltmeter between the red and black tilt relay leads. Use jumper leads to connect the light green relay lead to the positive battery terminal and the black relay lead to the negative battery terminal. The meter should read 12 volts.

5. Connect an ohmmeter or test lamp between the disconnected red and white relay leads. The meter should show continuity (test lamp should light).

6. Connect a voltmeter between the red and white tilt relay leads. Use jumper leads to connect the sky blue relay lead to the positive battery terminal and the black relay lead to the negative battery terminal. The meter should read 12 volts.

7. If the ohmmeter does not show continuity in Step 3 and Step 5 or if the voltmeter does not read 12 volts in Step 4 and Step 6, replace the tilt relay assembly.

Tilt Switch Test

The rocker-type tilt switch is mounted in the remote control box handle.

1. Disconnect the negative battery cable.

2. Remove the remote control box from its mounting bracket.

3. Remove the wire cover (**Figure 10**) and the back panels (**Figure 11**).

4. Disconnect the tilt switch leads (red, sky blue and light green).

5. Connect an ohmmeter between the sky blue and red switch leads. See **Figure 12**. The meter should show continuity when the switch is depressed in the UP direction.

6. Connect the ohmmeter between the red and light green switch leads. See **Figure 13**. The meter should show continuity when the switch is depressed in the DOWN direction.

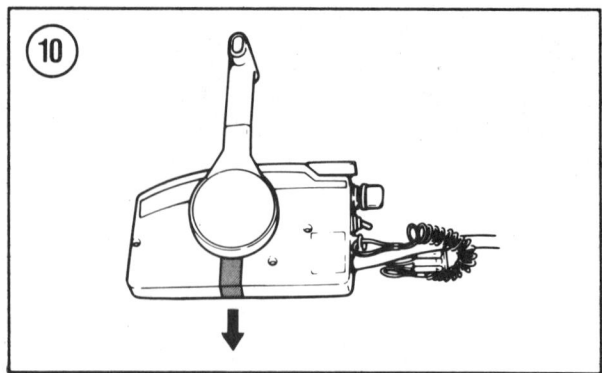

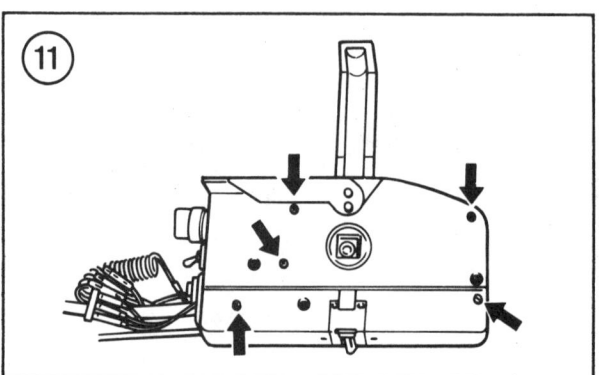

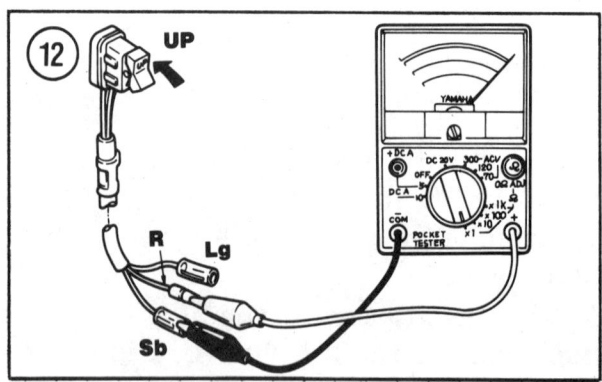

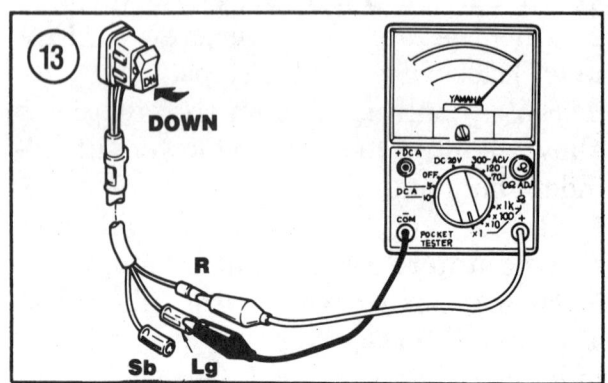

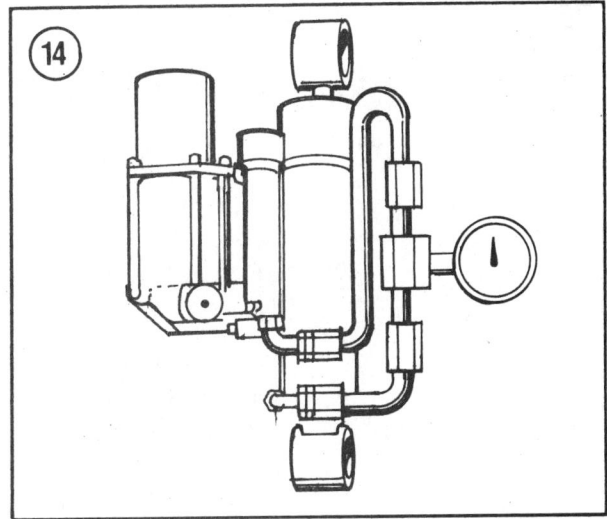

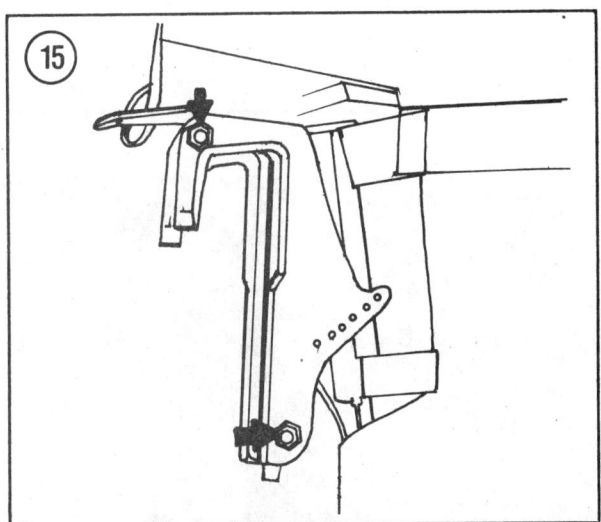

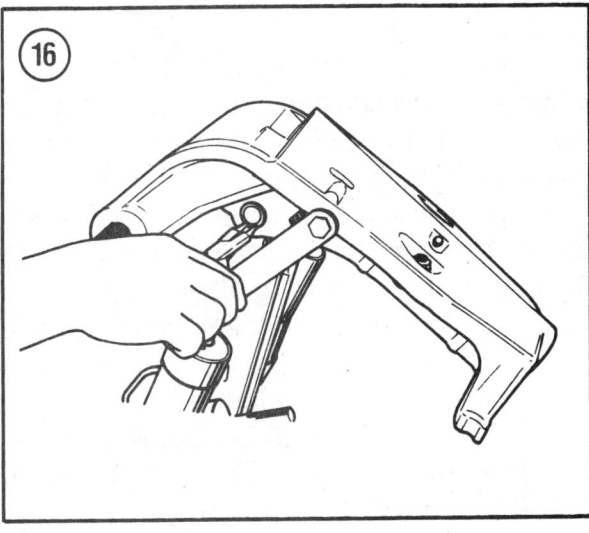

7. Connect the ohmmeter between the switch leads in the following order without depressing the switch:
   a. Red and sky blue.
   b. Red and light green.
   c. Sky blue and light green.

There should be no continuity between any of the leads with the switch in the OFF position.

8. If the switch does not perform as specified in Steps 5-7, replace it.

Hydraulic Test

1. Remove the tilt cylinder drain bolts and install pressure gauge part No. YB-6181 as shown in **Figure 14**.

2. Check the pump hydraulic fluid level as described in this chapter.

3. Move the tilt switch to the UP position and run the outboard up as far as it will go. The pressure gauge should read 500-570 psi.

4. Move the tilt switch to the DOWN position and run the outboard down as far as it will go. The pressure gauge should read 430-500 psi.

5. If the pressure gauge does not read as specified in Step 3 and Step 4, remove the power tilt unit as described in this chapter and have it serviced by a dealer or qualified specialist.

Removal/Installation

1. Disconnect the negative battery cable.

2. Remove the engine cover.

3. Disconnect the power tilt motor leads at the terminal assembly or junction box.

4. Unbolt and remove the outboard from the boat. Secure it to a repair stand.

5. Remove the upper and lower clamp bracket fastener on each side. See **Figure 15**. Remove the clamp brackets from the swivel bracket.

6. Remove the snap ring at each end of the tilt cylinder upper shaft. See **Figure 16** (typical). Carefully drive the upper shaft out.

12

7. Remove the power tilt assembly.

8. Installation is the reverse of removal. Check, bleed and refill the unit (as required) with hydraulic fluid as described in this chapter.

POWER TRIM/TILT SYSTEM

Components

This system consists of a manifold or trim/tilt housing containing a hydraulic pump (consisting of an electric motor, oil reservoir, oil pump and valve body) and 2 hydraulic trim pistons. A separate hydraulic tilt cylinder is attached to the housing by a shaft and bushings. See **Figure 17**. A trim/tilt switch mounted in the remote control box handle sends current to the up/down relays mounted on the bottom cowling to the rear of the starter relay on 70-90 hp (**Figure 5**) or the power head relay bracket on 115-220 hp (**Figure 18**). The relays connect to a terminal assembly or junction box on the power head, sending current to the power trim/tilt motor.

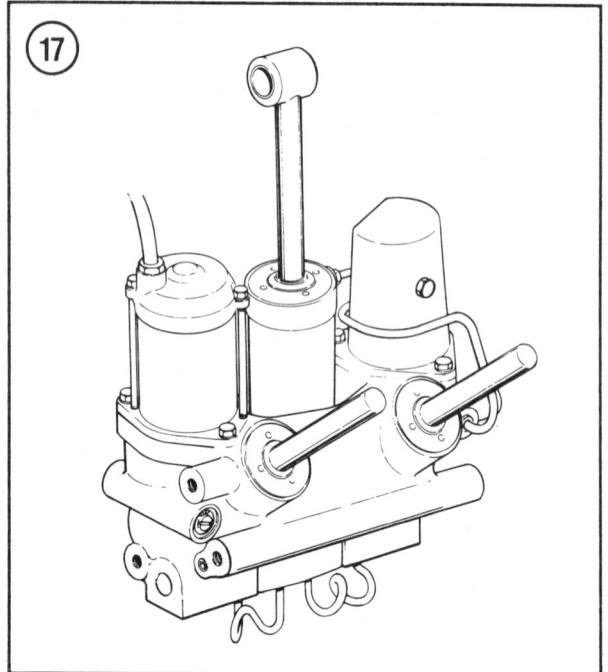

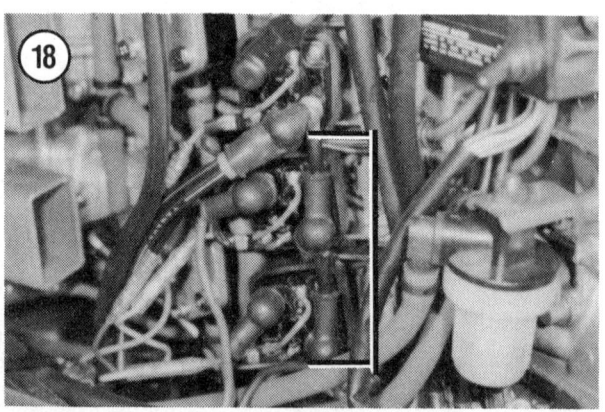

Operation

Moving the trim/tilt switch to the UP position closes the pump motor circuit. The motor drives the oil pump, forcing oil into the up side of the trim cylinders. The trim cylinder pistons push on the swivel bracket thrust pads to trim the engine upward. Once the trim cylinders are fully extended, the hydraulic fluid is diverted into the tilt cylinder, which now moves the engine throughout the remaining range of travel.

Moving the trim/tilt switch to the DOWN position also closes the pump motor circuit. The reversible motor runs in the opposite direction, forcing oil into the tilt cylinder and bringing the engine back to a position where the swivel brackets rest on the trim cylinder pistons. The pistons then lower the engine the remainder of the way.

The power trim/tilt system will temporarily maintain the engine at any angle within its range to allow shallow water operation at slow speed, launching, beaching or trailering.

To prevent damage if the outboard strikes an underwater object, a relief valve in the hydraulic pump opens to allow the outboard to pivot upward quickly and return slowly, absorbing the shock.

A trim gauge sending unit is located on the inside of the port clamp bracket (**Figure 19**). Access to the sending unit requires the engine to be fully tilted.

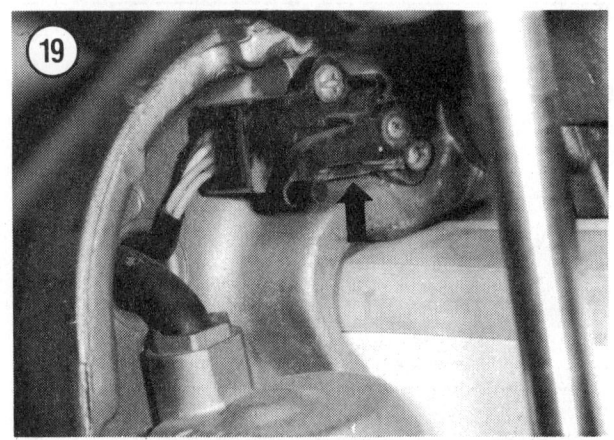

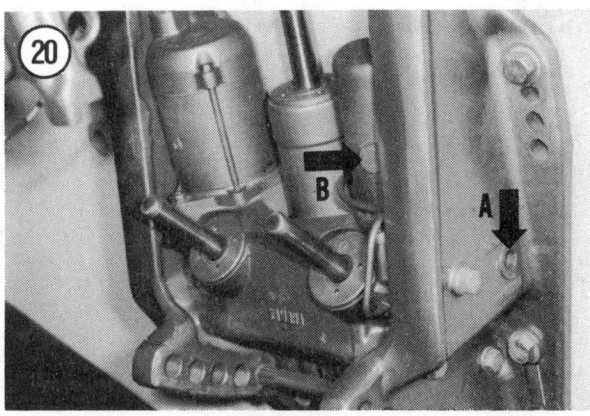

A manual release valve (A, **Figure 20**) is provided on the hydraulic pump. Opening this valve by turning its screw head permits manual raising and lowering of the engine if the electrical system fails.

Hydraulic Pump Fluid Check

1. Tilt the outboard to its fully UP position.
2. Clean area around pump fill plug. Remove the plug (B, **Figure 20**) and visually check the fluid level in the pump reservoir. It should be at the bottom of the fill hole threads.
3. Top up if necessary with Yamalube Power Trim/Tilt Fluid.

Hydraulic Pump Fluid Bleed and Refill

Follow this procedure when a large amount of fluid has been lost due to an overhaul of the system or a leakage that has been corrected.

1. Tilt the outboard to the fully UP position. Remove the fill plug (B, **Figure 20**) and top off the reservoir with Yamalube Power Trim/Tilt Fluid. Install the fill plug.
2. Operate the trim/tilt switch to completely lower, then raise the engine to its full tilt position, adding more fluid as required to keep its level at the bottom of the fill plug hole.
3. Operate the engine up and down several times.
4. Tilt the outboard to the fully UP position. Remove the fill plug and add fluid to the reservoir as required to keep the fluid level at the bottom of the fill plug hole.
5. Repeat Step 3 and Step 4 until the fluid level stabilizes at the bottom of the fill plug hole. Install and tighten the fill plug securely.
6. Recharge the battery. See Chapter Seven.

Troubleshooting

Whenever a problem develops in the power trim/tilt system, the initial step is to determine whether the problem is in the electrical or hydraulic system. Electrical and hydraulic tests are given in this chapter. If the problem appears to be in the hydraulic system, refer it to a dealer or qualified specialist for necessary service.

1. Make sure the plug-in connectors are properly engaged and that all terminals and wires are free of corrosion. Clean and tighten as required.
2. Make sure the battery is fully charged. Charge or replace as required.
3. Check the fluid level as described in this chapter. Top off if necessary.
4. Make sure the manual release valve (A, **Figure 20**) is fully closed.

12

Trim/Tilt Relay Test

1. Disconnect the negative battery lead, then the positive battery lead.

2. Disconnect the brown and black power trim/tilt relay leads.

3. Connect an ohmmeter or test lamp between the disconnected relay leads. The meter should show continuity (test lamp should light).

4. Connect a voltmeter or test lamp between the trim/tilt relay terminals. Use jumper leads to connect the brown relay lead to the positive battery terminal and the black relay lead to the negative battery terminal. See **Figure 21**.

5. The relay should click and the meter needle deflect or the test lamp light when the relay leads are connected to the battery in Step 4.

   a. If the relay clicks but the needle does not deflect (test lamp does not light) in Step 4, the relay contacts are defective. Replace the starter relay.

   b. If there is no relay click or needle deflection (test lamp does not light) in Step 4, the relay coil is defective. Replace the starter relay.

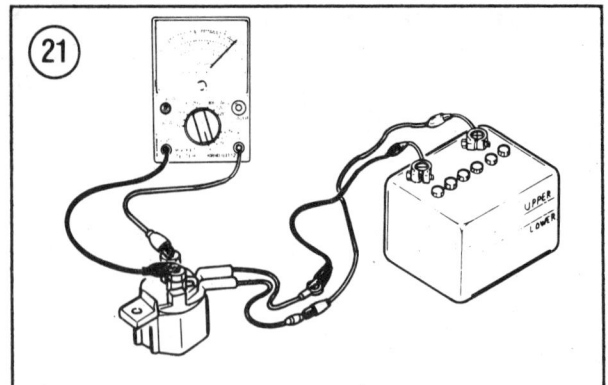

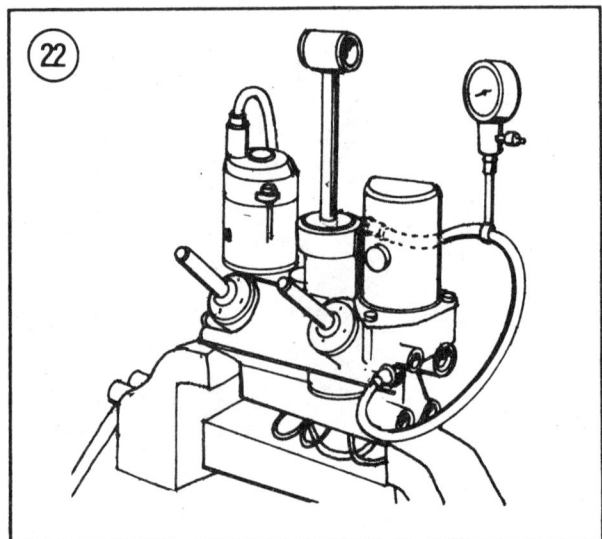

Trim/Tilt Switch Test

The rocker-type trim/tilt switch is mounted in the remote control box handle. A second switch is installed inside the bottom cowling of 1985-on V4 and V6 models to permit operation from the engine.

1. Disconnect the negative battery cable.

2. Remove the remote control box from its mounting bracket.

3. Remove the wire cover (**Figure 10**) and the back panels (**Figure 11**).

4. Disconnect the red, sky blue and light green tilt switch leads (the cowling switch on 1985-on V4 and V6 models has pink, sky blue and light green leads).

5. Connect an ohmmeter between the sky blue and red (pink) switch leads. See **Figure 12**. The meter should show continuity when the switch is depressed in the UP direction.

6. Connect the ohmmeter between the red (pink) and light green switch leads. See **Figure 13**. The meter should show continuity when the switch is depressed in the DOWN direction.

7. Connect the ohmmeter between the switch leads in the following order without depressing the switch:

   a. Red (pink) and sky blue.

   b. Red (pink) and light green.

   c. Sky blue and light green.

There should be no continuity between any of the leads with the switch in the OFF position.

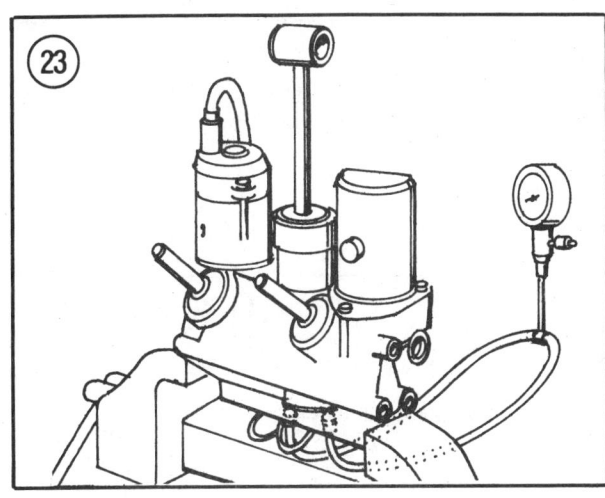

8. If the switch does not perform as specified in Steps 5-7, replace it.

Trim/Tilt Motor Test

1. Disconnect the pump motor electrical connector from the wiring harness.

2. Connect the red voltmeter lead to the blue wiring harness lead. Connect the black voltmeter lead to the black wiring harness lead.

3. Move the trim/tilt switch to the UP position. If the voltmeter does not read at least 12 volts, replace the up relay.

4. Move the red voltmeter lead to the green wiring harness lead.

5. Move the trim/tilt switch to the DOWN position. If the voltmeter does not read at least 12 volts, replace the down relay.

6. If the voltmeter readings are as specified in Step 3 and Step 5 and the motor still does not run, replace the motor.

Hydraulic Test

1. Turn the manual valve toward the manual tilt position until it stops.

2. Disconnect the hydraulic lines at the bottom of the reservoir and the upper chamber of the tilt cylinder. Install pressure gauge part No. YB-6181 to the reservoir and tilt cylinder fittings. See **Figure 22**.

3. Turn the manual valve to the power tilt position and immediately operate the trim/tilt switch to drive the outboard up and then down.

4. Turn the manual valve toward the manual tilt position until it stops (to bleed any air from the system), then return it to the power tilt position.

5. Check the hydraulic fluid level as described in this chapter.

6. Operate the trim/tilt switch to drive the outboard up, noting the pressure gauge readings. The gauge should read 0-71 psi during upward movement and 1,350-1,635 psi once the outboard has reached the full up position.

7. Return the outboard to the full down position, noting the pressure gauge readings. The gauge should read 85-156 psi once the outboard has reached the full down position.

8. Remove the pressure gauge and reconnect the hydraulic lines to the reservoir and tilt cylinder upper chamber.

9. Repeat Step 1.

10. Disconnect the hydraulic lines at the lower chambers of the trim cylinder and the tilt cylinder. Install pressure gauge part No. YB-6181 to the trim and tilt cylinder fittings. See **Figure 23**.

11. Repeat Steps 3-5.

12. Operate the trim/tilt switch to drive the outboard up, noting the pressure gauge readings. The gauge should read 0-71 psi during upward movment and 0 psi once the outboard has reached the full up position.

13. Return the outboard to the full down position, noting the pressure gauge readings. The gauge should read 583-782 psi once the outboard has reached the full down position.

14. Remove the pressure gauge and reconnect the hydraulic lines to the trim and tilt cylinder chambers.

15. Bleed and refill reservoir with hydraulic fluid as described in this chapter.

12

Trim/Tilt Housing
Removal/Installation

1. Disconnect the negative battery cable.
2. Remove the engine cover.
3. Disconnect the ground lead connected to the grease nipple on the starboard swivel bracket.
4. Disconnect the power trim/tilt motor leads at the terminal assembly or junction box. Disconnect the trim sender leads.
5. Unbolt and remove the outboard from the boat. Secure it to a repair stand.
6. Turn the manual valve screw to the manual tilt position, tilt the swivel bracket up and lock it in place with the tilt lock lever.
7. Remove the adjustment stud or trim pin from the clamp brackets.
8. Remove the nut from the starboard side clamp bracket.
9. Remove the bolts holding the trim/tilt housing to each clamp bracket. Separate the clamp brackets slightly.
10. Pull the trim sender and motor leads through the hole in the port clamp bracket.
11. Support the trim/tilt housing and remove the snap ring at the end of the tilt cylinder upper pin. See **Figure 24** (typical). Carefully drive the upper pin out.
12. Remove the power tilt assembly.
13. Installation is the reverse of removal. Check, bleed and refill the unit (as required) with hydraulic fluid as described in this chapter.

Trim Sender Removal/Installation

1. Raise the outboard to a full tilt position and support the unit in this position.
2. Disconnect the trim sender leads.
3. Remove the 2 screws holding the trim sender to the inside of the port clamp bracket (**Figure 19**).
4. Pull the trim sender leads through the hole in the port clamp bracket.

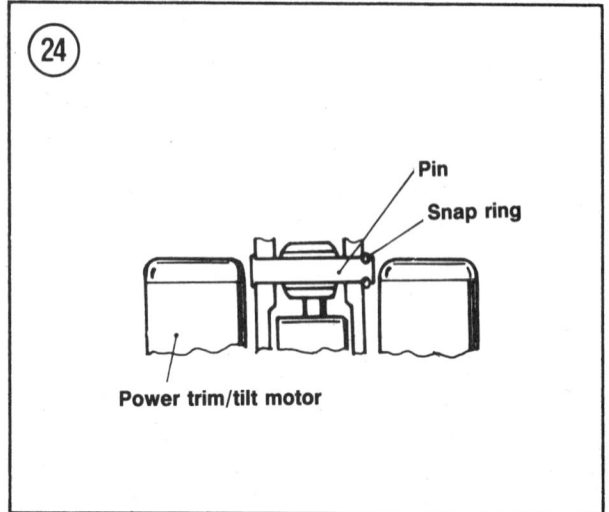

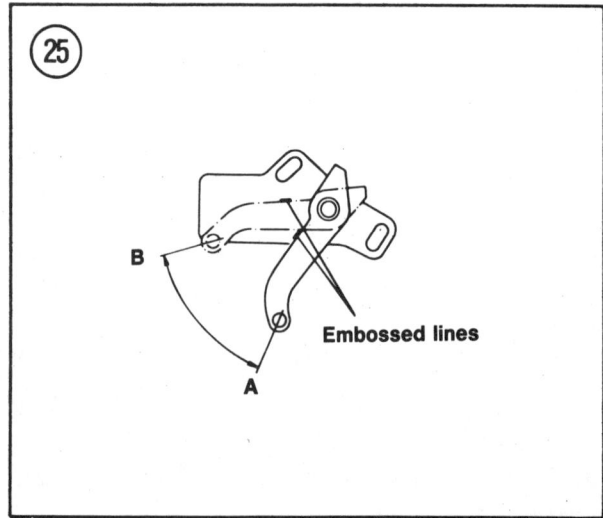

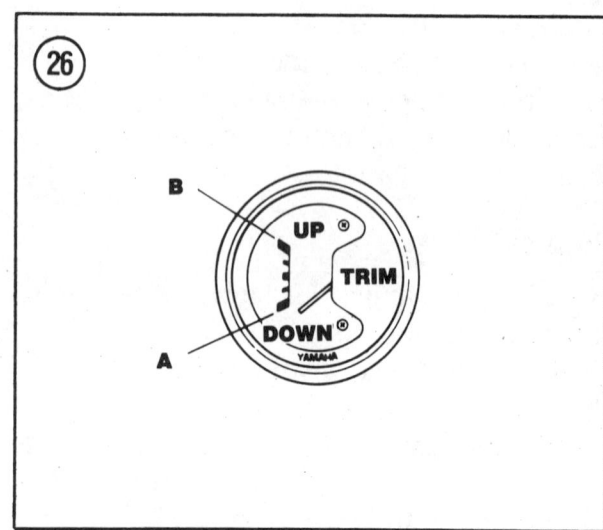

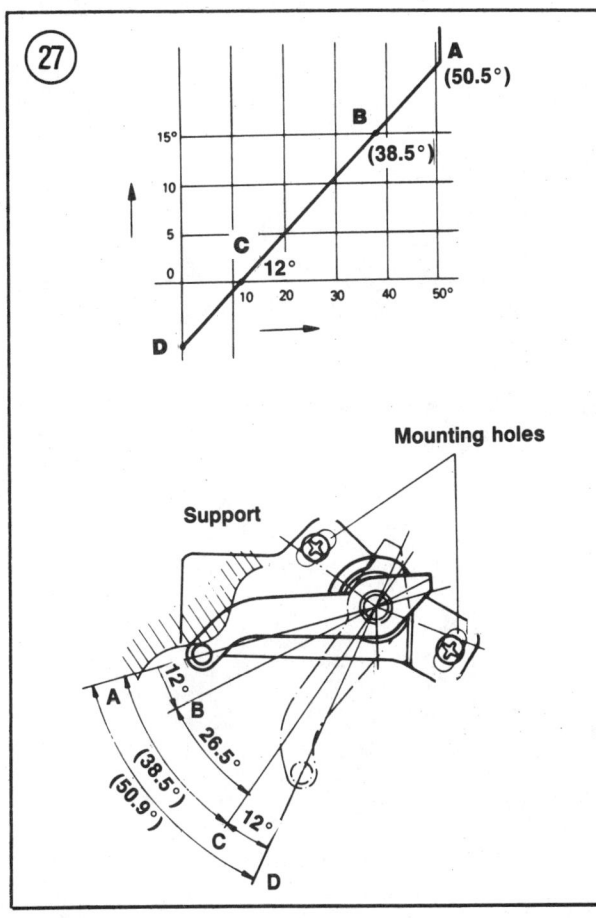

5. Installation is the reverse of removal. Adjust the trim sender as described in this chapter.

Trim Sender Adjustment

1. Raise the outboard to a full tilt position and support the unit in this position.

2. Manually move the trim sender lever to position A, **Figure 25**. The trim sender gauge should read full DOWN (A, **Figure 26**).

3. Manually move the trim sender lever to position B, **Figure 25**. The trim sender gauge should read full UP (B, **Figure 26**).

4. If the sender gauge is not in agreement with the lever position in Step 2 or Step 3, loosen the trim sender attaching screws.

5. Repeat Step 2 while repositioning the sender unit to achieve correct gauge alignment. When adjustment is correct, the outboard will trim upward a maximum 50.5°. See **Figure 27**.

12

Chapter Thirteen

Oil Injection System

The fuel-oil ratio required by outboard motors depends upon engine demand. Without oil injection, oil must be hand-mixed with gasoline at a predetermined ratio to assure that sufficient lubrication is provided at all operating speeds and engine load conditions. This ratio is adequate for high-speed operation, but contains more oil than required to lubricate the engine properly during idle and low-speed operation.

With oil injection, the amount of oil provided with the fuel sent to the engine cylinders can be varied instantly and accurately to provide the optimum ratio for proper lubrication at any operating speed or engine load condition.

All 1988-on 25 hp models, 1987-on 30 hp models, 1984-1986 40 hp electric start models and 1987-on 40 hp electric and manual start models, as well as larger outboards are equipped with the Precision Blend injection system. This is a mechanical

oil injection system using a crankshaft-driven injection pump and integral oil tank. V-block engines also have a remote reserve oil tank which automatically supplies the integral tank on the engine as required. The pump draws oil from the oil tank and supplies it under pressure to the intake manifold where it is sprayed into the air-fuel mixture. See **Figure 1** (typical). A linkage rod connects the pump to the throttle and varies the pump stroke according to throttle opening. The fuel-oil ratio ranges between 200:1 at idle to 100:1 at wide-open throttle on 25-50 hp models and 50:1 at wide-open throttle on 70-225 hp models.

The 25 and 30 hp models are not equipped with a control unit. When the oil level in the tank is reduced to 10 percent of its capacity, the oil level sensor circuit closes to activate the engine rpm reduction circuit to lower the engine rpm while activating the light in the motor pan and the buzzer on electric start models. On all other models, a 3-stage control unit monitors the oil level. The control unit informs the user of its status and automatically reduces engine rpm while sounding a warning buzzer if the oil level is low or the injection system is not functioning properly.

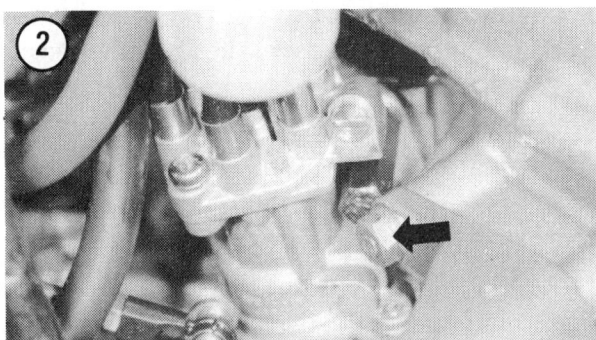

This chapter covers the operation, troubleshooting and component replacement required by the oil injection system.

SYSTEM COMPONENTS

The Yamaha Precision Blend system is a factory-installed standard feature. Models equipped with oil injection use a self-contained oil pump on the power head (**Figure 2**) in addition to the fuel pump. The oil pump is connected to the throttle by a linkage rod. An oil tank or reservoir is mounted on the power head. V-block models also have a remote reserve tank. When full, the tank(s) contain sufficient oil for approximately 5 hours of continuous wide-open throttle operation. A warning light or a series of warning lights mounted in the lower engine cover or on the instrument

13

panel and a warning buzzer located in the remote control box monitors oil level in the tank(s). **Figure 3A** shows the components of the 25 and 30 hp Precision Blend system; **Figure 3B** shows the components of the 3-cylinder (except 30 hp) Precision Blend system; **Figure 4** shows the V4 and V6 system components.

OPERATION

The Precision Blend injection system supplies oil to the engine separately from the fuel. A drive gear on the crankshaft engages a driven gear on the oil pump shaft. This driven gear transmits crankshaft rotation through a series of reduction gears inside the pump, controlling the stroke of the pump plunger according to crankshaft speed. See **Figure 5**. The plunger cam is designed to provide a 100:1 ratio with each full rotation of the crankshaft. Since the pump is mechanically linked to the throttle, it supplies the proper amount of oil according to engine speed and load.

NOTE
The oil injection pump is sealed at the factory and is serviced by replacement if

defective. Any attempt to disassemble the pump will void the factory warranty.

Oil travels to and from the oil pump through a series of lines. **Figure 6** shows the V6 configuration; others are similar. A transparent hose connects to the outlet of each oil delivery hose on 1985 and later models. This permits the user to check oil flow without removing a delivery hose.

Sensors in the integral and remote oil tanks are connected to the overheat warning buzzer in the remote control box and a series of 3 warning lamps mounted on the instrument panel.

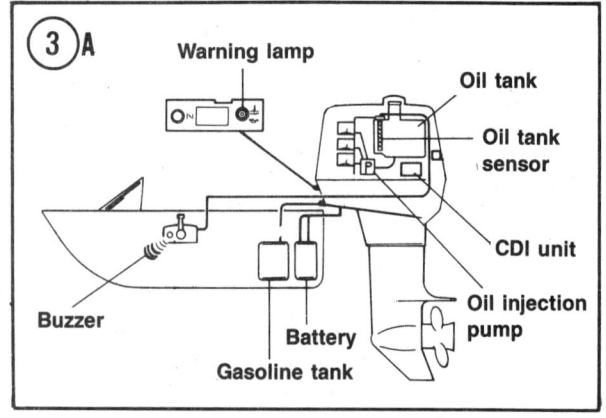

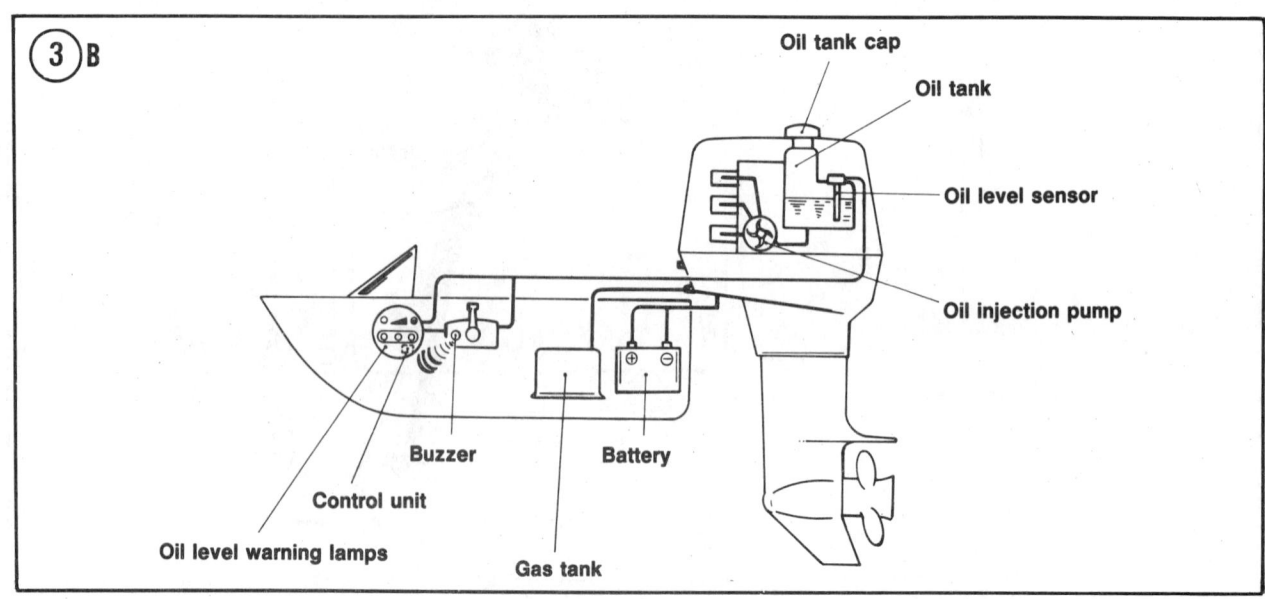

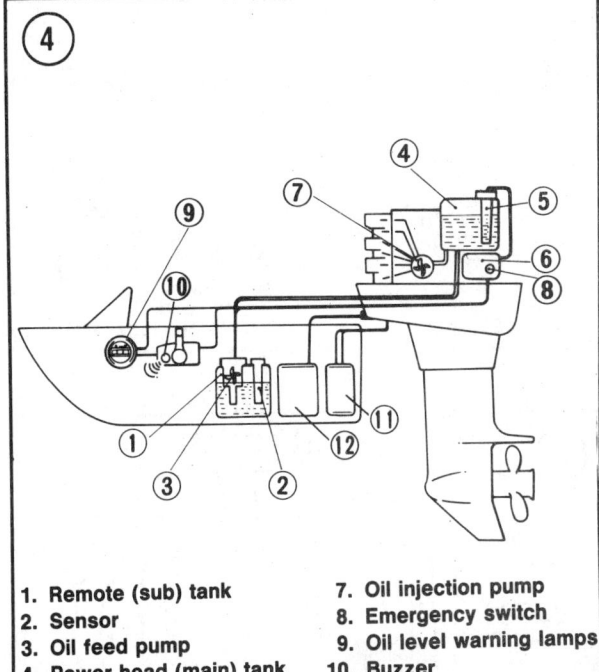

④

1. Remote (sub) tank
2. Sensor
3. Oil feed pump
4. Power head (main) tank
5. Sensor
6. Control unit
7. Oil injection pump
8. Emergency switch
9. Oil level warning lamps
10. Buzzer
11. Battery
12. Gas tank

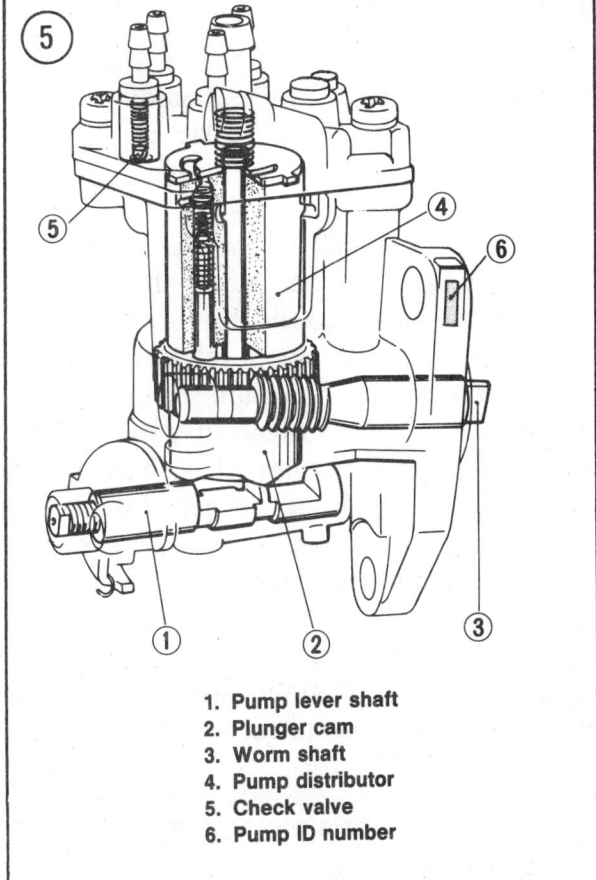

⑤

1. Pump lever shaft
2. Plunger cam
3. Worm shaft
4. Pump distributor
5. Check valve
6. Pump ID number

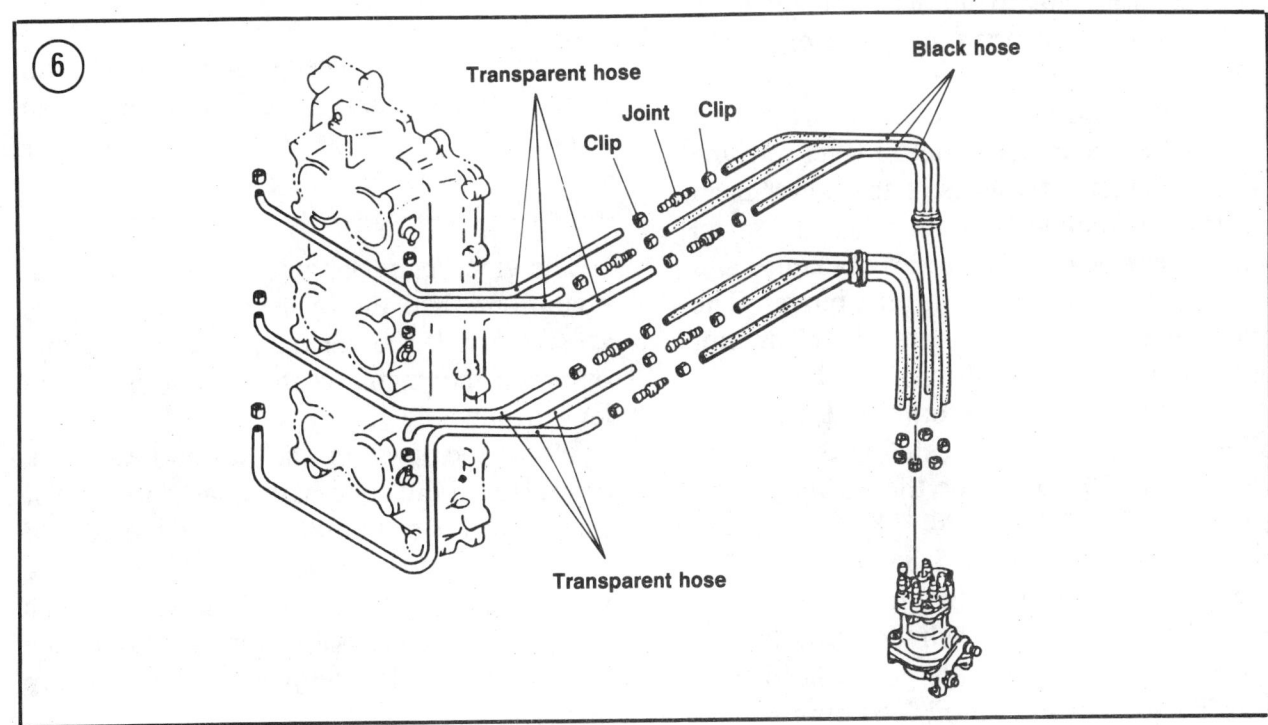

⑥

Transparent hose

Joint   Clip

Clip

Black hose

Transparent hose

13

Break-in Procedure

The first 10 hours of operation with a new or rebuilt oil-injected engine should be with a 50:1 fuel-oil mixture (see Chapter Four) *in addition* to the lubricant supplied by the injection pump.

Mark the oil level on the translucent oil tank mounted to the power head, then periodically check to make sure that the system is working (oil level diminishing) before switching over to plain gasoline at the end of the 10 hour break-in period. This applies both to engines that have been overhauled and new engines out of the box.

When an engine has not been used for several weeks or is just brought out of storage, make sure the remote oil tank on V-block models contains a minimum of 5.3 qt. oil or the oil feed pump chamber will not be filled and no oil will be supplied.

OIL LEVEL WARNING SYSTEM

Components

On 25 and 30 hp models, the oil level warning system consists of a sensor in the power head oil tank, one warning lamp and a buzzer on electric start models.

On all other models, the oil level warning system consists of sensors in the power head (main) and remote (sub) oil tanks, a control unit, emergency switch, 3 warning lamps and the remote control box warning buzzer. The emergency switch and remote oil tank are not used on 3-cylinder models.

The control unit monitors signals from the oil level sensor(s). It is located in the bottom cowling (40-50 hp), inside the warning lamp assembly (70-90 hp [1984-1985]) or on the power head (70-90 [1986-on] and 115-225 hp). **Figure 7** shows the location of the control unit on some 3-cylinder models.

On V-block models, the power head tank sensor also activates an oil feed pump when

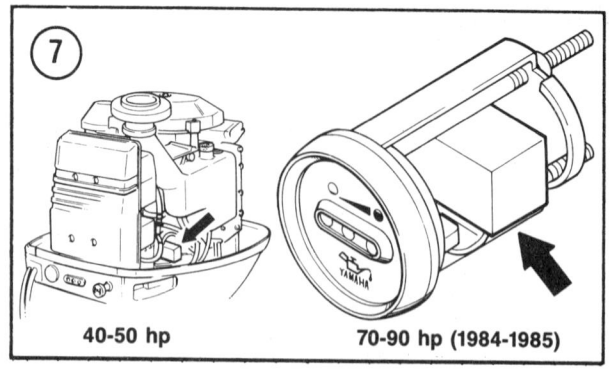
40-50 hp   70-90 hp (1984-1985)

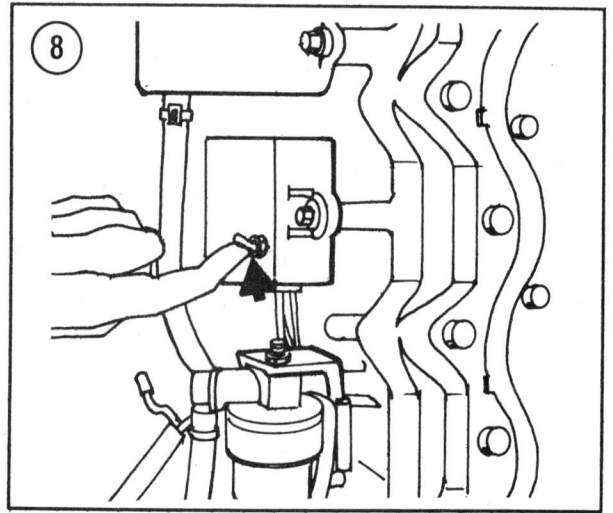

necessary. The oil feed pump supplies oil from the remote tank to refill the power head tank.

V-block Operation

The green light remains on during normal operation when the oil level in the tanks is satisfactory. If the green light does not come on when the engine is started, check the oil level immediately.

When the level in the power head tank drops to about one-half quart, its sensor signals the control unit to activate the oil feed pump and transfer additional oil from the remote tank to the power head tank. If the remote tank level is too low to supply the power head tank, the yellow warning lamp lights.

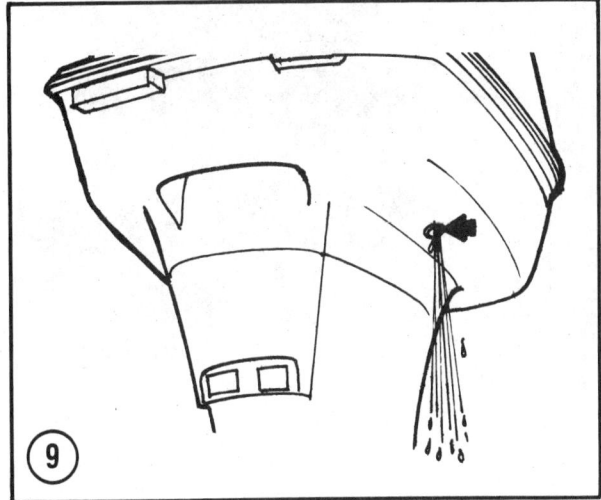

Continued operation with the yellow light on without adding oil will cause the red lamp to light, the warning buzzer to sound and a reduction in engine speed to approximately 2,000 rpm.

NOTE
The oil level and overheat warning systems on V4 models are designated Type A, B or C according to electrical wiring. Refer to the wiring diagram for your engine to determine which is applicable.

At this time, you should shut the ignition switch OFF to stop the engine. Use the emergency switch to pump the oil remaining in the remote tank to the power head tank (**Figure 8**). Turn the ignition switch back ON, make sure the red light is off and restart the engine. This will allow you to reach port where the tanks should be refilled immediately. Failure to turn the ignition switch OFF will prevent engine speed from exceeding 2,000 rpm, even after the oil tanks have been refilled.

CAUTION
If the emergency switch is used as described above, be sure to bleed the oil feed pump as described in this chapter when refilling the tank.

Three-cylinder Operation (Except 30 hp)

The green light remains on during normal operation when the oil level in the tank is satisfactory. If the green light does not come on when the engine is started, check the oil level immediately.

When the level in the power head tank drops to about one-half quart, its sensor turns on the yellow lamp. If engine operation continues without adding oil, the red lamp will light, the warning buzzer sounds and engine speed is reduced to approximately 2,000 rpm. At this time, you should shut the engine OFF, refill the power head tank and restart the engine. Failure to stop the engine will prevent it from exceeding 2,000 rpm even after the tank is refilled.

25 and 30 hp Operation

No light is illuminated during normal operation when the oil level in the tank is satisfactory.

When the level in the tank drops to 10 percent of capacity, the red lamp lights, the warning buzzer sounds (if so equipped) and the engine speed is reduced to approximately 2,000 rpm. At this time, you should shut the engine OFF, refill the tank and restart the engine. Failure to stop the engine will prevent it from exceeding 2,000 rpm even after the tank is refilled.

Overheat Protection

A temperature sending unit is installed in the cylinder head of all models and connected to the warning lamp on 25 and 30 hp models and to the warning buzzer through the key switch on 25 and 30 hp electric start models and all other models to warn of an overheat condition. If the power head temperature exceeds the sending unit's specified value, the warning lamp flashes on 25 and 30 hp models and the buzzer sounds continuously on 25

13

and 30 hp electric start models and all other models (the red lamp does not light on these models) and engine speed will automatically be reduced to approximately 2,000 rpm. Backing off on the throttle will shut the buzzer off as soon as power head temperature drops to a specified value, unless a restricted engine water intake is causing the overheat condition. If a steady stream of water does not flow from the tell-tale or pilot hole (**Figure 9**) or if the warning lamp continues to flash or the buzzer continues sounding after 2 minutes, the engine should be shut off immediately to prevent power head damage.

CAUTION
If the engine overheats and the warning buzzer sounds, retorque the cylinder head fasteners after the engine cools to minimize the possibility of power head damage from a blown head gasket.

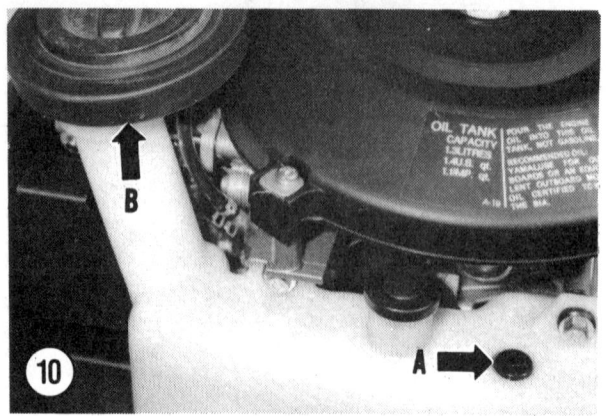

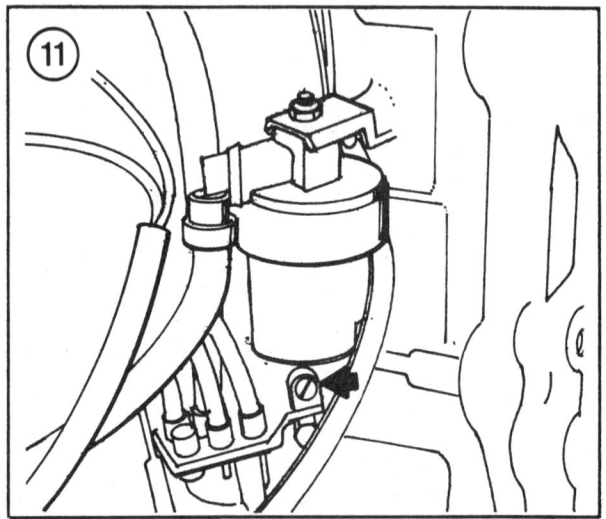

INJECTION SYSTEM SERVICE

Power Head Injection
Pump Bleeding (All Models)

The oil injection pump on 25 and 30 hp models is self-bleeding. The oil injection pump on all other models must be gravity-bled to remove any air whenever the system is serviced or when the outboard has not been used for a lengthy period of time. Proceed as follows:

1. Remove the engine cover.
2. Remove the oil tank air vent valve (A, **Figure 10**). If tank is not equipped with a vent valve, remove the oil tank cap (B, **Figure 10**).
3. Make sure that the engine is in an upright position. If tilted in a trailering position, tilt the engine upright.
4. If the oil pump has been removed from the power head, fill the oil lines with oil before reconnecting them to the pump fittings.

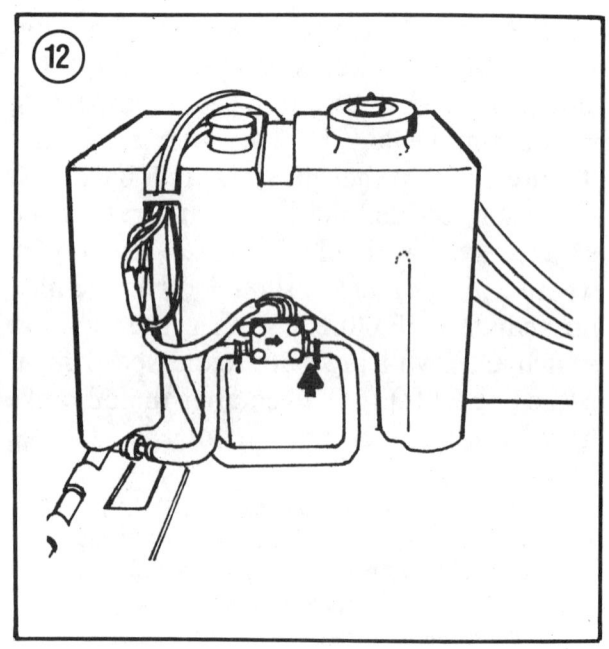

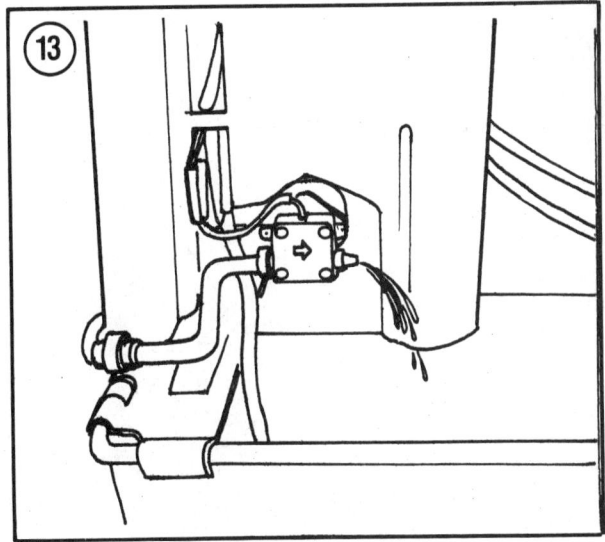

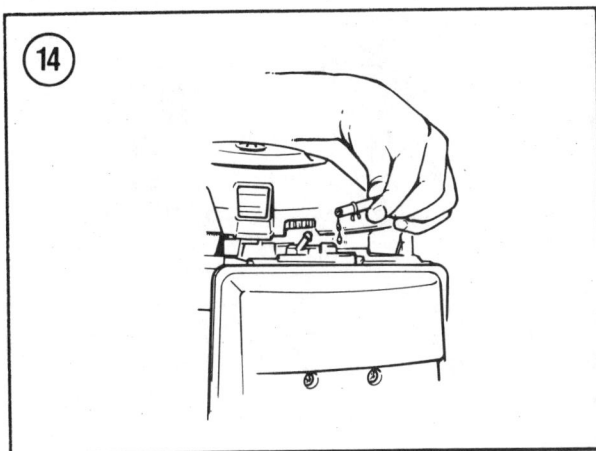

5. Open the bleed screw on the injection pump 3-4 turns counterclockwise. See **Figure 11**.

6. Wait several seconds, make sure that oil is flowing through the screw hole and then tighten the screw securely.

7. Reinstall the oil tank vent or cap and the engine cover.

Oil Feed Pump
(V-block Engines Only)

The remote oil tank oil feed pump must be gravity-bled to remove any air whenever the system is serviced or when the emergency switch has been used.

1. Remove the remote oil tank from its mounting bracket.

2. Add sufficient oil to bring the tank level up to a minimum of 5.3 qt.

3. Disconnect the oil feed pump outlet hose (**Figure 12**) and place a suitable container under the outlet.

4. Turn the ignition switch ON to operate the oil feed pump. If the pump does not operate with the ignition switch ON, use the emergency switch (**Figure 8**).

5. Operate the pump until oil runs out of the pump outlet (**Figure 13**), then reconnect the hose.

Oil Pump Discharge Adjustment

See *Oil Pump Link Adjustment*, Chapter Five.

Oil Pump Delivery Rate Test

> *CAUTION*
> *The engine should be operated with a 50:1 fuel-oil mixture in the fuel tank during this procedure.*

1. Start the engine and run at idle for 5 minutes or until it reaches operating temperature.

2. Remove the engine cover.

3. Install a tachometer according to manufacturer's instructions.

4. Run the engine at 1,500 rpm during Step 5 and Step 6.

5. Disconnect one of the pump injection lines at the intake manifold (**Figure 14**) and insert it into a suitable graduated container so that the oil will drip directly into the container, not onto the cylinder sidewall. Disconnect oil pump control rod from oil pump lever and rotate lever to wide open position.

6. Collect the oil discharged for 3 minutes, then reconnect the line to the intake manifold fitting.

7. Measure the amount of oil discharged. The pump should discharge 0.76 cc (25-300 hp), 0.7-1.1 cc (40-50 hp), 1.97-2.77 cc (70 hp), 2.70-3.73 cc (90 hp) or 2.6-4 cc (115-225 hp) in 3 minutes.

13

8. Repeat Steps 4-6 with each remaining injection line.

9. Shut the engine off and bleed the oil injection pump as described in this chapter.

10. Reconnect oil pump control rod to the oil pump lever.

11. If the pump does not perform as described in Step 7 with each injection line, check the line(s) for possible leakage or restrictions. If none are found, replace the pump.

Oil Pump Control Unit Troubleshooting

See Chapter Three.

COMPONENT REPLACEMENT

Oil Tank Removal/Installation

1. Remove the engine cover.

> *NOTE*
> *If a clean container is not available in Step 2, pinch the end of the line during removal and then plug it with an M6 bolt to prevent oil leakage.*

2. Place a clean shop cloth under the oil tank, then disconnect the line between the oil tank and the injection pump. Drain the contents of the oil tank into a suitable container.

3. Disconnect the oil sensor electrical leads.

4. Remove the bolts holding the tank to the power head. Remove the tank.

5. Installation is the reverse of removal. Refill the tank with Yamalube Two-cycle lubricant (for outboards). Bleed the oil injection pump as described in this chapter.

Oil Injection Pump Removal/Installation

> *CAUTION*
> *Proper injection line routing and connections are important. Injection lines must be installed between the pump and power head correctly and connected to the proper cylinder fitting*

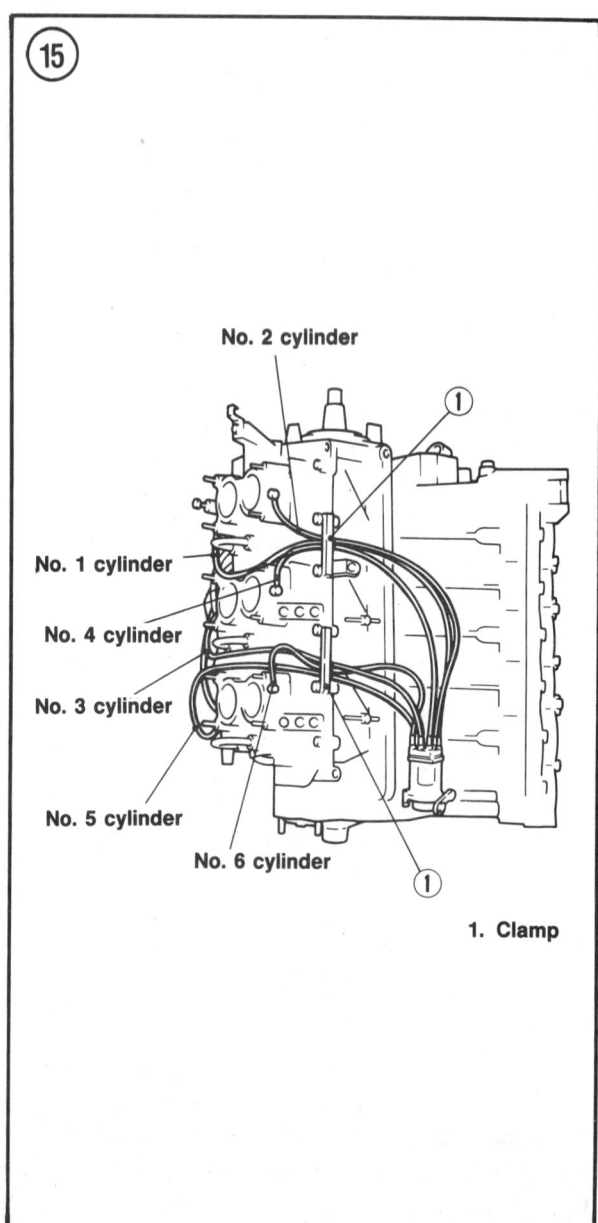

(15)

No. 2 cylinder

No. 1 cylinder

No. 4 cylinder

No. 3 cylinder

No. 5 cylinder

No. 6 cylinder

1. Clamp

(16)

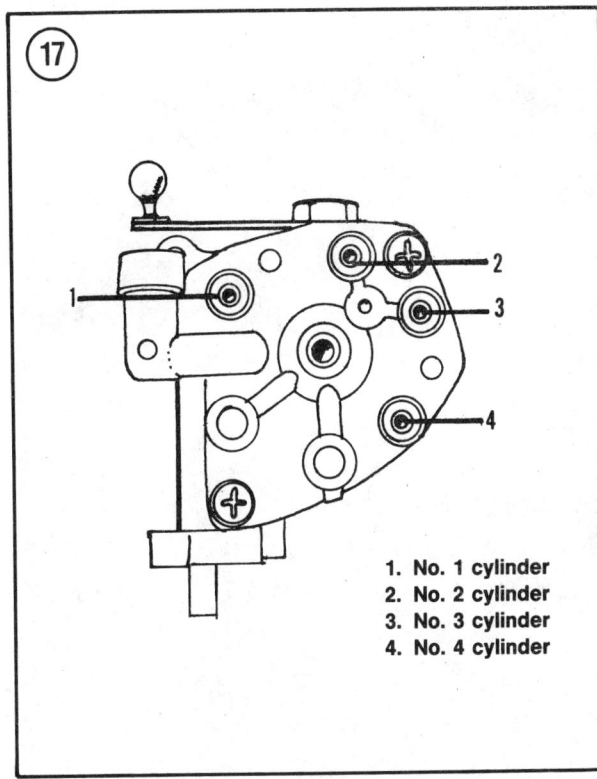

1. No. 1 cylinder
2. No. 2 cylinder
3. No. 3 cylinder
4. No. 4 cylinder

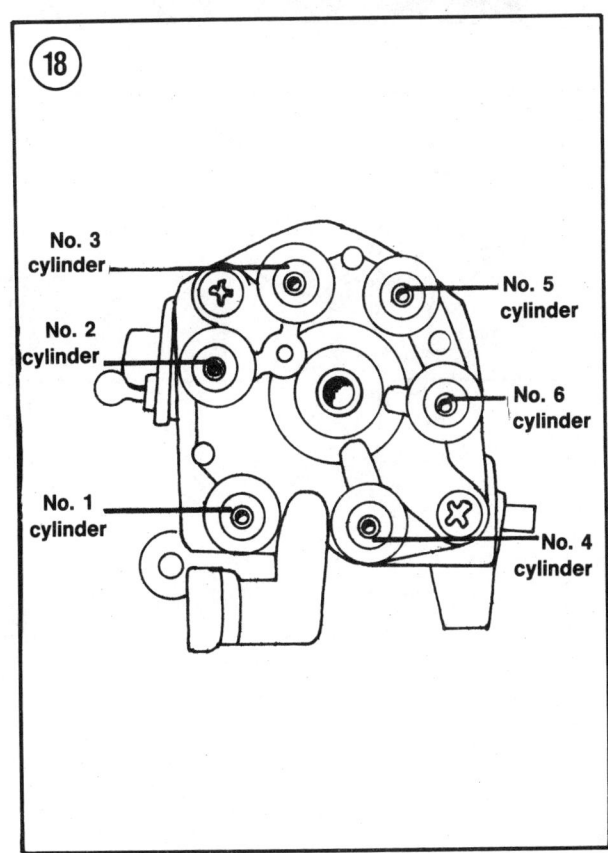

on the intake manifold. ***Figure 15*** *shows the V6 line routing.*

1. Disconnect the negative battery cable.
2. Remove the engine cover.
3. Remove the oil tank as described in this chapter.
4. Place a clean shop cloth under the injection pump to catch any oil that leaks when the lines are disconnected or the pump removed.
5A. If the pump is to be replaced but the injection lines reused, disconnect the lines at the pump.
5B. If the pump and injection lines will be reinstalled, label and disconnect the lines at the intake manifold.
6. V6—Remove the injection line retaining clamps from the power head. See **Figure 15**.
7. Disconnect the control link at the oil pump (**Figure 16**).
8. Remove the bolts holding the pump to the power head. Remove the pump.
9. Pull the driven gear from the power head.
10. Installation is the reverse of removal, plus the following:

a. Align the slit in the power head driven gear with the tang on the pump.
b. If the lines were removed from the pump, refer to **Figure 17** (V4) or **Figure 18** (V6) for proper reconnection.
c. Fill the injection lines with Yamalube Two-cycle lubricant (for outboards) before connecting them to the intake manifold.
d. Connect the control link at the pump and adjust as required. See Chapter Five.
e. Carefully check all lines and connections for signs of oil leakage before starting the engine.
f. Bleed the injection pump as described in this chapter.

13

Oil Pump Control Unit
Removal/Installation

The control unit receives signals from the oil tank sensor(s). It activates the warning lights and buzzer and reduces engine speed when necessary.

40, 50, 115-225 and 70-90 (1986-on) hp

1. Disconnect the negative battery cable.
2. Remove the engine cover.
3. Disconnect the control unit leads at their respective connectors.
4. Remove the bolts holding the control unit to the bottom cowl on 40-50 hp (**Figure 19**) or the power head on 70-90 (1986-on) and 115-225 hp (**Figure 20**). Remove the control unit.
5. Installation is the reverse of removal.

70-90 (1984-1985) hp

1. Disconnect the negative battery cable.
2. Disconnect the electrical connector at the warning lamp unit on the instrument panel.
3. Remove the nuts holding the warning lamp unit to the instrument panel. Remove the warning lamp unit.
4. Unbolt and remove the control unit from the warning lamp unit. See **Figure 21**.
5. Installation is the reverse of removal.

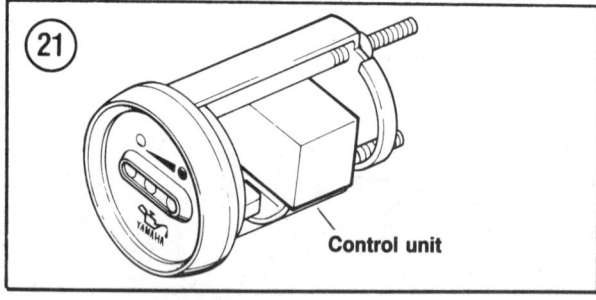

Control unit

Chapter Fourteen

Yamaha 703 Remote Control Box

The Yamaha 703 remote control box is provided with all 30-225 hp electric models and available as an option for 6-25 hp models. A light duty (701), a flush-mounted and single or dual binnacle remote controls are optional for some models.

The 703 remote control box houses the ignition or main switch, emergency stop, choke, tilt or trim/tilt and neutral start switches, and the overheat/oil level warning buzzer. The remote control box is connected to the engine gear shift handle and magneto control lever by cables, allowing the operator to shift the engine and control the throttle at a point away from the engine. See **Figure 1**.

This chapter covers disassembly and assembly of the 703 remote control box only—the 701, flush-mounted and binnacle models are not covered here. Testing of the 703 switches is covered in Chapter Three.

Disassembly/Assembly

Refer to **Figure 2** for this procedure.
1. Disconnect the wire harness coupler and remove the remote control box from its mounting bracket.
2. Remove the cover from the lower side of the box (**Figure 3**).
3. Remove the screws holding the 2 back panels (**Figure 4**). Remove the panels.
4. Disconnect the remote control cables at the throttle and shift arms (**Figure 5**).
5. Disconnect the power tilt or power trim/tilt switch lead, if so equipped.
6. Disconnect the electrical leads (**Figure 6**) and remove the warning buzzer.
7. Loosen the nuts holding the ignition, choke and emergency stop switches. Remove the switches.
8. Remove the neutral start switch.
9. Loosen the throttle friction screw (A, **Figure 7**). Pull out the throttle arm (B, **Figure**

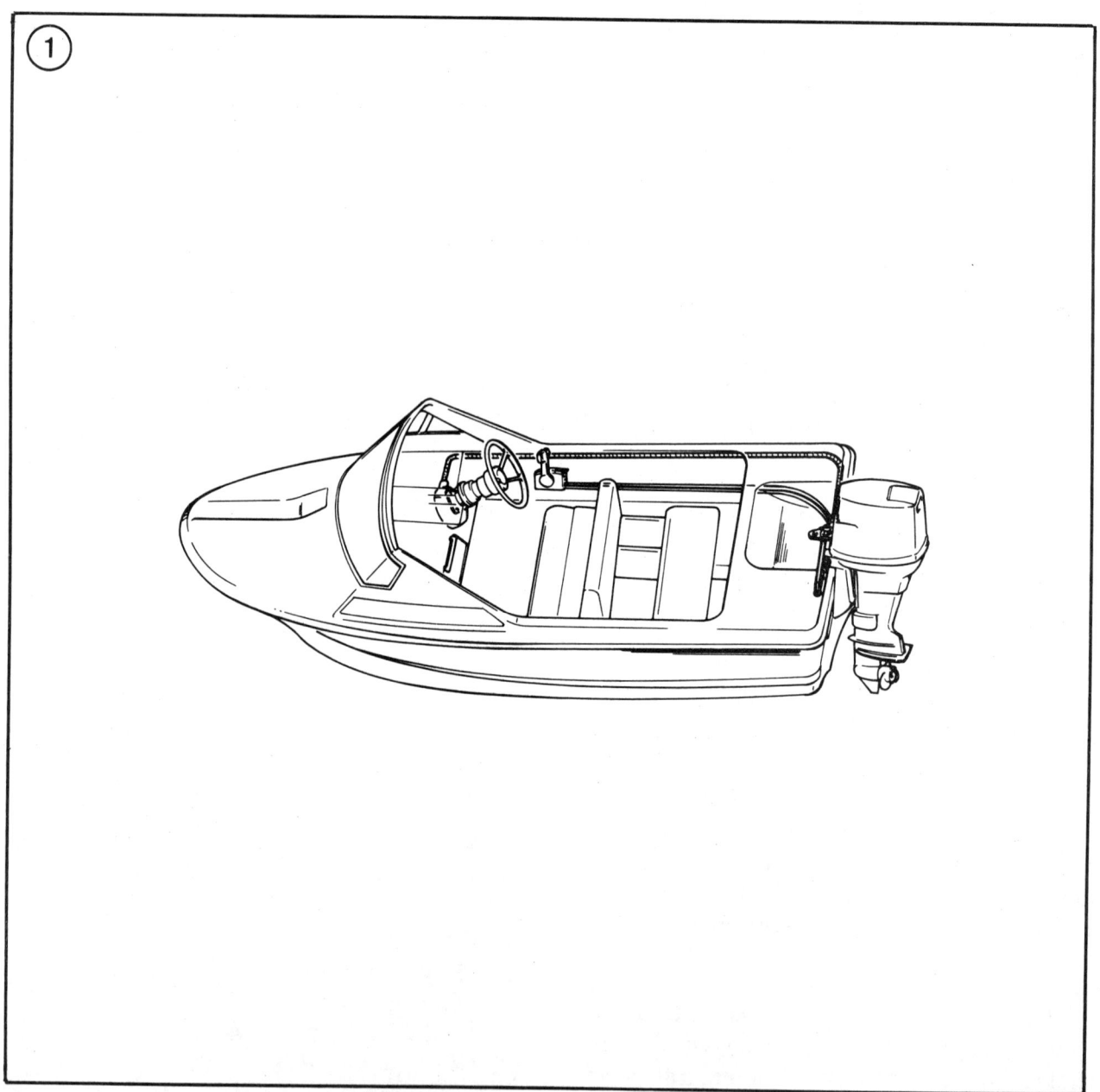

7) and remove the bushing, cam roller and washer from the control box or throttle arm. See **Figure 8**.

10. Remove the throttle friction screw circlip. Remove the throttle friction band and screw. See **Figure 9**.

11. Remove the neutral switch operating arm (**Figure 10**) from the gear.

12. Remove the leaf spring and detent roller. See **Figure 11**.

13. Loosen the center bolt until its head comes free of the gear, then tap the bolt with a plastic hammer and pull the control lever from the gear. See **Figure 12**.

14. Remove the gear and control cam (**Figure 13**).

15. Remove the shift lever arm (**Figure 14**).

16. Remove the neutral position lever from the control lever. See **Figure 15**. Remove the spring.

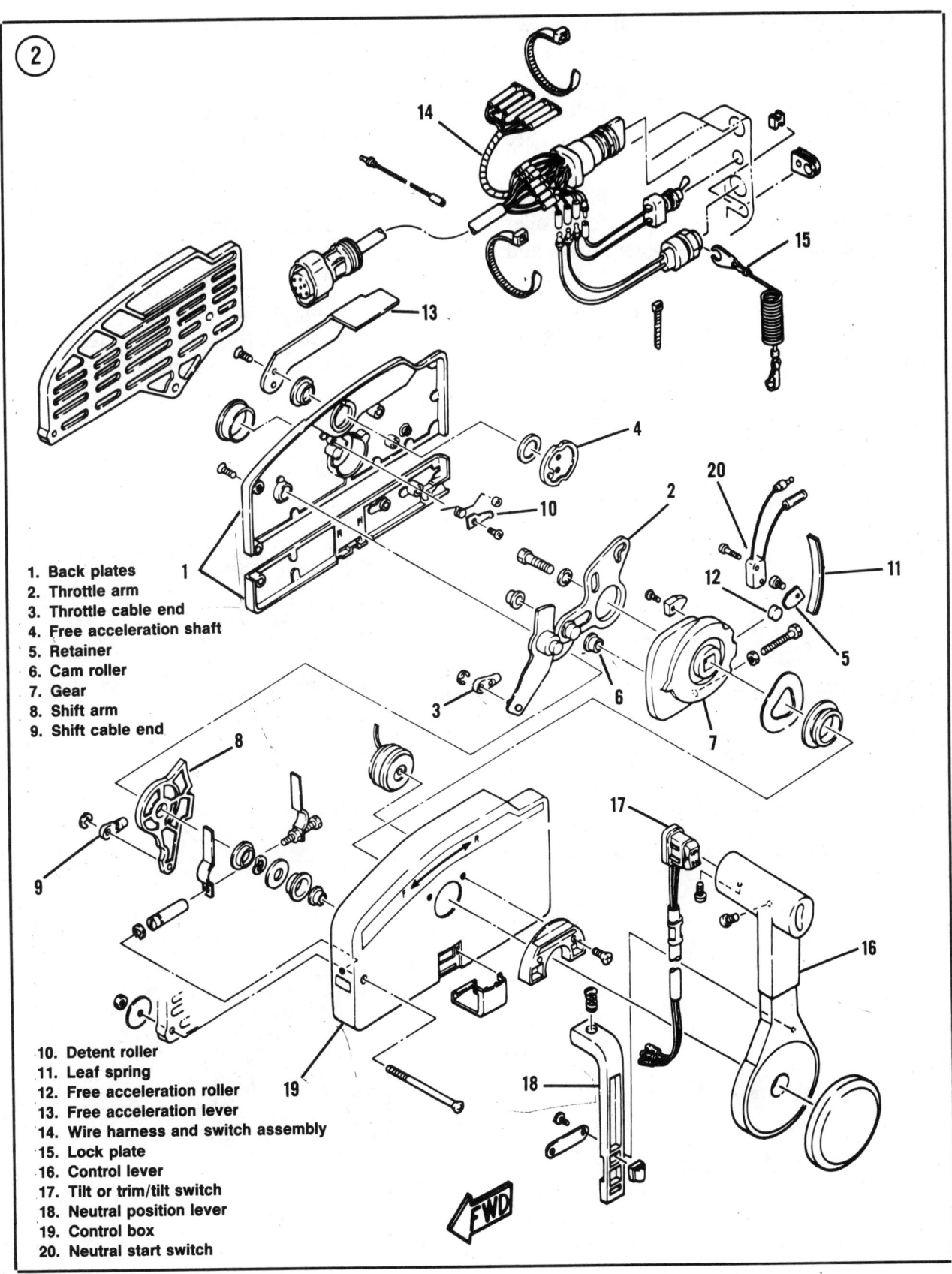

1. Back plates
2. Throttle arm
3. Throttle cable end
4. Free acceleration shaft
5. Retainer
6. Cam roller
7. Gear
8. Shift arm
9. Shift cable end
10. Detent roller
11. Leaf spring
12. Free acceleration roller
13. Free acceleration lever
14. Wire harness and switch assembly
15. Lock plate
16. Control lever
17. Tilt or trim/tilt switch
18. Neutral position lever
19. Control box
20. Neutral start switch

14

17. If equipped with a power tilt or power trim/tilt switch, remove the switch from the control lever as required. See **Figure 16**.

18. Remove the neutral lock holder or position plate from the control box. See **Figure 17**.

19. Carefully remove the retainer plate from the back cover to prevent the detent roller from flying out. Remove the detent roller and spring (**Figure 18**).

20. Clean the box and components with solvent and blow dry with low-pressure compressed air.

21. Check the box and all parts for excessive wear, cracks or other defects. Replace as required.

22. Assembly is the reverse of disassembly, plus the following:

    a. Lubricate the contact surface of all moving parts with Yamalube All-purpose Marine grease.

    b. Make sure the wire harness leads are properly reconnected to the switches according to color-coding.

    c. Attach the spring to the detent roller and install as an assembly.

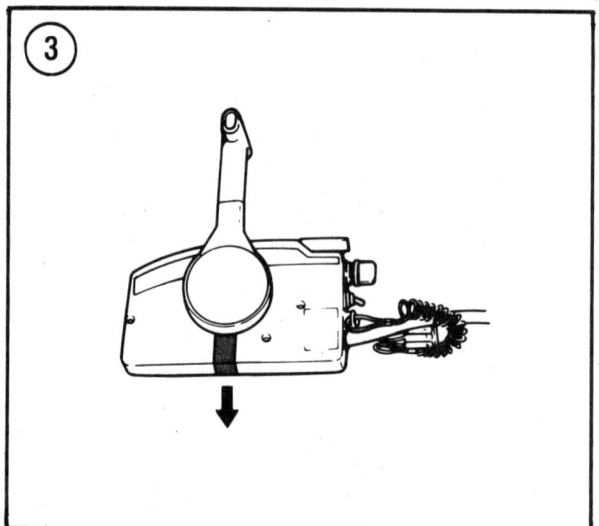

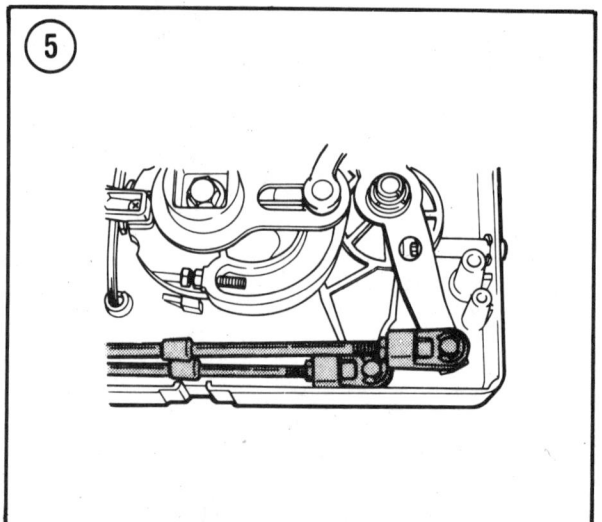

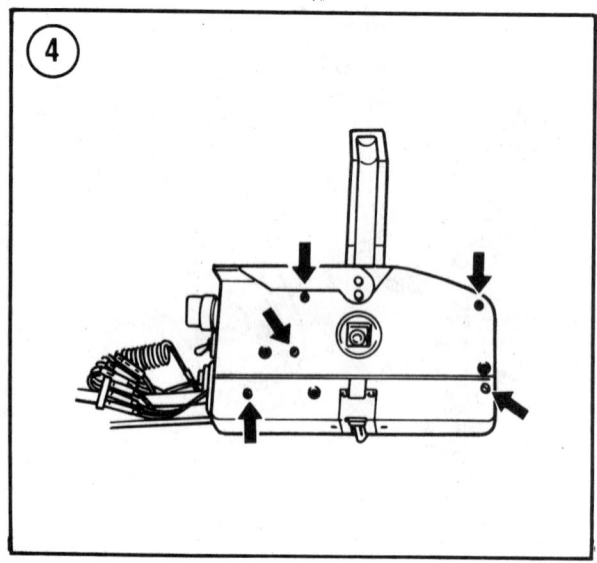

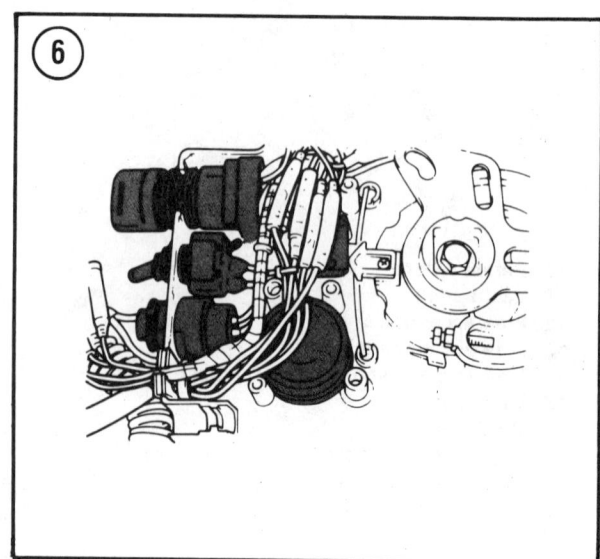

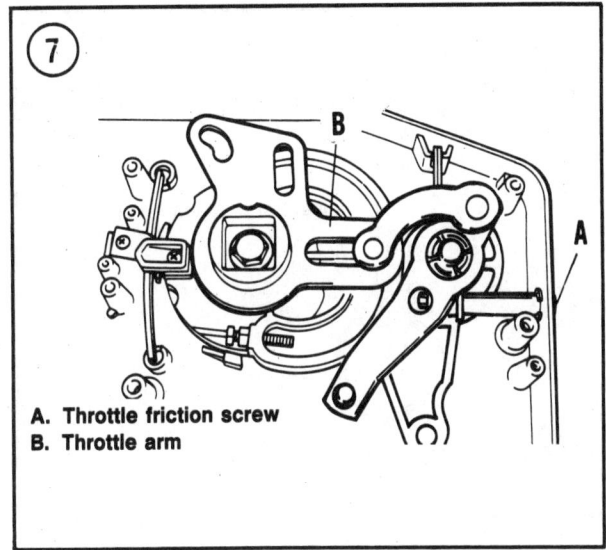

A. Throttle friction screw
B. Throttle arm

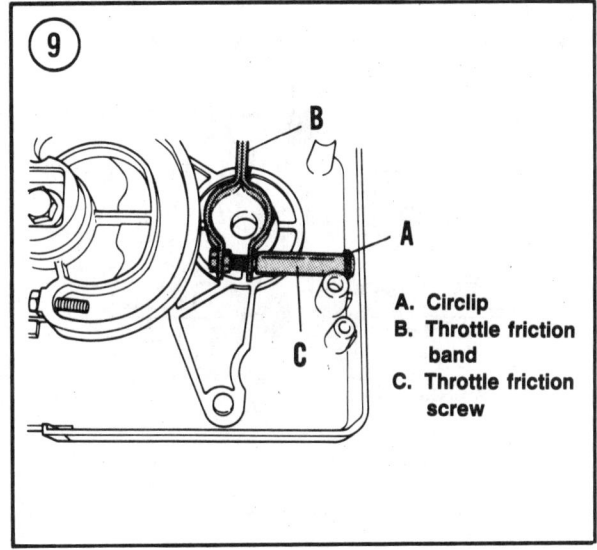

A. Circlip
B. Throttle friction band
C. Throttle friction screw

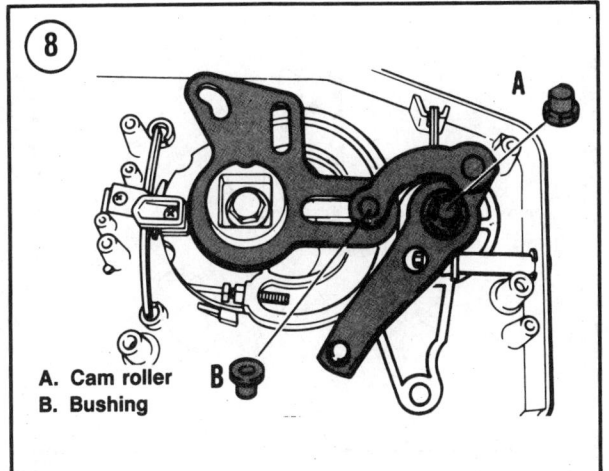

A. Cam roller
B. Bushing

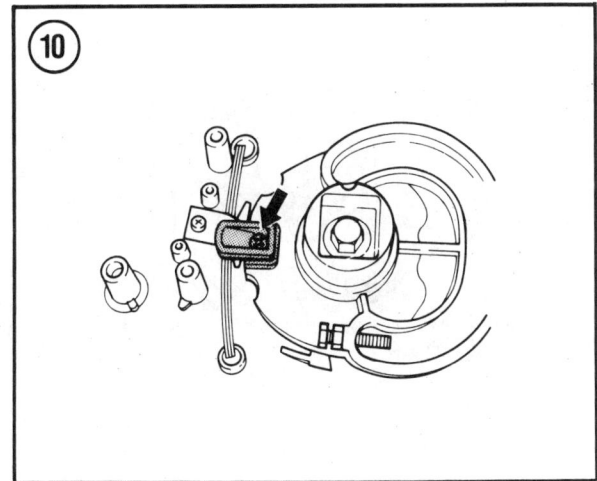

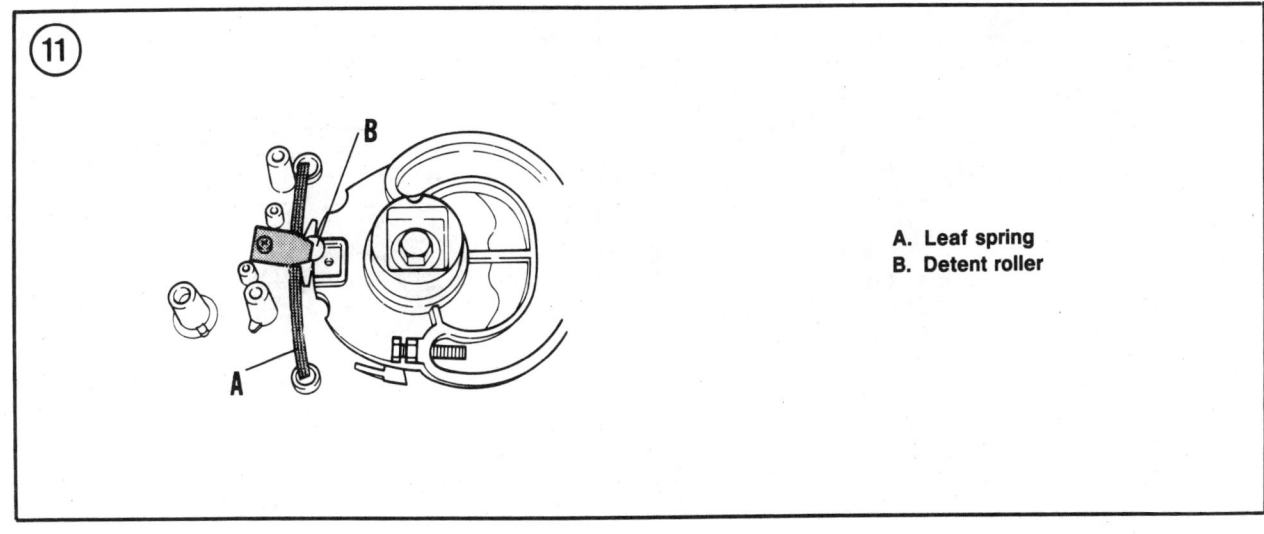

A. Leaf spring
B. Detent roller

14

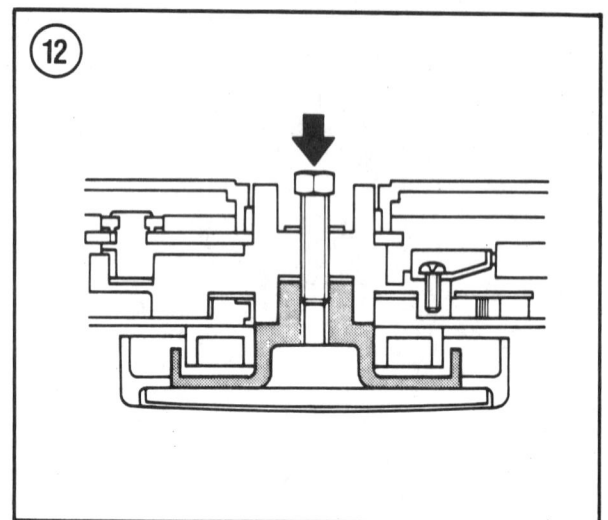

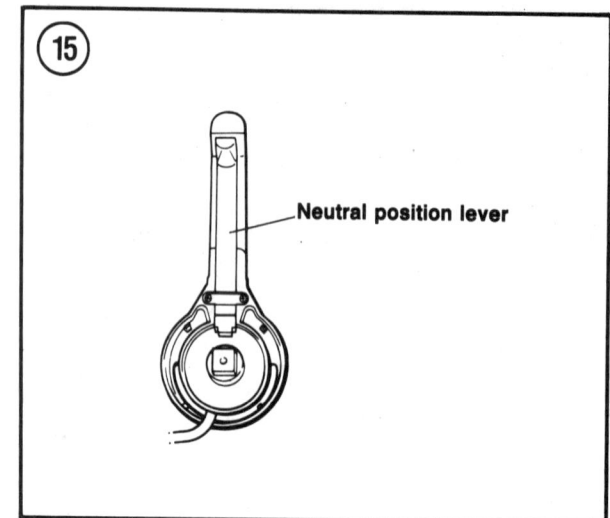

Neutral position lever

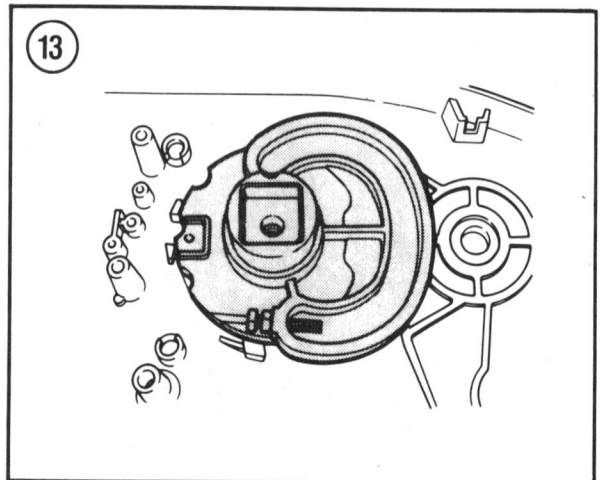

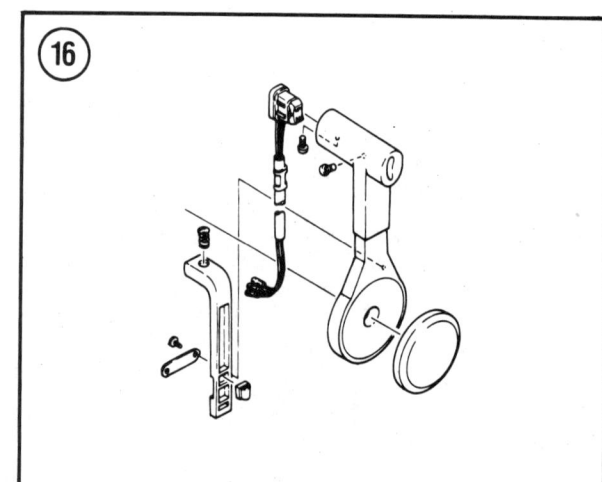

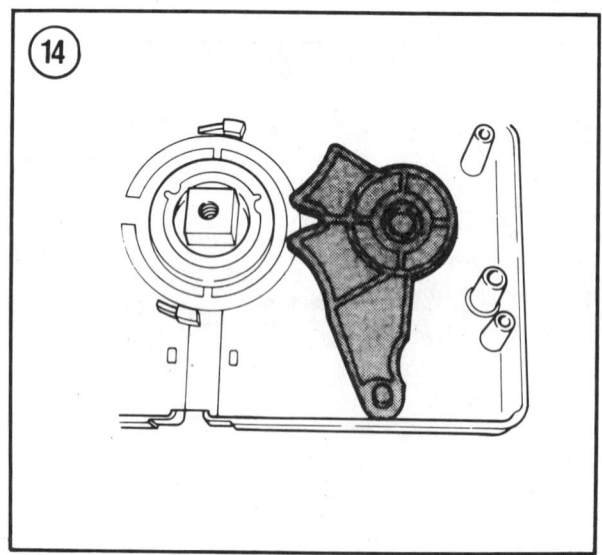

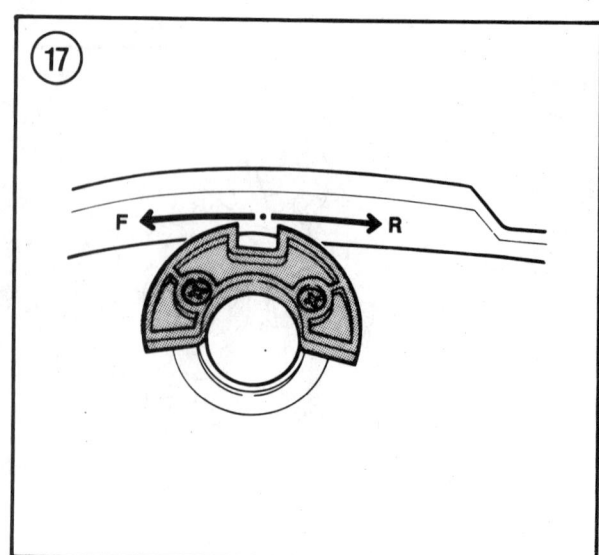

F ← → R

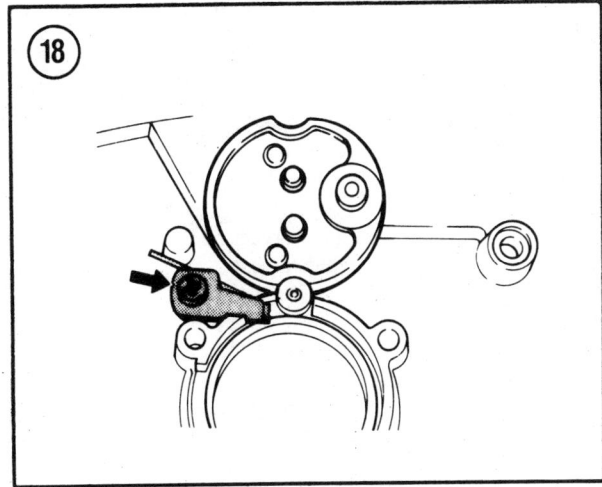

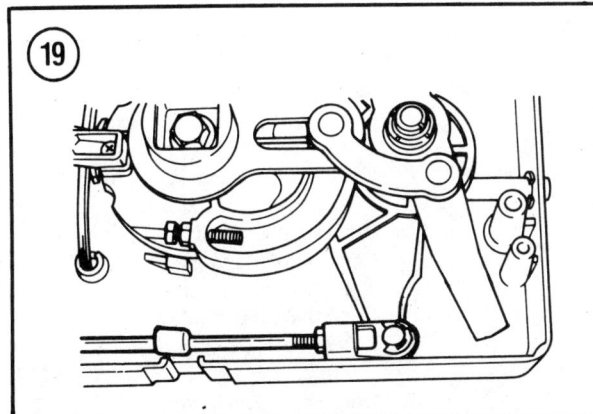

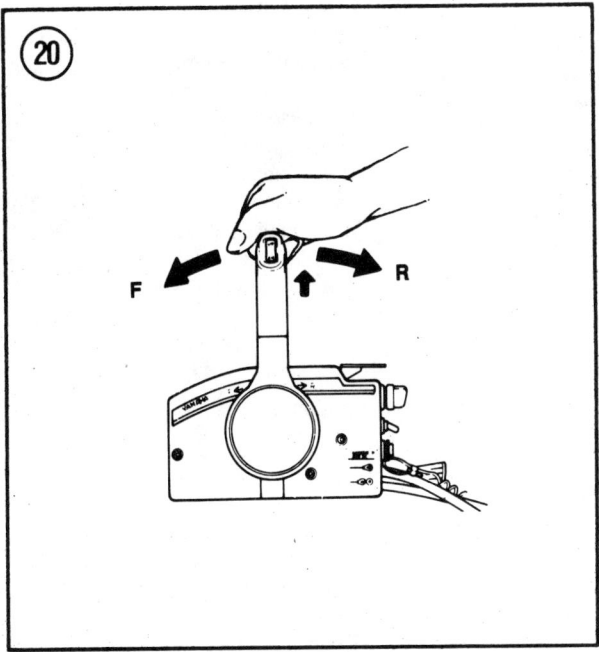

d. Route the wire harness leads properly during reassembly so they will not be pinched when the back plates are reattached.

e. Fit the shift cable on the shift lever arm (**Figure 19**) and secure it with the circlip. Install the cable anchor spacer, then connect the throttle cable to the throttle lever arm (**Figure 5**) and secure it with the circlip.

f. After reinstalling the back plates, operate the control lever to make sure the cables work properly.

g. Once the control box is reassembled and reinstalled on the boat, operate the control lever several times to make sure that the mechanism shifts properly (**Figure 20**). The shift and throttle operation should follow the pattern shown in **Figure 21**. If shift or throttle adjustments are necessary, adjust the cable length at the engine shift lever or carburetor.

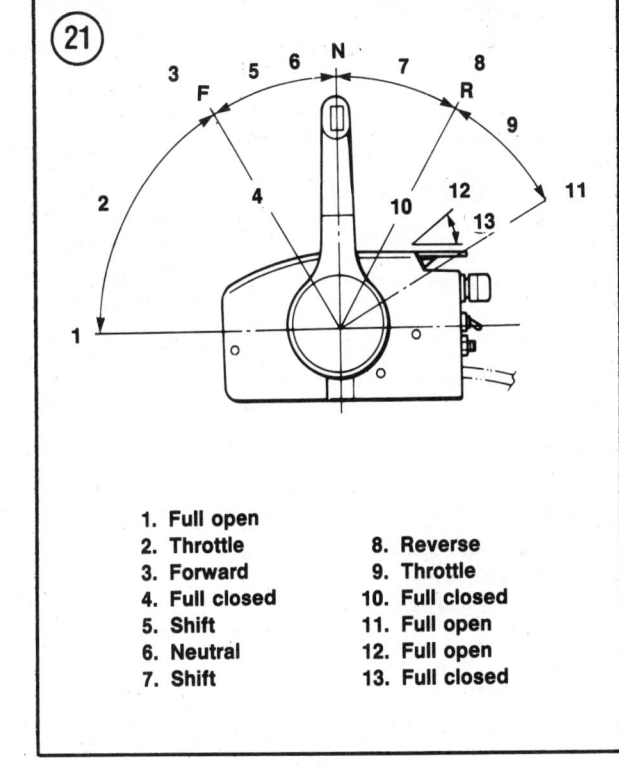

1. Full open
2. Throttle
3. Forward
4. Full closed
5. Shift
6. Neutral
7. Shift
8. Reverse
9. Throttle
10. Full closed
11. Full open
12. Full open
13. Full closed

14

Index

15

Wiring Diagrams

8N

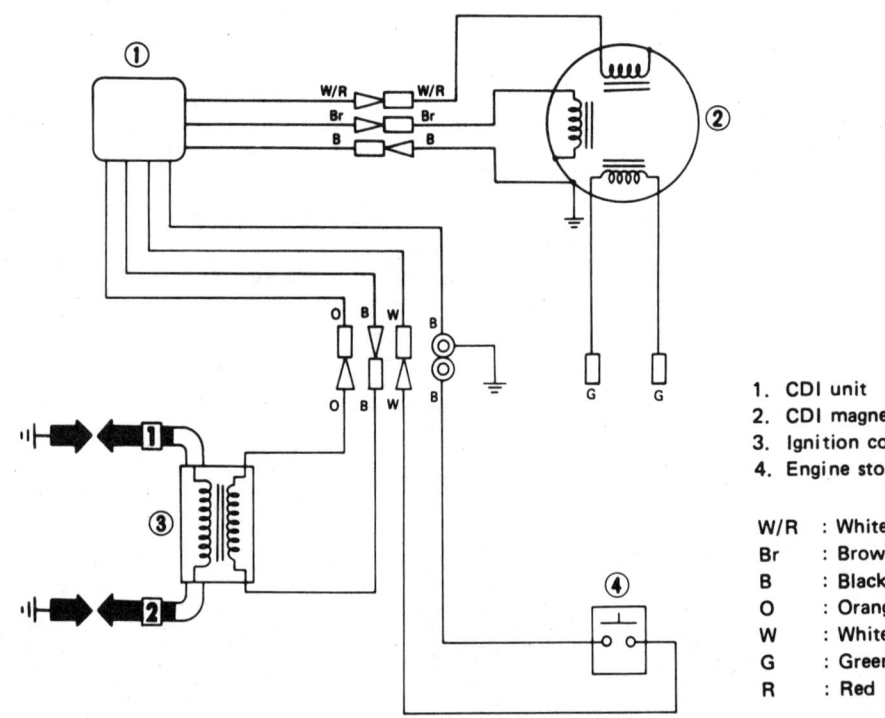

1. CDI unit
2. CDI magneto
3. Ignition coil
4. Engine stop button

W/R : White/Red
Br : Brown
B   : Black
O   : Orange
W   : White
G   : Green
R   : Red

9.9N

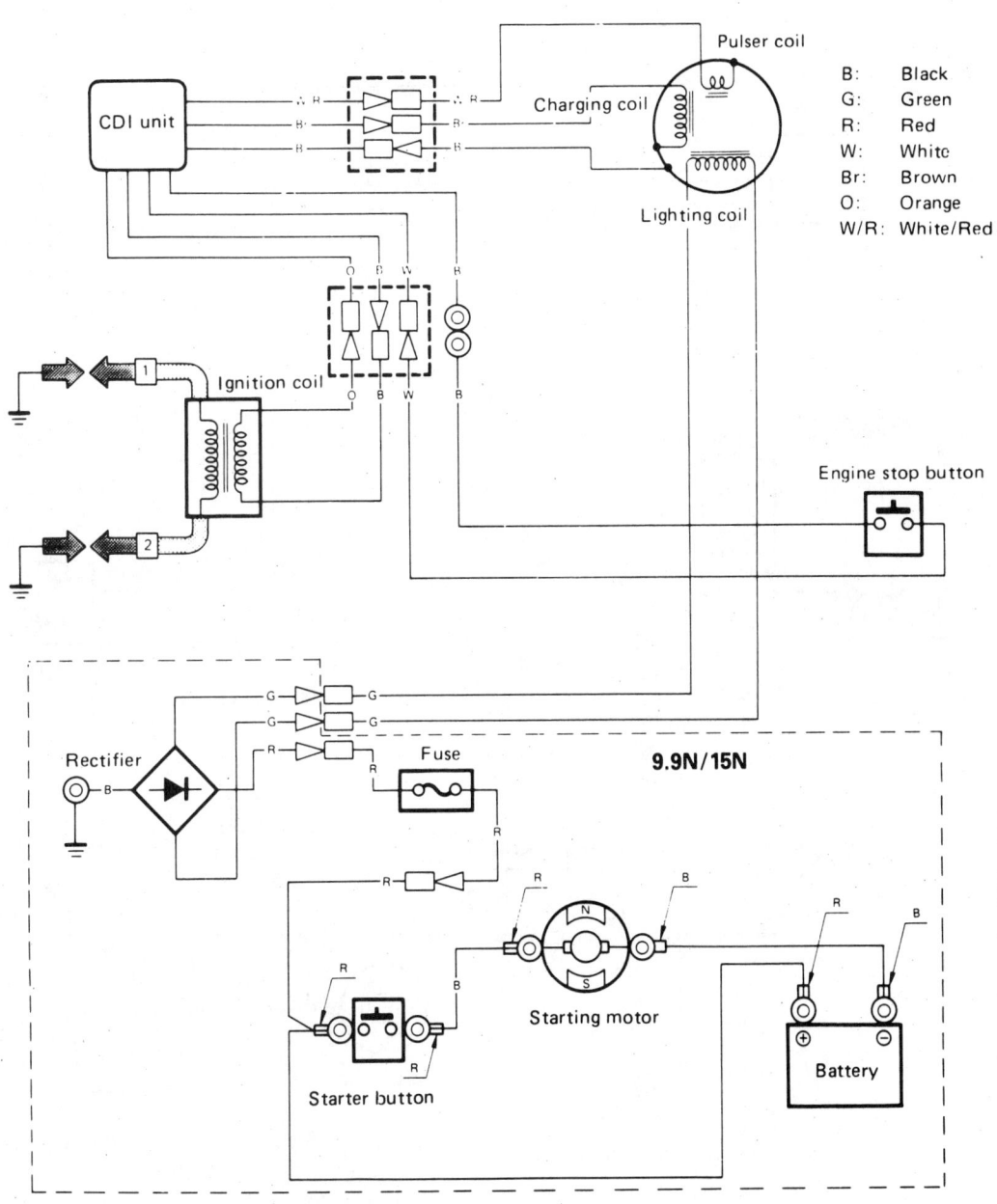

B: Black
G: Green
R: Red
W: White
Br: Brown
O: Orange
W/R: White/Red

16

15N

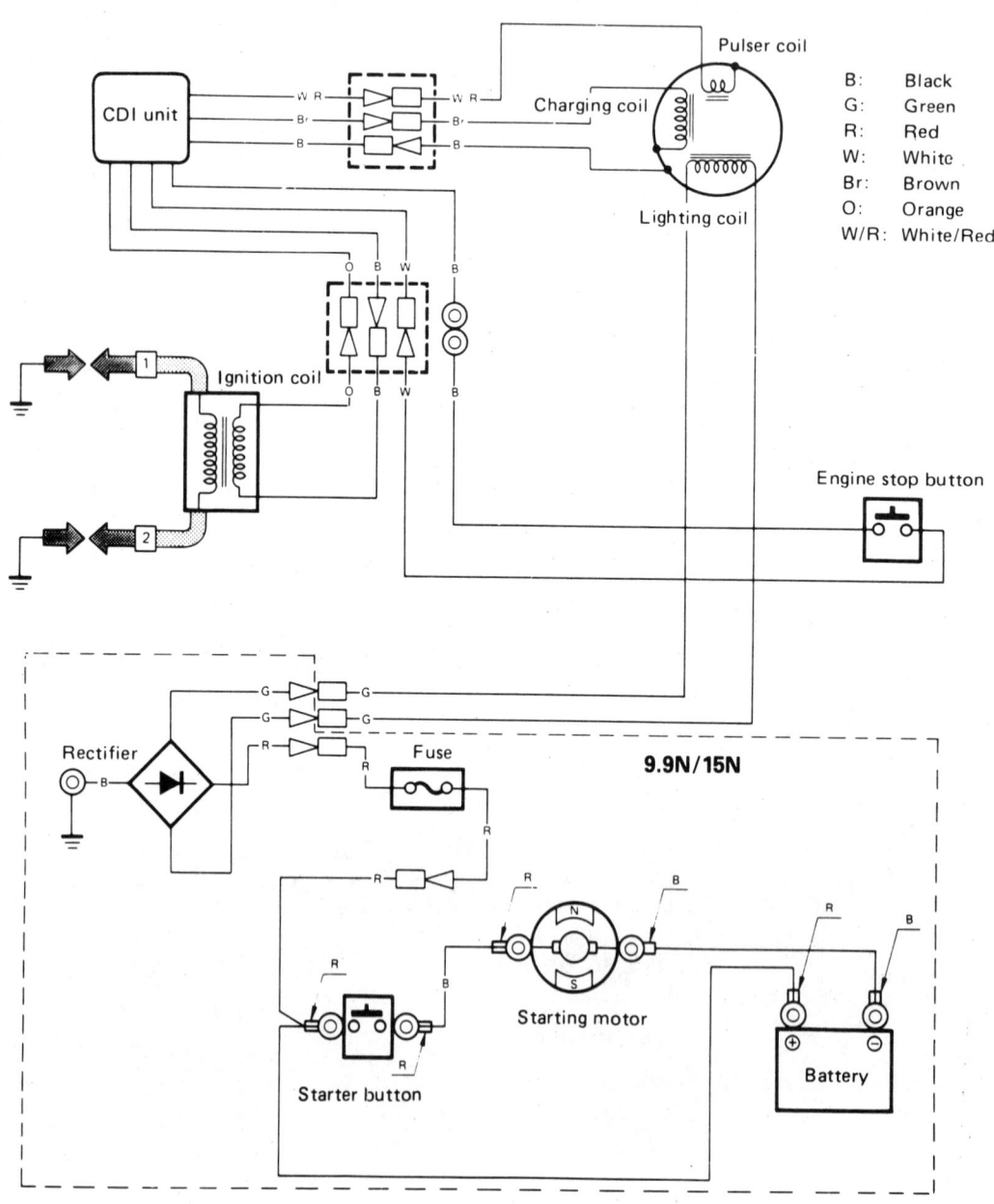

B: Black
G: Green
R: Red
W: White
Br: Brown
O: Orange
W/R: White/Red

Pulser coil

Charging coil

Lighting coil

CDI unit

Ignition coil

Engine stop button

Rectifier

Fuse

9.9N/15N

Starting motor

Battery

Starter button

30EN

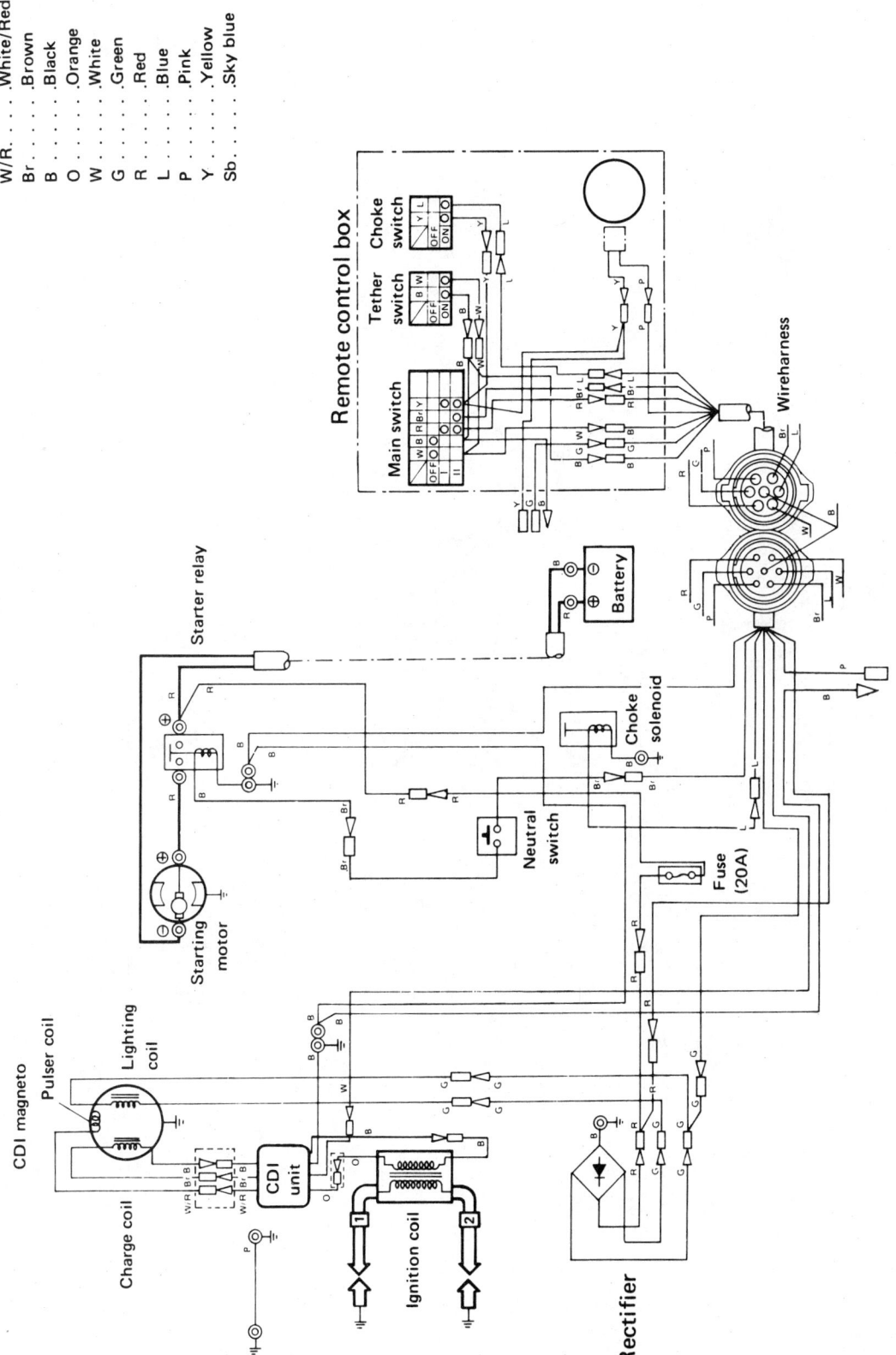

W/R	White/Red
Br	Brown
B	Black
O	Orange
W	White
G	Green
R	Red
L	Blue
P	Pink
Y	Yellow
Sb	Sky blue

16

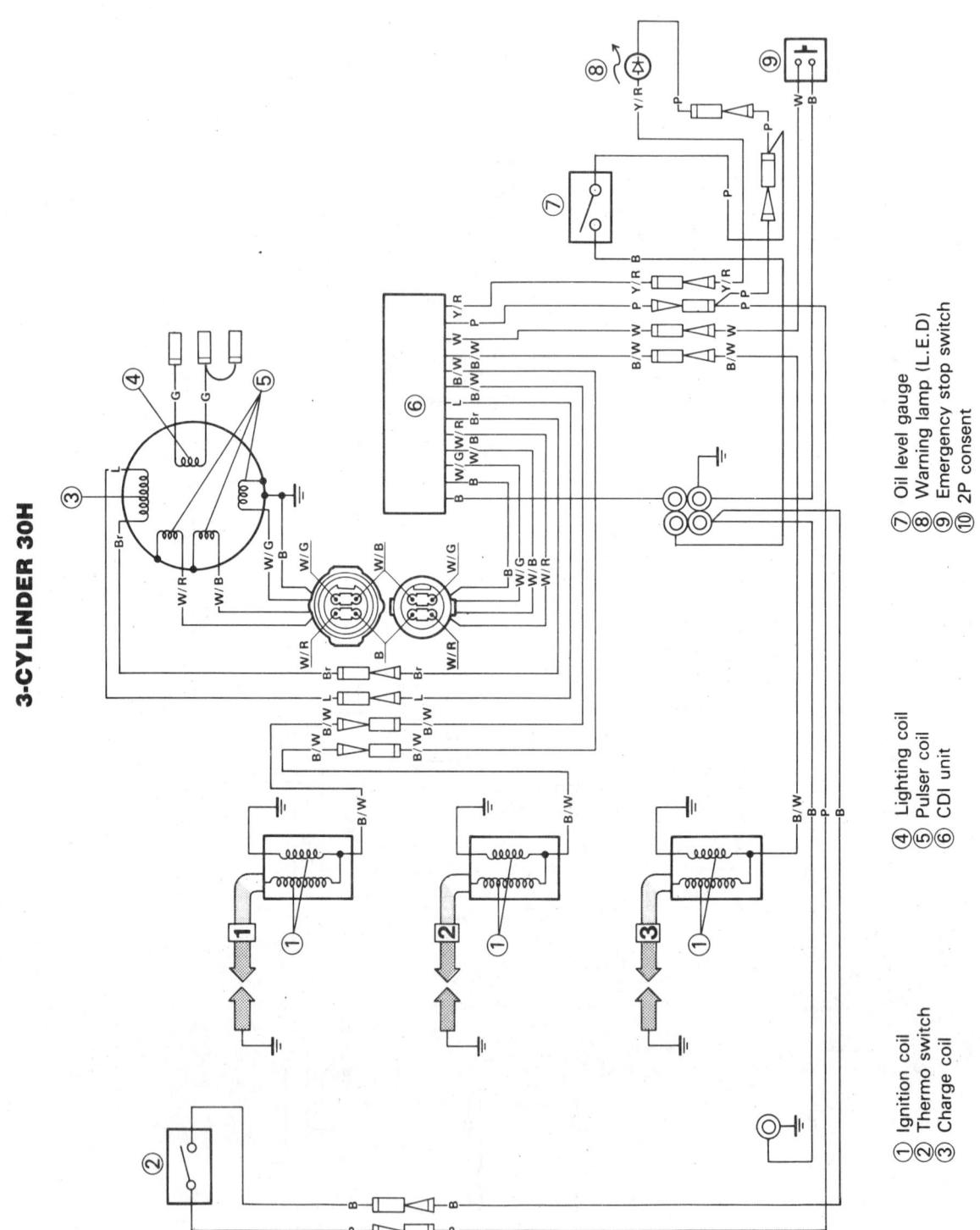

3-CYLINDER 30H

① Ignition coil
② Thermo switch
③ Charge coil

④ Lighting coil
⑤ Pulser coil
⑥ CDI unit

⑦ Oil level gauge
⑧ Warning lamp (L.E.D)
⑨ Emergency stop switch
⑩ 2P consent

3-CYLINDER 30EH

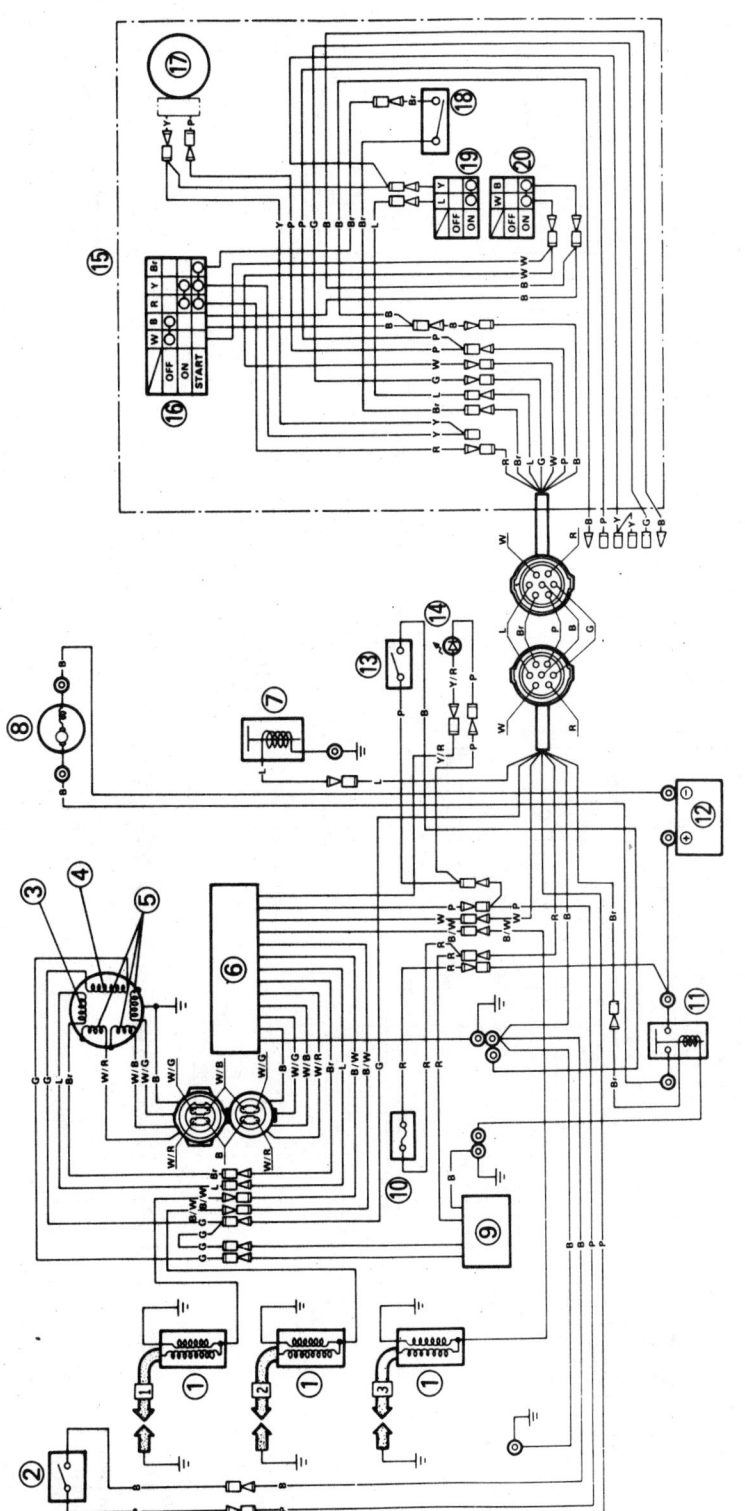

① Ignition coil
② Thermo switch
③ Charge coil
④ Lighting coil
⑤ Pulser coil
⑥ C.D.I. unit

⑦ Choke solenoid
⑧ Starting motor
⑨ Rectifier
⑩ Fuse (10A)
⑪ Starter relay
⑫ Battery

⑬ Oil level gauge
⑭ Warning light
⑮ 703 Remote control box
⑯ Main switch
⑰ Buzzer
⑱ Neutral switch
⑲ Choke switch
⑳ Emergency stop switch

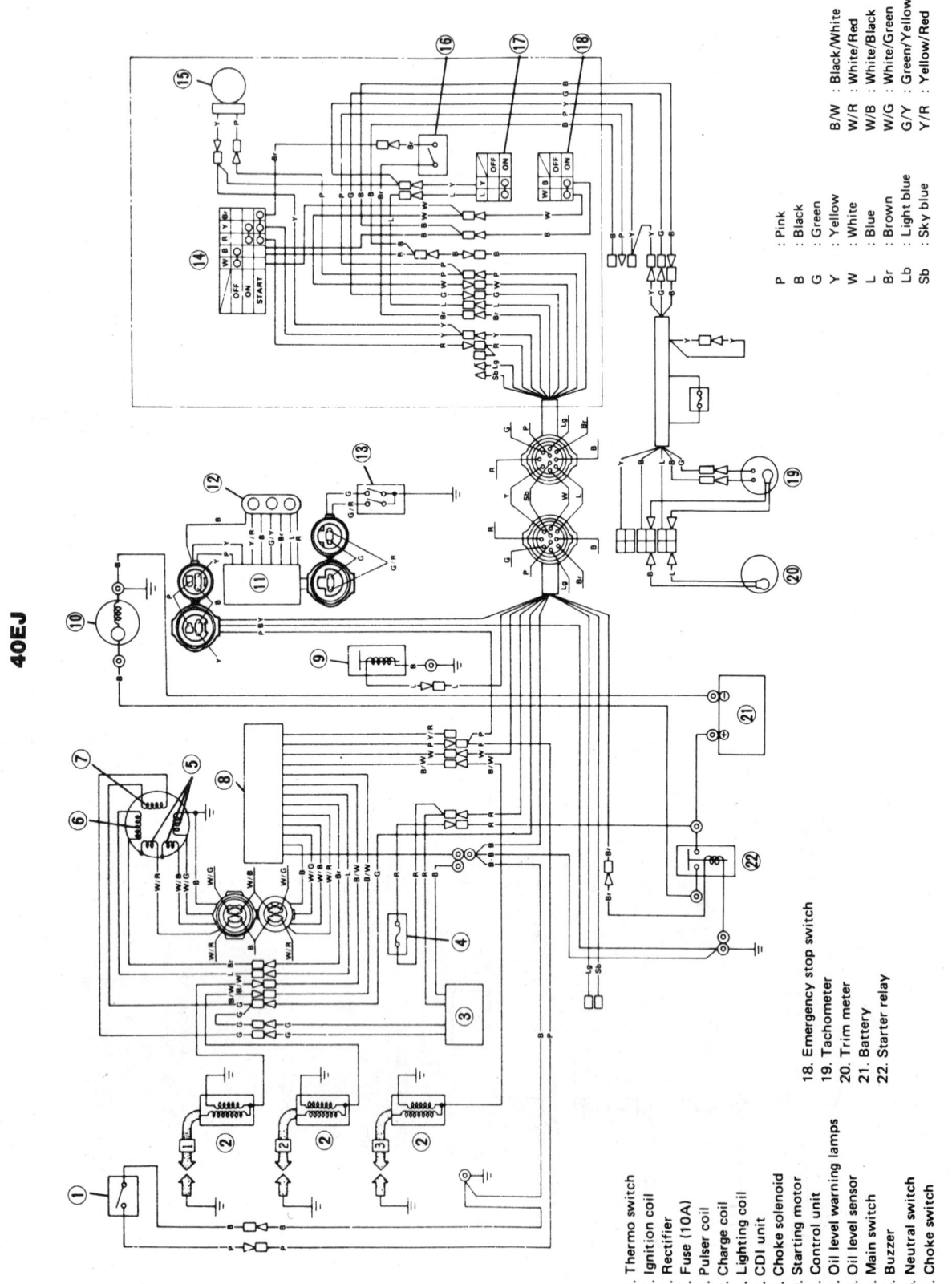

40EJ

1. Thermo switch
2. Ignition coil
3. Rectifier
4. Fuse (10A)
5. Pulser coil
6. Charge coil
7. Lighting coil
8. CDI unit
9. Choke solenoid
10. Starting motor
11. Control unit
12. Oil level warning lamps
13. Oil level sensor
14. Main switch
15. Buzzer
16. Neutral switch
17. Choke switch
18. Emergency stop switch
19. Tachometer
20. Trim meter
21. Battery
22. Starter relay

P : Pink
B : Black
G : Green
Y : Yellow
W : White
L : Blue
Br : Brown
Lb : Light blue
Sb : Sky blue

B/W : Black/White
W/R : White/Red
W/B : White/Black
W/G : White/Green
G/Y : Green/Yellow
Y/R : Yellow/Red

40N

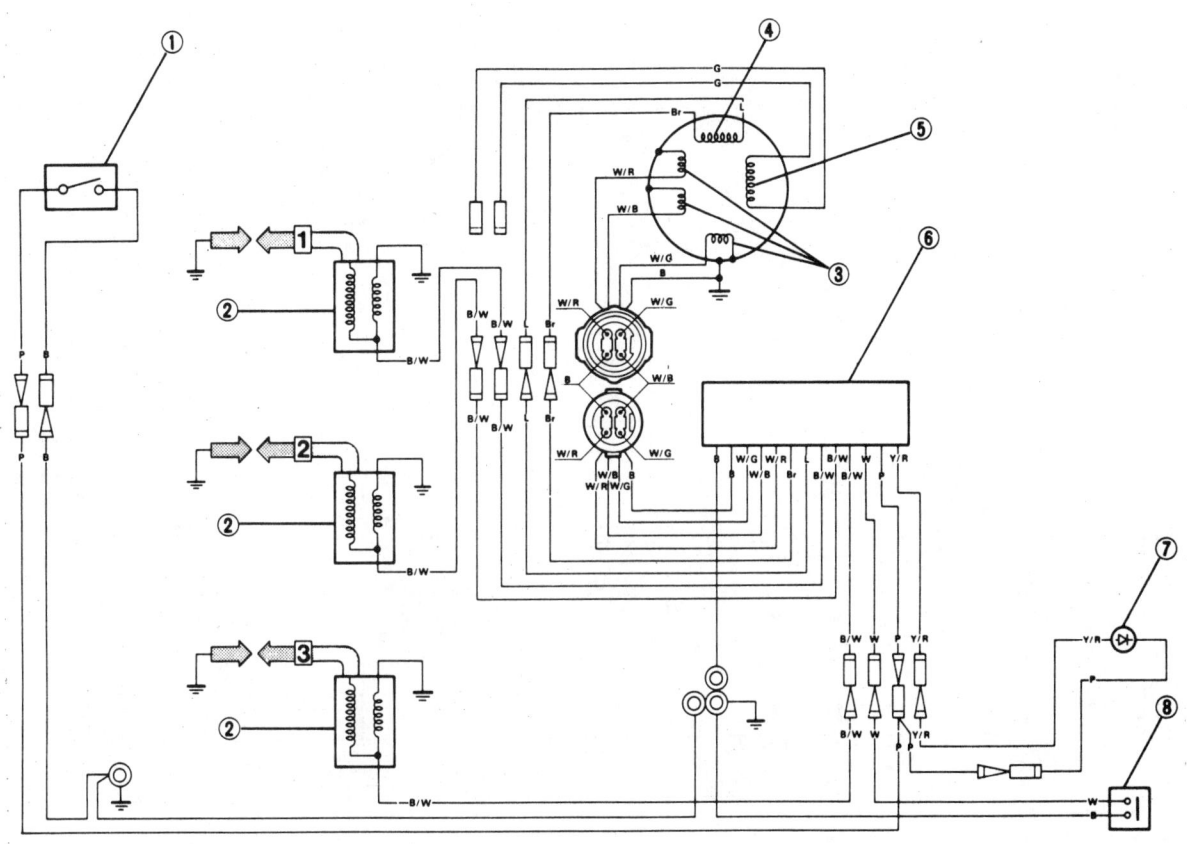

B	Black	L	Blue	Sb	Sky blue
Br	Brown	O	Orange	W	White
G	Green	P	Pink	Y	Yellow
Gy	Gray	R	Red	Lg	Light green

① Thermo switch
② Ignition coil
③ Pulser coil
④ Charge coil
⑤ Lighting coil
⑥ CDI unit
⑦ Overheat warning lamp
⑧ Emergency stop switch

16

40EN

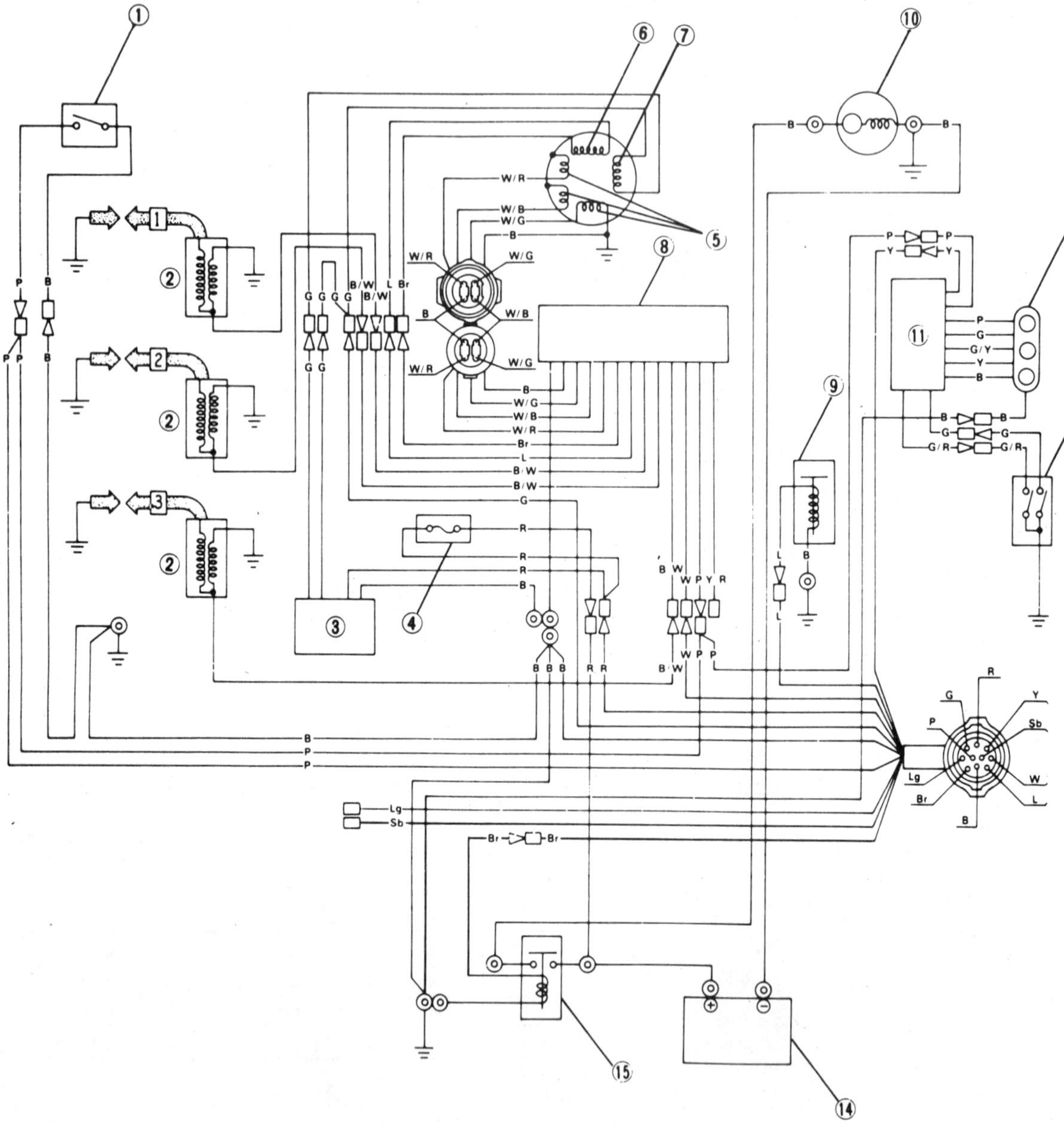

B	Black	L	Blue	Sb	Sky blue
Br	Brown	O	Orange	W	White
G	Green	P	Pink	Y	Yellow
Gy	Gray	R	Red	Lg	Light green

1. Thermo switch
2. Ignition coil
3. Rectifier
4. Fuse (10A)
5. Pulser coil
6. Charge coil
7. Lighting coil
8. CDI unit
9. Choke solenoid
10. Starting motor
11. Control unit
12. Oil level warning lamps
13. Oil level sensor
14. Battery
15. Starter relay

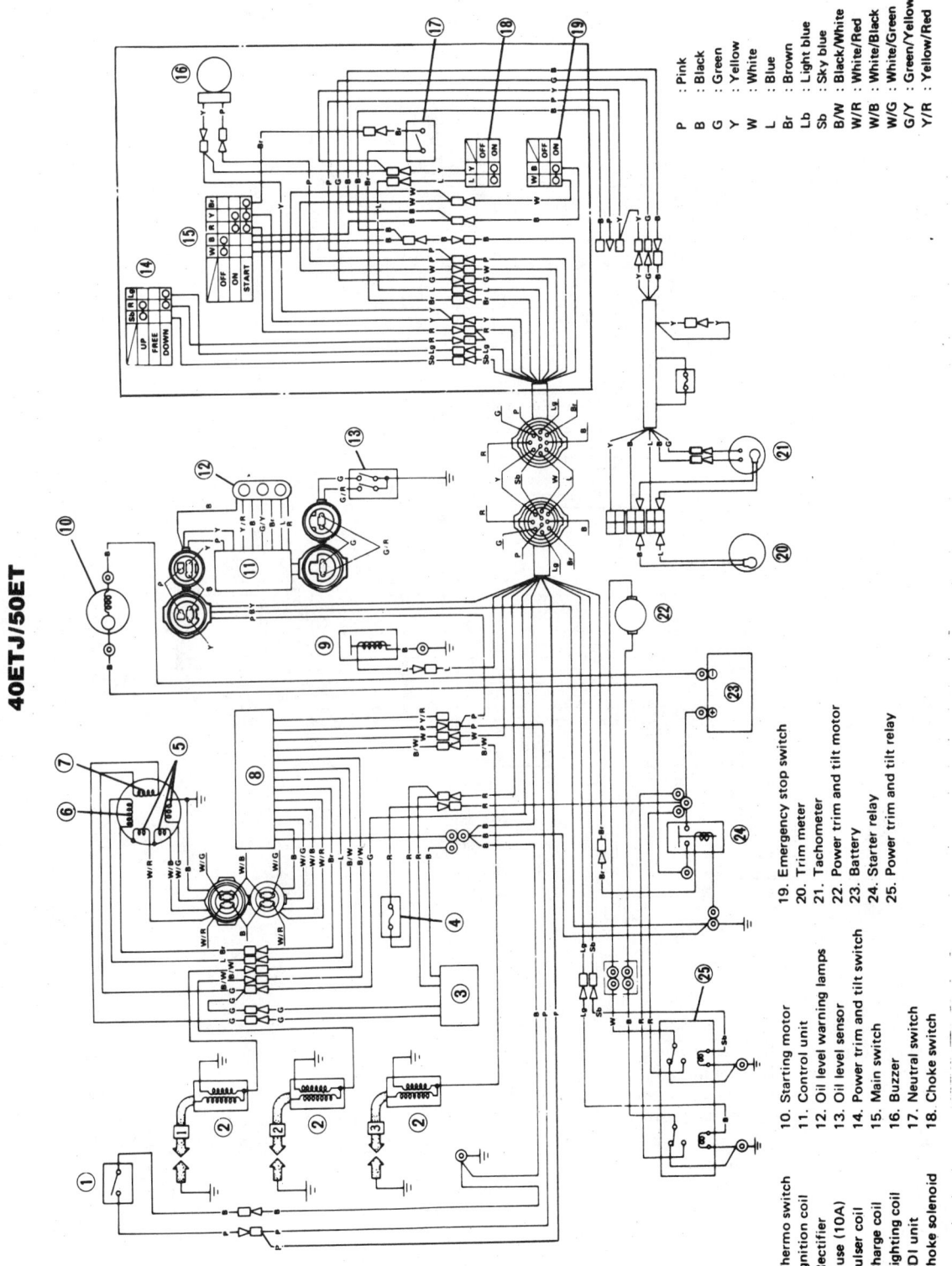

40ETJ/50ET

P	: Pink
B	: Black
G	: Green
Y	: Yellow
W	: White
L	: Blue
Br	: Brown
Lb	: Light blue
Sb	: Sky blue
B/W	: Black/White
W/R	: White/Red
W/B	: White/Black
W/G	: White/Green
G/Y	: Green/Yellow
Y/R	: Yellow/Red

1. Thermo switch
2. Ignition coil
3. Rectifier
4. Fuse (10A)
5. Pulser coil
6. Charge coil
7. Lighting coil
8. CDI unit
9. Choke solenoid

10. Starting motor
11. Control unit
12. Oil level warning lamps
13. Oil level sensor
14. Power trim and tilt switch
15. Main switch
16. Buzzer
17. Neutral switch
18. Choke switch

19. Emergency stop switch
20. Trim meter
21. Tachometer
22. Power trim and tilt motor
23. Battery
24. Starter relay
25. Power trim and tilt relay

16

50ETN

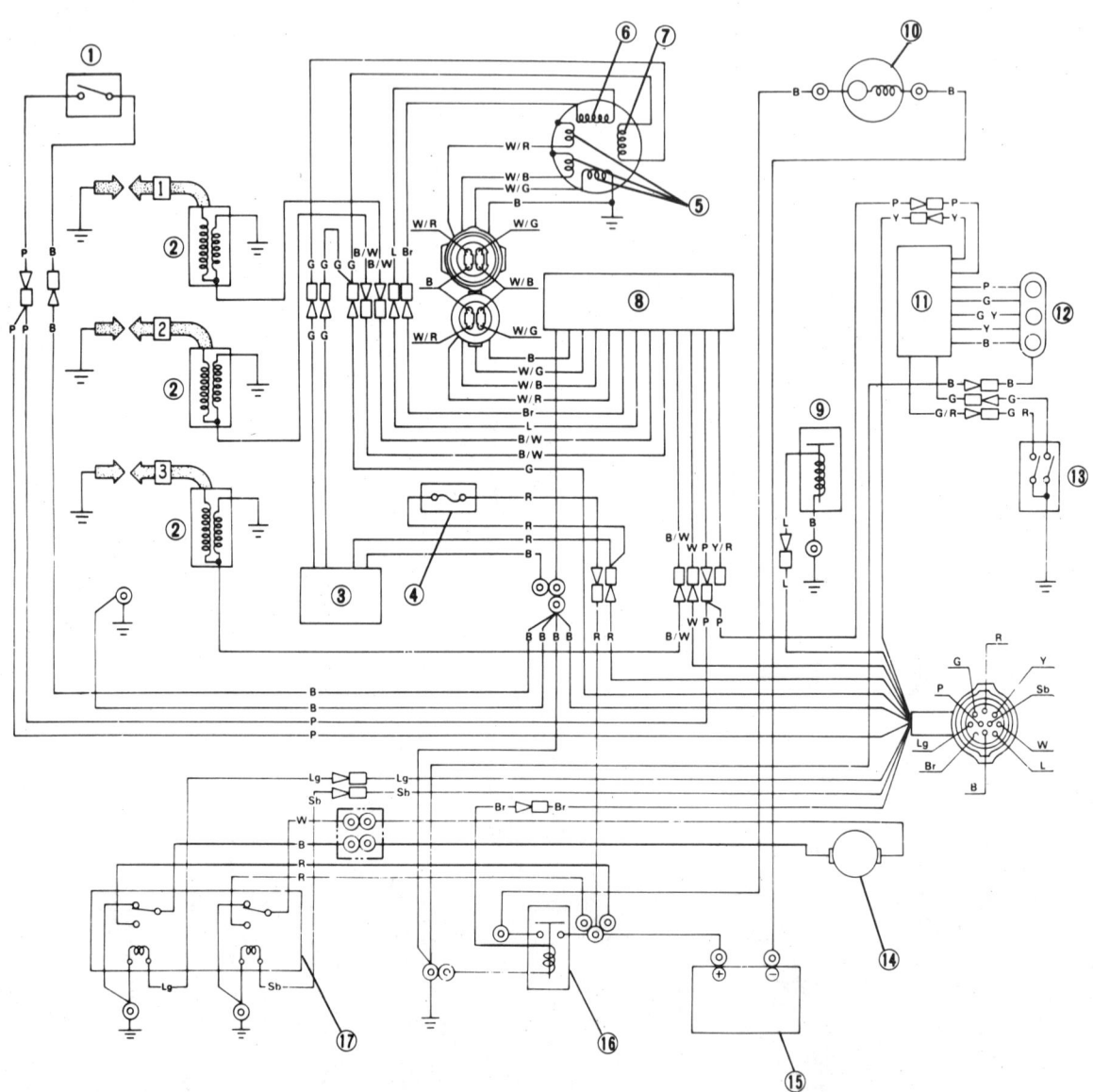

1	Thermo switch		10	Starting motor		
2	Ignition coil		11	Control unit		
3	Rectifier		12	Oil level warning lamps		
4	Fuse (10A)		13	Oil level sensor		
5	Pulser coil		14	Power tilt motor		
6	Charge coil		15	Battery		
7	Lighting coil		16	Starter relay		
8	CDI unit		17	Power tilt relay		
9	Choke solenoid					

B	Black	L	Blue	Sb	Sky blue
Br	Brown	O	Orange	W	White
G	Green	P	Pink	Y	Yellow
Gy	Gray	R	Red	Lg	Light green

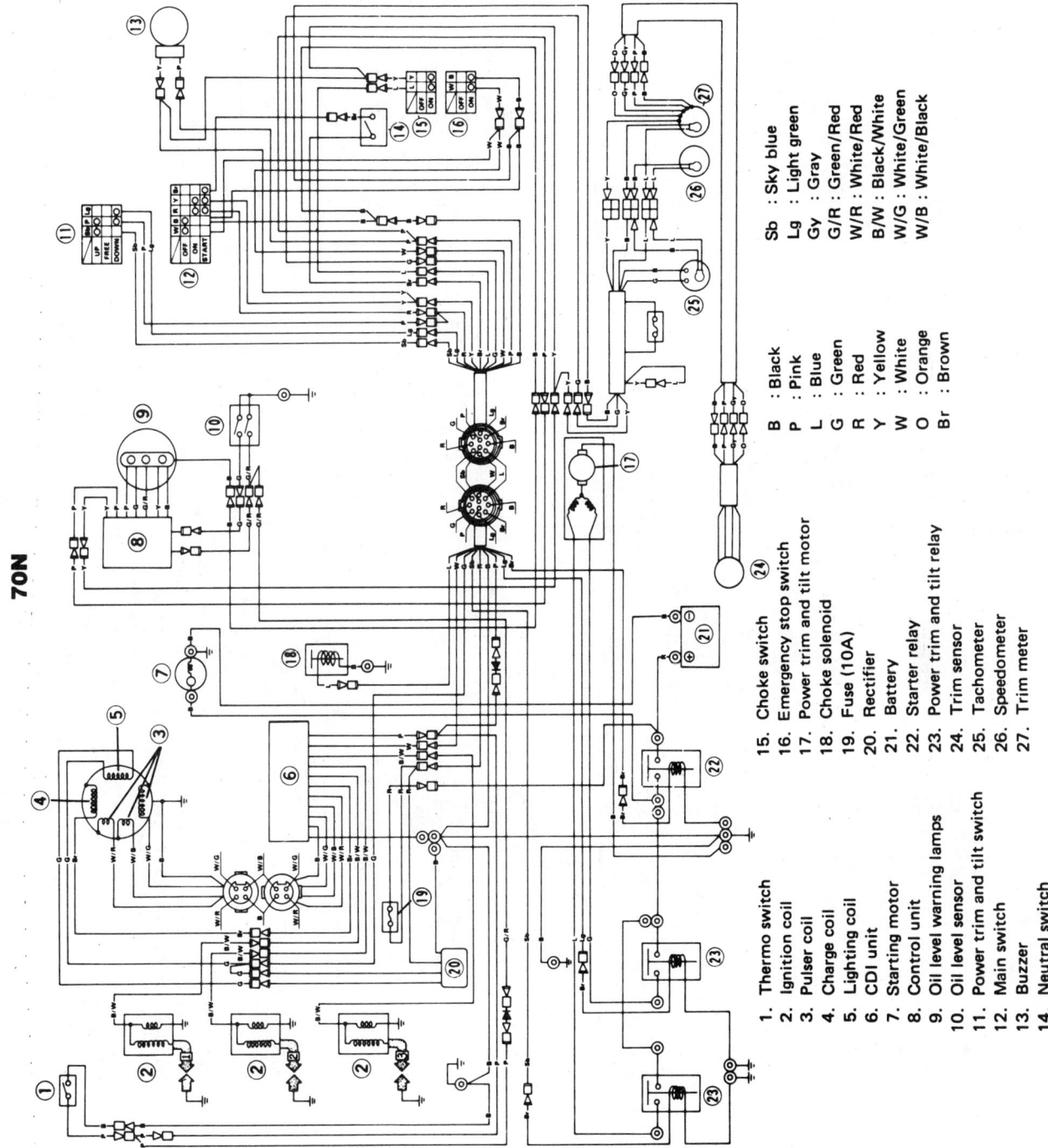

70N

Sb : Sky blue
Lg : Light green
Gy : Gray
G/R : Green/Red
W/R : White/Red
B/W : Black/White
W/G : White/Green
W/B : White/Black

B : Black
P : Pink
L : Blue
G : Green
R : Red
Y : Yellow
W : White
O : Orange
Br : Brown

1. Thermo switch
2. Ignition coil
3. Pulser coil
4. Charge coil
5. Lighting coil
6. CDI unit
7. Starting motor
8. Control unit
9. Oil level warning lamps
10. Oil level sensor
11. Power trim and tilt switch
12. Main switch
13. Buzzer
14. Neutral switch

15. Choke switch
16. Emergency stop switch
17. Power trim and tilt motor
18. Choke solenoid
19. Fuse (10A)
20. Rectifier
21. Battery
22. Starter relay
23. Power trim and tilt relay
24. Trim sensor
25. Tachometer
26. Speedometer
27. Trim meter

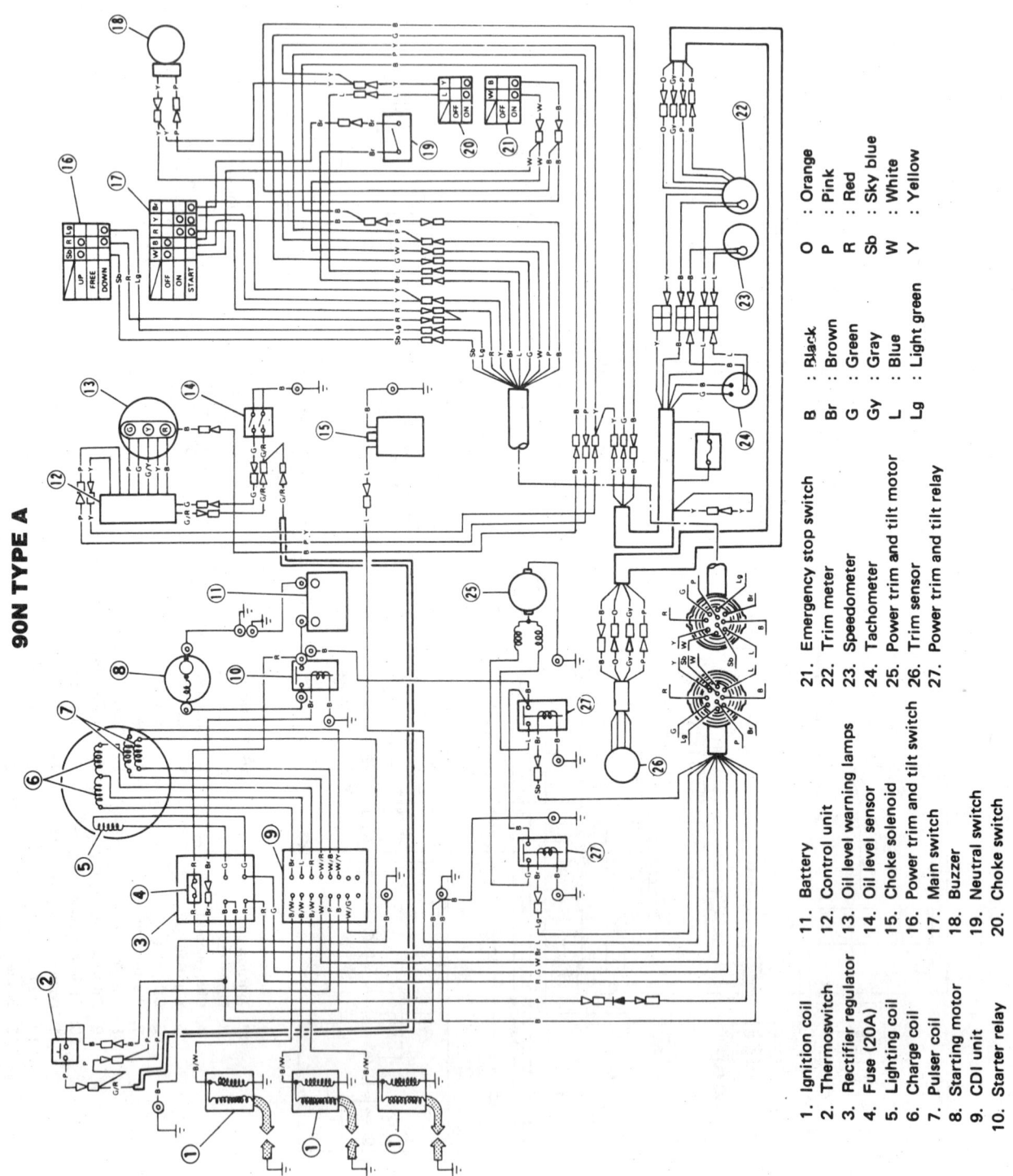

90N TYPE A

1. Ignition coil
2. Thermoswitch
3. Rectifier regulator
4. Fuse (20A)
5. Lighting coil
6. Charge coil
7. Pulser coil
8. Starting motor
9. CDI unit
10. Starter relay

11. Battery
12. Control unit
13. Oil level warning lamps
14. Oil level sensor
15. Choke solenoid
16. Power trim and tilt switch
17. Main switch
18. Buzzer
19. Neutral switch
20. Choke switch

21. Emergency stop switch
22. Trim meter
23. Speedometer
24. Tachometer
25. Power trim and tilt motor
26. Trim sensor
27. Power trim and tilt relay

B   : Black
Br   : Brown
G   : Green
Gy   : Gray
L   : Blue
Lg   : Light green

O   : Orange
P   : Pink
R   : Red
Sb   : Sky blue
W   : White
Y   : Yellow

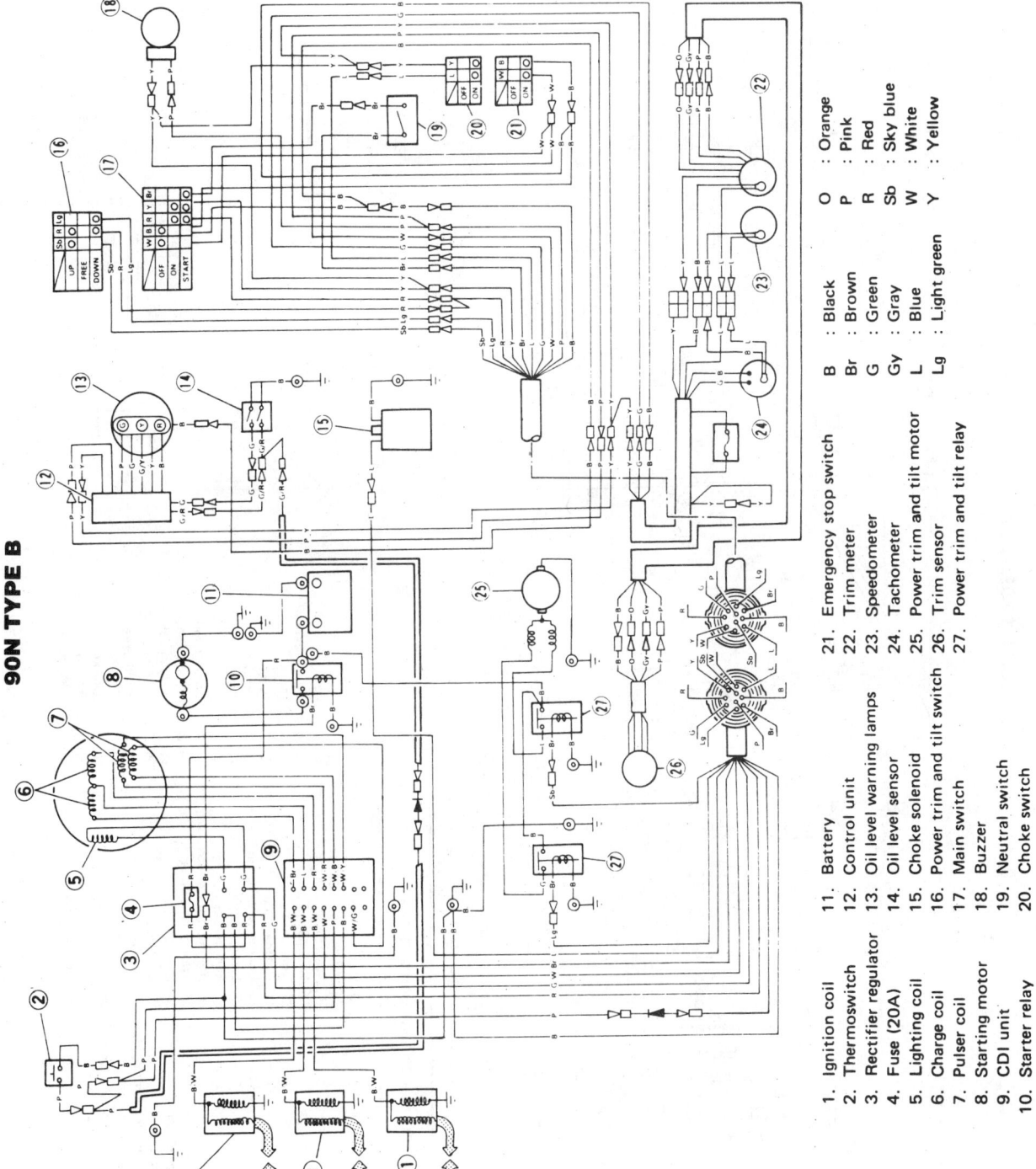

90N TYPE B

1. Ignition coil
2. Thermoswitch
3. Rectifier regulator
4. Fuse (20A)
5. Lighting coil
6. Charge coil
7. Pulser coil
8. Starting motor
9. CDI unit
10. Starter relay
11. Battery
12. Control unit
13. Oil level warning lamps
14. Oil level sensor
15. Choke solenoid
16. Power trim and tilt switch
17. Main switch
18. Buzzer
19. Neutral switch
20. Choke switch
21. Emergency stop switch
22. Trim meter
23. Speedometer
24. Tachometer
25. Power trim and tilt motor
26. Trim sensor
27. Power trim and tilt relay

B : Black	O : Orange
Br : Brown	P : Pink
G : Green	R : Red
Gy : Gray	Sb : Sky blue
L : Blue	W : White
Lg : Light green	Y : Yellow

1984 115N TYPE A

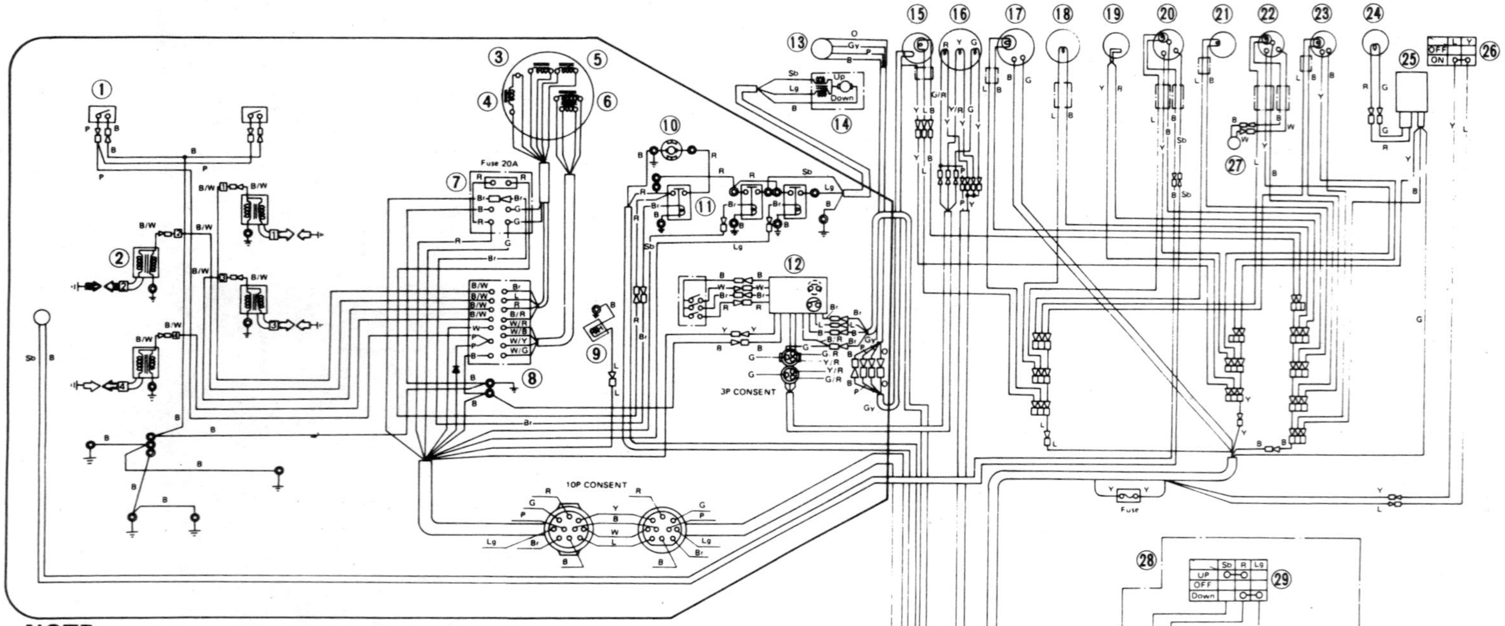

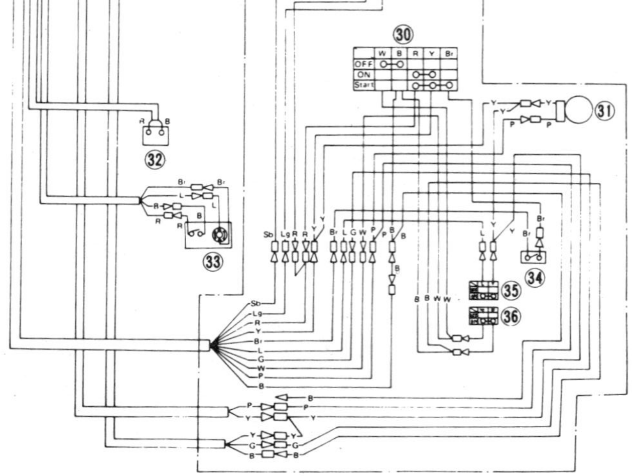

NOTE:

Be sure to re-connect wires firmly.

1. Thermo switch
2. Ignition coil
3. CDI magneto
4. Lighting coil
5. Charge coil
6. Pulser coil
7. Rectifier regulator
8. CDI unit
9. Choke solenoid
10. Starting motor
11. Starter relay
12. Control unit
13. Trim sensor
14. Trim & tilt motor
15. Trim gauge
16. Oil level warning lamps
17. Tachometer
18. Speedometer

19. Hour meter
20. Water temp. gauge
21. Water pressure meter
22. Fuel gauge
23. Voltmeter
24. Charge lamp
25. Charge warning unit
26. Lamp switch
27. Fuel sensor
28. Remote control box
29. Trim & tilt switch
30. Main switch
31. Buzzer
32. Battery
33. Oil level gauge
34. Neutral switch
35. Choke switch
36. Emergency stop switch

B	Black
P	Pink
B/W	Black/White
L	Blue
G	Green
Br	Brown
R	Red
W	White
W/R	White/Red
W/B	White/Black
W/Y	White/Yellow
W/G	White/Green
Lg	Light green
Gy	Grey
O	Orange
Y	Yellow
Sb	Sky blue

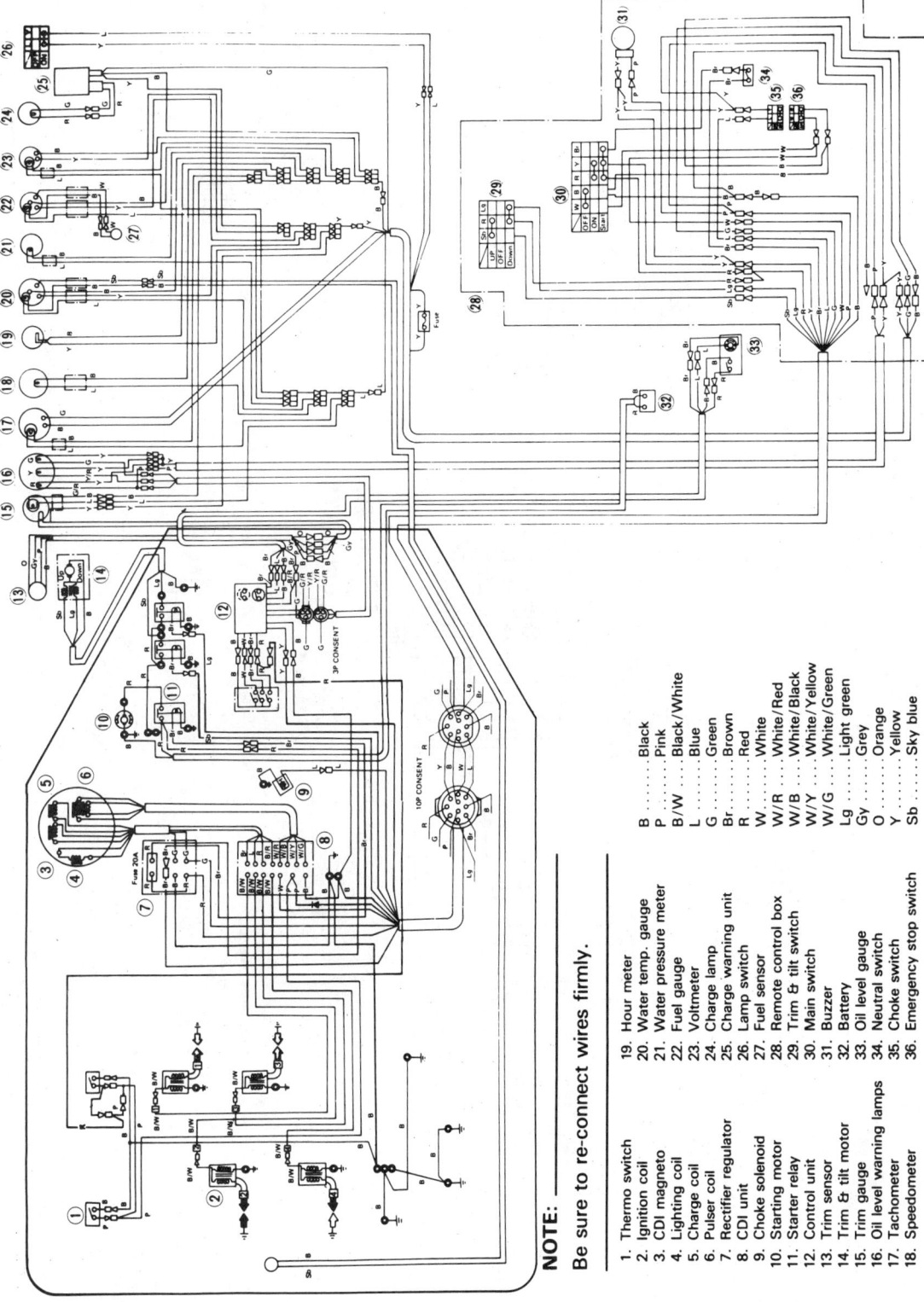

1984 115N TYPE B

NOTE:

Be sure to re-connect wires firmly.

1. Thermo switch	19. Hour meter	B	Black
2. Ignition coil	20. Water temp. gauge	P	Pink
3. CDI magneto	21. Water pressure meter	B/W	Black/White
4. Lighting coil	22. Fuel gauge	L	Blue
5. Charge coil	23. Voltmeter	G	Green
6. Pulser coil	24. Charge lamp	Br	Brown
7. Rectifier regulator	25. Charge warning unit	R	Red
8. CDI unit	26. Lamp switch	W	White
9. Choke solenoid	27. Fuel sensor	W/R	White/Red
10. Starting motor	28. Remote control box	W/B	White/Black
11. Starter relay	29. Trim & tilt switch	W/Y	White/Yellow
12. Control unit	30. Main switch	W/G	White/Green
13. Trim sensor	31. Buzzer	Lg	Light green
14. Trim & tilt motor	32. Battery	Gy	Grey
15. Trim gauge	33. Oil level gauge	O	Orange
16. Oil level warning lamps	34. Neutral switch	Y	Yellow
17. Tachometer	35. Choke switch	Sb	Sky blue
18. Speedometer	36. Emergency stop switch		

16

1985 115 HP

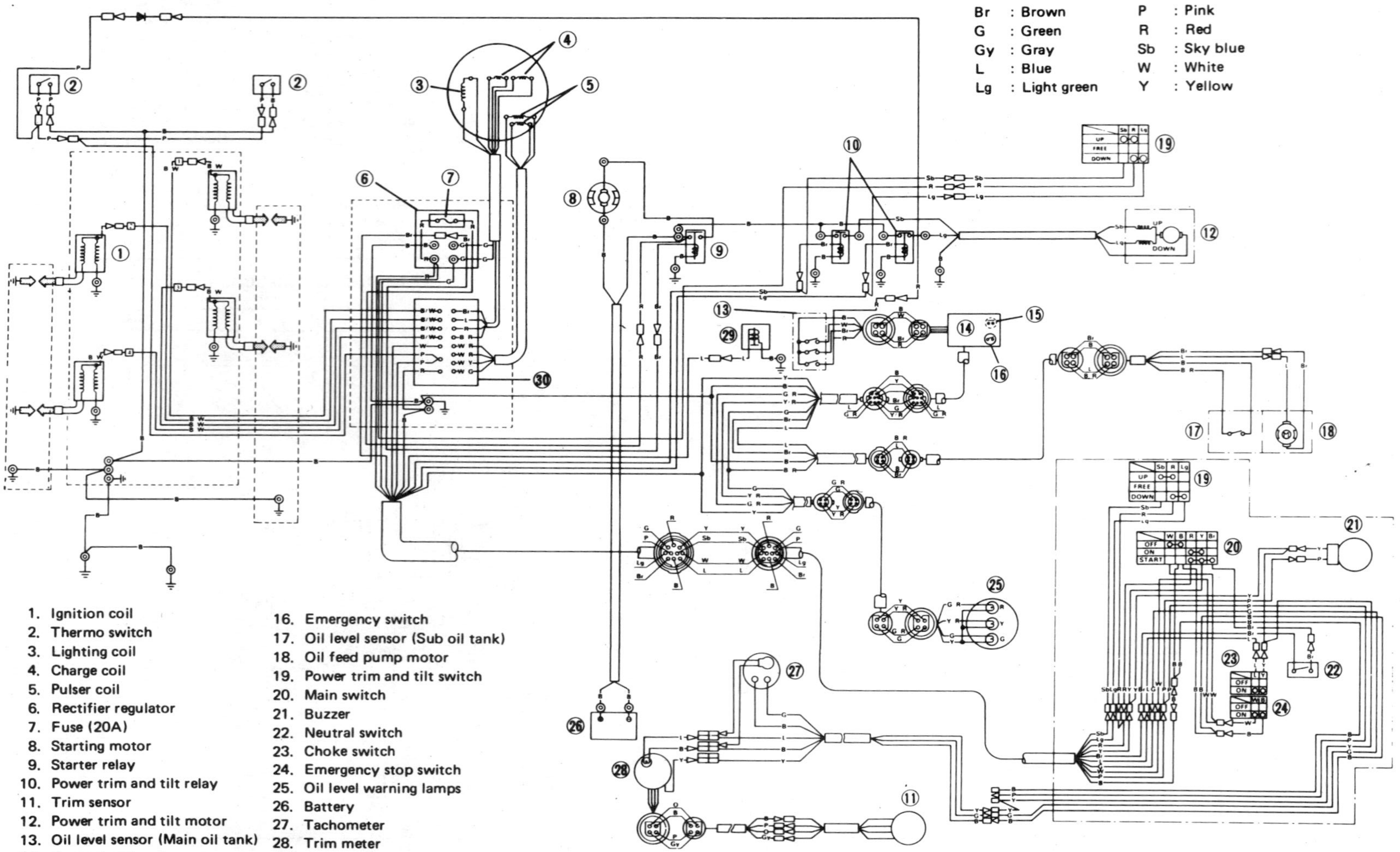

B : Black
Br : Brown
G : Green
Gy : Gray
L : Blue
Lg : Light green

O : Orange
P : Pink
R : Red
Sb : Sky blue
W : White
Y : Yellow

1. Ignition coil
2. Thermo switch
3. Lighting coil
4. Charge coil
5. Pulser coil
6. Rectifier regulator
7. Fuse (20A)
8. Starting motor
9. Starter relay
10. Power trim and tilt relay
11. Trim sensor
12. Power trim and tilt motor
13. Oil level sensor (Main oil tank)
14. Control unit
15. Tilt switch

16. Emergency switch
17. Oil level sensor (Sub oil tank)
18. Oil feed pump motor
19. Power trim and tilt switch
20. Main switch
21. Buzzer
22. Neutral switch
23. Choke switch
24. Emergency stop switch
25. Oil level warning lamps
26. Battery
27. Tachometer
28. Trim meter
29. Choke solenoid
30. CDI unit

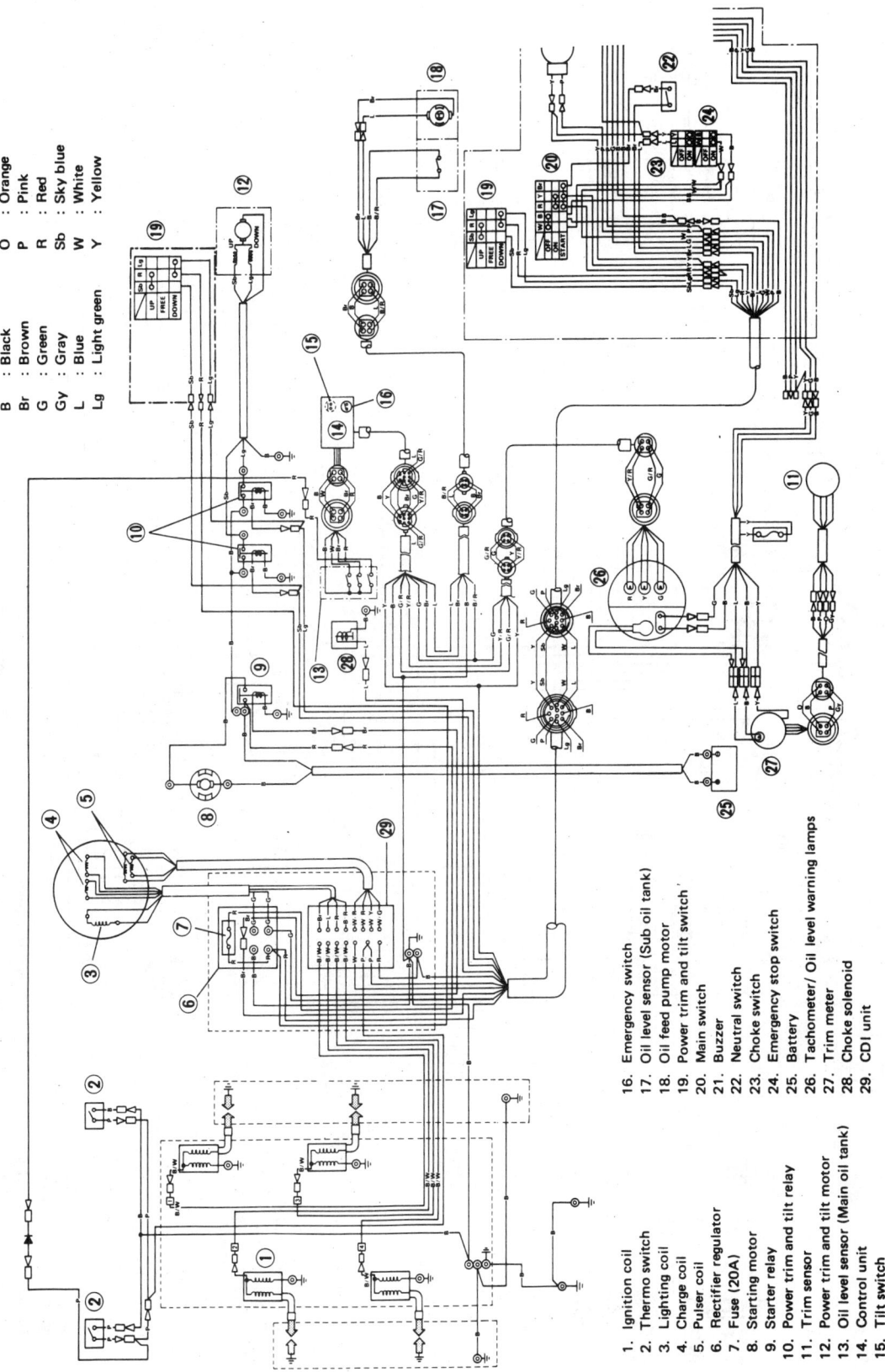

1986-ON 115 HP

B : Black
Br : Brown
G : Green
Gy : Gray
L : Blue
Lg : Light green

O : Orange
P : Pink
R : Red
Sb : Sky blue
W : White
Y : Yellow

1. Ignition coil
2. Thermo switch
3. Lighting coil
4. Charge coil
5. Pulser coil
6. Rectifier regulator
7. Fuse (20A)
8. Starting motor
9. Starter relay
10. Power trim and tilt relay
11. Trim sensor
12. Power trim and tilt motor
13. Oil level sensor (Main oil tank)
14. Control unit
15. Tilt switch

16. Emergency switch
17. Oil level sensor (Sub oil tank)
18. Oil feed pump motor
19. Power trim and tilt switch
20. Main switch
21. Buzzer
22. Neutral switch
23. Choke switch
24. Emergency stop switch
25. Battery
26. Tachometer/ Oil level warning lamps
27. Trim meter
28. Choke solenoid
29. CDI unit

16

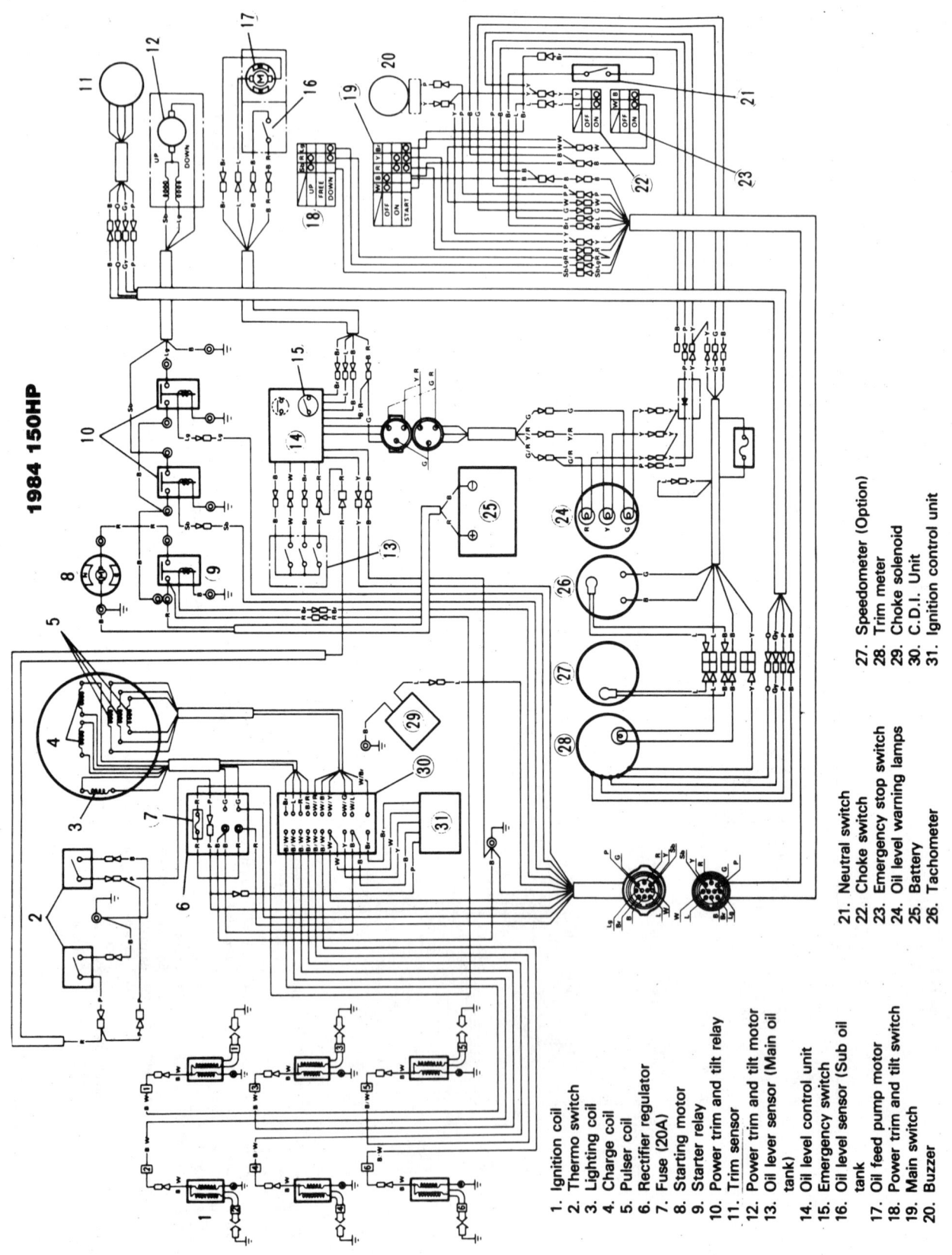

1984 150HP

1. Ignition coil
2. Thermo switch
3. Lighting coil
4. Charge coil
5. Pulser coil
6. Rectifier regulator
7. Fuse (20A)
8. Starting motor
9. Starter relay
10. Power trim and tilt relay
11. Trim sensor
12. Power trim and tilt motor
13. Oil lever sensor (Main oil tank)
14. Oil level control unit
15. Emergency switch
16. Oil level sensor (Sub oil tank)
17. Oil feed pump motor
18. Power trim and tilt switch
19. Main switch
20. Buzzer
21. Neutral switch
22. Choke switch
23. Emergency stop switch
24. Oil level warning lamps
25. Battery
26. Tachometer
27. Speedometer (Option)
28. Trim meter
29. Choke solenoid
30. C.D.I. Unit
31. Ignition control unit

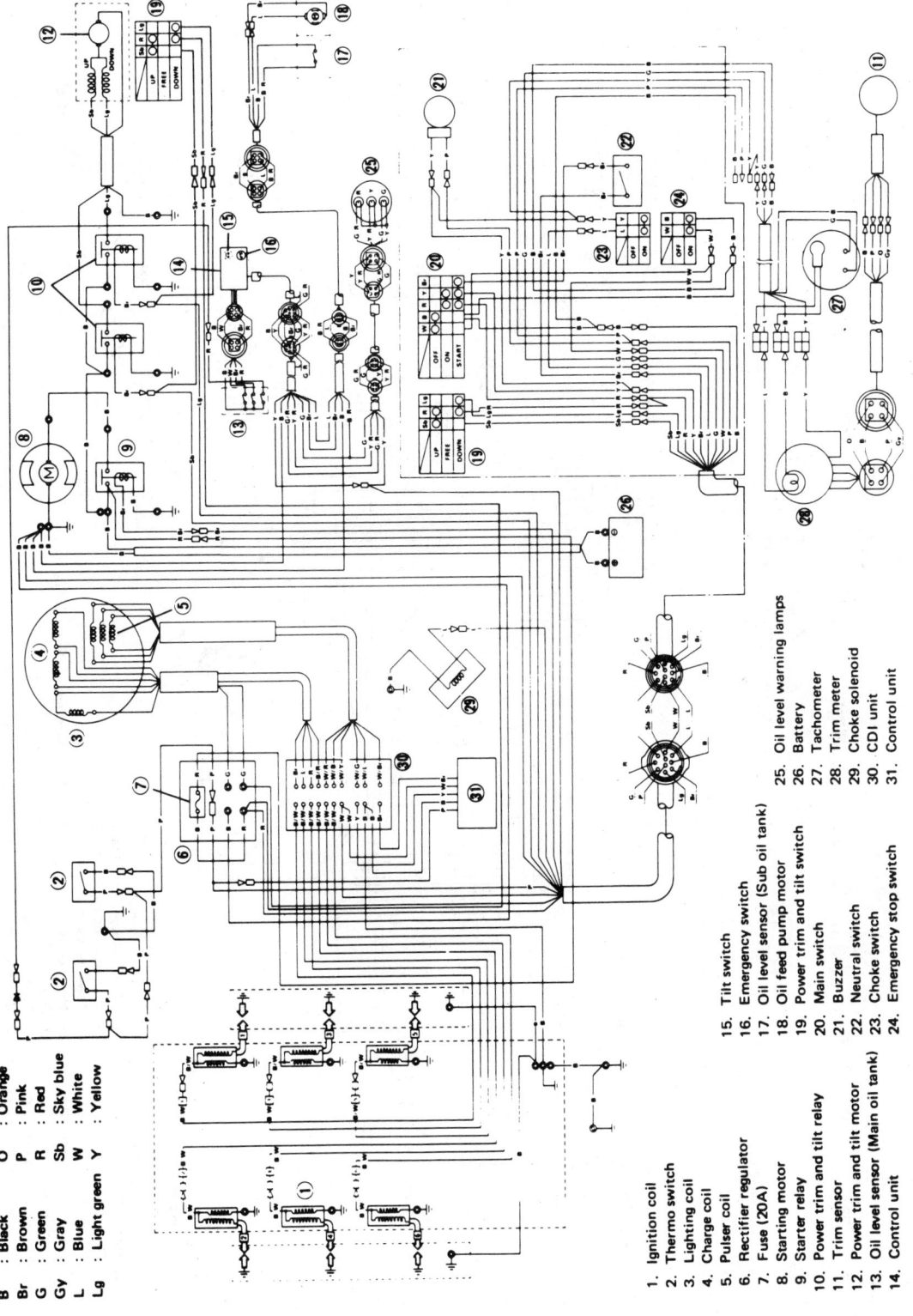

1985-ON 150HP

B : Black   O : Orange
Br : Brown   P : Pink
G : Green   R : Red
Gy : Gray   Sb : Sky blue
L : Blue   W : White
Lg : Light green   Y : Yellow

1. Ignition coil
2. Thermo switch
3. Lighting coil
4. Charge coil
5. Pulser coil
6. Rectifier regulator
7. Fuse (20A)
8. Starting motor
9. Starter relay
10. Power trim and tilt relay
11. Trim sensor
12. Power trim and tilt motor
13. Oil level sensor (Main oil tank)
14. Control unit
15. Tilt switch
16. Emergency switch
17. Oil level sensor (Sub oil tank)
18. Oil feed pump motor
19. Power trim and tilt switch
20. Main switch
21. Buzzer
22. Neutral switch
23. Choke switch
24. Emergency stop switch
25. Oil level warning lamps
26. Battery
27. Tachometer
28. Trim meter
29. Choke solenoid
30. CDI unit
31. Control unit

16

1986-ON 150/175/200 HP

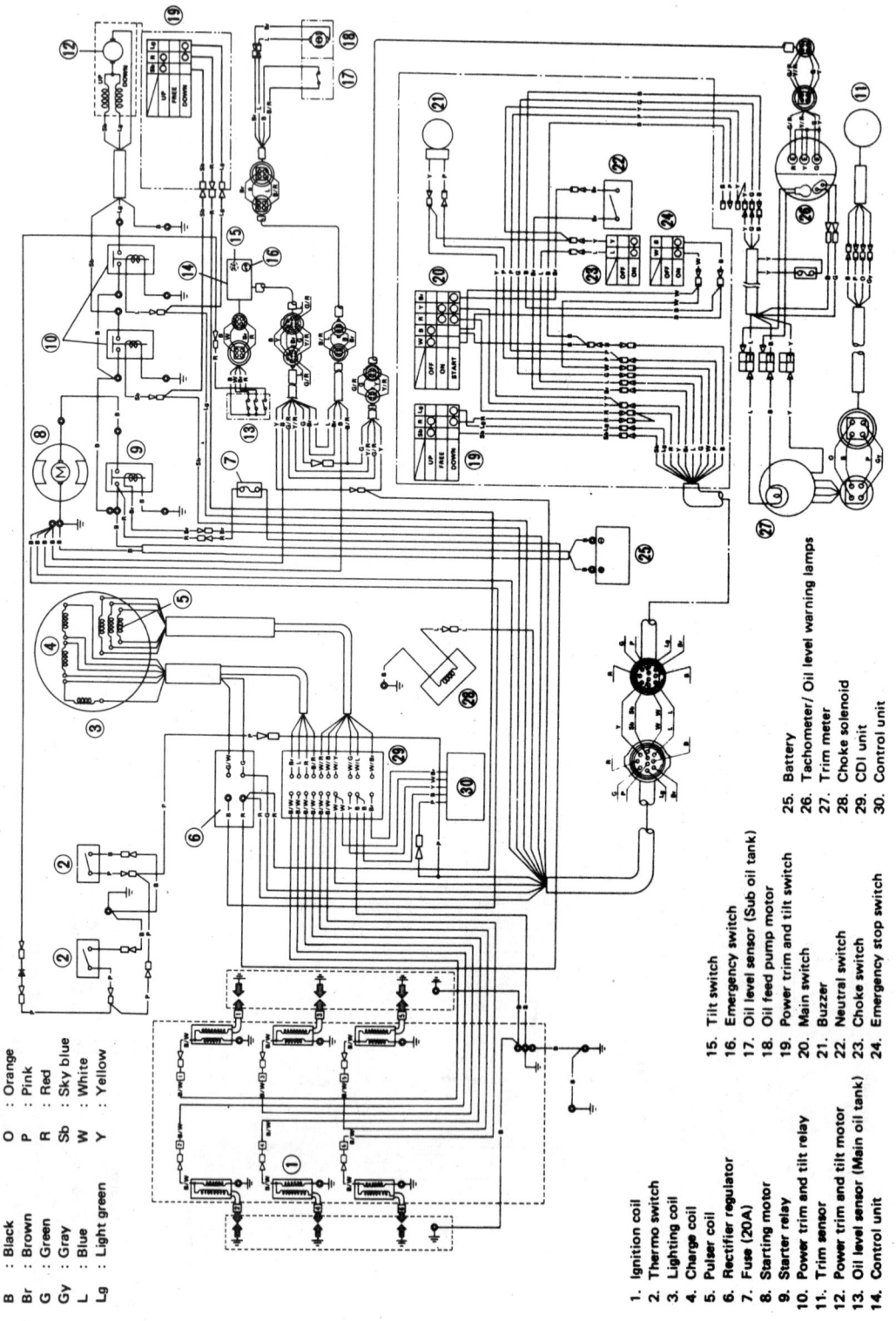

B : Black
Br : Brown
G : Green
Gy : Gray
L : Blue
Lg : Light green

O : Orange
P : Pink
R : Red
Sb : Sky blue
W : White
Y : Yellow

1. Ignition coil
2. Thermo switch
3. Lighting coil
4. Charge coil
5. Pulser coil
6. Rectifier regulator
7. Fuse (20A)
8. Starting motor
9. Starter relay
10. Power trim and tilt relay
11. Trim sensor
12. Power trim and tilt motor
13. Oil level sensor (Main oil tank)
14. Control unit

15. Tilt switch
16. Emergency switch
17. Oil level sensor (Sub oil tank)
18. Oil feed pump motor
19. Power trim and tilt switch
20. Main switch
21. Buzzer
22. Neutral switch
23. Choke switch
24. Emergency stop switch

25. Battery
26. Tachometer/ Oil level warning lamps
27. Trim meter
28. Choke solenoid
29. CDI unit
30. Control unit

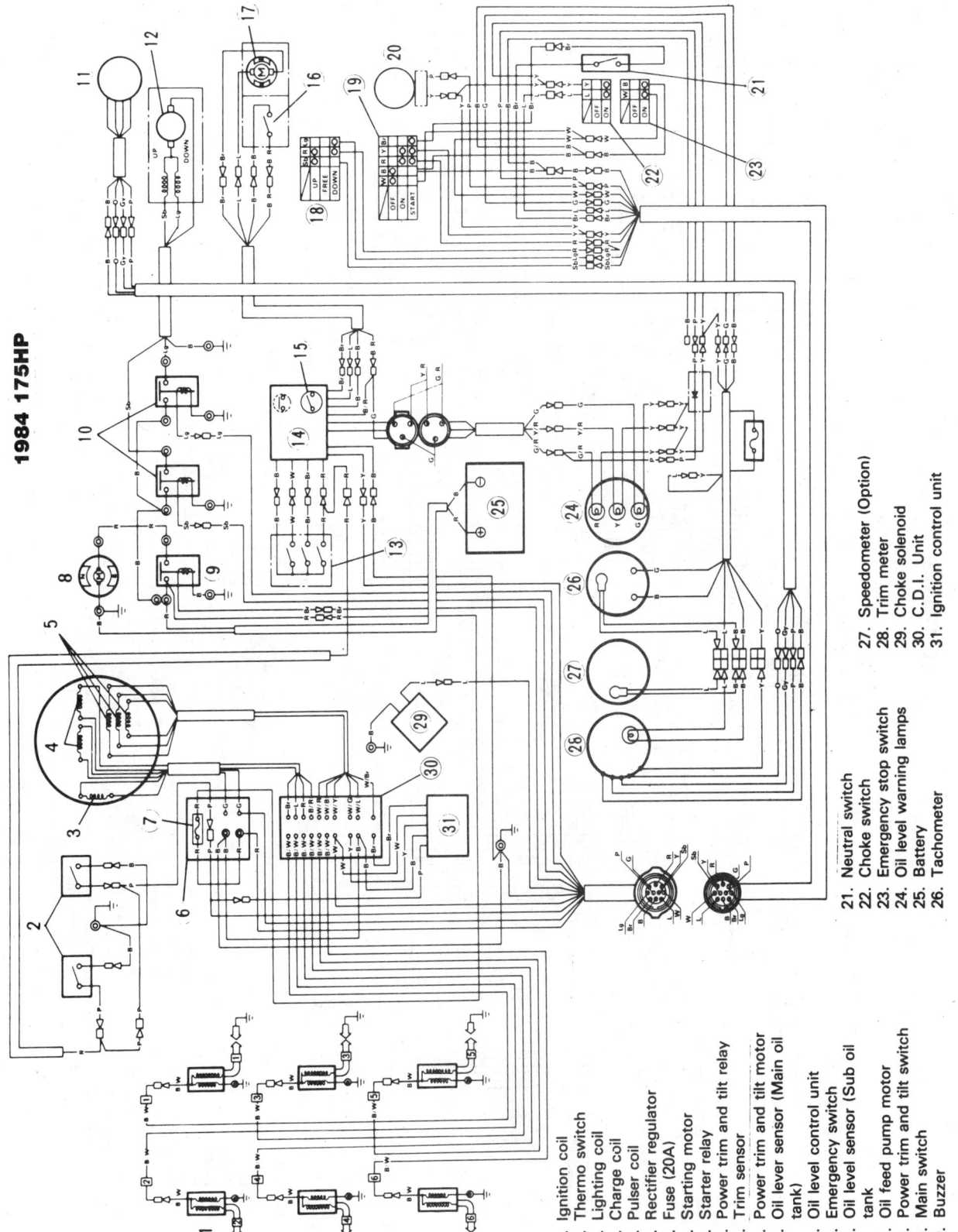

1984 175HP

1. Ignition coil
2. Thermo switch
3. Lighting coil
4. Charge coil
5. Pulser coil
6. Rectifier regulator
7. Fuse (20A)
8. Starting motor
9. Starter relay
10. Power trim and tilt relay
11. Trim sensor
12. Power trim and tilt motor
13. Oil lever sensor (Main oil tank)
14. Oil level control unit
15. Emergency switch
16. Oil level sensor (Sub oil tank)
17. Oil feed pump motor
18. Power trim and tilt switch
19. Main switch
20. Buzzer

21. Neutral switch
22. Choke switch
23. Emergency stop switch
24. Oil level warning lamps
25. Battery
26. Tachometer

27. Speedometer (Option)
28. Trim meter
29. Choke solenoid
30. C.D.I. Unit
31. Ignition control unit

1985 175 HP

B : Black
Br : Brown
G : Green
Gy : Gray
L : Blue
Lg : Light green

O : Orange
P : Pink
R : Red
Sb : Sky blue
W : White
Y : Yellow

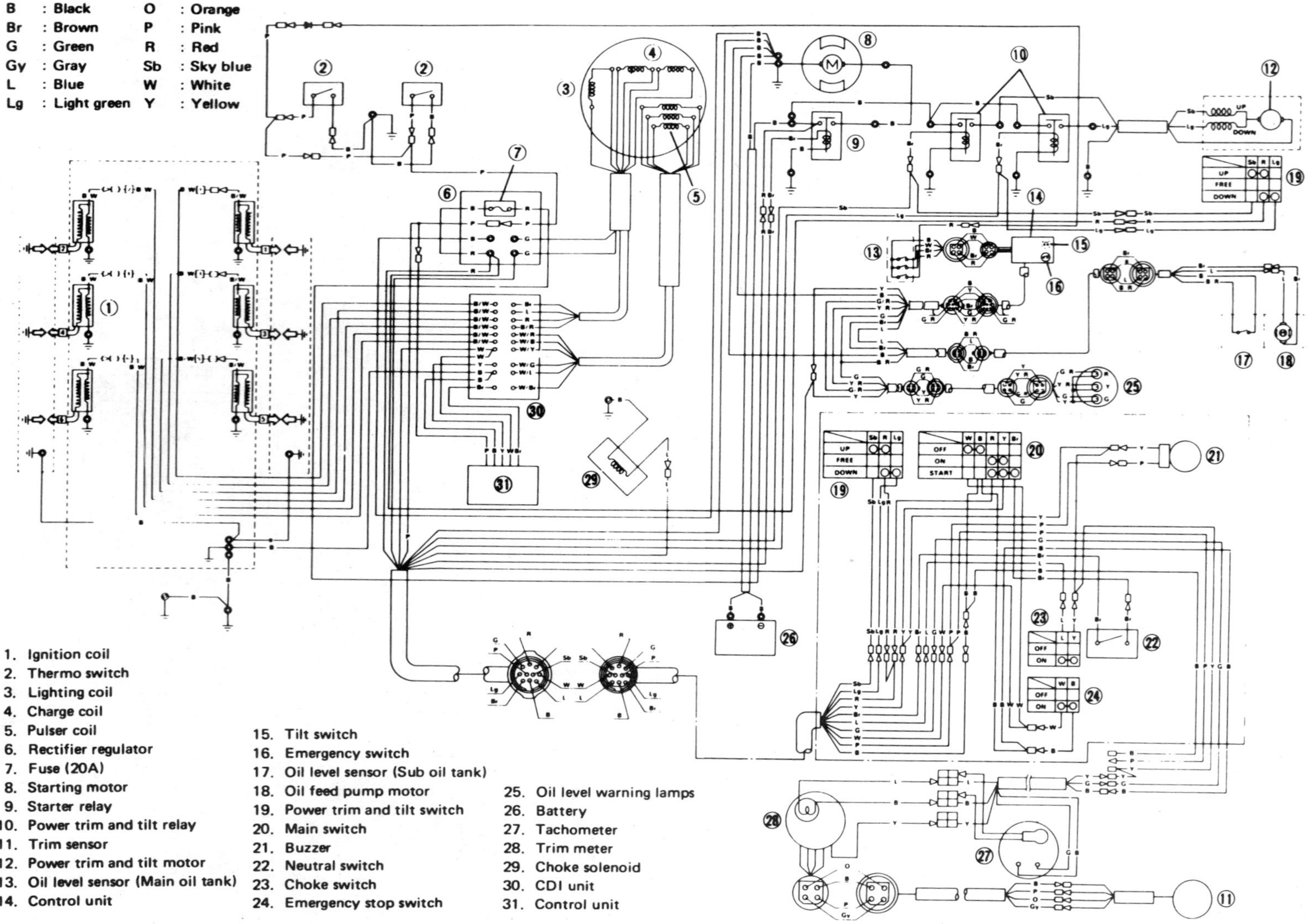

1. Ignition coil
2. Thermo switch
3. Lighting coil
4. Charge coil
5. Pulser coil
6. Rectifier regulator
7. Fuse (20A)
8. Starting motor
9. Starter relay
10. Power trim and tilt relay
11. Trim sensor
12. Power trim and tilt motor
13. Oil level sensor (Main oil tank)
14. Control unit

15. Tilt switch
16. Emergency switch
17. Oil level sensor (Sub oil tank)
18. Oil feed pump motor
19. Power trim and tilt switch
20. Main switch
21. Buzzer
22. Neutral switch
23. Choke switch
24. Emergency stop switch

25. Oil level warning lamps
26. Battery
27. Tachometer
28. Trim meter
29. Choke solenoid
30. CDI unit
31. Control unit

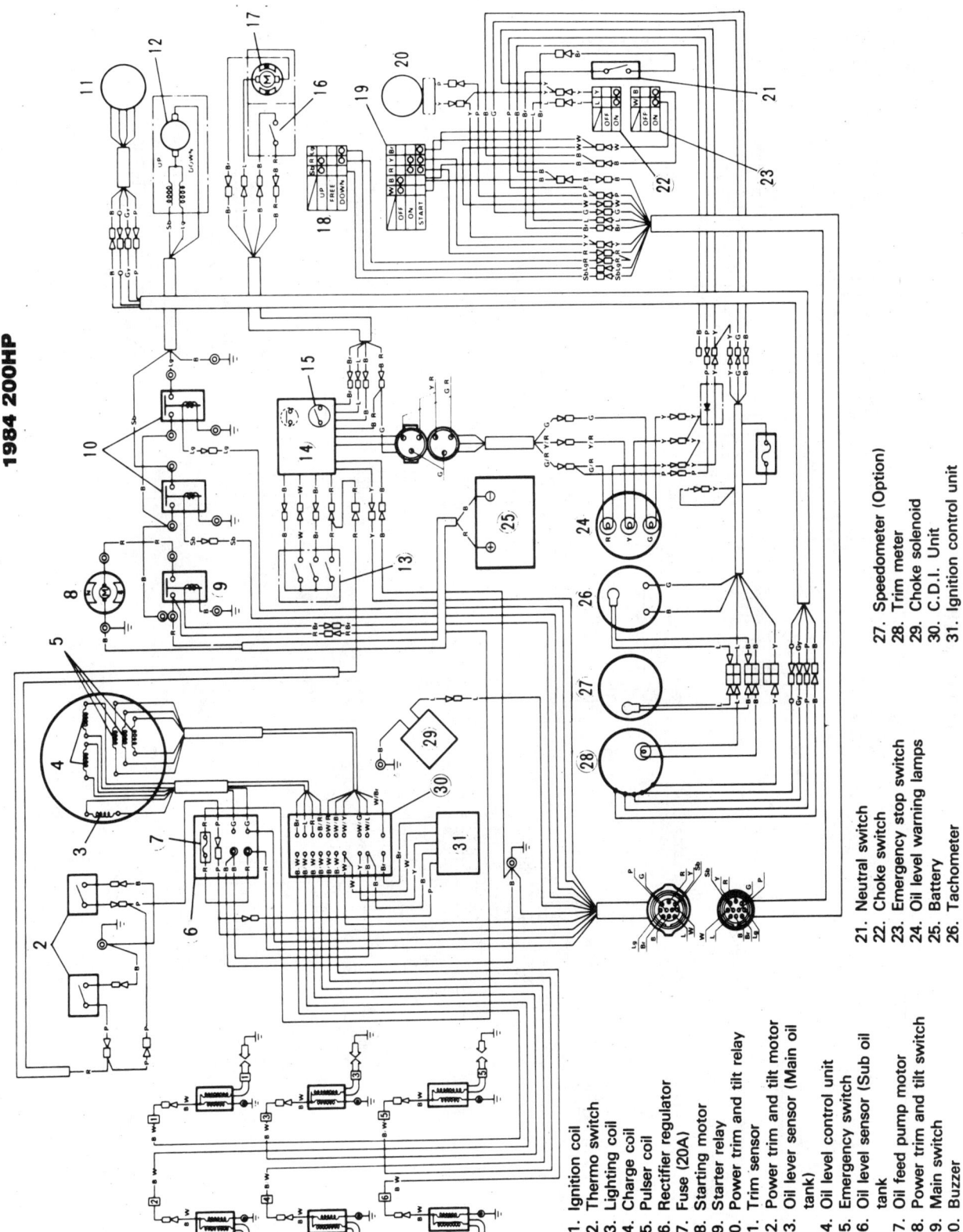

1984 200HP

1. Ignition coil
2. Thermo switch
3. Lighting coil
4. Charge coil
5. Pulser coil
6. Rectifier regulator
7. Fuse (20A)
8. Starting motor
9. Starter relay
10. Power trim and tilt relay
11. Trim sensor
12. Power trim and tilt motor
13. Oil lever sensor (Main oil tank)
14. Oil level control unit
15. Emergency switch
16. Oil level sensor (Sub oil tank)
17. Oil feed pump motor
18. Power trim and tilt switch
19. Main switch
20. Buzzer
21. Neutral switch
22. Choke switch
23. Emergency stop switch
24. Oil level warning lamps
25. Battery
26. Tachometer
27. Speedometer (Option)
28. Trim meter
29. Choke solenoid
30. C.D.I. Unit
31. Ignition control unit

16

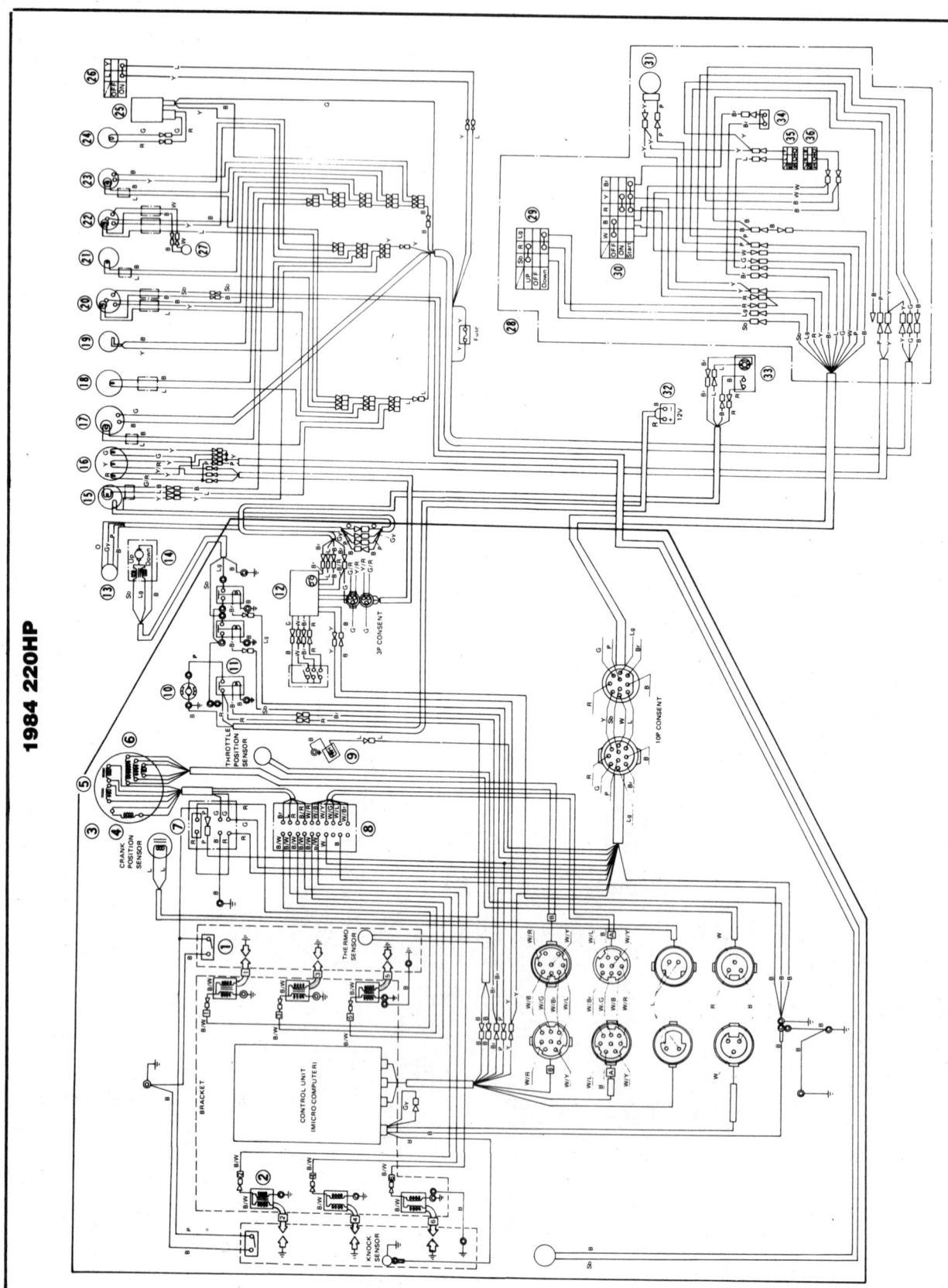

1984 220HP

Equipment

1. Thermo switch
2. Ignition coil
3. CDI magneto
4. Lighting coil
5. Charge coil
6. Pulser coil
7. Rectifier regulator
8. CDI unit
9. Choke solenoid
10. Starting motor
11. Starter relay
12. Control unit
   (Oil injection)

13. Trim sensor
14. Trim & tilt motor
15. Trim gauge
16. Oil level lamp
17. Tachometer
18. Speed meter
19. Hourmeter
20. Water temp. gauge
21. Water pressure meter
22. Fuel gauge
23. Voltmeter
24. Charge lamp

25. Charge warning unit
26. Lamp switch
27. Fuel sensor
28. Remote control box
29. Trim & tilt switch
30. Main switch
31. Buzzer
32. Battery
33. Sub oil tank
34. Neutral switch
35. Choke switch
36. Emergency stop switch

Wiring color

B	: Black	L	: Blue	R	: Red	W/B	: White/Black
P	: Pink	G	: Green	W	: White	W/Y	: White/Yellow
B/W	: Black/White	Br	: Brown	W/R	: White/Red	W/G	: White/Green
Lg	: Light green	Gy	: Grey	O	: Orange	Y	: Yellow
Sb	: Sky blue						

16

1985 200 HP

B	:	Black	
Br	:	Brown	
G	:	Green	
Gy	:	Gray	
L	:	Blue	
Lg	:	Light green	

O	:	Orange
P	:	Pink
R	:	Red
Sb	:	Sky blue
W	:	White
Y	:	Yellow

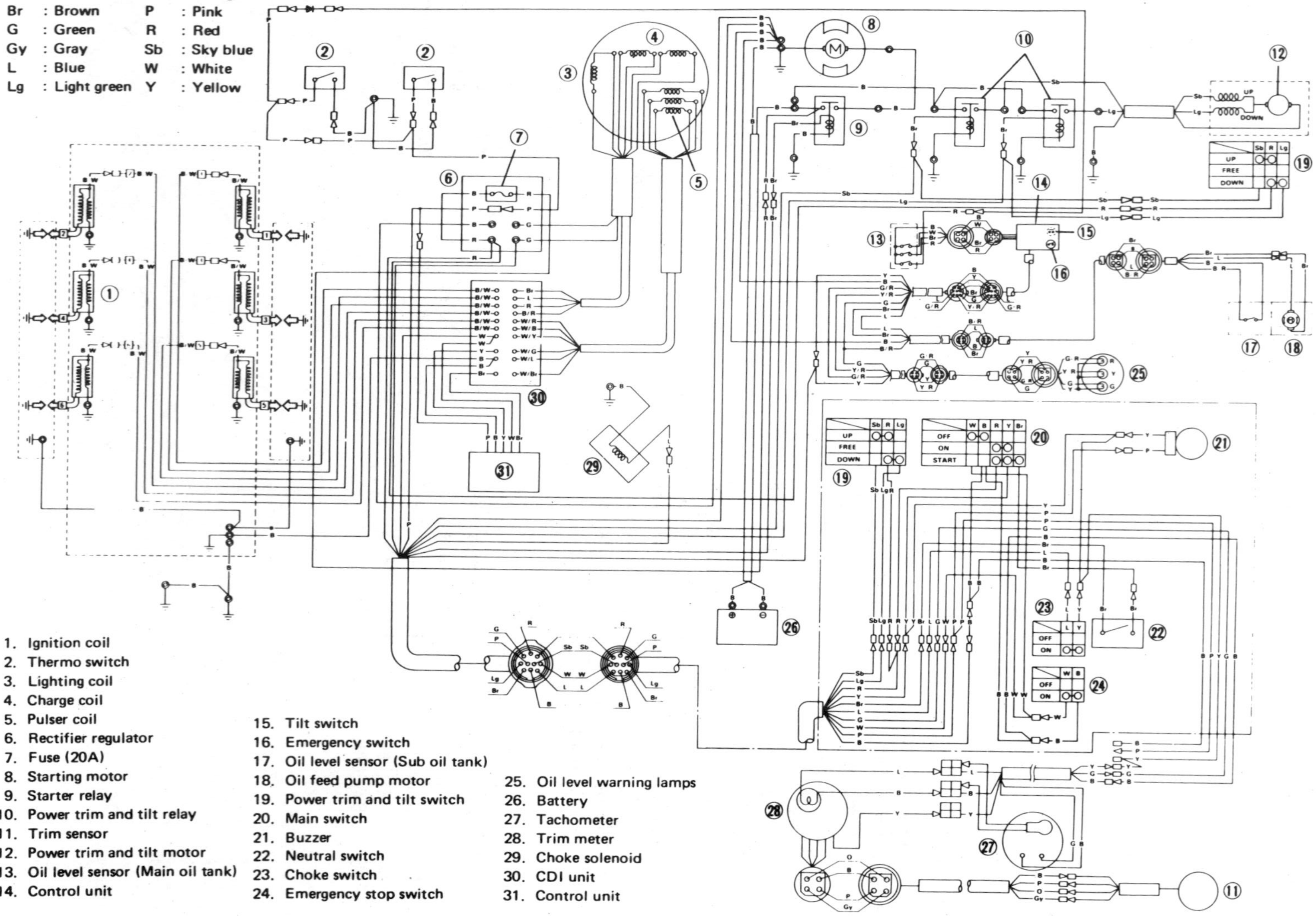

1. Ignition coil
2. Thermo switch
3. Lighting coil
4. Charge coil
5. Pulser coil
6. Rectifier regulator
7. Fuse (20A)
8. Starting motor
9. Starter relay
10. Power trim and tilt relay
11. Trim sensor
12. Power trim and tilt motor
13. Oil level sensor (Main oil tank)
14. Control unit

15. Tilt switch
16. Emergency switch
17. Oil level sensor (Sub oil tank)
18. Oil feed pump motor
19. Power trim and tilt switch
20. Main switch
21. Buzzer
22. Neutral switch
23. Choke switch
24. Emergency stop switch

25. Oil level warning lamps
26. Battery
27. Tachometer
28. Trim meter
29. Choke solenoid
30. CDI unit
31. Control unit

1985 220 HP V6 SPECIAL

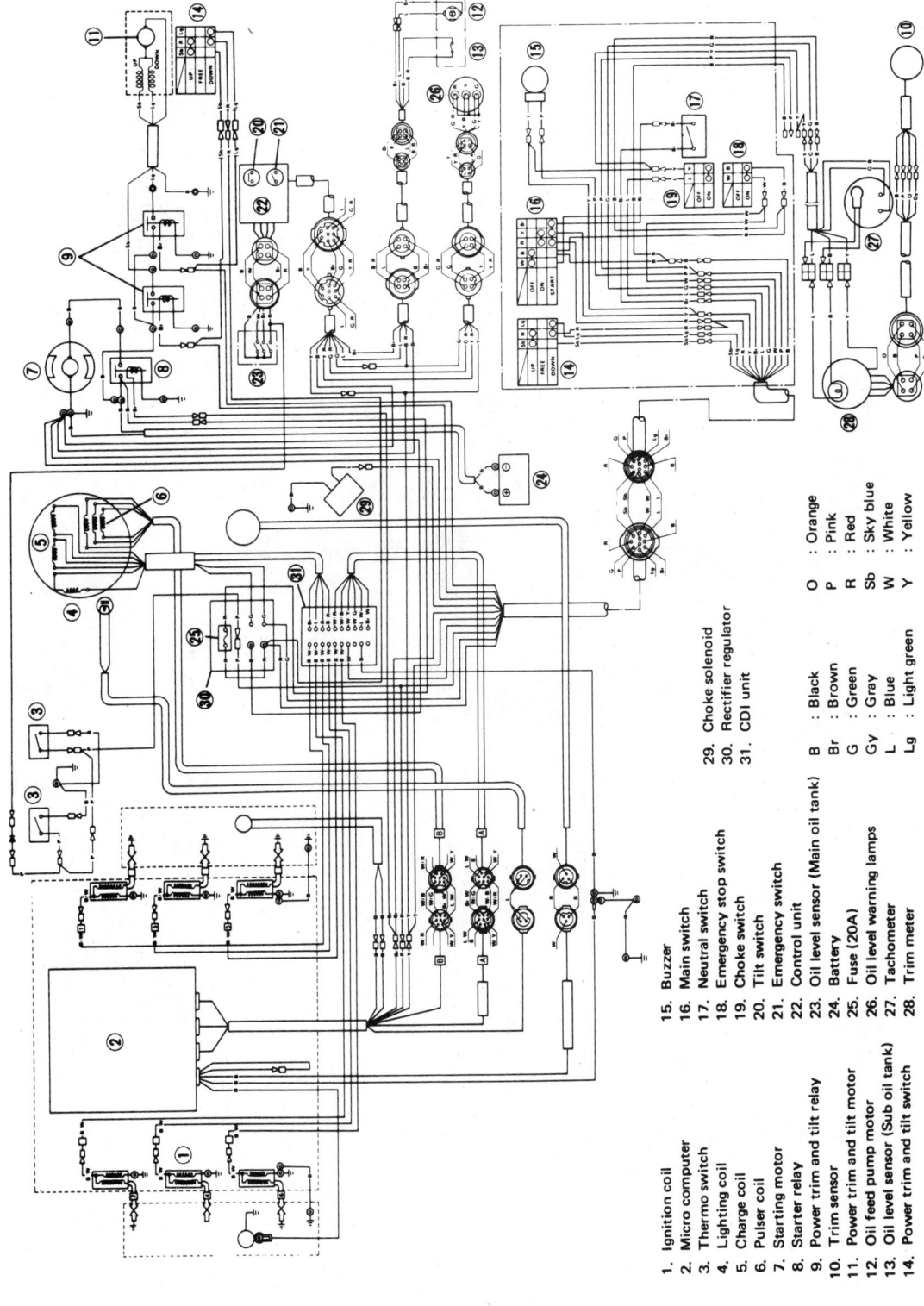

1. Ignition coil
2. Micro computer
3. Thermo switch
4. Lighting coil
5. Charge coil
6. Pulser coil
7. Starting motor
8. Starter relay
9. Power trim and tilt relay
10. Trim sensor
11. Power trim and tilt motor
12. Oil feed pump motor
13. Oil level sensor (Sub oil tank)
14. Power trim and tilt switch

15. Buzzer
16. Main switch
17. Neutral switch
18. Emergency stop switch
19. Choke switch
20. Tilt switch
21. Emergency switch
22. Control unit
23. Oil level sensor (Main oil tank)
24. Battery
25. Fuse (20A)
26. Oil level warning lamps
27. Tachometer
28. Trim meter

29. Choke solenoid
30. Rectifier regulator
31. CDI unit

B : Black
Br : Brown
G : Green
Gy : Gray
L : Blue
Lg : Light green

O : Orange
P : Pink
R : Red
Sb : Sky blue
W : White
Y : Yellow

16

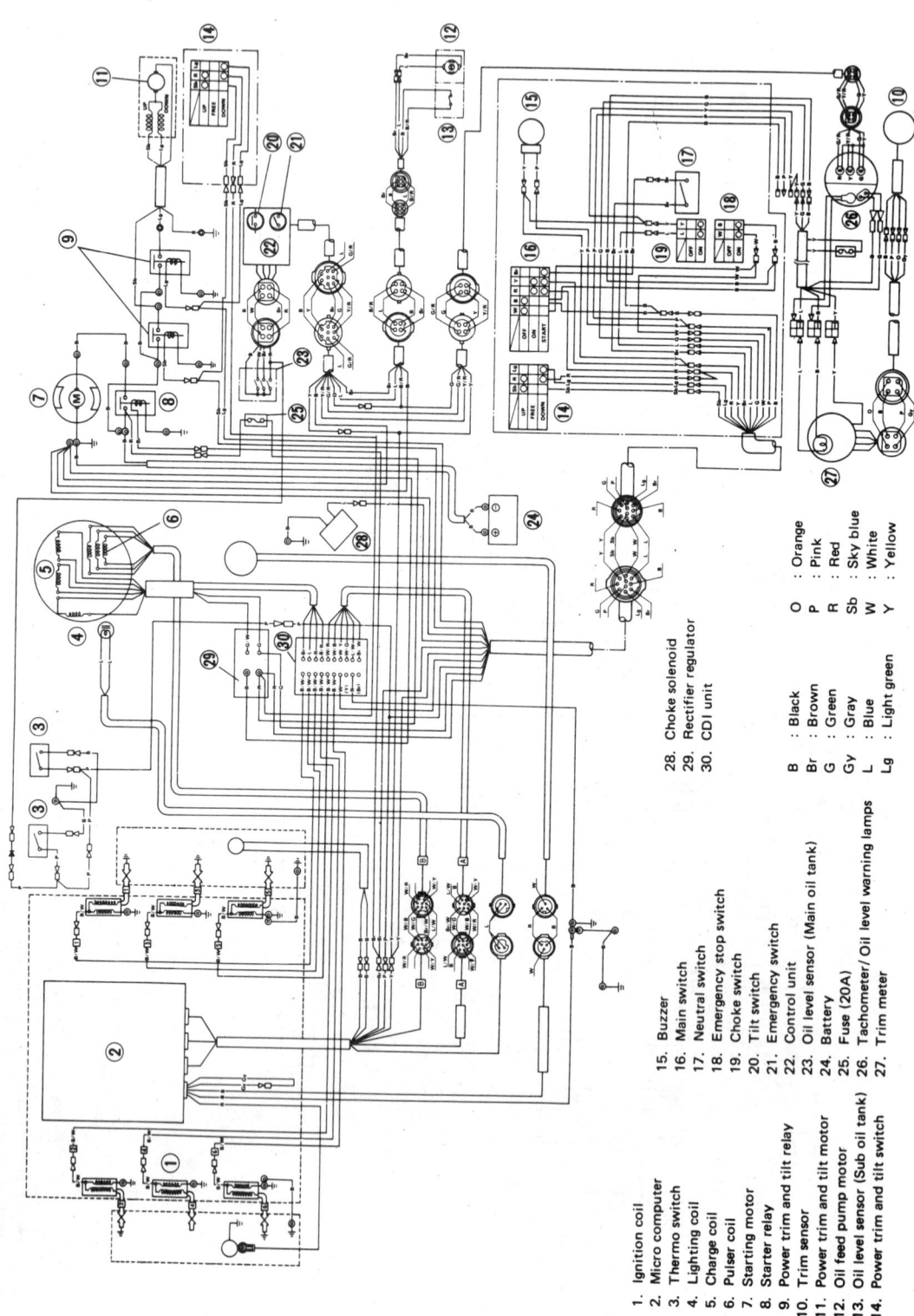

1986-on 220 HP V6 SPECIAL

1. Ignition coil
2. Micro computer
3. Thermo switch
4. Lighting coil
5. Charge coil
6. Pulser coil
7. Starting motor
8. Starter relay
9. Power trim and tilt relay
10. Trim sensor
11. Power trim and tilt motor
12. Oil feed pump motor
13. Oil level sensor (Sub oil tank)
14. Power trim and tilt switch

15. Buzzer
16. Main switch
17. Neutral switch
18. Emergency stop switch
19. Choke switch
20. Tilt switch
21. Emergency switch
22. Control unit
23. Oil level sensor (Main oil tank)
24. Battery
25. Fuse (20A)
26. Tachometer/ Oil level warning lamps
27. Trim meter

28. Choke solenoid
29. Rectifier regulator
30. CDI unit

B : Black
Br : Brown
G : Green
Gy : Gray
L : Blue
Lg : Light green

O : Orange
P : Pink
R : Red
Sb : Sky blue
W : White
Y : Yellow

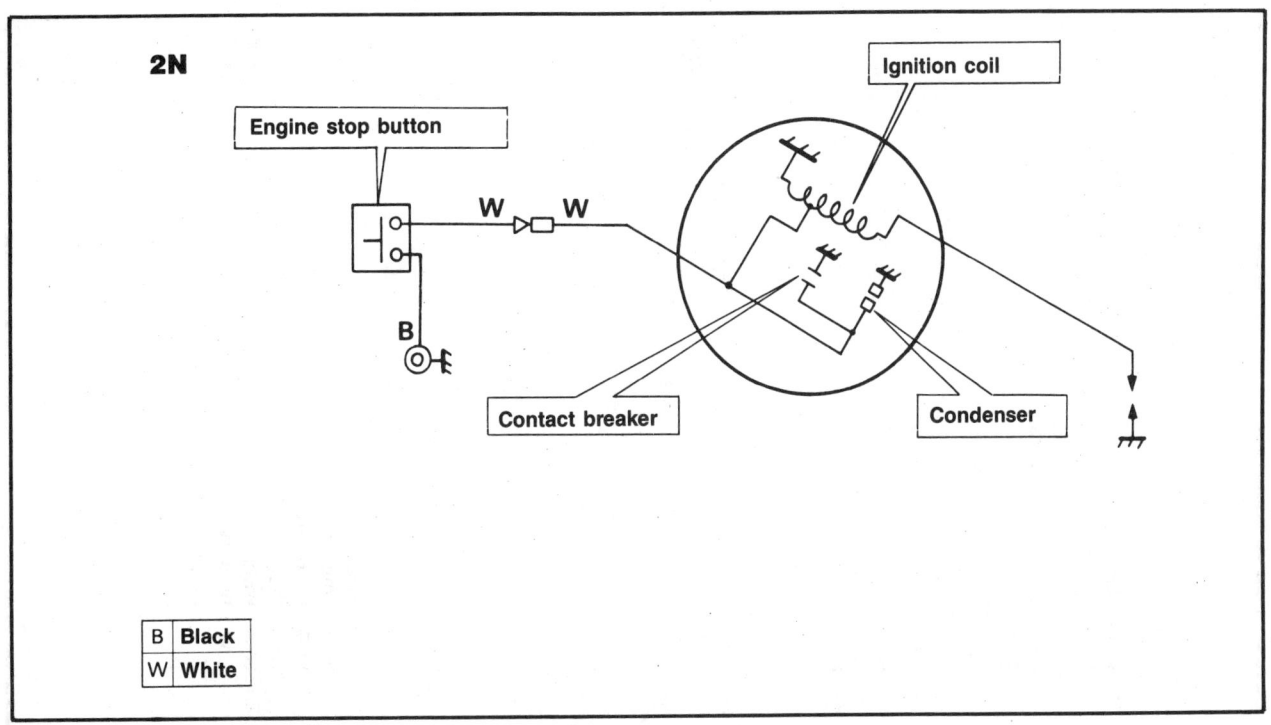

2N
Engine stop button
Ignition coil
W
W
B
Contact breaker
Condenser
B Black
W White

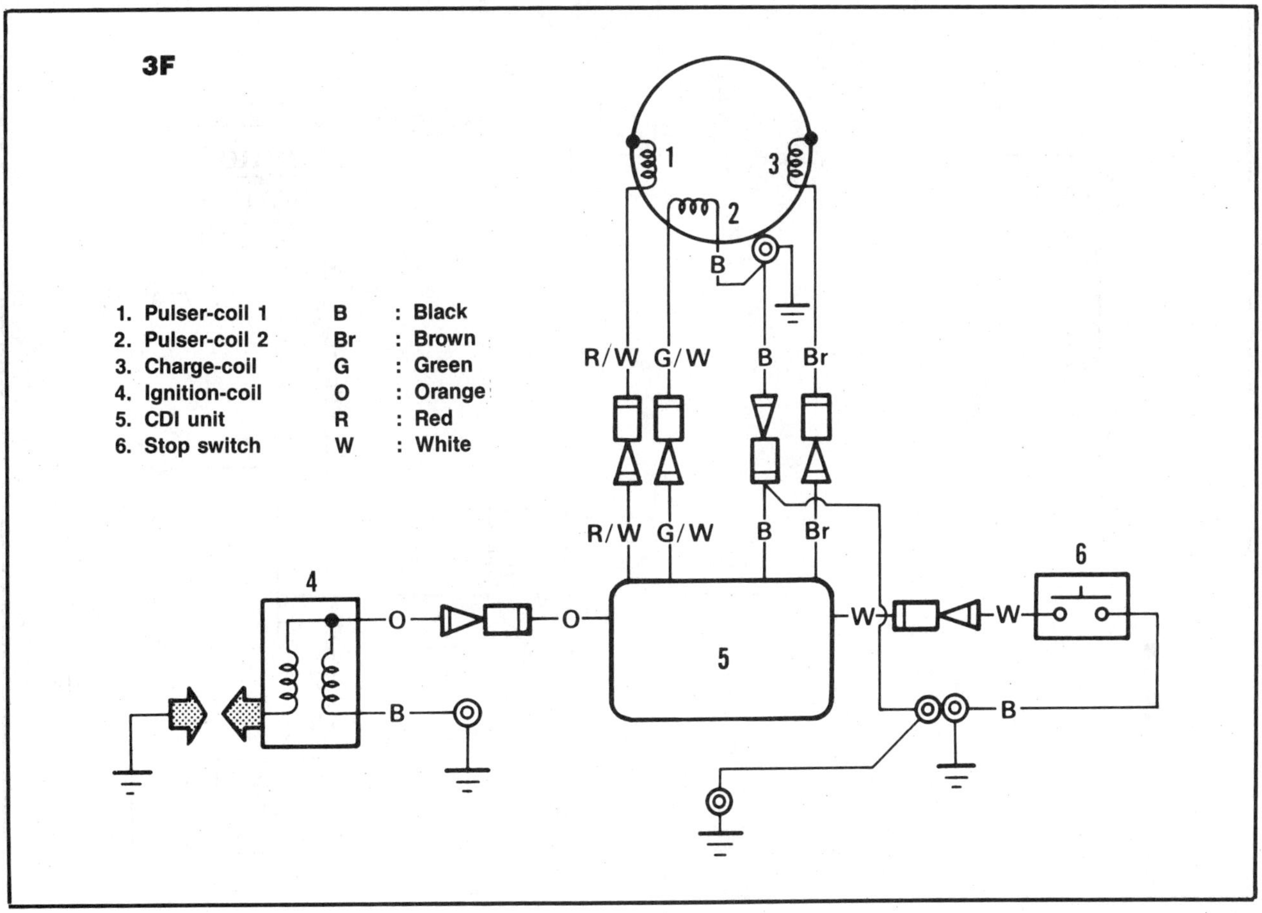

3F
1. Pulser-coil 1
2. Pulser-coil 2
3. Charge-coil
4. Ignition-coil
5. CDI unit
6. Stop switch
B : Black
Br : Brown
G : Green
O : Orange
R : Red
W : White
1
2
3
4
5
6
R/W G/W B Br
R/W G/W B Br
O
O
B
W
W
B

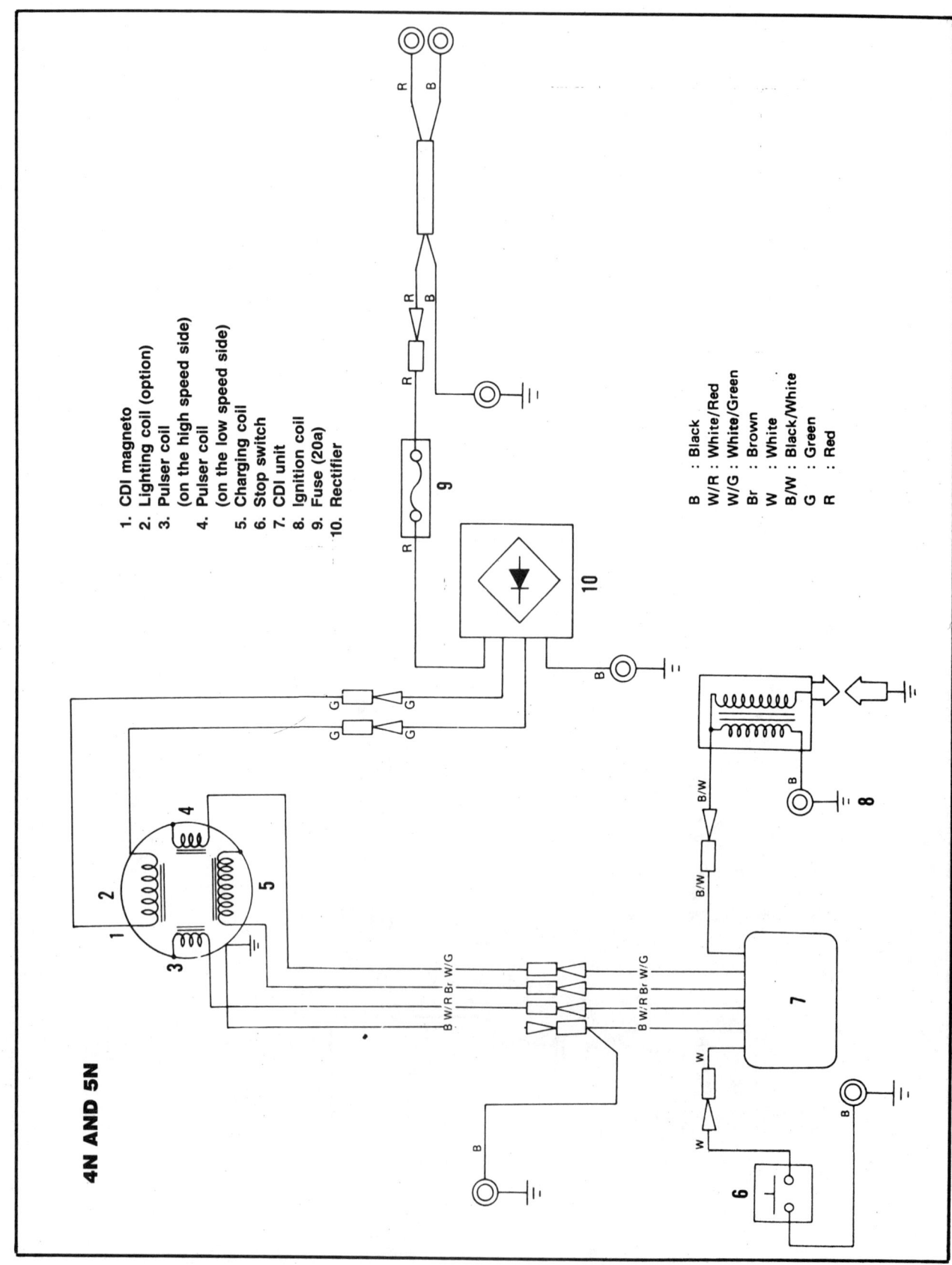

4N AND 5N

1. CDI magneto
2. Lighting coil (option)
3. Pulser coil
   (on the high speed side)
4. Pulser coil
   (on the low speed side)
5. Charging coil
6. Stop switch
7. CDI unit
8. Ignition coil
9. Fuse (20a)
10. Rectifier

B   : Black
W/R : White/Red
W/G : White/Green
Br : Brown
W   : White
B/W : Black/White
G   : Green
R   : Red

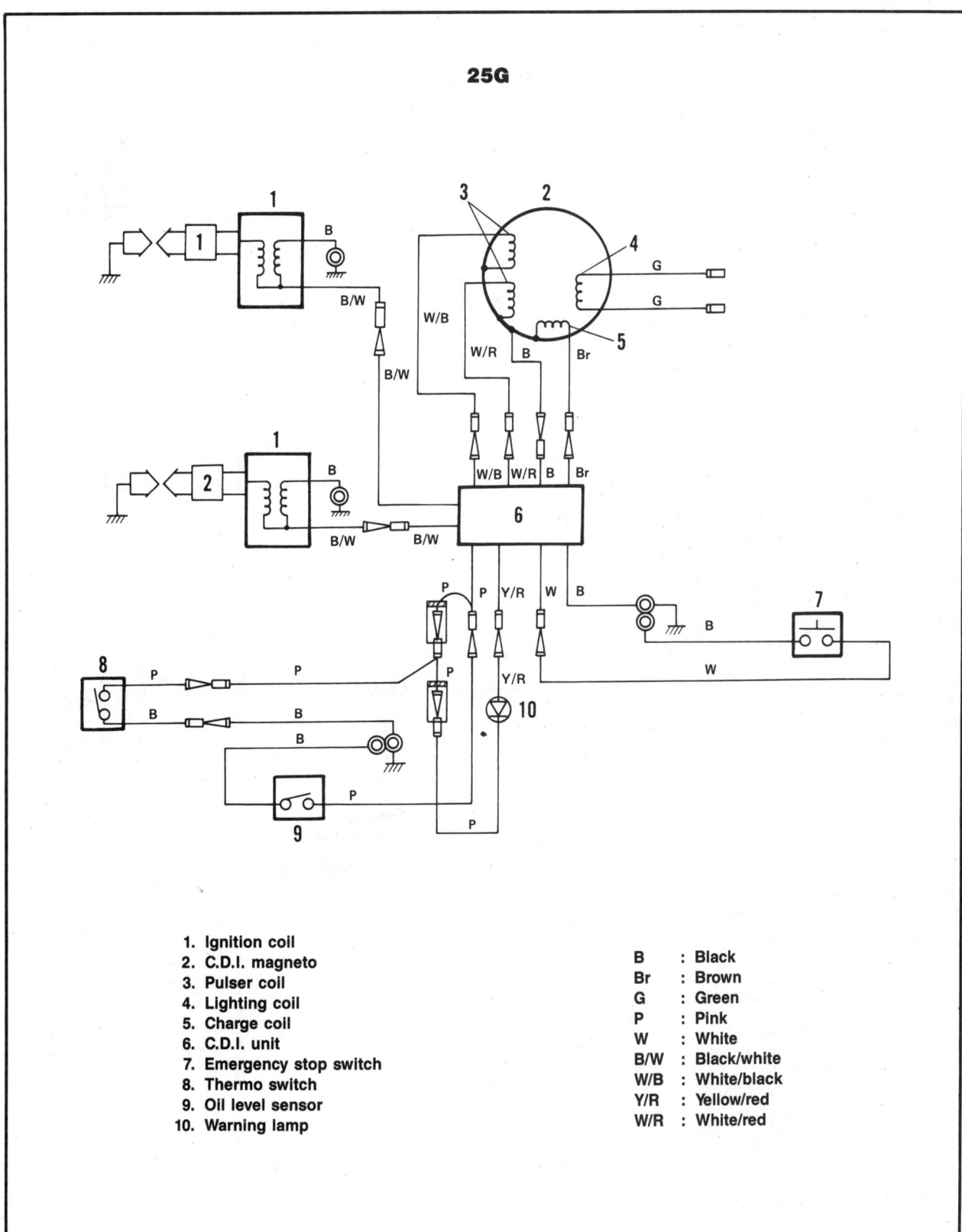

25G

1. Ignition coil
2. C.D.I. magneto
3. Pulser coil
4. Lighting coil
5. Charge coil
6. C.D.I. unit
7. Emergency stop switch
8. Thermo switch
9. Oil level sensor
10. Warning lamp

B	:	Black
Br	:	Brown
G	:	Green
P	:	Pink
W	:	White
B/W	:	Black/white
W/B	:	White/black
Y/R	:	Yellow/red
W/R	:	White/red

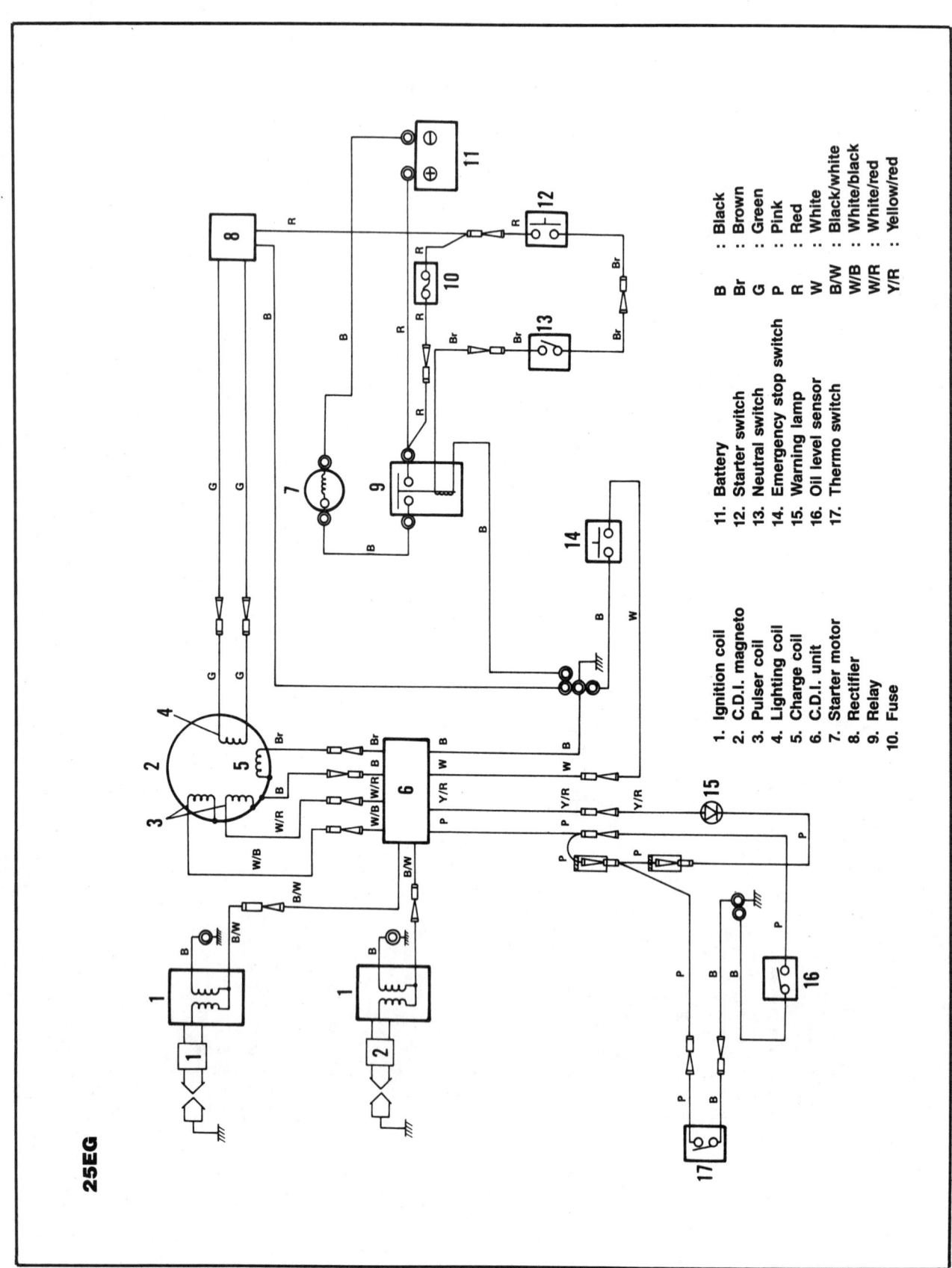

Black : B
Brown : Br
Green : G
Pink : P
Red : R
White : W
Black/white : B/W
White/black : W/B
White/red : W/R
Yellow/red : Y/R

1. Ignition coil
2. C.D.I. magneto
3. Pulser coil
4. Lighting coil
5. Charge coil
6. C.D.I. unit
7. Starter motor
8. Rectifier
9. Relay
10. Fuse
11. Battery
12. Starter switch
13. Neutral switch
14. Emergency stop switch
15. Warning lamp
16. Oil level sensor
17. Thermo switch

25EG

25 SN/LN

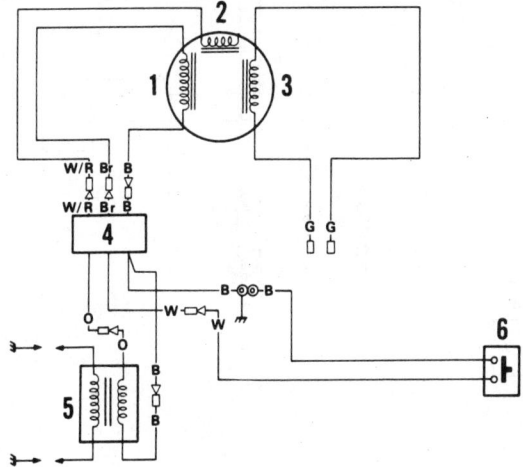

COLOR CODE

B . . . . Black
W . . . . White
R . . . . Red
O . . . . Orange
G . . . . Green
Br . . . . Brown
W/R . . . . White/red

1. Charge coil
2. Pulser coil
3. Lighting coil
4. CDI unit
5. Ignition coil
6. Stop switch

16

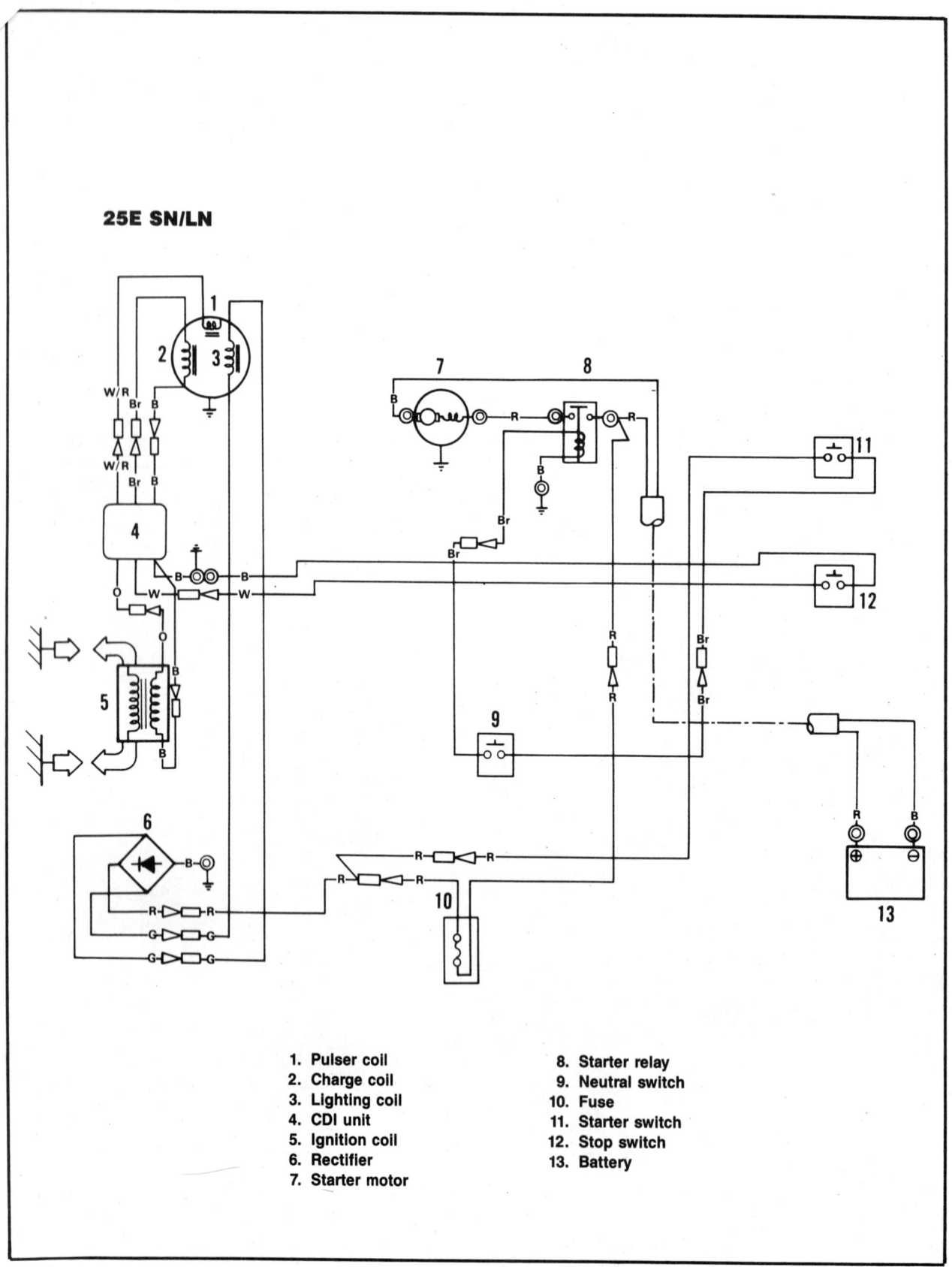

25E SN/LN

1. Pulser coil
2. Charge coil
3. Lighting coil
4. CDI unit
5. Ignition coil
6. Rectifier
7. Starter motor
8. Starter relay
9. Neutral switch
10. Fuse
11. Starter switch
12. Stop switch
13. Battery

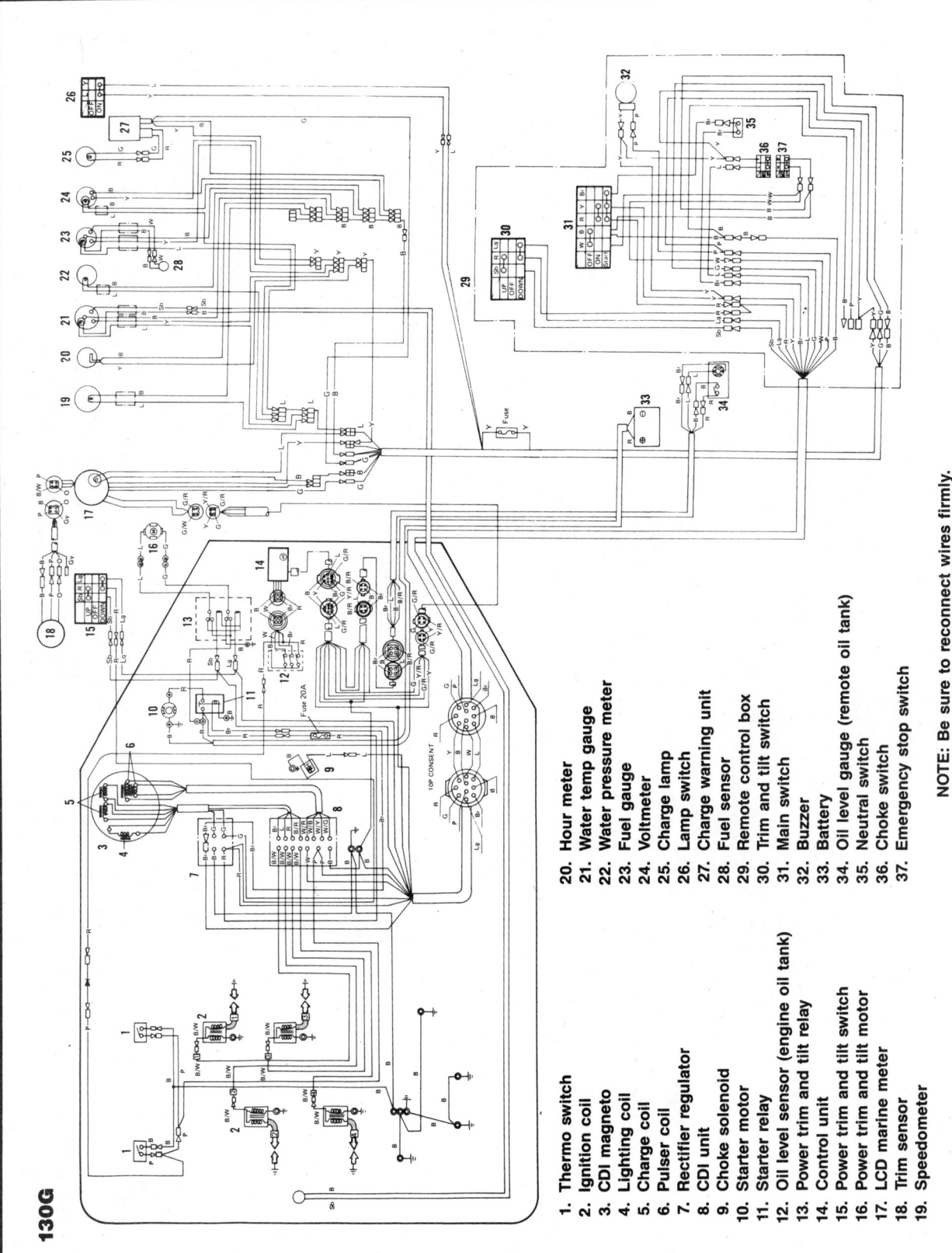

130G

1. Thermo switch
2. Ignition coil
3. CDI magneto
4. Lighting coil
5. Charge coil
6. Pulser coil
7. Rectifier regulator
8. CDI unit
9. Choke solenoid
10. Starter motor
11. Starter relay
12. Oil level sensor (engine oil tank)
13. Power trim and tilt relay
14. Control unit
15. Power trim and tilt switch
16. Power trim and tilt motor
17. LCD marine meter
18. Trim sensor
19. Speedometer
20. Hour meter
21. Water temp gauge
22. Water pressure meter
23. Fuel gauge
24. Voltmeter
25. Charge lamp
26. Lamp switch
27. Charge warning unit
28. Fuel sensor
29. Remote control box
30. Trim and tilt switch
31. Main switch
32. Buzzer
33. Battery
34. Oil level gauge (remote oil tank)
35. Neutral switch
36. Choke switch
37. Emergency stop switch

NOTE: Be sure to reconnect wires firmly.

16

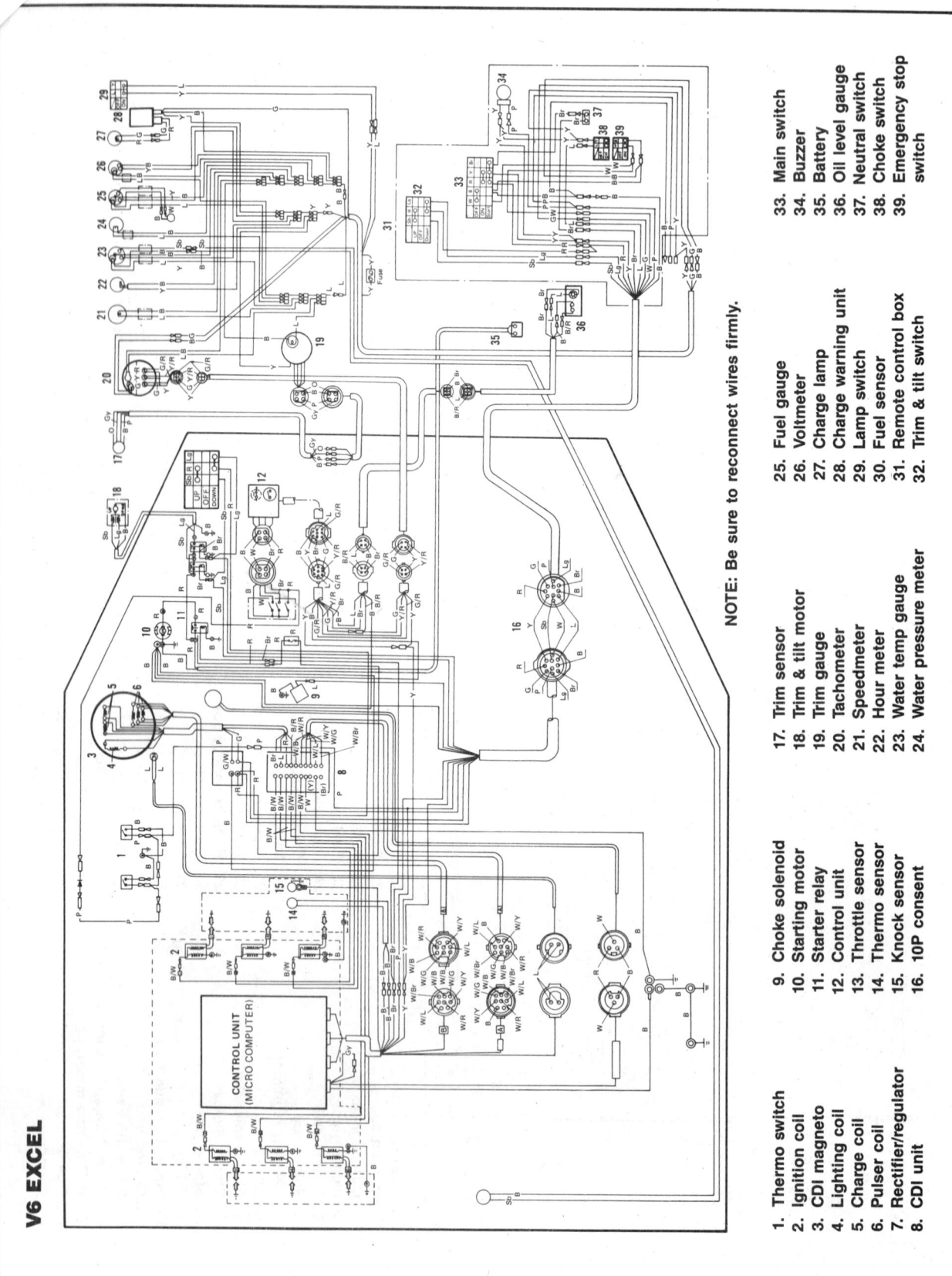

V6 EXCEL

NOTE: Be sure to reconnect wires firmly.

1. Thermo switch
2. Ignition coil
3. CDI magneto
4. Lighting coil
5. Charge coil
6. Pulser coil
7. Rectifier/regulator
8. CDI unit

9. Choke solenoid
10. Starting motor
11. Starter relay
12. Control unit
13. Throttle sensor
14. Thermo sensor
15. Knock sensor
16. 10P consent

17. Trim sensor
18. Trim & tilt motor
19. Trim gauge
20. Tachometer
21. Speedmeter
22. Hour meter
23. Water temp gauge
24. Water pressure meter

25. Fuel gauge
26. Voltmeter
27. Charge lamp
28. Charge warning unit
29. Lamp switch
30. Fuel sensor
31. Remote control box
32. Trim & tilt switch

33. Main switch
34. Buzzer
35. Battery
36. Oil level gauge
37. Neutral switch
38. Choke switch
39. Emergency stop
    switch

CONTROL UNIT
(MICRO COMPUTER)